Pharmaceutical Discovery, Development and Manufacturing Forum 2020

Held at the 2020 AIChE Annual Meeting

Online
16 – 20 November 2020

ISBN: 978-1-7138-2319-3

Printed from e-media with permission by:

Curran Associates, Inc.
57 Morehouse Lane
Red Hook, NY 12571

AIChE
120 Wall Street, FL 23
New York, NY 10005-4020

Phone: (800) 242-4363
Fax: (203) 775-5177

www.aiche.org

Additional copies of this publication are available from:

Curran Associates, Inc.
57 Morehouse Lane
Red Hook, NY 12571 USA
Phone: 845-758-0400
Fax: 845-758-2633
Email: curran@proceedings.com
Web: www.proceedings.com

TABLE OF CONTENTS

(25A) UNDERSTANDING AND PREDICTING THE EFFICACY OF TAILOR MADE ADDITIVES ON THE CRYSTALLISATION OF P-AMINOBENZOIC ACID 1
Peter L. Kaskiewicz, Ian Rosbottom, Robert B. Hammond, Nicholas J. Warren, Colin Morton, Peter J. Dowding, Neil George, Kevin J. Roberts

(25B) MULTISCALE COMPUTER MODELING TO INVESTIGATE POLYMERIC MICELLE BASED NANOCARRIERS FOR ANTICANCER DRUG DELIVERY 2
Tibo Duran, Antonio Costa, Xiaoming Xu, Diane Burgess, Bodhisattwa Chaudhuri

(25C) SIMULATION OF DRYING OF DROPLETS AND ITS EFFECT ON PARTICLE'S MORPHOLOGY IN SPRAY-DRYING 4
Siavash Zamani, Aaron Morris

(25D) POPULATION BALANCE MODEL COMPARISON OF GROWTH, SECONDARY NUCLEATION BY ATTRITION AND RIPENING 5
Luca Bosetti, Marco Mazzotti

(25E) SOLUBILITY AND NUCLEATION KINETICS OF A SERIES OF REPRESENTATIVE SINGLE AND MIXED DIESEL N-ALKANES IN DODECANE AND TOLUENE 7
Alexander Jackson, Kevin J. Roberts

(25F) MULTI-SCALE MODELING OF SOFT MATTERS: A QUALITY BY DESIGN OF DRUG DELIVERY CARRIERS BY A COMPUTATIONAL TOOL 8
Abdollah Koolivand, Panagiotis Dimitrakopoulos

(25H) INVESTIGATING MECHANISMS OF BIMETALLIC NANOCRYSTAL FORMATION WITH LIQUID PHASE TRANSMISSION ELECTRON MICROSCOPY 9
Mei Wang, Taylor J. Woehl

(45A) DISCOVERY OF LIGHT-DRIVEN HYDROGEN EVOLUTION CATALYSTS USING HIGH THROUGHPUT SIMULATION, EXPERIMENTS AND MACHINE LEARNING 10
John R. Kitchin, Zachary Ulissi, Jill E. Millstone, Stefan Bernhard

(45B) HIGH THROUGHPUT SCREENING OF ALLOY STRUCTURES FOR PROPANE DEHYDROGENATION REACTION 11
Ranga Rohit Seemakurthi, Siddharth Deshpande, Yinan Xu, Fabio H. Ribeiro, Jeffrey T. Miller, Jeffrey Greeley

(45C) THEORY-GUIDED, INTERPRETABLE MACHINE LEARNING FINDS PREDICTIVE GEOMETRIC STRUCTURE-PROPERTY RELATIONSHIPS FOR CHEMISORPTION ON ALLOYS 12
Jacques Esterhuizen, Bryan R. Goldsmith, Suljo Linic

(45D) TRENDS IN CATALYST STABILITY AND REACTIVITY: EXTRACTING PHYSICAL INSIGHTS USING SIMPLE DATA-DRIVEN APPROACHES 13
Tej Choksi, Lavie Rekhi, Verena Streibel, Paulami Majumdar, Jeffrey Greeley, Frank Abild-Pedersen

(45E) ORIENTEERING IN AN UNCHARTED CHEMICAL SPACE: SEARCHING FOR AN OPTIMAL BIMETALLIC NANOCATALYST 15
James Dean, Giannis Mpourmpakis

(45F) PHYSICS INFORMED MACHINE LEARNING OF CHEMISORPTION AT METAL SURFACES 16
Shih-Han Wang, Siwen Wang, Noushin Omidvar, Luke Achenie, Hongliang Xin

(45G) IMPROVED CATALYST PREDICTIONS WITH MACHINE LEARNING COUPLED ALCHEMICAL PERTURBATION DENSITY FUNCTIONAL THEORY 17
Charles Griego, Lingyan Zhao, Karthikeyan Saravanan, John A. Keith

(48A) TANDEM METAL NANOPARTICLE-OXIDASE CATALYSIS FOR SELECTIVE OXIDATION REACTIONS 18
Joseph Brindle, Chinmay Verma, Patrick Nelson, Michael Nigra

(48B) A REDOX-ACTIVE HYDROGEL MEDIATES ELECTRON TRANSFER FOR EFFICIENT ELECTROSYNTHETIC HYDROGEN EVOLUTION VIA THREE CLASSES OF HYDROGENASES 19
John C. Ruth, Ross D. Milton, Wenyu Gu, Alfred M. Spormann

(48C) CHARACTERIZING THE BIOCATALYTIC PROPERTIES OF CYTOSOLICALLY EXPRESSED PENICILLIN G ACYLASE FROM E. COLI FOR THE SYNTHESIS OF CEPHALEXIN AND COMPARISON WITH ßF24A MUTANT 20
Patrick R. Harris, Andreas Bommarius, Martha A. Grover, Ronald Rousseau

(48F) DYNAMICS OF THE UREA-UREASE REACTION IN CSTR – EXPERIMENTS AND DETAILED MODEL 21
Frantisek Muzika, Lenka Schreiberova, Igor Schreiber

(48G) CA@ZIF-L-STABILIZED PICKERING INTERFACIAL CATALYTIC SYSTEM FOR EFFICIENT CO2 CONVERSION 22
Jiafu Shi, Yang Zhao, Yiying Sun, Xueying Wang, Yu Chen

(48H) IMMOBILIZATION OF THREONINE ALDOLASE AND ITS APPLICATION IN THE SYNTHESIS OF CHIRAL ß-HYDROXY-A-AMINO ACID 23
Kaitong Chen

(98A) IMPACT OF DRY COATING ON THE AGGLOMERATION AND DISSOLUTION OF COHESIVE, POORLY WATER-SOLUBLE DRUG POWDERS 24
Sangah Kim, Rajesh Davé

(98B) USING THE CONICAL SCREEN MILL AS A PRACTICAL AND EFFECTIVE DRY-COATING PROCESS TO ENABLE DIRECT COMPACTION FOR HIGH DRUG LOADING FORMULATIONS 26
Maxx Capece, Christian Borchardt, Arun Jayaraman

(98C) EFFECT OF ALD AND MLD COATING ON WETTING AND FLOW PROPERTIES OF MICRON-SIZED INHALATION PHARMACEUTICAL POWDERS 27
Fuweng Zhang, Damiano La Zara, Feilong Sun, Mike J. Quayle, Gunilla Petersson, Staffan Folestad, J. Ruud Van Ommen

(98D) IMPACT OF THE MICRO/NANO- STRUCTURAL PROPERTIES OF MATERIALS ON TRIBO-CHARGING: A CASE STUDY ON SALBUTAMOL SULPHATE PARTICLES 28
Michela Beretta, Joana T. Pinto, Wen-Kai Hsiao, Amrit Paudel

(98E) INVERSE GAS CHROMATOGRAPHY (IGC) OF PHARMACEUTICAL POWDERS: HARNESSING THE POWER OF ADHESION-COHESION FORCES TO IMPROVE PHARMACEUTICAL DEVELOPMENT 31
Ricardo Goncalves, Luis Sousa, Maria Paisana

(98F) SPHERICAL AGGLOMERATION OF PARACETAMOL: A MECHANISTIC APPROACH 34
Kate Pitt, Jonathan D. Tew, James D. Litster, Rachel M. Smith

(98G) REDUCTION OF PUNCH-STICKING PROPENSITY OF CELECOXIB BY SPHERICAL CRYSTALLIZATION VIA POLYMER ASSISTED QUASI-EMULSION SOLVENT DIFFUSION 35
Hongbo Chen, Shubhajit Paul, Hongyun Xu, Kunlin Wang, Mahesh K. Mahanthappa, Changquan Sun

(106A) MOLECULAR DESIGN BLUEPRINTS: CATALYSTS AND PRINCIPLES FROM NEW SIMULATION AND MACHINE LEARNING TOOLS 37
Heather J. Kulik, Aditya Nandy, Daniel Harper, Michael Taylor, Chenru Duan, Fang Liu, Jon Paul Janet, Naveen Arunachalam

(106B) MACHINE LEARNING QUANTUM TUNNELING IN THE KINETICS OF CHEMICAL REACTIONS 38
Stephanie Valleau

(106C) IMPROVED CATALYST DESIGN AND PERFORMANCE THROUGH APPLICATION OF MACHINE LEARNING ALGORITHMS 39
Foster Agblevor, Luke Achenie

(106D) IMPROVING THE REUSABILITY OF MACHINE LEARNING MODELS FOR MATERIALS 40
Matthew M. Montemore, Gbolade O. Kayode, Chukwudi F. Nwaokorie

(106E) THE OPEN CATALYST PROJECT DATASET 41
Lowik Chanussot, Abhishek Das, Javier Heras-Domingo, Siddharth Goyal, Caleb Ho, Thibaut Lavril, Aini Palizhati, Devi Parikh, Morgane Riviere, Muhammed Shuaibi, Kevin Tran, Zachary Ulissi, Junwoong Yoon, C. Lawrence Zitnick

(106F) ACTIVE-LEARNING FRAMEWORK FOR THE DISCOVERY OF NOVEL MATERIALS FOR CATALYSIS 42
Kirsten T. Winther, Raul A. Flores, Ankit Jain, Michal Bajdich, Thomas Bligaard

(106G) IDENTIFYING INTERACTION TRENDS BETWEEN SINGLE METAL ATOMS AND MODIFIED MGO(100) SUPPORTS WITH STATISTICAL LEARNING 43
Chun-Yen Liu, Shijia Zhang, Daniel Martinez, Meng Li, Thomas P. Senftle

(106H) HOW TO FIX SOME OF OUR ERRORS VIA DATA SCIENCE 44
Dionisios G. Vlachos

(109A) OPTIMIZATION-DRIVEN QUASI-DETERMINISTIC MONTE CARLO SAMPLING FOR ACCURATE UNCERTAINTY QUANTIFICATION: APPLICATIONS TO THE PHARMACEUTICAL INDUSTRY 45
Francesco Rossi, Linas Mockus, Gintaras V. Reklaitis

(109B) TOPOLOGICAL DATA ANALYSIS: CONCEPTS, COMPUTATION, AND APPLICATIONS 47
Alexander Smith, Victor M. Zavala

(109C) COMPUTATIONAL FLUID DYNAMICS-BASED IN-SITU SENSOR ANALYTICS OF DIRECT METAL LASER SINTERING PROCESS USING MACHINE LEARNING 49
Yi Ming Ren, Yichi Zhang, Yangyao Ding, Panagiotis D. Christofides

(109D) A DATA-ENGINEERING APPROACH FOR ESTIMATING CHEMICAL RELEASES FROM INDUSTRIAL POLLUTION ABATEMENT TECHNOLOGIES 50
Jose D. Hernandez-Betancur, Shuyun Li, Fernando V. Lima, Mariano Martín, Gerardo Ruiz-Mercado

(109E) PHYSICS INFORMED MACHINE-LEARNING FOR STATIC SECURITY ANALYSIS OF OPTIMAL POWER FLOW SOLUTIONS 51
Zachary Kilwein, Logan Blakely, Anya Castillo, Michael Eydenberg, Carl D. Laird, Fani Boukouvala

(109F) STATISTICS PATTERN ANALYSIS-BASED MULTICLASS WOODCHIP MOISTURE CLASSIFICATION SYSTEM USING SHORT-RANGE IOT WI-FI 52
Kerul Suthar, Jin Wang, Zhihua Jiang, Q. Peter He

(109G) PROAD (PROCESS ADVISOR): HEALTH MONITORING DASHBOARD FOR CENTRIFUGAL PUMPS 54
Arnab Dutta, Iftekhar A. Karimi, Farooq Shamsuzzaman

(109H) DATA-DRIVEN APPROACHES TOWARDS REAL-TIME EQUIPMENT HEALTH MONITORING AND PREDICTIVE MAINTENANCE IN DRUG PRODUCT MANUFACTURING 57
Philipp Zürcher, Sara Badr, Hirokazu Sugiyama

(121A) MICROFLUIDICS-BASED BIOCATALYST-PRODUCT SEPARATION FOR CONTINUOUS ENZYMATIC PROCESSES. A CASE STUDY ON ß-LACTAM ANTIBIOTICS MANUFACTURING 58
Hossein Salami, Sajad Razavi Bazaz, Matthew A. McDonald, Martha A. Grover, Ronald Rousseau, Majid Ebrahimi Warkiani, Andreas Bommarius

(121B) NOVEL INTEGRATION OF CONTINUOUS CRYSTALLIZATION PLATFORMS FOR CONTROL OF CRYSTAL SIZE DISTRIBUTION AND POLYMORPHIC FORM 59
Wei-Lee Wu, Joseph Oliva, Shivani Kshirsagar, Christopher Burcham, Zoltan K. Nagy

(121C) A SCALABLE MEMBRANE PERVAPORATION APPROACH FOR CONTINUOUS FLOW RING CLOSING METATHESIS 61
Matthew Bio, Hannah Murnen, Ning Shangguan, Sudip Majumdar, Christopher P. Breen, Christine Parrish, Timothy Jamison

(121D) DEVELOPMENT OF CONTINUOUS HOMOGENEOUS HYDROGENATION PROCESS FOR PHARMACEUTICALS USING A FLOW REACTOR 62
Onkar Manjrekar, Duygu Gerceker, Greg Storer, Brian Kotecki, Kaid Harper, Elie Chaaya, Shashank Shekhar, Moiz Diwan

(121E) LIQUID PHASE PEPTIDE SYNTHESIS VIA NANOSTAR-SIEVING 63
Jet Yeo, Ludmila G. Peeva, Piers Gaffney, Carla Luciani, Fernando Albericio, Andrew G. Livingston

(121F) LIQUID-LIQUID EXTRACTIONS IN API MANUFACTURING PROCESSES: IS CONTINUOUS THE WAY TO GO? 65
Rui Pina Campos, Maria Inês Lopes, Ruth Morais, Susana Nascimento

(121G) DEVELOPMENT AND PROCESS INTENSIFICATION OF A THREE STEP CONTINUOUS FLOW PROCESS FOR TELESCOPED HIGH ENERGY REACTIONS 66
Christopher Lippelt, Pieter Otten, Shawn Conway

(130B) AN ACCURATE, PRECISE, AND AFFORDABLE LIGHT EMITTING DIODE SPECTROPHOTOMETER WITH A RASPBERRY PI USER INTERFACE FOR DIABETES TESTING IN HONDURAS 67
Tanner Kontny

(130C) EMBEDDED DROPLET PRINTING FOR SPHERICAL CRYSTALLIZATION OF PHARMACEUTICALS 68
Arif Z. Nelson, Saif A. Khan, Patrick S. Doyle

(130D) HEAT EXCHANGER MODEL FOR PREDICTING TEMPERATURE AND DISSOLUTION OF API FOR A HIGH-THROUGHPUT TEMPERATURE SHIFT SPRAY DRYING PROCESS 69
Molly Adam, Kimberly B. Shepard, John Baumann, Leon Z. Wang

(130E) ADVANCING SPRAY DRYING PROCESS CAPABILITIES FOR LOW SOLUBILITY DRUG COMPOUNDS BY LEVERAGING TEMPERATURE-SHIFT SOLUBILITY ENHANCEMENT: A CASE STUDY WITH ALECTINIB 71
Dillon Wright, John Baumann, Kimberly B. Shepard, Deanna Mudie, Molly Adam

(130G) NOVEL DISPENSING MECHANISM TO ACHIEVE CONTINUOUS FLOW OF HIGHLY COHESIVE POWDERS 73
Jon Hilden, Carl Hartmann III

(144A) PRODUCING PREDICTIVE CHEMICAL PROCESS MODELS FROM A STATE OF INCOMPLETE MECHANISTIC KNOWLEDGE AND SMALL DATA SETS 74
Daniel Griffin, Seth Huggins

(144C) IDENTIFYING THE PROCESSING SPACE OF CONTINUOUS GRANULATION: A CASE STUDY OF EXTENDED RELEASE TABLETS 75
Ahmed Zidan

(144D) DYNAMIC OPTIMIZATION STRATEGIES TOWARDS ADDRESSING THE CHALLENGES IN PROTEIN THERAPEUTICS 77
Chrysoula D. Kappatou, Adel Mhamdi, Alexander Mitsos

(144E) INTEGRATED SUPPLY CHAIN NETWORK DESIGN AND INVENTORY MANAGEMENT FOR AUTOLOGOUS CELL THERAPY. 79
Apoorva Katragadda, Iftekhar A. Karimi, Xiaonan Wang

(144F) TWO-STAGE CONTROL USING MODEL-BASED REINFORCEMENT LEARNING AND PREDICTIVE CONTROL FOR FED-BATCH BIOREACTOR 80
Tae Hoon Oh, Jong Min Lee

(144G) SUSTAINABLE API DEVELOPMENT: DEVELOPING A STRATEGY FOR MEASURING AND REDUCING ENVIRONMENTAL IMPACT IN THE PHARMACEUTICAL INDUSTRY 82
David Streather, Anna Parsons

(156B) AN INTEGRATED PROCESS AUTOMATION SYSTEM IN API MANUFACTURING: KEY CHARACTERISTICS AND CASE STUDY 83
Da Pan

(156E) A QUANTITATIVE ASSAY OF SODIUM TRIACETOXYBOROHYDRIDE 84
Michael Zacuto, Joseph Perona, Robert Dunn

(156F) DELIVERY STRATEGIES FOR LIVE THERAPEUTIC BACTERIA 85
Ava M. Vargason, Ryann Callaghan, Shruti Santhosh, Aaron C. Anselmo

(156G) IN VITRO BIOMIMETIC TUMOR MODEL WITH CANCER-SPECIFIC GENE REGULATORS AND GREEN FLUORESCENCE REPORTER FOR HIGH THROUGHPUT DRUG SCREENING 86
Shang-Tian Yang, You Li

(156I) PREPARATION AND USE OF FINE GRADE ENGINEERED EXCIPIENTS FOR DIRECT COMPRESSION OF BINARY BLENDS OF COHESIVE DRUG POWDERS 87
Zhixing Lin, Liang Chen, Kai Zheng, Kuriakose Kunnath, Sangah Kim, Rajesh Davé

(156J) APPLYING MACHINE LEARNING TO PREDICT THERAPEUTIC ANTIBODY VISCOSITY 88
Pin-Kuang Lai

(156K) DEVELOPMENT OF MULTI-ADJUVANT SYSTEMS FOR VACCINES AGAINST INFECTIOUS DISEASES 89
Julia E. Vela Ramirez, Lauren A. Austin, Heidi Ferguson, Pedro J. Cejas, Rob Saklatvala

(156L) DEVELOPMENT OF A HARVEST SMALL SCALE MODEL TO SIMULATE CONTINUOUS CENTRIFUGATION 90
Jianfa Ou, Qing Wong, Michael Kagan, Dominique Monteil, Daniel Bock

(156M) OPTIMIZING AND REDUCING CLEANING PRODUCTION LOSS IN PHARMA, AGROCHEMICAL AND SPECIALTY CHEMICAL COMPANIES USING A NEW EXPERT SYSTEM. 91
Joan Cordiner

(156N) REVOLUTIONIZE OF PHARMACEUTICAL PRODUCT CHANGEOVERS BY ENHANCED CLEANING WITH MICROBUBBLES 92
William B. Zimmerman, Joan Cordiner

(210A) (INVITED) SKIN-LIKE WEARABLE AND IMPLANTABLE SENSORS 93
Zhenan Bao

(210B) AN ADHESION-BASED ANTI-CORROSION STRATEGY FOR WEARABLE ELECTROCHEMICAL SENSING AND SYSTEM INTEGRATION 94
Bo Wang, Yichao Zhao, Sam Emaminejad

(210E) RAPID SCREENING PLATFORM FOR THE ISOTOPIC DETERMINATION OF RADIONUCLIDES IN WATER 95
Scott M. Husson, Abenazer W. Darge, James C. Foster, Valery Bliznyuk, Timothy A. Devol

(210F) ELECTROCHEMICAL DETECTION OF ENVIRONMENTAL POLLUTANTS 96
Ariel Furst

(210H) FABRICATION OF HIGH-RESOLUTION GRAPHENE-BASED FLEXIBLE ELECTRONICS VIA POLYMER CASTING AND MICROFLUIDIC APPROACHES 97
Metin Uz, Matthew Lentner, Kyle Jackson, Maxsam Donta, Juhyung Jung, John Hondred, Eric Mach, Jonathan Claussen, Surya K. Mallapragada

(231A) DISORDER AND ENTROPY IN MOLECULAR CRYSTALS 99
Michael R. Shirts, Nathan Abraham, Eric Dybeck, Marcus Hock

(231B) A SIMPLE FIX ADDRESSES SEVERE DENSITY FUNCTIONAL THEORY POLYMORPH RANKING PROBLEMS IN PHARMACEUTICALS 100
Gregory Beran

(231C) VIRTUAL COFORMER SCREENING BY CLOUD-COMPUTING CRYSTAL STRUCTURE PREDICTIONS 101
Yuriy Abramov, Guangxu Sun, Yingdi Jin, Sizhu Li, Zhuocen Yang, Baimei Shi, Chao Chang

(231F) MORPHOLOGY PREDICTION FOR ORGANIC SALT CRYSTALS 102
Yongsheng Zhao, Steven Landis, Michael F. Doherty

(231G) MITIGATING PUNCH STICKING PROPENSITY OF CELECOXIB BY COCRYSTALLIZATION – AN INTEGRATED COMPUTATIONAL AND EXPERIMENTAL APPROACH 103
Chenguang Wang, Shubhajit Paul, David Sun, Sten Nilsson Lill, Changquan Sun

(259A) DIFFUSION GROWTH MECHANISM OF FIVE-FOLD TWINNED AG NANOWIRE 110
Jianming Cui, Kristen Fichthorn

(259B) VOLUME EFFECTS ON ACETAMINOPHEN NUCLEATION THROUGH AN INDUCTION TIME APPROACH 111
Isaac Jerome C. Dela Cruz, Gerard Capellades, Jem Valerie D. Perez, Bryan G. Alamani, Allan S. Myerson

(259C) AMORPHOUS INTERMEDIATE STATES DURING THE PRECIPITATION OF CERIUM OXALATE: TOWARDS A NEW MODEL 112
Maxime Durelle, Sophie Charton, David Carrière

(259D) LANTHANIDE PRECIPITATION FROM SPENT NI-MH BATTERY LEACHATES: KINETIC STUDY AND PHASE IDENTIFICATION 113
Margot Zielinski, Laurent Cassayre, Philippe Destrac, Nicolas Coppey, Cédric Foulet, Béatrice Biscans

(259E) MOLECULAR MODIFIERS SUPPRESS STRUVITE FORMATION VIA UNIQUE MECHANISMS 117
Doyoung Kim, Jessica Moore, Nicola Irwin, Jeffrey D. Rimer

(259F) SOLVENT STRUCTURING AND VISCOSITY EFFECTS ON THERMODYNAMICS AND KINETICS OF CRYSTALLIZATION 118
Rajshree Chakrabarti, Lakshmanji Verma, Monika Warzecha, Jeremy C. Palmer, Peter G. Vekilov

(259G) IDENTIFYING CRYSTAL MORPHOLOGIES TO OPTIMIZE GROWTH KINETICS OF COVALENT ORGANIC FRAMEWORKS – AN EXAMPLE OF COF-5 119
Rajan Bhawnani, Santanu Chaudhuri, Meenesh R. Singh

(369A) ENZYMATIC REACTIVE CRYSTALLIZATION OF CEPHALEXIN USING IMMOBILIZED PENICILLIN G ACYLASE FROM E. COLI 120
Patrick R. Harris, Hossein Salami, Matthew A. McDonald, Grant D. Marshall, Ronald Rousseau, Martha A. Grover, Andreas Bommarius

(369B) RECRYSTALLIZATION DEVELOPMENT FOR SMALL MOLECULE API (ACTIVE PHARMACEUTICAL INGREDIENT) ISLATRAVIR 122
Rachel S. Bade, Richard Varsolona, Dirk Stueber, Kevin Sirk, Jacob Heltzel, Athanas Koynov, Eric Margelefsky

(369C) SEEDING STRATEGIES FOR THE SELECTIVE CRYSTALLISATION OF MODEL PROTEINS 123
Ian Rosbottom, Xiaoyu Li, Wenqian Chen, Jerry Y. Y. Heng

(369D) DEVELOPING A SCALABLE TEMPLATING-PLATFORM FOR ENHANCING CRYSTALLISATION OF BIO-ACTIVE PROTEINS 124
Frederik J. Link, Jerry Y. Y. Heng

(369E) SENSOR INTEGRATED CONTINUOUS MICROFLUIDIC MIXER FOR PROTEIN SCREENING AT CONSTANT SUPERSATURATION 126
Paria Coliaie, Manish S. Kelkar, Nandkishor K. Nere, Meenesh R. Singh

(369G) PARTICLE SIZE CONTROL OF A DRUG SUBSTANCE THROUGH PROCESS CRYSTALLIZATION 127
Otute Akiti, Tanya Wallace, Matteo Daldosso, Federico Scaravelli, Martina Saccani, Irene Parisini, Francesco Martinelli, Davide Comisso, Wayne Morley, Marina Galvani

(435A) REACTION KINETIC MODEL APPLICATION TO SPEED UP DEVELOPMENT 128
Filipe Ataíde, Elisabete Mateus, Lucia Volta E Sousa, Marta Mourão, Ana Ribeiro, Fabrice Odille, Georges Assaf, Ugo Cocchini, Patrick Lo Brutto

(435B) UTILIZING MECHANISTIC MODELLING TO SUPPORT SCALE-UP OF FED-BATCH REACTION PROCESS 129
Marcello Bosco, Leonor Rosa, Niall Mitchell, Sam Wilkinson, Meera Mahadevan

(435C) SCALING DOWN A REACTIVE CRYSTALLIZATION USING VARIOUS MIXING MODELS 131
Christopher Jordan, Christopher Marton, Yohanes Agustinus, Christopher Wilbert, Joseph Perona, Kira Ragazzo, Buck Holloway

(435D) CENTRIFUGATION AT SCALE: SCALING UP BASED ON BENCH TOP, PILOT PLANT AND MODELING DATA 132
Laura McKee, Molly Riccitelli, Adelia Carragher, Xiao Serenity Hou, Elie Chaaya

(435E) EXPERIMENTAL CONFIRMATION OF SCALE-DOWN MIXING MODELS FROM COMMERCIAL TO LAB SCALE WITH CFD FOR STERILE DRUG PRODUCTS 133
Matthew H. Flamm, Tracey Mascaro, Amy L. Quach, Craig Ikeda, Brian Marks, Francis Kwofie, Kyra Schwartz

(435F) COMPUTATIONAL APPROACHES TO SCALE-UP AND TECH-TRANSFER IN AGITATED FILTER BED DRYING 134
Eric Murphy, Kushal Sinha, Mothivel Mummudi, Rahul Bharadwaj, Viraj Vilas Belekar, Shankar Subramaniam, Nandkishor K. Nere

(204A) INDUSTRIAL CRYSTALLIZATION – THE INTERSECTION OF SOLID STATE CHEMISTRY AND PROCESS ENGINEERING 135
Thomas Vetter

(204B) CRYSTALLIZATION OPTIMIZATION BY ENZYMATIC CONTROL OF A REACTIVE CRYSTALLIZATION 136
Matthew A. McDonald, Grant D. Marshall, Hossein Salami, Andreas Bommarius, Martha A. Grover, Ronald Rousseau

(204C) CONTINUUM OF SOLUTE DIMERS DRIVES DISTINCT GROWTH MODES ON ANISOTROPIC FACES OF ETIOPORPHYRIN 137
Peter G. Vekilov, Rajshree Chakrabarti, Lakshmanji Verma, Monika Warzecha, Viktor Hadjiev, Jeremy C. Palmer

(249A) PHYSICOCHEMICAL MODELING OF DISSOLUTION AND STABILITY OF POORLY SOLUBLE AND UNSTABLE DRUG LOADED MULTILAYER SOLUBLE POLYMERIC FILMS 138
M. Arif Khan, Andie L. Macmillan, Aktham Aburub, Karthik Vaideeswaran, Sarah Clark, Mohamed Elsayed, Shekhar K. Viswanath, Patrick Marsac, Thomas D. Dziubla

(249C) A SEMI-THEORETICAL MODEL FOR SIMULATING THE TEMPORAL EVOLUTION OF MOISTURE-TEMPERATURE DURING INDUSTRIAL FLUIDIZED BED GRANULATION 139
Hossein Amini, Xiaorong He, Yni-Chao Tseng, Gulsad Kucuk, Robert Schwabe, Leon Schultz, Martin Maus, Daniela Schröder, Pavol Rajniak, Ecevit Bilgili

(249D) MODELLING AMORPHOUS SOLID DISPERSIONS: THE ROLE OF SUPERSATURATION, NUCLEATION AND CRYSTAL GROWTH 140
Patrícia D. Nunes, Inês Nande, João F. Pinto, Mafalda Paiva, João Henriques

(249F) DETERMINATION OF DRYING STRESSES AND FRACTURE ENERGY IN TABLET COATING FILMS 142
Bhawana S Tomar, Mahesh S. Tirumkudulu, Weili Yu, Bruce C. Macdonald, Alfred Berchielli, Pankaj Doshi

(249G) HEAT AND MASS TRANSFER IN A PHARMACEUTICAL COATING PROCESS 143
Stefan Madlmeir, Thomas Forgber, Martina Trogrlic, Adrian Kape, Alan Carmody, Avik Sarkar, Lourdes Contreras, Johannes G. Khinast, Dalibor Jajcevic

(291A) (INVITED TALK) EXPLOITING BARRIERS OF THE BODY FOR TARGETED NANOMEDICINE 144
Eun Ji Chung

(291B) SURFACTANT EFFECTS ON NANOTHERAPEUTIC FATE WITHIN THE BRAIN 145
Andrea Joseph, Georges Motchoffo Simo, Tora Gao, Norah Sulaiman Alhindi, Elizabeth Nance

(291C) MICRORHEOLOGICAL CHARACTERIZATION OF COVALENT ADAPTABLE HYDROGEL DEGRADATION IN RESPONSE TO PH CHANGES THAT MIMIC THE PH IN THE GASTROINTESTINAL TRACT 147
Nan Wu, Kelly Schultz

(291D) DUAL RESPONSIVE BACKBONE SHATTERING ORGANELLE TARGETED NANO-PRODRUG TO MODULATE OXYGEN CONSUMPTION FOR EFFECTIVE CHEMOPHOTOTHERAPY 148
Vishnu Sunil, Anbu Mozhi Thamizhchelvan, Wang Chi-Hwa

(291F) PH-RESPONSIVE, CELL-PENETRATING, MAGNETITE-SILVER NANOPARTICLES FOR DELIVERY OF PLASMID-BASED GENE THERAPIES: PREPARATION, CHARACTERIZATION, AND IN VITRO EVALUATION 149
Carlos Manuel Ramirez Acosta, Claudia Castellanos, Javier F Cifuentes, Juan C Cruz, Luis H. Reyes

(291G) ENZYMATIC PROTECTION AND BIOCOMPATIBILITY SCREENING OF ENZYME-LOADED POLYMERIC NANOPARTICLES FOR NEUROTHERAPEUTIC APPLICATIONS. 152
Rick Liao, Elizabeth Nance

(303A) RATIONAL DESIGN OF MODIFIED-RELEASE DOSAGE FORMS 154
Yihong Qiu

(303B) STRATEGIES TO DEVELOP IV FORMULATIONS OF POOR WATER-SOLUBLE COMPOUNDS AND THEIR IN VITRO EVALUATION 155
Elisa A. Torrico Guzmán, Kirk Overhoff, Katie Sokolowsky, Setu Roday

(303C) INTEGRATED PRODUCT & PROCESS DEVELOPMENT 156
Xiaorong He

(303D) PREDICTIVE AND MODEL BASED PRODUCT & PROCESS DESIGN 157
Xiaorong He, Dabing Chen

(303E) RATIONAL DESIGN OF AMORPHOUS SOLID DISPERSION 158
Yi Gao

(318A) STOCHASTIC SIMULATIONS OF MOTION OF SOLUTE MOLECULES DURING THE PROCESS OF CRYSTALLIZATION YIELDS THE RATE CONSTANTS FOR THE PROCESSES OF CRYSTALLIZATION 159
Anish V. Dighe, Meenesh R. Singh

(318B) IRREVERSIBLE INHIBITION OF BARITE CRYSTALLIZATION: A UNIQUE MECHANISM FOR TREATING SCALE FORMATION 160
Ricardo D. Sosa, Xi Geng, Michael A. Reynolds, Jacinta C. Conrad, Jeffrey D. Rimer

(318C) MULTI-STEP CRYSTALLIZATION OF SELF-ORGANIZED SPIRAL EUTECTICS 161
Saman Moniri, Ashwin J. Shahani

(318D) MODELING IMPURITY-MEDIATED CRYSTAL GROWTH 162
Tobias Mazal

(318E) MONTE CARLO SIMULATIONS OF HYDROTHERMAL METAL ORGANIC FRAMEWORK (MOF) CRYSTAL GROWTH 163
Andrew R. Garcia, Janani Sampath, Sergey Vasenkov, Kirk J. Ziegler

(318F) THE ROLE OF WATER IN THE CRYSTALLISATION AND NUCLEATION OF GLYCINE HOMOPEPTIDES 164
Mingxia Guo, Ian Rosbottom, Wenqian Chen, Jerry Y. Y. Heng

(322A) DATA AND INFORMATION MANAGEMENT VISION AND STRATEGY TOWARDS PHARMA 4.0 166
Kushal Sinha, Alex Ruggles, Eric Murphy, Moiz Diwan, Andreas Gryczke, Samrat Mukherjee

(322B) DATA DRIVEN PERFORMANCE MEASUREMENT & MANAGEMENT IN THE EXTERNAL API NETWORK 167
Daniel Perkins

(322C) REAL TIME DYNAMIC OPTIMISATION FOR ENHANCED PRODUCT QUALITY IN BIOPROCESSES 168
Maria M. Papathanasiou, Pavlos Kotidis, Cleo Kontoravdi, Iosif Pappas, Iosif Pappas, Styliani Avraamidou, Efstratios N. Pistikopoulos

(322D) CONDITION MONITORING OF PHARMACEUTICAL PROCESSES UNDER A PROBABILISTIC FRAMEWORK 170
Rexonni Lagare, Sudarshan Ganesh, Zoltan K. Nagy, Gintaras V. Reklaitis

(322E) PHARMA 4.0: ADVANCED CONTROL AND CYBER-PHYSICAL SECURITY OF CONTINUOUS PHARMACEUTICAL MANUFACTURING PILOT-PLANT 171
Ravendra Singh, Rohit Ramachandran, Fernando Muzzio

(322F) WEB DEPLOYMENT OF MECHANISTIC MODELS FOR PHARMACEUTICAL APPLICATIONS: CASE STUDIES IN SPRAY DRYING AND DISTILLATION 173
Dana Barrasso, Kerri Fischer, Christopher Burcham, Sean Bermingham

(327A) KINETICS OF FRIEDEL CRAFTS ACYLATION OF 2-METHYLFURAN WITH OCTANOIC ANHYDRIDE AND AL-MCM-41 174
Ankita Naik, Kristeen Joseph, M. Alexander Ardagh, Manish Shetty, Paul Dauenhauer

(327B) ADVANCES IN CATALYTIC HYDROPROCESSING FOR THE PRODUCTION OF HIGH QUALITY LUBE BASE STOCKS 175
Jason McMullan

(327C) CATALYTIC EVALUATION OF SULFATED METAL OXIDE CATALYSTS FOR THE SOLVENT-FREE TERT-BUTYLATION OF PHENOL 176
Adam Zuber, Aditya Khandare, Shreyas Acharya, Justin Marlowe, George Tsilomelekis

(327D) ARTIFICIAL NEURAL NETWORK AIDED SCALE-UP OF A MULTI-TUBULAR REACTOR FOR THE OXIDATIVE DEHYDROGENATION OF BUTYLENE TO BUTADIENE 177
Dela Quarme Gbadago, Sungwon Hwang

(327E) A PARADIGM SHIFT IN CATALYST STABILITY FOR TOLUENE ALKYLATION WITH METHANOL 178
Deependra Parmar, Seunghyeok Cha, Taha Salavati-Fard, Lars C. Grabow, Jeffrey D. Rimer

(327F) ACTIVITY ENHANCEMENT AND KINETICS OF NH3 OXIDATION ON PT/AL2O3:RATE ENHANCEMENT BY BALL MILLING AND STRUCTURE SENSTIVITY EFFECTS 179
Ghosh Rajat, Pritpal Singh Dhillon, Michael P. Harold, Di Wang

(327G) NOVEL MULTIFUNCTIONAL ORGANOCATALYSTS AND CONTINUOUS FLOW PLATFORM FOR SUSTAINABLE CHEMICAL TRANSFORMATION 181
Abdo-Alslam Alwakwak, Fateme Rezaei, Ali Rownaghi

(334A) SOLVENT -SOLUTE INTERACTION AND VISCOSITY EFFECTS ON THERMODYNAMICS AND KINETICS OF CRYSTALLIZATION 182
Rajshree Chakrabarti, Lakshmanji Verma, Monika Warzecha, Jeremy C. Palmer, Peter G. Vekilov

(334B) CHARACTERIZING AND MODELING PHARMACEUTICAL TWIN SCREW FEEDER MASS FLOW RATES USING STATISTICAL TIME SERIES ANALYSIS 183
Brad Johnson, Salvador García-Muñoz, Maitraye Sen, Joshua Hanson, David Slade, Nick Sahinidis

(334C) UNDERSTANDING SECOND-SHELL CATALYST INTERACTIONS FOR ENHANCING THE PERFORMANCE OF ARYLBORANE-CATALYZED EPOXIDE RING-OPENING 184
Mihir Bhagat, Charmaine Bennett, Ying Yu, Arjun Raghuraman, Matthew Belowich, Sonbinh Nguyen, Linda Broadbelt, Justin Notestein

(334D) MOLECULAR SIMULATIONS TO CHARACTERIZE THE ASSEMBLY AND TRANSPORT OF BIOMOLECULES IN SOLUTION AND AT LIPID INTERFACES 187
Samarthaben J. Patel

(334E) ENABLING SELECTIVE CONVERSION OF BIOMASS DERIVED OXYGENATES TO C4-C5 DIENES 188
Gaurav Kumar, Omar Abdelrahman, Michael Tsapatsis, Paul Dauenhauer

(334F) GAS-LIQUID SEPARATORS: CHALLENGES AND UNCERTAINTIES 191
Michael Miranda

(334H) RHEOLOGICAL CHARACTERIZATION OF CELLULOSE NANOMATERIALS FOR QUALITY CONTROL AND PROCESSING 193
Jianshan Liao

(334I) DATA-DRIVEN PARAMETER ESTIMATION OF HYBRID MODELS 194
William Bradley, Fani Boukouvala

(334J) PROCESS MODELING AND OPTIMIZATION FOR MONOCLONAL ANTIBODY PRODUCTION 195
Ou Yang

(334K) SINGLE ATOM CATALYST FOR OXIDATION - UNDERSTANDING FUNDAMENTALS OF SYNTHESIS AND REACTIVITY 196
Shyam Deo, Linxi Wang, Ahana Mukhopadhyay, Robert Rioux, Michael Janik

(334L) ELUCIDATING STRUCTURE-FUNCTION RELATIONSHIPS OF METAL OXIDES AND SULFIDES FOR RATIONAL CATALYST DESIGN 197
Emily Cheng, Justin Notestein

(334M) MUCIN GLYCOPOLYMERS PREVENT MICROBIAL VIRULENCE WHILE PREVENTING THE DEVELOPMENT OF ANTIBIOTIC RESISTANCE 198
Caroline Werlang

(334O) OPERANDO STUDY AND ANALYSIS OF CRYSTALLIZATION FOR HEALTH CARE AND ENERGY APPLICATIONS 199
Paria Coliaie

(334P) DEVICE-LEVEL ENGINEERING OF ELECTROCATALYTIC REACTORS UNDER PRACTICAL OPERATING CONDITIONS TO ENABLE SUSTAINABLE SMALL-MOLECULE TRANSFORMATIONS 201
Kindle Williams

(334Q) MODELING AND OPTIMIZATION APPROACHES FOR PROCESS DEVELOPMENT 203
Caroline Nielsen

(334R) CHEMICAL RESEARCH AND DEVELOPMENT: PROCESS MODELING AND MACHINE LEARNING 204
Shiyan Wang

(334S) MOLECULAR MECHANISMS OF THE AAA+ PROTEASE CLPXP CHARACTERIZED USING SINGLE-MOLECULE FLUORESCENCE QUENCHING 205
Harris W. Manning, Benjamin M. Stinson, Tristan A. Bell, Tania A. Baker, Robert T. Sauer, Matthew J. Lang

(334T) PAVING THE WAY FOR CYBERNETIC MODELING OF BIOLOGICAL PROCESSES IN MAMMALIAN SYSTEMS 206
Lina Aboulmouna

(334U) LIGNIN DEPOLYMERIZATION AND ESTERIFICATION BY CARBOXYLIC ACIDS TO PRODUCE BIODIESEL 208
Parinaz Hafezisefat, Marge Rover, Santanu Bakhshi, Robert C. Brown

(334V) AUTOMATED OPTIMIZATION AND CONTROL OF MODULAR CHEMICAL SYSTEMS 210
Anastasia Nikolakopoulou, Richard Braatz

(334Y) CROSS-LINKED NONWOVEN FIBERS BY ROOM-TEMPERATURE CURE BLOWING AND IN SITU PHOTOPOLYMERIZATION 212
Aditya Banerji, Mahesh K. Mahanthappa, Christopher J. Ellison

(334Z) MAGNETIC CHARACTERIZATION AND QUANTIFICATION OF MAGNETIC NANOPARTICLE ENDOCYTOSIS MECHANISMS BY PARTICLE TRACKING VELOCIMETRY 213
Abhinav Sannidhi, Paul W. Todd, Thomas R. Hanley

(334W) DESIGN OF A HIGH-PRESSURE CATALYTIC PHOTORREACTOR TO OPERATE AT SUPERCRITICAL CO2 CONDITIONS. 214
Estefany Pajaro, Victor Baldovino-Medrano, Fernando Martínez-Ortega

(334AA) TUNING CATALYSTS FOR EFFICIENT CHEMICAL AND ENERGY TRANSFORMATION 216
Joshua Gopeesingh, Omar Abdelrahman, Jesse Bond

(334AB) MOTILITY AND TRANSPORT BEHAVIOR OF SOFT COLLOIDS 218
Ambika Somasundar

(334AC) BULK PROPERTY PREDICTION OF PHARMACEUTICAL POWDERS VIA BOND NUMBER ESTIMATION 220
Kuriakose Kunnath

(334AD) INFLUENCE OF NATURAL CATALYTIC ENVIRONMENTS ON THE ACTIVATION OF CELLULOSE VIA FAST PYROLYSIS 221
Vineet Maliekkal, Paul Dauenhauer, Matthew Neurock

(334AE) AMPING UP THE ORGANIC SYNTHESIS IN THE PHARMACEUTICAL INDUSTRY: EXPLORE THE POTENTIAL OF STRUCTURE-PROPERTY RELATIONSHIP. 224
Sameh Elsaidi

(334AF) RADIOLYTIC DEGRADATION OF ORGANIC SOLVENTS IN THE ACTINIDE LANTHANIDE SEPARATION PROCESS (ALSEP) FOR NUCLEAR WASTE TREATMENT 227
Christian G. Bustillos, Randy O. Ngelale, Mikael Nilsson

(334AH) A SYSTEMATIC MULTISCALE COMPUTATIONAL APPROACH FOR ENGINEERING NOVEL ADAPTIVE MATERIALS FOR BIOLOGICAL APPLICATIONS 228
Sriramvignesh Mani

(334AI) HYBRID MONITORING METHODS FOR DETECTION, DIAGNOSIS, AND CLASSIFICATION 230
M. Ziyan Sheriff

(334AK) SMART BIOMANUFACTURING: A PROCESS SYSTEMS ENGINEERING AND ARTIFICIAL INTELLIGENCE APPROACH 231
Yu Luo

(334AP) DYNAMIC PROCESS MODELING AND SYSTEM ANALYSES FOR CONTINUOUS PHARMACEUTICAL MANUFACTURING 232
Pooja Bhalode

(334AQ) SELF-ASSEMBLY FOR COLLOIDAL CRYSTALLIZATION AND BIOMIMETIC STRUCTURAL COLOR 233
Tianyu Liu, Bryan Vansaders, Sharon C. Glotzer, Michael J. Solomon

(334AR) QUANTIFYING THE EFFECTS OF HYDROSTATIC PRESSURE ON THE BARRIER FUNCTION OF MAMMARY EPITHELIAL CELLS 234
Lena A. Barrett, Celeste M. Nelson

(334AS) COMPUTATIONAL APPROACHES FUELING CLEAN MATERIALS CHEMISTRY - FROM FIRST PRINCIPLES CALCULATIONS TO DATA DRIVEN DISCOVERY AND MATERIALS DESIGN 235
Luke Johnson, Aleksandra Vojvodic

(334AT) COMBINING MOLECULAR SIMULATIONS AND MACHINE LEARNING FOR NANOMATERIAL AND REACTION DESIGN 237
Alex K. Chew

(334AU) MATERIALS SIMULATION FOR MANUFACTURING 238
Andrew P. Santos

(334AV) DEVELOPING EXPERIMENTAL AND COMPUTATIONAL SINGLE-CELL TECHNIQUES TO UNCOVER BIOLOGICAL INSIGHTS 239
Chatarin Wangsanuwat

(334AX) NUCLEATION AND EARLY STAGE GROWTH OF LITHIUM ELECTRODEPOSITS 240
Prayag Biswal

(334AY) JET-MIXING REACTOR FOR NANOMATERIAL SYNTHESIS AND SCALE-UP CONSIDERATIONS 250
Pinaki Ranadive, Aamena Parulkar, Alexander Spanos, Nicholas Brunelli

(334AZ) LAYER-BY-LAYER IMPLANT COATINGS FOR CRANIOMAXILLOFACIAL BONE REGENERATION 251
Maylin Howard, John Martin, Sheryl Wang, Adam G. Berger, Paula T. Hammond

(334BA) FABRICATION OF MULTI-COMPARTMENT MICRO-CAPSULES FOR DRUG DELIVERY OF MULTIPLE DRUGS 252
Pompon Mputu Udipabu, Panagiotis Dimitrakopoulos

(334BC) TUNING AROMATICS HYDROGENATION BY SYSTEMATIC VARIATION OF THEIR SURFACE POLARITY IN SBA-15-TYPE PERIODIC MESOPOROUS CATALYSTS 253
Hyunjin Moon, Songi Han, Susannah L. Scott

(334BD) LEVERAGING MOLECULAR SIMULATIONS AND DATA SCIENCE TO IMPROVE THE DESIGN OF BIOMATERIALS WITH FINE-TUNED HYDROPHOBICITY 254
Bradley C. Dallin

(334BE) UNDERSTANDING THE FUNDAMENTALS OF ZEOLITE CRYSTALLIZATION FOR THE OPTIMIZATION OF THEIR PHYSICOCHEMICAL PROPERTIES 255
Rishabh Jain, Madhuresh K. Choudhary, Jeffrey D. Rimer

(334BF) EX VIVO MINIGUT MUCOSAL MODEL FOR DEVELOPMENT OF NEW ORAL VACCINE 263
Yijun Qi

(334BH) SUPERHYDROPHOBIC COATINGS SYNTHESIZED BY HIGH-TEMPERATURE FREE-RADICAL POLYMERIZATION 264
Hossein Riazi, Michael C. Grady, Masoud Soroush

(334BI) DETERMINING THE FOLDING LANDSCAPE OF A-SYNUCLEIN (35-97) USING REPLICA EXCHANGE MOLECULAR DYNAMICS 265
Karnesh Jain, Othman Ghribi, Jerome Delhommelle

(334BJ) HIGH-THROUGHPUT AND DATA-DRIVEN STRATEGIES FOR THE DESIGN OF DEEP EUTECTIC SOLVENTS 266
Jaime Rodriguez Jr., Shrilakshmi Bonageri, Maria Politi, Sage Scheiwiller, Lilo Pozzo

(334BK) REDESIGN AND DEVELOP EFFECTIVE LIPOSOMAL FORMULATION FOR PRODRUG DELIVERY THROUGH SYNCHROTRON X-RAY STUDIES OF MOLECULAR INTERACTIONS 267
Tiep Hoang Pham, Paola Leon Plata, Pin Zhang, Anand Vellara, Chang Liu, Wei Bu, Binhua Lin, Ying Liu

(334BL) SYSTEMS ENGINEERING APPROACHES AND CAREER ADVANCEMENT IN BIOPROCESS DEVELOPMENT AND BIOMANUFACTURING 268
Remil Aguda

(334BO) CADHERIN BINDING INTERACTIONS IN LIPID BILAYERS AT THE SINGLE-MOLECULE LEVEL 269
Connor J. Thompson, Zhaoqian Su, Vinh H. Vu, Yinghao Wu, Deborah E. Leckband, Daniel K. Schwartz

(334BQ) DEVELOPMENT OF PROCESS MODELS AND CONTROL STRATEGIES TO SUPPORT THE CONTINUOUS MANUFACTURING OF PHARMACEUTICALS 270
Andrew J. Maloney

(334BR) POLYELECTROLYTE SURFACE DIFFUSION IN A NANO-SLIT GEOMETRY 271
Gregory T. Morrin, Daniel Kienle, James S. Weltz, Jeremiah Traeger, Daniel K. Schwartz

(334BV) MULTI-SCALE MODELING OF HETEROGENEOUS CATALYSTS USING FIRST-PRINCIPLES AND DEEP LEARNING 272
Pushkar Ghanekar, Jeffrey Greeley

(334BW) DYNAMICS OF VESICLES IN STRONG FLOWS USING A STOKES TRAP 273
Dinesh Kumar, Charles M. Schroeder

(334BX) CATALYST DESIGN FOR OXYGEN REDUCTION REACTION USING MOLECULAR MODELING 275
Jiayi Xu, Bin Liu

(334BY) PHOTOTHERMAL AND IMMUNOMODULATORY NANOMATERIALS FOR RAPID TISSUE REPAIR AND WOUND HEALING 276
Deepanjan Ghosh, Russell Urie, Jordan Yaron, Suneel Kumar, David Dicaudo, Jacquelyn Kilbourne, François Berthiaume, Kaushal Rege

(334CB) SESEED 277
Tianyu Yan, Kristen Fichthorn

(334CC) FUNCTIONAL PARTICLES FOR CONTROLLED RELEASE AND STEM CELL ENGINEERING 278
Yu-Tong Hong, Hyunjoon Kong

(334CD) FLOW BEHAVIOR IN COMPLEX FLUIDS AND PARTICULATES SYSTEMS 279
Ehsan Akbari Fakhrabadi, Matthew Liberatore

(366A) CONTINUOUS MIXING TECHNOLOGY: THE IMPACT OF BLEND PROPERTIES AND PROCESS PARAMETERS ON MIXING PERFORMANCE 280
Kai Lee, James Kimber, Pankaj Doshi, Jenna Brandon, David Wilsdon, Daniel O. Blackwood, Ashwinkumar Jain

(366B) INVESTIGATION OF CRITICAL PROCESS PARAMETERS OF SCREW ELEMENTS OF A TWIN SCREW EXTRUDER BY AN IMPROVED SIMULATION METHOD 281
Vanessa Duephans, Isabella Putz, Julius Arntzen, Markus Thommes

(366C) DRUG PRODUCT CONTINUOUS POWDER FEEDING FOR LOW-PERCENTAGE FORMULATION COMPONENTS 285
Carl Hartmann III, Jon Hilden

(366D) STOCHASTIC ANALYSIS AND MODELING OF PHARMACEUTICAL SCREW FEEDER MASS FLOW RATES 286
Brad Johnson, Salvador García-Muñoz, Maitraye Sen, Joshua Hanson, David Slade, Nick Sahinidis

(366E) TABLET PAN COATING PERFORMANCE DURING SCALE UP AND BETWEEN BATCH AND CONTINUOUS PROCESSES 287
Michael Choi, Stuart C. Porter, Betsy Macht

(366F) DESIGN METHOD FOR INJECTABLE MANUFACTURING PROCESSES CONSIDERING CONTINUOUS AND BATCH COMPOUNDING 291
Masahiro Yamada, Yasuyuki Yoshioka, Hayao Nakanishi, Hirokazu Sugiyama

(366G) EFFICIENT PRODUCTION OF MULTI-COMPONENT AMORPHOUS SOLID DISPERSIONS BY SPRAY DRYING 292
Vojtech Klimsa, Gabriela Ruphuy Chan, František Štepánek

(367A) LEVERAGING OF MODELING AND AUTOMATION IN DEVELOPMENT OF A CONTROL STRATEGY FOR A CGMP BUCHWALD-HARTWIG CROSS-COUPLING 294
Brandon Reizman, Mark S. Kerr, Mark A. Pietz, Daniel J. Jarmer

(367C) MOVING TO A CHROMATOGRAPHY-FREE SYNTHESIS: ESTABLISHING PRODUCT-RELATED AND POLYMERIC IMPURITY CONTROL STRATEGY DURING PROCESS DEVELOPMENT 295
Geoffrey E. Purdum, Changxia Yuan, Guanghui Zhu, Benjamin M. Cohen, Miao Yu, Qinggang Wang, Amanda Rogers

(367D) PROCESS CONTROL STRATEGY DEVELOPMENT FOR AN INTEGRATED CONTINUOUS PLATFORM FOR MONOCLONAL ANTIBODY MANUFACTURING 296
Andrew J. Maloney, Elizabeth M. Cummings Bende, Dragana Bozinovski, Amos E. Lu, Jose Sangerman, Moo Sun Hong, Anastasia Artamonova, Rui Wen Ou, Weike Sun, Nili Persits, Rajeev J Ram, Jacqueline Wolfrum, Paul W. Barone, Stacy Springs, Anthony J Sinskey, Richard Braatz

(367E) MODULATING ANTIBODY GLYCOSYLATION USING TRADITIONAL AND PAT METHODS 297
Madiha Khurshid, Prince Bhebe, Li Zhang, Tony Wang, Richard Wu, Jack Chung-Jr Huang

(367F) MODEL PREDICTIVE IN VITRO DISSOLUTION TESTING FOR REAL-TIME RELEASE IN PHARMACEUTICAL END-TO-END INTEGRATED CONTINUOUS MANUFACTURING: AN EQUIVALENCE STUDY 298
Qinglin Su, Paul Hermant, Federica Casati, Bayan Takizawa, Wei Wu, Atul Dubey, Lucy L. Botros, Stephen Born, Bayan Takizawa, Salvatore Mascia

(402A) OR-GYM: A REINFORCEMENT LEARNING LIBRARY FOR OPERATIONS RESEARCH PROBLEMS 299
Christian D. Hubbs, Owais Sarwar, Hector Perez, Ignacio Grossmann, Nick Sahinidis

(402B) UPDATED LEAPS2 FOR SURROGATE RECOMMENDATION 301
Maaz Ahmad, Iftekhar A. Karimi

(402C) WHEN DEEP LEARNING MEETS SPARSE MODEL IDENTIFICATION: ONLINE ADAPTIVE SPARSE IDENTIFICATION OF SYSTEMS (OASIS) 303
Bhavana Bhadriraju, Faisal Khan, Joseph Kwon

(339G) APPLYING REINFORCEMENT LEARNING TO CONTROL BATCH PROCESSES 305
Wenbo Zhu, Zhenyu Wang, Ivan Castillo, Ricardo Rendall, Leo H. Chiang, Jose A. Romagnoli

(402F) INVERSE LINEAR OPTIMIZATION: CONCEPTUAL FRAMEWORK, TWO-PHASE ALGORITHM, AND ADAPTIVE SAMPLING 306
Rishabh Gupta, Qi Zhang

(402G) CONNECTIONS BETWEEN RESIDUAL NETWORKS AND EXPLICIT NUMERICAL INTEGRATORS, WITH APPLICATIONS TO IDENTIFICATION OF NONINVERTIBLE DYNAMICAL SYSTEMS 307
Tianqi Cui, Tom S. Bertalan, Yorgos M. Psarellis, Ioannis G. Kevrekidis

(409A) CRYSTAL MORPHOLOGY MODIFICATION IN A NUCLEATION DOMINANT CRYSTALLIZATION: A TOOL FROM TRADITIONAL CHEMISTRY 308
Tharanga Wijethunga, Michael Waldo, Sonja Sharpe

(409B) PROCESS DEVELOPMENT TO AVOID META-STABLE LIQUID-LIQUID PHASE SEPARATION DURING CRYSTALLIZATION OF API 309
Patricia Cho, John Ryan Coombs, Victoria Mbachu, Eric M. Saurer

(409C) DECOUPLING NUCLEATION AND GROWTH FOR CONTINUOUS CRYSTALLIZATION IN MICROFLUIDICS 310
Naghmeh Fatemi, Tom Van Gerven, Simon Kuhn

(409D) EFFECTS OF VOLUME REDUCTION ON GLYCINE NUCLEATION 313
Isaac Jerome C. Dela Cruz, Gerard Capellades, Jem Valerie D. Perez, Bryan G. Alamani, Allan S. Myerson

(409E) FORMATION OF SOLID SOLUTIONS AS A KEY IMPURITY RETENTION MECHANISM IN SOLUTION CRYSTALLIZATION IN THE PRESENCE OF STRUCTURALLY SIMILAR IMPURITIES 314
Fredrik Nordstrom

(409G) DESIGN OF A CONTINUOUS CRYSTALLIZATION FOR A DRUG SUBSTANCE USING A DYNAMICALLY MIXED REACTOR 315
Srividya Ramakrishnan, Ramana Susarla, Suhas Jawlekar, Ravi Kumar Gorle, Jagannadharao Velaga, Manohar Venkatachalam, Yakoob Sardar Mohammed

(469A) DIGITAL DESIGN VIA MECHANISTIC MODELLING OF A CONTINUOUS CRYSTALLIZATION PROCESS WITH PERIODIC WET MILLING 317
Niall Mitchell, Christopher Burcham, Filipe Calado, Simao Pereira, Steven Myers

(469B) CONTINUOUS GENERATION OF GLYCINE CRYSTALS WITH PURE FORM AND NARROW SIZE DISTRIBUTION IN SLUG FLOW 319
Mingyao Mou, Mo Jiang

(469C) CONTINUOUS CRYSTALLISATION OSCILLATORY FLOW PLATFORM FOR PROTEIN PURIFICATION AND BIO-SEPARATION 320
Ian Rosbottom, Huaiyu Yang, Wenqian Chen, Xiaoyu Li, Jerry Y. Y. Heng

(469D) TEMPERATURE CONTROL APPROACHES FOR COOLING CRYSTALLIZATION USING A CONTINUOUS MICROFLUIDIC MIXER 321
Paria Coliaie, Manish S. Kelkar, Nandkishor K. Nere, Meenesh R. Singh

(469E) "FILLING THE GAP" IN CONTINUOUS PHARMACEUTICAL TECHNOLOGIES: A NOVEL CONTINUOUS FILTRATION DRYING SYSTEM 322
Simon Coleman

(469F) INTEGRATED CONTINUOUS CRYSTALLIZATION AND SPRAY DRYING OF INSULIN 323
Ruipeng Chen, Jingwen Weng, Shing Fung Chow, Richard Lakerveld

(498A) REAL-TIME MODELING OF CRYSTAL NUCLEATION, GROWTH, AND BREAKAGE IN AN AGITATED BIOREACTOR 325
John A. Thomas

(498B) CFD PREDICTIONS OF POWER DISSIPATION AND RELATED HYDRODYNAMIC EFFECTS FOR A RETREAT-BLADED IMPELLER IN A STIRRED VESSEL UNDER DIFFERENT BAFFLING CONFIGURATION 326
Chadakarn Sirasitthichoke, Piero M. Armenante

(498C) PREDICTING FILTRATION PERFORMANCE BY STUDYING THE PORE NETWORK OF THE FILTER CAKE: THE CASE OF NEEDLE-LIKE CRYSTALS 327
Giulio Perini, Carlos Avendano, Arash Rabbani, Masoud Babaei, William Hicks, Anna Parsons, Thomas Vetter

(498D) MODELING CHO CELL GLYCOSYLATION PROCESS USING DYNAMIC KRIGING 330
Ou Yang, Marianthi Ierapetritou

(0) SPONSORED TECHNOLOGY WORKSHOP - MICROSTRUCTURE CHARACTERIZATION FOR DRUG DEVELOPMENT: NOVEL AI ANALYTICS AND RELEASE PREDICTION - DIGIM SOLUTION 332
Shawn Zhang, Joshua Lomeo, Lisa Ma

(403A) HIGH-THROUGHPUT APPROACH TO STUDYING ADSORPTION ON ZEOLITE SURFACES 333
Hassan Aljama, Martin Head-Gordon, Alexis T. Bell

(403B) MACHINE LEARNING-GUIDED FLOW SYNTHESIS OF INORGANIC METAL HALIDE PEROVSKITE QUANTUM DOTS 334
Robert Epps, Amanda A. Volk, Kameel Abdel-Latif, Kristopher G. Reyes, Milad Abolhasani

(403C) HIGH-THROUGHPUT AND DATA-DRIVEN STRATEGIES FOR THE DESIGN OF DEEP EUTECTIC SOLVENT ELECTROLYTES 336
Jaime Rodriguez Jr., Shrilakshmi Bonageri, Maria Politi, Sage Scheiwiller, Lilo Pozzo

(403D) DEVELOPMENT OF A HYBRID MODEL TO DESCRIBE THE DYNAMICS OF A PARTIALLY KNOWN SYSTEM 337
Dongheon Lee, Arul Jayaraman, Joseph Kwon

(403E) FFIBER: FRAMEWORK FOR FLUORESCENT NEUROIMAGING BASED EXPERIMENTAL ROUTINES 339
Hawley Helmbrecht, Elizabeth Nance

(403F) DATA-DRIVEN DISCOVERY OF POLYMERIC VEHICLES FOR GENE EDITING:SERENDIPITY-INSPIRED DESIGN DIRECTIONS. 341
Ramya Kumar, Ngoc Le, Zhe Tan, Theresa M. Reineke

(403G) ARTIFICIAL INTELLIGENCE-ASSISTED FULL-MAP UNDERSTANDING OF STRAIN SENSING DEVICES AND DESIGNATED PREDICTIONS FOR SOFT ROBOTIC SYSTEMS 342
Jiali Li, Haitao Yang, Kaizhuo Lim, Chuanji Pan, Xiaoli Liu, Xiaonan Wang, Po-Yen Chen

(407A) CALIBRATION-FREE APPROACH FOR END-POINT DETERMINATION OF A CHALLENGING POWDER BLENDING PROCESS: SCIENTIFIC CASE STUDY 343
Huiquan Wu, Koushik Sowrirajan, Alpa Patel

(407B) PREDICTION OF ENTIRE TABLET COMPOSITIONS FROM PURE POWDER SPECTRA USING A NONLINEAR ITERATIVE OPTIMIZATION METHOD 344
Yukteshwar Baranwal, Andrés D. Román-Ospino, Douglas Hausner, Fernando Muzzio, Rohit Ramachandran

(407C) REAL-TIME MONITORING OF POWDER MASS FLOWRATES FOR MPC/PID CONTROL OF A CONTINUOUS DIRECT COMPACTION TABLET MANUFACTURING PROCESS 346
Yan-Shu Huang, Sergio Medina-González, Yasasvi Bommireddy, Sunidhi Bachawala, Marcial Gonzalez, Zoltan K. Nagy, Gintaras V. Reklaitis

(407D) REGULATORY CASE STUDIES ON IN-PROCESS CONTROLS AND SCALE UP OF PHARMACEUTICAL POWDER BLENDING 348
Maotang Zhou

(407E) USE OF PROCESS DATA FOR GRANULE MOISTURE ESTIMATION IN THE CONSIGMA 25TM SEMI-CONTINUOUS FLUID BED DRYER 349
Jakob Rehrl, Stephan Sacher, Wen-Kai Hsiao, Martin Horn, Johannes G. Khinast

(407F) MONITORING LOSS-IN-WEIGHT PHARMACEUTICAL FEEDERS THROUGH STATE ESTIMATION 350
Francesco Destro, Salvador García-Muñoz, Pierantonio Facco, Fabrizio Bezzo, Massimiliano Barolo

(426A) TOWARDS AN UNDERSTANDING OF FLUID AND PARTICLE HISTORY IN A TURBULENT MIXING TANK USING LARGE EDDY SIMULATION: EXAMINING EFFECT OF DIFFERENT IMPELLER TYPES. 352
Matthew H. Flamm, Katherine Raudenbush

(426B) BUILDING THE PREDICTIVE PROCESS ENGINEERING TOOLBOX TO DEFINE MIXING PARAMETERS FOR EARLY-STAGE DRUG PRODUCT MANUFACTURING OF LARGE MOLECULES 353
Robert Kuo, Zbynek Kalal, Julia Eckhoff, Smeet Deshmukh, Matthew H. Flamm, Sachin Mittal, Adam Procopio

(426C) HOLD TANK MIXING PROFILE OPTIMIZATION FOR POTENCY CONTROL IN FILLING A SHEAR THINNING SUSPENSION FORMULATION 354
Richard Shields

(426D) EFFICIENT PROCESS DEVELOPMENT IN THE MICROMIXING-MESOMIXING SPACE: THE BOURNE PROTOCOL AND BEYOND 355
Aaron Sarafinas

(426E) DEVELOPMENT OF A MIXING INDEX PROVIDES A MEANS TO SOLVE MANY INDUSTRIAL MIXING PROBLEMS 357
David S. Dickey

(426F) IMPLEMENTATION OF A VAPOR-LIQUID EQUILIBRIUM AND VAPOR PURGE MODEL FOR SCALE-UP OF A TRANSFER HYDROGENATION 358
Kira Ragazzo, Federico Lora Gonzalez, Zhongping Shi, Ronald Carrasquillo-Flores, Shawn Brueggemeier

(426G) COUPLING REACTION KINETICS, CALORIMETRY AND MIXING STUDIES FOR SUCCESSFUL SCALE-UP OF EXOTHERMIC REACTIONS 359
Filipe Ataíde, João Sardinha, Ricardo Mendonça, Elsa Gonçalves, Ana Monteiro, Susana Nascimento

(459A) MOLECULAR RECOGNITION AND IN VIVO DETECTION OF TEMOZOLOMIDE AND 5-AMINOIMIDAZOLE-4-CARBOXAMIDE FOR GLIOBLASTOMA USING FLUORESCENT NANOSENSORS 360
Manki Son, Freddy T. Nguyen, Punit Mehra, Michael A. Lee, Naveed Bakh, Michael Strano

(459B) AN ENZYMATIC ELECTROCHEMICAL BIOSENSOR FOR REAL-TIME DETECTION OF PHYSIOLOGICALLY RELEVANT NICOTINE CONCENTRATIONS 361
Uros Kuzmanovic, Mingfu Chen, Margarita Tararina, Nicolas S. Shu, Prerana Sensharma, Anant Gupta, Andy Fan, Catherine M. Klapperich, Karen Allen, Mark Grinstaff, James Galagan

(459C) SIMULATIONS GUIDE OPTIMIZATION OF ELECTROENZYMATIC BIOSENSORS FOR NEUROTRANSMITTERS AND ENABLE PROPER INTERPRETATION OF SENSOR RESPONSE IN VIVO 362
Mackenzie Clay, Harold G. Monbouquette

(459D) THEORETICAL AND EXPERIMENTAL STUDIES ON AN ELECTROCHEMICAL ENZYME IMMUNOSORBENT BIOSENSOR 363
Neda Rafat, Paul Satoh, Robert M. Worden

(459E) HTL DERIVED BIOCHAR AND GRAPHENE NANOPLATELETS FOR BIOSENSOR APPLICATIONS 364
Bharath Maddipudi, Vinod S. Amar, Hope Dosch, Anuradha Shende, Rajesh Shende

(459F) CONSTRUCTION OF A RED EMISSION BODIPY-BASED PROBE FOR TRACING LYSOSOMALVISCOSITY CHANGES IN CULTURE CELLS 365
Baoxing Shen

(459G) SUPER SENSITIVE CERIUM OXIDE-BASED COMPOSITE SENSOR FOR THE DETECTION OF HYDROXYL RADICALS 366
Surachet Duanghathaipornsuk, Cheng-Han Li, Joerg Jinschek, Dong-Shik Kim, Ana Alba-Rubio

(459H) HIGH THROUGHPUT ANTIBIOTIC SUSCEPTIBILITY TESTING WITH OPTICAL NANOSENSORS 367
Megan Jewell, Samuel C. Saccomano, Alexa A. David, J. Kirk Harris, Edith Zemanick, Kevin J. Cash

(478B) DATA-DRIVEN PROCESS DEVELOPMENT AND SCALE-UP OF AN ACETAL PROTECTION REACTION AND CRYSTALLIZATION 368
Karthik Narsimhan, Joshua Lee, Jonathan McMullen, Eric Sirota, Anne Mohan

(478C) MODEL DEVELOPMENT FOR ENZYMATIC REACTIVE CRYSTALLIZATION OF ß-LACTAM ANTIBIOTICS: A REACTION-DIFFUSION-CRYSTALLIZATION APPROACH. 369
Hossein Salami, Matthew A. McDonald, Patrick R. Harris, Colton Lagerman, Martha A. Grover, Ronald Rousseau, Andreas Bommarius

(478E) OPTIMIZATION OF PROCESS RISK, COST, AND ENVIRONMENTAL SUSTAINABILITY DURING LATE STAGE DRUG SUBSTANCE CONTROL STRATEGY DEVELOPMENT 371
Patrick Sipple, Matthew Hickey, Nathaniel D. Kopp, Jia Zang, Kenneth Fraunhoffer, Michaël Fenster, Jose E. Tabora

(478F) DIGITAL DESIGN OF AN INTENSIFIED FILTRATION-DRYING UNIT FOR CONTINUOUS MANUFACTURING 372
Francesco Destro, Inyoung Hur, Daniel M. Casas-Orozco, Su-Lin Lee, Mesfin Abdi, Xin Feng, Erin Wood, Massimiliano Barolo, Zoltan K. Nagy

(478G) INTEGRATED FILTRATION AND WASHING MODELLING TO PREDICT CONTINUOUS ISOLATION PERFORMANCE 374
Sara Ottoboni, Cameron Brown, Ekaterina Gramadnikova, Jan Sefcik, Chris Price

(522A) COMPARISON OF SURROGATE MODELING TECHNIQUES FOR DESIGN SPACE APPROXIMATION AND SURROGATE-BASED OPTIMIZATION: EFFECT OF SAMPLING TECHNIQUE AND SAMPLE SIZE 376
Bianca Williams, Selen Cremaschi

(522C) INVESTIGATING PERFORMANCE METRICS AND MACHINE LEARNING MODELS TOWARDS AUTOMATING THE DIAGNOSIS OF MEMBRANOUS NEPTHROPATHY 379
Antonios Armaou, Yao Xue

(522D) HIGH-THROUGHPUT ADSORBENT SCREENING FOR GAS SEPARATION VIA ARTIFICIAL NEURAL NETWORKS 381
Akhil Arora, M M Faruque Hasan

(522E) A HYBRID MODELLING FRAMEWORK FOR ON-LINE DATA-DRIVEN MODELLING OF PARTICLE PROCESSES 382
Rasmus Fjordbak Nielsen, Nima Nazemzadeh, Martin Peter Andersson, Krist V. Gernaey, Seyed Soheil Mansouri

(522F) INTEGRATION OF CONTROL WITH PROCESS OPERATION THROUGH REINFORCEMENT LEARNING 384
Panagiotis Petsagkourakis, Vasileios Charitopoulos, Federico Galvanin, E. Antonio Del Rio Chanona

(522G) CONTROL LYAPUNOV-BARRIER FUNCTION-BASED PREDICTIVE CONTROL OF NONLINEAR PROCESSES USING REAL-TIME MACHINE LEARNING MODELING 386
Zhe Wu, David Rincon, Panagiotis D. Christofides

(522H) CONSTRAINED REINFORCEMENT LEARNING FOR PROCESS OPTIMIZATION AND CONTROL 387
Panagiotis Petsagkourakis, Eric Bradford, Ilya Orson Sandoval Cardenas, Dongda Zhang, Federico Galvanin, E. Antonio Del Rio Chanona

(528A) FINITE-SIZE EFFECTS ON THE CHEMICAL POTENTIALS OF MOLECULAR SOLIDS 389
Eric Dybeck, Geoffrey Wood

(528B) COMPUTER-AIDED DESIGN OF SOLVENT BLENDS FOR THE CRYSTALLISATION OF IBUPROFEN – REGULATING CRYSTAL MORPHOLOGY BY THE SPIRAL GROWTH MODEL 390
Oliver L. Watson, Suela Jonuzaj, George Jackson, Amparo Galindo, Claire S. Adjiman

(528C) MACHINE LEARNING FOR THERMODYNAMIC PROPERTIES OF MATERIALS: TOWARDS THE PREDICTION OF SOLUBILITY 392
Florence H. Vermeire, William H. Green

(528D) APPROACHING COUPLED CLUSTER ACCURACY WITH A GENERAL-PURPOSE NEURAL NETWORK POTENTIAL 395
Olexandr Isayev

(528E) KINETIC MODELLING AND RISK ASSESSMENT OF AN EXOTHERMIC FED-BATCH REACTION USING CALORIMETRY 396
Han Xia, Brandon Reizman, Meera Mahadevan, Niall Mitchell, Sam Wilkinson, Xiaohui Liu

(528F) KINETIC MODELING OF A NON-HOMOGENEOUS METAL-CATALYZED REACTION SYSTEM FOR DRUG SYNTHESIS 397
Brandon Reizman, Meera Mahadevan, Niall Mitchell, Sam Wilkinson, Xiaohui Liu

(542A) CONTINUOUS MIXING TECHNOLOGY: DEVELOPMENT OF DIGITAL TWIN 398
Peter Toson, Pankaj Doshi, Eva Siegmann, Daniel O. Blackwood, Kai Lee, James Kimber, Jenna Brandon, David Wilsdon, Ashwinkumar Jain, Johannes G. Khinast, Dalibor Jajcevic

(542B) INTEGRATED CONTINUOUS API PURIFICATION AND DIRECT FORMULATION OF DRUG PRODUCTS VIA CONFINED SUSPENSION AGITATED BED CRYSTALLIZATION AND FLUIDIZED BED COATING 399
Michael Stocker, Matthew Harding, Anne-Marie Healy, Steven Ferguson

(542D) DEVELOPING HME-BASED DRUG PRODUCTS USING QBD AND EMERGING SCIENCE 401
Josip Matic, Hannes Bauer, Amrit Paudel, Johannes G. Khinast

(542E) FIRST-PRINCIPLES DYNAMIC SIMULATION OF AN INTEGRATED CONTINUOUS BIOMANUFACTURING PLATFORM 403
Moo Sun Hong, Amos E. Lu, Andrew J. Maloney, Elizabeth M. Cummings Bende, Dragana Bozinovski, Jose Sangerman, Anastasia Artamonova, Rui Wen Ou, Paul W. Barone, Jacqueline Wolfrum, Stacy Springs, Anthony J Sinskey, Richard Braatz

(542F) A COMPREHENSIVE INVESTIGATION OF PROCESS DESIGN AND PARAMETERS OF SPRAY DRIED PROTEINS 405
Daniela Fiedler, Sonja Hartl, Tanja Gerlza, Christina Trojacher, Andreas Kungl, Eva Roblegg, Johannes G. Khinast

(542G) H2O2 DECONTAMINATION FOR STERILE MANUFACTRUING: IMPACT OF SORPTION BY POLYMERS ON THE DURATION OF AERATION 406
Keisho Yabuta, Haruka Futamura, Koji Kawasaki, Hirokazu Sugiyama

(561A) SIMULATION OF DRYING OF PHARMACEUTICAL WET GRANULES IN A FLUIDIZED BED DRYER USING COUPLED CFD-DEM APPROACH 407
Syed Ahsan, Hossain Aziz, Bodhisattwa Chaudhuri, Yijie Gao

(561C) CONTINUUM SIMULATIONS OF HEAT AND MASS TRANSFER IN WET GRANULAR MIXTURES 408
Viraj Vilas Belekar, Nandkishor K. Nere, Kushal Sinha, Theodore J. Heindel, Shankar Subramaniam

(561D) NEW INSIGHTS INTO LOCAL RESIDENCE TIME DISTRIBUTIONS OF TWIN SCREW ELEMENTS USED IN HOT MELT EXTRUSION VIA SMOOTHED PARTICLE HYDRODYNAMICS SIMULATION 409
Hannes Bauer, Josip Matic, Johannes G. Khinast

(561E) TOWARDS A GENERIC MODEL FOR TWIN-SCREW WET GRANULATION: CALIBRATION AND VALIDATION OF AN IMPROVED PBM FOR MULTIPLE APIS. 410
Ana Alejandra Barrera Jiménez, Daan Van Hauwermeiren, Michiel Peeters, Ingmar Nopens, Thomas De Beer

(561F) DEVELOPING A DIGITAL TWIN FOR THE L.B. BOHLE TABLET COATING PROCESS 412
Alfred Berchielli, Ismail Anifowoshe, Sandra Conway, Pankaj Doshi, Naomi Hodgins, Reza Kamyar, Peiyuan Liu, Bruce C. Macdonald, Zilong Wang, Gopal Kasat, Mothivel Mummudi, Parth Sinha, Andreas Altmeyer

(561G) IMPLEMENTATION OF A DYNAMIC MECHANISTIC MODEL TOWARDS OPTIMAL DESIGN AND SCALE-UP OF AN ATYPICAL FREEZE-DRYING PROCESS 413
Maitraye Sen, Lindsay Wegiel, Joao Moreira, Pavol Rajniak

(261A) NUMERICAL MODELING OF DEAGGLOMERATION OF COHESIVE PARTICLES BY TURBULENCE 414
Yuan Yao, Jesse Capecelatro

(261B) REGIMES OF IMPACT-INDUCED EROSION OF PARTICLES OF VARYING STRENGTH 415
Chanh Nguyen, Jiecheng Yang, Jennifer S. Curtis, Rose McCallen, Kambiz Salari

(261C) BREAKAGE OF STRUCTURED PARTICLES: A COMPARISON BETWEEN STRESS-STRAIN AND IMPACT EVENTS 416
Kerry Johanson

(261D) ABRADABLE PARTICLES IN DEM THROUGH SPHERICAL HARMONICS 417
Rosario Capozza, Kevin J. Hanley

(261E) INVESTIGATION INTO THE ROLE OF HOLDUP ON THE GAS AND PARTICLE FLOW PATTERN IN A SPIRAL JET MILL 419
Lewis Scott

(261F) BREAKAGE FUNCTIONS FOR HIGH ASPECT RATIO PARTICLES 420
Priscilla J. Hill

(261G) A NONLINEAR ELASTO-PLASTIC BONDED-SPHERE MODEL FOR THE DISCRETE ELEMENT SIMULATION OF BIOMASS PARTICLE BREAKAGE 421
Yuan Guo, Qiushi Chen, Yidong Xia, Feiyang Chen, Jordan Klinger

(261H) MATHEMATICAL MODELING OF CONTINUOUS MILLING PROCESSES USING A CELL-BASED PBM APPROACH: A PARAMETRIC STUDY 422
Nontawat Muanpaopong, Rajesh Davé, Ecevit Bilgili

(431A) SPHERICAL AGGLOMERATION: PROCESS MODELLING AND VALIDATION FOR KINETIC ESTIMATION 424
Bilal Ahmed, Victoria R. Kitching, Omid Arjmandi-Tash, Jonathan D. Tew, Kate Pitt, Rachel M. Smith, James D. Litster

(431B) UNDERSTANDING AUTO-AGGLOMERATION OF DRY ACTIVE PHARMACEUTICAL INGREDIENTS 426
Svetlana Bibiceva, Ali Hassanpour, Sven Schroeder

(431C) MODELING OF GRANULE SIZE DISTRIBUTION IN PHARMACEUTICAL ROLLER COMPACTION PROCESS 427
Hossein Amini, Venkata Bobba, Jaya Malladi, Ilgaz Akseli

(431D) INFLEUNCE OF NATURAL AND SYNTHETIC BINDERS ON PARACETAMOL GRANULES PROPERTIES 428
Sheena Reeves, Michael Aliado, Tyree Madison, Vatallio Abrams, Mahmoud Soliman

(431E) IDENTIFYING OPTIMUM LIQUID-TO-SOLID RATIO RANGE FOR MICROCRYSTALLINE CELLULOSE BASED FORMULATIONS IN TWIN SCREW WET GRANULATION 429
Shankali Pradhan, Joseph W. Bullard, Peter Ojakovo, Mark Costello, Steven Dale

(431F) PARAMETER SENSITIVITY ANALYSIS OF A HIGH-SHEAR WET GRANULATION MODEL FOR EXPERIMENTAL DESIGN AND PARAMETER ESTIMATION. 430
Aaron S. Yeardley, Stefan Bellinghausen, Robert A. Milton, James D. Litster, Solomon F. Brown

(431H) ANALYSIS AND EVALUATING DESIGN SPACE OF CONTINUOUS HIGH SHEAR GRANULATION AND DRYING PROCESS USING GPROMS 431
Mingzhe Yu, Jerome Castro, Jane Mathews, James D. Litster, Agba D. Salman

(438A) ROBOTIC PLATFORM FOR EXECUTION, ANALYSIS, AND OPTIMIZATION OF MULTISTEP REACTIONS IN CONTINUOUS FLOW 432
Anirudh M. K. Nambiar, Christopher P. Breen, Timothy M. Monos, Travis Hart, Tim Kulesza, Federico Florit, Timothy Jamison, Klavs F. Jensen

(438C) CONTINUOUS ENZYMATIC PRODUCTION OF ENANTIOPURE AMINES IN A PACKED BED REACTOR WITH AN AMINE DEHYDROGENASE 433
Robert D. Franklin, Joshua A. Whitley, Adam A. Caparco, Bettina Bommarius, Julie Champion, Andreas Bommarius

(493B) SCALE-UP OF CO- AND COUNTER-CURRENT IRRADIATED PHOTOCHEMICAL REACTORS USING COMMON LEDS FOR INDUSTRIAL APPLICATIONS 473
Glen Meir, M. Enis Leblebici, Simon Kuhn, Tom Van Gerven

(493C) DESIGN OF MULTICHANNEL PHOTOCATALYTIC REACTORS 474
Thomas Claes, Mumin E. Leblebici, Tom Van Gerven

(493D) VISIBLE LIGHT MEDIATED SWITCHABLE SELECTIVITY IN C-O, C-C AND C-S BOND FORMATION INDUCED BY DISULFIDE 475
Qingwei Meng, Jingnan Zhao

(567A) ELUCIDATING MOLECULAR ORIGINS OF DRUG SELECTIVITY FOR RATIONAL DESIGN OF SELECTIVE DRUGS 476
Diwakar Shukla

(567B) COMPUTER AIDED GENERATION OF DIVERSE AND SYNTHESIZABLE MOLECULAR LIBRARY 477
Hanyu Gao, Klavs F. Jensen

(567C) DE NOVO DESIGN OF ENGINEERED CYTOKINE MIMETICS AS TARGETED IMMUNOTHERAPEUTICS 478
Jamie B. Spangler

(567D) INTEGRATED PREDICTION OF CHEMICAL EFFECTS ON IMPORTANT PATHWAYS OF EMBRYO DEVELOPMENT BASED ON IN VITRO EMBRYONIC STEM CELL TEST AND MACHINE LEARNING 479
Fengli Zhang, Shang-Tian Yang

(567E) A COMPUTATIONAL APPROACH TO IDENTIFYING SYNERGISTIC COMPOUNDS FOR TREATING ANTIMICROBIAL-RESISTANT PATHOGENS 480
Fangyuan Zhang, Tianhua Zhai, Yanhong Liu, Shozeb Haider, Zuyi (Jacky) Huang

(567F) LIQUID PHASE MEMBRANE SUPPORTED CYCLICAL FLOW SYNTHESIS OF OLIGONUCLEOTIDES FOR THERAPEUTIC APPLICATIONS 482
Catalina Parga, Brian Glennon, Steven Ferguson

(567G) PREDICT NEW COCRYSTALS VIA MECHANOCHEMISTRY 484
Jan Gröls

(576A) INVESTIGATING INDIRECT ULTRASOUND ON PARTICLE SIZE, SHAPE AND NUCLEATION OF APIS THROUGH PROCESS INTENSIFICATION 485
Bilal Ahmed, Charles D. Papageorgiou, Yihui Yang, Justin L. Quon, Humera Siddique, Ian Houson, Alastair J. Florence

(576B) EXPERIMENTAL INVESTIGATION OF WET-MILL INTEGRATED CRYSTALLIZATION WITH TEMPERATURE CYCLES FOR SIZE AND ASPECT RATIO CONTROL OF A COMMERCIAL PHARMACEUTICAL INGREDIENT 487
Ayse Eren, Botond Szilagyi, Justin L. Quon, Charles D. Papageorgiou, Zoltan K. Nagy

(576C) BEYOND STRUCTURE DETERMINATION: CRYSTALLISATION AS A PURIFICATION UNIT PROCESS FOR PROTEINS AND PEPTIDES 489
Ian Rosbottom, Niall Mitchell, Frederik J. Link, Wenqian Chen, Jerry Y. Y. Heng

(576D) MANEUVERING AROUND LIQUID-LIQUID PHASE SEPARATION (LLPS) TO ENABLE DEVELOPMENT OF ROBUST CRYSTALLIZATION PROCESSES. 491
Moussa Boukerche, Timothy Grieme, Manish S. Kelkar, James C. Marek, Collin Morris, Nandkishor K. Nere

(576E) UTILIZING MECHANISTIC MODELLING TO SUPPORT CRYSTALLIZATION SEEDING DESIGN 492
Olivier Monnier, Lea Heraud, Clemence Valla, Guillermo Jimeno, Niall Mitchell

(576F) PROCESS OPTIMIZATION FOR CONTROLLING THE MORPHOLOGY OF COF-5 CRYSTALS 494
Rajan Bhawnani, Santanu Chaudhuri, Meenesh R. Singh

(576G) CRYSTALLIZATION MODEL DISCRIMINATION USING IN SITU CHORD LENGTH MONITORING AND A SIZE CONVERSION MODEL (CLD2PSD MODEL).. 495
Roberto Irizarry, Jochen Schoell

(576H) CONTINUOUS SOLID-STATE DERACEMIZATION VIA TEMPERATURE CYCLES – A MODEL-BASED PROCESS DEVELOPMENT 496
Brigitta Bodák, Marco Mazzotti

(591B) REAL-TIME PROCESS MONITORING IN CONTINUOUS MANUFACTURING OF ORAL SOLID DOSAGES USING RAMAN PROCESS ANALYTICAL TECHNOLOGY 499
Sumit Kumar

(591D) MULTIPURPOSE AUTOMATED OPTIMIZATION PLATFORM FOR REACTIONS INVOLVING SOLIDS 501
Kakasaheb Nandiwale, Yiming Mo, Maria Jose Nieves-Remacha, Martin Johnson, Pablo García-Losada, Carlos Mateos, Juan Antonio Rincón, Brandon Reizman, Klavs F. Jensen

(591E) IMPLEMENTATION OF MODEL-PREDICTIVE CONTROL IN CONTINUOUS CRYSTALLIZATION 502
Daniel Patience, Erwin Irdam

(591F) MODEL-BASED DYNAMIC OPTIMIZATION OF FEEDER REFILL STRATEGY USED IN CONTINUOUS PHARMACEUTICAL MANUFACTURING PROCESS.................................... 503
Ravendra Singh, Fernando Muzzio

(591G) REAL TIME, DYNAMIC (NON-STEADY STATE) PROCESS MODELS FOR OSD CONTINUOUS MANUFACTURING CONTROL 504
Keirnan Lamarche, Eric Jayjock

(603A) BIG DATA-BASED FAULT DETECTION WITH ADVANCED ANALYTICS IN THE PHARMA INDUSTRY.. 505
Deniz Koç, Daniel Castro-Rodriguez, Dimitrios I. Gerogiorgis

(603B) RECENT ADVANCES IN KINETIC PARAMETER ESTIMATION TOOLKIT (KIPET) WITH SPECTRA 507
Kevin McBride, Michael Short, Weifeng Chen, Kuan-Han Lin, David Thierry, Salvador García-Muñoz, Lorenz Biegler

(603D) TEACHING COMPUTERS TO INTERPRET MDSC THERMOGRAMS.. 508
Pedro Monteiro, Teresa Marta

(603E) DATA-SMART MACHINE LEARNING METHODS FOR PREDICTING YOUNGS MODULUS OF DIRECTLY COMPRESSED BLENDS OF PHARMACEUTICAL POWDERS 509
Stephen Thomas, Hossein Amini, Venkata Bobba, Hannah Palahnuk, Jaya Malladi, Ilgaz Akseli

(603G) ADVANCES AND NEW DEVELOPMENTS ON EIOT FOR PAT APPLICATIONS IN PHARMACEUTICAL DEVELOPMENT AND MANUFACTURING 510
Salvador García-Muñoz

(625A) SOLVENT MIXTURE DESIGN FOR THE INTEGRATED PURIFICATION IN PHARMACEUTICAL MANUFACTURING. 511
Suela Jonuzaj, Oliver L. Watson, Sara Ottoboni, Chris Price, Jan Sefcik, Amparo Galindo, George Jackson, Claire S. Adjiman

(625C) IN-LINE MONITORING OF LOW DRUG CONCENTRATION OF FLOWING POWDERS IN A NEW SAMPLER DEVICE 513
Nobel O. Sierra-Vega, Pedro A. Martínez-Cartagena, Bárbara B. Alvarado-Hernández, Rodolfo Romañach, Rafael Mendez

(625D) FABRICATION OF CURCUMIN-LOADED MICROPARTICLES FROM OIL-IN-WATER EMULSION BY SUPERCRITICAL ASSISTED ATOMIZATION 514
Zi-Dan Wang, Hu-Hong Peng, Yi-Xin Guan, Shan-Jing Yao

(625E) DEVELOPMENT OF TARGETED EXOSOMES FOR DRUG DELIVERY 515
Yingnan Si, Seulhee Kim, Lufang Zhou, Xiaoguang Liu

(625F) LINKING PROCESS & PERFORMANCE OF DRUG PRODUCTS WITH MICRO-IMAGE ANALYTICS 516
Liping Zhou, Joshua Lomeo, Shawn Zhang

(625G) SURROGATE MODELING OF DISSOLUTION BEHAVIOR TOWARD EFFICIENT DESIGN OF TABLET MANUFACTURING PROCESSES 517
Kensaku Matsunami, Tomohiro Miura, Keita Yaginuma, Shuichi Tanabe, Hirokazu Sugiyama

(642A) OPTIMIZATION OF HYDRIDE VAPOR PHASE EPITAXY (HVPE) DEPOSITION REACTOR MANUFACTURING III-V MATERIALS USING MULTIPLE LARGE-EDDY SIMULATIONS (LES) 518
Malik Hassanaly, Hariswaran Sitaraman, Kevin Udwary, Kevin L. Schulte, Heather Splawn, Aaron Ptak, John Simon

(642B) PROCESS INTENSIFICATION AND INTEGRATION POTENTIAL IN MANUFACTURING POLLUTION PREVENTION 521
Tapas K. Das

(642C) PROCESS INTENSIFICATION FOR CERAMIC PART PRODUCTION – FIELD ASSISTED SINTERING OF CORE/SHELL POWDERS FABRICATED BY PARTICLE ATOMIC LAYER DEPOSITION 522
Rebecca O'Toole, Bola Yoon, Christopher Gump, Sanjit Ghose, Rishi Raj, Alan W. Weimer

(642D) SINKER ELECTRIC DISCHARGE MACHINING AS AN INEXPENSIVE METHOD TO PROTOTYPE MICROSCALE-BASED LAMINA-PLATES WITH HIGH ASPECT RATIO FEATURES 523
Lucas Freiberg, Jad Touma, Matthew Coblyn, Goran Jovanovic

(642E) PROCESS DESIGN AND DEVELOPMENT OF A SMALL SCALE HYBRID MANUFACTURING SYSTEM FOR THE CANCER DRUG LOMUSTINE 524
Jaron Mackey, Ahmed Mufti, Su-Lin Lee, Mesfin Abdi, Xin Feng, Erin Wood, David H. Thompson, Zoltan K. Nagy

(642G) SELECTION OF FUNCTIONAL SOLVENT FACTORY FOR AUTOMATIC ALDEHYDE AND KETONE SEPARATION IN ALDOL REACTION: A COMBINED SIMULATION AND EXPERIMENTAL APPROACH 525
Chenyue Zhang, Zhen Song, Can Jin, Job Nijhuis, Teng Zhou, Timothy Noël, Harald Gröger, Kai Sundmacher, Jan Van Hest, Volker Hessel

(644A) NEXT GENERATION LIPID-BASED EXCIPIENTS: OVERCOMING THE CHALLENGES OF SPRAY DRYING APPLICATIONS ON THE DEVELOPMENT OF INHALABLE LIPID-MICROPARTICLES 528
Carolina Corzo, Agnes Fuchsbichler, Đana Crvenjak, Sebastian Reyer, Dirk Lochmann, Andreas Zimmer, Sharareh Salar-Behzadi

(644B) COMPARATIVE EVALUATION OF DISSOLUTION BEHAVIOR AND BIOEQUIVALENCE OF VARIOUS CAPSULES 530
Jin Zhao, Amie Gehris, Thomas Watson, Laura McReady, Tom Breshamer, Bahi Mahesan, Elizabeth Tocce, Yeli Zhang, Paul Zajac, Michael Baumann

(644C) DEVELOPMENT OF HIGH-SOLIDS, QUICK-SET PHARMACEUTICAL TABLET COATINGS........ 532
Thomas H. Kalantar, Mladen Ladika, Hui Shao, Stacey L. Dean, J. Keith Harris, Paul J. Sheskey, Karen Coppens, Karen M. Balwinski, Debora L. Holbrook

(644E) INFLUENCE OF EXTERNAL LUBRICANTS AND THE ENVIRONMENTAL CONDITIONS ON THE PUNCTURING PROPERTIES OF GELATIN AND HPMC CAPSULES FOR INHALATION 534
Thomas Wutscher, Milica Stankovic-Brandl, Sarah Zellnitz, Mirjam Kobler, Francesca Buttini, Veronica Chierici, Laura Andrade Benítez, Verónica Daza Fernández, Alberto Mercandelli, Stefano Biserni, Susana Ecenarro Probst, Laura Canalejas, Amrit Paudel

(644F) FORMULATION, STABILITY, AND SCALABILITY OF FAST-RELEASING LUMEFANTRINE NANOPARTICLES FOR THE TREATMENT OF MALARIA 537
Madeleine Armstrong, Jie Feng, Chester E. Markwalter, Chang Tian, Kurt D. Ristroph, Leon Z. Wang, Jiankai Yang, Hui Du, Huizhuan Lin, Fengjiao He, Shuhong Jiang, Santipharp Panmai, Hanu Ramachandruni, Yingyue Zhang, Simon A. McManus, Kai Gong, Claire White, Aditya Rawal, Robert K. Prud'Homme

(646A) MONITORING AND DETECTION OF BYPRODUCT CRYSTALLIZATION USING IN-SITU MICROSCOPY........ 538
Matthew A. McDonald, Hossein Salami, Patrick R. Harris, Andreas Bommarius, Martha A. Grover, Ronald Rousseau

(646B) INTEGRATION OF IN-SILICO TOOLS AND HIGH-THROUGHPUT AUTOMATION FOR ENHANCED CRYSTALLIZATION DEVELOPMENT 539
Kirsten Springer, Jie Chen, Manish S. Kelkar

(646C) UTILIZATION OF QUALITY-BY-CONTROL FOR RAPID PROCESS DESIGN OF AGROCHEMICAL CRYSTALLIZATION 540
Wei-Lee Wu, Jasson T Patton, Chris Chappelow, Paul A. Larsen, Zoltan K. Nagy

(646D) DUAL CHAMBER AND MEMBRANE EMBEDDED CONTINUOUS MICROFLUIDIC DEVICE FOR MOFS SCREENING 541
Paria Coliaie, Manish S. Kelkar, Nandkishor K. Nere, Meenesh R. Singh

(646E) IN SITU POLYMORPH TRANSFORMATION OF A PHARMACEUTICAL INTERMEDIATE 542
Thomas Vetter, Anthony J. Nockett, Charles A Goss

(646F) DETECTION OF MORPHOLOGY AND MODALITY CHANGES DURING CRYSTALLIZATION USING IN SITU CHORD LENGTH MONITORING AND INFERENCE SYSTEM MODEL. 543
Roberto Irizarry, Akshaya Nataraj, Jochen Schoell

(646G) INDUSTRIAL SCALE DOWNSTREAM PROCESS DEVELOPMENT FOR LOW CONCENTRATION AMINO ACID PRODUCT USING PAT 544
Iman Beheshti Tabar, Michael Taussig

(674A) HYBRID MODELING APPROACH INTEGRATING PLS MODELS WITH FIRST PRINCIPLES KNOWLEDGE 545
Debanjan Ghosh, Prashant Mhaskar, John Macgregor

(674B) INDIRECT PARAMETER ESTIMATION OF DYNAMIC MECHANISTIC MODELS VIA HYBRID AND NEURAL DIFFERENTIAL EQUATIONS. 546
William Bradley, Fani Boukouvala

(674C) BAYESIAN LEARNING FROM HYBRID MODELS AND THEIR APPLICATIONS IN THERMODYNAMICS AND REACTION ENGINEERING 547
Elvis Eugene, Xian Gao, Alexander Dowling

(674D) DEVELOPMENT OF A HYBRID FIRST PRINCIPLES – ARTIFICIAL INTELLIGENCE APPROACH FOR DYNAMIC MODELING OF COMPLEX SYSTEMS 549
Angan Mukherjee, Debangsu Bhattacharyya

(674E) A HYBRID SIMULATION-BASED AND MATHEMATICAL OPTIMIZATION FRAMEWORK: DEMONSTRATION ON A CARBON CAPTURE SYSTEM OPTIMIZATION 551
Miguel A. Zamarripa, Anuja Deshpande, Benjamin P. Omell, Michael S. Matuszewski, David C. Miller

(674F) COUPLING MECHANISTIC AND DATA-DRIVEN MODELS BY MEANS OF NEURAL DIFFERENTIAL EQUATIONS TO INCORPORATE UNMODELED DYNAMICS 554
Ward Quaghebeur, David Fernandes Del Pozo, Elena Torfs, Bernard De Baets, Ingmar Nopens

(674G) ENHANCING SYMBOLIC REGRESSION WITH GRADIENT INFORMATION 556
Marissa Engle, Nick Sahinidis

(674H) MIXED-INTEGER FORMULATIONS FOR FAIR CLASSIFICATION 557
Apoorva Sampat, Dilara Goreke, Victor M. Zavala

(375A) MODELING AND NONLINEAR STATE ESTIMATION FOR ADVANCED PROCESS CONTROL OF THE ENZYMATIC CONVERSION OF LACTOSE INTO VALUE-ADDED PRODUCTS 559
Ronald Alexander, Guilhermina Schultz, Marcelo P. A. Ribeiro, Fernando V. Lima

(375B) DESIGNING DATA-BASED AND MODEL-BASED METHODS FOR PROCESS MONITORING AND EQUIPMENT DEGRADATION TRACKING 561
M. Ziyan Sheriff, M. Nazmul Karim, Costas Kravaris, Hazem Nounou, Mohamed Nounou

(375C) OPTIMAL SENSOR NETWORK DESIGN AND NONLINEAR STATE ESTIMATION FOR IN-SITU CORROSION MONITORING IN COAL-FIRED BOILERS 562
Chandra Sekhar Somayajula, Debangsu Bhattacharyya, Xingbo Liu

(375D) CO2 INJECTION MONITORING WITH A SCALABLE, AUTOMATED SPARSE SEISMIC ARRAY 564
Cesar Barajas-Olalde, Donald Adams, Lu Jin, John A. Hamling, Nicholas Bosshart, Wesley D. Peck

(81A) CONTINUOUS REVOLUTIONS: DISTRIBUTED MANUFACTURING TO RE-SHORE THE PHARMACEUTICAL SUPPLY CHAIN 565
Luke Rogers

(81B) BUILDING RESILIENT SUPPLY CHAINS THROUGH THE MANUFACTURING USA INSTITUTE NETWORK 566
William Grieco

(81C) CHANGE CREATES OPPORTUNITY AND DISRUPTIVE ORGANIZATIONAL INNOVATION 567
Brendon G. Ricart

(81D) RESPONSES TO COVID-19 FROM DOW LIFE SCIENCES 568
Ida(Xue) Chen

(81E) SOCIAL DISTANCING VIA TELEPRESENCE FOR PRODUCT DEVELOPMENT AND MANUFACTURING 569
Jaquan Levons

(81F) INNOVATION IN CHALLENGING TIMES 570
Gregory Frank

(332D) (INVITED PLENARY TALK) WEARABLE ELECTROCHEMICAL SENSORS 571
Joseph Wang

(332C) (INVITED PLENARY TALK) AN INTEGRATED PARADIGM FOR PRECISION EXPOSURE TO AIRBORNE CHEMICAL AND BIOLOGICAL STRESSORS BASED ON PERSONAL SENSING 572
Dimosthenis Sarigiannis, Dimitrios Chapizanis, Marianthi Kermenidou, Ioannis Petridis, Spyros Karakitsios

(279B) HIGH THROUGHPUT PROCESS DEVELOPMENT FOR CEX CHROMATOGRAPHY 574
Nicholas Vecchiarello, Pranali Shah, Nitya Jacob, Glen Bolton

(279C) THE USE OF TANGENTIAL FLOW FILTRATION SYSTEMS IN SYNTHETIC PEPTIDE PURIFICATION 575
Cyrus Fisher, Kevin D. Seibert, Katerina Williams

(279E) IN-SITU MONITORING DURING MIXING FOR PROTEIN CRYSTALLISATION 576
Frederik J. Link, Ian Rosbottom, Jerry Y. Y. Heng

(279F) THE USE OF NANOFILTRATION SYSTEMS IN SYNTHETIC PEPTIDE PROCESSES 578
Nil Tandogan, Nicholas Klitzing, Frank Torres Torres, Kevin D. Seibert

(377A) DESIGN SPACE VIA SYMBOLIC COMPUTATION 579
Fei Zhao, Ignacio Grossmann, Salvador García-Muñoz, Stephen D. Stamatis

(377B) A MULTI-SCALE HYBRID MODELING APPROACH FOR CONTINUOUS PHARMACEUTICAL UNIT OPERATIONS 581
Pooja Bhalode, Marianthi Ierapetritou

(377C) INTEGRATED USE OF DIGITAL TWIN-BASED FRAMEWORK FOR THE DEVELOPMENT OF A LOMUSTINE MANUFACTURING PATHWAY 583
Daniel Laky, Jaron Mackey, Ahmed Mufti, Daniel M. Casas-Orozco, Gintaras V. Reklaitis, Carl Laird, Su-Lin Lee, Mesfin Abdi, Xin Feng, Erin Wood, Zoltan K. Nagy

(377D) DIGITAL DESIGN OF END-TO-END MANUFACTURING PROCESS FOR MEFENAMIC ACID USING MECHANISTIC MODELING 584
Cameron Brown, Niall Mitchell, Frederik Doerr, John McGinty, Magdalene Chong, Murray Robertson, Elke Prasad, Sara Ottoboni, Wei Li, Guillermo Jimeno, Patricia Marce Villa, Chris Rielly, Chick Wilson, Alastair J. Florence

(377E) DIGITAL DESIGN OF AN END-TO-END DRUG MANUFACTURING PROCESS USING MECHANISTIC MODELING 585
Jonathan Meyer, Amit Goda, Filipe Calado, Renato Wong, Ananya Chowdhury, Dana Barrasso, Edward Close, Niall Mitchell, Sean Bermingham, Sélim Douieb, Ugo Cocchini

(377F) DIGITAL DESIGN OF A DRY GRANULATION-BASED DRUG PRODUCT THROUGH INTEGRATED MANUFACTURING PROCESS AND PRODUCT PERFORMANCE MODELING 587
Ananya Chowdhury, Parind Desai, Rachel Apfelbaum, Tim Boung-Wook Lee, Yash Melkeri, Dana Barrasso, Edward Close

(377G) A MECHANISTIC MODEL-BASED DIGITALISATION FRAMEWORK FOR CONTINUOUS MANUFACTURING PROCESSES 588
Li Ge Wang, David Slade, Chalak Omar, James D. Litster, Agba D. Salman

(414A) DATA-RICH EXPERIMENTATION AT MERCK. A RETROSPECTIVE ON A 6YEAR JOURNEY. 589
Shane T. Grosser

(414C) EMBEDDING AUTOMATION IN THE PHARMA RESEARCH LABORATORY TO ENABLE DRUG SUBSTANCE AND PRODUCT DEVELOPMENT 590
Christopher Nunn

(414E) DEVELOPMENT A WORKUP FOR REMOVING TRIPHENYLPHOSPHINE OXIDE FROM A MITSUNOBU COUPLING 591
Eric G. Moschetta, Benoit Cardinal-David, Moiz Diwan

(414F) DRYING PROCESS CHARACTERIZATION OF A NOVEL CONTINUOUS VACUUM DRYING TECHNOLOGY 592
Manuel Zettl, Isabella Aigner, Thomas Mannschott, Peter Van Der Wel, Hartmuth Schröttner, Johannes G. Khinast, Markus Krumme

(414G) PREDICTING OPTIMAL SALT FORMS FOR ACTIVE PHARMACEUTICAL INGREDIENTS USING THE SAFT-γ MIE EQUATION OF STATE 593
Malak Wehbe, Andrew J. Haslam, Claire S. Adjiman, George Jackson, Amparo Galindo

(460A) ELUCIDATION OF LOCAL DRUG RELEASE DYNAMICS IN COMPLEX FORMULATIONS WITH MULTI-NUCLEAR NMR MICRO-IMAGING 594
Mick Mantle

(460B) PHARMACY ON DEMAND: AN END-TO-END APPROACH TO DEVELOPING A HIGH-DOSE, POORLY FLOWING API INTO A ROBUST CONTINUOUS MANUFACTURED FORMULATION 595
Sarah Aboelela, Gregory Hammersmith, Kersten Rapp, Anna Forzano, Annessa Burnett, Naomi Briggs, Thomas D. Roper

(460C) PREPARATION OF SOLID DISPERSIONS BY SPRAY DRYING AND ELECTROSTATIC PRECIPITATION 596
Anna Justen, Adrian Dobrowolski, Damian Pieloth, Gerhard Schaldach, Helmut Wiggers, Markus Thommes

(460D) MODELING AND SIMULATION OF MAGNETOPHORESIS OF NANOPARTICLES FOR MAGNETIC TARGETING APPLICATIONS 601
Ayankola Ayansiji, Anish V. Dighe, Andreas Linninger, Meenesh R. Singh

(460E) RATIONAL DESIGN AND MATHEMATICAL MODELING OF INTRATHECAL THERAPEUTICS: BIODISPERSION AND PHARMACOKINETICS 602
Daniel Gehrke, Ayankola Ayansiji, Meenesh R. Singh, Andreas Linninger

(460F) COMPUTER AIDED DRUG FORMULATION DESIGN WITH IMAGE-BASED MICROSTRUCTURE MODELING 603
Shawn Zhang, Athanas Koynov, Joshua Lomeo

(460G) MECHANISTIC MASS TRANSPORT ANALYSIS OF PREDICTIVE IMPROVED DISSOLUTION AND BIOAVAILABILITY USING PH MODIFIER IN DRUG FORMULATIONS 605
Niloufar Salehi, Gislaine Kuminek, Robert M. Ziff, Gregory E. Amidon

(645A) HYBRID NANOCRYSTAL-AMORPHOUS SOLID DISPERSIONS (HYNASDS): A NEW CLASS OF MATERIALS FOR BIOAVAILABILITY ENHANCEMENT OF POORLY SOLUBLE DRUGS 606
Ecevit Bilgili, Mahbubur Rahman, Faustin Arevalo, Alexander Coelho

(645B) LEVERAGING HOLLOW FIBRE MEMBRANES AND HIGH BOILING POINT ORGANIC SOLVENTS IN THE PRODUCTION OF ENTERIC AMORPHOUS SOLID DISPERSIONS 607
Rute F. Mota, Catarina M. Portugal, Paulo Lino, Rui C. Silva, Ana-Aguiar Ricardo, João Vicente

(645D) INSIGHT INTO AMORPHOUS SOLID DISPERSION DISSOLUTION BEHAVIOR: LASER-DIFFRACTION AND RAMAN SPECTROSCOPY HOLDING HANDS 609
Maria Paisana, Patricia Nunes, Mafalda Paiva, João Henriques

(645E) MULTI-FACTOR MODELING TO PREDICT SIGNIFICANT FACTORS IN TABLET DISINTEGRATION AND DISSOLUTION USING THE FBRM 613
Catherine Metzler, Gneya Harish Pandya, Joseph W. Bullard

(645F) MELT-COATING OF APIS WITH SURFACTANT FOR DRUG DISSOLUTION ENHANCEMENT 614
Zhanjie Liu, Fernando Muzzio

(645G) COCRYSTALLIZATION PROCESS DEVELOPMENT OF BENZOIC ACID – APREMILAST COCRYSTAL 615
Jan Jirát, Miroslav Soos, Vít Zvonícek, Ludek Ridvan

(692A) A SYSTEMATIC STUDY OF STATE-OF-THE-ART METHODS IN CRYSTAL STRUCTURE PREDICTION FOR ORGANIC HYDRATES 616
Yizu Zhang, Isaac J. Sugden, Susan M. Reutzel-Edens, Constantinos C. Pantelides, Claire S. Adjiman

(692C) EFFECTS OF INORGANIC SALTS AND VOLUME REDUCTION ON GLYCINE POLYMORPHISM 618
Isaac Jerome C. Dela Cruz, Gerard Capellades, Jem Valerie D. Perez, Bryan G. Alamani, Allan S. Myerson

(692D) USING FLUID DYNAMICS AND INTERFACE ENGINEERING TO CONTROL POLYMORPHISM AND MORPHOLOGY OF ORGANIC MOLECULES. 619
Stephanie Guthrie, Ashley Conley, Gaurav Giri

(692E) ACETAMINOPHEN POLYMORPHISM AND TEXTURE CONTROL USING NOVEL COATING REGIMES 620
Stephanie Guthrie, Gaurav Giri

(692F) TERNARY PHASE DIAGRAM AND PHASE TRANSFORMATION BEHAVIORS OF VARIOUS CREATINE PHOSPHATE SODIUM HYDRATES 621
Jiayu Dai, Ling Zhou, Qiuxiang Yin

(195A) TOWARDS THE DEVELOPMENT OF DIGITAL TWIN FRAMEWORK FOR CONTINUOUS PHARMACEUTICAL MANUFACTURING: REAL-TIME MODEL MAINTENANCE AND SYSTEM ANALYSES 622
Pooja Bhalode, Yingjie Chen, Marianthi Ierapetritou

(195B) DIGITAL TRANSFORMATION OF UNIT OPERATIONS LABORATORY 624
David Henthorn

(195C) DEVELOPMENT OF A JOINT INFRASTRUCTURE FOR CHEMICAL ENGINEERING APPLICATIONS RELATED TO INDUSTRY 4.0 AND DIGITAL TRANSFORMATION 625
Selorme Agbleze, Willy Araujo, Heleno Bispo, Fernando V. Lima

(195D) ROBUST LARGE-SCALE STATE-ESTIMATE PREDICTION 627
Leila Samandari Masooleh, Ulku Oktem, Warren Seider, Jeffrey E. Arbogast, Masoud Soroush

(264B) IS GLOBAL HARMONIZATION ACHIEVABLE – AN INDUSTRY PERSPECTIVE 629
Timothy J. Watson

(264C) QUALITY BY DESIGN AND EXPEDITED DEVELOPMENT PROGRAMS 630
Kevin Fitzpatrick

(264D) QBD IN DEVELOPMENT AND COMMERCIAL MANUFACTURE 631
Anita Lalloo

(264E) THE HANDLE AND THE BRIEFCASE: MODEL QUALITY FOR QUALITY BY DESIGN 632
Pablo Rolandi

(264F) QBD AT VERTEX: BACK TO THE FUTURE 633
Billie Kline, Mike Filios, Cristian Harrison, Adam Looker, David Moon, Brian Phenix, Michael Ryan, David Willcox

(264G) QBD IN MOBILE PRODUCTION SCENARIOS: PROGRESS TOWARD A MULTISTAGE PAT STRATEGY FOR PHARMACY ON DEMAND 634
Thomas D. Roper

(264A) QBD (QUALITY BY DESIGN) AN ENABLER FOR PHARMACEUTICAL PRODUCT DEVELOPMENT: FDA PERSPECTIVE 635
Sharmista Chatterjee

Author Index

609420 Understanding and Predicting the Efficacy of Tailor Made Additives on the Crystallisation of p-Aminobenzoic Acid

Abstract:

Unwanted crystallization is a major concern in a number of chemical industries, affecting fuels, agrochemicals, pharmaceuticals and food. The ability to overcome this issue by disrupting the crystallization process, through the use of additives, has been a long studied area of crystallization science and engineering [1]. Since the work of Lahav, Leiserowitz and colleagues at the Weizmann Institute of Science[2,3], a large number of research studies have focussed on tailor-made additives (TMAs), whereby an additive with a very similar molecular structure to a target crystallization compound is used to inhibit aspects of the crystallization process. The majority of work has focussed on crystal growth inhibition, however, more recently the ability of TMAs to inhibit nucleation has been a growing area of research.

This work focuses on the nucleation of alpha-p-aminobenzoic acid (pABA) from ethanol solution, in the presence of TMAs. pABA in ethanol solution is known to nucleate through carboxylic group H-bonding dimer formation, with the crystal structure formed also containing strong amine-carboxyl group H-bonding and π-π stacking interactions [4]. TMAs that could potentially disrupt pABA-pABA dimer formation as well as the formation of the previously described synthonic interactions were chosen to study their effect on pABA nucleation inhibition form ethanol solution.

Seven TMAs were chosen and molecular modelling using an intermolecular grid search was undertaken to screen and understand the dominant molecular interactions between pABA and the TMAs, in terms of synthonic interaction energies. Two TMAs were found to have stronger interactions to pABA than pABA itself, forming competitive carboxylic H-bonding dimers, these were 4-amino-2-methoxybenzoic acid (AMBA2), 4-amino-3-nitrobenzoic acid (ANBA3). These TMAs had the same molecular structure of pABA, but with methoxy or nitro groups attached, respectively. From calculated atomic charge energies, the nitro group of ANBA3 was assessed to be able to disrupt the subsequent binding of pABA molecules into the crystal structure through electrostatic interactions.

A polythermal experimental screen was performed to assess the intermolecular grid search's ability to predict TMA efficacy. The experimental results corroborated findings from the modelling results, with ANBA3 being the most effective nucleation inhibitor followed by AMBA2.

Nucleation kinetics were assessed for pABA vs. pABA-ANBA3 in ethanol solution using the KBHR approach. It was found that pABA nucleates by instantaneous nucleation, whereas the pABA-ANBA3 nucleates by progressive nucleation. The interfacial tension was found to also increase in the presence of ANBA3.

References

1) J. W. Mullin, Crystallization, Butterworth-Heinemann, Oxford, 4th ed., 2001.

2) L. Addadi, Z. Berkovitch Yellin, N. Domb, E. Gati, M. Lahav and L. Leiserowitz, Nature, 1982, 296, 21–26.

3) I. Weissbuch, L. Addadi, M. Lahav and L. Leiserowitz, Science, 1991, 253, 637–645.

4) D. Toroz, I. Rosbottom, T. D. Turner, D. M. C. Corzo, R. B. Hammond, X. Lai and K. J. Roberts, Faraday Discuss., 2015, 179, 79–114.

606384 Multiscale Computer Modeling to Investigate Polymeric Micelle Based Nanocarriers for Anticancer Drug Delivery

Abstract:

Purpose:

Typical micelle-based drug delivery carriers are formed from amphiphilic block copolymers which self-assemble into core-shell nanoarchitectures with a hydrophobic core being able to encapsulate drug molecules and a biocompatible hydrophilic shell. An innovative turbulent co-flow platform of water and ethanol (with drug and polymers) was previously developed to continuously process drug loaded micelle nanocarriers. Polymer aggregation is the result of the intermolecular forces among molecules and along with the jet flow characteristics impact to the formation of micelles. However, the underlying mechanism and the detailed effects from material attributes are only partially understood. In order to investigate the underlying mechanism and quantitative estimation of the effect of material attributes and process parameters on the quality of the polymeric micelle formulation, we implemented a multiscale computational approach to study micelle formation using a coaxial turbulent jet flow.

Methods:

All-atom (AA), coarse-grained molecular dynamics (CG-MD), as well as computational fluid dynamics (CFD) simulations have been conducted to not only reveal the effects to material attributes and processing parameters during the micelle formation, but also as parametric case studies for this process. The initial conditions applied into the simulations were based on polymeric micelle experiments in which PEG-PLA (2kD-1.7kD) was used as model drug carrier. The CFD simulation was implemented using Large Eddy Simulation (LES) model in COMSOL Multiphysics incorporating energy equation and high-resolution mesh in mixing area. The MD simulation trajectories and their analysis were carried out using GROMACS package. We applied both CHARMM and MARTINI force-fields for all-atom as well as coarse-grained simulations, respectively. The initial coordinates of models and their force field parameters were generated by CHARMM software and MARTINI forward mapping approach. The steepest decent algorithm was used in energy minimization with a 20 fs time step followed by 10 ns equilibration step in isothermal–isochoric NVT ensemble at temperature of 300 K, then the production runs were performed beyond 1 μs in the NPT ensemble using the Nosé–Hoover thermostat and the Parrinello–Rahman barostat at pressure of 1 atm with constant temperatures at every 20 fs time steps. The simulations were carried out using periodic boundary conditions. Computations were performed in High Performance Center Supercomputer Cluster at the University of Connecticut.

Results:

The self-assembled process of polymeric micelle was studied in the system with and without drug molecules, and in each case, block copolymer was randomly packed in the simulation box with explicit water and ethanol for all simulations. With optimized CG-MD forward mapping strategy and force field parameters incorporating MARTINI standard beads and recent developed S-beads, we observed that polymeric micelles formed successfully and fell into the experimental size range. The simulation of micelles formation and the size distribution the micelles indicate that CHARMM, MARTINI force fields (FFs) can effectively capture the experiment behavior and results . By adjusting Smagorinsky coefficient (Cs) of LES model companied with energy and mass transfer equations, CFD simulations successfully modeled flow patterns and formation temperatures of co-axial turbulent jet flow from experiments, as well as successful comparison and verification. Our CG-MD and CFD simulations are in good agreement with experimental results.

Conclusions:

In this present work, the model predictions provide an understanding of the impact of drug-polymer interactions, drug-polymer ratio, initial organic phase polymer concentration, as well as maintaining the agreement between the computational predictions and experimental findings highlight the power of the multiscale approach to be able to cover the long and short length scale predictions. Moreover, the details of the formation process was revealed in micro, macro, and meso scales by connecting discrete and continuum computational modeling work which are found to be complementary to one another. The multiscale approach used in this work can be effectively utilized as a powerful tool in the process of discovery, development, and optimization of new drug delivery systems in the co-flow continuous processing.

Acknowledgements: FDA Grant# 1U01FD005773-01.

Disclaimer: This article reflects the views of the authors and should not be construed to represent FDA's views or policies.

609574 Simulation of Drying of Droplets and Its Effect on Particle's Morphology in Spray-Drying

Abstract:

Spray-drying is a manufacturing process utilized by a wide range of industries such as food, agricultural, and pharmaceutical for converting a liquid feed containing insoluble solid particles into particulate form. By removing the moisture, it is intended to achieve a final product with desired chemical composition, morphology (hollow particles, solid cores, agglomerates, etc.), and particle size distribution. There are many variables that affect the drying process, and due to its complexity, traditional trial-and-error methods have been the primary technique in achieving the desired product.

The drying dynamics affect the internal flow patterns within a droplet in spray-drying. The flow patterns are dominated by surface tension and evaporative effects which ultimately disperse the suspended particles in the droplet and finally determine the final product morphology. In the presented work, the drying and internal transport of insoluble particles within a droplet were simulated using a volume of fluid method coupled with a Lagrangian solver for tracking particle movement. The simulations were performed within the open-source framework, OpenFOAM. The solver has been heavily modified to include interfacial mass transfer by employing a coupled Eulerian-Lagrangian framework capable of simulating drying of a droplet and the solid particles suspended inside. Modeling droplet's internal flow created through evaporative effects, and tracking movement of solid particles, drying and the morphological evolution of the product is simulated and analyzed.

606888 Population Balance Model Comparison of Growth, Secondary Nucleation By Attrition and Ripening

Abstract:

Introduction

Secondary nucleation is a phenomenon present everywhere in nature and of fundamental importance for crystallization processes [1,2]. There are two families of mechanisms of secondary nucleation: the first occurs at the interface of the seed crystal and the solution, and it is a combination of an activated process and the effect of shear, i.e. surface- or shear secondary nucleation. The second family results from mechanical collision happening to the seed crystal, i.e. secondary nucleation by attrition. Attrition is the mechanism through which fragments form after an impact of a crystal with a stirrer, other crystals or the wall of the reactor. Those fragments, if small enough, can be considered secondary nuclei. For both families, there is not a clear understanding of the governing physics, and secondary nucleation rates are usually given by an empirical relationship depending on the extent of mixing, on the amount of crystals suspended and on the supersaturation level [1].

Methods and Models

In this contribution, we have extended the well-established mechanistic attrition model developed by Gahn and Mersmann [3] to two different population balance equation (PBE) models, which are used to simulate secondary nucleation processes. The traditional approach describes the formation of attrition fragments due to collisions as a secondary nucleation rate, which is included in the model as a boundary condition (we call it nucleation model). In the alternative approach, the formation of attrition fragments is described as a breakage expression and the growth rate is the result of size-dependent solubility (we call it breakage model) [4,5]. Conversely to the traditional model, this formulation takes into account the size distribution of attrition fragments and their evolution due to growth, see the figure for a schematic representation of the two models. The two approaches have specific challenges and they have different pros and cons.

In both models, the number of parameters to estimate has been limited by using expressions, where the physical and mechanical properties of the system, together with operative conditions, have been employed. In particular, for the breakage frequency, a physical model suitable for PBE has been derived and used in the simulations. The traditional attrition distribution of Mersmann has been adapted to a mechanistic breakage daughter distribution, that we were able to use in a PBE. For the secondary nucleation rate, a new expression has been derived and then compared with standard ones[5].

Results

The physical-based model of Mersmann was successfully integrated into two different population balance models to describe secondary nucleation by attrition for a population of crystals. In order to compare the models, two simulations of a batch, isothermal system have been performed. In both cases, the simulated time was very long, thus resulting in an almost complete depletion of supersaturation. The two models lead to very similar results in the region where supersaturation is high enough to ensure growth and secondary nucleation. However, the breakage model can describe the mechanism of Ostwald ripening, as a result of the embedded size-dependent solubility, when supersaturation is very close to 1, and the simulated time is very long.

The two models are solved in two different numerical ways since the population balance in the case of the breakage model is more complex and numerically intensive than the implementation of the boundary condition, which is the only necessary thing to solve for the nucleation model. This required the implementation of two different numerical schemes: the nucleation model has been solved by following a finite volume scheme with boundary conditions for nucleation. The attrition model has been solved by discretizing the growth part with a finite volume scheme and the breakage term with the fixed pivot technique [6,7].

Conclusions

The two models are very similar in the growth regime, thus where secondary nucleation and growth are the dominant phenomena. At extremely low values of supersaturation, thanks to size-dependent solubility, the first model yields to further development of the crystal population, e.g., Ostwald ripening and aging. The main result is that secondary nucleation by attrition can be described as a birth/death term or as a source term according to the final application of the model. Since the two approaches have very different computational intensities, one can choose the right model based on the objective of the simulation study.

Acknowledgments

This project has received funding from the European Research Council (ERC) under the European Union's Horizon 2020 research and innovation program under grant agreement No 2-73959-18.

References

1. Agrawal, S. G., & Paterson, A. H. J. (2015). *Chemical Engineering Communications*, 202(5), 698–706.
2. Mersmann, A. (Ed.). (2001). *Crystallization technology handbook*. CRC Press.
3. Gahn, C., & Mersmann, A. (1999). *Chemical Engineering Science*, 54(9), 1273–1282.
4. Iggland, M., & Mazzotti, M. (2011). *Crystal Growth and Design*, 11(10), 4611–4622.
5. Bosetti, L., & Mazzotti, M. (2019) *Crystal Growth & Design*1, 307-319.
6. LeVeque, R. J. (2002). *Finite volume methods for hyperbolic problems* (Vol. 31). Cambridge university press.
7. Kumar, S., & Ramkrishna, D. (1996) *Chemical Engineering Science*, 51(8), 1311–1332.

Figure. Results of an isothermal desupersaturation experiment. On the l.h.s., schematic representation of the two models. On the center, the evolution of supersaturation (green), and the increase of the zeroth moment (blue for breakage model, red for nucleation model) corresponds to the formation of attrition fragments: in this region, the two models are almost indistinguishable. On the r.h.s., evolution of the initial population with a zoom on the population of fines, where the difference in the mechanisms of formation of nuclei is visible.

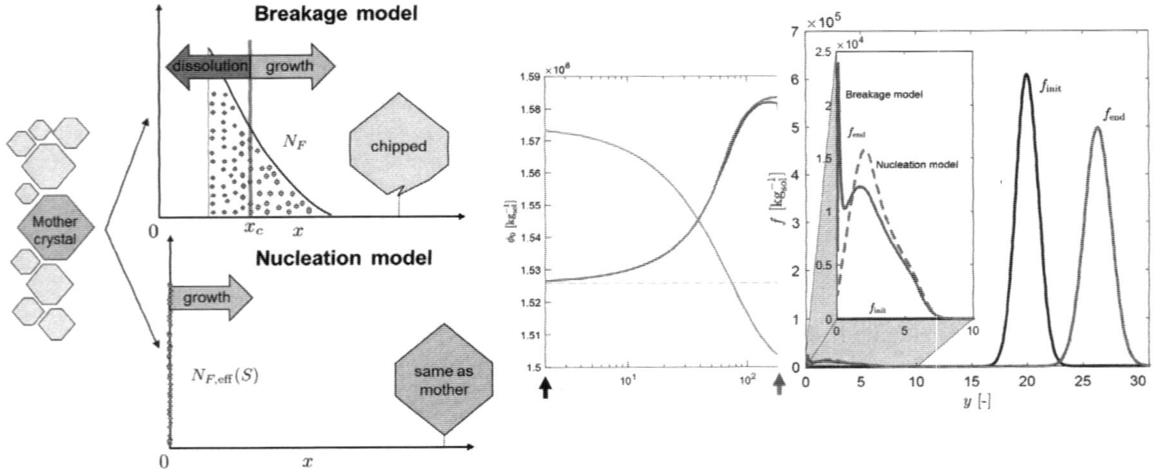

610983 Solubility and Nucleation Kinetics of a Series of Representative Single and Mixed Diesel *N*-Alkanes in Dodecane and Toluene

Abstract:

The formation of *n*-alkane waxes in solution during cold weather is a recurrent problem in the automotive industry. In diesel fuels, crystallisation of long chains *n*-alkanes can result in the formation of large flat plate-like crystals leading to flow impairment, blockage of filters and engine fuel starvation. Clearly the unwanted formation of these waxes and their associated problems can cause significant issues for the manufacturer and end-users.

Current wax mitigation strategies involve the development of chemical additives to control or inhibit the nucleation and growth of crystals so that they no longer form large deposits or block filters. However, the variable nature of fuel composition, specifically the fractionation of the *n*-alkanes, results in a complex challenge for the additive provider to meet operational needs[1]. In order to understand crystallisation of waxes in fuels, it is important to understand the crystallisation of *n*-alkanes in mixtures and singularly.

Here we present the results from an experimental study on the nucleation kinetics and solubility behaviour single and mixtures of 8 *n*-alkanes representative of diesel fuel, ranging from $C_{16}H_{34}$ to $C_{23}H_{48}$ in toluene and dodecane. Analysis of the experimental polythermal data making, use of the analytically derived KBHR approach[2–4] reveals, that in all bar two cases the preferred nucleation mechanism for single n-alkane solutions is instantaneous. i.e. all the nuclei are formed at once at a certain level of supersaturation, to further grow . Furthermore, the influence of chain length on solubility was observed between odd and even alkanes in both toluene and dodecane. By analysing the nucleation and solubility behaviour within the set of n-alkanes tested, suggestions of the factors controlling wax crystallisation in fuel systems are presented and discussed from first thermodynamic principles.

References

[1] A. R. Gerson; K. J. Roberts; J. N. Sherwood, *Powder Technol.*, **65** (1991) *243.*

[2] D. Kashchiev; A. Borissova; R. B. Hammond; K. J. Roberts, *J. Cryst. Growth*, **312** (2010) *698.*

[3] D. Kashchiev; A. Borissova; R. B. Hammond; K. J. Roberts, *J. Phys. Chem.*, **114** (2010) *5441.*

[4] D. M. Camacho Corzo; A. Borissova; R. B. Hammond; D. Kashchiev; K. J. Roberts; K. Lewtas; I. More, *CrystEngComm*, **16** (2014) *974.*

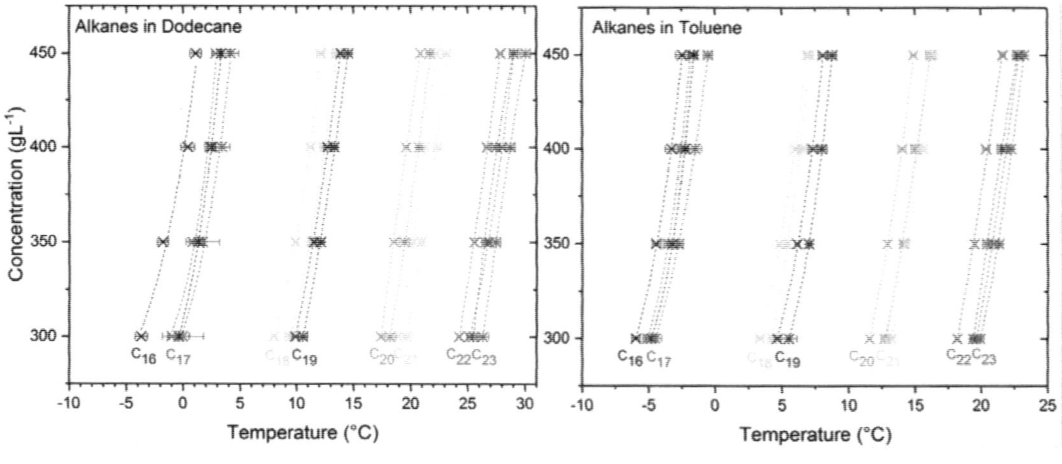

Figure 1 – Solubility curves of 8 *n*-Alkanes C_{16} – C_{23} in toluene and dodecane showing their metastable zones

611896 Multi-Scale Modeling of Soft Matters: A Quality By Design of Drug Delivery Carriers By a Computational Tool

Abstract:

In this work we investigate computationally the relationship between transient motion of multi-compartment microcapsules and their mechanical properties by flowing them in a microfluidic device. Different active ingredients can be encapsulated separately in inner (daughter) capsules that are surrounded by a larger (mother) capsule. The capsule shell (membrane) can be fabricated with materials that are stimuli (e.g. pH, temperature) responsive. Despite significant experimental advance in the fabrication of multi-compartment capsules [1-4], their deformation mechanisms and rheological behavior are poorly understood. Multi-scale modeling is a tool that facilitates the quality by design (QBD) for the fabrication of pharmaceutical capsules (e.g. Chitosan–alginate microcapsules [5]). QBD of microcapsules increases the probability in targeted delivery and the release of encapsulated materials at desired time and location, and mitigates the risk of cross-contamination between the ingredients. Here, we present the effects of different physical and geometrical parameters such as viscosity ratio, number of inner capsules, their sizes and arrangements on the overall dynamical behavior of multi-compartment capsules, by utilizing our membrane spectral boundary element method [6-9]. We further establish the link between the mechanical properties of these capsules and their deformational behavior, which leads to enhanced manufacturing of these class of drug carriers.

[1] Kim, Shin-Hyun, et al. "Multiple polymersomes for programmed release of multiple components." Journal of the American Chemical Society 133.38 (2011): 15165-15171.

[2] Adams, L. L. A., et al. "Single step emulsification for the generation of multi-component double emulsions." Soft Matter 8.41 (2012): 10719-10724.

[3] Dams, Stanley S., and Ian M. Walker. "[5] Multiple emulsions as targetable delivery systems." Methods in enzymology. Vol. 149. Academic Press, 1987. 51-64.

[4] Vasiljevic, Dragana, et al. "An investigation into the characteristics and drug release properties of multiple W/O/W emulsion systems containing low concentration of lipophilic polymeric emulsifier." International journal of pharmaceutics 309.1-2 (2006): 171-177.

[5] Ghaffarian, Rasa, et al. "Chitosan–Alginate Microcapsules Provide Gastric Protection and Intestinal Release of ICAM-1-Targeting Nanocarriers, Enabling GI Targeting In Vivo." Advanced functional materials 26.20 (2016): 3382-3393.

[6] Dodson, W. R., and P. Dimitrakopoulos. "Dynamics of strain-hardening and strain-softening capsules in strong planar extensional flows via an interfacial spectral boundary element algorithm for elastic membranes." Journal of fluid mechanics 641 (2009): 263-296.

[7] Koolivand, A., and P. Dimitrakopoulos. "Deformation of an elastic capsule in a microfluidic T-junction: settling shape and moduli determination." Microfluidics and Nanofluidics 21.5 (2017): 89.

[8] Koolivand, Abdollah, and Panagiotis Dimitrakopoulos. "Dynamics of Elastic Multi-Compartment Capsules in Microfluidic Channels." 2019 AIChE Annual Meeting. AIChE, 2019.

[9] Koolivand, Abdollah, and Panagiotis Dimitrakopoulos. "An Advanced 3D Computational Model for Designing Next Generation Drug Carriers." 2019 AIChE Annual Meeting. AIChE, 2019.

605435 Investigating Mechanisms of Bimetallic Nanocrystal Formation with Liquid Phase Transmission Electron Microscopy

Abstract:

Liquid phase transmission electron microscopy (LP-TEM) has been widely used to investigate the formation mechanisms of metal nanoparticles in liquid, providing guidance to synthesize nanocrystals with desired structures and functional properties.[1, 2] During LP-TEM imaging, the electron beam acts both as the imaging tool and the reducing agent to reduce either metal precursors to form nanocrystals, which enables visualizing nanocrystal formation in real time with nanometer scale spatial resolution. Despite numerous insights gained from these experiments, it is unclear how electron beam induced nanocrystal formation compares to nanocrystal formation during flask-based synthesis. Indeed, a systematic study comparing the chemistry and kinetics of nanocrystal formation during LP-TEM and flask-based colloidal synthesis has not been undertaken.

Here we establish ranges of LP-TEM experimental conditions that are representative of flask-based nanocrystal synthesis. We investigate Au/Cu bimetallic nanoparticle synthesis as a model system due to its promising application as a CO_2 reduction electrocatalyst.[3] Controllable synthesis of alloyed bimetallic nanocrystals is difficult due to differing kinetics and thermodynamics of precursor reduction of different metal species.[4] We utilized poly(ethylene glycol) methyl ether thiol (PEG-thiol, Mw=800 g/mol) to complex Au/Cu metal salts into prenucleation complexes (PNC) prior to nanoparticle synthesis.[5] Our results showed that flask-based synthesis using sodium borohydride to reduce PNCs formed ~2-3 nm Au/Cu alloy nanocrystals. LP-TEM synthesis, which utilizes radiolysis to produce aqueous electrons and hydrogen radicals as reducing agents,[6] formed 3 - 7 nm nanocrystals at low electron dose rates. Imaging at high electron dose rates (high image magnification and beam current) formed irregular aggregated particles. High electron dose rates increase the nucleation rate and create more oxidizing radical species,[7] which oxidizes the sulfur group on the PEG-thiol ligand, leading to formation of uncapped aggregated particles. Our study identified key differences between flask-based and LP-TEM synthesis methods, giving important guidance on how to set experimental conditions during LP-TEM to study nanocrystal formation mechanisms at conditions relevant to flask-based synthesis.

1. Zheng, H.; Smith, R. K.; Jun, Y.-w.; Kisielowski, C.; Dahmen, U.; Alivisatos, A. P., Observation of Single Colloidal Platinum Nanocrystal Growth Trajectories. *Science* **2009,** *324* (5932), 1309-1312.
2. Loh, N. D.; Sen, S.; Bosman, M.; Tan, S. F.; Zhong, J.; Nijhuis, C. A.; Král, P.; Matsudaira, P.; Mirsaidov, U., Multistep nucleation of nanocrystals in aqueous solution. *Nature Chemistry* **2016,** *9*, 77.
3. Kim, D.; Resasco, J.; Yu, Y.; Asiri, A. M.; Yang, P., Synergistic geometric and electronic effects for electrochemical reduction of carbon dioxide using gold–copper bimetallic nanoparticles. *Nature Communications* **2014,** *5*, 4948.
4. Wang, D.; Li, Y., Bimetallic Nanocrystals: Liquid-Phase Synthesis and Catalytic Applications. *Advanced Materials* **2011,** *23* (9), 1044-1060.
5. Marbella, L. E.; Chevrier, D. M.; Tancini, P. D.; Shobayo, O.; Smith, A. M.; Johnston, K. A.; Andolina, C. M.; Zhang, P.; Mpourmpakis, G.; Millstone, J. E., Description and Role of Bimetallic Prenucleation Species in the Formation of Small Nanoparticle Alloys. *Journ American Chemical Society* **2015,** *137* (50), 15852-15858.
6. Schneider, N. M.; Norton, M. M.; Mendel, B. J.; Grogan, J. M.; Ross, F. M.; Bau, H. H., Electron–Water Interactions and Implications for Liquid Cell Electron Microscopy. *The Journal of Physical Chemistry C* **2014,** *118* (38), 22373-22382.
7. Wang, M.; Park, C.; Woehl, T. J., Quantifying the Nucleation and Growth Kinetics of Electron Beam Nanochemistry with Liquid Cell Scanning Transmission Electron Microscopy. *Chemistry of Materials* **2018,** *30* (21), 7727-7736.

599612 Discovery of Light-Driven Hydrogen Evolution Catalysts Using High Throughput Simulation, Experiments and Machine Learning

Abstract:

The renewable production of hydrogen is an important technology. Current approaches to catalyst discovery for thermo/electro and photochemical methods for hydrogen production have been difficult to optimize. This difficulty arises in part because of the separation of the synthesis and characterization of catalysts from their performance under operating conditions. It is challenging to build models that relate catalyst performance to synthesis conditions. We have taken a different approach to this problem. We use a photo-driven process that couples a photoabsorber with the synthesis of multi-component metal particles and hydrogen production. This approach is readily parallelized in a multi-well plate, enabling us to explore a wide range of synthesis conditions that are directly connected to catalyst performance measurements. We will show how we used this approach to screen thousands of possible catalyst synthesis conditions to identify promising combinations. The high-throughput experiments along, however, do not tell us what metals to test. For this we augment the approach with a parallel high-throughput computation approach where hundreds of thousands of DFT calculations are performed on multi-component alloy surfaces to screen for promising candidates with desired hydrogen and CO adsorption energies. Finally, we combine these efforts to drive focused studies of the most promising candidate synthesis conditions to develop a more conventional understanding of how the synthesis conditions led to the observed activity. We will show how this approach has led to confirmation of some known results from the electrochemical production of hydrogen from water, as well as some new results we may not have discovered with any single approach.

601035 High Throughput Screening of Alloy Structures for Propane Dehydrogenation Reaction

Abstract:

Non-oxidative propane dehydrogenation (PDH) on intermetallic alloys is a promising technology to convert large amounts of propane available in the shale gas to propylene. However, the high temperatures required for the reaction lead to reduced selectivity towards propylene formation and deactivation of the catalyst surface.[1] Even though Pt and Pd alloys have shown promise in terms of improved selectivity in comparison to pure metals, they ultimately experience loss of reactivity under reaction conditions. This difficulty motivates development of a more fundamental understanding of why Pt and Pd alloys have improved performance compared to pure metals and, further, elucidation of how such insights can be leveraged to find other alloy combinations with simultaneously high selectivity, stability, and activity for PDH.

Through our previous analysis on PdIn alloys, we have demonstrated that, as compared to terraces, the step surfaces of 1:1 alloys are highly active under PDH conditions (5 orders of magnitude higher rates). Furthermore, a comprehensive microkinetic analysis pointed to useful descriptors for activity, selectivity and stability.[2] With these descriptors as a starting point, the current study focuses on high-throughput screening of intermetallic alloy structures, including both steps and terraces, to identify reactivity trends across Pt and Pd alloys for PDH and, ultimately, find improved catalysts for this chemistry. First, we deploy a simple automated framework for adsorption and activation energy calculations on large number of chosen terrace and step alloy structures. The Materials Project is used to identify the most stable bulk structure of the alloy composition, while a surface energy analysis points to the most stable surface terminations, and CatKit is employed to identify all the distinct adsorption sites [3]. Subsequently, an in-house algorithm identifies all unique adsorbate configurations and finds the most stable adsorption site. These data are organized with a python-based databasing tool, leading to identification of BEP-type correlations between bond breaking barriers and the binding energies of various adsorbates. Further the binding energies of key descriptors were correlated with electronic and geometric properties of the alloy surfaces, using Random Forest techniques. This, therefore, allowed for a deeper understanding of the surface features that impacts the binding of descriptors the most.

References:

[1] J.J.H.B. Sattler, J. Ruiz-Martinez, E. Santillan-Jimenez, B.M. Weckhuysen, Catalytic dehydrogenation of light alkanes on metals and metal oxides, Chem. Rev. 114 (2014) 10613–10653. https://doi.org/10.1021/cr5002436.

[2] R.R. Seemakurthi, et al. Structure Sensitivity and Microkinetic Analysis of Propane Dehydrogenation on PdIn alloy and Pd, In Preparation.

[3] J.R. Boes, O. Mamun, K. Winther, T. Bligaard, Graph Theory Approach to High-Throughput Surface Adsorption Structure Generation, J. Phys. Chem. A. 123 (2019) 2281–2285. https://doi.org/10.1021/acs.jpca.9b00311.

609929 Theory-Guided, Interpretable Machine Learning Finds Predictive Geometric Structure-Property Relationships for Chemisorption on Alloys

Abstract:

Chemisorption, the direct chemical bonding of an atomic or molecular species to a solid-state material, is central in the fields of catalysis, corrosion, and electrochemistry, among many others. Despite the fundamental nature and importance of chemisorption, linking the geometry (i.e., the structure and composition) of different materials to their chemisorption properties remains a critical challenge. Developing physically transparent and quantitatively accurate models that can relate the chemisorption strength between an adsorbate and a solid surface to the adsorption site's geometry is critical to advance our understanding of chemisorption. In this talk, we discuss our efforts to use a theory-guided machine learning approach, which uses an interpretable class of machine learning models called generalized additive models (iGAM models),[1] to discover predictive structure-property models that can quantify the chemisorption strength of O, OH, S, and Cl on Pt-metal and Au-metal alloy surfaces subject to various strain- and ligand-induced changes in the local geometric structure. The iGAM models show a strong degree of predictive accuracy, with an average root-mean-square-error (RMSE) of 0.046 eV for samples in the test set. Through quantification of the relative importance of the features used to construct the models, we identified three important geometric features of the adsorption site that impact the relative chemisorption strength on metal alloys: the strain in the surface layer, the number of *d*-electrons in the ligand metal, and the size of the ligand atom. The interpretable functional form of the iGAM models allows us to analyze the model behavior with respect to each of these geometric features. Comparison between the iGAM chemisorption models and established electronic-structure models shed light on the critical physical concepts that control the chemisorption process on metal surfaces.

References

(1) Lou, Y.; Caruana, R.; Gehrke, J. Intelligible Models for Classification and Regression. In *Proceedings of the 18th ACM SIGKDD International Conference on Knowledge Discovery and Data Mining*; KDD '12; ACM: New York, NY, USA, 2012; pp 150–158. https://doi.org/10.1145/2339530.2339556.

608786 Trends in Catalyst Stability and Reactivity: Extracting Physical Insights Using Simple Data-Driven Approaches

Abstract:

The catalytic turnovers of a typical reaction pathway are determined by the energies of between 10 to 100 reaction intermediates.[1] Discovering optimally performing catalysts entails maximizing catalytic turnovers across vast spaces of energy and materials. One way to simplify this formidable optimization problem is to parametrize the reaction mechanism using energy correlations.[2] These energy correlations, often linear functions, are mappings between binding energies of reaction intermediates (e.g. $C_xH_yO_z$*) and one to two catalytic descriptors (e.g. binding energies of CH* and O*). Such energy correlations have enabled the high-throughput screening of metals, oxides, and zeolites. The slopes of these linear correlations (e.g. 0.5 for correlations between OH* and O*) across metals, oxides, and zeolites are determined by the ratios of bond orders of the corresponding species. Recent advances notwithstanding, these correlations are limited by three key features. First, since these correlations do not include catalyst stability metrics, they are unable to describe catalyst dynamics under reaction conditions. Second, within the existing formalism it is difficult to reverse-engineer active sites through property (e.g. rate) → structure (e.g. chemical composition) "inverse design" approaches. Third, their linearity, together with rigid constraints on the slope, imposes severe limitations in the accessible regions of materials space. There has been a concerted effort in the computational catalysis community to bridge these three knowledge gaps using data-driven statistical learning models.[3]-[5] Herein, we illustrate our contributions towards reverse-engineering realistic active ensembles sites that possess targeted catalytic turnovers.[6]-[9] We place a special emphasis on extracting physical insights from regressed parameters within our simple data-driven approach.[6], [10]

In the first part, we present accelerated schemes for determining the energies of metal atoms that, in turn, influence catalyst stability.[6] Predicting energies of metal atoms with atomic site specificity enables expeditious evaluations of surface energies, cohesive energies of nanoparticles having generic shapes, and binding energies of metal adsorption sites. We calculate the energies of metal atoms by melding quadratic and linear interpolations across the space of coordination numbers, composition, and external strain.[7] These new interpolations shrink the training set size for 12 fcc p- and d-block metals from > 204 to as few as 24 Density Functional Theory (DFT) calculations, without sacrificing accuracy on the test set.[6] We will specifically highlight physical insights regarding metal-metal bonding that are obtained from the regressed parameters. As a proof of concept, we employ this method to reverse engineer thermodynamically stable active site motifs of nanoparticles that catalyze C_1 conversion.

In the second part, we investigate the existence of energy correlations on supported nanoparticles using Au nanorods on doped MgO supports as a prototypical model.[10] Using > 1700 DFT calculated binding energies, we show that although DFT-derived energy correlations between 31 reaction intermediates of CO oxidation, water gas shift, and methanol synthesis do exist, their slopes depart from bond order conservation constraints. This departure from bond order conservation at bifunctional metal/oxide interfaces enhances the flexibility in materials space during catalyst optimization. We build a conceptual framework using the multipole expansion in electrostatics to semi-quantitatively explain trends within the slopes. Through these trends, we elucidate the electrostatic origin of the departure in slopes from bond order conservation.

Taken together, these two studies reveal new functional forms for energy correlations that determine metal nanoparticle stabilities, and reactivity trends at bifunctional interfaces. More crucially, we show that parameters of simple data-driven models contain a wealth of intriguing physical insights that enhance model interpretability.

References

1. Norskov et al., *Proc. Natal. Acad. Sci. USA,* (2011), **108,** 937.
2. Abild-Pedersen et al., *Phys. Rev. Lett.* (2007), **99,** 016105.
3. Tran et al., *Nature Catal.* (2018), **1**, 696.
4. O'Connor et al., *Nature Catal.* (2018), **1**, 531.
5. Goldsmith et al., *AICHE J.* (2018), **64,** 2311.
6. Choksi, Streibel, and Abild-Pedersen., *J. Chem. Phys.* (2020), **152,** 094702.
7. Streibel, Choksi and Abild-Pedersen. *J. Chem. Phys.* (2020), **152,** 094701.
8. Choksi et al. *J. Phys. Chem. Lett.* (2019), **10,** 1852.
9. Roling, Choksi, and Abild-Pedersen. *Nanoscale* (2019), **11,** 4438.
10. Choksi et al. *Angew. Chem. Int. Ed.* (2018), **57,** 15410.

604578 Orienteering in an Uncharted Chemical Space: Searching for an Optimal Bimetallic Nanocatalyst

Abstract:

Nanoparticles (NPs) have received tremendous attention as catalysts due to their high specific surface area and abundance of unique active sites. When designing such a catalyst, the Sabatier principle tells us that it must be tuned to adsorb key intermediates neither too strongly nor too weakly. For this reason, it is imperative to screen candidate NPs for their adsorption behavior. This task is conventionally accomplished either through time-consuming experiments, or computationally-expensive *ab-initio* calculations, which are primarily limited to highly idealized periodic surfaces. This presents a bottleneck in materials discovery, one which can be overcome through the development of rapid methods to predict adsorption behavior. To this end, we present a universal model of small molecule adsorption to NPs [1], which is able to accurately capture the adsorption behavior of metal NPs of any size, shape, or composition with minimal computational cost. By combining this adsorption model with our new genetic algorithm approach to NP structure prediction, we are now able to efficiently traverse the vast chemical space of bimetallic NPs and identify possible catalytic NPs to feed experiments. Overall, this work leverages cutting-edge chemical modeling techniques, such as first principles calculations, machine learning and statistical regression to significantly accelerate materials discovery and enable the rapid identification of promising bimetallic nanocatalysts.

1. Dean, J.; Taylor, M.; Mpourmpakis, G. Unfolding Adsorption on Metal Nanoparticles: Connecting Stability with Catalysis. *Science Advances* 5 (9), eeax5101.

604822 Physics Informed Machine Learning of Chemisorption at Metal Surfaces

Abstract:

The formation and breakage of chemical bonds at active sites is the molecular basis of catalysis. Being able to rapidly compute interaction strengths between bonding entities and understand their trends holds the key to the design of improved catalysts. Despite recent advances, machine learning (ML) faces a tremendous challenge for catalysis applications due to its poor transferability and explainability. Here we present a physics informed machine learning (PIML) approach that integrates convolutional neural networks with the *d*-band theory of chemisorption for predicting the chemical reactivity of metal surfaces. With *OH and *CO as two representative adsorbates, we demonstrated that the hybrid ML models outperform the purely data-driven ones in both data scarce and rich regions, especially for out-of-sample systems. More importantly, the architecture design enables its physical interpretability, shedding light on the nature of chemical bonding at metal surfaces.

606955 Improved Catalyst Predictions with Machine Learning Coupled Alchemical Perturbation Density Functional Theory

Abstract:

First principles Kohn-Sham density functional theory (KS-DFT) calculations hinder large-scale searches for new and sustainable heterogeneous catalysts due to significant computational expense. Alchemical perturbation density functional theory (APDFT) is a promising method that provides rapid predictions of catalyst descriptors based on perturbations of electrostatic potentials that arise from isoelectronic transmutations made to a reference catalyst. While straightforward and simple, first-order APDFT approximations are largely inaccurate when large and/or many transmutations are made. Here, we first discuss systematic trends in errors related to these factors for binding energy (BE) prediction of CH_x, NH_x, and OH_x adsorbates on hypothetical alloy variations of fcc Pt(111) surfaces. Based on these observations, we demonstrate our approach fingerprinting locations of transmutations in these hypothetical alloys, and we show how to correct APDFT errors with machine learning models trained using about 3600 BE data points. Our models provide improved prediction accuracy with errors decreased by as much as an order of magnitude, and this approach extends the breadth of APDFT predictions, allowing rapid and accurate BE predictions on many more alloys made by numerous transmutation combinations.

599751 Tandem Metal Nanoparticle-Oxidase Catalysis for Selective Oxidation Reactions

Abstract:

There is great potential in combining heterogeneous metallic nanoparticle catalytic active sites and enzymatic active sites for sequential chemical transformations. This work combines enzymes, namely oxidases, and metallic nanoparticles in one material for tandem catalytic activity, analogous to substrate channeling. We investigate the cooperativity between the metallic and oxidase catalytic active sites in two selective oxidation reaction systems.

Gold nanoparticles and gold alloy nanoparticles are very selective oxidation catalysts using mild oxidants such as hydrogen peroxide. Oxidases convert sugars to carboxylic acids and hydrogen peroxide. This work demonstrates tandem oxidative activity using glucose oxidase and gold and gold alloy nanoparticles. The glucose oxidase will oxidize glucose to gluconic acid and hydrogen peroxide, and the gold/gold alloy nanoparticle will perform a second oxidation reaction.

In the first reaction sequence, glucose is oxidized to gluconic acid and hydrogen peroxide using glucose oxidase, and the gluconic acid is further oxidized to saccharic acid using hydrogen peroxide as the oxidant on the gold nanoparticle surface. We demonstrate that glucose oxidase is bound to the gold nanoparticle surface and that the hybrid system produces saccharic acid, while the glucose oxidase on its own does not produce saccharic acid, only gluconic acid. The selectivity of the hybrid system can be adjusted by changing the pH. Higher pH values promote oxidation to saccharic acid.

The second reaction sequence uses the same glucose oxidase hybrid gold/gold alloy nanoparticle system for the oxidation of glucose by glucose oxidase, and the hydrogen peroxide product is used to partially oxidize benzyl alcohol to benzaldehyde. Cooperativity is demonstrated between the glucose oxidase and the gold nanoparticles through increased production of benzaldehyde in the hybrid system versus the gold nanoparticles on their own. The effect of the alkalinity of the solution on the catalytic performance is also demonstrated in this system.

599905 A Redox-Active Hydrogel Mediates Electron Transfer for Efficient Electrosynthetic Hydrogen Evolution Via Three Classes of Hydrogenases

Abstract:

Molecular hydrogen is an important high-energy carrier for future energy technologies if produced from renewable electrical energy. Hydrogenase enzymes offer a pathway for bioelectrochemically producing hydrogen that is advantageous over traditional platforms for hydrogen production due to low overpotentials and ambient operating conditions. However, catalysis is often limited by the rate of electron transfer from the electrode surface to the enzyme's active site. We demonstrate the ability of a cobaltocene-functionalized polyallylamine (Cc-PAA) redox polymer to immobilize and mediate rapid electron transfer to three different hydrogenases from *Clostridium pasteurianum* and *Methanococcus maripaludis* for hydrogen evolution at a cathode. Furthermore, Cc-PAA-mediated hydrogenases can operate at high faradaic efficiency (80-100%) and low apparent overpotential (-0.578 to -0.593 V vs. SHE). Specific activities of these hydrogenases electrosynthetically evolving hydrogen via Cc-PAA were comparable to their respective activities in traditional methyl viologen assays, indicating that Cc-PAA mediates electron transfer at high rates, to most of the embedded enzymes.

605458 Characterizing the Biocatalytic Properties of Cytosolically Expressed Penicillin G Acylase from *E. coli* for the Synthesis of Cephalexin and Comparison with βF24A Mutant

Abstract:

Beta-lactam antibiotics have made large contributions in the fight against bacterial infections ever since their discovery in 1928. Currently the industrial production of beta-lactam antibiotics is conducted via batch-wise chemical synthetic pathways consisting of multiple reaction steps, use of protecting groups, harsh solvents and very low temperatures, generating large amounts of hazardous waste. To approach this, a great deal of research has been conducted in developing a more efficient enzymatic synthetic processing pathway. The biocatalyst of interest is penicillin G acylase from *E. coli* (EcPGA). The combination of distinct acyl-donor moieties and nucleophilic beta-lactam moieties results in the synthesis of different semi-synthetic beta-lactam antibiotics. While wild-type EcPGA has been shown to have poor synthetic properties due to high product affinity, protein engineering on the binding pocket has altered the binding properties of substrates and products to enhance specificity.

We have improved the expression of wild -type EcPGA 20-fold to 45 mg/L culture. Additionally, we have characterized the synthetic properties of wild type EcPGA, βF24A, and Assemblase®, a commercial variant of EcPGA for cephalexin synthesis. We found that Assemblase® possessed marginally improved synthetic parameters over βF24A and wild-type enzymes. Higher product accumulation during long-term reaction experiments confirmed this observation. Next, wild type EcPGA was immobilized on various epoxy-functionalized supports to enable recycling of the biocatalyst between reactions. Approximately 30% of free enzymatic activity was retained after immobilization with decreased apparent activity attributed to diffusion limitations, as suggested by the decreasing apparent specific activity with increasing particle sizes. The synthesis to hydrolysis ratio decreased slightly, possibly due to internal product accumulation due to diffusion limitations. We are currently modeling the reaction and diffusion within the carrier pores using a reaction diffusion model to predict optimum enzyme loading to achieve high productivity while minimizing reduction of the synthesis to hydrolysis ratio.

601077 Dynamics of the Urea-Urease Reaction in CSTR – Experiments and Detailed Model

Abstract:

We present experimental evidence of pH oscillations and bistability in the urea-urease enzyme reaction carried out in a continuous stirred tank reactor (CSTR) and compare these results with dynamics predicted by a detailed model of urea-urease reaction.

The CSTR is open to atmosphere with three inlet streams delivering solutions of urease, urea and sulphuric acid and one outlet stream. Sulfuric acid is used as a second substrate to form a feedback loop controlling production of ammonia making it possible to display nonlinear dynamical effects. The pH in the reaction mixture is measured via a pH electrode. Concentration of the inflow reactants are taken as variable parameters as well as the flow rate k_0. We show experimental k_0 - pH hysteresis curves for a wide range of urease concentrations and various inlet concentrations of sulfuric acid. More importantly, we also report pH oscillatory behavior with amplitudes of ~1 pH units occurring in a region of parameters adjacent to the region of hysteresis.

The model is based on a full scheme for protonation/deprotonation of the enzyme and its activated complex and, in addition, includes inhibition by the ammonium ions that are produced in weakly acidic conditions. A closer analysis based on the reaction network theory reveals that it is the inhibition that is responsible for the oscillatory behavior observed in the experiments. The model also shows an asymmetric double-bell-shaped curve of steady state reaction rate vs external pH in agreement with prior work. Near the higher peak there is a region of instability, where stable oscillations occur via Hopf bifurcation. Set of kinetic parameters for which the oscillatory behavior occurs was determined by the constrained stoichiometric network analysis, which uses linear optimization to select proper network parameters consistent with experimental observation of oscillations.

608773 CA@ZIF-L-Stabilized Pickering Interfacial Catalytic System for Efficient CO_2 Conversion

Abstract:

Enzymatic catalysis offers a green and potent alternative for efficient CO_2 conversion due to its high selectivity and specificity, high efficiency, and mild operational conditions. Carbonic anhydrase (CA) enzyme-based absorption technology for CO_2 capture has been intensively investigated. The main issue related to this method is the activity, stability and reusability of the CA enzyme in vitro. To address this issue, CA@ZIF-L-stabilized Pickering interfacial catalytic system was constructed. In detail, CA enzymes were embedded into zeolitic imidazolate framework-L (ZIF-L) particles through biomimetic crystallization. The as-synthesized CA@ZIF-L particles were then spontaneously assembled into capsule at the interface between the oil phase (hexadecane) and water phase during the formation of Pickering emulsions. CA@ZIF-L-stabilized Pickering interfacial catalytic system was finally constructed to convert CO_2 to bicarbonates. The performance of the system was evaluated by adding calcium ions to form calcium carbonate.

The results showed that CA could rapidly initiate the crystallization of ZIF-L to form CA@ZIF-L particles under mild aqueous conditions. The immobilization efficiency of CA could be above 90%. The synthesized CA@ZIF-L, which was served as a solid emulsifier to stabilize Pickering emulsion with its self-assembly properties at the oil-water interface, exhibited excellent catalytic activity in Pickering interfacial reaction and the corresponding cyclic performance. In detail, CA@ZIF-L-stabilized Pickering interfacial catalytic system exhibited higher catalytic efficiency than that of free CA@ZIF-L mainly because of the shortened diffusion path of CO_2 from gas phase to enzyme active center. The production rate of $CaCO_3$ through Pickering interfacial catalysis reached 2.73 mg min^{-1}, which is 1.14 times higher than free CA@ZIF-L. In the cycling experiment, about 96.2% of conversion efficiency could be retained for our system after 6 times recycling. These findings above implies that the system we constructed exhibits high activity and stability, particularly, superior recyclability for conversion of CO_2 into $CaCO_3$.

595910 Immobilization of Threonine Aldolase and Its Application in the Synthesis of Chiral β-Hydroxy-α-Amino Acid

Abstract:

Abstract: Threonine aldolase can synthesize β-hydroxy-α-amino acids from glycine and aldehyde in one step, which are important intermediate with two chiral centers. The free threonine aldolase is difficult to be separated and reused, and its stablity is not satisfied under industrial condition. Enzyme immobilization technology can effectively solve these problems. Based on the evaluation of enzyme activity recovery and reaction batch stability, immobilization methods of threonine aldolase such as adsorption, entrapment, covalent binding and crossing-linking were studied. After extensive screening, the covalent binding using amino resin with further glutaraldehyde cross-linking appeared to be the most promising method. Then we optimized the immobilization conditions such as the amount of carrier, glutaraldehyde concentration, the activation time for carrier, immobilization time, pH, and temperature. After optimizations, the immobilized enzyme activity recovery was up to 119.3%, and the enantiomeric excess (*ee*) value of products was 98%. After continuous immersion in the reaction solution for 18 days(enzyme activity is measured daily), the enzyme activity remained 82.4%. And the enzyme activity retention basically unchanged after being stored in a refrigerator at 4 °Cfor 62 days. Compared with free enzyme, immobilized threonine aldolase has better batch reaction stability, storage stability and stereoselectivity. This work also contributed to the economic flexibility of β-hydroxy-α-amino acids biosynthesis.

610124 Impact of Dry Coating on the Agglomeration and Dissolution of Cohesive, Poorly Water-Soluble Drug Powders

Abstract:

For fine powders, in particular those below 30μm, interparticle cohesion is the dominating force, inducing significant agglomerate formation that delays their dissolution rate[1,2]. For such fine powders, dry coating with glidants, typically silica nanoparticles, is very effective in reducing cohesion and hence the extent of agglomeration[3,4]. Previous papers have reported on the impact of dry coating on enhancements in drug powder bulk density, flow, and dissolution[5,6,7,8,9]. However, those papers did not quantify the extent of agglomerate size reduction and its impact on the dissolution[3,4]. In addition to presumed agglomerate size reduction after dry coating, the competing relationship with the surface hydrophobicity on the drug dissolution rate was not examined. In this work, the investigation is carried out to fill these gaps and develop a more quantitative understanding of the relationship between cohesion reduction and agglomerate size reduction. The latter objective will be based on proposing a predictive model that is based on the dimensionless parameter that captures the effect of interparticle cohesion, in the form of the granular Bond number. Griseofulvin (d_{50} of 10 μm) and ibuprofen (d_{50} of 20 μm) were selected as model drugs. Each drug was dry coated with either hydrophilic (A200) or hydrophobic (R972P) fumed silica via LabRAM, while varying theoretical surface area coverage (SAC) of silica. Agglomerated and primary particle sizes were evaluated by two different particle sizers, a gravity-driven dispersion method, and a compressed air dispersion method, respectively. Following USP <711> protocol, USP IV dissolution testing was conducted in de-ionized water. Surface hydrophilicity was measured via the modified Washburn method[10,11]. Inverse gas chromatography was employed to measure the surface energy of the dry coated powders. The results show that the dissolution rate significantly influenced by the changed surface hydrophilicity and the reduced agglomerate size. The multi-asperity model was employed to calculate the bond number for this work[12], utilizing the measured surface energy of the dry coated drug particles. The applicability of the previously presented mathematical correlation between a bond number to an agglomerate ratio[13] was tested by comparing the predicted agglomerate size to the measured agglomerate size keeping in mind that it does not account for the polydispersity of the powders and process shear effect from the dry coating. Overall, the outcomes from this study will help the design of solid oral dosages through the understanding of the impacts of the surface hydrophobicity and the agglomerate size after dry coating on the dissolution of poorly water-soluble drugs.

References

1. Yang, J.; Sliva, A.; Banerjee, A.; Dave, R. N.; Pfeffer, R., Dry particle coating for improving the flowability of cohesive powders. *Powder Technology* **2005,** *158* (1-3), 21-33.
2. de Villiers, M. M., Influence of agglomeration of cohesive particles on the dissolution behaviour of furosemide powder. *International Journal of Pharmaceutics* **1996,** *136* (1-2), 175-179.
3. Han, X.; Ghoroi, C.; To, D.; Chen, Y.; Davé, R., Simultaneous micronization and surface modification for improvement of flow and dissolution of drug particles. *International Journal of Pharmaceutics* **2011,** *415* (1-2), 185-195.
4. Han, X.; Ghoroi, C.; Davé, R., Dry coating of micronized API powders for improved

dissolution of directly compacted tablets with high drug loading. *International Journal of Pharmaceutics* **2013,** *442* (1-2), 74-85.
5. Jallo, L. J.; Ghoroi, C.; Gurumurthy, L.; Patel, U.; Davé, R. N., Improvement of flow and bulk density of pharmaceutical powders using surface modification. *International Journal of Pharmaceutics* **2012,** *423* (2), 213-225.
6. Huang, Z.; Xiong, W.; Kunnath, K.; Bhaumik, S.; Davé, R. N., Improving blend content uniformity via dry particle coating of micronized drug powders. *European Journal of Pharmaceutical Sciences* **2017,** *104*, 344-355.
7. Capece, M.; Huang, Z.; To, D.; Aloia, M.; Muchira, C.; Davé, R. N.; Yu, A. B., Prediction of porosity from particle scale interactions: Surface modification of fine cohesive powders. *Powder Technology* **2014,** *254*, 103-113.
8. Chen, L.; Ding, X.; He, Z.; Huang, Z.; Kunnath, K. T.; Zheng, K.; Davé, R. N., Surface engineered excipients: I. improved functional properties of fine grade microcrystalline cellulose. *International Journal of Pharmaceutics* **2018,** *536* (1), 127-137.
9. Kunnath, K.; Huang, Z.; Chen, L.; Zheng, K.; Davé, R., Improved properties of fine active pharmaceutical ingredient powder blends and tablets at high drug loading via dry particle coating. *International Journal of Pharmaceutics* **2018,** *543* (1-2), 288-299.
10. Thakker, M.; Karde, V.; Shah, D. O.; Shukla, P.; Ghoroi, C., Wettability measurement apparatus for porous material using the modified Washburn method. *Measurement Science and Technology* **2013,** *24* (12).
11. Washburn, E. W., The dynamics of capillary flow. *Physical Review* **1921,** *17* (3), 273-283.
12. Chen, Y.; Yang, J.; Dave, R. N.; Pfeffer, R., Fluidization of coated group C powders. *AIChE Journal* **2008,** *54* (1), 104-121.
13. Castellanos, A., The relationship between attractive interparticle forces and bulk behaviour in dry and uncharged fine powders. *Advances in Physics* **2005,** *54* (4), 263-376.

599608 Using the Conical Screen Mill As a Practical and Effective Dry-Coating Process to Enable Direct Compaction for High Drug Loading Formulations

Abstract:

Direct compaction is the simplest and most preferred process route in the manufacturing of tablets. Unfortunately, direct compaction cannot often be developed without severely restricting drug loading due to the poor flow and compaction properties typical of the active pharmaceutical ingredient (API). Dry-coating processes, which coat API or excipients with glidant (nano-sized silicon dioxide), have been shown effective to improve the manufacturability of formulations containing APIs with poor flowability and tabletability. Accordingly, dry-coating has been considered an enabling technology for direct compaction. Despite the advantages of dry-coating at the lab-scale using a wide variety of equipment such as powder blenders/mixers and mills, manufacturing-scale processes have yet to gain significant acceptance. This may be due to the poor scalability of dry-coating processes or presumably due to the aversion of risk associated with adopting new technology in the pharmaceutical industry.

One particularly attractive dry-coating device which has high relevance to the pharmaceutical industry is the conical screen mill (comill). The comill is a continuous process which is frequently employed in the production of tablets as a screening/de-lumping or particle size reduction process. Several lab-scale studies have shown that the comill is capable of dry-coating, but typically not as effective or efficient as batch processing equipment due to the very short residence time.

As a major novelty, this study makes a simple and easy modification to the comill screen in order to improve the effectiveness and efficiency of the comill used as a dry-coating device. The effect of screen properties and processing conditions are evaluated for a model cohesive material, Avicel PH 105. It is shown that the flowability of Avicel PH 105 can be improved equally as well as a benchmark batch dry-coating device known as the LabRAM (Resodyn Acoustic Mixer). The comill process with the modified screen is applied to a micronized grade of acetaminophen as well as several in-house APIs. The API's are formulated into 50% drug loading blends which are all shown to have adequate flowability and tabletability. The industry's familiarity with the comill combined with its improved effectiveness may allow it to be adopted as a dry-coating process in order to enable direct compaction for high drug loading formulations.

AbbVie funded and participated in the study design, research, data collection, analysis and interpretation of data, as well as writing, reviewing, and approving the publication. Maxx Capece, is an AbbVie employees and may own AbbVie stock/options. Arun Jayaraman and Christian Borchardt are former employee of AbbVie and have no conflicts of interest to report.

600636 Effect of ALD and MLD Coating on Wetting and Flow Properties of Micron-Sized Inhalation Pharmaceutical Powders

Abstract:

Budesonide powder is an important pharmaceutical ingredient, used to treat asthma or chronic obstructive pulmonary disease by respiratory inhalation. As an important physical property of Budesonide, wettability has a significant influence on dissolution rates and drug release characteristics. Good flowability of Budesonide powders is vital in the re-suspending of powder in inhalers. However, micron-sized Budesonide powders generally exhibit poor flowability as the inter-particle cohesive forces play a dominant role in the bulk powder behavior when the size of individual particles is less than 30 microns.

Coating of powder material can be a solution to modify surface properties of individual particles, so as to improve wettability and flowability. Therefore, we experimentally coated the Budesonide powders (size range of 0.1 to 10 μm) with SiO_2, TiO_2, Al_2O_3 and PET (Polyethylene terephthalate) in a vibrated fluidized bed reactor. Nano-sized films of SiO_2, TiO_2, and Al_2O_3 was coated on the surface of Budesonide particles via atomic layer deposition (ALD, inorganic coatings). Nano-scaled layers of PET was deposited onto Budesonide via molecular layer deposition (MLD, organic coatings). The effects of different coating materials, coating cycles (thickness) and coating process (ALD and MLD) on the two important properties of Budesonide powders, wettability and flowability were investigated.

Uncoated Budesonide powder is classified as a very hydrophobic material with poor flowability. Improvements both in the wettability and flowability of Budesonide after the ALD coating process are observed. The coated Budesonide becomes less hydrophobic and its flow property improves; the degree of improvement depends on the coating material. The positive effect can be further enhanced by increasing the number of coating cycles. In particular, Budesonide coated with TiO_2 shows the largest increase in wettability (smallest water contact angle). In contrast to ALD coated Budesonide, PET coated budesonide via MLD becomes more hydrophobic compared to the uncoated Budesonide.

608882 Impact of the Micro/Nano- Structural Properties of Materials on Tribo-Charging: A Case Study on Salbutamol Sulphate Particles

Abstract:

INTRODUCTION

Powder tribo-charging occurs due to the charge exchange between two solid surfaces that are brought into contact and then separated. During powder manufacturing, particles can acquire electrostatic charges from their frequent collisions within the powder and the surfaces of the processing equipment. The attained electrostatic charges can alter powder behaviours and affect its processability, i.e., reducing flow, inducing particle agglomeration and/or segregation, causing adhesion to surfaces, etc. Besides the type of processing operation (e.g., sieving, blending, conveying, etc.) and environmental conditions (i.e., temperature and relative humidity) used during production, the charging propensity of powders also depends on their multiple material attributes[1], particularly, their solid-state characteristics (e.g., crystallinity), presence of impurities, etc. For instance, crystalline salbutamol sulphate (SS) has been reported to charge distinctly, compared to its amorphous counterpart[2]. The presence of surface and bulk impurities can also contribute to powder tribo-electrification[3]. In this context, the present work aims at evaluating the impact of solid-state, including different amorphous contents and impurity contents, on the tribo-charging tendency of SS powders.

MATERIALS AND METHODS

Materials

Salbutamol sulphate (SS) was purchased from Fagron GmbH Co.KG (Barsbüttel, Germany) and engineered to obtain particles with different solid-state properties.

Methods

Powder engineering via ball-milling

Particle engineering was performed via ball-milling on a PM 100 planetary mill (Retsch, Germany), equipped with a 250 ml agate jar. Milling was performed at 400 rpm for different milling times (i.e., 10, 20, 40, 120 min) at room temperature.

Solid-state characterization of the salbutamol sulphate powders

The crystallinity of SS particles was analysed via wide angle X-ray scattering (WAXS) and attenuated total reflectance-Fourier transformed infrared (ATR-FTIR) spectroscopy. Information on the structural heterogeneity of the samples at nano-scale was obtained by the analysis of the small angle X-ray scattering (SAXS) signal. Karl Fischer (KF) titration was performed to determine the water content of the samples, whereas particle size and specific surface area were characterized by laser diffraction and gas adsorption, respectively. The impurities arising from the milling process were quantified via high performance liquid chromatography (HPLC). Results of the powder characterization of the engineered samples were compared to the starting material (i.e., crystalline powder) and to spray-dried SS (i.e., amorphous powder). For spray-drying process, the same parameters described elsewhere were used[4].

Charge evaluation

The electrostatic charge density of the milled powders and raw material was measured using the GranuCharge™ (GranuTools, Belgium) apparatus at 22 ± 2 °C and 38 ± 3% RH. Approx. 2 g of powder were used for the analysis by measuring the powder charge before and after flow in contact with the stainless-steel tubes of the GranuCharge™ device and the charge density of the samples expressed as difference between the two charge values. Charge density data was normalized by the specific surface area of the powders to allow a proper comparison between the samples.

Statistical analysis

Principal component analysis (PCA) was selected as multivariate data analysis method with the primary aim of identifying material attributes contributing to powder tribo-electrification. Analysis was performed using Simca 16.0 software (Umetrics, Sweden).

RESULTS AND DISCUSSION

The room temperature ball-milling process induced a progressive amorphization of the SS powders with increasing milling time as confirmed from the changes noticed in both FTIR and WAXS patterns. A completely amorphous sample was obtained after 120 min of milling, as shown by the absence of Bragg peaks in X-ray pattern of the material. Similarly, a broad halo was found for the spray-dried SS, further confirming the complete amorphicity of the 120 min sample. In addition to powder amorphization, the milling energy applied on the material also led to the drug degradation in solid-state and resulted in the appearance of impurities, particularly at longer milling times. All the engineered powders showed similar particle size distribution (i.e., x_{50}~7.2 µm; span~3.7), whereas a narrower distribution was obtained for the starting material (i.e., x_{50}=8.3 µm; span=2.7) and the spray-dried sample (i.e., x_{50}=7.0 µm; span=2.2). Since tribo-charging is particularly sensitive to the presence of water, the water content of the samples was quantified and found to be in the range between 3.6-4.7 wt% for all the produced samples. These amounts were considerably higher, when compared to the starting material (i.e., 0.2 wt%) and the spray-dried sample (i.e., 1.7 wt%). Regarding charge evaluation, all the milled powders and the starting material were found to charge positively before and after contact with the stainless-steel surface of the GranuCharge tubes. However, distinct charging trends were obtained for the different milling times. Charge density was found to initially decrease with milling time, reach a minimum at 20 min and thereafter, progressively increase up to 120 min (i.e., fully amorphous sample). The tribo-charging behaviour of fully crystalline and amorphous SS has also been explored in earlier studies[2,5], where, likewise to our current results, amorphous SS was found to charge to a higher extent (compared to its crystalline form). However, the effect of different degree of amorphicity and impurities on the SS tribo-electrification have not been investigated before. Both factors could potentially impact tribo-charging by inducing variations to surface chemical composition of the materials, thus altering their effective work function (i.e., driving force of the tribo-charging phenomenon)[2,3,5,6]. Statistical analysis of the results presented in this study revealed underlying correlations between the various material attributes (Figure 1). From the arrangement of the powder properties in the loading plot (i.e., plot displaying the variables in the principal components planes), the main source of variability among the samples was attributed to the particle size of the materials and their amorphous and water content. Those variables were in fact located along the first principal component (PC), explaining 53.3% of variability in the data, whereas the surface charge density (Δq), impurities and nano-heterogeneity were contributing to the second principal component of the PCA model. Surprisingly, powder tribo-charging was found to be not only correlated to solid-state properties, such as the amorphous content and impurity content, but a negative correlation was also revealed in respect to the structural heterogeneity of the samples at the nano-scale level. This observation suggests that the presence of heterogeneous domains at the nano-scale might disrupt charge transfer within the material, resulting in charge mitigation.

CONCLUSION

The present study showed that changes in solid-state properties and impurity level of SS powders can alter their tribo-charging behaviour. Statistical analysis revealed a correlation between powder tribo-electrification and the amorphous and impurity contents. Also, a relationship between the structural heterogeneity of the particles at the nano-scale and tribo-charging was found. This suggests an inter-dependence between amorphous and impurity contents and the nano-heterogeneity of materials and its consequent effect on tribo-charging.

REFERENCES

1. Kaialy, K. On the effects of blending, physicochemical properties, and their interactions on the performance of carrier-based dry powders for inhalation, Adv. Colloid Interface Sci. 235, 70-89 (2016).
2. Wong, J. Effect of crystallinity on electrostatic charging in dry powder inhaler formulations, Pharm Res. 31, 1656-1664 (2014).
3. Mukherjee, R. Effects of particle size on the triboelectrification phenomenon in pharmaceutical excipients: experiments and multi-scale modelling. Asian J. Pharm. Sci. 11, 603-617 (2016).
4. Littringer E. M. Spray Drying of Aqueous Salbutamol Sulfate Solutions Using the Nano Spray Dryer B-90—The Impact of Process Parameters on Particle Size. Dry. Technol. 31, 1346–1353 (2013).
5. Zellnitz S. Tribo-Charging Behaviour of Inhalable Mannitol Blends with Salbutamol Sulphate Pharm Res. 36, 80 (2019).
6. Mirkowska, M. Principal Factors of Contact Charging of Minerals for a Successful Triboelectrostatic Separation Process – a Review. Berg Huettenmaenn Monatsh 161, 359–382 (2016).

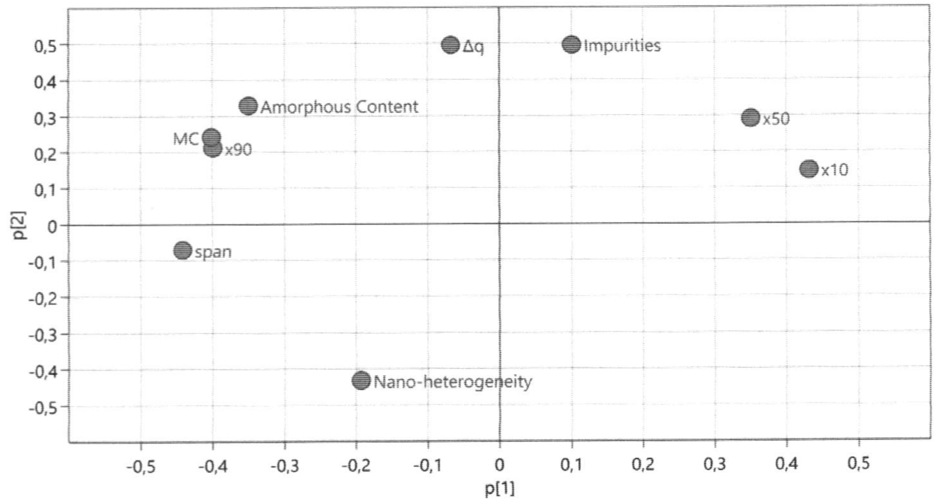

Figure 1 Loading plot of the PCA model.

610037 Inverse Gas Chromatography (IGC) of Pharmaceutical Powders: Harnessing the Power of Adhesion-Cohesion Forces to Improve Pharmaceutical Development

Abstract:

The surface energy of pharmaceutical powders can affect their processing behavior including flowability and miscibility with other powders. Traditionally, surface energies are measured by classical contact angle measurements. However, the surface roughness, the presence of pores, and surface energy gradients of some materials make the contact angle measurements less appropriate for surface energetics determinations [1]. Surface energies has also been reported as the property of pharmaceutical powders that has the most impact on aerosol formation and powder aerosol performance, in inhalation formulations. Therefore, the measurement of these physicochemical properties is considered critical to describe and predict the aerodynamic performance of inhalation products [2].

An alternative technique that can be used for surface energy measurements is inverse gas chromatography (IGC). This technique has been previously applied to the measurement of important physicochemical properties of pharmaceutical materials such as powder surface energies, acid/base/polar functionality of surfaces, diffusion kinetics, solubility parameters and surface heterogeneity. In IGC the roles of the stationary (solid) and mobile (gas or vapor) phases are inverted with respect to traditional analytical gas chromatography (GC).

This work aims at describing the development and application of a home-made inverse gas chromatograph for the determination of surface properties in different pharmaceutical powders at infinite dilution regimen. In this study, IGC was used to explore different properties of formulations used in dry powder inhalers (DPI) and oral dosage formulations and the results obtained were compared to those obtained using next generation impactor (NGI) and contact angle measurements.

Two different case studies where devised to demonstrate the applicability of IGC as a tool for the characterization of pharmaceutical powders during the development stages of formulation design and development. This study was conducted not only to demonstrate the suitability of IGC for surface energetics measurement but also to benchmark the results obtained using the home-made equipment and the results obtained by the well-established classical methodology.

For the first case study DPI formulations were used and a correlation between IGC results and emitted dose (ED) and fine particle fraction (FPF) obtained by NGI was evaluated. Lactose carrier-based formulations were prepared with different active pharmaceutical ingredient (API) content (2% 3.5% and 5%) and different proportions of fine lactose (5%, 7.5% and 10%). The aerodynamic performance of each formulation was determined using NGI performed in duplicate for each formulation. The dispersive (apolar) component of the surface energy, as well as the specific (polar) component were measured for each formulation in triplicate. By calculating the specific work of adhesion (W_a^S) of tetrahydrofuran (slightly basic) and chloroform (slightly acid) to the surface particles it was possible to chemically characterize the surface of the particles for all formulations, regarding their acid/base properties. The ratio between the works of adhesion calculated is a good indicator of the chemical behavior of the surface. The results obtained for the ratio of the work of adhesion varied between 1.63 and 1.71 for all formulations tested, demonstrating that the surface chemistry does not change significantly with changes in the formulation variables. This observation is consistent to the fact that the API content is always much lesser than the Lactose content. The average surface chemistry is not significantly changed by varying the API content between 5% (w/w) and 10%(w/w).

The results obtained for the dispersive component of the surface free energy (g^d) did not correlate well with the % Emitted Dose (ED) (r^2=0.199) neither with the Fine Particle Fraction (FPF) (r^2=0.548) measured by NGI. Only when the contribution of the specific interactions (Acid/Base) was accounted for, the determination of the surface interaction energy (SIE) correlation with FPF was observed with good agreement (r^2=0.808; Figure 1 a) confirming that the polar contributions of the surface free energy measurements of particles play a crucial role in understanding and determine particle-particle interactions as previously described [3]. It was also observed that the increase in the % of fine lactose present in the formulation increases the SIE (fic. 1 c), which is in agreement with previous observations [2] and counter intuitive to observations made for other materials since higher SIE are typically associated to more particle to particle interaction. However, it is proposed that for DPI formulations it is required some additional energy to ensure that micronized API which is very cohesive to be easily deagglomerated to originate better aerodynamic performance. The fine lactose increases the total surface area of the powder and competes with the API fort the high energy spots present in the coarse lactose therefore the increase in FPF observed when fine lactose is added to the formulation.

In addition to the IGC experiments, the traditional liquid wetting angle approach, for surface energy determination was also applied to surface characterization of four excipients commonly used in oral dosage forms, namely two diluents (lactose monohydrate, LM, and microcrystalline cellulose, MCC), one glidant (silicon dioxide, SD) and one lubricant (magnesium stearate, MgSt). Throughout these wetting angle studies, the surface energy, work of cohesion, work of adhesion and spreading coefficient between different excipients were determined.

Results showed that both SD and MgSt present a lower work of cohesion between their particles (28 mJ/m^2 for MgSt and 31 mJ/m^2 for SD) compared to the diluents (120 mJ/m^2 for LM and 105 mJ/m^2 for MCC). Since MgSt and SD present lower work of cohesion than adhesion to the different excipients (45 mJ/m^2 for LC-MgSt, 43 mJ/m^2 for LC-SD; 45 mJ/m^2 MCC-MgSt and 37 mJ/m^2 MCC-SD) they are likely to adhere and spread over both diluents surfaces. Raman imaging confirmed the surface coverage of diluents by both MgSt and SD when physical mixtures between the diluents and glidant/lubricant were prepared. As a result of the ability of MgSt or SD to spread over diluent particles during blending, these additives showed to be able to reduce the angle of repose, cohesion, compressibility and specific energy of the individual diluents.

IGC measurements were conducted in the same four excipients to determine the surface energy as well as the work of adhesion and cohesion and correlations between the IGC and liquid wetting angle results were explored, allowing to understand the benefits of each technique and major inconvenient.

The work conducted demonstrated that IGC can be used as an alternative methodology to support both oral dosage forms and DPI formulation development. The proposed methodology was successfully used to estimate the aerodynamic performance of a DPI formulation as demonstrated by the good correlation between NGI data and surface energy obtained using IGC. Preliminary work with excipients used in oral dose formulations also allowed identifying that the IGC data can be translated to this field of science, allowing insights on blending efficiency, predicting the impact of using excipients in formulations regarding final blend flowability and performance.

References:

[1] B. Riedl and P. D. Kamdem, "Estimation of the dispersive component of surface energy of polymer-grafted lignocellulosic fibers with inverse gas chromatography," *J. Adhes. Sci. Technol.*, vol. 6, no. 9, pp. 1053–1067, 1992.

[2] D. Cline and R. Dalby, "Predicting the quality of powders for inhalation from surface energy and area," *Pharm. Res.*, vol. 19, no. 9, pp. 1274–1277, 2002.

[3] D. Traini, P. Rogueda, P. Young, and R. Price, "Surface energy and interparticle forces correlations in model pMDI formulations," *Pharm. Res.*, vol. 22, no. 5, pp. 816–825, 2005.

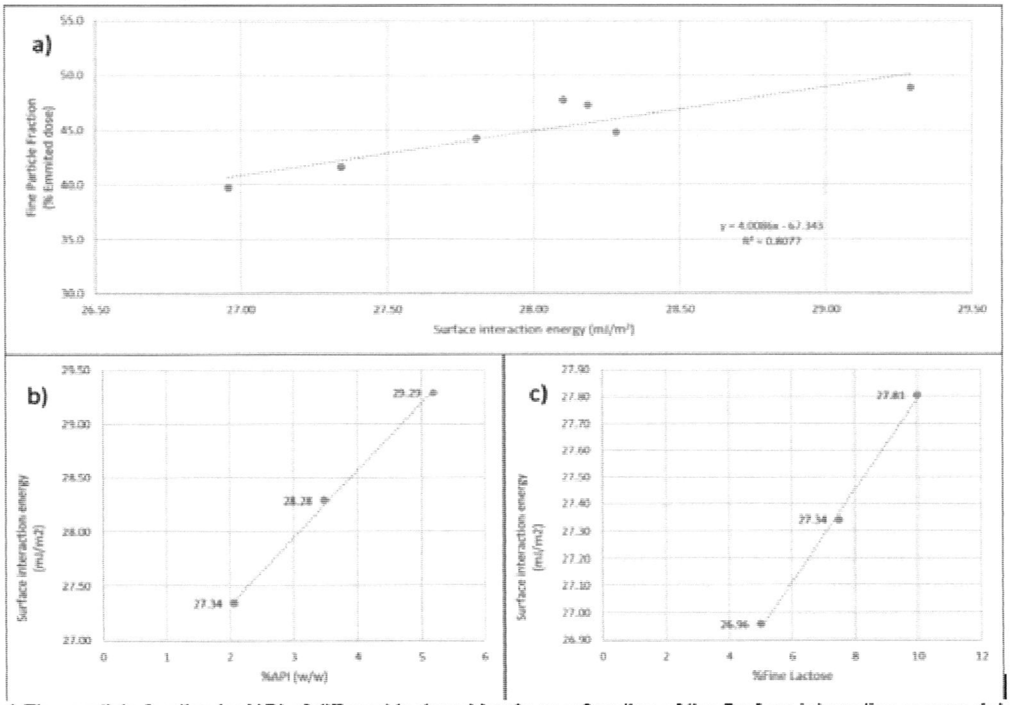

Figure 1. a) Fine particle fraction by NGI of different lactose blends as a function of the Surface interaction energy determined by IGC; **b)** Example of surface interaction energy as a function of API load in formulations containing 7.5% fine lactose; **c)** Example of surface interaction energy as a function of %fine Lactose in formulations containing 2% API load.

597399 Spherical Agglomeration of Paracetamol: A Mechanistic Approach

Abstract:

Spherical agglomeration is a size enlargement technique which offers many advantages such as ease of active pharmaceutical ingredient (API) handling and improved tabletability. Consequently, this reduces the need for further downstream processing during the manufacture of pharmaceuticals. In the spherical agglomeration process, an immiscible bridging liquid is added to agglomerate a suspension of primary particles. There are many process and formulation parameters that influence the final agglomerate properties, and the bridging liquid to solid ratio (BSR) is considered to be highly important. Although there has been a considerable amount of research regarding spherical agglomeration, there are a lack of studies that focus specifically on mechanistic understanding and the rate processes involved [1].

In this work, experiments have been designed to study the spherical agglomeration mechanism via immersion nucleation, *i.e.* where the bridging liquid droplets are larger than the primary particles to be agglomerated. Here, particles cover the droplets and subsequently penetrate the droplets forming agglomerate nuclei. A model system using paracetamol crystals suspended in heptane with water as the bridging liquid is used in this study.

A range of bridging liquid to solid ratios was investigated by adding the appropriate amount of bridging liquid directly to a stirred suspension of paracetamol crystals in heptane, at different solid loadings using a novel methodology. The product was characterized using image analysis (optical microscopy and digital photography). To establish the kinetics of the process, residence time studies were conducted. Samples were taken from the reactor at different time intervals and analysed. Agglomerate size and shape were determined via image analysis. Agglomerate density was measured using a combination of mass measurement and image analysis. The extent of agglomeration as a function of time was determined using a wet sieving technique. In certain experiments, the bridging liquid was dyed with acid red. This allowed visualization of the droplets, and confirmed that the spherical agglomeration process was via the immersion nucleation mechanism.

An optimum BSR range (0.7-0.8) has been identified where 100% agglomeration occurs. Here, stable, dense and highly spherical agglomerates with excellent flow properties are formed. This optimum BSR is independent of solid loading. Below the optimum BSR, a mixture of primary particles and agglomerates is produced. Above the optimum BSR, 100% agglomeration occurs, followed by formation of a paste; the timescales of which are dependent on both the BSR and the solid loading, as are the properties of the agglomerates prior to paste formation.

Changes in the extent of agglomeration and agglomerate properties (density, size) as a function of residence time also give insights into the mechanisms of spherical agglomeration via immersion nucleation. These kinetic studies will be used to validate a recently developed immersion nucleation mathematical model [2].

[1] K. Pitt et al. Powder Technol. 326 (2018) 327-343

[2] O. Arjmandi-Tash et al. Chem. Eng. Sci. X 4 (2019) 100048

610358 Reduction of Punch-Sticking Propensity of Celecoxib By Spherical Crystallization Via Polymer Assisted Quasi-Emulsion Solvent Diffusion

Abstract:

Purpose

Punch sticking is a common problem in tablet manufacturing that leads to poor tablet quality, such as tablet weight variation. This work was aimed at overcoming punch sticking problems using spherical crystallization achieved through quasi-emulsion solvent diffusion (QESD). The hypothesis is that a suitable polymer containing both hydrophilic and hydrophobic moieties facilitates the formation of emulsions because of its simultaneous interactions with water and celecoxib (CEL). Such a polymer is expected to remain largely at the water-drug interface, which results in CEL spherulites coated with polymer. Both the size enlargement and surface coating reduce punch sticking propensity by reducing direct contact between CEL and punch during compaction.

Methods

Celecoxib, a COX-2 selective nonsteroidal anti-inflammatory drug for treating pain and inflammation associated with osteoarthritis or rheumatoid arthritis, was selected as a model drug because it exhibits high punch sticking propensity. HPMC was used as an emulsion stabilizer. To prepare CEL spherical crystals, a CEL solution in ethyl acetate was added dropwise into HPMC aqueous solutions at different concentrations (0.1 %, 0.3 %, 0.5 %, w/w) while mixed by an overhead stirrer at 600 rpm. The growth process of CEL spherulites was monitored by observing samples under an optical microscope at 0.5, 1, 3, 6, 10, and 15 min. The appropriate concentration of HPMC sufficient to eliminate punch sticking was determined based on the cleanness of punch surface after compressing CEL spherulites at 50 MPa. The phase purity of the CEL spherulite powders was verified by DSC and PXRD. The particle size and shape were analyzed by laser diffraction and microscopy, respectively. Sticking kinetics of direct compression formulations, consisting of 20% of CEL 79.5% Avicel PH102 and 0.5% magnesium stearate, were assessed using a punch with a removable tip on a compaction simulator. The amount of material transferred to punch was determined gravimetrically every 10 compressions up to 50 compactions. Powder flow properties were characterized using a ring shear tester and powder tabletability was determined on a Zwick universal material testing machine. HPMC content was measured via size-exclusion chromatography (SEC). Surface coating of HPMC on CEL spherulites was studied by scanning electronic microscopy (SEM) and X-ray photoelectron spectroscopy (XPS). A solution 1D ^{1}H NMR was carried out to understand the mechanism of the QESD process.

Results

Spherical CEL was successfully prepared through a QESD process. No detectable punch sticking was observed when the HPMC concentration was 0.5 % (w/w), which was selected for further study. The CEL crystals were spherulites when HPMC was used but irregular in shape without using HPMC. Upon adding the CEL ethyl acetate solution into the HPMC aqueous solution, emulsions form with size of drug rich droplets gradually grew and eventually solidified. The PXRD, DSC and TGA of the spherical CEL matched well with the CEL as-received, suggesting the phase purity and negligible residual solvents. The amount of HPMC in the CEL QESD powder was determined 1.1 ± 0.1 %

(w/w). Both tabletability and flowability of the CEL spherical crystals were much better than the as-received CEL. The spherical CEL crystals exhibited significantly lower punch sticking propensity compared to that of as-received CEL. SEM images showed that the spherulites were covered by materials with irregular morphology distinct from that of the CEL crystals. The significantly reduced signals of fluorine and nitrogen atoms on the top surface of a compressed tablet than that of a CEL tablet suggested that the spherulites were coated by HPMC. Addition of HPMC into a 3 mg/mL CEL DMSO solution shifted the peaks within 7.50 ~ 7.60 ppm, 7.16 ~ 7.24 ppm and 2.28 ~ 2.36 ppm, to the upfield, corresponding to the H atoms on aromatic ring and methyl group of a CEL molecule.

Conclusion

This was the first work that reported polymer assisted QESD process to reduce the punch sticking propensity, attributed to the polymer on the surface of CEL particles, higher bonding strength and enlarged particle size of CEL QESD. Polymer coating was confirmed through: (1) surface observation by SEM, where the rough and irregular surface of QESD instead of sharp and smooth surface of as-received was observed, and (2) elemental composition analysis by XPS, where the peak intensity of feature elements F and N of CEL significant reduced in QESD compared to as-received. Mechanism investigation on polymer coating on SA surface through solution NMR suggested that the strong interactions between the hydrophobic groups on CEL and HPMC played a crucial role on the surface coverage of SA by HPMC but not HPC and PVP, whose intermolecular interactions with CEL were much weaker due to the lack of hydrophobic functional groups. Collectively, QESD is a powerful API crystal engineering approach to enable the successful development of direct compression tablets of sticky drugs.

597813 Molecular Design Blueprints: Catalysts and Principles from New Simulation and Machine Learning Tools

Abstract:

Many highly selective catalysts have been discovered that are defined by well-isolated sites with tailored metal-organic bonding. The rational design of such sites in de novo transition metal complexes, metal-organic frameworks (MOFs), or single atom catalysts, however, remains challenging. First-principles (i.e., with density functional theory, or DFT) high-throughput screening is a promising approach but is hampered by high computational cost, particularly in the brute force screening of large numbers of materials. In this talk, I will outline our efforts over the past few years to accelerate the design of single-site catalysts. I will describe our software and machine learning (ML) models that both simplify and accelerate the screening of new materials. These tools have enabled us to uncover new design rules and exceptions in both molecular and periodic materials (e.g., MOFs). We have paired ML models with multiobjective optimization strategies, robust uncertainty quantification, and highly parallel accelerated computing to discover optimal materials in weeks instead of decades. Time permitting, I will describe our recent efforts in autonomous computational chemistry by developing ML models as decision engines capable of predicting when calculations will fail and when methods beyond DFT are needed.

600080 Machine Learning Quantum Tunneling in the Kinetics of Chemical Reactions

Abstract:

The computation of reaction rate constants has been tackled using a multitude of theories ranging from classical transition state theory[1] to the fully quantum flux-flux correlation function approach.[2] The main cost of these methods is related to the need to explore potential energy surfaces, PES, either at the level of geometry optimizations, minimum energy path searches or to describe dynamics in time. In this context, machine learning algorithms which accelerate the search for reactive pathways or the prediction of rate constants are of great interest.[3,4] We have constructed a database of over 130000 reaction rate constants[5] obtained via quantum calculations of the reaction rate constants for particles crossing one dimensional single and double symmetric and asymmetric barriers. A deep neural network, DNN, was trained on the database with the ADAM optimizer and used to predict reaction rate constants. The most significant input features were identified using the Pearson correlation coefficient, and a grid search analysis was carried out for a coarse optimization of the DNN model hyperparameters. The network was trained for hundreds of epochs. The predicted results from the test set are in good agreement with the exact values. This approach shows promise for the application of machine learning to predict reaction rate constants.

1. Pechukas, P. Transition State Theory. *Annu. Rev. Phys. Chem.* **32**, 159–177 (1981).
2. Tromp, J. W. & Miller, W. H. The reactive flux correlation function for collinear reactions H + H2, Cl + HCl and F + H2. *Faraday Discuss. Chem. Soc.* **84**, 441–453 (1987).
3. Amabilino, S. *et al.* Training neural nets to learn reactive potential energy surfaces using interactive quantum chemistry in Virtual Reality. *J. Phys. Chem. A* **123**, 4486–4499 (2019).
4. Ulissi, Z. W. *et al.* Machine-learning methods enable exhaustive searches for active Bimetallic facets and reveal active site motifs for CO2 reduction. *ACS Catal.* **7**, 6600–6608 (2017).
5. Valleau, S. Machine learning quantum tunneling in the kinetics of chemical reactions. *In preparation.* (2020).

611092 Improved Catalyst Design and Performance through Application of Machine Learning Algorithms

- Download/View Extended Abstract

Abstract:

Conversion of biomass feedstocks to hydrocarbon fuels is often fraught with catalyst deactivation due to deposition of coke on catalyst surface or deactivation due to inorganic components of the biomass. Acidic zeolite catalyst get easily deactivated because of the basic oxides in the biomass feedstock. The high aromatic content and macromolecules in the biomass pyrolysis oils also tend to promote carbon formation on the active sites, which eventually lead to their deactivation. We developed a nickel supported on mixed oxide (Ni/RM) catalyst that is more robust than the commercial nickel on silica-alumina catalyst. This new catalyst is regenerable and can be used effectively for hydrodeoxygenation of biomass pyrolysis oils to long chain hydrocarbon fuels. The new catalyst is effective in opening both furan and benzene rings and thus producing hydrocarbon fuels with linear and branched carbon chain length between C6 to C15. The catalyst was not deactivated by inorganic components in the biomass, however, the catalyst activity decreased with deposition of coke on the active sites and required frequent catalyst regeneration through combustion and reduction to restore activity. We developed a machine learning (ML) algorithm which enabled us to reformulate the catalyst and reduce the carbon deposition on the active site. The *signature* of each nickel supported mixed oxide catalyst (Ni/RM)) was defined as the weight percent of *eleven elements* (Al, Ca, Fe, K, Mg, Na, Ni, P, S, Si, Ti). These together with the *reaction temperature formed the feature vector* (input to ML). The output of the ML was coke formation, which was also a performance measure for the designed catalyst. The network relating the ML inputs to the output was either a simple Neural Network or a Convolutional Neural Network (CNN). Using this ML model as a basis, we inverted the model (inverse modeling) to obtain a new Ni/RM catalyst, which was then validated through experiments. The machine learning approached reduced catalyst development time and improved the robustness of the catalyst. Thus, the carbon deposition rate was reduced and catalyst regeneration frequency was also reduced.

603620 Improving the Reusability of Machine Learning Models for Materials

Abstract:

Machine learning models have proven useful in a variety of chemical and materials applications. However, one drawback of these models is that they are often geared to a particular application, and thus a new model must be created for a new application. This is particularly noticeable in applications that feature guest-host interactions, broadly defined, as there is a huge combinatorial challenge in considering many possible guests and many possible hosts. For example, this combinatorial challenge applies to catalysis, adsorption, doping, and intercalation. In this work, we develop and apply strategies to improve the generality and reusability of machine learning models.

We have developed a machine-learning architecture that takes advantage of inherent aspects of chemical systems, and is particularly geared towards those that feature guest-host interactions. This framework partially decouples the guests from the hosts by introducing intermediate variables, which are a set of implicit host properties that control how the host interacts with all guests. These intermediate variables are learned during the fitting process. We create separate sub-models for different host elements for predicting these intermediate variables, and separate sub-models for how different guests respond to these host variables. The sub-models are all fit simultaneously. Our framework takes advantage of the fact that elements are discrete entities, and greatly simplifies the huge combinatorial challenge of considering many possible guests and many possible hosts. We apply this framework to multiple datasets and show how it can be used to improve the efficiency and accessibility of initial screening in a variety of applications.

610067 The Open Catalyst Project Dataset

Abstract:

The Open Catalyst Project aims to develop new ML methods and models to accelerate the catalyst simulation process for renewable energy technologies and improve our ability to predict activity/selectivity across catalyst composition. To achieve that in the short term we need participation from the ML community in solving key challenges in catalysis. One path to interaction is the development of grand challenge datasets that are representative of common challenges in catalysis, large enough to excite the ML community, and large enough to take advantage of and encourage advances in deep learning models. Similar datasets have had a large impact in small molecule drug discovery, organic photovoltaics, and inorganic crystal structure prediction. We present the first open dataset from this effort on thermochemical intermediates across stable multi-metallic and p-block doped surfaces. This dataset includes full-accuracy DFT calculations across 53 elements and their binary/ternary materials, various low-index facets. Adsorbates span 56 common reaction intermediates with relevance to carbon, oxygen, and nitrogen thermal and electrochemical reactions. Off-equilibrium structures are also generated and included to aid in machine learning force field design and fitting. Collectively, this dataset represents the largest systematic dataset that bridges organic and inorganic chemistry and will enable a new generation of catalyst structure/property relationships. Fixed train/test splits that represent common chemical challenges and an open challenge website will be discussed to encourage competition and buy-in from the ML community.

607970 Active-Learning Framework for the Discovery of Novel Materials for Catalysis

Abstract:

The discovery of abundant and affordable materials for sustainable energy applications is a significant challenge of this decade. The space of possible material compositions and crystal structures is so vast that brute-force screening approaches are unfeasible. However, density functional theory (DFT) combined with machine-learning (ML) techniques have a huge potential to speed up the search and hold a great promise for the discovery of novel materials.

Here, we present a ML-based active-learning framework to search for stable and meta-stable inorganic materials. The workflow includes the generation of candidate structures, which is followed by automated DFT job submission, ML-training, and acquisition. When using a search space of experimentally observed crystal structures, this approach was recently applied to the discovery of new stable polymorphs of IrO_x, predicted to have an enhanced stability and catalytic activity for the oxygen evolution reaction (OER). In this talk, we extend the search to hypothetical crystal structures that are generated from a bottom-up prototype enumeration scheme, based on spacegroup and Wyckoff positions. This is applied to the search for cheaper and more abundant materials. Last, we envisioned how the screening approach can be extended to surface and chemisorption properties, on order to extend the search to catalytic applications.

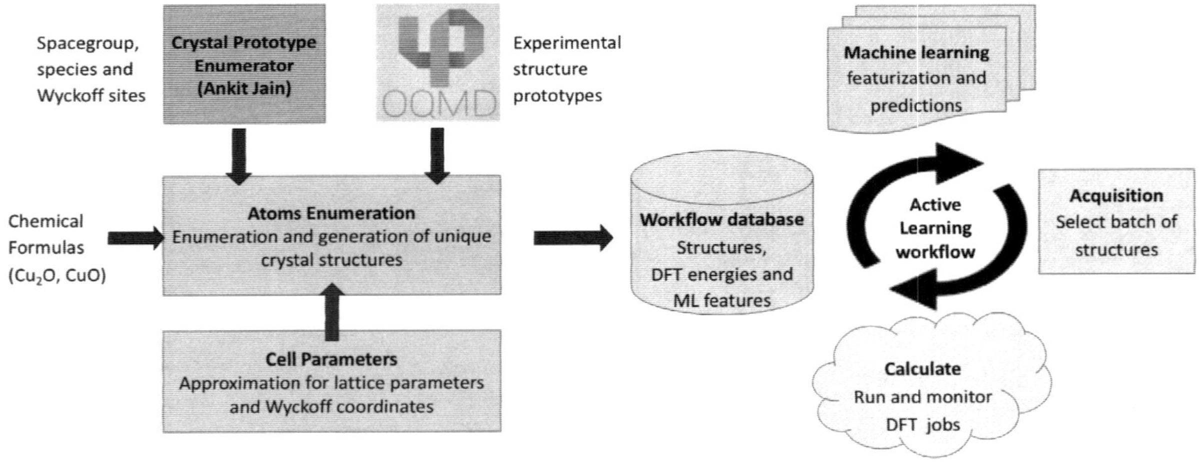

606321 Identifying Interaction Trends between Single Metal Atoms and Modified MgO(100) Supports with Statistical Learning

Abstract:

In this work, we apply statistical learning (SL) to identify physical descriptors for predicting electronic metal-support interactions (EMSI) in catalysis. Charge transfer between metal atoms and their underlying oxide support influences both the binding strength of the metal to the support and the binding strength of reaction intermediates to the metal. Tuning charge transfer by introducing surface-modifying dopants or co-adsorbates is an important strategy for controlling catalytic behavior. We use SL to build models for predicting metal atom binding energy to the oxide surface as a function of readily available physical descriptors containing properties of the metal, the support, and the surface-modifiers. Models are trained against DFT datasets consisting of metal adsorption energies on modified MgO(100) surfaces featuring various aliovalent dopants (e.g., Al, B, Li, and Na) and adsorbates (e.g., F, H, OH, and NO_2). These modifications generate both electron-rich and electron-poor MgO surfaces, which in turn enhance the extent of charge transfer between the metal and the support. We also test multiple SL procedures, including LASSO, Horseshoe prior, and Dirichlet-Laplace prior, and find that features derived by Dirichlet-Laplace prior construct the most accurate predictive models (i.e., achieving RMSE of ~0.2 eV for predicted binding energies). Moreover, we find that the features selected by SL using MgO training data generally are transferable to similar oxides, such as CaO(100), BaO(100), and ZnO(100), where data for these oxides are excluded from the feature selection process. This demonstrates the robust nature of features selected by the Dirichlet-Laplace prior procedure, both in terms of model accuracy and transferability to oxide surfaces beyond those included in the training set.

611372 How to Fix Some of Our Errors Via Data Science

Abstract:

Over the past two decades, multiscale modeling has advanced tremendously, and several algorithms currently exist. Yet, our ability to apply first principles modeling to process design is seriously limited due to multiple challenges. One of the challenges relates to error of first principles calculations, reaction mechanisms, active sites, etc. and its impact on catalyst discovery and reactor design. In this talk we will discuss sources of errors and new statistical learning methods that are transferable and explainable to quantify and correct some of the errors. We will discuss data quality and probabilistic graph models as a framework that integrates data, knowledge, and physical models to enable uncertainty quantification.

611147 Optimization-Driven Quasi-Deterministic Monte Carlo Sampling for Accurate Uncertainty Quantification: Applications to the Pharmaceutical Industry

Abstract:

Numerical algorithms for estimation of probability distributions are the backbone of advanced process design, monitoring and optimization techniques, e.g. stochastic dynamic optimization (Rossi et al., 2016), robust state estimation (Mondal et al., 2010) and real-time risk-based decision making (Si et al., 2012). The latter have recently attracted the attention of both industry and academia, due to the emergence of the industry 4.0 paradigm and the steady increase in the availability of readily accessible experimental and process data. In view of these considerations, the development of new and more efficient methods for PDF estimation could facilitate the systematic implementation of the industry 4.0 guidelines, thus improving process automation, economics, reliability and safety.

Most of the conventional algorithms for estimation of probability density functions (PDFs), e.g. Metropolis-Hastings, Gibbs sampling and Hamiltonian Monte Carlo (Gamerman and Lopes, 2006), consist of Monte Carlo sampling strategies, which rely on acceptance/rejection criteria to draw samples from the high probability regions of the PDF of interest. These algorithms are well-established, reasonably accurate and generally reliable but suffer from some important drawbacks: (I) poor performance in the presence of high parameter correlation; (II) limited applicability to complex PDF estimation problems (these state-of-the-art algorithms may not properly capture the features of multimodal PDFs, heavily skewed PDFs and/or PDFs with non-convex level sets); (III) considerable (often unaffordable) computational cost; and (IV) poor scalability and limited intrinsic concurrency. Unfortunately, probability distributions with complex shapes and high parameter correlation are very common in engineering applications, thus there is a need for new algorithms for estimation of probability distributions, which can mitigate some of the aforementioned drawbacks.

This contribution proposes a new strategy for PDF estimation, which combines an adaptive space partitioning scheme, inspired by multidimensional integration algorithms (Hahn, 2005), with a quasi-deterministic sampling method, based on optimization. Adaptive space partitioning alternates with optimization-driven sampling in an iterative fashion, until appropriate convergence conditions are met, e.g. the sample moments of the probability distribution of interest no longer vary between consecutive iterations. The adaptive cubature step serves three principal purposes (Figure 1): (I) identification of those regions of the uncertainty space (the space spanned by the parameters of the probability distribution of interest), in which the current estimate of the PDF requires further refinement; (II) detection of very low probability density regions, which can be safely ignored; and (III) improvement of the concurrency and overall computational efficiency of the algorithm (sampling can be conducted in parallel within different subregions). The optimization-driven sampling phase complements the former, in that it allows selection of optimal samples, which uniformly span every single subregion of the uncertainty space, without the need for acceptance/rejection schemes, which often accept only a small fraction of all the candidate samples analyzed. As consequence of the combination of adaptive space partitioning and quasi-deterministic, optimization-driven sampling, the proposed PDF estimation method can generate accurate estimates of complex PDFs at reasonable computational cost, is generally insensitive to parameter correlation, and exhibits better scalability features than similar state-of-the-art algorithms.

The new PDF estimation strategy, proposed in this contribution, has been demonstrated on several different case studies, namely, the estimation of several known probability distributions as well as the computation of the PDF of the parameters of the mathematical model of a continuous drug product manufacturing plant (the scale of these PDF estimation problems ranges from 2 to 8 dimensions). As a basis for comparison, all of these applications have also been solved with conventional Monte Carlo methods. These case studies confirm that the new PDF estimation strategy, proposed in this work, is more accurate and robust than state-of-the-art Monte Carlo methods.

References

Gamerman, D., Lopes, H.F. (2006). Markov Chain Monte Carlo - Stochastic simulation for Bayesian inference. Taylor & Francis Group, New York (NY).

Hahn, T. (2005). Cuba – a library for multidimensional numerical integration. Computer Physics Communications, 168, 78-95.

Mondal, S., Chakraborty, G., Bhattacharyy, K. (2010). LMI approach to robust unknown input observer design for continuous systems with noise and uncertainties. International Journal of Control, Automation and Systems, 8, 210-219.

Si, H., Ji, H., Zeng, X. (2012). Quantitative risk assessment model of hazardous chemicals leakage and application. Safety Science, 50, 1452-1461.

Rossi, F., Reklaitis, G., Manenti, F., Buzzi-Ferraris, G. (2016). Multi-scenario robust online optimization and control of fed-batch systems via dynamic model-based scenario selection. AIChE Journal, 62, 3264-3284.

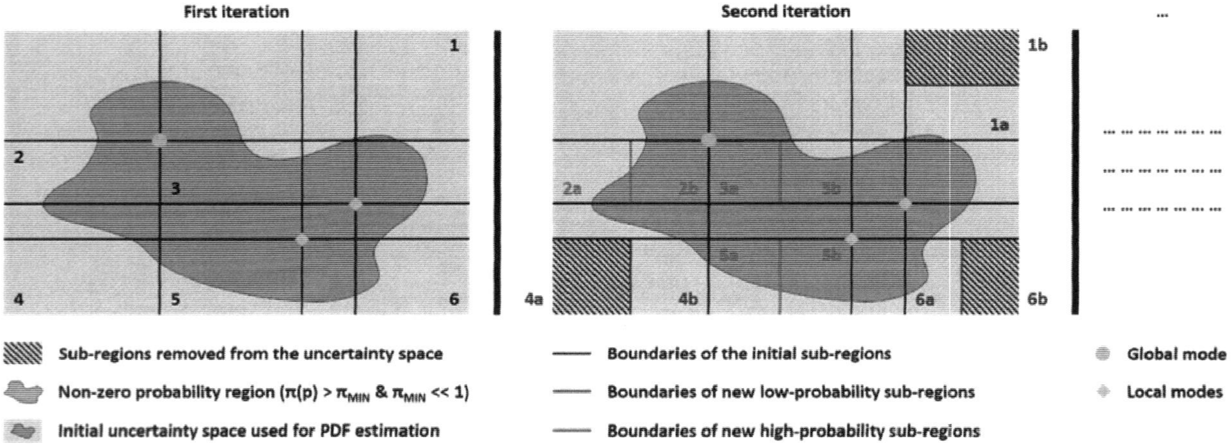

Figure 1: Rationale of the adaptive space partitioning scheme.

596368 Topological Data Analysis: Concepts, Computation, and Applications

Abstract:

Statistics, machine learning, and signal processing are the dominant paradigms used to analyze data; unfortunately, such techniques provide limited capabilities to analyze certain types of datasets. A couple of interesting examples that illustrate this limitation are the anscombe quartet [1] and the datasaurus dozen [2] datasets. These datasets are visually distinct (define a different geometrical object) but they have the exact same descriptive statistics (e.g., mean, standard deviation, and correlation).

The recent application of algebraic and computational topology to data science has led to the development of a new field known as Topological Data Analysis (TDA) [3]. TDA techniques are based on the observation that data can be interpreted as elements of a geometrical object; as the name suggests, TDA utilizes techniques from computational topology to quantify the geometry of data [4]. Fundamentally, topology studies geometric and spatial relations that are persistent (are stable) in the face of continuous deformations of an object (e.g., stretching, twisting, and bending). This perspective provides multiple advantages over other techniques [3, 5]:

• Topology studies the geometry of the data in a manner that is independent of the chosen coordinates.

• Topology studies the geometry of the data in a way that minimizes sensitivity to the metric chosen.

• Topology generalizes well to high-dimensional spaces.

The main focus of this talk is a technique in the field of TDA that is known as persistence homology [6, 7]. The goal of persistent homology is to extract topologically dominant features within the data in the form of basic features such as connected components, holes, loops, and voids. This feature information can be quantified and leveraged by statistical and machine learning techniques to perform regression, classification, hypothesis testing, and clustering tasks [8, 9, 10, 11, 12, 13].

TDA can be seen as dimensionality reduction technique that maps data from its original high-dimensional space to a low-dimensional space that it is easier to understand and visualize. This is similar in spirit to principal component analysis (PCA), which projects the data into a low-dimensional space by extracting latent variables (principal components) that contain information in terms of variance. In TDA, the latent variables are homologies that contain information in terms of topological features.

In this talk, we review orelevant concepts and computational methods of TDA from the perspective of chemical engineering applications. We show how to apply persistent homology to analyze datasets described by point clouds and functions in high dimensions and we discuss fundamental stability results of topological features in the face of perturbations. We present multiple case studies with complex synthetic and experimental datasets to demonstrate the concepts and advantages of TDA. Specifically, we show that TDA extracts informative features from complex datasets that correlate strongly with emerging features of practical interest. For instance, we demonstrate the application of these techniques in the analysis of time series and state space geometry, and in the geometric analysis of 2-dimensional diffusion scalar fields. Our work seeks to open new research directions and applications of TDA in chemical engineering.

[1] Francis J Anscombe. Graphs in statistical analysis. The american statistician, 27(1):17–21, 1973.

[2] Justin Matejka and George Fitzmaurice. Same stats, different graphs: generating datasets with varied appearance and identical statistics through simulated annealing. In Proceedings of the 2017 CHI Conference on Human Factors in Computing Systems, pages 1290–1294, 2017.

[3] Gunnar Carlsson. Topology and data. Bulletin of the American Mathematical Society, 46(2):255–308, 2009

[4] Herbert Edelsbrunner and John Harer. Computational topology: an introduction. American Mathematical Soc., 2010.

[5] Afra Zomorodian. Topological data analysis. Advances in applied and computational topology, 70:1–39, 2012.

[6] Robert Ghrist. Barcodes: the persistent topology of data. Bulletin of the American Mathematical Society, 45(1):61–75, 2008.

[7] Gunnar Carlsson, Afra Zomorodian, Anne Collins, and Leonidas J Guibas. Persistence barcodes for shapes. International Journal of Shape Modeling, 11(02):149–187, 2005

[8] Peter Bubenik. Statistical topology using persistence landscapes. arXiv preprint arXiv:1207.6437, 3, 2012.

[9] Peter Bubenik, Gunnar Carlsson, Peter T Kim, and Zhi-Ming Luo. Statistical topology via morse theory persistence and nonparametric estimation. Algebraic methods in statistics and probability II, 516:75–92, 2010.

[10] Peter Bubenik and Paweł Dłotko. A persistence landscapes toolbox for topological statistics. Journal of Symbolic Computation, 78:91–114, 2017.

[11] Andrew J Blumberg, Itamar Gal, Michael A Mandell, and Matthew Pancia. Robust statistics, hypothesis testing, and confidence intervals for persistent homology on metric measure spaces. Foundations of Computational Mathematics, 14(4):745–789, 2014.

[12] Henry Adams, Tegan Emerson, Michael Kirby, Rachel Neville, Chris Peterson, Patrick Shipman, Sofya Chepushtanova, Eric Hanson, Francis Motta, and Lori Ziegelmeier. Persistence images: A stable vector representation of persistent homology. The Journal of Machine Learning Research, 18(1):218–252, 2017.

[13] Jan Reininghaus, Stefan Huber, Ulrich Bauer, and Roland Kwitt. A stable multi-scale kernel for topological machine learning. In Proceedings of the IEEE conference on computer vision and pattern recognition, pages 4741–4748, 2015.

596537 Computational Fluid Dynamics-Based in-Situ Sensor Analytics of Direct Metal Laser Sintering Process Using Machine Learning

Abstract:

The wide utilization of the AM method lies in its advantages including rapid prototyping, faster manufacturing, reduced time and operating cost, and also the possibility to create almost any geometry. For complicated, high yield strength and high precision metal part construction, laser powder bed fusion (LPBF) has been extensively researched and prototyped [1]. To identify errors in real-time to avoid detrimental damage to the machine and material and energy waste, in-situ sensor monitoring technologies are developed to record the manufacturing information [2]. Nevertheless, the overwhelming amount of image based data produced by the optical sensors poses challenges in the data analysis and storage [3].

To address the aforementioned problem, we develop an automated data-flow that integrates the simulation model and machine learning network for real-time process simulation and sensor data analytics. In particular, the simulation models, such as thermal history analysis, are tailored to mimic the manufacturing platform characteristic and to incorporate potential process disturbances. From the simulation result, the heat map images are reconstructed with corresponding classified disturbances categories, according to the industrial sensor technologies. Advanced machine learning algorithms like convolutional neural networks (CNN) are designed and trained with transfer learning to achieve a robust yet computationally acceptable in-situ process analysis.

[1] Frazier, W.E., 2014. Metal additive manufacturing: a review. Journal of Materials Engineering andPerformance 23, 1917–1928.

[2] Everton, S.K., Hirsch, M., Stravroulakis, P., Leach, R.K., Clare, A.T., 2016. Review of in-situ process monitoring and in-situ metrology for metal additive manufacturing. Materials & Design 95.

[3] Scime, L., Beuth, J., 2019. Using machine learning to identify in-situ melt pool signatures indicative of flaw formation in a laser powder bed fusion additive manufacturing process. AdditiveManufacturing 25, 151–165.

598188 A Data-Engineering Approach for Estimating Chemical Releases from Industrial Pollution Abatement Technologies

Abstract:

Chemical substances available for industrial and commercial activities or uses may present an unreasonable risk to human health and the environment. In the U.S., the chemical risks should be evaluated according to the Toxic Substances Control Act, while in Europe, this is under the Registration, Evaluation, Authorization, and Restriction of Chemicals (REACH). Therefore, knowing what occurs with a chemical at each stage of its life cycle is essential for its evaluation, selection, use, and regulatory decision-making. However, gathering information to evaluate chemical risk is a time-consuming task, and especially at the end-of-use (EoU) stage because of the high uncertainty about chemical fate and exposure pathway. The contribution of this work is a data engineering approach for gathering information about pollution control units (PCUs) from publicly-available regulatory databases and filling data gaps based on technical information. The data collected include the type of waste stream having the chemical of interest (e.g., liquid waste), composition, and PCUs (e.g., absorber) and their efficiencies, which the framework transforms into a machine-readable structure for future automation. Thus, the developed approach can rapidly streamline chemical release estimations from PCUs and allocate potential effluents. Besides, it could be applied to identify and recommend the application of industrial PCUs for managing chemicals at generator facilities. The developed approach will be applied for case studies with the resulting identification of PCUs and release estimates. This approach may support the risk evaluation process by developing learning-from-data models to select PCUs and estimate releases.

The views expressed in this abstract are those of the authors and do not necessarily represent the views or policies of the U.S. Environmental Protection Agency.

611106 Physics Informed Machine-Learning for Static Security Analysis of Optimal Power Flow Solutions

Abstract:

In power grid control centers, optimal power flow (OPF) problems are solved several times per day to find optimal setpoints given load demands. However, an even more challenging problem is to have the ability to quickly assess the security of the optimal operating points in the case of contingency events which can result in operating at insecure or suboptimal points. Static grid security analysis can be done through N-1 contingency enumeration, where each contingency event is simulated and the resulting OPF solution is checked for voltage and line flow violations. An operating point that has no voltage or line flow violations is N-1 secure, but this becomes intractable to solve online for large grids that have hundreds of contingencies and thousands of variables. Deep machine learning models have had success in predicting system security under contingency [1, 2] or functioning as a quick OPF solver [3], but operators are wary of these fully black-box models because they require a lot of reliable training data, they ignore well-established power-flow physics, and have no interpretability.

Physics-informed machine learning models have been shown to approximate complex, nonlinear functions while retaining physical generalizability in the areas of physics, differential equations, and systems engineering [4, 5]. In this work, deep neural networks with embedded physics have been trained to predict OPF solutions under contingency and the resulting system security. Various approaches are considered for the generation of balanced and realistic training data, using offline simulations of varying load profiles and rigorous optimization for all contingencies using power flow software package Egret [6]. Different techniques for developing the physics-embedded NN models will be presented, including an augmented loss term that considers Kirchoff's Current Law at every node in the system, which allows for fewer training points and better performance outside of the training set. This framework is demonstrated on IEEE case studies of increasing size and complexity, and all solutions are compared to fully black-box approaches with respect to accuracy, training data requirements and complexity. Finally, integration of dimensionality reduction techniques with physics-informed NN training is presented for large case studies, which allows for maintaining model accuracy and tractability in the case of high-dimensional input spaces.

References:

1. Sunitha, R., R.S. Kumar, and A.T. Mathew, *Online static security assessment module using artificial neural networks.* IEEE transactions on power systems, 2013. **28**(4): p. 4328-4335.
2. Donnot, B., et al., *Fast Power system security analysis with Guided Dropout.* arXiv preprint arXiv:1801.09870, 2018.
3. Hu, X., et al., *Physics-Guided Deep Neural Networks for PowerFlow Analysis.* arXiv preprint arXiv:2002.00097, 2020.
4. Raissi, M., P. Perdikaris, and G.E. Karniadakis, *Physics-informed neural networks: A deep learning framework for solving forward and inverse problems involving nonlinear partial differential equations.* Journal of Computational Physics, 2019. **378**: p. 686-707.
5. Lutter, M., C. Ritter, and J. Peters, *Deep lagrangian networks: Using physics as model prior for deep learning.* arXiv preprint arXiv:1907.04490, 2019.
6. Knueven, B., et al., *Egret v. 0.1 (beta).* 2019: ; Sandia National Lab. (SNL-NM), Albuquerque, NM (United States). Python.

Disclaimer: Sandia National Laboratories is a multimission laboratory managed and operated by National Technology and Engineering Solutions of Sandia LLC, a wholly owned subsidiary of Honeywell International Inc. for the U.S. Department of Energy's National Nuclear Security Administration under contract DE-NA0003525. This paper describes objective technical results and analysis. Any subjective views or opinions that might be expressed in the paper do not necessarily represent the views of the USDOE or the United States Government.

610737 Statistics Pattern Analysis-Based Multiclass Woodchip Moisture Classification System Using Short-Range Iot Wi-Fi

Abstract:

The pulp and paper industry is the third-largest consumer of energy in the US industrial sector thereby leading to tremendous opportunities to improve its energy efficiency and productivity. The pulping process, which converts woodchips into pulp by displacing lignin from cellulose fibers, is one of the most important operations in a pulp and paper mill. For the pulping process in a pulp & paper plant that uses wood as raw material, it is important to have real-time knowledge about the moisture content of the woodchips so that the process can be optimized and/or controlled correspondingly to achieve satisfactory product quality while minimizing the consumption of energy and chemicals. Both destructive and non-destructive methods have been developed for estimating moisture content in woodchips, but these methods are often lab-based that cannot be implemented online due to time constraint [1], relatively complex and costly to install and operate[2], or too fragile to stand the harsh manufacturing environment. Due to these limitations, at present, the moisture level is measured only a few times throughout the year in the pulp and paper industry.

To address these limitations, we propose a non-destructive and economic approach based on the Internet-of-Things (IoT) based Wi-Fi and use channel state information (CSI) to estimate the moisture content in woodchips. We extract CSI by modifying the open-source device drivers for Intel Wi-Fi link 5300 Network Interface Card based on CSITool [3]. CSI contains information about the channel in the form of individual data sub-carriers capturing indoor channel features like the effect of scattering, fading, and power decay with distance. We hypothesize that when a Wi-Fi signal passes through wood chips, the water or moisture content in woodchips would cause a corresponding change in the Wi-Fi signal which is then captured in CSI. The goal is to develop a data-driven model to extract the relationship between the change in CSI and the woodchip moisture content. CSI has been successfully used for moisture detection in wheat [4]. However, compared to wheat moisture detection, woodchip moisture detection is much more challenging due to the significant heterogeneity in the shape, structure and arrangement of the woodchips. For example, the woodchip arrangement in the container is expected to have a significant impact on the CSI data.

In this work, our goal is to develop a robust approach that is insensitive to the heterogeneity of the woodchips, and in particular the effect of shuffling, to achieve a consistent and accurate moisture estimation. To achieve this goal, our proposed approach is based on statistics pattern analysis (SPA) framework [5][6] we developed previously, where SPA is integrated with different classification approaches to build robust multivariate statistical models for multiclass moisture classification in woodchips. Specifically, SPA is used to extract features from extremely noisy raw Wi-Fi CSI data thereby converting the messy big data into so called smart data for accurate sensing. These features are then used to build multiclass classification models using machine learning techniques including linear discriminant analysis (LDA), subspace discriminant classification (SDC), support vector machines (SVM), and artificial neural networks (ANN). We demonstrate an extensive study on the above-mentioned approaches for multiclass classification which, when combined with SPA, provide potential solutions for estimating woodchip moisture content for the pulp and paper industry. Our results show highly accurate classification of more than 20 moisture levels even when two moisture classes are separated by a very small margin of $<1\%$ (wet basis). Our approach is not only economic and non-destructive but can also be implemented online, making it highly desirable for the pulp and paper manufacturing environment.

[1] R. Govett, "A PRACTICAL GUIDE FOR THE DETERMINATION OF MOISTURE CONTENT OF WOODY BIOMASS A Practical Handbook of Basic Information, Definitions, Calculations, Practices and Procedures for Purchasers and Suppliers of Woody Biomass."

[2] S. O. F. Sensing, W. H. O. Uses, and E. Method, "Learn the Six Methods For Determining Moisture."

[3] D. Halperin, W. Hu, A. Sheth, and D. Wetherall, "Tool Release: Gathering 802.11n Traces with Channel State Information," *ACM SIGCOMM Comput. Commun. Rev.*, 2011.

[4] W. Yang, X. Wang, S. Cao, H. Wang, and S. Mao, "Multi-class wheat moisture detection with 5GHz Wi-Fi: A deep LSTM approach," in *Proceedings - International Conference on Computer Communications and Networks, ICCCN*, 2018.

[5] Q. P. He and J. Wang, "Statistics pattern analysis: A new process monitoring framework and its application to semiconductor batch processes," *AIChE J.*, 2011.

[6] J. Wang and Q. P. He, "Multivariate Statistical Process Monitoring Based on Statistics Pattern Analysis," *Ind. Eng. Chem. Res.*, 2010.

PROAD (Process Advisor): Health monitoring dashboard for centrifugal pumps

Abstract ID: 601927

Arnab Dutta[a,b], Iftekhar A Karimi[a], Shamsuzzaman Farooq[a]

[a] Department of Chemical and Biomolecular Engineering
National University of Singapore, 4 Engineering Drive 4, Singapore 117585

[b] Department of Chemical Engineering
Birla Institute of Technology and Science (BITS) Pilani, Hyderabad Campus
Jawahar Nagar, Kapra Mandal, Hyderabad 500 078, Telangana, India

In the era of IoT and digitalization, developing user-friendly tools to facilitate the maintenance of critical assets within a chemical plant is an area of growing industrial interest. We have integrated concepts of data analytics along with information and insights obtained from our industrial collaborator to develop a tool that will aid engineers to monitor the health of centrifugal pumps using plant data. In the proposed talk, we will report the development of PROAD – a health monitoring dashboard for centrifugal pumps, a critical and common equipment in a range of process industries spanning from oil and gas, chemicals, and petrochemicals, to food, agriculture, paper and pulp, and pharmaceuticals.

In our work, we have developed a comprehensive methodology that allows health monitoring, diagnosis, and prescriptive maintenance of centrifugal pumps. We adopt a multi-pronged approach where we use process operational data for health monitoring, and equipment-specific data for prescriptive maintenance and fault diagnosis. Process operational data are used to compute and monitor the deviation of the pump head as a function of flow rate (pump head curve) from the vendor-supplied head curve for a fresh pump. Recent trends of these deviations are then used to predict the remaining useful life of the pump w.r.t. a given threshold head deviation. On the other hand, continuous readings from different vibration sensors on a pump are analyzed for trends to prescribe more detailed spectral checks for potential incipient faults. In addition, a diagnostic tool is also developed that analyzes spectral data and recognizes known patterns linked to various mechanical faults (such as imbalance, misalignment, bearing damage, cavitation, etc.) common in centrifugal pumps. Our diagnostic tool also has the ability to add new and unknown faults that may arise during the pump operation and get diagnosed by maintenance engineers.

Besides these process/mechanical health indicators, we estimate the energy equivalent economic loss and carbon emissions associated with the pump over its operational period.

The above methodologies are implemented within a standalone plug and play desktop application (PROAD) easily implementable in any chemical plant without requiring any external software license. The app revolves around a graphical user interface (GUI) – based dashboard built using the MATLAB app designer feature, which displays all pumps on a single screen. Data input to PROAD is via well-directed, interactive, and robust MS-EXCEL templates. Based on different performance and diagnostic metrics, health status of each pump is indicated by specific color codes. Detailed health status and historical performance of each pump are conveyed via reports, and visualized via head, efficiency, vibration, and spectral plots. PROAD maintains a database of all reports and plots for future access. PROAD also has a feature that summarizes the health status of all pumps, thus providing an easy way to assess their performance at once. The GUI of PROAD is developed in consultation with engineers from our industrial collaborator, thus any information obtained from PROAD can be easily interpreted by any industry personnel. An overview of health monitoring of centrifugal pumps via PROAD is illustrated in Figure 1.

We have tested PROAD on multiple centrifugal pumps with about 20 months of real data obtained from our industrial collaborator. The health status and diagnoses inferred from PROAD were consistent with the performance observed and confirmed by the plant engineers. The tool is currently being tested by our industrial collaborator and refined with their feedback.

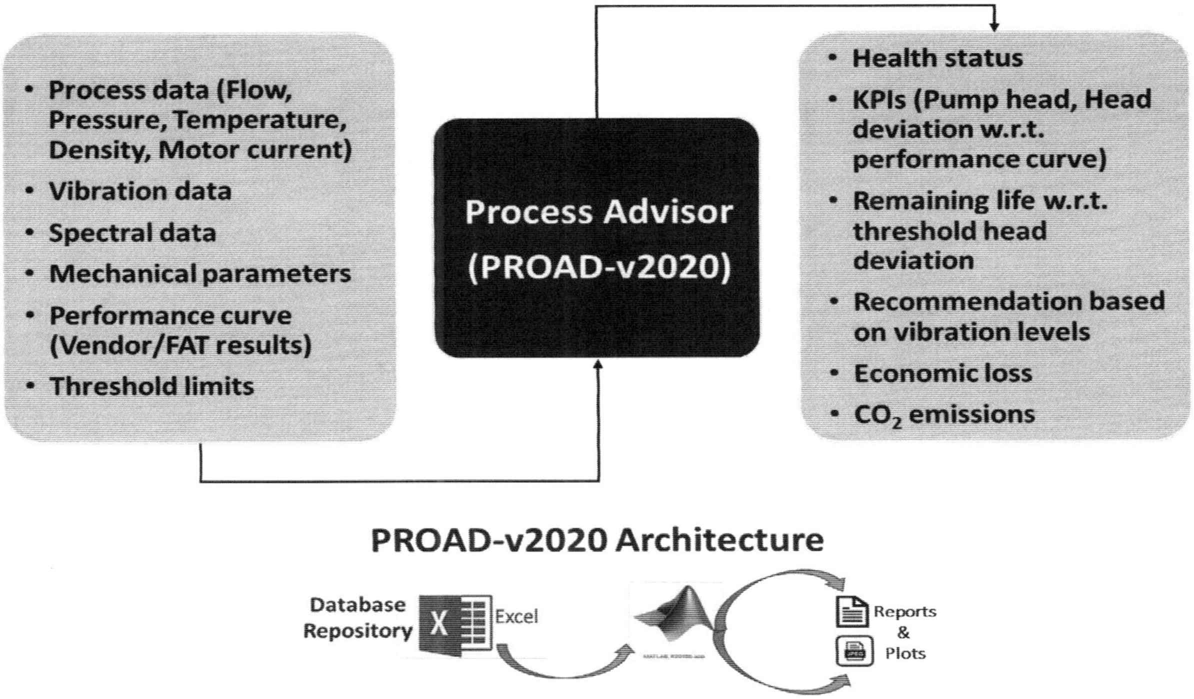

Figure 1: Overview of health monitoring of centrifugal pumps using PROAD.

Acknowledgement

This project is funded under Energy Innovation Research Programme (EIRP) Award NRF2017EWT-EP003-020, administrated by the Energy Market Authority (EMA). The EIRP is a competitive grant call initiative driven by the Energy Innovation Programme Office and funded by the National Research Foundation (NRF) of Singapore. We would also like to thank our industrial collaborator for providing plant data and feedback on PROAD's development.

605925 Data-Driven Approaches Towards Real-Time Equipment Health Monitoring and Predictive Maintenance in Drug Product Manufacturing

Abstract:

Digitalization is the central pillar towards the introduction of Industry 4.0 in manufacturing. However, the pharmaceutical industry lags behind in this regard and sophisticated approaches in terms of data analysis are still seldom used despite the availability of an abundance of recorded process data. Highly regulated fields such as drug manufacturing require manufacturers to store process data over many years for backtracking purposes. The analysis of such stored data highlights the potential to use it for predictive maintenance purposes.

This work presents a data-driven approach for equipment monitoring and the implementation of predictive maintenance (PdM) in an aseptic filling line in drug product manufacturing. Multiple years' worth of industrial data are used for this purpose. The data mainly consists of Supervisory Control and Data Acquisition (SCADA) variables which incorporate parameters such as pressure, temperature or vibration data. The goal of this work is to replace currently implemented time-based maintenance strategies with more efficient predictive and condition-based ones. The data obtained from the production facility varies in their source and structure based on their main intended purposes. The data is therefore first preprocessed and prepared. In this work, different machine learning algorithms are compared to identify the most suitable given the structure of the available data (i.e. the installed sensors at different locations in the line), the information content as for certain areas specific measurements are recorded (e.g. changeover processes), and the resulting prediction accuracy. Random forest algorithms are the most commonly applied in this regard due to their high performance potential with limited information incorporated in the data and an avoidance for over-fitting. Artificial neural networks are also popularly applied for a diverse set of PdM problems followed by support vector machines and k-means models. Ensemble learning or combinations of the different algorithms are also tested. A detailed analysis of the results obtained by the different algorithms is provided. The best strategies for each equipment are determined in terms of maintenance frequency.

A decision support tool is proposed to determine if an equipment is expected to fail in the foreseeable future. Expert knowledge is also then incorporated in the tool to determine the appropriate course of action identified with each type of failure or countermeasures needed to prevent failure. Different maintenance strategies are compared based on the expected downtime, uncertainty of the decision and the associated costs. Such a tool can be easily integrated with other facility scheduling decision tools.

604021 Microfluidics-Based Biocatalyst-Product Separation for Continuous Enzymatic Processes. a Case Study on β-Lactam Antibiotics Manufacturing

Abstract:

Immobilizing an enzyme on a support often is essential for downstream separation of the biocatalyst from a product and retaining the valuable enzyme in many enzymatic processes. This is particularly true in the case of continuous enzymatic manufacturing processes where the enzyme needs either to be retained in the reactor or separated from the product downstream and recycled back to the reaction vessel. This becomes challenging when the product is in solid form, e.g., crystals, necessitating the separation of two types of solid particles, biocatalyst and the crystalline product. This is the case for the continuous synthesis and crystallization of β-lactam antibiotics using the enzyme Penicillin G acylase (PGA) [1].

In this contribution, we propose a 3D-printed microfluidics-based device for downstream separation of the biocatalyst from the crystalline antibiotic product. Microfluidics is a technique characterized by the precise manipulation of fluid dynamics at the microscale and has shown a great promise for industrial processing applications. Among microfluidic devices, inertial microfluidics has experienced massive growth due to their high throughput, automation, operational simplicity, and low cost [2]. In such a device, particles with different sizes are separated by taking advantage of inherent inertial lift and drag forces. Inertial microfluidics can be employed by various channel shapes and structures; spiral channels are frequently used for the aim of high-throughput particle focusing by reduction of lateral migration positions and they can be an ideal candidate for this particular application. Here, we focus on the separation of high aspect ratio crystals from spherical biocatalyst particles. The efficiency of the separation for different product and catalyst sizes, and potentials/challenges for scaling up the proposed separation strategy will be discussed.

[1] McDonald, M. A., Bommarius, A. S., Grover, M. A., & Rousseau, R. W. (2019). Continuous reactive crystallization of β-lactam antibiotics catalyzed by penicillin G acylase. Part II: Case study on ampicillin and product purity. *Computers & Chemical Engineering*, *126*, 332-341.

[2] Moloudi, R., Oh, S., Yang, C., Teo, K. L., Lam, A. T. L., Warkiani, M. E., & Naing, M. W. (2018). Inertial-based filtration method for removal of microcarriers from mesenchymal stem cell suspensions. *Scientific reports*, *8*(1), 1-10

605697 Novel Integration of Continuous Crystallization Platforms for Control of Crystal Size Distribution and Polymorphic Form

Abstract:

Continuous crystallization has been in the focus of pharmaceutical industry as an appealing method to obtain drug substances of consistent critical quality attributes (CQAs), which meet strict regulatory guidelines. Traditionally, continuous crystallization is done in mixed suspension mixed product removal (MSMPR) crystallizers, but one main disadvantage of MSMPR crystallizers is the broad crystal size distribution (CSD) generated due to broad residence time distribution (RTD).[1] On the other hand, the new emerging continuous oscillatory baffled crystallizers (COBCs) offer plug-flow conditions with narrow RTD properties to generate product with narrow CSD.[2] Unlike traditional plug flow crystallizers that use high volumetric flow rate to achieve plug flow conditions, the COBCs use an oscillating piston to generate turbulent mixing within its concentric baffles without a high volumetric flow rate. However, a main disadvantage of COBCs is fouling and encrustation due to nucleation on the crystallizer walls which can lead to clogging and process failure.[3] To remedy fouling and prolong the operating time of the COBC, seeding is a common methodology to not only reduce the supersaturation in the system for nucleation but also provide polymorphic form control. Although seeding is very common method in the pharmaceutical industry, seeding for continuous crystallization has proven to be challenging, especially in the case of the COBC. Consistent manual preparation of seeds is labor intensive and economically infeasible, and, furthermore, any fluctuations of seed quality can impact startup dynamics.[4] In addition, seeds can be washed away after a few residence times, which no longer provide the benefits of seeding.

In this work, a novel integration of the MSMPR crystallizer with a COBC demonstrated in-situ seed generation capabilities while demonstrating CSD and polymorphic control in continuous combined cooling antisolvent crystallization (CCAC). Firstly, the 3D solubility surfaces of the metastable and stable form were generated to provide an operating regime for the continuous CCAC. It was discovered that the continuous generation of the metastable form required higher concentration of solvent than that of the stable form. Secondly, the MSMPR crystallizer was integrated into the COBC as a separate temperature zone to generate seeds via primary nucleation. The seeds generated were then fed into the COBC without pump transfer to allow for further growth. While compared to a single stage MSMPR crystallizer, the integrated system showed little signs of fouling and encrustation while generating product of uniform CSD. While both platforms are capable of polymorphic control using a temperature switch, the integrated system was able to maintain the narrower CSD while compared to the MSMPR, which makes this combined system a novel next generation continuous crystallization platform for systems with polymorphism and seeding requirements.

References:

(1) Ferguson, S.; Morris, G.; Hao, H.; Barrett, M.; Glennon, B. Characterization of the Anti-Solvent Batch, Plug Flow and MSMPR Crystallization of Benzoic Acid. *Chem. Eng. Sci.* **2013**, *104*, 44–54. https://doi.org/10.1016/j.ces.2013.09.006.

(2) Oliva, J. A.; Pal, K.; Barton, A.; Firth, P.; Nagy, Z. K. Experimental Investigation of the Effect of Scale-up on Mixing Efficiency in Oscillatory Flow Baffled Reactors (OFBR) Using Principal

Component Based Image Analysis as a Novel Noninvasive Residence Time Distribution Measurement Approach. *Chem. Eng. J.* **2018**, *351* (June), 498–505. https://doi.org/10.1016/j.cej.2018.06.029.

(3) Pena, R.; Oliva, J. A.; Burcham, C. L.; Jarmer, D. J.; Nagy, Z. K. Process Intensification through Continuous Spherical Crystallization Using an Oscillatory Flow Baffled Crystallizer. *Cryst. Growth Des.* **2017**, *17* (9), 4776–4784. https://doi.org/10.1021/acs.cgd.7b00731.

(4) Wood, B.; Girard, K. P.; Polster, C. S.; Croker, D. M. Progress to Date in the Design and Operation of Continuous Crystallization Processes for Pharmaceutical Applications. *Org. Process Res. Dev.* **2019**, *23*, 122–144. https://doi.org/10.1021/acs.oprd.8b00319.

606175 A Scalable Membrane Pervaporation Approach for Continuous Flow Ring Closing Metathesis

Abstract:

Pervaporation membranes can be used to avoid azeotropic distillation, enable flow chemistry technology, degas organic solvents, dehydrate solvents, drive reactions towards product formation, and more. Compact Membrane Systems (CMS) has demonstrated the effectiveness of their fluoropolymer pervaporation membranes to degas and dewater organic solvents for the pharmaceutical and specialty chemical industries in recent years. These membranes are applicable to a wide variety of solvents and operating conditions including difficult solvents such as NMP, DMSO or THF. In this talk, CMS will describe their newest pervaporation case study, performed in collaboration with Snapdragon Chemistry. A CMS membrane was used to selectively remove ethylene from a ring closing metathesis reaction in toluene to drive product formation and enable the use of flow chemistry.

Using traditional batch techniques, implementing olefin metathesis reactions on the commercial scale has proven challenging, if not impossible. Ring closing reaction mechanisms are commonly used during the preparation of small molecules for the pharmaceutical and fragrance industries. Membranes are a modular, scalable solution that can continuously remove the ethylene byproduct and enable the use of commercial scale continuous ring closing metathesis. This case study details the advantage of using a sheet-in-frame membrane reactor as opposed to a stainless steel tubular reactor. The fluoropolymer membrane has excellent chemical and thermal stability which allows for an extended range of thermal conditions without sacrificing selectivity or flux. Pure ethylene flux of the membrane was also investigated to ensure that the mass transport across the membrane would always be greater than the reaction's ethylene production rate. Scale up considerations including residence time, required ethylene removal rate and module design will be addressed. The feasibility of using a membrane reactor to remove byproducts shown in this case study is only one example of the reactions that could benefit from the implementation of membrane reactors.

604827 Development of Continuous Homogeneous Hydrogenation Process for Pharmaceuticals Using a Flow Reactor

Abstract:

In past decade there has been significant interest in the development of continuous processes in the pharmaceutical industry. Key drivers for continuous processing are product quality improvement, yield and process robustness, footprint reduction of processing area and increased process safety. Hydrogenations are commonly used in synthesis of pharmaceutical intermediates and active pharmaceutical ingredients. Types of hydrogenations include removal of protecting groups such as benzyl groups, reduction of nitro compounds and asymmetric hydrogenations. In this work we have demonstrated:

- Application of a laboratory scale tubular reactor for development and scale-up of a continuous asymmetric hydrogenations process.
- The effect of operating parameters such as pressure, gas to liquid ratio, temperature, catalyst loading and residence time of the substrate on enantioselectivity and product quality, are studied to develop a scalable and robust process.

A ‘Fit for purpose’ model was developed and found to be very effective in efficiently identifying the optimal reaction conditions, assessing risk to product CQAs, and increasing overall robustness.

Disclosures:

All authors are employees of AbbVie and may own AbbVie stock. AbbVie sponsored and funded the study; contributed to the design; participated in the collection, analysis, and interpretation of data, and in writing, reviewing, and approval of the final publication.

593483 Liquid Phase Peptide Synthesis Via Nanostar-Sieving

Abstract:

Peptides, an important class of polymers, mainly used for therapeutic purposes, are generally synthesised using solid phase methods. While the rapid synthesis characteristic of solid phase peptide synthesis (SPPS) benefits the drug discovery, this technology faces challenges for large scale synthesis, where its non-quantitative coupling leads to error sequences. In contrast liquid phase peptide synthesis (LPPS), offers high peptide purity and scalability but its development is hampered by inefficient intermediate separations.

Liquid phase peptide synthesis via nanostar-sieving (LPPS-NS) is a platform which synthesises peptides in solution with facile intermediate separations. Amino acids (AA) are coupled iteratively onto a 3-armed, star-shaped macromolecule, forming peptide-nanostar intermediates. After coupling, the unreacted AA is quenched and subsequently proceeded to N-terminal deprotection. The bulky intermediates are then 'sieved' out from the debris and quenched AA all together via organic solvent nanofiltration (OSN) thus omitting the post-coupling isolation step. This synthetic cycle is repeated until the desired peptide length is achieved, as shown in Figure 1. Standard Fmoc peptide chemistry is applied throughout the synthetic cycle. The use of nanostar greatly enhances the molecular sieving efficiency by the >3-fold mass difference between the nanostar and the unreacted building blocks. Most importantly, real-time reaction monitoring can be undertaken by LC-MS with high accuracy because of the monodispersity of the nanostar. OSN plays a pivotal role in synthesis efficiency. A solvent resistant membrane made of crosslinked polybenzimidazole (PBI) polymer was chosen for OSN. The polymer-based membrane was proven to be durable and to have a high separation factor which remained consistent throughout many synthesis cycles. We speculate that this new concept of LPPS is a truly one-pot synthesis where no transfer of liquids between cycles occurs and has a good potential for full-automation.

In this work we demonstrate the successful synthesis of Enkephalin-type model peptides (~5-10 mers) via nanostar-sieving technology. The products are of higher purity than, or at least comparable to, peptides produced by a reliable solid phase vendor, while using less equivalents of AA during coupling. To further validate this technology, high purity linear Octreotide (8AA) was synthesised. Our ambition is to develop this platform into a robust technology for large-scale, fully automated and high purity peptide synthesis.

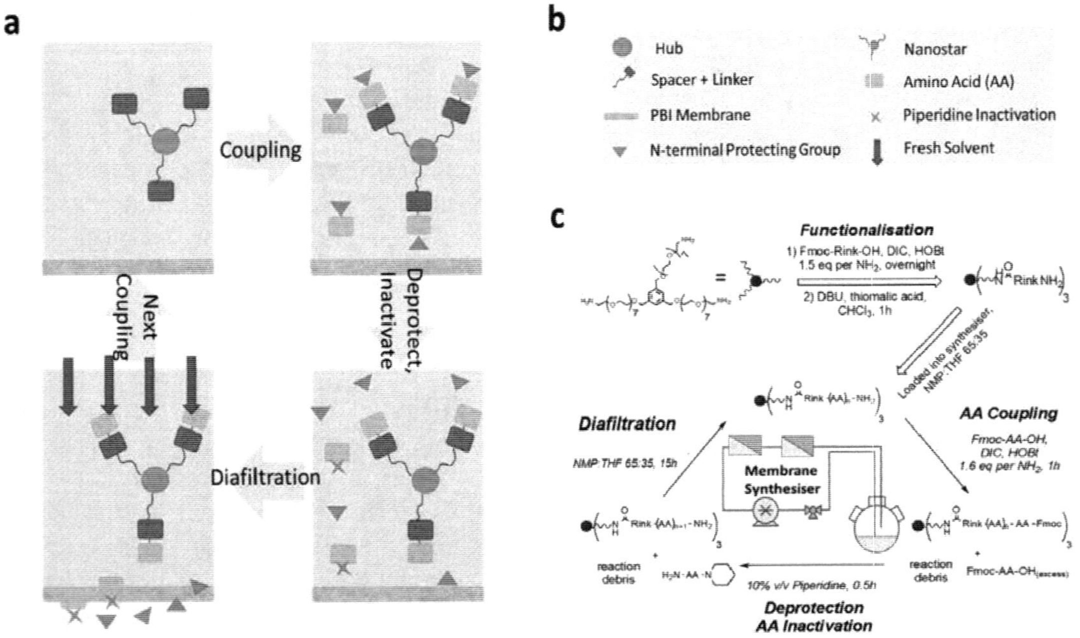

Figure 1: *(**a**) Flow diagram of Liquid Phase Peptide Synthesis via Nanostar-sieving (LPPS-NS) and (**b**) diagram legend. (**c**) The synthetic cycle of LPPS-NS. Prior to peptide synthesis, the star-shaped and 3-armed nanostar is functionalised with rink-amide linker. The functionalised nanostar then undergoes repeated cycle of coupling, deprotection / AA inactivation and diafiltration in the membrane synthesiser until target peptide length is reached. No liquid transfers between cycles ie. genuine one-pot synthesis.*

600743 Liquid-Liquid Extractions in API Manufacturing Processes: Is Continuous the Way to Go?

Abstract:

Extraction and separation operations are present in most API processes, in different stages, and with different purposes. Both unit operations are traditionally carried out under batch mode in multi-purpose reactors and consist in two/three sequential steps of mixing immiscible liquids with different densities and then separate them by gravity, which results in a time-consuming stage. Moreover, sometimes emulsions are obtained leading to difficulties in phase separation with impact in cycle time, yield and, potentially, product quality. Another bottleneck from these operations is related to the reactor occupancy at this stage, considering that usually one phase is carried forward and the other is discarded, which results in a limitation of batch size and production capacity.

To tackle all these operation deficiencies, continuous manufacturing is an attractive approach. The main drivers associated when going from batch to continuous highlight the opportunity to improve reproducibility, capacity, faster development, among others. For extraction and separation processes, several alternatives are available in the market with different characteristics and advantages.

In this work, we demonstrate our approach in selecting technologies for continuous process of extraction and separation, and case studies with different purposes such as pH adjustments, product/API transfer between phases and washings. The examples will show characteristics of the selected technologies (membrane based separator and centrifugal extractor) and types of liquid-liquid extraction and separation process evaluated along with the impacts resulting from going from batch to continuous manufacturing.

608996 Development and Process Intensification of a Three Step Continuous Flow Process for Telescoped High Energy Reactions

Abstract:

Continuous manufacturing (CM) offers significant advantages over traditional batch manufacturing for chemistries that generate or require high energy (usually in the form of heat). The comparatively large surface-area-to-volume ratio of flow reactors makes them particularly attractive for chemical reactions such as lithiation-carboxylation, nitro reduction, debenzylation, and thermal cyclization. Continuous processing also allows the reaction rates for these transformations to be orders of magnitude higher than would be possible with batch manufacturing. Additionally, performing these reactions in flow allows them to be telescoped to reduce operator handling and processing downtime.

This presentation will focus on development and process intensification of a three step continuous flow process toward synthesis of an active pharmaceutical ingredient (API). It will detail how construction of fit-for-purpose continuous reactors improves safety, cost, and process mass intensity while enabling use of preferred reagents and unlocking otherwise unattainable operating conditions.

611798 An Accurate, Precise, and Affordable Light Emitting Diode Spectrophotometer with a Raspberry Pi User Interface for Diabetes Testing in Honduras

Abstract:

The spectrophotometric determinations of glucose in blood serum and urine are commonly used for diabetes testing in Honduras. However, the high cost of spectrophotometers often limits access to diabetes testing in this and other low-income countries. To make this testing more accessible, a highly affordable light-emitting diode (LED) spectrophotometer was designed, built, and tested for 3 years using best engineering practices and state-of-the-art electrical components. In addition, this LED spectrophotometer used a Raspberry Pi credit-card sized computer for a user interface, and for data processing and storage. The cost and performance of this LED spectrophotometer/Raspberry Pi system was compared against 3 common commercial spectrophotometers. More specifically, the performance of these instruments was evaluated from the upper limits of linear range, upper limits of operational range, calibration sensitivities, R^2 values, estimated limits of detection, and percent calibration check standard recoveries for the determination of glucose using o-toluidine in glacial acetic acid. The parts for our LED spectrophotometer/Raspberry Pi system costs $168 United States dollars. The 3 commercial spectrophotometers ranged in price from $2,382 to $3,382. Our results suggest that the LED spectrophotometer/ Raspberry Pi system has upper limits of linear range, upper limits of operational range, calibration sensitivities, R^2 values, estimated limits of detection, and percent calibration check standard recoveries that are generally comparable to those of the 3 commercial spectrophotometers used in this study. In conclusion, this LED spectrophotometer/Raspberry Pi system is highly affordable and could be used to make diabetes testing in Honduras and other low-income countries more widely accessible.

601744 Embedded Droplet Printing for Spherical Crystallization of Pharmaceuticals

Abstract:

We present the application of a newly developed technique, embedded droplet printing in yield-stress fluids (1), to the formation of uniform spherical pharmaceutical particles. Yield-stress fluids are rheologically complex materials that are solid-like below a critical stress but undergo a dramatic drop in viscosity and flow above their yield stress. Printing droplets that are embedded inside a yield-stress fluid enables their processing in the absence of exterior influences including convection and solid boundaries. This makes embedded droplet printing ideal for performing sensitive processes like pharmaceutical crystallization.

In this study, we crystallize various drug-excipient systems using embedded droplet printing methods. We demonstrate the advantages of our technique in obtaining particles that are as spherical and uniform as possible, key targets in the manufacturing of pharmaceutical materials. The powders we produce show significant improvement over conventional materials in terms of their flowability and processability, and we believe embedded droplet printing can help pave the way for the manufacturing of more exquisite and targeted pharmaceutical products.

1. A. Z. Nelson, B. Kundukad, W. K. Wong, S. A. Khan, P. S. Doyle, Embedded droplet printing in yield-stress fluids. *Proc. Natl. Acad. Sci.* **117**, 5671–5679 (2020).

599475 Heat Exchanger Model for Predicting Temperature and Dissolution of API for a High-Throughput Temperature Shift Spray Drying Process

Abstract:

Purpose:

The temperature-shift spray-drying process enables high-throughput, commercially relevant manufacturing of poorly soluble active pharmaceutical ingredients (APIs) in organic solvents by increasing API solubility by at least 10-fold relative to room temperature operation. In the temperature-shift technique, an in-line tube-in-tube heat exchanger rapidly increases the temperature of a spray suspension to fully dissolve an API immediately before entering the spray dryer. API dissolution kinetics are a key factor when designing the process and optimizing operating parameters. The temperature and residence time must be selected to create an operating space to allow for full dissolution without causing API degradation. A first-principles model was developed to provide guidance on the heat and mass transfer requirements to aid in experimental design. Temperatures of the solution flowing through the heat exchanger were measured, and the model was found to accurately predict the heat transfer. Additional experiments were done with a model compound to assess the dissolution model and were also found to be in good agreement. This tool will help improve thermal shift spray drying process development and process parameter optimization to achieve the desired product quality more efficiently.

Methods:

A scale-independent model to describe both heat and mass transfer was developed using first principles approaches. To develop a heat transfer model, we employed an energy balance across the heat exchanger and then evaluated typical heat transfer correlations and dimensionless numbers. These correlations from literature were found to be applicable to our physical situation. This approach allowed calculation of the temperature profile of both the hot and cold sides of the tube-in-tube heat exchanger as a function of length.

Dissolution of an API was modeled using the Noyes Whitney Equation. By knowing the temperature profile of the heat exchanger, the concentration of dissolved drug was determined based on the changing solubility of the API as a function of temperature. Additionally, fundamental parameters like the diffusion coefficient and solution viscosity were also allowed to vary as a function of temperature leading to a good approximation of the physical properties of the suspension during dissolution in the heat exchanger.

An experimental set-up was fabricated to measure the temperature of the inlet and outlets of both the hot and cold streams. Initial trials were completed testing ranges of temperatures, flowrates, and heat exchanger geometries using water as a test fluid. To further test the heat transfer model, placebo spray drying experiments using the dispersion polymer HPMCAS were conducted on a clinical scale dryer with a water-acetone mixture. HPMCAS was tested at two solids loadings to see how solution temperatures with different viscosities were predicted by the model.

To evaluate the dissolution of an API through the heat exchanger, Riboflavin was used as a model compound. An in-line UV probe was used to determine the effective concentration in solution as a function of heat exchanger residence time across a range of flowrates.

Results:

The temperatures of the tube side are predicted with less than 5% error compared to experimental data spanning small scale feasibility to large commercial scale for water, solvent, and polymer solutions. Dissolution of Riboflavin in water was predicted with good accuracy over the flowrates tested at small scale.

Conclusions:

The model was found to accurately describe the heat and mass transfer for this novel rapid heating spray drying process. This can be used for better designing DOE studies to understand the operating space with respect to drug dissolution and degradation. Future work will focus on better understanding this sensitivity of API dissolution kinetics during the temperature shift process.

599585 Advancing Spray Drying Process Capabilities for Low Solubility Drug Compounds By Leveraging Temperature-Shift Solubility Enhancement: A Case Study with Alectinib

Abstract:

Purpose: Amorphous solid dispersions can provide bioavailability enhancement to active pharmaceutical ingredients with low aqueous solubility, which enables oral delivery. A spray drying process is a scalable technique for manufacture of amorphous solid dispersions. Manufacturing spray dried dispersions (SDDs) provides opportunities for particle engineering to optimize quality attributes such as flowability, compactibility, dissolution rate and drug release profiles that can be essential to drug product development. One of the limiting factors that can impact the SDD formulation and process is a drug's solubility in organic solvents. Low drug solubility in organic solvents can have several process and product impacts that can make an SDD challenging as a lead formulation technology, including: low throughput, large batch size and non-optimal powder properties (e.g. particle size and bulk density). One novel technique to overcome the organic solubility limitation is to take advantage of the temperature-shift solubility enhancement. Achieving rapid drug dissolution from short term heating of a spray drying solution requires consideration to the drug's chemical stability and modification of the processing equipment to solubilize the drug in an inline heat exchanger followed by flash atomization, which utilizes vaporization and solvent cavitation for droplet formation. Here, we discuss the technical considerations of developing such a process for an SDD of Alectinib, a BCS class II oral dosage oncology drug.

Methods: The solubility of Alectinib was first determined as a function of temperature. Two small-scale techniques were compared: 1) High temperature NMR, and 2) thermogravimetric analysis (TGA). The solubility improvement at the selected temperature was then investigated using a high pressure solution vessel with a sight glass to visually confirm solubility. Lastly, two internally-developed models were used to optimize a PSD-1 spray drying process: 1) an in-line heat exchanger model to tune flow rate and temperatures required for heat and mass transfer and 2) a spray drying thermodynamic model to predict drying gas inlet and outlet temperatures. The temperature-shift SDD was then characterized using DSC, XRPD and dissolution testing to compare the performance relative to crystalline drug. A control SDD was manufactured by a standard spray drying process using a dilute, ambient-temperature spray solution for further comparison.

Results: High-throughput solubility screening using a TGA method at 25°C and 37°C showed that a 90/10 methanol/water solvent system provided the highest temperature-shift advantage with a thermodynamic solubility of 0.4wt% at 25°C and 0.5wt% at 37°C. High temperature NMR enabled solubility measurements up to 120°C and demonstrated an increase to 1.2wt% solubility at that temperature. Extrapolating the two-point TGA method to higher temperatures with the van't Hoff relationship resulted in excellent agreement with the NMR solubility measurements in the anticipated operating range of 100-130°C, which was also confirmed by using a larger volume pressurized solution vessel. Improvements to the NMR solubility was also observed with the addition of dissolved HPMCAS-H. Successful manufacture of amorphous SDDs was achieved on a PSD-1 spray dryer through a model-based process optimization strategy and demonstrated a 6x increase in API throughput using the temperature-shift process. Dissolution in gastric and intestinal media with varying micelle concentrations showed significant dissolution rate, extent and AUC improvements against pure crystalline material. No performance difference was observed between SDD manufactured using the inline heating process when compared to a dilute solution sprayed at room temperature

Conclusion: Utilizing the described temperature-shift process with a challenging low organic solubility drug compound provided significant throughput improvements while reducing spray drying energy requirements by enabling higher spray solution concentrations and reducing batch sizes. Further application of this process can provide continued cost reduction in the spray drying manufacturing process while promoting the advancement of research in critical areas including atomization, drying kinetics and drug/polymer interactions.

602085 Novel Dispensing Mechanism to Achieve Continuous Flow of Highly Cohesive Powders

Abstract:

Under the best of circumstances, powder flow is difficult to control. Powders can bridge in hoppers, rat hole, avalanche, segregate, and adhere to surfaces through electrostatic or other interactions. Cohesive powders are especially challenging, and a significant body of research spanning many decades has been devoted to their study. These flow challenges are amplified in continuous pharmaceutical manufacturing processes where powders must not only be fed reliably, but also at precisely controlled feed rates. In these cases, feed screws are typically employed to meter powders at a desired mass flow rate. While the screw feeder design provides excellent results for many powders, they can fail to provide adequate results when feeding very low levels (mass flow rates) of cohesive powders or powders with very low bulk density. These low-level ingredients may include lubricants such as sodium stearyl fumarate and magnesium stearate, glidants such as silica, or various active pharmaceutical ingredients which are often milled to mean particle sizes below 20 microns. In this work, a novel continuous dispensing mechanism was explored to achieve continuous feeding of pharmaceutical powders. Initial results suggest that powders can be fed as low as $1/10^{th}$ of the mass flow rate and with improved precision compared to twin feed screws.

601987 Producing Predictive Chemical Process Models from a State of Incomplete Mechanistic Knowledge and Small Data Sets

Abstract:

ABSTRACT: In this work, we explore the possibility of developing a quantitative and predictive model for chemical processes, without having to first gain a complete understanding of the underlying mechanisms. We find that this can be done efficiently combining: (1) expectations for the functional expression of the process—extracted from a general, *qualitative* understanding on how the chemical process ought to behave—and, (2) small data sets providing examples of the process outputs resulting from varying process inputs. The strategy discussed is labeled knowledge-constrained machine learning. The main example we take is on reaction kinetic models—and, here, the expectations are written in terms of constraints on the model predictions over time, and the data is a series of time-point measurements for the reaction outcome under different conditions. But, the intent of the presentation is to discuss a general methodology that may be used to develop predictive models across the range of unit operations found in the processes used to produce synthetic drug substance API's.

594384 Identifying the Processing Space of Continuous Granulation: A Case Study of Extended Release Tablets

Abstract:

While continuous manufacturing (CM) has been employed for directly compressible tablets, the continuous wet granulation has also been used as an alternative in continuous tableting lines. To date, no matrix-based extended release tablets manufactured by a CM technology has been approved. Compared to batch granulation, continuous twin screw granulation exerts higher shear to produce denser granules with broader size distribution within very short time (seconds). Changing the screw configuration, process, and formulation parameters can alter the physical properties of the granules, e.g., shape, flow, compressibility, density, size distribution, etc. In twin screw granulation, complex interactions among these parameters manage the granule properties. This study aimed at identifying the critical material attributes and process parameters and their interactions to be controlled for switching a batch process to a CM granulation process.

Using batch high shear granulation, we evaluated the formulation and process parameters that can produce metoprolol succinate extended release tablets. These formulation and process parameters were successfully transferred to continuous twin screw granulation to produce the same dissolution properties of the finished product. The effects of the screw configuration parameters on the characteristics of the granules were further investigated. The interaction between the processing, e.g., liquid to solid ratio, screw speed, barrel fill volume, etc., and the significant screw parameters was studies to unveil the processing space and to identify the failure modes.

Similar granule quality attributes were obtained for most of the formulations; although, the conditioned bulk density and compressibility of the granules determined the dissolution properties of the tablets. The number of kneading and sizing elements had the most significant impact on granule characteristics. The effects of the staggering angle and distance between the kneading zone of the screws were negligible.

This presentation will then highlight the processing space of screw granulation to produce extended release tablets. Some of the failure modes will be discussed along with the feasibility of implementing PAT tools to monitor the granulation process and granules properties.

DISCLAIMER

This presentation reflects the views of the presenter and should not be construed to represent FDA's views or policies.

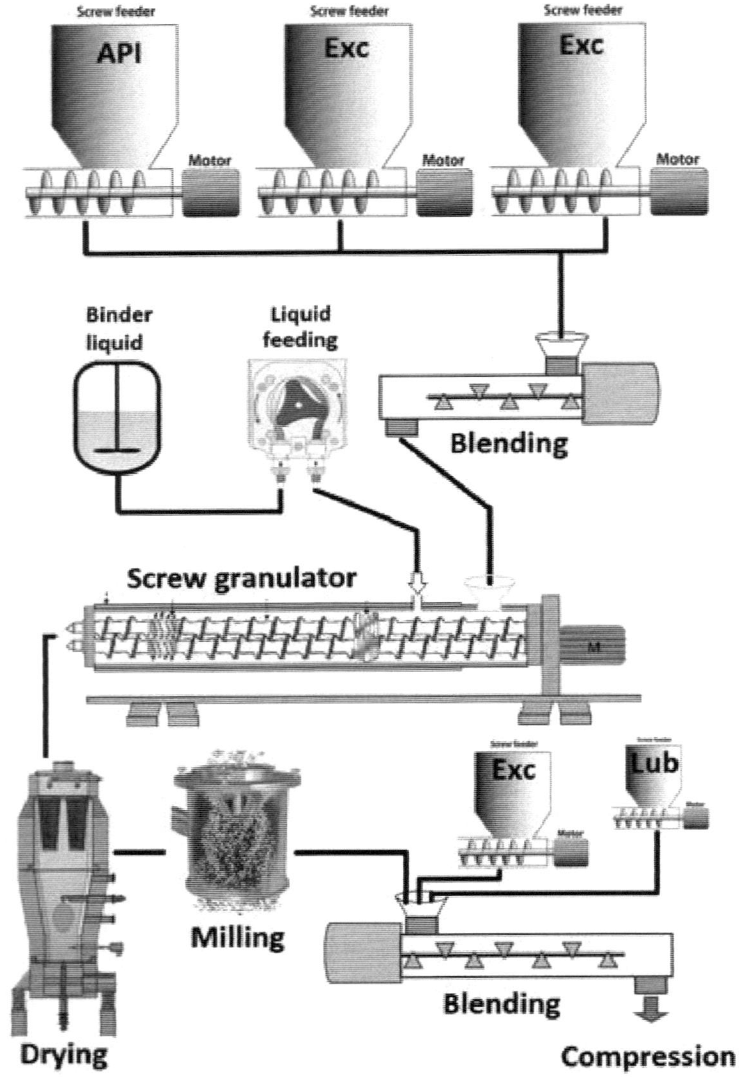
Screw feeder
API
Motor
Screw feeder
Exc
Motor
Screw feeder
Exc
Motor
Binder liquid
Liquid feeding
Blending
Screw granulator
M
Exc
Lub
Milling
Blending
Drying
Compression

608665 Dynamic Optimization Strategies Towards Addressing the Challenges in Protein Therapeutics

Abstract:

Protein therapeutics is an important field of the pharmaceutical industry. Current challenges in this field include meeting the increasing market demands, handling the entrance of biosimilar products and complying with quality by design (QbD) principles [1-3]. To tackle these challenges effectively, efficient use of existing tools, methods and solution approaches together with the development of novel ones is necessitated. This work provides an overview of our recent efforts to address some of the aforementioned challenges using model-based dynamic optimization strategies.

Model-based approaches are powerful tools towards facilitating process development and advancing process performance. They offer a promising alternative to costly and time-consuming experimentation [1], and can have a profound effect on process sustainability and profitability. Recent advances in cell culture modeling (with cell cultures representing a typical production platform for many of the products of this class) have enabled significant insight to the system's dynamics and the underlying production mechanisms, and have paved the way towards developing dynamic optimization strategies for process intensification (cf., [4-6]).

Existing numerical methods and optimization software for solving the resulting dynamic optimization problems rely almost exclusively on local techniques. Due to the noncovexity of the described complex biological systems, the vast majority of these problems exhibits the presence of multiple local minima. Nevertheless, deriving suboptimal local solutions can have direct quality and cost impacts. Despite the significant progress in deterministic global optimization [7], it is not yet tractable for the considered systems. Hence, alternative methods to overcome the hazards of suboptimal local optimality are required. Stochastic and heuristic approaches can be advantageous to this end. In our previous work [8], we have shown how incorporating biological process knowledge into the optimization problem formulation can facilitate the derivation of superior local solutions. In a similar direction, we illustrate how proper decomposition of the optimization problem can outweigh certain advantages [9].

Beyond increasing product yield, monitoring and controlling product quality is essential. Regulatory authorities including FDA are increasingly recommending consideration of QbD to develop processes that consistently meet quality specifications [3, 10]. This requires further attempts in modeling, as well as reconsiderations in the formulation of the dynamic optimization problem to effectively account for direct quality attributes. Yet, current methods in this direction are still in an early stage. The recent work in [11] demonstrates a successful example of model-based optimization to maximize galactosylation (a glycosylation related metric, with glycosylation presenting a critical quality attribute for many therapeutic proteins [12, 13]) content for an antibody case study. In our recent study [14], we have combined model simulations and design of experiments to investigate the effects of nutrients and nucleotide sugars feeding in protein glycosylation. Building on this basis, we have developed an optimization framework to maximize process performance while meeting quality specifications, and thus provided a successful example for utilizing dynamic optimization to apply the QbD paradigm in protein therapeutics [15].

Acknowledgements:
This work has received funding from the European Union's Horizon 2020 research and innovation programme under the Marie Skłodowska-Curie grant agreement no.675251.

References:

[1] Farzan, P., Mistry, B., & Ierapetritou, M. G. (2017). Review of the important challenges and opportunities related to modeling of mammalian cell bioreactors. AIChE Journal, 63(2), 398-408.
[2] Shukla, A. A., Wolfe, L. S., Mostafa, S. S., & Norman, C. (2017). Evolving trends in mAb production processes. Bioengineering & Translational Medicine, 2(1), 58-69.
[3] Collins, P. C. (2018). Chemical engineering and the culmination of quality by design in pharmaceuticals. AIChE Journal, 64(5), 1502-1510.
[4] Dhir, S., Morrow Jr, K. J., Rhinehart, R. R., & Wiesner, T. (2000). Dynamic optimization of hybridoma growth in a fed-batch bioreactor. Biotechnology and Bioengineering, 67(2), 197-205.
[5] Kiparissides, A., Pistikopoulos, E. N., & Mantalaris, A. (2015). On the model-based optimization of secreting mammalian cell (GS-NS0) cultures. Biotechnology and Bioengineering, 112(3), 536-548.
[6] Kappatou, C. D., Mhamdi, A., Campano, A. Q., Mantalaris, A., & Mitsos, A. (2018). Model-based dynamic optimization of monoclonal antibodies production in semibatch operation—Use of reformulation techniques. Industrial & Engineering Chemistry Research, 57(30), 9915-9924.
[7] Chachuat, B., Singer, A. B., & Barton, P. I. (2006). Global methods for dynamic optimization and mixed-integer dynamic optimization. Industrial & Engineering Chemistry Research, 45(25):8373–8392.
[8] Kappatou, C. D., Mhamdi, A., Campano, A. Q., Mantalaris, A., & Mitsos, A. (2017). Dynamic optimization of the production of monoclonal antibodies in semi-batch operation. In Computer Aided Chemical Engineering (Vol. 40, pp. 2161-2166). Elsevier.
[9] Kappatou, C. D., Altunok, O., Mhamdi, A., Mantalaris, A., & Mitsos, A. (2019). Sequential and Simultaneous Optimization Strategies for Increased Production of Monoclonal Antibodies. In Computer Aided Chemical Engineering (Vol. 46, pp. 1021-1026). Elsevier.
[10] Pais, D. A., Carrondo, M. J., Alves, P. M., & Teixeira, A. P. (2014). Towards real-time monitoring of therapeutic protein quality in mammalian cell processes. Current Opinion in Biotechnology, 30, 161-167.
[11] Kotidis, P., Jedrzejewski, P., Sou, S. N., Sellick, C., Polizzi, K., del Val, I. J., & Kontoravdi, C. (2019). Model-based optimization of antibody galactosylation in CHO cell culture. Biotechnology and Bioengineering, 116(7), 1612-1626.
[12] Hossler, P., Khattak, S. F., & Li, Z. J. (2009). Optimal and consistent protein glycosylation in mammalian cell culture. Glycobiology, 19(9), 936-949.
[13] Blondeel, E. J., & Aucoin, M. G. (2018). Supplementing glycosylation: A review of applying nucleotide-sugar precursors to growth medium to affect therapeutic recombinant protein glycoform distributions. Biotechnology Advances, 36(5), 1505-1523.
[14] Ehsani, A., Kappatou, C. D., Mhamdi, A., Mitsos, A., Schuppert, A., & Niedenfuehr, S. (2019). Towards model-based optimization for quality by design in biotherapeutics production. In Computer Aided Chemical Engineering (Vol. 46, pp. 25-30). Elsevier.
[15] Kappatou, C. D., Ehsani, A., Niedenführ, S., Mhamdi, A., Schuppert, A., & Mitsos, A. Quality-targeting dynamic optimization of monoclonal antibody production. (Submitted)

608965 Integrated Supply Chain Network Design and Inventory Management for Autologous Cell Therapy.

Abstract:

Personalized medicine is gaining a lot of attention in the pharmaceutical industry as it can treat diseases that earlier had no cure. In the past few years, we witnessed exponential growth in the development of cell and gene therapy products. Cancer Research Institute reported that 753 cancer cell therapies are currently under development all over the world, of which 375 are in clinical trials [1]. There has been an increase in the growth and demand for autologous therapy due to the recent approval from the FDA [2]. Therefore, autologous therapy is undergoing the transformation from laboratory to commercial scale.

The autologous therapy starts with a collection of cells from the patient, which are cryopreserved and sent to the manufacturing site. The cells are then formulated and enriched. The formulated cells are again cryopreserved and transferred back to the patient, where they are injected back. Unlike traditional pharmaceutical drugs that are produced in bulk, these drugs are patient-centric and have to be manufactured for each patient separately. In addition, the process also deals with living cells that go around different nodes in the supply chain. Therefore, the cells have to be cryopreserved in storage device and transported across the supply chain.

An integrated approach is considered for supply chain network design and inventory management of the raw materials. We consider a network with a set of hospitals where the patients arrive. The facilities to be set up are distribution centers and manufacturing centers. The manufacturing center receives the samples from the patient and engineers the cells to create the drug. The distribution centers have the inventory for the storage containers and are installed for the intermediate halt for the sample transportation between the hospital and manufacturing center. To account for the transit times for the material transfer between facilities, we define the parameter accounting for both transportation time and waiting time for different transportation modes as well as the processing times at each of the node. We also simultaneously optimize inventory policy. We use periodic-review policy for the raw materials inventory at the distribution center [3]. Under this policy, if the inventory is below the defined level, the order is placed to restore the inventory to level . Since the storage devices are reusable, the policy is modified to account for the returning devices after the delivery of the drug.

Given the demand distribution, the potential locations of distribution and manufacturing centers, costs and time for processing, inventory, and transportation, we determine the location of different centers, inventory levels at the distribution centers, response time, and cost of delivery for each sample, with objective function of minimizing the cost and response time. The problem is formulated as a single objective optimization where response time is multiplied by a factor and added to the cost function. A case study of China is considered, with 20 hospitals situated in different provinces and a total of 130 patients over a period of 1 month. Also, the location of 20 potential locations of the distribution centers and 20 manufacturing centers are given. We found out that the majority of the cost is incurred in setting up manufacturing and distribution centers. The cost accounting for the response time depends on the multiplicative factor we choose. In addition, we also explored the importance of batching the samples at the distribution centers and transporting to the manufacturing center subjected to the response time constraint. Decisions related to inventory of raw material are critical to have an efficient supply chain. Also, sensitivity analysis is conducted to study the effect of different parameters on the optimal network design and overall cost.

References:

1. Tang, J., et al., *The global landscape of cancer cell therapy.* Nat Rev Drug Discov, 2018. **17**(7): p. 465-466.
2. Bach, P.B., S.A. Giralt, and L.B. Saltz, *FDA approval of tisagenlecleucel: promise and complexities of a $475 000 cancer drug.* Jama, 2017. **318**(19): p. 1861-1862.
3. Brunaud, B., et al., *Inventory policies and safety stock optimization for supply chain planning.* AIChE Journal, 2019. **65**(1): p. 99-112.

600490 Two-Stage Control Using Model-Based Reinforcement Learning and Predictive Control for Fed-Batch Bioreactor

Abstract:

As the demand for biomass and pharmaceutical materials is increasing in the market, various genetically modified cells are tested in a batch bioreactor in order to increase production rate [1]. At the same time, a robust and efficient control strategy is required to successfully produce the product with uncertainty of cellular behaviors [2]. However, limitation of the information on the system dynamics makes it challenging problems to optimize the operating conditions of a bioreactor and to implement an automatic controller. In general, the biosystem is less reproducible and shows higher nonlinearity compared to other engineering systems. In particular, most of the parameters in the first-principle model are lumped type and these values would be varied not only in a single batch operation with changes of the states but in a batch-to-batch fashion. Therefore, implementing an automatic control to track the optimal control trajectories obtained off-line is not a robust approach to produce fine products for a large number of batch reactors.

In this work, we propose a two-stage control strategy to optimally control a fed-batch type bioreactor that produces penicillin. The proposed strategy utilizes model-based Reinforcement Learning (RL) off-line to optimize the operating trajectories of state and input. In on-line, the Moving Horizon Estimation (MHE) is applied to estimate the parameter values, and an adaptive Model Predictive Control (MPC) is utilized to simultaneously modify and track the trajectories obtained in off-line. First, the first-principle model of the system dynamics is represented in Stochastic Differential Equations (SDEs) form which could theoretically handle the white noise in a continuous system. In addition, the variance of the process noise could shift a mean value of drifting terms of the system so that the changes in parameter values could be presented by adjusting the variance of the process noise [3]. These SDEs are numerically integrated and served as a virtual plant. The next step is obtaining the operating trajectories of the reactor off-line. The optimal control method could be classified in three classes, direct, indirect and RL (dynamic programming-based method). The direct method is the one mostly applied in practice since the problem formulation and the procedure of obtaining the solution is intuitive. On the other hand, RL has several advantages over the direct method. RL provides a closed-loop solution of the optimal control problem so that the control policy could be obtained instead of just a control trajectory [4]. This is a more robust approach since RL evaluates the control cost for the case where the states have deviated from the optimal trajectories. In addition, value based RL provides not only the optimal trajectories but also the value of state which is an expected total cost of the control policy starting from the state [4]. Although this value is just a by-product to find the optimal control policy, it is useful information when the trajectories are modified on-line.

The trajectories suggested by RL could serve as a baseline of optimal control. However, due to the change in parameter values and disturbances, tracking those trajectories might not be feasible which requires a modification of trajectories. For this purpose, MHE and adaptive MPC could be utilized to estimate the parameters and modify the trajectories on-line. However, even though changes in parameter values are successfully estimated by MHE, the modification of trajectories by MPC for the entire horizon could not be feasible by the lack of computational time. This forces MPC to solve the optimal control problem with a limited length of prediction horizon where the terminal state of the bioreactor is not considered. This becomes a problem because most of the important cost is imposed only at the terminal state of the bioreactor. Therefore, providing a proper terminal cost of MPC becomes a crucial point to properly modify the trajectories. At this point, the value calculated from the RL could be used as the terminal cost of MPC. Although this value would not be exact by the changes of parameter values, the simulation results suggest that this approximation is valid for MPC to stably calculate the new trajectories.

The key to this combination of MPC and RL is selecting the proper length of the prediction horizon. Under this scheme, the length of the prediction horizon is not only determined by the computational limits, but it is related to the reliability of the parameter values. The long prediction horizon implies that MPC is more rely on the newly estimated parameter values than the one used for RL. The presence of the parameter changes is one support on using the longer prediction horizon. However, the reliability of the estimated parameter values is related to both the quality and quantity of measurement data. Therefore, the choice of prediction horizon length should be carefully determined by evaluating the reliability of the parameter values. We propose one method for evaluating such reliability and propose a method of adjusting the prediction horizon length.

[1] C. Larroche, Sanromán Maria Ángeles, G. Du, and A. Pandey, Current developments in biotechnology and bioengineering. Amsterdam: Elsevier, 2017.

[2] Shuler and Kargi, Bioprocess engineering. United States: Academic Internet Publishers, 2007.

[3] H.-H. Kuo, Introduction to stochastic integration. New York: Springer, 2006.

[4] R. S. Sutton, Reinforcement learning: an introduction. Cambridge, Mass: MIT Press, 2018.

608716 Sustainable API Development: Developing a Strategy for Measuring and Reducing Environmental Impact in the Pharmaceutical Industry

Abstract:

In January 2020, AstraZeneca announced its ambition to eliminate greenhouse gas emissions from its sites and fleet by 2025 and to become carbon negative across the entire value chain by 2030. The development and manufacture of active pharmaceutical ingredients (API) contributes significantly to AstraZeneca's carbon footprint and process engineers have a responsibility to play a leading role in achieving the zero-carbon target.

This presentation will share the approach taken to implement a strategy for measuring and reducing environmental impact during the development and manufacture of APIs. This approach comprises of influencing ways of working during API process development, implementing robust measurement of process environmental impact and identifying opportunities for innovation through science and technology to minimise AZ's carbon footprint.

This approach describes developing a procedure to appropriately consider sustainability alongside other key drivers for process design such as product quality, manufacturability and regulatory factors. Similarly, creating a culture to deliver continuous improvement against established metrics for sustainability in every aspect of API development and manufacturing.

This case also describes initiation of novel technology investigations to significantly reduce the carbon footprint of AstraZeneca's API manufacturing processes. This will be a leading factor in achieving a 50% reduction in carbon footprint for API development and the zero-carbon ambition of the company.

609303 An Integrated Process Automation System in API Manufacturing: Key Characteristics and Case Study

Abstract:

Safety is always first priority and a critical concern in the chemical industry. Jiangsu Provincial Emergency Management Administration (JSEMA) in China released <Basic Requirements for Diagnosis and Improvement of Instinct Safety> (Su Emer. NO.53) in June 2019, which specifies the requirements for the main aspects of chemical production. The concept of "integrated process automation" in Su Emer. No. 53 poses a great challenge to API manufacturing. It emphasizes mandatory safety and intensified automatic implementation. This paper illustrates the key characteristics in the integrated process automation of API manufacturing and gives some case studies.

Such automation system includes sophisticated solution and cutting-edge tools. According to the recipe and process of the product, the consultant optimizes the control strategic with expertise to enhance the safety and meet the regulation requirements. The process control modules or batch control program are deployed in DCS system for the purpose of multifunctional and flexible manufacturing. In case study it gives example of the mixing control strategic in the reactors which is critical and should consider all concerned factors. Usually the geared motor electricity is ignored to monitor, the absence may interfere the cooling system activation in time, its real time monitoring and interlock with cooling media are necessary; another example is the utility facilities configuration and interlock with automatic systems which should be tuned duly to eliminate the hidden trouble. The integrity of safety instrumented functions (SIFs) should be determined through risk assessment tools and rating tools, such as HAZOP (Hazard and Operability Analysis) and LOPA (Layer of Protection Analysis). The related case study reports sharing control components between SIS and DCS will increase the possibility of failure, which should be validated. The allocation of center control room and controller assignment should be concerned with HSE guideline. It also discusses the roadmap of information technology and MES (Manufacture Execution System) for API manufacturing. Our implementation shows that the integrated process automation system can help to reach instinct safety, which increase the productivity and improve the quality of products with regulatory compliance.

605455 A Quantitative Assay of Sodium Triacetoxyborohydride

Abstract:

Sodium triacetoxyborohydride (STAB) is a common reducing agent for reductive amination reactions. Due to sensitivities to water and air, the potency of this reagent tends to degrade over time. Traditional analytical assays for STAB fail to accurately assess the content of the active borohydride species, and often report assay purities of 97% or greater, when the true potency is lower. An incorrect or unknown potency of the STAB being used may lead to significant challenges when scaling up reductive aminations. An excess of STAB may be required and reactions may not achieve full conversion with the amount of STAB charged, requiring additional charges. The use of an excess of STAB comes with greater raw material costs and the thermal and H_2 evolution hazards associated with quenching the excess reagent.

This presentation outlines a simple assay, based on an aldehyde reduction, that was developed to determine the active borohydride content of STAB. The HPLC assay yield of a salicylaldehyde reduction has been shown to accurately determine this potency and has been validated against an H_2 evolution method, as well as yields obtained from a reductive amination.

607708 Delivery Strategies for Live Therapeutic Bacteria

Abstract:

The microbiome is a collection of bacteria, viruses and fungi that perform essential functions for human health. Dysbiosis, or the dysregulation of the microbiome, can contribute to chronic inflammation, metabolic disorders, neurological diseases, and cancer. Clinically, one of the only methods for altering the composition of the microbiome is with broad-spectrum antibiotics, which contribute to dysbiosis and create selective pressure for antibiotic resistant pathogens. As such, live therapeutic bacteria (LTB) have been identified as an attractive alternative to antibiotics, as they can modulate the composition of the microbiome without eliminating commensal species. However, LTBs are commonly delivered orally and experience significant challenges during gastrointestinal (GI) transit, including survival in the harsh conditions of the stomach, adherence to the intestinal lumen, and navigation of interpatient variability in the microbiome due to factors such as diet. Few engineered drug delivery platforms have been rationally developed to address these challenges. In this work, we describe two complementary platforms that separately adapt LTBs to improve viability during gastric transit and modify LTB surfaces to aid in adhesion along the GI tract. In the first system, we sought to understand the influence of factors such as diet on LTB adaptation and colonization. We delivered an LTB to female BALB/c mice on two distinct diets and found that diet significantly influenced both the appearance of a small colony variant of the LTB, as well as its distribution along the GI tract. As adaptation appeared to significantly affect LTB colonization, we developed an approach to control the adaptation of LTBs to specific environmental insults during GI transit. We found that growth in an *in vitro* adaptation media can prime LTBs for the acidic conditions of the stomach, significantly improving their viability during a challenge in simulated gastric fluid. Following transit through the stomach, LTBs must adhere to the intestinal lumen to persist and colonize in the GI tract. Therefore, for our second system, we developed a platform to chemically conjugate targeting ligands to the surface of LTBs to improve their adhesion to the GI mucosa. This platform significantly improves both the rate of colonization, as well as the concentration of LTBs in the GI tract of female BALB/c mice. Collectively, this work represents a simple approach to adapt LTBs to the intestinal environment during *in vitro* growth, while enabling subsequent modification of the LTB surface for improved intestinal adherence. Together, these delivery techniques for LTBs may help overcome challenges associated with GI transit, colonization and adaptation.

609030 In Vitro Biomimetic Tumor Model with Cancer-Specific Gene Regulators and Green Fluorescence Reporter for High Throughput Drug Screening

Abstract:

Although much scientific effort has been expended for cancer treatment, this disease is still one of the leading causes of death around the world. The limitations and shortcomings of current therapies highlight the urgent need for more effective therapeutic strategies. So far, most of the chemotherapy drugs act negatively on the cell division without selection, leading to serious adverse effect on normal cells. The current mainstream of anti-cancer drug development is to find the drugs that have specific targets (DNA, mRNA or proteins). Those targets are overexpressed in cancerous cells and are critical for carcinogenesis. In this study, we engineered breast and lung cancer cells to express the enhanced green fluorescent protein (EGFP) under cancer-specific gene promoters for three known carcinogenesis genes - survivin, Bcl-2 and hTERT, which enable online high-throughput screening of drugs that may up- or down-regulate these cancer genes as indicative of efficacy in inhibiting breast and lung cancers. Additionally, the mechanism of drugs and the crosstalk between cancer genes can be elucidated at the early stage of drug discovery. Another issue in the pharmaceutical industry is the high drug failure rate owing to poor predictability of early drug assays in a two-dimensional cell culture system. In such traditional assay, multiple cell-cell and cell-matrix interactions are deficient, leading to incomprehensive cellar responses to drug agents. To improve the drug assay predictability, our assay has focused on the construction of three-dimensional biomimetic cancer cell models with the recapitulation of multiple elements of tumor microenvironment. While developing the new drug screening assay, we also combined the rapid cell detection approach to achieve the high-throughput capability. Our optimized drug screening process can help to discover more anti-cancer therapies.

609939 Preparation and Use of Fine Grade Engineered Excipients for Direct Compression of Binary Blends of Cohesive Drug Powders

Abstract:

Novel, surface modified excipients are prepared using a fluid energy mill (FEM) to simultaneously mill and dry coat microcrystalline cellulose (MCC) with nano-silica to produce different sizes of fine, well flowing MCC excipient [1]. The purpose of these new class of excipients is to enable direct compression (DC). DC route is one of the most efficient and economic methods for tablet manufacturing since only blending and compression operations are required, greatly simplifies the process-train. However, the popularity of direct compression is limited by its stringent requirements, such as good flowability, high packing density, and excellent compaction for excipient-active pharmaceutical ingredient (API) blends. When the blend has high drug loading with a poorly flowing API, the flowability and tabletability of blends is generally decreased sharply. In a recent work, it has been shown that surprisingly, fine excipients of 20 mm size that have improved flow and bulk density after silica dry coating, can greatly improve direct compression performance [2]. Unfortunately, it could not work for very fine cohesive APIs such as 10 mm acetaminophen (APAP) at higher drug loading. In this work, binary blends of several fine sized dry coated MCC-based excipients are considered with three different fine, cohesive APIs at high drug loadings. Bulk density, FFC, and tablet compactability for all these blends are measured and compared to determine which excipient sizes provide the best overall performance. The results are analyzed for the relative sizes of the excipients and APIs to better understand the interplay between the components of the blends and assessment of the extent of silica transfer from coated excipient powders to API powders and its effect on blend bulk density and flowability. It is expected that using such novel surface engineered fine excipients in blends with fine APIs may also help in reduced segregation tendency during overall processing. Hence, these results are expected to provide industry relevant guidelines for the use of finer surface modified excipients in DC tableting.

596229 Applying Machine Learning to Predict Therapeutic Antibody Viscosity

Abstract:

A general machine learning feature selection protocol to predict antibody stabilities is proposed. These features only require information from sequences and molecular structures of antibodies. In addition, the physically relevant features are easily interpretable and help to design experiments to understand the underlying mechanisms of antibody stabilities. Two case studies are demonstrating using this protocol. First, we applied this framework to predict aggregation behaviors of 21 marketed monoclonal antibodies at high concentration (150 mg/mL), yielding an accuracy of 97% on validation tests with only three features. Models trained using structural features, rather than sequence features, correlated better with aggregation. Second, the concentration dependent viscosity behavior of 27 FDA approved antibody drugs was measured in concentrated solutions at high concentration (150 mg/mL). Combining molecular modeling and machine learning feature selection, we found that the net charge in the antibodies and the amino acid composition in the variable region are key features which govern the viscosity behavior. We presented two predictive models for aggregation and viscosity based on machine learning trained models, facilitating early stage design.

602107 Development of Multi-Adjuvant Systems for Vaccines Against Infectious Diseases

Abstract:

The ability to elicit potent humoral and cellular immune responses is a key element required for the development of efficacious prophylactic and therapeutic vaccine formulations. For vaccines against intracellular pathogens, the induction of a balanced, long-lived, and potent $CD4^+$ and $CD8^+$ T cell activation and proliferation is critical, combined with robust antibody responses. One strategy is to activate innate immunity and proinflammatory responses through Toll-like receptor-independent cytosolic pathways. Importantly, engagement of these mechanisms has demonstrated its ability to provide protection against bacterial challenge, and several small molecules have been evaluated in preclinical and clinical studies as vaccine adjuvants against infectious diseases.

Small molecules can bind to cytoplasmic receptors triggering signaling cascades that activate innate immunity. Despite their potential as adjuvants, small molecules have some limitations that need to be overcome. For example, there may be low exposure in target tissues due to their size and charge, and are prone to enzymatic biodegradation; with some off-target toxicities. In order to address these challenges, the use of drug delivery systems is being explored.

In this work, three different polymeric nanoparticle platforms were evaluated for small molecule delivery and for their capability to act synergistically as co-adjuvants and elicit robust immune responses. In these studies, nanoparticles based on natural and synthetic polymers were fabricated using water/oil/water and oil/water emulsions. Particle characterization was performed using dynamic light scattering, scanning electron microscopy, and zeta potential measurements that confirmed the size, morphology, and charge of the resulting nanoparticles. Small molecule compounds were loaded into each of the platforms, and their loading and release efficiency were determined using liquid chromatography. Finally, these systems were evaluated using *in vivo* murine models following parenteral administration to assess their ability to induce cytokine secretion, antibody and cell-mediated immunity. The results obtained demonstrate the potential of polymeric particulate systems as dual delivery and co-adjuvant platforms for the development of small molecule-adjuvanted prophylactic and therapeutic vaccines.

603577 Development of a Harvest Small Scale Model to Simulate Continuous Centrifugation

Abstract:

Disk stack centrifugation is one of the main clarification steps used in large scale mammalian cell culture processes. During process development, pilot scale disc stack centrifuge is normally used to develop and characterize the harvest process for new cell culture processes due to the lack of smaller scale model for disc stack centrifugation. In order to increase speed of development, a small scale model (SSM) of a disk stack centrifuge is of high demand to enable harvest development in an earlier stage of process development. A small scale model can also improve throughput and reduce cost. In this study, we have developed a SSM for disc stack centrifugation using a capillary shear device combined with benchtop centrifugation. Capillary tube size, length, and flow rate were optimized to generate energy dissipation rates comparable to large scale disk stack centrifuges. Additionally, shear force from multiple centrifuges at different scales were standardized and compared. Mathematical correlations were established for energy dissipation, viable cell density, cell viability, and lactate dehydrogenase (LDH) activity. Our SSM offers an approach to understand how the separation process, such as centrifuge designs, sizes, and configurations affect filtration performance and product quality. This work can speed up harvest development during process scale-up in biopharmaceutical mammalian cell culture processes.

599615 Optimizing and Reducing Cleaning Production Loss in Pharma, Agrochemical and Specialty Chemical Companies Using a New Expert System.

Abstract:

Pharmaceutical, Biopharmaceutical, Agrochemical and Specialty chemical companies make many different products in their plants. This is true both for the active ingredient stages as well as the secondary of formulation processing. Cleaning and set up between products is critical and reduce productivity time up to 50% in some cases. The presentation will show the critical areas important in the modelling development in this expert system and how it will aid companies in scheduling their productions to reduce cleaning time and to reduce issues with cleaning.

599656 Revolutionize of Pharmaceutical Product Changeovers By Enhanced Cleaning with Microbubbles

Abstract:

Pharmaceutical, Biopharmaceutical, Agrochemical and Specialty chemical companies make many different products in their plants. This is true both for the active ingredient stages as well as the secondary of formulation processing. Cleaning and set up between products is critical and reduce productivity time up to 50% in some cases. Microbubbles have been shown to enhance many unit operations e.g. distillation. The benefits of microbubbles in cleaning will be presented along with the impact this will enable in regulatory/ registration processes.

603892 (Invited) Skin-like Wearable and Implantable Sensors

Abstract:

In this talk, I will present our recent work on BodyNet concept for stretchable wearable and implantable sensors. Examples include, full body sensing, blood flow sensor and morphing electronics - neurostimulator.

600113 An Adhesion-Based Anti-Corrosion Strategy for Wearable Electrochemical Sensing and System Integration

Abstract:

The large-scale deployment of biomonitoring sensors through integration in wearable consumer electronics enables population-level health and wellness monitoring, and thus can play a critical role in transforming personalized medicine. Currently, physical sensors have been widely incorporated within commercialized wearable platforms, but they cannot capture biomarker information that lie in molecular levels and provide highly specific information about the body's dynamic chemistry. Electrochemical biosensing interfaces can be positioned to target biomarker molecules in a wearable format, as they can be constructed on miniaturized/flexible footprints/substrates and can convert (bio)chemical signals into electrical signals measurable by electronic readouts. The contact pads on the printed circuit board (PCB), which is a core component of wearable consumer electronics, can directly serve as electrodes and as a substrate for electrochemical sensors. However, we found that the commercialized PCB metal contact pads (*e.g.*, gold) were prone to corrosion and unsuitable for electrochemical modifications in solution environments. Additionally, electrochemical biosensors often are not reusable and need to be replaced frequently, while PCB units are not positioned for disposable applications.

To this end, we devise an adhesion-based anti-corrosion strategy for PCB-interfaced electrochemical sensing, which enables seamless sensor-system integration. The strategy is based on decoupling the sensing interface and PCB metal contact pads by an intermediary ultrathin (~50 μm), vertical conductive, and double-sided adhesive film. This strategy has dual benefits of making electrochemical sensors disposable and insulating the contact pads from sounding solutions. We first establish the suitability of the adhesion film to serve as an electrode and as an interconnection film simultaneously. One side of the double-sided adhesive film was patterned with noble metal electrodes, followed by the modifications with recognition layers, such as permselective membrane and enzymes, to render biosensing functionality. The electrochemical biosensors were transferred and adhered to the PCB contact pads with the aid of water-soluble tape. As compared with bare PCB contact pads, this fabrication strategy allows for realizing a corrosion-resistant and stable electrochemical biosensing interface on PCB for operation in biofluid (*e.g.*, sweat). Amperometry and open circuit potential characterization experiments validated the electrochemical corrosion resistance properties in saline solution environments. The excellent anti-corrosion attributes of the modified film could be ascribed to: (1) the polymer glues on both sides of the adhesive film acting as corrosion-protective layers; (2) bypassing the need for oxidation prone-adhesion metal (*e.g.*, chromium), achieved by e-beam evaporation (at the noble metal patterning step); and (3) the electrodeposited permselective membrane (here, phenylenediamine), which has been reported to presented anti-corrosion properties. We specifically tailored the enzymatic layer of the electrochemical interface to target biomarker molecules such as glucose, lactate, and choline as example of informative metabolites and nutrients. By modifying different (bio)chemical layers various other molecules can be targeted. In this way, our strategy serves as a foundation for electrochemical sensor development and integration with wearable consumer electronics.

607021 Rapid Screening Platform for the Isotopic Determination of Radionuclides in Water

Abstract:

Traditional approaches for the detection of trace-level radioisotopes in water require lengthy offsite sample preparations and do not lend themselves to rapid screening. Therefore, a novel sensor platform that combines onsite purification, concentration, and isotopic screening with a fieldable detection system will be an invaluable tool.

In this presentation, we will discuss findings on the development of a sensor platform based on reactive polymer film coatings for the rapid isotopic screening of waterborne radionuclides. Nanothin coatings were prepared by grafting uranium-selective polymers from polyethersulfone (PES) surfaces via UV-initiated polymerization, and by introducing uranium-selective functional groups to polyacrylonitrile (PAN) surfaces by chemical reaction. Ellipsometry was used to study film growth kinetics on PES films. X-ray photoelectron spectroscopy of modified PAN films revealed conversion of nitrile groups to amidoxime groups to be as high as 40% and showed that the extent and depth of reaction could be varied precisely. Static uptake experiments with solutions of depleted uranium spiked with ^{233}U were conducted to determine uranium binding capacities and kinetics of the modified polymer films at different pH values from 4 to 8. Sorption isotherm data were fitted to the Langmuir model, and the highest sorption capacities were obtained at pH 6 for modified PAN (M-PAN) and PES (M-PES) films. Capacities at pH 4 and 8 were lower and could be explained by differences in sorption mechanisms. Uranium batch uptake kinetics followed a pseudo-second order rate model. Alpha spectroscopy pulse height spectra were analyzed to study the role of selective layer film thickness on peak energy resolution. Excellent resolution with full width at half maximum values from 29 to 41 keV were recorded for M-PAN film and from 26 to 45 keV for M-PES film. Whereas uranium uptake increased with selective layer film thickness and varied with polymer chemistry/extent of modification, the peak energy resolution was independent of layer thickness and polymer chemistry within the experimental measurement uncertainties.

Results from this work were used to modify PES and PAN ultrafiltration membranes with the same experimental conditions to develop uranium selective membranes. High-throughput uptake experiments were performed to study the effect of residence time on detection efficiency and isotopic resolution. This membrane-based detection method enables the rapid, fieldable screening analysis of radionuclides in water for environmental monitoring studies and nuclear forensics investigations.

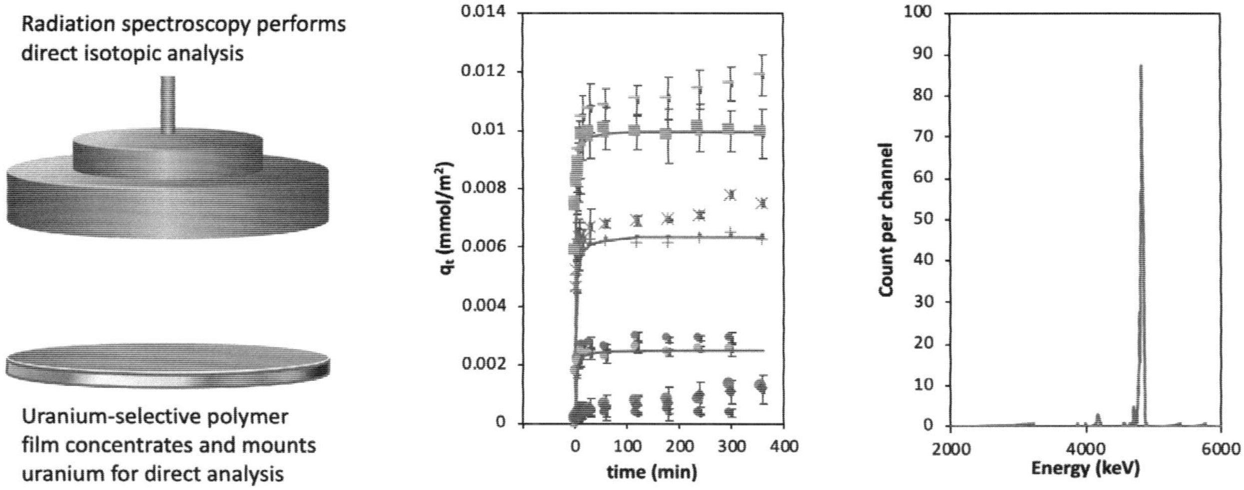

600875 Electrochemical Detection of Environmental Pollutants

Abstract:

Many small molecule pollutants interact with human hormonal systems. These compounds, known as endocrine disruptors, are ubiquitous, as they are found in plastics, medications, and pesticides. Such chemicals are especially problematic because they have been linked to severe health problems, including cancer, infertility, and diabetes. We have developed two complementary electrochemical methods to rapidly detect and quantify these compounds. One technique involves the specific monitoring of Bisphenol A (BPA) using a DNA aptamer immobilized on an electrode. The second method involves the measurement of the biological activity of chemically dissimilar endocrine disruptors using a human hormone receptor as a component of an electrochemical sandwich assay. These electrochemical platforms require no specialized skills to implement and have enabled very sensitive detection (sub-ppb levels) of these environmental pollutants from complex solutions, such as baby formula.

604144 Fabrication of High-Resolution Graphene-Based Flexible Electronics Via Polymer Casting and Microfluidic Approaches

Abstract:

Graphene-based flexible electronic devices for sensor applications has been receiving growing interest due to unique properties of graphene.[1] Various fabrication methods, such as photolithography, printing and chemical vapor deposition (CVD), has been used to produce high resolution and low feature size graphene-based devices.[2, 3] However, most of these methods are complex and require multiple processing steps (i.e. stamping, vacuum drying and etching) as well as additional post treatments (i.e. high temperature (300-1000 °C) baking or laser annealing) that can thermally or chemically degrade polymer-based substrates limiting substrate material selection.[2, 3] In addition, these methods can only be applied to 2D planar substrates and are not feasible to fabricate 3D circuits and some of them, such as sticky/adhesive tape peeling method, require high amounts of graphene consumption.[4] Despite the recent progress in the field, there is still a need to develop novel fabrication approaches that can address the current limitations of the existing methods.

In this study, we present two flexible electronics device fabrication methods based on simple polymer casting and microfluidics approaches.[5, 6] The first method focuses on simple polymer casting based graphene transfer at room temperature that does not require any post-processing. Briefly, our method consists of two main steps; (i) the formation of graphene patterns/films on substrates/molds via conventional methods such as CVD, channel filling or ink-jet printing and (ii) direct casting of target substrate polymer solution on the substrates/molds with graphene micropatterns and direct graphene transfer to the target substrate via peeling off upon drying and film formation. This method simply relies on the differences in the surface energies and adhesive forces between the graphene/mold and graphene/target polymer substrate.[5] Alternatively, the second method concentrates on a novel microfluidic approach to fabricate high resolution and low feature size graphene-based circuits/devices.[6] This method involves controlled and selective filling of microchannels on substrate surfaces with a conductive binder-free graphene nanoplatelets (GNP) solution. The ethanol-thermal reaction of GNP solution at low temperatures (~75 °C) prior to microchannel filling (as called "pre-heating") potentially reduces and homogenize GNP, which in turn enhances conductivity and reduces sheet resistance (down to ~0.05 kΩ sq^{-1}). Therefore, harsh post-processing of GNP is eliminated, and room temperature fabrication is enabled, which allows the use of degradable substrates, including biodegradable and natural polymers. This simple process is followed by selective administration of conductive GNP solution to the predetermined microfluidic channels on flexible substrates in a controlled manner via a syringe pump to create high resolution and low feature size graphene circuits/devices. This microfluidic approach can also be used to create 3D circuits with different geometries, using combined approaches with 3D printing, which is difficult to obtain with other methods (i.e. inkjet, screen or aerosol printing) that work mainly with 2D planar substrates.[6]

We demonstrated that using these fabrication methods, we obtained high resolution and low feature size graphene circuits with different geometries, dimensions and 3D superficial microstructures. We investigated the effect of temperature and microfluidic parameters (such as concentration, volume and flow rate) as well as the type of substrate materials on the graphene structure and formed micropatterns. The XPS, Raman, SEM and TEM results validated the structural changes of graphene along with confirming the stable presence of graphene patterns on the target substrates.[5, 6] The feature sizes of the graphene patterns could range from a few micrometers (down to ~15 μm in width and ~5 μm in depth) to a few millimeters with very small amounts of GNP used (~2.5 mg of graphene to obtain ~0.1 kΩ sq^{-1} of sheet resistance for 1 cm^2). [5, 6] The generated graphene patterns demonstrated significant stability after multiple washing and bending cycles. These methods were also implemented using other conductive inks, such as conductive graphene-silver based composite ink.[5, 6] Room temperature processing also provided precise control of 3D microstructural and mechanical properties (such as film porosity, pore size, elasticity etc.) of the target substrate materials, as demonstrated by SEM images.[5, 6] Using these methods, we successfully fabricated pressure sensors and near field communication antennas. [5, 6] Our results demonstrated that the devices worked with high sensitivity and stability.

In conclusion, the circuits/devices fabricated using this method can easily be implemented as wearable electronic, sensors and batteries, robotic components or motion detectors. In a broader sense, these simple, environmentally friendly and low-cost methods could pave the way for the fabrication of 2D and 3D electronic circuits/devices on different degradable flexible substrates using various graphene-based conductive inks.

References:

1. Jang, H.; Park, Y. J.; Chen, X.; Das, T.; Kim, M. S.; Ahn, J. H., Graphene-Based Flexible and Stretchable Electronics. *Advanced Materials* **2016,** *28* (22), 4184-4202.
2. Das, S. R.; Nian, Q.; Cargill, A. A.; Hondred, J. A.; Ding, S.; Saei, M.; Cheng, G. J.; Claussen, J. C., 3D nanostructured inkjet printed graphene via UV-pulsed laser irradiation enables paper-based electronics and electrochemical devices. *Nanoscale* **2016,** *8* (35), 15870-15879.
3. Das, S. R.; Srinivasan, S.; Stromberg, L. R.; He, Q.; Garland, N.; Straszheim, W. E.; Ajayan, P. M.; Balasubramanian, G.; Claussen, J. C., Superhydrophobic inkjet printed flexible graphene circuits via direct-pulsed laser writing. *Nanoscale* **2017,** *9* (48), 19058-19065.
4. Oren, S.; Ceylan, H.; Schnable Patrick, S.; Dong, L., High-Resolution Patterning and Transferring of Graphene-Based Nanomaterials onto Tape toward Roll-to-Roll Production of Tape-Based Wearable Sensors. *Advanced Materials Technologies* **2017,** *2* (12), 1700223.
5. Uz, M.; Jackson, K.; Donta, M. S.; Jung, J.; Lentner, M. T.; Hondred, J. A.; Claussen, J. C.; Mallapragada, S. K., Fabrication of High-resolution Graphene-based Flexible Electronics via Polymer Casting. *Scientific Reports* **2019,** *9* (1), 10595.
6. Uz, M.; Lentner, M. T.; Jackson, K.; Donta, M. S.; Jung, J.; Hondred, J.; Mach, E.; Claussen, J.; Mallapragada, S. K., Fabrication of Two-Dimensional and Three-Dimensional High-Resolution Binder-Free Graphene Circuits Using a Microfluidic Approach for Sensor Applications. *ACS Applied Materials & Interfaces* **2020,** *12* (11), 13529-13539.

611518 Disorder and Entropy in Molecular Crystals

Abstract:

Many computational methods for predicting polymorphic behavior generate many possible crystal structures and rapidly evaluate the lattice energies of these candidate structures to determine the most likely to be observed experimentally. However, lattice energies alone do not fully represent the stability of systems that are at room temperature and sample a range of configurations due to the thermally-induced motions at non-zero temperatures, which have both harmonic and anharmonic character.

In this talk, I survey our work using molecular dynamics to examine the free energy differences between enantiotropic crystalline systems, where the approximation that the lowest lattice energy crystal is the most stable must necessarily break down. We have found that although small rigid crystals are well-described by quasiharmonic approximations, even moderately conformationally flexible molecules can have extremely complex potential energy surfaces, making it hard to find appropriate structures from which to compute quasiharmonic approximations. We also find that many of these minima are accessible at room temperature, and proper annealing can help identify appropriate ensembles of structures to use for quasiharmonic analysis, and that bulk crystals differ thermodynamically in important ways from single unit cells. We also describe the extent to which there is moderate entropy-enthalpy compensation between different models and computational methods, potentially opening the door for using lower levels of computational detail to determine stability at room temperature even if classical potentials do not provide sufficient accuracy to determine lattice energies.

599827 A Simple Fix Addresses Severe Density Functional Theory Polymorph Ranking Problems in Pharmaceuticals

Abstract:

Knowledge of the solid form landscape is key during the pharmaceutical development process. Crystal structure prediction is increasingly used to help survey these landscape, and density functional theory (DFT) has established a successful track record of predicting the most stable crystal structures for a variety of species. However, a few significant limitations of the widely used functionals have recently become apparent. Typical density functionals predict the polymorph stabilities poorly in certain conformationally flexible pharmaceuticals, for example. In other cases, these functionals incorrectly predict salt formation in are actually neutral co-crystals. More accurate correlated wavefunction methods can overcome these limitations, albeit at a much higher computational cost. The origins of these deficiencies will be discussed, and a simple, computationally inexpensive approach that can overcome the difficulties in conformational polymorphs will be presented. Examples will include prolific polymorph formers ROY, Axitinib, and Galunisertib.

605426 Virtual Coformer Screening by Cloud-Computing Crystal Structure Predictions

Abstract:

One of the most popular strategies of the optimization of drug properties in the pharmaceutical industry appears to be a solid form changing into a cocrystalline form. A number of virtual screening approaches have been previously developed to allow a selection of the most promising cocrystal formers (coformers) for an experimental follow-up. A significant drawback of majority of those methods is related to the lack of accounting for the crystallinity contribution to cocrystal formation. To address this issue, we propose a virtual coformer screening approach based on a modern cloud-computing crystal structure prediction (CSP) technology at a dispersion-corrected density functional theory (DFT-D) level.[1] The CSP-based methods were validated on challenging cases of indomethacin and paracetamol cocrystallization, for which the previously developed approaches provided poor predictions. The calculations demonstrated a dramatic improvement of the virtual coformer screening performance relative to the other methods. It is demonstrated that the crystallinity contribution to the formation of paracetamol and indomethacin cocrystals is a dominant one and, therefore, should not be ignored in the virtual screening calculations.

1. GX Sun, Y Jin, S Li, Z Yang, B Shi, C Chang, YA Abramov. *Phys. Chem. Lett. 2020,* 11, 8832−8838.

600506 Morphology Prediction for Organic Salt Crystals

Abstract:

Crystal morphology plays a significant role in many industrial applications. For example, it has a great impact on the end-use properties of the materials and the downstream performance of the manufacturing process. However, due to the large number of crystals, such as more than 1,000,000 crystals in the Cambridge Structural Database (CSD) at present [1], coupled with numerous growth conditions including solvent, additive, temperature, supersaturation, and crystallization, it is a great challenge to the industry to obtain specific crystals with particular product functionality and morphology.

The above challenge inspired our group at UCSB to develop the first version of our automated crystal morphology prediction software ADDICT (Advanced Design and Development of Industrial Crystallization Technology) [2]. Very recently, we have upgraded ADDICT's software framework to enable it to implement crystal growth models for any crystal complexity. [3] and rewritten its code using object-oriented programming in Matlab as well as addition of new features in the third version of ADDICT (ADDICT3) [4]. Therefore, based on these recent advances we can start attempting to predict the morphology for organic salt crystals.

Organic salt crystals are an important class of materials in the chemical, pharmaceutical, and nonlinear optical (NLO) industries. It is challenging to use mechanistic growth models to predict the morphology of such crystals, and in fact there are no other papers on the prediction of their morphology grown from solvents using such models. Organic salt crystals generally have stronger interactions than organic crystals, resulting in more complex periodic bond chains (PBCs) that are difficult to handle.

In this study, the concepts of a neutral growth unit, a building unit, and new PBC rules are proposed to predict the morphology of organic salts. We use two examples to study the morphology of such crystals grown from solvent. They are: (1) $_{\text{L}}$-leucine hydrobromide and (2) glycinium trifluoroacetate. The predicted crystal shapes grown from water solution are consistent with the experimental results, which provides evidence for the reliability of the proposed method.

References

[1] https://www.youtube.com/watch?v=4G6cFXtbwNo (accessed Jul 2, 2019).

[2] Li, J., Tilbury, C. J., Kim, S. H., & Doherty, M. F. (2016). Prog. Mater. Sci., 82, 1-38.

[3] Landis, S., Zhao, Y., & Doherty, M. F. (2020). Comput. Chem. Eng., 133, 106637.

[4] Zhao, Y., Tilbury, C. J., Landis, S., Sun, Y., Li, J., Zhu, P., & Doherty, M. F. (2020). Cryst. Growth Des., (Accept)

CRYSTAL GROWTH & DESIGN

ACS Publications

pubs.acs.org/crystal

Communication

Mitigating Punch Sticking Propensity of Celecoxib by Cocrystallization: An Integrated Computational and Experimental Approach

Chenguang Wang,# Shubhajit Paul,# David J. Sun, Sten O. Nilsson Lill, and Changquan Calvin Sun*

Cite This: *Cryst. Growth Des.* 2020, 20, 4217−4223

Read Online

ACCESS | Metrics & More | Article Recommendations | Supporting Information

ABSTRACT: An L-proline cocrystal of celecoxib was identified based on a computational approach for predicting crystal mechanical properties in order to significantly reduce the punch sticking propensity of celecoxib. The reduced punch sticking propensity of this cocrystal was attributed to both reduced plasticity through deactivating slip planes and minimizing exposure of high electronegative functional groups to the punch tip during compression. This material-sparing and reliable approach of integrated computational and experimental investigation of punch sticking holds promise in crystal engineering of drugs for crystal forms suitable for developing tablet formulations.

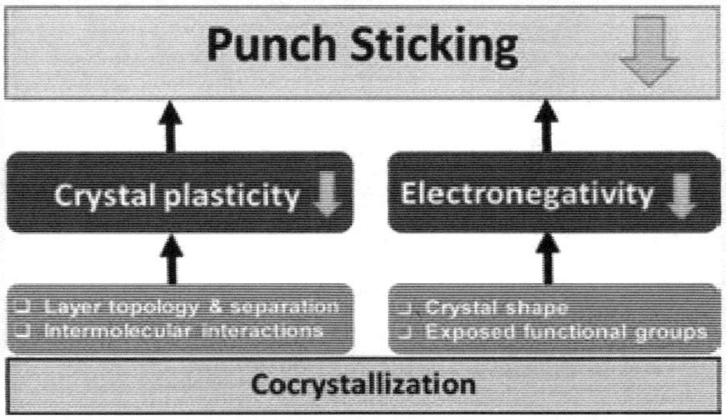

Cocrystallization is a supramolecular heterosynthon synthesis process to crystallize two or more different molecular and/or ionic compounds in a stoichiometric ratio through hydrogen bonds, halogen bonds, $\pi-\pi$ interactions, and weak intermolecular interactions.[1,2] This technology has been used to discover new materials with desirable properties, e.g., organic optics,[3,4] semiconductors,[5] energetic materials (explosives, propellants, and pyrotechnics),[6] and pharmaceuticals.[7,8] Cocrystals with demonstrated advantageous material properties also offer potential intellectual property opportunities.[9]

Pharmaceutical cocrystal screening is commonly performed during the preformulation stage to overcome the deficient properties of active pharmaceutical ingredients (API), such as solubility, bioavailability, stability, and tabletability.[10−12] Drug products enabled by cocrystals, including an antidiabetic drug Suglat (ipragliflozin and L-proline), an anticancer drug Odomzo (sonidegib and phosphoric acid), and a cardiovascular drug Entresto (valsartan and sacubitril), attest to the potential value of the cocrystallization technology in the pharmaceutical development. A rational pharmaceutical cocrystal design starts from understanding intermolecular interactions between functional groups in known cocrystals based on a structural informatics approach.[13] Many new structural insights into the solubility, chemical/physical stability, pharmacokinetics, and mechanical properties of cocrystals have been gained in recent years.[10,14−17] Among them, the structural understanding of the crystal mechanical behavior (elastic or plastic bending and brittle fracture) and properties (hardness and elastic moduli),[18−20] facilitates the design of cocrystals with desirable mechanical properties to enable the more efficient development of solid oral drug products.

Punch sticking is an outstanding common problem that hinders commercial tablet manufacturing. The occurrence of punch sticking is evident when powder mass adheres onto tooling surfaces during the course of tablet compression. It causes poor tablet quality, such as rough tablet surface and inconsistent drug content, and results in low manufacturing efficiency because of the required intermittent process stoppage for cleaning.[21] Previous works showed that a number of factors, such as crystal habit, surface energy, particle shape, size, choice of excipients, manufacturing conditions, and environment, can affect punch sticking.[21−26]

According to the Paul−Sun punch sticking model,[27] the punch sticking propensity of an API depends on the interplay among three forces, i.e., F_1 (API-punch adhesion), F_2 (API-API cohesion), and F_3 (API-excipient adhesion). API sticking occurs if F_1 is greater than both F_2 and F_3. The sticking ceases after a monolayer of API has coated the punch tips if $F_2 < F_3$, while continuous buildup of API is expected when $F_2 > F_3$. Thus, an API with a high F_2, due to strong bonding strength

Received: April 10, 2020
Revised: June 9, 2020
Published: June 10, 2020

among API particles, could lead to severe punch sticking. Accordingly, punch sticking can be reduced by (1) lowering F_1 with a lubricant, polymer coating,[23] or changing tooling surface[28] or (2) increasing F_3 by using suitable excipients in a formulation.[26] The use of plastic and strongly bonding excipients in the tablet matrix can reduce sticking propensity because it leads to higher F_3 without affecting F_1 and F_2.[26]

A highly plastic API tends to exhibit greater punch sticking because of the more facile increase in API-punch contact areas, which leads to higher F_1, and API–API contact area, which corresponds to higher F_2.[26] Therefore, we hypothesized that the punch sticking propensity of an API can be reduced by using a harder crystal form to reduce both F_1 and F_2. Although not yet tested, we are encouraged by the observation that punch sticking propensities among solid forms differ.[29,25] In this work, we test this hypothesis using celecoxib (CEL, Scheme 1a) and one of its less plastic cocrystals. The commercial CEL (form III) is an API with known high punch sticking propensity.[27]

Scheme 1. Molecular Structures of (a) Celecoxib and (b) L-Proline

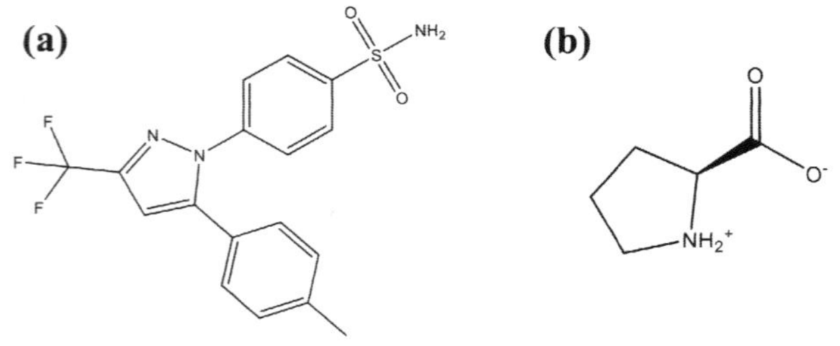

A CSD search returned 21 unique structures that contain CEL (20 neutral and one anionic CEL) (Table S1). To select a crystal harder than CEL form III for testing the hypothesis, we first performed a computational screening.[25] Crystal plasticity depends on the mobility of dislocations, while crystal stiffness is determined by the molecular packing pattern and the strength of intermolecular interactions.[20,30] Low plasticity is expected for a crystal meeting the following criteria: (1) strong 3D intermolecular interactions, such as the presence of 3D hydrogen bonding network, and (2) absence of active slip planes due to strong interlayer bonding energy and rough topology of layer surface. A quantitative digital topology analysis tool was employed to predict the slip plane and estimate the layer roughness.[31] The degree to which atoms from one layer interpenetrate into the adjacent layers was calculated and ranked. Miller planes with the least physical impediment are predicted as slip planes. A positive interlayer distance suggests complete separation between atoms in the two layers while negative interlayer distance implies the interdigitation between layers. Thus, potential molecular layers in a crystal structure were identified and analyzed for packing topological rugosity and interlayer hydrogen bonds. The topology analysis identified a total of 15 crystal structures that are likely hard because all possible molecular layers are interlocked (i.e., negative interlayer distance), which hinders slip among layers. The other 6 crystal structures have positive interlayer separation distance, suggesting easier slip (Figure 1).

The three most interlocked structures are bipyridine derivative solvates (Table S1). However, they are not suitable for pharmaceutical applications because of safety concerns.

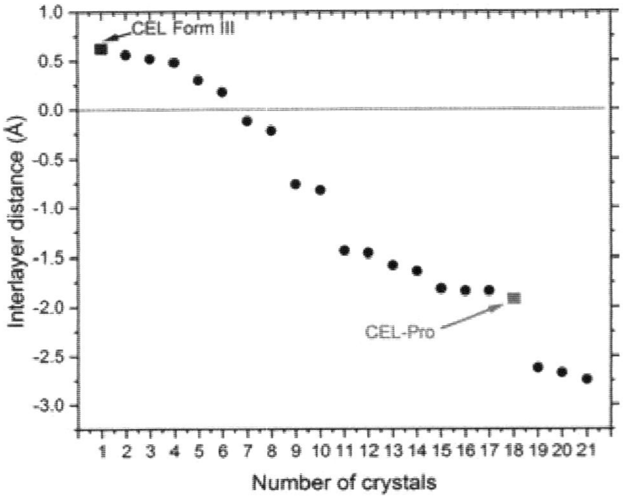

Figure 1. Celecoxib-containing crystals rank order based on the interlayer distance between the most likely slip planes in each crystal. The more negative separation distance corresponds to larger extent of interdigitation between adjacent layers.

Hence, we selected the likely next least plastic 1:2 CEL cocrystal with L-proline (Pro, Scheme 1b).

The structures of CEL-Pro (BEJNAI) and CEL (DIBBUL) were first analyzed to assess the potential hardness difference between them. For triclinic CEL (space group $P\bar{1}$, $Z = 2$, and $Z' = 1$), the asymmetric unit consists of a single CEL molecule. CEL molecules are connected through N–H⋯O=S ($d_{(D\cdots A)}$: 2.952 Å) and N–H⋯N (3.087 Å) hydrogen bonds, arranged in $C_1^1(4)$ and $R_2^2(18)$ motifs, to form one-dimensional CEL chains running along the *c*-direction. Adjacent chains are connected through S=O⋯H–C (3.294 Å) and $\pi\cdots\pi$ stacking to form 2D layers parallel to the *ac* plane, i.e., the (010) plane. Only weak C–F⋯π interactions are present between the neighboring layers. The energy framework (Figure 2 and Table S2) shows the much lower total interlayer bonding energies (−25.1 kJ/mol) compared with intralayer interaction energies (−256.6 kJ/mol) for (010) (see the Supporting Information). This is consistent with the lowest attachment energy and largest surface area of (010) in the predicted crystal morphology (Figure S2 and Table S2). Hence, the slip of the (010) planes is energetically favorable. The interlayer spacing of 0.20 Å and the absence of interplanar hydrogen bonds corroborate the possible slip of the (010). It is noteworthy that the most separated (1−1 0) planes, with a spacing of 0.62 Å (Figure S3c), are less energetically favorable for slip to occur because of strong dispersive interactions. Therefore, CEL is expected to be plastic with (010) as the primary slip plane.[32] The predicted plasticity of CEL here should not be confused with the high 1D elasticity of CEL observed before.[33] While CEL crystals can undergo reversible elastic bending on the major facet up to at least 3.56% of strain,[33] the stress required to initiate plastic deformation is low. That is to say that the elastically bending CEL crystals can still be plastic when the external stress deforms the crystal beyond its elastic limit.

The asymmetric unit of the orthorhombic CEL-Pro cocrystal (space group $P2_12_12_1$, $Z = 8$ and $Z' = 2$) is much more complex than CEL, containing two CEL and four L-proline molecules. The crystal consists of alternating layers of Pro and

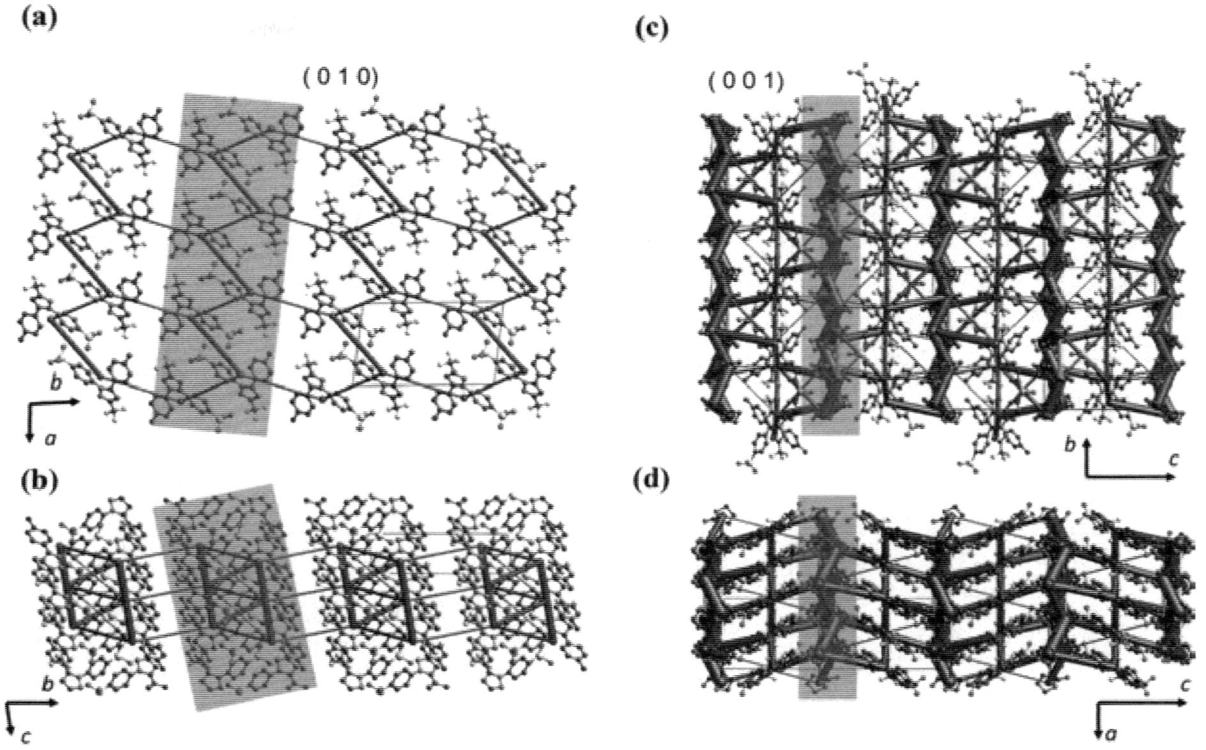

Figure 2. Crystal energy framework of CEL form III (a and b) and CEL-Pro (c and d). The molecular layers identified are shaded in pink. The thickness of each cylinder (in blue) represents the relative strength of interaction. The energy threshold for the energy framework is set at −10 kJ/mol.

CEL parallel to the (001) plane. The zwitterionic Pro molecules in the Pro layer form a 2D network of charge-assisted hydrogen bond N−H···O⁻ (2.736 Å, 2.819 Å, 2.856 Å, 2.878 Å, and 2.960 Å). This structural feature was correlated to high hardness in some amino acid crystals.[20] Within the CEL layer, CEL molecules interact through several hydrogen bonds, including S=O···H−N (2.856, 2.948 Å) and S=O···H−C (3.302, 3.399 Å). Despite the clearly layered structure, the overall packing pattern corresponds to a relatively isotropic energy framework, suggesting comparable interlayer and intralayer interaction strength (Figure 2c,d). This may be attributed to the charge-assisted N−H···O⁻ hydrogen bonds (2.849 Å, 3.022 Å) between the adjacent layers. The predicted crystal morphology of CEL-Pro exhibited a lower aspect ratio (2.609) than CEL (7.093) (Figure S3), which is consistent with the energetically more isotropic packing of CEL-Pro. The predicted CEL crystal has the major facet of (010); the CEL-Pro crystal has a few comparable facets, but the (002) facet is the largest. In both crystals, the electrostatic contribution dominates the intermolecular interaction (Figure S1). Although both crystal structures exhibit layered packing, the bonding energy between (001) layers of CEL-Pro (−152.6 kJ/mol) is much higher than that between (010) layers of CEL (−25.1 kJ/mol). In addition, the (001) planes are interlocked (interlayer distance of −1.92 Å) in CEL-Pro but not between (010) planes in CEL. Thus, the CEL-Pro cocrystal is expected to be less plastic.

From the Heckel analysis of in-die compressibility data of bulk powders (Figure 3a, see the Supporting Information), the plasticity parameter, P_y, of CEL (52 MPa) is lower than that of CEL-Pro (84 MPa). Therefore, CEL is more plastic than CEL-Pro by this measure.[34] The stiffness of CEL and CEL-Pro was quantified by their elastic tensors calculated from their crystal structures (Figure S2). The slightly higher anisotropy index of CEL (3.339) than CEL-Pro (3.235) based on Young's modulus, *E*, distributions, implied that CEL is likely more plastic than CEL-Pro. The *E* (averaging scheme of Hill) of CEL (9.5 GPa) is higher than that of CEL-Pro (7.7 GPa). The elastic modulus determined by nanoindentation was 10.19 ± 0.45 GPa for the (002) facet of CEL-Pro and 16.27 ± 0.43 GPa for the (010) facet of CEL, which are close to the calculated *E* values of 9.7 GPa for CEL-Pro and 14.0 GPa for CEL of corresponding facets, respectively. It should be pointed out that the (010) facet of CEL in this work is identical to the (001) facet in a previous report because of the different assignments of unit cell axes in the two crystal structures.[32] According to the *E* values, CEL and CEL-Pro fall in the intermediate stiffness category of the mechanical properties of organic crystals.[35]

The nanoindentation data (Figure 3b) suggested that contact hardness (H_c) of CEL on the major face (010), 0.45 ± 0.02 GPa,[33] is not significantly different from that on the (002) facet of CEL-Pro (0.46 ± 0.02 GPa). However, the total deformation during the nanoindentation test is a sum of elastic and plastic components, which depend on both elastic modulus and plastic deformation resistance (*H*, a true measure of crystal plasticity), respectively.[36] The percent plastic deformation in CEL (60%) is higher than that in CEL-Pro (50%), which corresponds to the higher *E* of CEL. After correcting for the contribution from elastic deformation (see the Supporting Information), the *H* of CEL-Pro (1.84 ± 0.11 GPa) is much higher than CEL (1.21 GPa).[36] Additionally, the

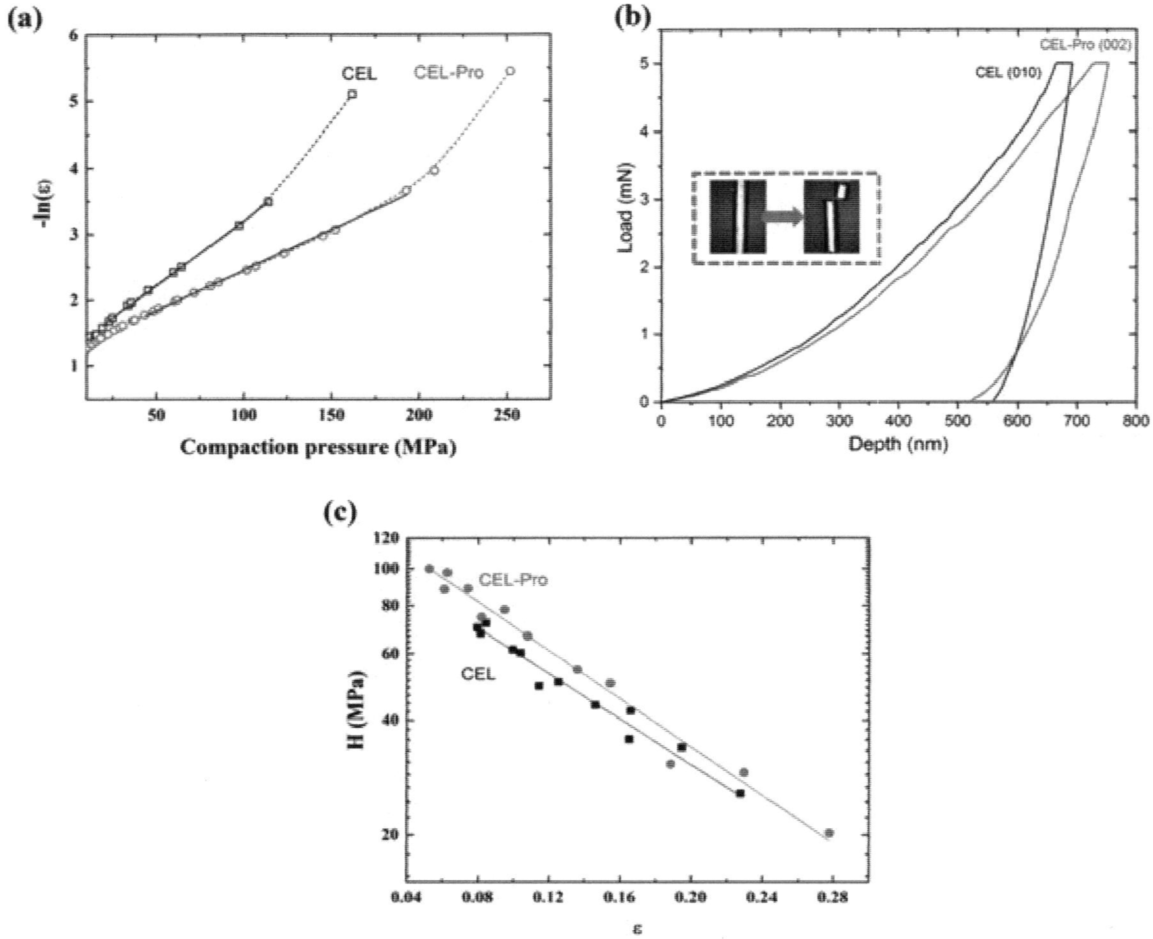

Figure 3. Mechanical properties of CEL form III and CEL-Pro assessed by (a) in-die Heckel analysis, (b) nanoindentation on single crystals (inset shows the fracture of a CEL-Pro crystal upon bending), and (c) macroindentation on tablets of 1:1 mixtures with MCC.

CEL-Pro crystals fractured with a sharp fracture plane when bent, suggesting a limited extent of plastic deformation (Figure 3b).

Because nanoindentation could not be used to quantify H of all crystal facets, the powder plasticity of the two crystals was also assessed by the macroindentation method using compressed tablets.[37] Because intact tablets of pure crystals could not be prepared as a result of extensive punch sticking by CEL and fracture of CEL-Pro tablet during indentation, 1:1 binary mixtures with microcrystalline cellulose were used for compression. Results show that tablet hardness extrapolated to zero porosity (H_0) of the CEL mixture (121.1 $\pm$ 6 MPa) is lower than that of the CEL-Pro mixture (146 $\pm$ 6 MPa) (Figure 3c), suggesting the plasticity of CEL is greater than that of CEL-Pro. In summary, the reduced plasticity of CEL-Pro is supported by computational structure analysis, in-die Heckel analysis, nanoindentation, and macroindentation data.

Plasticity plays an important role in powder compression according to the bonding area–bonding strength model, which suggests that more plastic particles tend to develop larger bonding areas in a tablet.[38] The observation of higher tabletability of CEL than CEL-Pro (Figure 4) is consistent with the higher plasticity of CEL (Figure 3), which leads to larger bonding area among particles in a tablet compressed at the same pressure. The reduction in tablet tensile strength (σ) at 250 MPa for CEL-Pro and 300 MPa for CEL suggest the phenomenon of overcompression.[39] Compared to CEL-Pro, the delayed occurrence of overcompression at a higher pressure for CEL is in agreement with its lower extent of

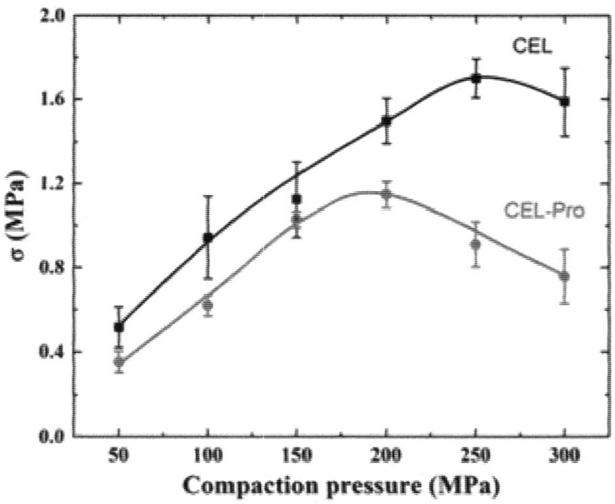

Figure 4. Tabletability of CEL form III and CEL-Pro crystals.

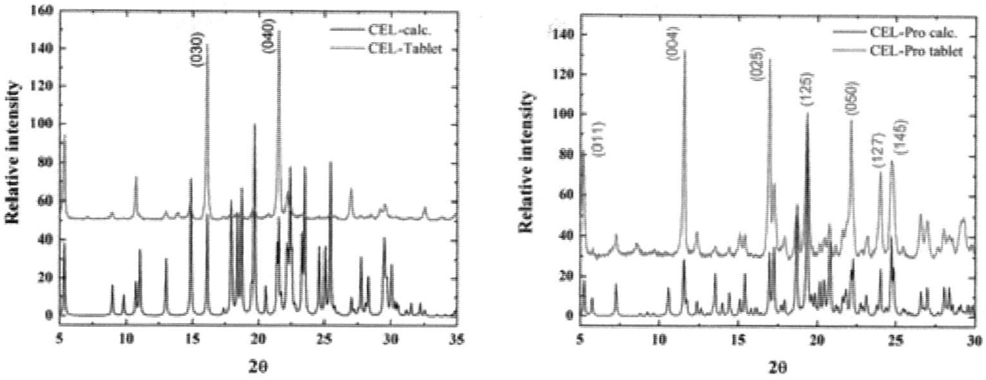

Figure 5. XRD patterns of tablets and calculated PXRD patterns of (a) CEL form III and (b) CEL-Pro. Corresponding crystallographic (*hkl*) planes of main diffraction peaks are indicated.

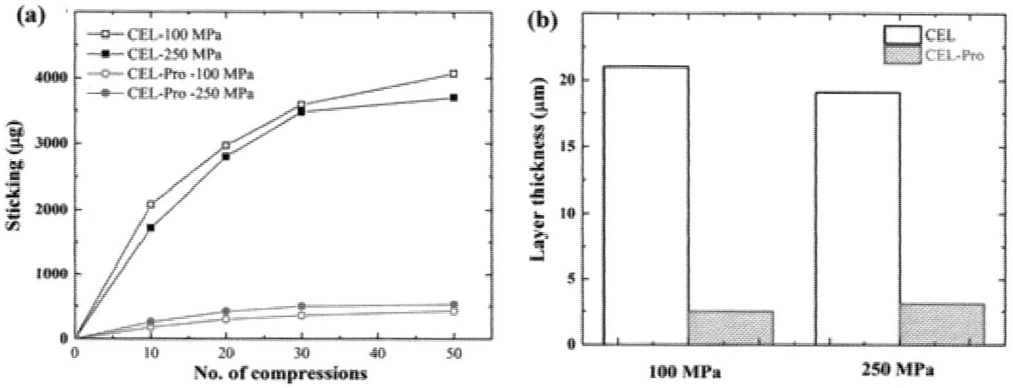

Figure 6. (a) Sticking propensity and (b) sticking layer thickness of CEL form III and CEL-Pro formulations (20% active in MCC matrix) at compaction pressures of 100 and 250 MPa.

elastic recovery and its higher plasticity. The former corresponds to its higher E, suggested by the nanoindentation data, and the latter affords greater resistance against reduction of bonding area induced by elastic recovery. Regardless of the mechanism, the lower tabletability of CEL-Pro than CEL suggests that the F_2 of CEL-Pro is lower than that of CEL, i.e., interparticle bonding force between CEL-Pro crystals is weaker, which favors lower sticking propensity according to the punch sticking model.[27] Hence, both the lower plasticity and the consequently lower F_2 favor a sticking propensity of CEL-Pro that is lower than that of CEL.

As demonstrated previously, the large electrostatic potential difference of exposed functional groups and punch surface (Fe) facilitates the initiation of sticking.[25] The PXRD diffractograms of tablets showed that the (010) plane is preferably exposed to the surface of the CEL tablet while the (004) and (025) planes are preferably exposed to the CEL-Pro tablet surface (Figure 5). Out of the possible scenarios, the 4-(trifluoromethyl)-1H-pyrazole rings of the (010) plane is more likely exposed to the CEL tablet surface while the pyrrolidine rings of Pro are more likely exposed to the CEL-Pro tablet surface because their interlayer bonding energies are lower than those of the alternative surface exposure (Figure S3d,e). The molecular electrostatic potentials maps (Figure S3f) suggested the carbon–fluorine functional group in CEL is more electronegative than the exposed portion of the pyrrolidine group of Pro. Therefore, the initiation of sticking for CEL-Pro is expected to be also slower than that of CEL.[24,25]

The consistently predicted lower sticking propensity of CEL-Pro by the various predictive tools (crystal structure analysis, computational calculation, and bulk powder mechanical properties) was experimentally confirmed by punch sticking assessment, where the sticking propensity of CEL-Pro was only ~10% of that of CEL (Figure 6a). After 50 compressions, a thin film was developed in the case of CEL-Pro while a much thicker layer of sticking mass was observed for CEL (Figure 6b).

The significantly lower punch sticking propensity of CEL-Pro supports the hypothesis that deactivating the slip plane in CEL through cocrystallization with L-proline reduces crystal plasticity, which lowers σ by limiting bonding area among particles. The effects of both the lower plasticity and weaker API–API interaction collectively reduced punch sticking propensity of CEL significantly. The less exposed electronegative functional groups to punch during compression of CEL-Pro reduces the onset of sticking. Therefore, CEL-Pro appears to be a suitable candidate for developing a CEL tablet formulation. By demonstrating the ability to reduce punch sticking, this example expands the list of potential applications of cocrystallization in facilitating drug development.

■ ASSOCIATED CONTENT

Supporting Information

The Supporting Information is available free of charge at https://pubs.acs.org/doi/10.1021/acs.cgd.0c00492.

Experimental details of material and methods, reference codes of crystals containing celecoxib in Cambridge Structural Database (CSD), topology analysis, predicted crystal morphology, attachment energy, energy framework, and elastic constants (PDF)

■ AUTHOR INFORMATION

Corresponding Author

Changquan Calvin Sun – *Pharmaceutical Materials Science and Engineering Laboratory, Department of Pharmaceutics, College of Pharmacy, University of Minnesota, Minneapolis, Minnesota 55455, United States;* orcid.org/0000-0001-7284-5334; Phone: 612-624-3722; Email: sunx0053@umn.edu; Fax: 612-626-2125

Authors

Chenguang Wang – *Pharmaceutical Materials Science and Engineering Laboratory, Department of Pharmaceutics, College of Pharmacy, University of Minnesota, Minneapolis, Minnesota 55455, United States;* orcid.org/0000-0003-1926-5722

Shubhajit Paul – *Pharmaceutical Materials Science and Engineering Laboratory, Department of Pharmaceutics, College of Pharmacy, University of Minnesota, Minneapolis, Minnesota 55455, United States*

David J. Sun – *Pharmaceutical Materials Science and Engineering Laboratory, Department of Pharmaceutics, College of Pharmacy, University of Minnesota, Minneapolis, Minnesota 55455, United States*

Sten O. Nilsson Lill – *Early Product Development and Manufacturing, Pharmaceutical Sciences, R&D, AstraZeneca Gothenburg, Mölndal SE-431 83, Sweden*

Complete contact information is available at: https://pubs.acs.org/10.1021/acs.cgd.0c00492

Author Contributions

#C.W. and S.P. contributed equally to this work.

Notes

The authors declare no competing financial interest.

■ ACKNOWLEDGMENTS

We thank the Minnesota Supercomputing Institute (MSI) at the University of Minnesota for providing resources that contributed to the research results reported in this paper. URL: http://www.msi.umn.edu

■ REFERENCES

(1) Aitipamula, S.; Banerjee, R.; Bansal, A. K.; Biradha, K.; Cheney, M. L.; Choudhury, A. R.; Desiraju, G. R.; Dikundwar, A. G.; Dubey, R.; Duggirala, N.; Ghogale, P. P.; Ghosh, S.; Goswami, P. K.; Goud, N. R.; Jetti, R. R. K. R.; Karpinski, P.; Kaushik, P.; Kumar, D.; Kumar, V.; Moulton, B.; Mukherjee, A.; Mukherjee, G.; Myerson, A. S.; Puri, V.; Ramanan, A.; Rajamannar, T.; Reddy, C. M.; Rodriguez-Hornedo, N.; Rogers, R. D.; Row, T. N. G.; Sanphui, P.; Shan, N.; Shete, G.; Singh, A.; Sun, C. C.; Swift, J. A.; Thaimattam, R.; Thakur, T. S.; Kumar Thaper, R.; Thomas, S. P.; Tothadi, S.; Vangala, V. R.; Variankaval, N.; Vishweshwar, P.; Weyna, D. R.; Zaworotko, M. J. Polymorphs, Salts, and Cocrystals: What's in a Name? *Cryst. Growth Des.* **2012**, *12*, 2147–2152.

(2) Duggirala, N. K.; Perry, M. L.; Almarsson, Ö.; Zaworotko, M. J. Pharmaceutical Cocrystals: Along the Path to Improved Medicines. *Chem. Commun.* **2016**, *52*, 640–655.

(3) Yan, D.; Delori, A.; Lloyd, G. O.; Friščić, T.; Day, G. M.; Jones, W.; Lu, J.; Wei, M.; Evans, D. G.; Duan, X. A Cocrystal Strategy to Tune the Luminescent Properties of Stilbene-Type Organic Solid-State Materials. *Angew. Chem., Int. Ed.* **2011**, *50*, 12483–12486.

(4) Christopherson, J. C.; Topić, F.; Barrett, C. J.; Friščić, T. Halogen-Bonded Cocrystals as Optical Materials: Next-Generation Control over Light–Matter Interactions. *Cryst. Growth Des.* **2018**, *18*, 1245–1259.

(5) Zhang, J.; Geng, H.; Virk, T. S.; Zhao, Y.; Tan, J.; Di, C.-a.; Xu, W.; Singh, K.; Hu, W.; Shuai, Z.; Liu, Y.; Zhu, D. Sulfur-Bridged Annulene-Tcnq Co-Crystal: A Self-Assembled "Molecular Level Heterojunction" with Air Stable Ambipolar Charge Transport Behavior. *Adv. Mater.* **2012**, *24*, 2603–2607.

(6) Bolton, O.; Simke, L. R.; Pagoria, P. F.; Matzger, A. J. High Power Explosive with Good Sensitivity: A 2:1 Cocrystal of Cl-20:Hmx. *Cryst. Growth Des.* **2012**, *12*, 4311–4314.

(7) Sun, C. C. Cocrystallization for Successful Drug Delivery. *Expert Opin. Drug Delivery* **2013**, *10*, 201–213.

(8) Wang, C.; Perumalla, S. R.; Lu, R.; Fang, J.; Sun, C. C. Sweet Berberine. *Cryst. Growth Des.* **2016**, *16*, 933–939.

(9) Kapczynski, A.; Park, C.; Sampat, B. Polymorphs and Prodrugs and Salts (Oh My!): An Empirical Analysis of "Secondary" Pharmaceutical Patents. *PLoS One* **2012**, *7*, e49470.

(10) Battini, S.; Mannava, M. K. C.; Nangia, A. Improved Stability of Tuberculosis Drug Fixed-Dose Combination Using Isoniazid-Caffeic Acid and Vanillic Acid Cocrystal. *J. Pharm. Sci.* **2018**, *107*, 1667–1679.

(11) Shan, N.; Perry, M. L.; Weyna, D. R.; Zaworotko, M. J. Impact of Pharmaceutical Cocrystals: The Effects on Drug Pharmacokinetics. *Expert Opin. Drug Metab. Toxicol.* **2014**, *10*, 1255–1271.

(12) Sun, C. C.; Hou, H. Improving Mechanical Properties of Caffeine and Methyl Gallate Crystals by Cocrystallization. *Cryst. Growth Des.* **2008**, *8*, 1575–1579.

(13) Wood, P. A.; Feeder, N.; Furlow, M.; Galek, P. T. A.; Groom, C. R.; Pidcock, E. Knowledge-Based Approaches to Co-Crystal Design. *CrystEngComm* **2014**, *16*, 5839–5848.

(14) Good, D. J.; Rodríguez-Hornedo, N. Cocrystal Eutectic Constants and Prediction of Solubility Behavior. *Cryst. Growth Des.* **2010**, *10*, 1028–1032.

(15) Krishna, G. R.; Shi, L.; Bag, P. P.; Sun, C. C.; Reddy, C. M. Correlation among Crystal Structure, Mechanical Behavior, and Tabletability in the Co-Crystals of Vanillin Isomers. *Cryst. Growth Des.* **2015**, *15*, 1827–1832.

(16) Childs, S. L.; Kandi, P.; Lingireddy, S. R. Formulation of a Danazol Cocrystal with Controlled Supersaturation Plays an Essential Role in Improving Bioavailability. *Mol. Pharmaceutics* **2013**, *10*, 3112–3127.

(17) Kuminek, G.; Rodríguez-Hornedo, N.; Siedler, S.; Rocha, H. V. A.; Cuffini, S. L.; Cardoso, S. G. How Cocrystals of Weakly Basic Drugs and Acidic Coformers Might Modulate Solubility and Stability. *Chem. Commun.* **2016**, *52*, 5832–5835.

(18) Krishna, G. R.; Devarapalli, R.; Lal, G.; Reddy, C. M. Mechanically Flexible Organic Crystals Achieved by Introducing Weak Interactions in Structure: Supramolecular Shape Synthons. *J. Am. Chem. Soc.* **2016**, *138*, 13561–13567.

(19) Saha, S.; Mishra, M. K.; Reddy, C. M.; Desiraju, G. R. From Molecules to Interactions to Crystal Engineering: Mechanical Properties of Organic Solids. *Acc. Chem. Res.* **2018**, *51*, 2957–2967.

(20) Azuri, I.; Meirzadeh, E.; Ehre, D.; Cohen, S. R.; Rappe, A. M.; Lahav, M.; Lubomirsky, I.; Kronik, L. Unusually Large Young's Moduli of Amino Acid Molecular Crystals. *Angew. Chem., Int. Ed.* **2015**, *54*, 13566–13570.

(21) Chattoraj, S.; Daugherity, P.; McDermott, T.; Olsofsky, A.; Roth, W. J.; Tobyn, M. Sticking and Picking in Pharmaceutical Tablet Compression: An Iq Consortium Review. *J. Pharm. Sci.* **2018**, *107*, 2267–2282.

(22) Hooper, D.; Clarke, F. C.; Docherty, R.; Mitchell, J. C.; Snowden, M. J. Effects of Crystal Habit on the Sticking Propensity of Ibuprofen—a Case Study. *Int. J. Pharm.* **2017**, *531*, 266–275.
(23) Chen, H.; Paul, S.; Xu, H.; Wang, K.; Mahanthappa, M. K.; Sun, C. C. Reduction of Punch-Sticking Propensity of Celecoxib by Spherical Crystallization Via Polymer Assisted Quasi-Emulsion Solvent Diffusion. *Mol. Pharmaceutics* **2020**, *17*, 1387–1396.
(24) Paul, S.; Taylor, L. J.; Murphy, B.; Krzyzaniak, J. F.; Dawson, N.; Mullarney, M. P.; Meenan, P.; Sun, C. C. Toward a Molecular Understanding of the Impact of Crystal Size and Shape on Punch Sticking. *Mol. Pharmaceutics* **2020**, *17*, 1148–1158.
(25) Paul, S.; Wang, C.; Wang, K.; Sun, C. C. Reduced Punch Sticking Propensity of Acesulfame by Salt Formation: Role of Crystal Mechanical Property and Surface Chemistry. *Mol. Pharmaceutics* **2019**, *16*, 2700–2707.
(26) Paul, S.; Sun, C. C. Modulating Sticking Propensity of Pharmaceuticals through Excipient Selection in a Direct Compression Tablet Formulation. *Pharm. Res.* **2018**, *35*, 113.
(27) Paul, S.; Taylor, L. J.; Murphy, B.; Krzyzaniak, J.; Dawson, N.; Mullarney, M. P.; Meenan, P.; Sun, C. C. Mechanism and Kinetics of Punch Sticking of Pharmaceuticals. *J. Pharm. Sci.* **2017**, *106*, 151–158.
(28) Uchimoto, T.; Iwao, Y.; Yamamoto, T.; Sawaguchi, K.; Moriuchi, T.; Noguchi, S.; Itai, S. Newly Developed Surface Modification Punches Treated with Alloying Techniques Reduce Sticking During the Manufacture of Ibuprofen Tablets. *Int. J. Pharm.* **2013**, *441*, 128–134.
(29) Al-Karawi, C.; Leopold, C. S. A Comparative Study on the Sticking Tendency of Ibuprofen and Ibuprofen Sodium Dihydrate to Differently Coated Tablet Punches. *Eur. J. Pharm. Biopharm.* **2018**, *128*, 107–118.
(30) Mishra, M. K.; Ramamurty, U.; Desiraju, G. R. Solid Solution Hardening of Molecular Crystals: Tautomeric Polymorphs of Omeprazole. *J. Am. Chem. Soc.* **2015**, *137*, 1794–1797.
(31) Bryant, M. J.; Maloney, A. G. P.; Sykes, R. A. Predicting Mechanical Properties of Crystalline Materials through Topological Analysis. *CrystEngComm* **2018**, *20*, 2698–2704.
(32) Wang, C.; Sun, C. C. Identifying Slip Planes in Organic Polymorphs by Combined Energy Framework Calculations and Topology Analysis. *Cryst. Growth Des.* **2018**, *18*, 1909–1916.
(33) Wang, K.; Mishra, M. K.; Sun, C. C. Exceptionally Elastic Single-Component Pharmaceutical Crystals. *Chem. Mater.* **2019**, *31*, 1794–1799.
(34) Heckel, W. An Analysis of Powder Compaction Phenomena. *Trans. Metall. Soc. AIME* **1961**, *221*, 671–675.
(35) Wang, C.; Sun, C. C. The Landscape of Mechanical Properties of Molecular Crystals. *CrystEngComm* **2020**, *22*, 1149–1153.
(36) Oyen, M. L. Nanoindentation Hardness of Mineralized Tissues. *J. Biomech.* **2006**, *39*, 2699–2702.
(37) Patel, S.; Sun, C. C. Macroindentation Hardness Measurement—Modernization and Applications. *Int. J. Pharm.* **2016**, *506*, 262–267.
(38) Osei-Yeboah, F.; Chang, S.-Y.; Sun, C. C. A Critical Examination of the Phenomenon of Bonding Area - Bonding Strength Interplay in Powder Tableting. *Pharm. Res.* **2016**, *33*, 1126–1132.
(39) Shi, L.; Sun, C. C. Overcoming Poor Tabletability of Pharmaceutical Crystals by Surface Modification. *Pharm. Res.* **2011**, *28*, 3248–3255.

596992 Diffusion Growth Mechanism of Five-Fold Twinned Ag Nanowire

Abstract:

Five-fold twinned nanowires can be produced with high aspect ratios from diverse face-centered cubic (fcc) metals. However, the mechanism of their anisotropic growth is still poorly understood. We develop a model combining deposition and surface diffusion to predict the diameters and the aspect ratios of Ag nanowires. We describe two aspects of nanowire growth, denoted as the seed process and the wire process. In the seed process, the deposition time is much longer than the surface diffusion time, so that the seed shape only depends on the difference between interfacet diffusion times. The structure of nanowire seeds is similar to Marks decahedra with {111} notches and {110} steps that accelerate interfacet diffusion leading to wire growth. The result exhibits that the decahedron seeds could lead to three kinds of products with different strucutures as nanowires, decahedron particles and giant decahedron seeds. We focus on the seeds growing to nanowires and introduce the wire process, which occurs subsequently and continues until the diffusion time is equal to the deposition time. We predict the aspect ratio by contrasting the diffusion rate from {100} to {111} to the deposition rate on {100}. The aspect ratios predicted in our simulations are in the experimental range.

597237 Volume Effects on Acetaminophen Nucleation through an Induction Time Approach

Abstract:

Confined crystallization is an approach to develop nano-crystalline active pharmaceutical ingredients which improves dissolution and bioavailability properties of a formulated drug. To elucidate on this, the effect of volume reduction to the nucleation of acetaminophen in ethanol was explored. Using a supersaturated solution of acetaminophen in ethanol at volumes of 1000 μL to 25 μL, induction times for the primary nucleation event were obtained experimentally. The probability distribution of induction times showed that as sample volume is reduced, the induction time increases exponentially. The induction time at the volume of 25 μL is obtained to be at 4214 minutes, which is at a 5400% increase from the observed induction time of 78 minutes when the volume is set at 1000 μL.These results demonstrate the effect of volume reduction to nucleation induction times. Confinement may relate to very low probabilities of a nucleation event.

600993 Amorphous Intermediate States during the Precipitation of Cerium Oxalate: Towards a New Model

Abstract:

The precipitation is a common step in industry which is relatively ill-understood. In the last decades, many efforts were made to make numerical simulations of this step to avoid using the trial and error method, especially for the precipitation of active compound like plutonium oxalate. Many parameters are required to handle these numerical simulations, including the nucleation rate (the number of particles which appears in the reaction media by unit of time and of volume).

The nucleation rate is usually rationalized according to the Classical Nucleation Theory (CNT). This theory predicts the nucleation rate, assuming that thermal fluctuations of the reaction media lead to the formation of some "clusters" of monomers with the same composition and (?) symmetry as the crystal. These clusters tend to immediately dissolve into the reaction media unless they reach a particular size, called the critical size, from which they can grow fast to a macroscopic size. The so-called nucleation rate is the rate at which clusters reach this critical size. However, the CNT is more and more questioned, as it has been evidenced for many systems, including cerium oxalate, that nucleation occurs through amorphous states which are currently ignored by the CNT and which lead to an significant underestimation of the nucleation rate, up to 400 orders of magnitude.

In this work, the nucleation of cerium oxalate, a non-radioactive surrogate for plutonium oxalate, has been studied using cryo-microscopy and small and wide x-ray scattering (SAXS and WAXS). Here we show that cerium oxalate has two possible non-classical nucleation mechanisms: at low concentrations, nucleation in presence of amorphous nanoparticles, and at higher concentration, nucleation involving both a transient dense liquid phase and amorphous nanoparticles.

601001 Lanthanide Precipitation from Spent Ni-MH Battery Leachates: Kinetic Study and Phase Identification

Abstract:

Securing access to critical metals required for high-performance technologies, particularly the light rare earth elements (REEs = La, Ce, Nd, Pr), has become a major challenge for import-dependent economies such as the EU. The recycling of spent nickel metal hydride (Ni-MH) batteries from hybrid electric vehicles (HEV) serves as an increasingly-attractive secondary source of REEs thanks to the recent development of hydrometallurgical processes.[1] In order to recover REEs, precipitation from pregnant leach solutions (PLS) in sulfate media, using Na_2SO_4 is often reported in the literature. However, little consideration is given as to whether and how sodium ions influence the precipitation efficiency and selectivity, and detailed phase characterization of the products is rarely reported. This work focuses on a better understanding of the precipitation phenomenon and mass balances in this complex system, by coupling pilot-scale experiments on industrially-sourced PLS, in-depth precipitated particles identification, and thermodynamic calculations.

The initial PLS used in this study originates from the pilot-scale leaching of industrial samples of spent Ni-MH battery powders in H_2SO_4 media, according to a process described in [1]. The metallic content of the battery waste leachate consists of 71%mol Ni, 10 %mol REEs and 19%mol of other metals like Co, Mn and Fe. In the view of metal recycling, it is of great interest to find a selective precipitation pathway to recover REEs from such a complex solution. For each precipitation experiment, 1.5 L of PLS was added to a double-jacketed batch reactor; solution temperature was regulated using a cryostat and pH was monitored. In a standard experiment, a 2.9 M Na_2SO_4 solution was maintained at 40 °C and added to the PLS at constant flow rate, in simple jet, using a membrane pump. Mechanical agitation was implemented with a 3-blade Teflon marine propeller at 400 rpm. The influence of three operating parameters on precipitation efficiency and kinetics was studied: temperature (25 – 60 °C), Na:REEs molar ratio (1:1 – 4:1, by increasing the total addition time of Na_2SO_4 solution at a constant flow rate) and Na_2SO_4 solution addition (all at once or slowly at 7 $g.min^{-1}$ implying an initial semi-batch mode). Suspension volumes were regularly sampled, filtered on 0.45 µm syringe filters, and solutions were analyzed by ICP-OES. Precipitation efficiency was calculated as the proportion of the REEs initially present in solution that have precipitated. After 1 h, the suspension was filtered on a P3 Büchner filter and rinsed with demineralized water at room temperature. The obtained crystals were dried in an oven at 80 °C overnight before in-depth characterization using multiple techniques: SEM-EDX, powder XRD, TGA, microwave digestion in aqua regia and ICP-OES.

Results compiled in Figure 1) a) illustrate the evolution of the proportions of Ni, Co, Mn, Fe, La, Ce, Nd and Pr remaining in solution when precipitation is carried out at 60°C, for a Na:REEs molar ratio of 4:1 and a Na_2SO_4 solution flow rate of 7 $g.min^{-1}$. The simultaneous drop of pH when REE concentrations start to decrease (Figure 1) b) demonstrates that pH monitoring serves as an indicator of precipitation initiation and therefore allows determining the induction period, which is of about 3 min in this case. Curves in Figure 1) a) flatten out after approximately 30 min, suggesting that equilibrium has been reached. After 1 h, the precipitation efficiencies of La, Ce, Nd and Pr reach 99.3 %, 100.0 %, 99.7 % and 87.6 %, respectively, corresponding to 70 ppm of La, 0 ppm of Ce, 4 ppm of Nd and 52 ppm of Pr remaining in solution. REE precipitation is highly selective in this configuration since 99 % of Ni, 99 % of Co, 98 % of Mn, and 99 % of Fe remain in solution for further recovery. The molar ratios of REEs in the precipitate during the semi-batch and batch modes are reported in Figure 1) c). These ratios remain relatively constant during the 1 h run, suggesting that the same phases form throughout the whole precipitation period.

Precipitation efficiencies and aqueous REE concentrations are summarized in Table 1 for all experimental conditions. This compilation highlights the influence of operating parameters. For a fixed Na:REEs molar ratio of 4:1, increasing the temperature from 25 to 60 °C greatly increases precipitation efficiency of all REEs and yields faster precipitation kinetics; for example, 74.4 % of La is precipitated at 25 °C while at 60 °C, 99.3 % is precipitated. Similarly, increasing the Na:REEs molar ratio from 1:1 to 4:1 at 60 °C, strongly increases the precipitation efficiencies of all REEs. Furthermore, adding the equivalent volume of Na_2SO_4 solution at once yields the same precipitation efficiencies than when the solution is added at a 7 $g.min^{-1}$ flow rate but greatly increases precipitation kinetics; however, rapid injection favors nucleation generating very small particles and consequently, after 1 h, filtration was more difficult (low filtration rates). It is interesting to note that, for a given temperature or for a given Na:REEs molar ratio, precipitation efficiencies of the REEs is always as: Ce > Nd > La > Pr.

Powder XRD analyses of the precipitates yield diffraction peaks that lie between those of pure $NaLa(SO_4)_2.H_2O$ and $NaCe(SO_4)_2.H_2O$ double sulfates. Crystal lattice parameters were calculated and exhibit intermediate values between those of pure double sulfates suggesting that precipitates are solid solutions of lanthanide-alkali double sulfates. Structures are assigned to the hexagonal crystal system[2], which can be observed on a representative SEM micrograph provided in Figure 2. The chemical composition of the solids was determined by TGA (number of water molecules n) and ICP-OES analysis (elemental content of S, Na, La, Ce, Nd and Pr). The resulting solid compositions are gathered in Table 2 as $Na_x(REE)_1(SO4)_y.nH_2O$, where x and y are the respective Na and S proportions relative to the total REEs content. The water content (n) is close to 1 for each solid, which proves the formation of monohydrate-type crystals. The mass and electrical (3+x–2y) balances yield that both Na and SO_4 contents are lower than the stoichiometric ratios (x<1 and y<2), and that the overall charge of such compounds is about +0.35 – +0.40. This charge balance inconsistency, which is currently investigated, is most likely due to the incorporation of additional ions in the structure. As indicated in Table 2, REE molar ratios in the mineralized precipitates are relatively equivalent for all experiments. Therefore, it is concluded that the same lanthanide-alkali solid solution forms regardless of the temperature or the Na:REEs molar ratio.

In parallel, equilibrium calculations were performed using the mixed-solvent electrolyte model implemented in the OLI software, which includes an accurate thermodynamic description of rare earth sulfates recently developed by Anderko et al.[3] Calculations were carried out with the various initial PLS conditions (pH, temperature, and ion concentrations). They foresee the precipitation of the compounds $NaLa(SO_4)_2.H_2O$, $NaCe(SO_4)_2.H_2O$, $NaNd(SO_4)_2.H_2O$ and $NaPr(SO_4)_2.H_2O$ in all conditions, which is in good agreement with XRD characterizations. The calculated REEs aqueous concentrations, detailed in Table 1, are also rather close to experimental measurements, which indicates that the precipitation reactions are close to thermodynamic equilibrium after 1 h. More specifically, REE aqueous concentrations strictly meet calculated concentrations at 60 °C and for a 4:1 Na:REEs molar ratio. At 60 °C, calculations show that the solubilities of all lanthanide-alkali double sulfates decrease with increasing Na^+ concentration, which explains why increasing the Na:REEs molar ratio improves precipitation efficiencies at constant pH. Moreover, for a Na:REEs molar ratio of 4:1, increasing the temperature decreases the double sulfates solubilities (apart from $NaCe(SO4)2.H2O$ which shows the opposite trend). Therefore, based on both experimental results and thermodynamic trends, we conclude that the configuration (60 °C and 4:1 Na:REEs molar ratio) yields the best precipitation efficiencies for La, Ce, Nd and Pr (99.3 %, 100.0 %, 99.7 % and 87.6 %, respectively).

Key takeaways of this study are a straightforward approach to perform a grouped extraction of REEs contained in complex solutions originating from the leaching of industrial samples of spent Ni-MH batteries. Using a concentrated Na_2SO_4 solution, highly selective precipitation was obtained after 1 h at 60 °C and for a Na:REEs molar ratio of 4:1, whereby only 70 ppm La, 0 ppm Ce, 4 ppm Nd and 52 ppm Pr remain in solution and more than 98 % of the major elements do not co-precipitate. Thanks to multi-analytical characterization backed by thermodynamic calculations, the precipitated crystals were determined to be lanthanide-alkali double sulfate solid solutions of overall composition $Na_{0.79}La_{0.59}Ce_{0.20}Nd_{0.07}Pr_{0.14}(SO_4)_{1.72}.H_2O$. Results suggest that the same lanthanide-alkali solid solution forms regardless of the experimental conditions.

[1] M. Zielinski, L. Cassayre, P. Destrac, N. Coppey, G. Garin, B. Biscans, Leaching mechanisms of industrial powders of spent nickel metal hydride batteries in a pilot-scale reactor, ChemSusChem. 4 (2020) 616–628.

[2] A.C. Blackburn, R.E. Gerkin, Sodium lanthanum(III) sulfate monohydrate, NaLa(SO4)2.H2O, Acta Crystallogr. C. 50 (Pt 6) (1994) 833–838.

[3] G. Das, M.M. Lencka, A. Eslamimanesh, P. Wang, A. Anderko, R.E. Riman, A. Navrotsky, Rare earth sulfates in aqueous systems : Thermodynamic modeling of binary and multicomponent systems over wide concentration and temperature ranges, J. Chem. Thermodyn. 131 (2019) 49-79.

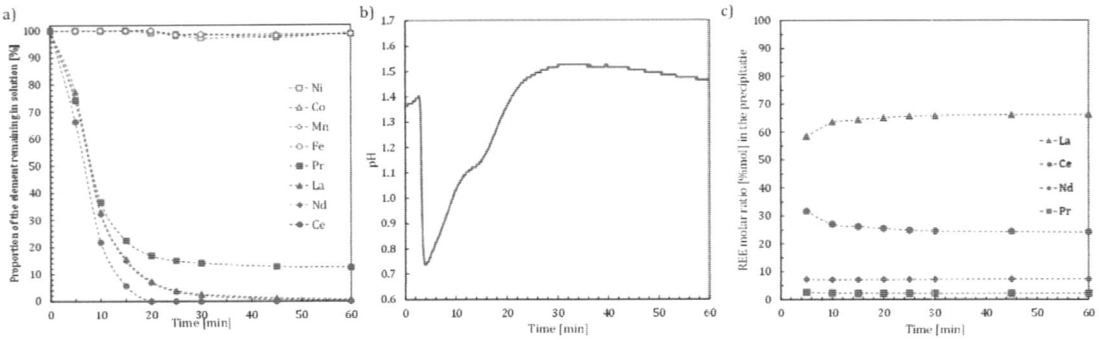

Figure 1 (a) Evolution of the proportions of Ni, Co, Mn, Fe, Pr, La, Nd and Ce remaining in solution [%] at 60 °C, for a Na:REEs molar ratio of 4:1 and a 7 g.min^{-1} Na_2SO_4 solution flow rate, (b) corresponding pH evolution and (c) evolution of the REE molar ratio [%mol] in the precipitate calculated from ICP analyses of the solution

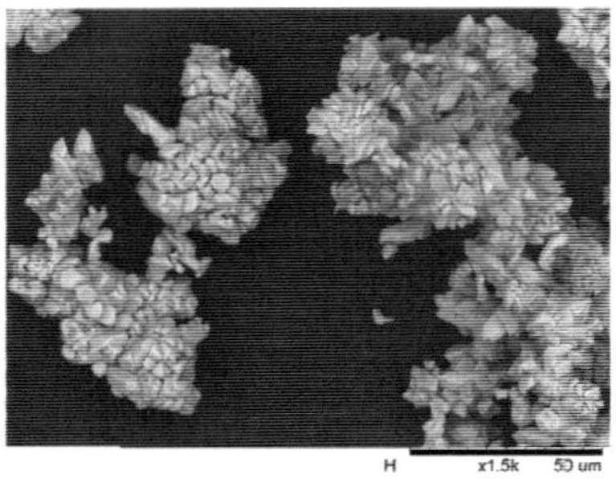

Figure 2 SEM micrograph of the precipitate obtained after 1 h of precipitation at 60 °C, a Na:REEs molar ratio of 4:1 and Na_2SO_4 solution flow rate of 7 g.min^{-1}.

Operating conditions										
	Influence of temperature			**Influence of Na:REEs ratio**			**Influence of Na_2SO_4 flow rate**			
Temperature [°C]	25	40	60	60	60	60	60	60	60	60
Na:REEs molar ratio	4:1	4:1	4:1	1:1	2:1	4:1	2:1	2:1	4:1	4:1
Na_2SO_4 flow rate	7 g.min^{-1}	7 g.min^{-1}	7 g.min^{-1}	7 g.min^{-1}	7 g.min^{-1}	7 g.min^{-1}	7 g.min^{-1}	At once	7 g.min^{-1}	At once
Precipitation efficiencies after 1 h of precipitation [%]										
Pr	69.3	84.3	87.6	69.1	80.0	87.6	80.0	82.7	87.6	14.4
La	74.4	92.9	99.3	71.7	92.4	99.3	92.4	93.0	99.3	1.2
Nd	77.5	94.4	99.7	74.1	93.5	99.7	93.5	93.7	99.7	1.2
Ce	87.5	100.0	100.0	78.7	95.6	100.0	95.6	94.4	100.0	1.9
Concentrations in solution after 1 h of precipitation [mol/L]										
Pr	0.0009	0.0005	0.0004	0.0011	0.0006	0.0004	0.0006	0.0005	0.0004	0.0004
La	0.0198	0.0056	0.0005	0.0224	0.0059	0.0005	0.0059	0.0058	0.0005	0.0010
Nd	0.0019	0.0005	0.0000	0.0024	0.0006	0.0000	0.0006	0.0006	0.0000	0.0001
Ce	0.0035	0.0000	0.0000	0.0059	0.0012	0.0000	0.0012	0.0015	0.0000	0.0005
Equilibrium concentrations according to OLI modelling [mol/L]										
Pr	0.0003	0.0003	0.0003	0.0034	0.0008	0.0003	0.0008	0.0008	0.0003	0.0003
La	0.0043	0.0016	0.0005	0.0069	0.0014	0.0005	0.0014	0.0014	0.0005	0.0005
Nd	0.0002	0.0001	0.0001	0.0011	0.0002	0.0001	0.0002	0.0002	0.0001	0.0001
Ce	0.0012	0.0021	0.0043	0.0263	0.0114	0.0043	0.0114	0.0114	0.0043	0.0043

Table 1 Precipitation efficiencies of La, Ce, Nd and Pr [%], corresponding concentrations [mol/L] remaining in solution after 1 h of precipitation and equilibrium concentrations [mol/L] as calculated by OLI modelling, with respect to operating conditions.

Parameter studied	**Temperature [°C]**	**Na:REEs molar ratio**	**Na_2SO_4 flow rate**	**Precipitate composition after 1 h of precipitation**
Influence of temperature	25	4:1	7 g.min^{-1}	$Na_{0.90}La_{0.58}Ce_{0.21}Nd_{0.07}Pr_{0.14}(SO_4)_{1.76}.1.1H_2O$
	40	4:1	7 g.min^{-1}	$Na_{0.87}La_{0.58}Ce_{0.21}Nd_{0.07}Pr_{0.14}(SO_4)_{1.74}.1.0H_2O$
	60	4:1	7 g.min^{-1}	$Na_{0.79}La_{0.59}Ce_{0.20}Nd_{0.07}Pr_{0.14}(SO_4)_{1.72}.1.0H_2O$
Influence of Na:REEs molar ratio	60	1:1	7 g.min^{-1}	$Na_{0.74}La_{0.58}Ce_{0.21}Nd_{0.07}Pr_{0.14}(SO_4)_{1.68}.1.0H_2O$
	60	2:1	7 g.min^{-1}	$Na_{0.78}La_{0.59}Ce_{0.20}Nd_{0.07}Pr_{0.14}(SO_4)_{1.70}.1.0H_2O$
	60	4:1	7 g.min^{-1}	$Na_{0.79}La_{0.59}Ce_{0.20}Nd_{0.07}Pr_{0.14}(SO_4)_{1.72}.1.0H_2O$
Influence of Na_2SO_4 flow rate	60	4:1	7 g.min^{-1}	$Na_{0.79}La_{0.59}Ce_{0.20}Nd_{0.07}Pr_{0.14}(SO_4)_{1.72}.1.0H_2O$
	60	4:1	At once	$Na_{0.78}La_{0.59}Ce_{0.20}Nd_{0.07}Pr_{0.14}(SO_4)_{1.70}.1.0H_2O$

Table 2 Precipitate composition after 1 h of precipitation with respect to operating conditions, deduced from microwave digestion of the precipitates, ICP-OES and TGA analyses.

604083 Molecular Modifiers Suppress Struvite Formation Via Unique Mechanisms

Abstract:

The crystallization of struvite ($MgNH_4PO_4 \cdot 6H_2O$) is relevant for nutrient (nitrogen and phosphorus) recovery in water purification and scale formation in pipelines, and it is a primary component of so-called infection stones arising from urinary tract infections. Research efforts have focused on understanding the effects of different parameters (e.g., supersaturation, pH, inhibitors) on crystallization of struvite; however, prior studies have only focused on the use of bulk techniques and have not provided mechanistic insights into these controls.

Herein, we present a comprehensive approach to evaluate struvite formation at both macroscopic and microscopic length scales in the absence and presence of various inhibitors. We conducted *in situ* kinetic measurements to estimate the extent of inhibition posed by each inhibitor on struvite formation. To this end, we employ a microfluidic platform to track the anisotropic growth rates of struvite crystals under various conditions, including comparisons of solutions with and without the presence of modifiers[1]. Furthermore, we elucidate the mechanism of inhibitory action by probing the surface dynamics in real time via *in situ* atomic force microscopy (AFM), which has proven to be a powerful technique to elucidate molecular level details of crystallization[2,3]. The results reveal unique mechanisms facilitating complete inhibition of growth and in select cases the unprecedented suppression of nucleation, which is not commonly observed for even the most potent modifiers of mineralization. Collectively, our findings identify highly efficient and commercially available modifiers that inhibit struvite formation. This study also uncovers a new class of inhibitory mechanisms that have potentially broader applicability to other minerals, thus establishing new paradigms for the prevention of commercial scale.

1. Kim, D., Olympiou, C., McCoy, C. P., Irwin, N. J., & Rimer, J. D. (2020). Time-Resolved Dynamics of Struvite Crystallization: Insights from the Macroscopic to Molecular Scale. *Chemistry–A European Journal*, *26*(16), 3555-3563.
2. Chung, J., Granja, I., Taylor, M. G., Mpourmpakis, G., Asplin, J. R., & Rimer, J. D. (2016). Molecular modifiers reveal a mechanism of pathological crystal growth inhibition. *Nature*, *536*(7617), 446-450.
3. Farmanesh, S., Ramamoorthy, S., Chung, J., Asplin, J. R., Karande, P., & Rimer, J. D. (2014). Specificity of growth inhibitors and their cooperative effects in calcium oxalate monohydrate crystallization. *Journal of the American Chemical Society*, *136*(1), 367-376.

610225 Solvent Structuring and Viscosity Effects on Thermodynamics and Kinetics of Crystallization

Abstract:

The structuring of the solvent at the crystal-solvent interface largely regulates the activation barrier for incorporation of solute into the crystal. Depending on the groups exposed on the crystal surface, the solvent at the interface may be ordered or disordered [1,2]. In general, the consequences of solvent structuring on the molecular level interactions comprising crystallization are poorly understood. Organic solvents, in which the bonds that could support a defined fluid structure are weak and varied present a particular challenge to understand the effect of solvent structuring on thermodynamics and kinetics of crystallization.

We explore the thermodynamic parameters of crystallization of a group of supramolecules known as porphyrins in organic solvents. Thermodynamic characteristics of crystallization were used as probes for the presence of solvent structuring which reveals that the interactions that govern the enthalpy and entropy of crystallization are dominated by the van der Waals contact between the solvent aliphatic tails and the non-polar groups of the etioporphyrin I (a free base porphyrin).

Time-resolved *in situ* atomic force microscopy monitoring of the step dynamics on (010) face of etioporphyrin I in two types of solvents, alcohols (octanol, butanol and hexanol) and DMSO, of varying aliphatic chain length and subsequent solvent viscosity demonstrates the impact of solvent structuring, solute-solvent interactions and solvent viscosity on the barrier of solute incorporation into kinks. The relation between step kinetic coefficient and solvent viscosity reveals that the crystallization of organic small molecules is not governed by diffusion limitation but rather by solute-solvent interaction at the kinks. The correlation between the step velocity and the solute concentration on the (010) face of etioporphyrin I for different solvents reveal that for solvents belonging to the same homologous series, solute incorporation scales with the solute diffusion coefficient. All-atom MD simulation for solvent structuring at the crystal interface emphasizes the importance of solute-solvent binding at the crystal interface.These emerging insights offer guidance for polymorph selection, chiral separations, structure design and numerous other open questions relevant to myriad crystallization systems in the pharmaceutical and chemical industries.

References

1. Vácha, R., Zangi, R., Engberts, J. B. F. N. & Jungwirth, P. Water Structuring and Hydroxide Ion Binding at the Interface between Water and Hydrophobic Walls of Varying Rigidity and van der Waals Interactions. *The Journal of Physical Chemistry C (2008)* **112**, 7689-7692.

2. Meister, K., Strazdaite, S., DeVries, A. L., Lotze, S., Olijve, L. L. C., Voets, I. K., & Bakker, H. J. Observation of ice-like water layers at an aqueous protein surface. *Proceedings of the National Academy of Sciences (2014)* **111**, 17732-17736.

611564 Identifying Crystal Morphologies to Optimize Growth Kinetics of Covalent Organic Frameworks – an Example of Cof-5

Abstract:

Covalent Organic Frameworks (COFs) have gained significant attention in the field of material science over last two decades [1]. 2D boronate-ester linked frameworks like COF-5 are porous, crystalline, organic polymers with low density, that have numerous applications such as gas storage and as membranes for various industrial purification applications. COF-5 exhibits high thermal stability, high surface area and permanent porosity, widening the scope of its applicability. Controlling crystal morphology is the most critical variable to maximize the applicability of these polymeric frameworks. Computing steady state and dynamic morphologies of COF-5 in a specific environment is essential for gaining comprehensive understanding of ways to control it. Face-specific growth rates have been previously utilized to understand these morphologies. The faces of these crystals are divided into three classes: flat, stepped and kinked faces, according to the Hartman-Perdok theory. This theory also states that layer by layer growth occurs on the F-face of the crystal lattice, which is the basis of studies conducted on COF-5 framework. Attachment energy calculations have been carried out along with kink sites distribution calculations to evaluate growth rates. The predicted rates have been used to understand the morphology of COF-5 crystals and the results are compared with experimental values, providing a deeper insight on optimizing process conditions for morphology control.

607359 Enzymatic Reactive Crystallization of Cephalexin Using Immobilized Penicillin G Acylase from *E. coli*

Abstract:

Beta-lactam antibiotics have been used extensively to treat bacterial infections for nearly a century. The two largest produced beta-lactam antibiotic classes, semi-synthetic penicillins (e.g. amoxicillin) and cephalosporins (e.g. cephalexin), constitute the largest amount of sales of antibiotics for the treatment of bacterial infections. Currently the industrial production of beta-lactam antibiotics is conducted largely via batchwise chemical synthetic pathways consisting of multiple reaction steps, use of protecting groups, harsh solvents, and very low (-40°C) temperatures. Using these methods, the chemical synthesis route produces a large amount of chemical waste with an E factor of roughly 50 and is highly energy intensive due to the operation of low temperature refrigeration equipment. Additionally, due to increases in labor and raw material costs, much of the production of beta-lactam antibiotics has moved to China and India [1]. Considering recent events regarding COVID-19, the disruption international supply lines could lead to widespread shortages of highly important antibiotics in Western countries [2]. Improving the economic viability and environmental sustainability of beta-lactam antibiotic production is paramount to globalizing the production of these important drugs to decrease the chance of shortages.

Recently, we have been working to develop a continuous enzymatic beta-lactam antibiotic production process to address many of the issues of the current batchwise chemical synthesis processes to produce cephalexin and amoxicillin. These drugs were selected because they are the two highest volume manufactured and prescribed antibiotics worldwide and have been included in the world health organization (WHO) list of essential medicines as key access antibiotics [3]. Next, the move from chemical synthesis to enzymatic synthesis decreases waste production and increases yield due to high enantioselectivity of enzymes and the ability to operate in a mild temperature aqueous environment. We have expressed, purified, and engineered penicillin G acylase from *E. coli* (EcPGA), an industrially relevant enzyme used to produce precursors for semi-synthetic beta-lactam antibiotics, to synthesize cephalexin and amoxicillin. Additionally, we have immobilized EcPGA on various solid supports via covalent attachment, allowing for the recycle of the biocatalyst and decreasing the overall cost of the catalyst for production. Next, the induction of crystallization of the API within the reaction vessel protects the desired product from hydrolysis by EcPGA into a minimally soluble byproduct. The *in-situ* crystallization of the API enables the downstream separation of the solid crystals from the mother liquor into a filtration, washing, and drying protocol.

By combining these elements, we have conducted several successful seeded batch reactive crystallization experiments to produce solid cephalexin crystals. We found that after three crystallization experiments of three hours each, the catalyst suffered a 20% decrease in specific activity. This decrease may be due to crystallization within the pores of the solid catalyst, as the catalyst experienced no measurable deactivation under reaction conditions without crystallization. We have also studied the effect of different sized support particles as well as different support materials (e.g. agarose, methacrylate, polypropylene) on the diffusive properties of cephalexin and its precursors as well as on-line stability of the immobilized biocatalyst. Additionally, during reactive crystallization, after approximately four hours, the scarcely soluble byproduct, phenylglycine, generated enough supersaturation to precipitate and contaminate the solid phase. The progress of cephalexin crystallization and growth as well as phenylglycine precipitation were detected via *in-situ* microscopy and focused beam reflectance measurement (FBRM). We are currently working to implement these same elements and use these experiments to inform the design and operation of continuous MSMPR. The shift from batch to continuous production will improve the capacity and decrease the overall footprint of the production process as well as decrease the amount of off-specification product due to more robust control during steady state operation. The implementation of process analytical technology (PAT) such as Fourier transform infrared spectroscopy (FTIR) will allow for the on-line monitoring of the liquid phase components while FBRM and *in-situ* microscopy will allow for the on-line monitoring of the solid phase components. Additionally, reverse phase HPLC will be used to confirm final product purity. With the information of the components' concentrations in both phases, we will study and develop a control strategy to maintain a "state-of-control" operation of the reactive crystallization process for the production of cephalexin and amoxicillin.

References:

[1] Elander, R. P. (2003). Industrial production of β-lactam antibiotics. *Applied Microbiology and Biotechnology*, *61*(5-6), 385–392. doi: 10.1007/s00253-003-1274-y

[2] Oehler, R. L., & Gompf, S. G. (2020). Shortcomings in the US Pharmaceutical Supply Chain. *Jama*. doi: 10.1001/jama.2020.1634

[3] WHO Model Lists of Essential Medicines. (n.d.). *The SAGE Encyclopedia of Pharmacology and Society*. doi: 10.4135/9781483349985.n433

607528 Recrystallization Development for Small Molecule API (Active Pharmaceutical Ingredient) Islatravir

Abstract:

Islatravir is a nucleoside reverse transcriptase inhibitor small molecule API (active pharmaceutical ingredient) targeted for HIV PrEP (pre-exposure prophylaxis). Islatravir has multiple crystal forms and a complex phase diagram. This talk will focus on the conversion from the monohydrate to an anhydrous form. To produce anhydrous API, we have designed a slurry conversion process, during which the less stable monohydrate form dissolves and a more stable anhydrous form crystallizes simultaneously. The solution composition also comes into play, as the process requires operating within a specific window of water-activity for effective form conversion. The slurry conversion process is also designed to proceed in a controlled manner to reduce crystallization of another kinetically favored phase. In addition, Merck has recently published a novel biocatalytic synthetic scheme for the synthesis of Islatravir[*]. This presentation will discuss differences in anhydrous crystallization performance (morphology, particle size, and milling performance) when starting with API from a traditional chemical synthetic route vs. the novel biocatalytic route. In this talk, we discuss the factors we can manipulate to design the most robust process, ultimately producing anhydrous API with consistent properties across various batches.

[*]Huffman et al. "Design of an in vitro biocatalytic cascade for the manufacture of islatravir." Science, 2019, 366 (6470), 1255-1259.

609555 Seeding Strategies for the Selective Crystallisation of Model Proteins

Abstract:

The recognition of the selective nature of biopharmaceutical drugs and the advances in biotechnology have led to over 246 approved products with cumulative revenues reaching $140 billion [1]. Upstream biomanufacture of proteins has advanced, such that the time and expense bottleneck in biopharmaceutics lies in the chromatographic purification steps [3-5]. Crystallisation is a more economic and energy efficient alternative purification step, producing crystalline proteins with higher purity and stability [3, 6]. However, this technology is yet to mature to manufacturing scale and robustness.

Here, we report our optimised crystallisation, additive and seeding strategies to crystallise model proteins and demonstrate selective crystallisation of target proteins from a binary mixture.

Mesoporous silica nanoparticles (NPs) are shown to be effective heterogenous nucleants for the selective crystallisation of lysozyme from a binary mixture of lysozyme and thaumatin. From hanging drop solution conditions which can crystallise both proteins, the 'hard template' SBA-15 NPs are found to not only promote the nucleation of lysozyme, but also suppress the crystallisation of thaumatin. These conditions are also effectively scaled up to the ml level. Further to this, the effectiveness of graphene oxides and DNA polymeric additives are also examined as seeds for protein crystallisation promotion.

Here, we report on the use of several templating strategies for the crystallisation of proteins demonstrating the feasibility for selective crystallisation of proteins for downstream separations.

ACKNOWLEDGEMENT

This study acknowledges by the SCoBiC project funded by the UK's EPSRC (EP/N015916/1).

References

1. Demain, A.L., REVIEWS: The business of biotechnology. Industrial Biotechnology, 2007. 3(3): p. 269-283.
2. Yang, H. and J.H. ter Horst, Crystal nucleation of small organic molecules, in New Perspectives on Mineral Nucleation and Growth. 2017, Springer. p. 317-337.
3. Basu, S.K., et al., Protein crystals for the delivery of biopharmaceuticals. Expert opinion on biological therapy, 2004. 4(3): p. 301-317.
4. Parambil, J.V., et al., Effects of Oscillatory Flow on the Nucleation and Crystallization of Insulin. Crystal Growth & Design, 2011. 11(10): p. 4353-4359.
5. Delmas, T., M.M. Roberts, and J.Y. Heng, Nucleation and crystallization of lysozyme: Role of substrate surface chemistry and topography. Journal of Adhesion Science and Technology, 2011. 25(4-5): p. 357-366.
6. Harrison, R.G., et al., Bioseparations science and engineering. 2015: Topics in Chemical Engineering.
7. Shah, U.V., et al., Heterogeneous nucleants for crystallogenesis and bioseparation. Current Opinion in Chemical Engineering, 2015. 8: p. 69-75.
8. Shah, U.V., D.R. Williams, and J.Y.Y. Heng, Selective Crystallization of Proteins Using Engineered Nanonucleants. Crystal Growth & Design, 2012. 12(3): p. 1362-1369.

609008 Developing a Scalable Templating-Platform for Enhancing Crystallisation of Bio-Active Proteins

Abstract:

Over the last decade, bio-active peptides have emerged as the next generation of bio therapeutics with a rapidly growing demand [1]. A lack of cost-effective and reliable processes for downstream purification is a main industrial challenge for meeting the growing demand of bio-active peptides [2]. Crystallisation of peptides is a novel strategy to debottleneck downstream purification, offering a more economical and efficient technique than expensive traditional techniques [3].

Due to their low solubility and large molecular size, one of the main challenges in peptide crystallisation is the lack of controllability of the nucleation. However, very little has been done in using templates for controlling nucleation to achieve a consistent and reproducible crystallisation process. Chen et al. recently reported that highly mesoporous inorganic surfaces, referred to as "hard templates", can remarkably induce significantly faster and more controlled nucleation, even at low supersaturation [4]. In our recent work, we have successfully demonstrated a "soft templating" strategy by using soft, organic molecules, such as amino acids, to enhance nucleation. In buffer solution, amino acids form various interactions with peptides, whereby electrostatic interactions have been identified as most important. Therefore, amino acids that exhibit strong electrically charged residual groups for the relevant buffer conditions were investigated. In this presentation, the impact of types, amount and surface charge of the amino acids on the crystallisation performance of insulin at various supersaturation will be reported. With the aim of providing a strategy suitable for high-throughput peptide crystallisation, the possibility of scaling this templating strategy up, from the microlitres to the tens of millilitres, will be evaluated.

Insulin was used as model peptide and comprises of two chains, A and B, with a combined 51 amino acids and an overall molecular weight of 5.8 kDa. The impact of amino acids on crystallisation is evaluated by means of measuring induction time, crystal yield, and size distribution as a function of time. Citric buffer with a pH between 5 and 7 was used with zinc sulphate as the precipitant. The impact of amino acids on crystallisation is evaluated by monitoring the peptide concentration with UV-vis spectroscopy every 20-30 minutes. Using this, the induction time and crystal yield are obtained. Furthermore, the pH is measured at the start and the end of the crystallisation. Images obtained with a microscope are analysed to evaluate the crystal sizes and shapes. The experiments were carried out in batch crystallisers with volumes up to 15 millilitres to demonstrate the scalability.

Our recent results show that amino acids reduce the induction time of insulin crystallisation by up to 40% when compared to the induction time with no amino acids. The faster decrease in insulin concentration when amino acids are added also implies an increase in the crystal growth rate. As amino acids accelerate the crystallisation process, a yield of 90% can be achieved in a shorter period of time in crystallisers with volumes of up to 15 millilitres. Preliminary investigations show that the type and amount of amino acid added have only a minor impact on solubility, leading to a slight increase. Instead of impacting the solubility significantly, the charged residual-groups of the added amino acids increase the pH significantly. This increase in pH leads to an increase in stability of the tertiary structure of insulin by altering its physio-chemical surface properties which result in a faster nucleation. As a result of the more controlled crystallisation with amino acids, a narrower crystal size distribution can be obtained while achieving the same mean crystal size (~20 μm) when compared to the mean crystal size obtained with no amino acids. The insulin crystals obtained with amino acids show a more uniform crystal size distribution with a smaller span for crystalliser volumes reaching from the scale of microliters to the tens of millilitres. The soft templating strategy could be successfully demonstrated for crystalliser volumes of up to 15 millilitres while identifying mixing as one of the main limiting factors of scaling-up.

It has been demonstrated that amino acids as organic templates have the potential of enhancing the controllability of insulin crystallisation even at low supersaturations and to the scales of tens of millilitres. These includes a significant reduction of residence time needed in the crystalliser to obtain a desired crystal yield by maintaining a high level of consistency. Although different polar amino acids were investigated, the magnitude of advancing crystallisation differs dramatically between them as well as the ratio of amino acids concentration to insulin supersaturation.

Bibliography

1. Lau, J.L. and M.K. Dunn, *Therapeutic peptides: Historical perspectives, current development trends, and future directions.* Bioorg Med Chem, 2018. 26(10): p. 2700-2707.
2. Yang, H., et al., *Development and Workflow of a Continuous Protein Crystallization Process: A Case of Lysozyme.* Crystal Growth & Design, 2019. 19(2): p. 983-991.
3. Roque, A.C.A., et al., *Anything but Conventional Chromatography Approaches in Bioseparation.* Biotechnol J, 2020.
4. Chen, W., et al., *High Protein-Loading Silica Template for Heterogeneous Protein Crystallization.* Crystal Growth & Design, 2019. 20(2): p. 866-873.

609313 Sensor Integrated Continuous Microfluidic Mixer for Protein Screening at Constant Supersaturation

Abstract:

Sensors are great assets to quick and sensitive detection inside miniaturized devices were the amount of sample is small. Connecting an integrated microfluidic system to a control unit increases the accuracy of the experiments due to the automated adjustments of parameters. One of the examples for the potential application of a sensor integrated microfluidic system is screening crystalline materials where there is a high demand for accurate measurement of the concentration within the crystallization zone. In our previous studies, we developed a continuous microfluidic mixer for screening crystalline materials at the constant supersaturation, where the homogeneous concentration was calculated using COMSOL Multiphysics. However, it was not possible to perform an in-situ measurement for the actual concentration to experimentally verify that. Here we have coupled the microfluidic mixer design with an electrochemical sensor to enable in-situ measurement of the actual concentration inside the mixing zone. Moreover, integration of a sensor within the continuous microfluidic platform we can control the imposed supersaturation by manipulation of the flowrates. To validate the sensor integrated platform experimentally, we have done the protein crystallization of lysozyme and NaCl. Products of this crystallization process can exhibit different morphologies according to the ratio of the starting solutions. Several different experimental conditions are explored to study the morphologies obtained at assigned initial concentration as well as the growth kinetics.

606008 Particle Size Control of a Drug Substance through Process Crystallization

Abstract:

The solid state characteristics of an Active Pharmaceutical Ingredient (API) have great impact on the Drug Product (DP) performances and its formulation development approach.

One Critical Quality Attribute (CQA) that has to be controlled in order to ensure the success of a given formulation strategy in terms of manufacturability is the API Particle Size Distribution (PSD). In this case study, the API (PSD) had to be controlled to produce a blend which was homogeneous and flowable once combined with the chosen excipients and, in parallel, to reduce the risk of segregation during the DP manufacturing process.

The initial drug substance batch available for the formulation prototype production and selection showed a D(v,0.9) around 350mm. A mechanical PSD reduction via FitzMill was required in order to tune the drug substance PSD to match the PSD of the chosen excipients to ensure proper blend flowability and homogeneity attributes as well as to mitigate the risk of segregation during the encapsulation. As the alternative, the control of the PSD via process crystallization was investigated. Ultimately an anti-solvent mediated seeded crystallization process was developed. This process was able to deliver drug substance in the desired PSD range and avoided the need for post isolation milling.

608673 Reaction Kinetic Model Application to Speed up Development

Abstract:

The present work describes the journey of a reaction co-development between a pharma company and a CDMO since the transfer from UCB to large scale production at Hovione. It comprises the transfer of a kinetic model developed by UCB to Hovione, the experimental confirmation of the model by Hovione, a screening of conditions for a better understanding of the design space and, finally, the optimization and scale-up of the reaction to a >100 kg scale.

To initially confirm the kinetic model provided by UCB (both the UCB and Hovione used Dynochem for modelling) it was necessary to execute experimental work that can provide enough variability regarding the reaction profiles. At Hovione, we have compiled a workflow/set of experimental conditions that is suitable to a wide range of reactions types and can capture most critical physical and chemical transformation rates during a reaction (e.g. product converted into impurities, reagents reacting with other species (aside from starting raw material), mass transfer limitations). This set of experiments was used to confirm the kinetic model and afterwards we were able to explore the design space using a design of experiments approach.

The DoE was a Plackett-Burman design with 4 targeted parameters namely reagent equivalents (upper and lower boundaries set by economic and quality targets), solvent volumes, reaction temperature and acid equivalents. In order to minimize for scale-up issues, the DoE used heating rates similar to the ones used at production scale. Additionally, a maximum reaction time of 33h was used, which enabled to reach a 99% yield for all experimental conditions of the DoE. The kinetic model was able to accurately describe all the experiments carried out during the DoE and was used to define the critical variables proven acceptable ranges (PAR) and normal operating ranges (PAR). The isolated product showed a 99.0 % w/w assay as demonstration trial in the lab, while at a >100 kg scale it was possible to achieve a 99.7 % w/w assay.

It was proven that by resorting to a mechanistic modelling approach and a reduced number of experiments we were able to save resources (time and materials) and generate more knowledge than the traditional approach and have a more straightforward way to reach production scale with Right-First-Time batches.

608750 Utilizing Mechanistic Modelling to Support Scale-up of Fed-Batch Reaction Process

Abstract:

Digital design of manufacturing processes using mechanistic models is fast becoming an essential tool during Active Pharmaceutical Ingredient (API) process development. It enables rapid and effective exploration of the decision space for Critical Process Parameters (CPP), helping to reduce risk and product time-to-market, thus aiding in the effective and safe production of high-quality pharmaceutical products.

This work involves the application of mechanistic modelling of a fed-batch reaction process to help support process scale-up. The reaction kinetic modelling was combined with mixing and heat transfer effects in order to describe the reaction process at laboratory scale using a RC1 reactor. Scale-up calculations were performed, which included optimization of process parameters to ensure that the exothermic reaction was performed within the temperature limits recommended for the reactor. The main of objective of this activity was to assess the process of developing and scaling up a reaction process in a CSTR from reaction kinetics regression to inclusion of mixing and heat transfer effects as a function of scale. Additionally, this work intended to develop an industrial workflow to document how mechanistic reaction modelling could be utilised in this case. This workflow involved the following steps:

1. Regress the kinetic parameters of a CSTR that combines the reaction kinetic modelling with mixing and heat transfer
2. Assess the scaling up capability of the model
3. Optimize the process parameters to obtain the desired qualities

Upon completion of step 2, the validated mechanistic reactor model was utilised to assess the impact of dosing flowrate on the maximum temperature observed during the reaction process with increasing scale. In this process, the feed rate is used to control the temperature in the reactor. To study this impact, a number of uncertainty analyses were performed to evaluate the reactor temperature with varying feed flowrate. The dynamics of the reactor temperature can be seen in Figure 1 below across three productions scales, namely 80 kg, 400 kg and 800 kg scales. These predictions can be used to define the minimum and maximum reactant feed rates for each scale to ensure operation of the reaction within the desired range of temperatures, shown by the highlighted region in the plots below.

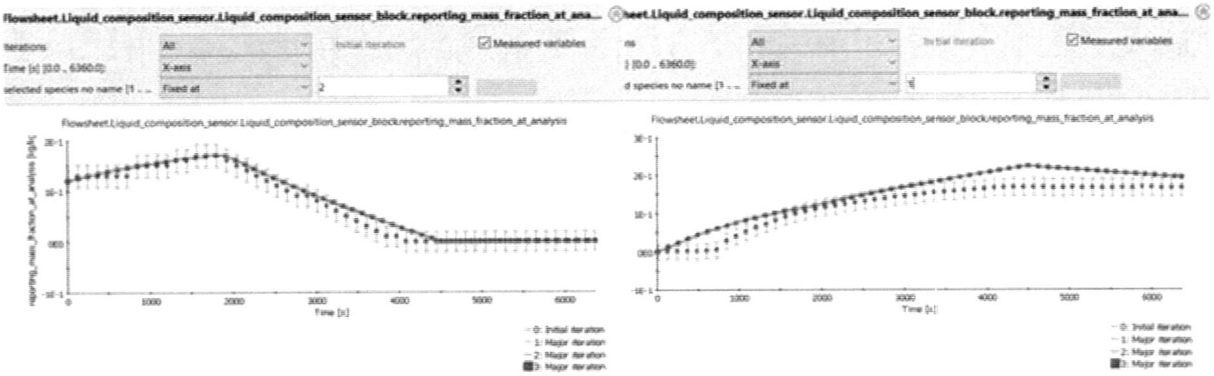

Figure 1: Plots representing the overlays between the measured data and the model prediction. Plot in the left shows the change in concentration of the limiting reactant. Similarly, plot in the right shows the increasing concentration of the product in the reactor.

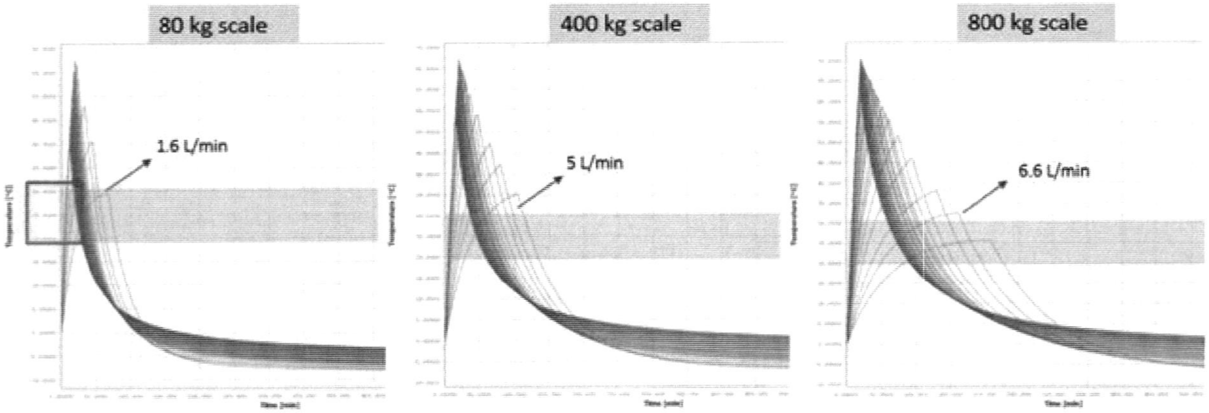

Figure 2: Scenario analysis of the impact of reactant addition rate on the temperature profile observed during a fed-batch reaction across three scales. The desired region of operation for process temperature is highlighted in the above figures.

601871 Scaling Down a Reactive Crystallization Using Various Mixing Models

Abstract:

Multiple commercially available mixing modeling tools were utilized to scale-down a reactive crystallization from the planned process performance qualification (PPQ) scale to the laboratory scale. Differences in the historical crystallization batch data with results in standard laboratory crystallization set-up drove the need to fully characterize the lab scale setup to ensure it was representative of the PPQ equipment train.

The mixing modeling tools were used to model mixing at both the PPQ and laboratory scales system and compared with experimental and batch data for physical properties (particle size and surface area) of the product. These tools encompassed both correlation-based and computational fluid dynamics based modeling (DynoChem, M-Star, and MixIT). In this presentation, the modeling results of numerous laboratory scale conditions and setups as well as large scale batch data were used to understand micro-mixing, meso-mixing, energy dissipation rate and strain rates. The modeling results were compare with experimental physical property data to give insight into the strengths and weaknesses of using these tools in identifying laboratory conditions and setup of the reactive crystallization.

604831 Centrifugation at Scale: Scaling up Based on Bench Top, Pilot Plant and Modeling Data

Abstract:

Selecting the recommended centrifugation parameters for full scale manufacturing before the first site run is key to a successful technical transfer. Centrifugation scale-up poses unique challenges due to the differences in centrifuge geometry, rotational speed, and automation levels at different manufacturing sites and lab set-ups. Using a combined approach modeling, lab experiments, and pilot plant data, starting parameters for manufacturing could be selected for loading rate, loading spin speed, and final speed. The recommended starting values were then used to shape the optimization at full scale.

604547 Experimental Confirmation of Scale-Down Mixing Models from Commercial to Lab Scale with CFD for Sterile Drug Products

Abstract:

The ability to model a large, commercial-scale process at a smaller scale in the laboratory is a valuable tool to increase the efficiency of process development. Characterizing the mixing process at commercial scale can be particularly challenging with sterile products where process monitoring tools and sampling can be a risk to the integrity of the sterile boundary. Having a lab-scale mixing model predictive of the large scale operation will enable more effective small scale batches to decrease the number of commercial scale batches needed during process development, consequently saving time, materials, and other costs. Additionally, the use of computational fluid dynamics (CFD) to create a virtual model can further reduce the number of small scale batches needed as well.

A scale-down model is being developed using CFD between a commercial scale 700L tank and a development scale 10L tank. These tanks are of stainless steel construction with bottom-mounted agitators used for mixing and formulation of sterile drug products. The primary responses investigated in this liquid-liquid mixing study was time to homogeneity as impacted by varying viscosities and densities of two equal-volume constituent components, mixing speed, and tank fill level. Raman and conductivity probes were utilized to collect real-time data throughout the mixing process, providing value in characterizing liquids that mix too quickly to be conducive to offline sampling. Experiments were designed considering statistical principles in conjunction with ranges of Richardson and Reynolds numbers to interrogate the boundaries between safe operating spaces and regions of failure. These experiments are executed physically in the tank and compared with corresponding CFD simulations to validate the model.

606470 Computational Approaches to Scale-up and Tech-Transfer in Agitated Filter Bed Drying

Abstract:

The agitated filter dryer (AFD) is a critical technology in the pharmaceutical industry for isolating potent active pharmaceutical ingredients (API) post crystallization. There the filtration and drying unit operations take place in a single contained vessel minimizing operator exposure. The central challenge in utilizing AFDs effectively is designing protocols to produce API powders with residual solvent content uniformity and targeted particle size distributions that affect the performance of drug product processes through flowability, blendability, compatibility, etc..

Scale-up/tech-transfer of AFDs is made difficult by the variety of different geometries available at the lab, pilot and plant scales. The differences in geometry and scale affect both the heat transfer and stresses experienced by API, and which in turn affects the physical properties of the resultant material, namely through agglomeration and attrition. Moreover, the isolation inherent to AFDs makes direct probing of plant and pilot scale vessels challenging.

In this work, we present several computational approaches to investigate the physics present in the spatially varying local environments within AFDs, such as the heat transfer, gas-liquid phase change, mixing, agglomeration and attrition. These investigations utilize a variety of computational tools such as discrete element method, computational fluid dynamics, and machine learning. We will present heuristics and metrics for designing and optimizing AFD protocol at a given scale and considerations for scale-up and tech-transfer.

Disclosures:

EM, KS and NN are AbbVie employees and may own AbbVie stock. MM is an employee of Tridiagonal Solutions. RB is an employee of Engineering Solutions and Scientific Software. SS is a Professor and VB is a PhD student at Iowa State University and have no additional conflicts to disclose. AbbVie has provided research funding to Tridiagonal Solutions and Iowa State University. AbbVie, Tridiagonal Solutions and Iowa State University contributed to the design; participated in the collection, analysis, and interpretation of data, and in writing, reviewing, and approval of the final abstract.

613144 Industrial Crystallization – the Intersection of Solid State Chemistry and Process Engineering

Abstract:

Crystallization is a complicated molecular recognition and self-assembly process that is completely managed by manipulating macroscopic process variables. Oftentimes, the crystallographic state of the resulting product is critial to its performance. The efficacy of products as diverse as pharmaceuticals and pesticides is often very sensitive to polymorph, for example. In addition, the particle size distribution depends on multiple process parameters in complicated, often nonlinear ways. As many crystallized products have become more advanced, with increasing demands on product performance, industrial crystallization has had to adapt. Pratitioners tend to come to the field from either process engineering or solid state chemistry. While it is possible to stay in one's own "lane" and contribute to the field, most successful practioners today are at least conversant with both sides of the chemistry/engineering "divide". Cross-training and collaboration often contribute greatly to successful crystalline product development and scale-up.

I come to industrial crystallization from process engineering, but have come to appreciate the necessity of understanding solid-state chemistry to adequately develop processes and products. I'll illustrate the interplay between process engineering and solid-state chemistry with several examples from several industries, and review where I think continued collaboration can lead to crystalline product control and improvement.

602066 Crystallization Optimization By Enzymatic Control of a Reactive Crystallization

Abstract:

Tunable reactions can be used to control precisely the generation of supersaturation in reactive crystallization systems. Enzymes, with high reaction specificity and predictable rate enhancement, enable fine control over a number of reactions, and have been demonstrated in several reactive crystallization systems. In this study, penicillin G acylase (PGA) is used to synthesize and simultaneously crystallize cephalexin, starting from two cephalexin precursors, in a controlled manner. It was recently shown that cephalexin, which typically forms high-aspect-ratio needle-shaped crystals, can form lower-aspect-ratio crystals when supersaturation is generated in a slow, controlled manner (Li et al., 2019). These lower-aspect-ratio crystals have many desirable properties including faster filtration (by an order of magnitude) and improved powder flowability and handling.

This study aims to show that enzyme control of cephalexin reactive crystallization can produce crystals with a more desirable aspect ratio while also providing the benefits of combined synthesis and separation (greater productivity, conversion). The mechanism of PGA synthesis of beta-lactam antibiotics such as cephalexin is known (McDonald et al. 2017); here an immobilized form of PGA is used to enable recycling of the enzyme. Use of process analytical technology (PAT) combined with a reaction-diffusion-crystallization model (Salami et al. 2020, *in preparation*) lends insight into the phenomena critical to successful implementation of this novel means of supersaturation control. *In situ* microscopy and focused beam reflectance measurement (FBRM) show that the amount of crystal surface area is important to both aspect ratio and secondary nucleation. The model suggests that the amount of enzyme, and unexpectedly the immobilized bead size, is critical in dictating the reaction timescale. In general, this study demonstrates how catalysts can lend control over the crystallization process, giving process designers an additional knob to turn in the synthesis of new routes to important pharmaceutical products.

Li, M., et al. (2019). "Optimizing the Aspect Ratio of Cephalexin in Reactive Crystallization by Controlling Supersaturation and Seeding Policy." Transactions of Tianjin University **25**(4): 348-356.

McDonald, M. A., et al. (2017). "Enzymatic reactive crystallization for improving ampicillin synthesis." Chemical Engineering Science **165**: 81-88.

Salami, H., et al. (2020). “Model development for enzymatic reactive crystallization of β-lactam antibiotics, a reaction-diffusion-crystallization modeling approach.” Computers & Chemical Engineering **in preparation**

610666 Continuum of Solute Dimers Drives Distinct Growth Modes on Anisotropic Faces of Etioporphyrin

Abstract:

One of the basic tenets of the classical theories of crystal nucleation and growth is the Szilard postulate, which states that molecules from the supersaturated phase join a nucleus or a growing crystal individually. Over the years, numerous crystals have been found to nucleate by assembly of monomers and grow by sequential association of monomers to growth sites. Recent experiments in bio- and geomimetic environments have spotlighted a partial deviation from the Szilard rule, a bimodal distribution of growth species that arises from the assembly of ordered or amorphous structures in the solution, which then incorporate into the crystal as a whole. Here we report a complete violation of the Szilard rule. In octanol solutions of etioporphyrin, a planar molecule whose crystals have attractive electronic and optical properties, the solute populates a continuum of quasi-parallel dimers that differ from the conformation in the crystal by shift and orientation. Molecular absorption and luminescence spectroscopies and all-atom molecular dynamics (MD) simulations reveal that the stabilities of the dimers are within the thermal energy and enforce free conversion between the distinct structures. The monomer concentration in the solution in below 0.1 %. Despite the structural complexity of the solute, the velocity of the steps on the (010) faces of etioporphyrin crystals (measured using time resolved in situ AFM) increases linearly with the total solute concentration. The observed linearity indicates that the molecular reaction at the growth sites in monomolecular and the majority of solute dimers seamlessly adopt the crystallographic conformation. By contrast, step growth on the other major face, (001), scales with the square of the solute concentration. We eliminate three potential mechanisms that may manifest as super-linear growth: inaccurate solubility determination, escalating kink density at elevated supersaturation, and adsorption of step pinners on the crystal surface. The disqualification of the three scenarios enforces the conclusion that the quadratic kinetic law indicates a bimolecular process of incorporation. We demonstrate that in contrast to numerous other solution grown crystals, solute molecules directly incorporate in the kinks and do not reach the growth sites after first adsorbing on the terraces between steps. The paucity of solute monomers precludes a collision between two monomers to form a crystallographic dimer as the mechanism of the crystallization reaction. MD evaluation of solute interactions with kinks on the (001) face reveal a molecular trap selective for solute dimers located on the approach to all kinks, which necessitates a collision with a second dimer to ease the captured solute towards the kink. The results with etioporphyrin demonstrate the lack of correlation between the solute state and the incorporation kinetics driven by the solute. In a broader context, these observations illuminate the immense diversity of unexplored crystallization scenarios that pave the long road to crystallization control.

604208 Physicochemical Modeling of Dissolution and Stability of Poorly Soluble and Unstable Drug Loaded Multilayer Soluble Polymeric Films

Abstract:

Biopharmaceutical Classification System (BCS) Class-II and Class-IV drugs (active pharmaceutical ingredients, APIs) have low aqueous solubility and potentially low bioavailability. To overcome the limitations, soluble polymeric drug dispersion as films by either solvent casting or hot-melt extrusion is emerging as a feasible technique, where lab-scale tablet can be made by 3D printing. Film based solid dispersion of drug formulation can side-step and replace current legacy-based process design through iterative trial-and-error in pharmaceutical industries and facilitate a step-by-step highly scalable and rapidly implementable predictive physicochemical modeling based on drug release, storage stability and mechanical structure of the film. However, product design requirements need to be transformed to mathematical design instructions (dosage, release, stability, aesthetics, multiple APIs) for optimized tuning of the drug release rate, stabilization and dosage flexibility with continuous manufacturing to be predicted in advance. Predictive modeling is greatly simplified for a film-based approach that provide dimensional control by regulating transport and phase behavior to a 1D layer, thus streamlining manufacturing and scaling process. It also provides the opportunity for a multilayer film-based formulation for faster transition from bench to clinical trials to commercial production.

A mathematical modeling framework for the design and parameter optimization during the formulation of hydrophobic API-loaded soluble polymeric film formulations was implemented for the foundation for customer specific drug design in terms of dosage, release rate, storage stability (self-life) and mechanical properties (bending/cutting). Overall optimization initially involves two components: dissolution and stability models, which have to be solved simultaneously for targeted design requirements in terms of thickness, geometry, order, composition and number of layers. Unsteady-state Fickian diffusion describes both dissolution and stability modeling. During dissolution, the relative rate of water intrusion (swelling controlled) and polymer erosion from surface coupled with API diffusion primarily govern the observed release rates, correlated with expected different release mechanisms (immediate/slow). For stability against water/oxygen, barrier polymer layer is modeled by an unsteady-state diffusion of water/oxygen along with mass transfer (partition) in the surface and interface to achieve $< 1\%$ impurity formation within 12 months considering impurity formation kinetics. Polymer diffusivity, solubility, active layer and barrier layer thickness, area, mass transfer coefficients and relative rate constants can be estimated and optimized using the model. Physical properties are determined from thermodynamics. Prototype creation and characterization, experiments will provide physical parameters for release, uniformity, stability and dose scaling. Generic algorithm for optimization/parameter fitting is pursued, where inner loops consisting smaller modules (release, stability, structure) are encircled by outer overall optimization loop (onion model) until any of the constraints (total mass < 600 mg, total thickness < 17 mm, area <153 mm^2, and volume <1071 mm^3) are met. The model is adaptive to progressive complexity such as presence of plasticizer/binders, glass-rubbery transition, phase separation, polymer diffusion, aqueous boundary layer, barrier dissolution and multi-dimensional transport, which can be independently adjusted, adapted and improved. Crystal or particle dissolution kinetics can be incorporated in case of crystalline/particulate drugs. Finally, film flexibility testing using 3-points flexural bending to express required mechanical properties (flexural stress, strain and modulus) in terms of force, deflection and dimension can also be incorporated in the overall optimization scheme.

598363 A Semi-Theoretical Model for Simulating the Temporal Evolution of Moisture-Temperature during Industrial Fluidized Bed Granulation

Abstract:

Moisture plays a major role in determining the attributes of granules prepared by fluidized bed granulation (FBG). In this talk, we are presenting a semi-theoretical droplet-based evaporation rate model that was developed and incorporated into moisture mass-enthalpy balances to simulate the temporal evolution of bed moisture-temperature. Experimental data from a GPCG30 unit were used to fit the model parameters. With only two fitting parameters, the model demonstrated very good capability to describe the moisture-temperature evolution for a wide range of operating conditions. Then, in a global process model (GPM) approach, the evaporation parameters were fitted to multi-linear functions of inlet air temperature, binder concentration, and spray rate. The GPM was validated successfully by simulating a different data set which was not used in its calibration. As the GPM demonstrated a good predictive capability, it was further used to investigate the impacts of process parameters. Numerical simulations suggest that the proposed GPM predicts the experimentally well-established trends of moisture-temperature profiles in previously published data, proving the applicability of the GPM approach. This study has demonstrated the capabilities of simple process models as a practical approach to predict time-wise evolution of bed moisture-temperature profiles in industrial FBG modeling, while also pointing out their limitations.

601130 Modelling Amorphous Solid Dispersions: The Role of Supersaturation, Nucleation and Crystal Growth

Abstract:

Amorphous solid dispersion (ASDs) are a rising strategy to overcome the poor solubility of drugs in water, since they can maintain supersaturation concentrations of the drug and thus increase its oral bioavailability. Many approaches are reported to predict the oral absorption of drugs, namely the combined use of biorelevant dissolution testing with *in silico* modeling.[1] There are not many publications regarding the prediction of *in vivo* ASDs behavior[2] and the current mathematical models are simplistic on describing supersaturation as the ratio between the bulk concentration and the concentration at saturation of the drug. This approach seems inadequate to describe supersaturation and can be a roadblock to accurate physiologically based pharmacokinetic (PBPK) modelling of ASDs.

This work aims at describing the behavior of ASDs when dissolving into a biorelevant medium taking into consideration a combination of four events: dissolution, supersaturation, nucleation and crystal growth. These events were described by differential equations and constants with physical meaning.

ASDs were produced by spray-drying following a full-factorial design of experiments, varying the drug load, the ratio of atomization and the outlet temperature of spray-drying. The behavior of these ASDs was evaluated by dissolution in biorelevant medium including a pH shift from FaSSGF medium (pH 1.2) to FaSSIF medium (pH 6.5) to mimic the conditions in the gastrointestinal tract.

The first attempt combined models describing dissolution and precipitation by using differential equations but describing supersaturation as a simple ratio between the concentration of the drug and its concentration at saturation. However, this attempt failed to be a good descriptor of supersaturation, as the model did not reflect the experimental data from ASDs presenting significant burst release.

In order to have the drug's concentration at saturation described by differential equations, rather than by the ratio between concentrations, the Noyes-Whitney equation was modified taking into consideration the maximum concentration for each ASD, rather than being restricted by the concentration at saturation of the drug. Then, the behavior of ASDs was modeled by combining four differential equations: the Noyes-Whitney equation for dissolution, the modified Noyes-Whitney equation for supersaturation, and the nucleation and crystal growth equations for precipitation. This approach was able to describe ASDs with different behaviors (error between experimental and simulated data < 2%). This outcome was possible because the dynamic process of dissolution was based on dissolution, supersaturation and precipitation equations of the drug.

Figure 1 shows the experimental and simulated data for ASDs produced by spray-drying with different behaviors: spring and parachute, absence of precipitation, absence of supersaturation, and delayed and incomplete drug release (produced by hot melt extrusion, HME).

This work contributes for a better understanding of an ASD behavior describing the supersaturation mechanism, which directly impacts on the drug's bioavailability and plasmatic concentration.

References:

1. Kaur, N., Narang, A. & Bansal, A. K. Use of biorelevant dissolution and PBPK modeling to predict oral drug absorption. *Eur. J. Pharm. Biopharm.* **129**, 222–246 (2018).

2. Mitra, A., Zhu, W. & Kesisoglou, F. Physiologically based absorption modeling for amorphous solid dispersion formulations. *Mol. Pharm.* **13**, 3206–3215 (2016).

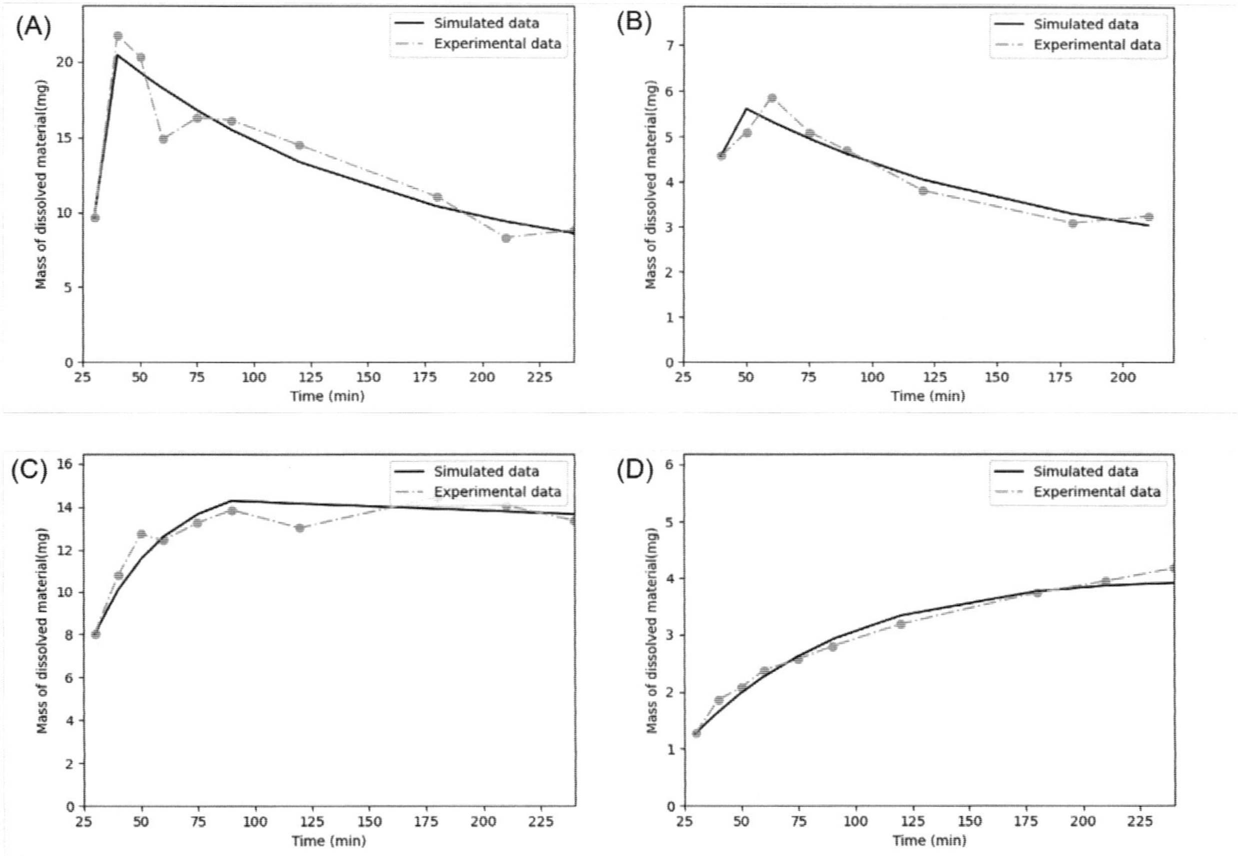

Figure 1: Experimental and simulated data for ASDs produced by spray-drying with different behaviors: (A) spring and parachute effect, (B) absence of precipitation, (C) absence of supersaturation, and (D) delayed and incomplete drug released from HME-ASDs.

600914 Determination of Drying Stresses and Fracture Energy in Tablet Coating Films

Abstract:

Thin coatings of polymer films find numerous technological applications in various industrial sectors related to protective and functional coatings. Tablets in the pharmaceutical industry may be coated for branding, to improve appearance or taste, increase stability or to achieve sustained or delayed release of the drug. In this study, these coatings are initially cast as thin films of polymer solutions, which dry to form a solid film. The solidification process due to the removal of the solvent phase is often accompanied by shrinkage of the film. If the polymer adheres to the substrate, the film can only shrink in the film-normal direction since the substrate prevents contraction in the transverse plane. In such cases, drying is accompanied by the development of transverse tensile stress in the film. If the tensile stress exceeds a critical value, the polymer film may develop cracks, thereby compromising the integrity of the film. Alternatively, if the adhesion between the film and the substrate is weak, the film may debond instead of cracking. It is essential to understand these mechanisms to obtain defect-free films.

In the present work, transverse stress was measured during the drying of the film using glass as a cantilever substrate for solutions of 10% and 20% total polymer solids (Figure 1, left panel). Two polymer systems used in this study were CA (cellulose acetate): PEG (polyethylene glycol) and CA (cellulose acetate): HPC (hydroxypropyl cellulose). It was found that final drying stress remains the same for both the polymer systems irrespective of concentration and coating layers except for 10% CA: HPC films. Transverse stresses in the latter films were found to be high and suggest a higher propensity for cracking (Figure 1, right panel). Cross-sectional SEM images were taken to relate the microstructure to the stress. Ultimate tensile strength of polymeric films was also determined using a tensile testing instrument. The tensile strength decreases with flaw size, in accordance with the Griffith's criterion. A simple mathematical model was developed to predict the critical thickness and critical stress beyond which the film would crack.

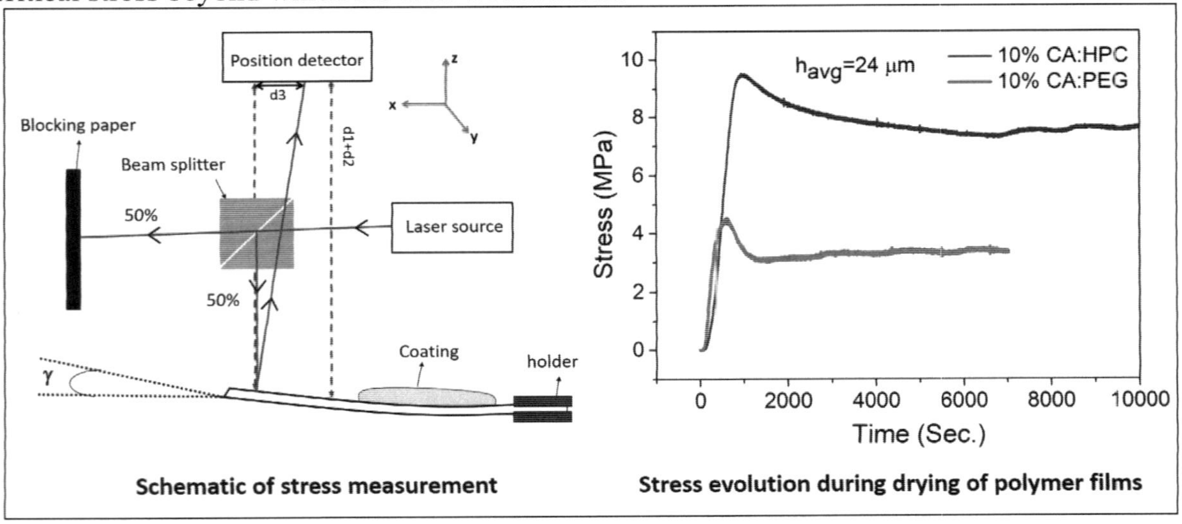

Figure 1: The left panel shows a schematic of the setup used to measure tensile stress in drying polymer films. The panel on the right shows the stress evolution for CA:HPC and CA:PEG films. In both cases, the average dry film thickness was 24 μm. Clearly, CA:HPC film shows higher drying stress indicating a higher propensity for cracking.

599153 Heat and Mass Transfer in a Pharmaceutical Coating Process

Abstract:

The Wurster process is commonly used to modify the release rate or the taste of granulated beads in the pharmaceutical industry. Hereby, beads are coated with one or multiple functional films that control the release of drug substances or mask a specific taste. The coating process takes place in the Wurster tube, in which the particles are transported centrally upwards followed by a subsequent downwards movement outside the tube. Thus, the particle flow resembles a torus-shaped recirculation pattern.

From former studies, it is known that small droplets, a diluted coating solution and low spray rate improve the coating layer homogeneity[1]. On the other hand, such a process is energy and time consuming. Low spray rates and small droplets promote spray drying (premature droplet drying, inhibits spray deposition on particles), reducing the process efficiency. Additionally, the coating quality increases with higher spray rates[1,2]. From product quality point of view, the spray rate is capped by the amount of spray drying on the lower boundary and by overwetting, which would lead to agglomerations, on the upper boundary. Therefore, investigating the heat and mass transfer inside the coating device can help to find optimal process parameters.

In our work, a fully integrated CFD-DEM environment is used to model the transport phenomena in a lab-scale Wurster coater. The commercial DEM code XPS is coupled to AVL-FIRE for the CFD simulation. State-of-the-art models for momentum, heat and mass transfer between continuous and discrete phase are integrated in a fully coupled manner[3]. Temperatures, vapor mass fractions and evaporation rates of a coating process with a multicomponent spray solution are calculated. The amount of spray drying is quantified and related to the input parameters of the process.

1. van Kampen A, Kohlus R. Statistical modelling of coating layer thickness distributions: Influence of overspray on coating quality. *Powder Technol.* 2018;325:557-567. doi:10.1016/j.powtec.2017.11.031
2. van Kampen A, Kohlus R. Systematic process optimisation of fluid bed coating. *Powder Technol.* 2017;305:426-432. doi:10.1016/j.powtec.2016.10.007
3. Forgber T, Toson P, Madlmeir S, Kureck H, Khinast JG, Jajcevic D. Extended validation and verification of XPS/AVL-Fire^{TM}, a computational CFD-DEM software platform. *Powder Technol.* 2020;361:880-893. doi:10.1016/J.POWTEC.2019.11.008

597652 (Invited Talk) Exploiting Barriers of the Body for Targeted Nanomedicine

Abstract:

Peptide-based, amphiphilic micelles are small, organic nanoparticles that can provide a tool for the detection and targeted delivery of therapeutics to diseases including cardiovascular and chronic kidney disease. Through rational design and molecular engineering of their physicochemical and biological properties, these nanoparticles can uniquely harness physiological and cellular barriers for diagnostic and therapeutic strategies. Herein, we discuss the development of multimodal, peptide micelles that can exploit clearance mechanisms and bind to specific markers for enhanced drug delivery.

For chronic kidney diseases such as autosomal dominant polycystic kidney disease (ADPKD), while small molecule drugs have been proposed as a therapy to manage disease progression, high dosages are often required to achieve therapeutic efficacy, generating off-target side effects, some of which are lethal. To address these limitations, our lab has designed kidney-targeting peptide micelles (KM) within a range of surface properties and sizes, and found KMs can cross the glomerular filtration barrier and bind to megalin, a multiligand cell surface receptor present on renal tubule cells. When incubated with human kidney proximal tubule cells, KMs were found to be biocompatible in vitro and showed higher accumulation in kidneys compared to nontargeted controls in vivo. Moreover, upon incorporation of a library of ADPKD drug candidates, KMs efficiency decreased cyst burden compared to free drugs using several routes of administration favorable for chronic diseases.

In addition, due to the modularity of micelles and their potential for theranostic applications, we will also discuss the development of nanoparticles delivering novel microRNAs as well as clinically relevant imaging agents to inhibit calcification and several pathological cell types in atherosclerosis. Throughout this presentation, we will discuss the potential of nanoparticles to harness barriers found in vivo to generate nanomedicine strategies with capabilities to bind to specific disease markers of interest, deliver a therapeutic, and monitor the progression and regression of the disease in real-time.

597031 Surfactant Effects on Nanotherapeutic Fate within the Brain

Abstract:

Background: Drug delivery to the brain is limited by poor transport of pharmaceutical agents across the blood-brain barrier (BBB), extracellular space (ECS), and into specific cells of interest. Nanoparticle drug carriers, due to their small size and highly-tailorable surfaces, have emerged as one strategy to overcome these barriers. In particular, polymeric nanoparticles made of poly(lactic-*co*-glycolic) acid (PLGA) can cross the BBB and, with a dense surface layer of poly(ethylene glycol) (PEG), can diffuse through the brain parenchyma. However, the formulation of PLGA-PEG nanoparticles often incorporates additional stabilizing molecules, called surfactants, whose effects have been understudied. Previous studies of surfactant influence on nanoparticle behavior have been limited to a single biological barrier. For example, cellular uptake in some immune and tumor cells is promoted by incorporation of the surfactant Pluronic F127 (F127) compared to poly(vinyl alcohol) (PVA). Polysorbate 80 (P80), another surfactant, can induce transcytosis across the BBB. Since the *in vivo* brain microenvironment exhibits many barriers in combination, the goal of this study was to determine the effect of surfactant choice on nanoparticle transport at multiple length and time scales.

Methods: AlexaFluor 647-labeled PLGA-PEG nanoparticles were formulated with the surfactants F127 (PLGA-PEG/F127), PVA (PLGA-PEG/PVA), or P80 (PLGA-PEG/P80). Control nanoparticles were formulated without surfactant (PLGA-PEG) and without PEG (PLGA). Nanoparticles were characterized by dynamic light scattering for size, surface charge, and polydispersity. To evaluate *in vivo* nanoparticle transport, we administered nanoparticles intravenously via tail vein to postnatal (P) day 9 female Sprague-Dawley (SD) rats and extracted all major organs four hours later. Uptake within neurons and microglia was observed in fixed, sectioned brains using immunohistochemistry and confocal microscopy. For all other organs, nanoparticle accumulation was measured by fluorescence spectrometry after tissue homogenization. Accumulation in the brain was also measured by fluorescence, but in order to more precisely capture nanoparticle extravasation across the BBB, we first used a capillary depletion technique to separate brain microvessels from the brain parenchyma. Finally, nanoparticle diffusion through the ECS was measured using multiple particle tracking. Freshly extracted brains from P9 female SD rats were sliced into 300 mm sections and 0.5 nL of nanoparticles was injected after a brief rest period. Videos of nanoparticles were acquired at 100 frames per second for 6.5 seconds and nanoparticle trajectories were extracted with TrackMate (ImageJ). Trajectory analysis was completed with a lab-developed Python package available on GitHub.

Results: The nanoprecipitation formulation technique resulted in nanoparticles of small size (59-65 nm in diameter), low polydispersity (0.06-0.2), and near-neutral surface charge (-11.4 to -9.86 mV) independent of surfactant choice. After intravenous administration, all formulations except PLGA-PEG/P80 were strongly associated with vascular structures, demonstrating the significant challenge posed by the BBB. However, the P80 formulation was able to demonstrate low endothelial association as well as internalization in microglia and neurons. These results were supported by quantification of fluorescent nanoparticles in brain microvessels versus the parenchyma: the average ratio of nanoparticle concentration in the parenchyma to capillaries was 8.5 for PLGA-PEG/P80 compared to 4.3, 1.9, and 0.2 for PLGA-PEG/F127, PLGA-PEG/PVA, and PLGA-PEG, respectively. Surfactant differences were also apparent at the microscopic scale of diffusion. Inclusion of a surfactant significantly decreased the diffusive ability of PLGA-PEG nanoparticles, which had an ensemble-averaged effective diffusion coefficient (D_{eff}) of 12.9 $\mu m^2/s$. The PLGA-PEG/PVA, PLGA-PEG/F127, and PLGA-PEG/P80 nanoparticles had diffusion coefficients of 8.48, 7.50, and 4.72 $\mu m^2/s$, respectively. The PLGA formulation had the lowest diffusion coefficient (D_{eff}=1.02 $\mu m^2/s$), likely due to the absence of a PEG layer to reduce interactions with cellular and protein components of the tissue environment. Analysis of trajectory features confirmed a comparatively large portion of PLGA nanoparticle trajectories as "bounded" by circular boundaries, consistent with cellular internalization.

Conclusions: In this study, we evaluated the effect of surfactant choice on nanoparticle transport behavior in the brain at multiple length and time scales. Characterization across transport scales is important because effective drug carriers must be able to both diffuse through a congested brain ECS in milliseconds and uptake into endothelial and target cells over hours. By more fully understanding the role of surfactants in determining nanoparticle fate, we can engineer nanoparticles that leverage biological interactions for enhanced therapeutic delivery to the brain.

595811 Microrheological Characterization of Covalent Adaptable Hydrogel Degradation in Response to pH Changes That Mimic the pH in the Gastrointestinal Tract

Abstract:

Covalent adaptable hydrogels (CAHs) are composed of network bonds that can reversibly break and reform in response to external stimuli, including addition of external force or change in environmental conditions. This stimuli-responsive nature is due to the incorporation of dynamic covalent chemistry into the polymeric scaffold. This chemistry results in stimuli-responsive structural evolution, which makes CAHs ideal for biological applications ranging from bio-ink for 3D-printing to pH responsive drug delivery vehicles. To design a CAH as a drug delivery vehicle, specifically for oral drug delivery, scaffold degradation in response to pH changes that mimic the native pHs in the gastrointestinal (GI) tract must be characterized. Our work focuses on characterization of the degradation of a self-assembled CAH consisting of 8-arm star poly(ethylene glycol) (PEG)-hydrazine and 8-arm star PEG-aldehyde, which create dynamic covalent hydrazone bonds. To mimic changes in pH through the digestive tract and simultaneously characterize scaffold degradation, we use μ^2rheology. μ^2rheology enables multiple particle tracking microrheological characterization in a microfluidic device. In multiple particle tracking microrheology (MPT), fluorescent probe particles are embedded in a sample and their Brownian motion is measured to extract material rheological properties. To mimic changes in the pH environment in each part of digestive tract, we use a two-layer microfluidic device. Our device allows the fluid environment to be exchanged around a single sample with minimal sample loss, regardless of the state of the sample. Using μ^2rheology, we characterize CAH degradation at a single pH (pH 4.3, 5.5 and 7.4), with a single pH exchange (pH 4.3 to 7.4 and pH 7.4 to 4.3) and during temporal pH changes that mimic the pH in the entire GI tract. To characterize the critical phase transition between a gel and a sol, the critical relaxation exponent is calculated from single pH and single pH exchange experiments. This value quantitatively identifies the state of the material, pinpoints where the phase transition occurs and identifies the material microstructure at the phase transition. The critical relaxation exponent for this scaffold is independent of incubation pH, indicating that the material microstructure and phase transition are also independent of incubation pH. From measurements of single pH exchange and temporal pH changes through the whole GI tract, we determine that degradation history has no impact on CAH degradation kinetics and material property evolution. However, the initial cross-link density of the scaffold at each pH exchange can be reduced by previous degradations, which reduces the time to reach the gel-sol transition. For applications of molecular delivery, this change in degradation time will impact release. These results indicate degradation can be tuned by changing scaffold cross-link density, which can be done by changing polymer concentration or the ratio of functional groups. This work will inform design of this CAH for site-specific oral drug or nutrient delivery.

611280 Dual Responsive Backbone Shattering Organelle Targeted Nano-Prodrug to Modulate Oxygen Consumption for Effective Chemophototherapy

Abstract:

The clinical translation of nanomedicine for glioblastoma multiforme (GBM) faces four key challenges: (1) lack of precise control over carrier architectures, drug composition and batch-to-batch variations in drug loading content/ release profile, (2) lack of a technique to monitor the fate of the payload without using any substitute labelling molecule, (3) over dependence on monotherapy to treat heterogeneous solid tumours thereby contributing to therapeutic resistance, (4) suboptimal therapeutic concentrations in the brain tumour due to the presence of Blood Brain Barrier (BBB). Photodynamic therapy is a promising clinical alternative to conventional treatments such as chemotherapy and radiotherapy. However, its dependence on oxygen limits its therapeutic potential in hypoxic tumours. To address this issue, several strategies have been studied including direct oxygenation of the tumour using components such as enzymes and perfluorocarbon, usage of red blood cells and haemoglobin as oxygen carriers. However, the realization of a satisfactory photodynamic therapeutic effect remains challenging due to irregularity in tumour blood vessels which leads to heterogeneous oxygen supply and uneven photosensitizer distribution. Recently, modulation of oxygen consumption has received considerable attention as an approach that reduces oxygen partial pressure gradients within tumours, thereby alleviating hypoxia in poorly vascularized regions. In this study, we report the design and synthesis of a dual responsive nano-prodrug (DRnp) that not only efficiently generates reactive oxygen species (ROS), but also specifically releases its chemotherapeutic module on demand in the tumour microenvironment. DRnp is self-assembled from an amphiphilic polymer brush, which comprises of a redox responsive photodynamic backbone grafted with poly(ethylene glycol) and conjugated with a pro-apoptotic peptide through a light responsive bond. DRnp is used to encapsulate an oxygen regulator, atovaquone. Furthermore, the cell penetrating peptide CGKRK targets the angiogenic blood tumour blood vessels to significantly improve the vascular damage through photodynamic therapy, while the proapoptotic D-amino acid peptide KLAKLAKKLAKLAK disrupts the mitochondrial membrane thereby promoting apoptotic cell death. A photosensitizer Protoporphyrin IX (PpIX) was polymerized with a redox responsive linker cystamine dihydrochloride which was further conjugated to PEG and a cell penetrating pro-apoptotic peptide ($CGKRK_D(KLAKLAK)_2$) using a ROS responsive thioketal linker. The synthesized material was characterized using NMR and GPC. DRnp was obtained by a nanoprecipitation method. The dual responsive ability of DRnp were studied in different pH and GSH environments. The ability of DRnp to pass the BBB was studied using an in-vitro BBB model. Finally, the penetration and cytotoxicity of DRnp were studied in an U87 and 3T3 co-culture tumour spheroid model. GPC data showed that the number average molecular weight of DRnp was circa 120,000 with good polydispersity (PDI). TEM showed DRnp has a uniform spherical morphology with an average diameter of ~120 nm, consistent with the results from dynamic light scattering. Significant penetration and cytotoxicity of DRnp was observed in multicellular spheroids after passing through the in-vitro BBB. The distinctive features of DRnp include (1) photosensitizer and chemotherapeutic peptide were combined at a fixed ratio for precise nanomedicine, (2) the photodynamic backbone was site specifically shattered at high GSH concentrations along with ROS mediated cleavage of the therapeutic peptide in the acidic tumour microenvironment, (3) Can pass through the BBB, and (4) minimizes the oxygen consumption rate within solid tumours. Overall, DRnp enables site specific integrated PDT and chemotherapy for synergistically amplified cancer treatment.

600944 pH-Responsive, Cell-Penetrating, Magnetite-Silver Nanoparticles for Delivery of Plasmid-Based Gene Therapies: Preparation, Characterization, and *in Vitro* Evaluation

Abstract:

Over the past few years, gene therapies have attracted much attention for the development of therapies for various conditions, including cancer, neurodegenerative diseases, protein deficiencies, and autoimmune disorders[1]. Despite the benefits of this approach, several challenges are yet to be solved to reach clinical implementation eventually. Some of these challenges include low transfection rates, limited stability under physiological conditions, and inadequate specificity towards the target cells[1]. An avenue to overcome such issues is to deliver the therapies with the aid of potent cell-penetrating vectors, which include viral and non-viral vehicles. The use of viral vectors has been criticized due to issues concerning genome integration, a potential for oncogenicity, and induced mutagenesis in the host cell[1]. Alternatively, non-viral vectors, such as nanostructured materials, have been successfully tested in drug and gene delivery without the concerns of the viral ones[2]. This has been attributed to their high surface area to volume ratio, ease of synthesis, and the possibility for surface modification to incorporate various surface chemistries and thereby a tunable multifunctional response. This has led to an increasing number of applications where the delivered molecules have shown a considerable increase in bioavailability in both *in vitro* and *in vivo* applications[3]. Among the most successful nanostructured vehicles, magnetite nanoparticles stand out due to their superior biocompatibility and the possibility for manipulation with the aid of magnetic fields[4].

Here, we propose to develop a nanostructured core-shell cell-penetrating vehicle composed of magnetite at the core and surrounded by a silver shell. A subsequent superficial conjugation of a pH-responsive polymer (poly(2-dimethylamino) ethyl methacrylate, pDMAEMA) was proposed to assure that the vehicle is capable of carrying and releasing circular DNA. Finally, Buforin II was conjugated to the vehicle to facilitate translocation and endosomal escape as well as reaching the nucleus[5]. The prepared core-shell system exhibited colloidal stability such that agglomerate formation and thereby precipitation was avoided. Besides the role of DNA carrier, the pDMAEMA was conjugated to attempt to avoid endosome formation, which is a cell internalization mechanism that traps exogenous material and consequently might lead to a significant reduction in transfection rate[4]. Buforin II was immobilized via the organosilane molecule aminopropyl 3-ethoxy silane and the surface spacer polyether amine (PEA) (Jeffamine 600) by following a protocol described by Cuellar et al5. Finally, a 10kb circular DNA vector was loaded on the modified nanoparticles at pH values in the range of 6 to 8.5 to estimate the loading efficiency and the loading maximum capacity. The designed vector contained information for a Cas9 enzyme, a CRISPR gRNA, and a mCherry reporter. This is an attempt to explore the suitability of the vehicle in the delivery of gene editing systems. The release of DNA was also tested *in vitro*, by exposing the loaded systems to a medium with pH values in the range of 8.5 to 6. The obtained nanobioconjugates were characterized by UV-Vis spectrophotometer, dynamic light scattering (DLS), thermogravimetric analysis (TGA), Fourier transform infrared spectroscopy (FTIR), scanning electron microscope equipped with energy dispersive spectroscopy (SEM+EDS), Transmission Electron Microscopy (TEM), and DLS. Moreover, the conjugates were tested in neuroblastoma (SH-SY5Y) and Vero cells to evaluate endosomal escape and cytotoxicity via LDH. Finally, hemolytic tendency and platelet aggregation assays were conducted to complete the biocompatibility analysis and therefore estimate the clinical potential of the vehicle.

The results indicate that the average hydrodynamic diameter of the obtained nanobioconjugate is 478 nm. SEM and TEM images revealed some unevenness of the Silver coating on nanoparticles but the crystalline structure corresponding to magnetite at the core. According to the TEM images, individual nanobioconjugates approached 10.4 nm in diameter. TGA analysis confirmed a conjugation efficiency of pDMAEMA of about 8.8% and of about 3% for Buforin II. The FTIR confirmed the characteristic bands of the polymer as well as those of Buforin II. The vehicle was capable of loading 35 ng of DNA per microgram of nanoparticles and it loads almost 16% of the added DNA while releasing around 8% of that amount. Figure 1 shows that cell viability remains above 95% for the pDMAEMA conjugated to the core@shell system (magnetite@silver-pDMAEMA) as well as for the vehicle with the additional conjugation of Buforin II (magnetite@silver-pDMAEMA-BUFII). This was also the case for the hemolysis assay where the release of hemoglobin was below 5% in both cases. To evaluate endosomal escape, the nanobioconjugates were fluorescently-labeled with Rhodamine B (magnetite@silver-pDMAEMA-BUFII-RhodB), delivered to SH-SY5Y, and subsequently incubated for 30 minutes and 4 hours. Pearson's correlation coefficients were calculated to evaluate the colocalization of the nanobioconjugates with the endosomes (labeled with Lysotracker Green). The results suggest an endosomal escape of about 80% for SH-SY5Y and 60% for Vero cells. Further evidence of escape and even cytosol distribution was provided by a calculated surface area coverage of about 50% for both cell lines.

Our findings regarding the possibilities of the vehicle to load and release of DNA, cell-penetrating and endosomal escape capacity as well as high biocompatibility, confers it with important chances for further *in vivo* testing. Also, it highlights a route to enable gene therapy applications that require large-size plasmid vectors such as those involved in CRISPR/Cas9 editing systems.

References:

1. Dai, W.-J. et al. CRISPR-Cas9 for in vivo Gene Therapy: Promise and Hurdles. Mol. Ther. - Nucleic Acids 5, e349 (2016).
2. Ho, B. X., Loh, S. J. H., Chan, W. K. & Soh, B. S. In vivo genome editing as a therapeutic approach. *International Journal of Molecular Sciences* **19,** (2018).
3. Jin, S. & Ye, K. Nanoparticle-mediated drug delivery and gene therapy. in *Biotechnology Progress* **23,** 32–41 (2007).
4. Varadan, V. K. *Nanomedicine : design and applications of magnetic nanomaterials, nanosensors, and nanosystems LK - https://univdelosandes.on.worldcat.org/oclc/916159742. TA - TT* - (Wiley, 2008).
5. Cuellar, M. *et al.* Novel BUF2-magnetite nanobioconjugates with cell-penetrating abilities. *Int. J. Nanomedicine* **Volume 13,** 8087–8094 (2018).

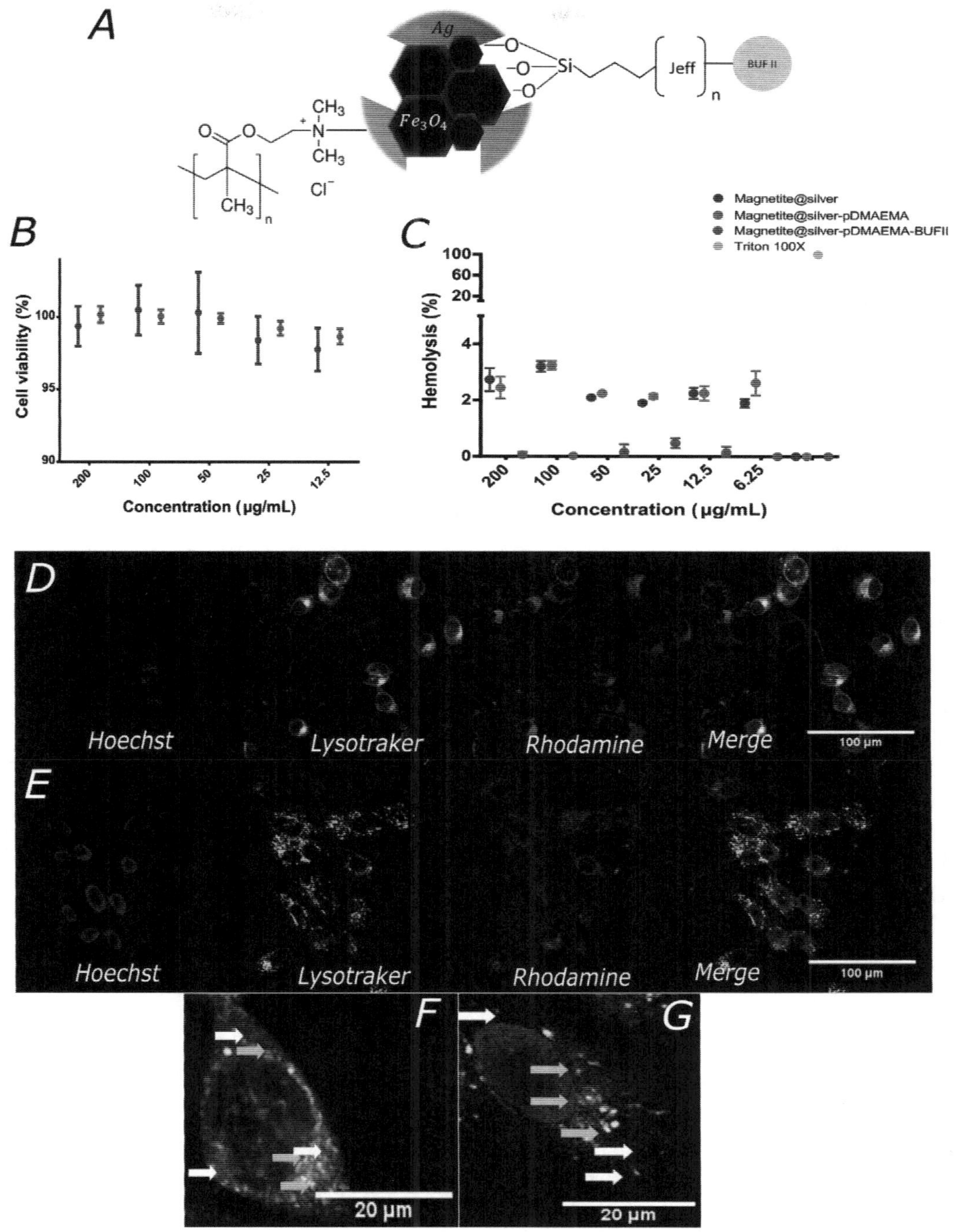

Figure 1.(A) Chemical schematic structure of nanobioconjugate magnetite@silver-pDMAEMA-BUFII (B) Cytotoxicity assay via LDH, in Vero cells after 24 and 48 hours for different concentrations of core@shell nanoparticles, silver@magnetite-pDMAEMA, and silver@magnetite-pDMAEMA-BUFII nanobioconjugates. The results indicate low cytotoxicity (C) Hemolysis assay indicates a low hemolytic tendency for evaluated conjugates (D) Cell internalization of magnetite@silver-pDMAEMA-BUFII-RhodB after 4 hours in SH-SY5Y. Nuclei were stained with Hoechst and endosomes were labeled with lysotracker Green (E) Cell internalization of magnetite@silver-pDMAEMA-BUFII-RhodB after 4 hours in Vero cells. (F)-(G) shows individual SH-SY5Y and Vero cells, respectively, where white arrows point to regions of possible endosomal escape while orange arrows point to regions of colocalization, which likely indicates endosomal entrapment.

605728 Enzymatic Protection and Biocompatibility Screening of Enzyme-Loaded Polymeric Nanoparticles for Neurotherapeutic Applications.

Abstract:

Introduction: Drug delivery of enzyme therapeutics encounters many obstacles, including systemic clearance, protease degradation, endothelial barriers to target organ entry, and poor penetration in local tissue parenchyma. The tightly regulated blood-brain barrier (BBB) is a further hindrance specifically for neurological disease and inhibits 98% of small molecule and ~100% of enzyme and other large-molecule therapeutics from entering the brain from the bloodstream. Leveraging polymeric nanoparticles for encapsulating and delivering therapeutic cargo can help overcome several of these barriers. However, nanoparticle platforms must be carefully engineered to ensure biocompatibility. An important consideration that could influence drug delivery and efficacy is how disease conditions alter the tissue microenvironment. The presence of inflammation or oxidative stress could affect the susceptibility of cells to inadvertent damage. Using organotypic whole hemisphere brain slice models, we can screen nanoparticles for biocompatibility under different disease conditions in high-throughput fashion.

Materials and Methods: We formulated enzyme-encapsulating poly(lactic-co-glycolic acid)-block-poly(ethylene glycol) (PLGA-PEG) polymeric nanoparticles with catalase as a model enzyme. Comparing double emulsion and nanoprecipitation formulation methods, and cholic acid (CHA) and poly(vinyl alcohol) (PVA) surfactants, we evaluated the nanoparticles' ability to protect encapsulated catalase from extracellular protease degradation to extend enzymatic activity. To study nanoparticle biocompatibility under neurological disease conditions, we developed organotypic whole hemisphere brain slice models of neuroinflammation with lipopolysaccharide (LPS) exposure, and excitotoxicity with monosodium glutamate (MSG) or oxygen glucose deprivation (OGD) exposure. Upon application of blank nanoparticles without catalase on our different disease models, we evaluated cytotoxicity via lactate dehydrogenase (LDH) release to assess nanoparticle biocompatibility. To explore the potential role of oxidative stress in cell death, we determined reduced glutathione (GSH) concentrations.

Results: All formulated nanoparticles exhibited a near-neutral surface charge and sub-100 nm size, requisite characteristics for effective diffusion in brain parenchyma. Regardless of surfactant, nanoprecipitation provided no extension of enzyme activity compared to free catalase, while CHA-emulsified double emulsion nanoparticles provided significant increases in catalase activity. For our brain slice models of neurological disease, we observed that MSG and OGD exposure elicited a 71.1% and 41.1% increase in cell death, respectively, and application of superoxide dismutase (SOD) on both MSG and OGD models fully inhibited cell death. LPS exposure did not induce an increase in cell death but did elicit an increase in inflammatory cytokine mRNA expression. We used our *ex vivo* slice models to screen our nanoparticles for toxicity to develop a biocompatible formulation. Surprisingly, double emulsion nanoparticles induced significant toxicity when applied to MSG- and OGD-exposed treated slices, but not for non-treated or LPS-exposed brain slices, while nanoprecipitation nanoparticles showed no toxicity for all slice treatment conditions. Double emulsion nanoparticle toxicity was consistently observed on excitotoxicity-induced slices for specifically PLGA-PEG nanoparticles made with dichloromethane as the organic solvent during formulation. Toxic nanoparticle application to MSG-exposed slices did not elicit a lower GSH concentration compared to MSG exposure alone, indicating nanoparticles did not exacerbate cell death through affecting the oxidant/antioxidant. Toxicity was eliminated when chloroform was used as the organic solvent, or with the addition of free PEG during the sonication steps of the double emulsion formulation process.

Conclusions: Our work demonstrates that polymeric nanoparticles provide promise for enzyme delivery. Double emulsion formulated polymeric nanoparticles can provide protection of enzymatic cargo from degradative *in vivo* conditions, but also showed toxicity in slices. Double emulsion PLGA-PEG nanoparticles are capable of enzyme protection from proteases, and are biocompatible with diseased brain slices, when free PEG is present during sonication. However, these findings demonstrate that nanoparticle therapeutics must be stringently evaluated and optimized for biocompatibility for effective progression into preclinical *in vivo* studies. Additionally, our results encourage the systematic approach of screening therapeutics on *ex vivo* brain slice models to isolate, understand, and overcome neurological disease processes to aid in preclinical neurotherapeutic development.

598340 Rational Design of Modified-Release Dosage Forms

Abstract:

Oral modified release (MR) dosage forms are developed by altering the kinetics and/or site of drug release/absorption in order to achieve specific clinical objectives, such as improved patient compliance, optimized efficacy and reduced adverse events. This presentation introduces basic design principles and different MR delivery technologies. It also discusses the importance of understanding material, formulation, process and their interplays in rational product and process development along with case studies.

602132 Strategies to Develop IV Formulations of Poor Water-Soluble Compounds and Their in Vitro Evaluation

Abstract:

Although a drug might be developed and formulated for oral delivery, the same compound is commonly administered as a solution via an IV injection/infusion in order to determine absolute bioavailability to support early development. A common challenge to develop IV formulation is the low solubility of drug candidates, risking phlebitis due to precipitation upon injection. Other parameters that need attention include the pH variation (too high or too low) that can cause irritation, and osmolality differences that may cause significant fluid shifts between the intracellular and extracellular spaces. Thus, the development of a stable and safe IV formulation requires an adequate selection of solubilizers and evaluation of the safety of the solution prior to administration.

In this study, a weak acid drug with poor water-solubility required a minimum of 40-fold increase in aqueous solubility for IV delivery. To achieve increased aqueous solubility, multiple strategies were assessed including screening of solvents and co-solvents, in situ salt, and pH control. Phlebitis risk was evaluated by in vitro static dilution using surrogate plasma and observation of flocculation in rat plasma by optical microscopy. In conclusion, a pH-buffered solution containing in situ sodium salt allowed for up to 133-fold increase in aqueous solubility of the compound showing no risk of precipitation while complying with pH and tonicity requirements for IV administration.

602618 Integrated Product & Process Development

Abstract:

During drug product development, science-based design of formulation and manufacturing process is essential for achieving and assuring consistent quality products. This presentation discusses rational product development based on the integrated product and process knowledge. It addresses the importance of fundamental understanding of physicochemical, mechanical and biopharmaceutical characteristics of API, formulation, process and their interactions during product design, development, scale-up and commercial manufacturing with case studies in ensuring consistent quality, robust drug delivery performance and process.

602619 Predictive and Model Based Product & Process Design

Abstract:

Use of predictive technologies and in silico modeling can significantly enhance speed, efficiency and effectiveness of drug product and process design. This presentation will discuss several case studies, in which multidisciplinary sciences and engineering principles, coupled with high power computing, are used to guide formulation development, manufacturability assessment, and drug substance solid form risk assessment.

605621 Rational Design of Amorphous Solid Dispersion

Abstract:

Amorphous solid dispersion (ASD) is a highly effective drug delivery technology for enabling oral absorption of insoluble APIs. This presentation discusses basic principles of formulation design & characterization, and importance of understanding material, formulation, dissolution (LLPS) and stability in rational product development using case studies

607865 Stochastic Simulations of Motion of Solute Molecules during the Process of Crystallization Yields the Rate Constants for the Processes of Crystallization

Abstract:

Theoretical and computational understanding of crystallization processes can significantly replace the trial-and-error methods in place for crystallization at different pharmaceutical, chemical and fertilizer industries. However, the self-assembly of molecules during the process of crystallization spans multiple timescales which makes full computational modeling of crystallization process difficult. Recently, with the help of molecular simulations, it was shown that dynamics of solvation shell determines the crystal structure and growth rates. However, more appropriate computational approach is required for calculation of rate constants for nucleation and growth of crystals. In this study, stochastic simulations of glutamic acid molecules were performed. The molecular features of glutamic acid were approximated using the already established models to retain its molecular feature while reducing the computational time. Such approximation allows simulating higher number of glutamic acid molecules in a simulation box for longer times. Findings of this study yield in rate constants for glutamic acid crystallization which reasonably agree with experimental observations. The approach also shows potential to be applied to other crystallization systems.

604531 Irreversible Inhibition of Barite Crystallization: A Unique Mechanism for Treating Scale Formation

Abstract:

The mineralization of sparingly soluble compounds in process equipment may result in corrosion, equipment damage, and cause reduction or loss of finished product for oil and gas production, wastewater treatment, and manufacturing processes. Conventional treatments rely on the use of caustic solutions that pose a negative environmental impact. Designing effective biodegradable chemical treatments to reduce or eliminate scale formation requires an understanding of the molecular-scale interactions of chemical additives in supersaturated environments at the crystal interface. Here we implement a cooperative approach to design and test environmentally friendly chemical treatments for scale formation using a combination of surface science techniques and microfluidic technology. Crystallization can be suppressed in highly supersaturated solutions using dilute quantities of modifiers. These studies focus on barium sulfate (barite), a highly insoluble and common scaling mineral. In this presentation we will describe the effect of organic acids on barite crystallization in bulk assays under flow and in quiescent conditions. Using *in situ* scanning probe microscopy, we explore the interfacial interactions between the modifiers and different surfaces of barite over a range of supersaturated conditions to elucidate the role of these inhibitors during crystal growth. Our findings identify unique crystal-modifier interactions that suppress barite crystallization via an irreversible inhibition mechanism. Collectively, these studies provide new fundamental understanding of crystal growth modifier action towards the design of improved scale treatments.

611669 Multi-Step Crystallization of Self-Organized Spiral Eutectics

Abstract:

A method for the solidification of metallic alloys involving spiral self-organization is presented as a new strategy for producing large-area chiral patterns with emergent structural and optical properties, with attention to the underlying mechanism and dynamics. This study reports the discovery of a new growth mode for metastable, two-phase spiral patterns from a liquid metal. Crystallization proceeds via a non-classical, two-step pathway consisting of the initial formation of a polytetrahedral seed crystal, followed by ordering of two solid phases that nucleate heterogeneously on the seed and grow in a strongly coupled fashion. Crystallographic defects within the seed provide a template for spiral self-organization. These observations demonstrate the ubiquity of defect-mediated growth in multi-phase materials and establish a pathway toward bottom-up synthesis of chiral materials with an inter-phase spacing comparable to the wavelength of infrared light. Given that liquids often possess polytetrahedral short-range order, our results are applicable to many systems undergoing multi-step crystallization.

600638 Modeling Impurity-Mediated Crystal Growth

Abstract:

Crystallization is commonly used in industrial processes to convert solute molecules dissolved in solvent to a structured solid state. Pharmaceutical companies often crystallize APIs in the form of organic molecules to selectively formulate specific crystal habits for optimal bioperformance. Crystal engineering is also of importance for developing catalysts with tailored surfaces to maximize active sites. Furthermore, tuning crystallization is desirable for varying electrical and optical properties in the field of electronic materials and for altering the impact sensitivity of energetic materials. Given the ubiquity of crystal growth in industrial processes, there is substantial demand for predictive and mechanistic modelling of crystallization. Crystallization of organic molecules is well understood for ideal systems (i.e., Kossel crystals with a single centrosymmetric growth unit). There is interest in studying crystal systems in which non-idealities are introduced, as these are more representative of realistic conditions. One such non-ideality involves the presence of impurities or 'imposter molecules.' There are numerous notable examples of crystal growth systems in which the presence of impurities significantly influences crystallization processes. Examples include "antifreeze" proteins and antimalarial medications. While these systems are prevalent and of importance in many fields, there is currently no comprehensive model that captures the full effect of impurities on crystal growth processes.

The goal of this research is to investigate the effect of impurities on the crystal growth process and to develop theoretical models for the mechanisms by which impurities influence crystal growth and hence affect crystal morphology and size. Impurities affect growth kinetics at the scale of kink attachment and detachment events, which are too fine to examine experimentally in real time. Thus, we use simulations to study the proposed mechanisms for growth inhibition. We employ Kinetic Monte Carlo (KMC) to simulate the time evolution of centrosymmetric organic crystal growth. Various mechanisms have been proposed to explain the growth-inhibiting effect of impurities, such as Cabrera and Vermilyea's *step-pinning* model and Doherty and Sizemore's *spiral-pinning* models. Model equations have been developed in accordance with these mechanisms to quantify the effect of inhibition on the crystal face. We use KMC simulations to examine the limits of these equations under different growth regimes and to elucidate the form of a more general model. The simulations serve to quantify the degree of growth inhibition of a simple crystal growth system populated with impurities. The results suggest a monotonic decrease of the step velocity with increasing impurity concentration, though the form of this dependence is based on the *nature of the impurity distribution*. Furthermore, the simulations suggest a critical impurity coverage at which crystal growth is halted completely.

611583 Monte Carlo Simulations of Hydrothermal Metal Organic Framework (MOF) Crystal Growth

Abstract:

Metal organic frameworks (MOFs) are crystalline materials which are well-known due to their applications in separations, catalysis, and chemical capture, among others, brought forth by their large surface areas and high porosity. These materials are structurally versatile as they are composed of metal cluster and organic ligand pairs, which can be strategically switched to make MOFs with certain functionalities. A subset of MOFs with a "rod-packing" motif ("rod-MOFs) are promising as breakthrough materials for size-selective, passive chemical separation processes due to their fine unidirectional channels. The main challenge for these applications, however, rests on the optimization of the crystallization process of these MOFs to make them larger than their conventional micron scales and with negligible intergrowths.

Crystallization through the hydrothermal method is a common process used to synthesize certain classes of Metal organic frameworks (MOFs), including MIL-53, which falls under the rod-MOF classification. Though the process is ubiquitous to the formation of a variety of crystals, control of the dynamics of the growth process itself remains a challenge as the closed nature of the batch system typically used for these reactions is reminiscent of a "black box". A wealth of literature exists on the exploration of crystal growth dynamics through Monte Carlo (MC) and Molecular Dynamics (MD) simulations, as well as powerful models which incorporate both. However, these models have seldomly been applied to investigate the crystal growth kinetics of MOFs, and more specifically, the MOF crystallization phenomena for special cases where the inputs to the system are changing periodically.

Herein we apply our understanding of the physics of crystallization, statistics and well established kinetic Monte Carlo algorithms (kMC) based on the solid-on-solid (SOS) model and Metropolis MC to develop a deeper insight on MOF crystallization phenomena under hydrothermal conditions.

609413 The Role of Water in the Crystallisation and Nucleation of Glycine Homopeptides

Abstract:

The combination of amino acids into peptides and proteins, through peptide bonds, are the building blocks of life on earth. The last decade has seen a significant increase in the use of these materials in the treatment of chronic and metabolic diseases (e.g. cancer, obesity and diabetes).[1-2] However, the general peptides product process still has severe problems, like low solubility, proteolytic degradation and physiochemical instability.[3] Peptide crystallisation, as a good alternative, can help us get the peptide structure and get ideal physicochemical products. Water as a most popular solvent used in peptide and protein crystallisation will not only support the bioactivity of these materials, but also can stabilise the more complex molecular conformations that arise from long chain biomolecules, through satisfying the multiple hydrogen bonding sites of these materials.[4-5] Solving how the water molecule effect the peptides crystal structure can help us to uncover the details of the peptide conformation and waters role in their crystallisation, giving insight into their roles in the crystallisation of highly hydrated protein structures.

Glycine is an important and well-studied amino acid, being abundant in proteins and utilized as a pharmaceutical excipient. Its simple hydrogen side chain makes this an ideal case study candidate for biopharmaceutical research.[6] Here we investigate the crystallisation of pure and hydrated forms of mono-, di- and tri-glycine. However, only mono- and tri-glycine are known to crystallise in hydrated forms. The crystallisation of mono, di- and tri-glycine from aqueous solutions was compared in this research. The single crystal structure of triglycine dihydrate (TGDH) form was solved for the first time, which can help us to analysis the interaction between water and peptide molecules. For Glycine homopeptides, crystal forms of up to five glycine residues in a peptide chain are known, and the antiparallel β sheet structure was found is the most stable crystalline form for triglycine anhydrate. However, from the crystal structure of triglycine dihydrate, the backbone of it trends to bend in water to find a more stable form.

The activation free energy (Gc) of nucleation was calculated based on induction time measurement at different supersaturation and different temperature. Compare with mono- and di-glycine, triglycine has the greatest decrease in Gc at 283.15K when the form is dihydrate. Molecular modelling was also used to investigate the delicate balance of conformation and packing forces, as a function of chain length.

In summary, we solved the first TGDH crystal structure and explored the water role in the hydration conformation of peptides. The modeling work help us to find several competing effects which are driving TGDH. It also gives protein crystallisation a new insight about the interaction between water and protein, this discovery has an important implication for pharmaceutical engineering and biology research.

Reference

1. Cicero, A. F. G.; Fogacci, F.; Colletti, A., Potential role of bioactive peptides in prevention and treatment of chronic diseases: a narrative review. *Br J Pharmacol* **2017,** *174* (11), 1378-1394.
2. Marqus, S.; Pirogova, E.; Piva, T. J., Evaluation of the use of therapeutic peptides for cancer

treatment. *J Biomed Sci* **2017,** *24* (1), 21.
3. Otvos, L., Jr.; Wade, J. D., Current challenges in peptide-based drug discovery. *Front Chem* **2014,** *2*, 62.
4. Scoppola, E.; Sodo, A.; McLain, S. E.; Ricci, M. A.; Bruni, F., Water-peptide site-specific interactions: a structural study on the hydration of glutathione. *Biophys J* **2014,** *106* (8), 1701-9.
5. Wyttenbach, T.; Liu, D.; Bowers, M. T., Hydration of small peptides. *International Journal of Mass Spectrometry* **2005,** *240* (3), 221-232.
6. Koleva, B. B.; Kolev, T. M.; Spiteller, M., Structural and spectroscopic analysis of hydrogensquarates of glycine-containing tripeptides. *Biopolymers* **2006,** *83* (5), 498-507.

607099 Data and Information Management Vision and Strategy Towards Pharma 4.0

Abstract:

Historical perspective of data in pharmaceutical industry has always been the by-product of scientific experiments and business processes required to maintain for regulatory reasons. With the rapid advancements in the field of machine learning (ML) and artificial intelligence (AI), there is a growing view of data as the new oil that could be mined to make quality decisions early and accelerate programs to make life-saving drug reach patients much faster.

To get to ML/AI, we need to build a robust data and information infrastructure which by design allows for these end use cases effortlessly. In this regard, a widely accepted philosophy is FAIRization of data which means data should be **F**indable, **A**ccessible, **I**nteroperable and **R**eusable. By making data FAIR, one can pass two litmus tests: a) data can be consumed without the presence of data owner and b) data is machine actionable.

In this talk, we will present vision and strategies that could help in making our laboratory and plant data FAIR and guide efficient process development and robust manufacturing across pipeline projects in pharmaceutical R&D space. We will talk about challenges, pitfalls and learning in this journey. We will also present the case-study of our in-house built platform that makes research data FAIR in the backend while making scientist's data journey simple in the frontend.

Disclosures: *All authors are AbbVie employees and may own AbbVie stock. AbbVie sponsored and funded the study; contributed to the design; participated in the collection, analysis, and interpretation of data, and in writing, reviewing, and approval of the final abstract.*

599249 Data Driven Performance Measurement & Management in the External API Network

Abstract:

Over the past two and a half years, Bristol Myers Squibb (BMS) has shifted from internal scale up and production of clinical small molecule Drug Substance, to external supply from contract manufacturing organizations (CMO`s). This paradigm shift has created a need for BMS to develop new strategies to measure, understand, and manage the performance of our technology transfer & scale up. Our team has developed a culture of performance data collection, gathered across our network of vendors for each project, to better understand performance metrics, such as yield variability, quality variability, key cost drivers, and supply delays. The aggregated, contextualized data enables the creation of fit-for-purpose analysis and reports to support future site selection decisions, vendor management, and internal continuous improvement efforts. Predictive models for business drivers such as project cost and vendor capacity can also be created from the aggregated data. The predictive models can then be used to drive tactical and strategic decisions across the vendor network and project portfolio. These data-driven products are used for multiple business purposes, including high-level performance metrics for API delivery, targeted feedback to vendors, budget estimation, and to support individual strategic decisions around network capacity and capabilities. This presentation will show examples of the workflow used to capture this data, build data-analysis products, and drive decisions and continuous improvement. In particular, we will describe how we measure vendor performance against various business metrics and provide data-driven feedback to partners. We will also cover the use of probabilistic models and Monte-Carlo simulation to understand future expected demand on our supplier network.

608857 Real Time Dynamic Optimisation for Enhanced Product Quality in Bioprocesses

Abstract:

Biopharmaceutical processes are prone to many challenges including highly nonlinear dynamics, batch-to-batch variability and high sensitivity to changes in operating parameters. Such complexities can introduce heterogeneity in biopharmaceutical products. At the same time, consistent product quality remains a top priority for this industry to ensure patient safety. Antibody-producing platforms consist of two main operations: upstream (USP) and downstream (DSP) production. In USP cells are inoculated into a series of progressively larger bioreactors for the growth of mammalian cells, typically Chinese hamster ovary (CHO) cells in the case of monoclonal antibodies (mAbs). These are then transferred to the production bioreactor for protein expression. At the end of this step DSP begins with a series of filtration and purification stages to remove impurities and other unwanted components from the final drug formulation.

The performance of such processes is assessed both for productivity and product quality. The latter is a broad concept that includes various aspects, amongst which product functionality. In mAb production, glycosylation percentage is one of the key indicators to measure cell culture quality performance. Glycosylation is a post-translational modification of mAbs that affects their stability and efficacy as therapeutics. As such, it is considered a Critical Quality Attribute (CQAs) under the Quality by Design (QbD) framework and needs to be controlled within a pre-specified range [1]. There have been various studies on the improvement of mAb glycosylation intervening both at process- [2], [3] and genetic- level [4], [5]. However, the complexity of the system and its underlying interactions make process optimisation difficult, with optimality unlikely to be achieved solely by experimental studies. This is because strategies to improve mAb glycosylation can enhance mAb glycosylation but, at the same time, reduce cell growth and, thus, mAb concentration at harvest.

In this work we use an Ordinary Differential Equation (ODE) model as digital-twin for an IgG-producing CHO cell culture system [6]. The model [7] comprises three main components; namely an unstructured model to describe cell growth, a reduced kinetics model for the intracellular nucleotide sugar donor (NSD) synthesis [8]and a mAb glycosylation model. The mAb glycosylation model as considered in this work, comprises four continuous stirred tank reactors (CSTRs) using the reaction network presented in del Val et al.[9]. We formulate and solve an exhaustive dynamic optimisation experiment to identify feeding regimes for galactose and uridine under which the amount of galactosylated antibody is maximised. The performance of every optimisation formulation is assessed based on: (a) achieved galactosylation percentage, (b) productivity, (c) culture viability and (d) experimental applicability. Towards an effort to enhance USP automation, measured process disturbances are introduced in the problem formulation. In this respect, the most successful scenarios are used as a basis for the design of a real time dynamic optimisation strategy as a means to achieve the optimal operational policy [10]. The system is re-optimised online and based on the effect of the disturbances to the process, adapts and calculates the optimal input trajectories. In this work these are validated in closed-loop against the process model.

References

[1] A. Eon-Duval, H. Broly, and R. Gleixner, "Quality attributes of recombinant therapeutic proteins: An assessment of impact on safety and efficacy as part of a quality by design development approach,"

Biotechnol. Prog., vol. 28, no. 3, pp. 608–622, May 2012.

[2] R. K. Grainger and D. C. James, "CHO cell line specific prediction and control of recombinant monoclonal antibody *N* -glycosylation," *Biotechnol. Bioeng.*, vol. 110, no. 11, pp. 2970–2983, Nov. 2013.

[3] M. J. Gramer *et al.*, "Modulation of antibody galactosylation through feeding of uridine, manganese chloride, and galactose," *Biotechnol. Bioeng.*, vol. 108, no. 7, pp. 1591–1602, Jul. 2011.

[4] C. Chung, Q. Wang, S. Yang, B. Yin, H. Zhang, and M. Betenbaugh, "Integrated Genome and Protein Editing Swaps α -2,6 Sialylation for α -2,3 Sialic Acid on Recombinant Antibodies from CHO," *Biotechnol. J.*, vol. 12, no. 2, p. 1600502, Feb. 2017.

[5] N. Yamane-Ohnuki *et al.*, "Establishment ofFUT8 knockout Chinese hamster ovary cells: An ideal host cell line for producing completely defucosylated antibodies with enhanced antibody-dependent cellular cytotoxicity," *Biotechnol. Bioeng.*, vol. 87, no. 5, pp. 614–622, Sep. 2004.

[6] P. Kotidis *et al.*, "Constrained global sensitivity analysis for bioprocess design space identification," *Comput. Chem. Eng.*, vol. 125, pp. 558–568, Jun. 2019.

[7] P. Kotidis *et al.*, "Model-based optimization of antibody galactosylation in CHO cell culture," *Biotechnol. Bioeng.*, vol. 116, no. 7, pp. 1612–1626, Jul. 2019.

[8] S. N. Sou, P. M. Jedrzejewski, K. Lee, C. Sellick, K. M. Polizzi, and C. Kontoravdi, "Model-based investigation of intracellular processes determining antibody Fc-glycosylation under mild hypothermia," *Biotechnol. Bioeng.*, vol. 114, no. 7, pp. 1570–1582, Jul. 2017.

[9] I. Jimenez del Val, J. M. Nagy, and C. Kontoravdi, "A dynamic mathematical model for monoclonal antibody N-linked glycosylation and nucleotide sugar donor transport within a maturing Golgi apparatus," *Biotechnol. Prog.*, vol. 27, no. 6, pp. 1730–1743, Nov. 2011.

[10] J. V. Kadam, M. Schlegel, B. Srinivasan, D. Bonvin, and W. Marquardt, "Dynamic optimization in the presence of uncertainty: From off-line nominal solution to measurement-based implementation," *J. Process Control*, vol. 17, no. 5 SPEC. ISS., pp. 389–398, Jun. 2007.

606263 Condition Monitoring of Pharmaceutical Processes Under a Probabilistic Framework

Abstract:

Condition-based maintenance in pharmaceutical processes requires the continuous monitoring of process parameters and asset performance features to enable predictions of when corrective operational or maintenance action would prove beneficial. Although this data does reflect the current-condition of the process and the equipment/facility, it is never sufficient to provide the complete picture, particularly, of the future state.1 This missing information, coupled with the variability in sensor data, contributes to uncertainty that needs to be explicitly quantified and incorporated into predictive models. Furthermore, identifying parameters and extracting features that best represent the state of the process require the use of training data,2 which may or may not completely represent the state of the process under normal operating conditions or under recurring abnormal conditions.

Even with the abundance of data brought about by the digitalization of processes, uncertainty is still prevalent. Aside from the inability to directly measure the condition of a process, the variability of operating parameters, e.g. raw materials, environment, personnel tendencies, can directly affect a process and hence affect the accuracy of the process model.

The explicit expression of uncertainty, of predictions and beliefs deduced from a model, may be reduced to calculation of the probability of that prediction/belief. With condition monitoring, this means using the data to estimate the probability of the condition of the system, and then updating these estimates to reflect new data in real time. Bayes' theorem fits well into this paradigm, since it provides a mechanism to correctly update probabilities (as a measure of uncertainty) in light of new data, hopefully reducing uncertainty in predictions.3

In this work, we demonstrate the use of Bayesian methods and the advantages of implementing condition monitoring under a probabilistic framework. We illustrate this approach using cases drawn from the operation of the Continuous Pharmaceutical Tableting Pilot Plant at Purdue University, which is a state-of-the-art high-bay facility that can be configured to produce tablets via direct compaction, dry granulation, or wet granulation. Moreover, the leading software for real-time manufacturing management would be used for these cases, particularly PI System (OSIsoft) and SmartFactory Rx (Applied Materials). The former will be mainly the data historian4 while the latter provides an environment for extracting the data from the historian and processing it for the purposes of operations productivity, analytics and control, knowledge management, and predictive maintenance.5 This environment would be ideal for implementing Bayesian methods for condition monitoring.

1 R. Flage, D.W. Coit, J.T. Luxhøj, and T. Aven, Reliability Engineering & System Safety **102**, 16 (2012).

2 C.M. Bishop, *Pattern Recognition and Machine Learning* (Springer, 2016).

3 O. Martin, *Bayesian Analysis with Python: Introduction to Statistical Modeling and Probabilistic Programming Using PyMC3 and ArviZ, 2nd Edition* (Packt Publishing Ltd, 2018).

4 OSIsoft, OSIsoft PI System (n.d.).

5 Applied Materials, Applied Materials Automation Software (n.d.).

607853 Pharma 4.0: Advanced Control and Cyber-Physical Security of Continuous Pharmaceutical Manufacturing Pilot-Plant

Abstract:

Currently, Industry 4.0 concepts are being applied to pharma industry to achieve Pharma 4.0 paradigm. Pharma 4.0 reduces the time and resources needed for continuous pharmaceutical manufacturing and also improves the product quality and production consistency. It has many advantages but also have bigger challenges on the advanced process control and cyber-physical security side since the process and CQA's are not only needed to controlled but also need to be protected from any vulnerability, in real time [1-4]. The quality of the pharmaceutical products can be improved significantly via implementing the advanced model predictive control (MPC) system coupled with an RTD based control system if an appropriate cyber-physical security defense is in place. However, much less attention has been paid to implement the integrated control and cyber-physical security system into the continuous pharmaceutical manufacturing plant.

In this work, an advanced model predictive control (MPC) system coupled with an RTD based control system has been implemented in the continuous pharmaceutical manufacturing (CPM) pilot-plant. The CPP's and CQA's are controlled in real time using advanced model predictive control (MPC) system while the none-confirming products are diverted in real time in waste to assure the final CQA's of qualified tablet lots. This bi-layer coupled control strategy assures the final product quality, improves the production efficiency, minimizes the need of off line testing, and facilitated the real time release. The critical control variables that have been controlled in first layer using model predictive control (MPC) system are drug concertation, powder level before tablet press, main and pre compression forces, tablet weight and hardness. A novel control strategy for powder level control in a chute placed in between blender and tablet press unit operation of continuous tablet manufacturing process has been developed, implemented and evaluated. A noninvasive technique based on change in electric field concept has been used for real time monitoring of powder level in continuous manufacturing pilot-plant. The APC has been used to control feeder, blender and tablet press as well [1]. A systematic framework including the methods and tools for real time diversion of tablets have been developed and implemented into continuous pharmaceutical manufacturing process as a second layer assurance of product quality. In CM, the drug concentration is measured in real time before the tablet compaction (chute &/or feed frame) using PAT sensor. The proposed control strategy then uses this inlet concentration to determine a signal for the diversion strategy that can accurately be used to reject tablets that are out of tolerance limits at the outlet of the tablet press. It is also providing a suitable platform to handle the pandemics such as Covid-19 via rapid production of suitable drugs and its real time release, with less time and resources.

A systematic framework including the methods and tools have been also developed for proactive identification and mitigation of potential cyber-physical attack risk on CPM. The cyber-physical security relevant software tools such as Snap 7, Wireshark, and Tripwire have been applied to CPM. A novel software tool named CPS (Cyber-Physical Security) has been developed for cyber-physical security of the continuous pharmaceutical manufacturing. The integrated commercially available and developed (in house) cyber-physical security tools have added an extra layer of security of our continuous pharmaceutical manufacturing pilot-plant for any unexpected attacks. All the relevant data generated during continuous manufacturing has been systematically collected, stored and organized in a data hub (OSI PI) and cloud system as per industry 4.0 standard.

The objective of this presentation is to demonstrate the performance of integrated MPC and RTD based bi-layer control strategy together with cyber-physical security system implemented into our continuous pharmaceutical manufacturing pilot-plant facility.

References

1. Bhaskar, A., Barros, F. N., Singh, R. (2017). Development and implementation of an advanced model predictive control system into continuous pharmaceutical tablet compaction process. International Journal of Pharmaceutics, 534 (1-2), 159-178.
2. Bhaskar, A., Singh, R. (2018). Residence time distribution (RTD) based control system for continuous pharmaceutical manufacturing process. Journal of Pharmaceutical Innovation. DOI: 10.1007/s12247-018-9356-7.
3. Singh, R. (2019). Systematic framework for implementation of RTD based control system into continuous pharmaceutical manufacturing pilot-plant. Pharma. Issue 34, 43-46.
4. Singh, R. (2020). The cyber-physical security of pharmaceutical manufacturing processes. Pharma., Issue 38, 53-57.

600809 Web Deployment of Mechanistic Models for Pharmaceutical Applications: Case Studies in Spray Drying and Distillation

Abstract:

In recent years, mechanistic models have been applied to a broad range of applications in pharmaceutical R&D and Engineering, improving efficiency through reduced data requirements, facilitating scale-up and tech transfer, and enabling virtual design space exploration. Despite their successes in these areas, mechanistic models have typically been utilized by a small group of modeling experts, who are not necessarily the key stakeholders in the answers that the models provide. As a result, scientists and engineers who are not themselves modelers have limited access to these tools and are less likely to take advantage of their full potential. Some common barriers to adoption include a lack of training, support and know-how, a high amount of effort required to establish a fit-for-purpose model-based solution, and a limited access to software for installation. Some recent efforts in the modeling community have aimed at "democratizing" models – reducing the barriers to adoption for those without modeling expertise. Web-based deployment of simple modeling tools is proposed as a potential solution to this problem.

In this work, model deployment for pharmaceutical applications to scientists and engineers will be discussed in the context of two case studies performed at Eli Lilly: a spray drying operation and a batch distillation and collection process. The modelling workflow requirements of scientists and engineers without prior modeling experience will be compared with those of modeling experts. For each application, a fit-for-purpose model was developed and deployed as a simple web interface to answer targeted questions of interest to scientists and engineers. Outcomes of the deployment exercise will be presented, and future perspectives on democratization of mechanistic models for pharmaceutical applications will be discussed.

601121 Kinetics of Friedel Crafts Acylation of 2-Methylfuran with Octanoic Anhydride and Al-MCM-41.

Abstract:

Friedel Crafts acylation reaction of aromatic substrates with different acylating agents produces ‘aryl alkyl ketones’. The acylation reaction of biomass derived furans and long-carbon chain (C_{12}-C_{20}) fatty carboxylic acids or anhydrides obtained by hydrolysis of plant -oil based triglycerides produces ‘alkyl furan ketones’. These ketone molecules are important precursors for the synthesis of renewable oleo-furan sulphonate (OFS) surfactants as demonstrated by Park et al. [1]. The exceptional stability of these OFS surfactants for use as detergents in hard water as well as other comparable detergent characteristics makes them a useful alternative to the petroleum-based surfactants. In this work, we elucidate the reaction kinetics of Friedel Crafts acylation reaction of 2-methylfuran and n-octanoic anhydride as a model system with Al-MCM-41 as a solid Brønsted acid catalyst in the presence of a non-protic solvent heptane. The apparent activation energy and the reaction rate orders are obtained in a flow reactor set-up. The experimental observations are explained using a potential mechanism and a kinetic model. The understanding of this C-C coupling reaction using long chain fatty acid derivatives will contribute to the scale-up of OFS surfactants as well as to the design of heterogeneous Friedel Crafts acylation catalysts.

Reference:

1. Park, D. S., Joseph, K. E., Koehle, M., Krumm, C., Ren, L., Damen, J. N., Shete, M. H., Lee, H. S., Zuo, X., Lee, B., Fan, W., Vlachos, D. G., Lobo, R. F., Tsapatsis, M. and Dauenhauer, P. J. ACS Cent. Sci. 2, 820–824 (2016).

604926 Advances in Catalytic Hydroprocessing for the Production of High Quality Lube Base Stocks

Abstract:

The lube manufacturing industry has benefited from remarkable advances in catalytic hydroprocessing technology. ExxonMobil and its affiliates worldwide have considerable technical know-how and experience in lubes manufacturing, and more than 75 years of rich history in applied catalysis. At the heart of the MSDW™/MAXSAT™ process for catalytic hydroisomerization and hydrofinishing, and the MWI™ process for wax conversion, are systems of proprietary, shape-selective catalysts containing zeolites and noble metals. These catalysts display superior activity and selectivity, and unsurpassed tolerance to various contaminants.

ExxonMobil has been a leader in applying state of the art technology in recent and current projects to meet the growing need for high quality basestocks. Recent ExxonMobil EHC™ Group II base stock investments include start-up of the Rotterdam hydrocracker unit early this year, and an upgrade to Group II quality to be completed in Singapore in the second quarter of this year. Together, these projects utilize the latest proprietary ExxonMobil catalyst and process technologies. Earlier this year, ExxonMobil also indicated that it had made a final investment decision on a multi-billion dollar expansion of its integrated manufacturing complex in Singapore that will produce higher-value products and expand lubricant base stocks production to meet growing market demand. That project will utilize a unique combination of proprietary ExxonMobil catalyst and process configuration technologies to upgrade and convert fuel oil into higher-value products. As part of the expansion, ExxonMobil will introduce a new, unique high viscosity Group II base stock into the marketplace.

This presentation will overview the latest ExxonMobil catalyst and process technologies being deployed in these projects, including state of the art, high activity Celestia hydrotreating catalyst. It will also highlight how these technologies are combined to capture very significant value from their superior productivity, selectivity, run length, and ability to handle more challenged feedstocks.

Catalytic Evaluation of Sulfated Metal Oxide Catalysts for the Solvent-free tert-Butylation of Phenol

***Adam Zuber**, Aditya Khandare, Shreyas Acharya, Justin Marlowe and George Tsilomelekis**

Department of Chemical and Biochemical Engineering, Rutgers, The State University of New Jersey, 98 Brett Road, Piscataway, NJ 08854

* Corresponding Author (Email: g.tsilo@rutgers.edu)

The transition from classical homogeneous catalysis to environmentally friendly, heterogeneous alternatives presents a unique challenge. Tert-butylation of phenol is a characteristic example due to its importance in industrial applications. An estimated 450,000 tonnes per year of tert-butyl phenols (TBP) are manufactured industrially for production of innumerable chemical commodities [1]. 2-TBP is utilized in production of pesticides and fragrances; 4-TBP functions as a flavoring agent and petroleum additive, among other compounds. However, the reaction is classically carried out in the presence of liquid acid catalysts [2], inducing economic and environmental concerns [3].

Numerous research groups have carried out this reaction using heterogeneous catalysts in both vapor and liquid phases and in batch and continuous flow systems, though few have managed to perform tert-butylation of phenol under conditions relevant to industrial applications. As phenol alkylation with tert-butyl alcohol (TBA) is catalyzed by Bronsted acidity, it is believed that solid superacid metal oxide catalysts with surface-bound sulfate species may be appealing.

At low pressure and temperature, sulfated metal oxide catalysts as well as Amberlyst ® 15 exhibited relatively high activity for tert-butylation of phenol. The observed activity for these catalysts follows the trend Amberlyst ® 15 > SnO_2/SO_4^{2-} > TiO_2/SO_4^{2-} > ZrO_2/SO_4^{2-}. At complete conversion, selectivity was observed to follow the same trend: ~60-65% 2-TBP, ~30-35% 4-TBP, 3-5% 2,4-TBP, and minor amounts of 2,6-TBP and 2,4,6-TBP. Production of TBPE, the O-alkylated product, was observed at very short reaction times. Catalytic regeneration and reuse were also evaluated. Remarkably, it was found that Amberlyst ® 15 could be regenerated simply by washing with ethanol and drying at 110°C; reuse was achieved 3 times without loss of conversion, quite contrary to the sulfated metal oxides. Given the high conversion of the alkylating agent and selectivity towards mono-alkylated products, the catalysts presented here offer an environmentally-friendly alternative to traditional homogeneous catalysts.

References:

[1] Modrogan E, Valkenberg MH, Hoelderich WF. J Catal 2009;261:177.
[2] Adam, F., and C. Kueh. J Taiwan Inst Chem Eng, vol. 45, no. 2, 2014, pp. 713–723.
[3] Matloubi MF et al. Trans C Chem Chem Eng 2009;16:81.
[4] Li, Huaju, et al. Journal of Energy Chemistry, vol. 24, no. 2, 2015, pp. 239–244.
[5] Patel, A. et al. Journal of the Chemical Society - Faraday Transactions 1997, 93, 347-353.

606907 Artificial Neural Network Aided Scale-up of a Multi-Tubular Reactor for the Oxidative Dehydrogenation of Butylene to Butadiene

Abstract:

Butadiene, a key raw material for the production of synthetic rubbers, and intermediate chemicals is conventionally manufactured via an energy intensive endothermic conversion of Butane. This process results in catalyst deactivation and low selectivity. Hence, an alternative, exothermic butadiene production pathway was proposed and is being studied extensively. In this study, an Artificial Neural Network (ANN) aided scale-up of a multi-tubular shell and tube reactor for the synthesis of Butadiene via the oxidative dehydrogenation of butylene was executed. A three-step solution strategy was adopted for the study. The step 1 comprised rigorous mathematical modelling of a lab-scale multi-tubular reactor using OpenFOAM. The developed model was used to generate data under varying process conditions (Feed composition and flowrates, coolant velocity, and temperature) for training an ANN surrogate model. The CFD results were validated against experimental data with minimal errors of less than 5%. In step 2, the generated data was fed to the ANN model for training and optimization. The optimized operating conditions were then used as inputs in the CFD model for simulation and cross-validation. Subsequently, the third and final step was implemented. In this step, the lab-scale multi-tubular reactor was scaled-up using the theory of dimensionality (π-theory), conventional strategies, and domain expert knowledge. Several parametric studies were carried-out by varying the baffle height, number and spacing, tube arrangement, number, length and diameter as well as the shell side length and diameter. Each of these case studies were simulated for data generation using the CFD model by leveraging expertise in advanced meshing techniques to reduce the computational mesh density without compromising the accuracy and numerical stability of the simulation results. The generated data was again fed to an ANN model for optimization. The optimized reactor configuration was then designed and simulated with the CFD model for comparison and validation.

601115 A Paradigm Shift in Catalyst Stability for Toluene Alkylation with Methanol

Abstract:

One of the commercially most important aromatic compounds is *p*-xylene, which is used for the production of various fine chemicals. Due to its cheap feedstocks, toluene alkylation with methanol is one of the most commercially attractive routes to produce *p*-xylene; however, challenges such as low catalyst stability and single pass *p*-xylene yield need to be addressed. Here we present a study of the catalytic properties of MWW-type zeolite catalyst, for toluene alkylation with methanol under high pressure conditions (600 psia).

To understand the catalytic behavior of MWW, we deconvoluted structure-function relationships for different topological features (supercages, sinusoidal channels, and external surface pockets) having profound impact on catalyst performance. Deconvolution was achieved by selectively passivating the active sites in specific pore systems, and then comparing their activity with the parent MWW catalyst.

Under high pressure operating conditions, the catalyst life increases multi-fold at high toluene conversion with side reactions such as methanol-to-hydrocarbon reactions being suppressed in the absence of water or hydrogen co-feed. Interestingly, the selectivity of xylene isomers changes significantly during the course of reaction, from thermodynamic equilibrium mixture to 55% *p*-xylene selectivity.

The combination of structure-deconvoluting experiments and DFT calculations reveals that the active sites in surface pockets are unselective and fast isomerization on external surface leads the reaction towards the thermodynamically determined xylene product distribution, while microporous sinusoidal channels and/or supercages preferentially form *p*-xylene. Overall, this study introduces a commercially viable route for *p*-xylene production as well as new insight into the reaction mechanism over MWW zeolite catalysts to aid in the development of optimized processes for various alkylation reactions.

604652 Activity Enhancement and Kinetics of NH_3 oxidation on Pt/Al_2O_3:Rate Enhancement By Ball Milling and Structure Senstivity Effects

Abstract:

Ammonia-oxidation(AMOX) over Pt-containing catalysts is an important reaction for eliminating NH_3 in diesel-engine exhaust. Kinetic models have been developed for Pt-single crystals under ultra-high vaccum[1] and for Pt-gauze for HNO_3 production[2,3] but these are of limited value for NH_3-oxidation on Pt-supported catalysts for emission control. Here we describe a kinetic study for Pt/Al_2O_3 at atmospheric pressure and the development of a microkinetic model. We measured reaction-rate(TOF) dependence over wide range of NH_3 concentration from 10 – 40,000ppm for Pt/Al_2O_3 powder and washcoated-monoliths. The data in Figs.1(a) and (b) show the presence of rate-maximum at 500ppm for unmilled-Pt/Al_2O_3 but at 10,000ppm for ball-milled Pt/Al_2O_3 powder and washcoated-monolith. Another unexpected observation is the enhanced activity of Pt/Al_2O_3 samples upon ball-milling, which resulted in lowering of light-off temperature(T_{50}) by 50°C[Fig.1(c)]. Ball-milling reduces particle size, in-turn decreases the extent of diffusion-limitations. However, steady-state AMOX done over ball-milled Al_2O_3 followed by Pt addition did not show the same enhanced activity ruling out diffusion-limitations as the main cause of the activity enhancement. Further characterization of Pt/Al_2O_3 samples using XRD[Fig.1(e)] reveals an alteration in the distribution of Pt crystalline-planes; notably, Pt<211> increased and Pt<111> decreased. These data suggest that the higher activity of ball milled-Pt/Al_2O_3 is a result of this crystalline-plane transformation. Microkinetics by Rebrov[1] assumed Pt<111>plane which predicted no rate variation with NH_3 concentration. This is due to NH_3 and O_2 adsorbing on independent sites 'b' and 'a' respectively. Contrary, experiments showed '+' to '-' NH_3 reaction-order variation over wide range of NH_3 concentration indicates site-competition[Fig-1(d)]. The model was modified by introducing site-competition between NH_3 and O_2 on site-a in addition to NH_3 adsorption on site-b which best described the experimental data and kinetics. These findings show the close dependence of kinetics and reactivity on Pt crystal-planes.

References:

[1]Rebrov.E.V et al.,Chem.Eng.J,90,61-76(2002)
[2]Hass.M et al.,ISCRE-25(2018)
[3]Baerns.M et al.,J.Catal,232,226-238(2005)

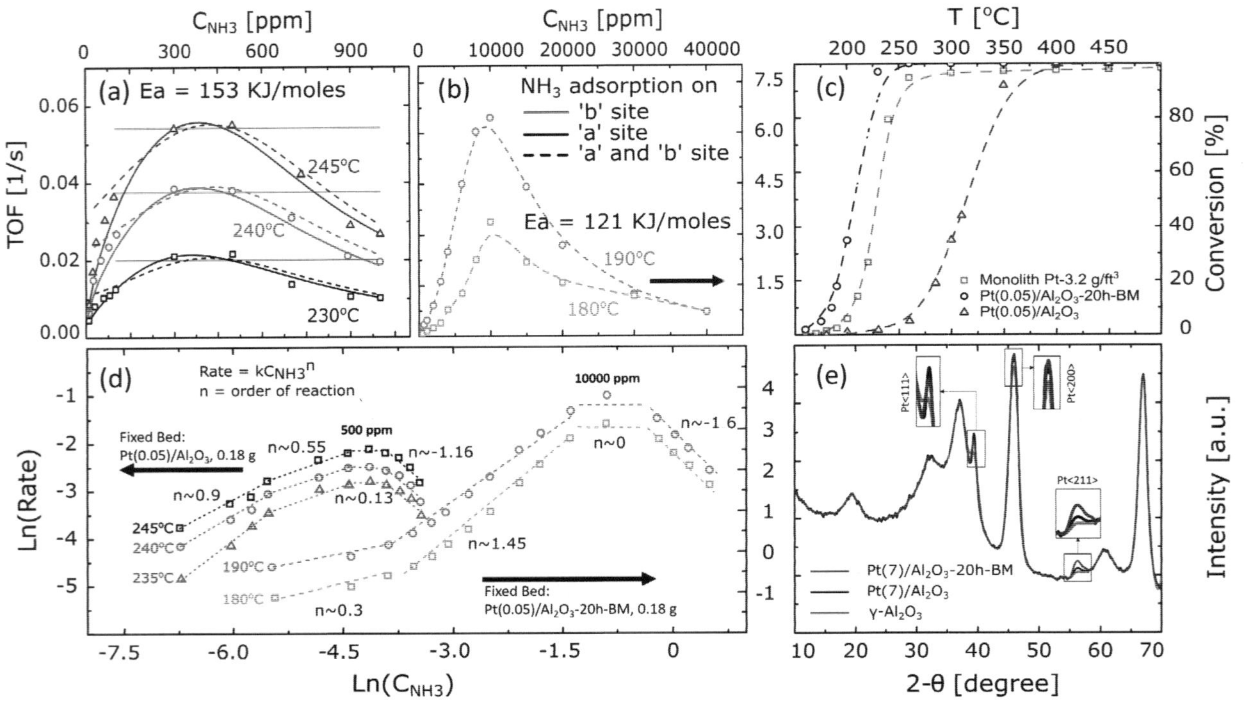

Figure 1: Reaction rate (TOF) against NH_3 concentration over (a) Un-milled Pt(0.05)/Al_2O_3 (b) Ball-milled Pt(0.05)/Al_2O_3 for 20h [Pt(0.05)/Al_2O_3-20h-BM] and Pt-3.2g/ft^3 washcoated monolith, (c) Steady state NH_3 conversion over Pt(0.05)/Al_2O_3, Pt(0.05)/Al_2O_3-20h-BM, Pt-3.2g/ft^3 washcoated monolith with model simulation based on NH_3 adsorbing on site 'a' and 'b'(d) Reaction order variation with respect with NH_3 for un-milled and ball milled Pt/Alumina (e) Powder XRD patterns for Al_2O_3 [support], Pt(7)/Al_2O_3 and Pt(7)/Al_2O_3-20h-BM. [Feed: NH_3 ppm + 10% O_2 + balance Argon, 1000 sccm, Fixed bed: 0.18 g of catalyst, Monolith: 400 cpsi – 0.5 cm, Pt(wt%)/Al_2O_3]. *Lines: Simulation in Figures 1(a, b,c), Symbols: Experimental data*

611565 Novel Multifunctional Organocatalysts and Continuous Flow Platform for Sustainable Chemical Transformation

Abstract:

The use of multifunctional organocatalysts and continuous flow platform are commonplace in modern chemical transformation.[1-4] Herein, we describe a method for immobilization of trifunctional organocatalysts on porous composite hollow fiber and demonstrate their application as heterogeneous catalyst and continuous-flow microfluidic reactor for chemical transformation. The polyamide imide hollow fibers (PAIHF's) are functionalized with aminosilanes and bromine source to immobilize covalently hydrogen-bond donor groups (-OH and -NH), and nucleophilic [Br^-] species upon the fiber surface and provide trifunctional acid-base-nucleophilic organocatalysts and microfluidic reactors. The cooperative effects of Br/APS/PAIHF trifunctional organocatalysts are elucidated in the CO_2 cycloaddition and hydroxyalkylation of aniline under batch and continuous flow synthesis. Our results indicated that the synergistic cooperative effect of trifunctional organocatalyst on PAIHF leads to a maximum 1-(phenylamino)propan-2-ol selectivity of 97.1% at 61% aniline conversion and 0.02 cm^3/min flow rate. While knowledge of the acid-base-nucleophilic trifunctional cooperativity is still limited, these findings demonstrate useful structure-property trends that can be used to design more efficient organocatalysts for sustainable chemical transformation.

References:

1. A.-A. Alwakwak, Y. He, A. Almuslem, M. Senter, A. K. Itta, F. Rezaei, A. Rownaghi, Metal- and solvent-free synthesis of aminoalcohols under continuous fl ow conditions, *React. Chem. Eng.* 5 (**2020**) 289-299.

2. Y. He, F. Rezaei, S. Kapila, A.A. Rownaghi, Engineering Porous Polymer Hollow Fiber Microfluidic Reactors for Sustainable C-H Functionalization, *ACS Appl. Mater. Interfaces.* 9 (**2017**) 16288–16295.

3. Y. He, A. Jawad, X. Li, M. Atanga, F. Rezaei, A.A. Rownaghi, Direct aldol and nitroaldol condensation in an aminosilane-grafted Si/Zr/Ti composite hollow fiber as a heterogeneous catalyst and continuous-flow reactor, *J. Catal.* 341 (**2016**) 149–159.

4. Y. He, A.K. Itta, A. Alwakwak, M. Huang, F. Rezaei, Aminosilane-Grafted SiO2-ZrO2 Polymer Hollow Fibers as Bifunctional Microfluidic Reactor for Tandem Reaction of Glucose and Fructose to 5-Hydroxymethylfurfural, *ACS Sustain. Chem. Eng.* 6 (**2018**) 17211–17219.

614721 Solvent -solute interaction and viscosity effects on thermodynamics and kinetics of crystallization

Abstract:

Research Interests

The structuring of solvent at the crystal-solvent interface largely regulates the activation barrier for incorporation of solute into the crystal. Depending on the groups exposed on the crystal surface, solvent at the interface may be ordered or disordered. In general, consequences of solvent structuring on the molecular level interactions comprising crystallization are poorly understood. Organic solvents, in which the bonds that could support a defined fluid structure are weak and varied, present a particular challenge to understand the effect of solvent structuring and solvent-solute interactions on thermodynamics and kinetics of crystallization.

We explore the thermodynamic parameters of crystallization for etioporphyrin I (a planar free base porphyrin) in a range of organic solvents from alcohols (octanol, hexanol and butanol), through a polar aprotic solvent (DMSO), to a carboxylic derivative (caprylic acid). Thermodynamic characteristics of crystallization along with absorption spectroscopy were used as probes to decipher the presence of solvent structuring and solute-solvent interactions for non-solvate crystals crystallizing from different organic solvents.

Time-resolved *in situ* atomic force microscopy monitoring of the step dynamics on (010) face of etioporphyrin in two types of solvent, alcohols (octanol, butanol and hexanol) and DMSO, having varying aliphatic chain lengths and subsequent solvent viscosities demonstrate the impact of solute-solvent interactions and solvent viscosity on the barrier for solute incorporation into kinks. The relationship between step kinetic coefficient and solvent viscosity reveals that the crystallization of organic small molecules is not governed by diffusion limitation but rather by solute-solvent interaction at the kinks. The correlation between the step velocity and the solute concentration on the (010) face of etioporphyrin I for different solvents reveal that for solvents belonging to the same homologous series, solute incorporation scales with the solute diffusion coefficient. All-atom MD simulation for solvent structuring at the crystal interface emphasizes the importance of solute-solvent interaction at the crystal interface.

These emerging insights offer guidance for polymorph selection, chiral separations, structure design and numerous other open questions relevant to myriad crystallization systems in the pharmaceutical and chemical industries.

613766 Characterizing and Modeling Pharmaceutical Twin Screw Feeder Mass Flow Rates Using Statistical Time Series Analysis

Abstract:

Screw feeders are the critical first unit operation in continuous manufacturing of drug product (CMDP) processes, influencing the mass flow rate of pharmaceutical powders downstream. Thus, industrial leaders are keen on accurately modeling feeders. Existing flowsheet models simulate the average mass flow rate [1]–[3], neglecting the stochastic behavior of the mass flow. Custom Discrete Element Method models realistically simulate particles' behavior but require prohibitively long computation times to simulate minutes of operation [4]. To better design processes and controllers, a quick-to-solve flowsheet model that can simulate the stochastic nature of real screw feeders is necessary. This is precisely the gap addressed by this work.

This work describes the novel characterization and modeling of the stochastic nature of a screw feeder's mass flow rate using statistical time series analysis and a deterministic flowsheet model. First, experimental data was used to estimate the parameters of a hybrid mechanistic-empirical screw feeder model, based on Bascone et al. 2020 [3]. Next, the stochastic residual of the mass flow was isolated by subtracting the flowsheet model's deterministic mass flow from the feeder-reported mass flow. Then, each experiment's stochastic residual was fit to an autoregressive moving average model (ARMA) [5], characterizing the mass flow variation. Finally, a predictive model mapping powder properties and operating conditions to ARMA model parameters was developed. This predictive model was integrated with our deterministic feeder model, yielding a novel mechanistic-empirical-stochastic flowsheet model that simulates realistic, high-variance mass flows and is suitable for the development of CMDP processes and controllers.

Research Interests: Dynamic Systems, Multivariate Analysis, Constrained Regression, Optimization, and Technical Software Development

[1] Y. Yu, "Theoretical modelling and experimental investigation of the performance of screw feeders," PhD thesis, 1997.

[2] F. Boukouvala, V. Niotis, R. Ramachandran, F. J. Muzzio, and M. G. Ierapetritou, "An integrated approach for dynamic flowsheet modeling and sensitivity analysis of a continuous tablet manufacturing process," *Computers & Chemical Engineering*, vol. 42, pp. 30–47, 2012.

[3] D. Bascone, F. Galvanin, N. Shah, and S. Garcìa-Muñoz, "A hybrid mechanistic-empirical approach to the modelling of twin screw feeders for continuous tablet manufacturing," *Industrial & Engineering Chemistry Research*, 2020.

[4] P. Toson and J. G. Khinast, "Particle-level residence time data in a twin-screw feeder," *Data in brief*, vol. 27, p. 104672, 2019.

[5] G. E. P. Box, G. M. Jenkins, G. C. Reinsel, and G. M. Ljung, *Time series analysis: Forecasting and control*. John Wiley & Sons, 2015.

613987 Understanding Second-Shell Catalyst Interactions for Enhancing the Performance of Arylborane-Catalyzed Epoxide Ring-Opening

Abstract:

Research Interests

- Homogeneous and Heterogeneous Catalysis
- Catalyst Synthesis
- Catalytic Mechanisms and Chemical Reaction Engineering

Introduction

Epoxide ring-opening is a key reaction for synthesizing a range of lab-scale and commercial compounds. A major application is the industrial production of polyols, which are primary raw materials for polyurethane plastics. This process involves the catalytic addition of an epoxide to an alcohol-based initiator, resulting in the formation of a β-alkoxy alcohol, which can further react with additional epoxide molecules to yield the desired polyol. Industrially utilized epoxides such as propylene oxide usually create a mixture of 1°-OH and 2°-OH ring-opening products (figure 1). The latter, however, exhibit poor reactivity for certain polyurethane applications and require substantial reprocessing prior to usage. This tedious step can be avoided by employing regioselective catalysts that predominantly yield 1°-OH products. However, nearly all conventional catalysts are unselective except for $B(C_6F_5)_3$, which exhibits high selectivity towards 1°-OH products. This work explains its unique behavior in epoxide ring-opening, through the formation of catalytically active adducts with H_2O[1] and alcohols[2] due to its strong Lewis acidity. Further on, we describe the influence of hydrogen bonding additives on the regioselectivity of the $B(C_6F_5)_3$–H_2O adduct and demonstrate a route to enhance its selectivity through the addition of α, β-diols.[3]

Methodology

Batch runs were performed in 20 mL vials on a hot shaker. 1 mL o-Xylene (G.C. I.S.) was added to 0.1 mol%/1 eq. $B(C_6F_5)_3$, with 5-20 eq. of an additive. The contents were heated, and the reaction started by adding 1 mL/1000 eq. 1,2-epoxyoctane (test epoxide) and 2 mL/4000 eq. of the corresponding alcohol. Progress was monitored using GC-FID and H_2O levels were determined via Karl Fisher titrations.

Selected Results

After studying this system at different conditions, we observed two unusual trends – reaction rates significantly varied with alcohol type (1°,2°,3°)/H_2O levels and reaction selectivities increased with conversion. After rejecting deactivation as a factor, we used DFT predictions to propose the catalytic activity $B(C_6F_5)_3$–H_2O/ROH adducts. Then, combining experimental data with theory, we developed microkinetic models which supported the simultaneous existence of Lewis acid, alcohol-mediated and water-mediated catalytic pathways (figure 2). Additionally, figure 3 also depicts the interesting trend where selectivity increases as a function of conversion in the absence of any additives. Ruling out product consumption as a factor, we proposed that reaction products P1&P2, were interacting with the $B(C_6F_5)_3$–H_2O adduct via hydrogen bonding and enhancing selectivity of the species. After verifying this hypothesis by adding co-catalytic amounts of P1 and P2, we sought to find external additives which could be used to deliberately enhance regioselectivity. Of all our target molecules, certain planar α, β-diols were most effective and presented a hydrogen-bonded complex which directed the incoming nucleophile in a manner that gave high selectivities.

Key Conclusions

This work demonstrates the existence of novel alcohol and water-mediated catalytic pathways for $B(C_6F_5)_3$. As these have not been well-studied for similar highly Lewis acidic catalysts, this merits further work. Moreover, having demonstrated the influence of hydrogen bonding on reaction selectivity, we were able to enhance selectivity through the judicious addition of α, β-diols which could also extend to other such highly Lewis acidic catalysts.

Literature Cited

1. Yu, Y., et al., "Mechanism of regioselective ring-opening reactions of 1, 2-epoxyoctane catalyzed by tris (pentafluorophenyl) borane: A combined experimental, DFT, and microkinetic study," ACS Catalysis, 8 (12), pp. 11119-11133 (2018).
2. Bennett, C.K., et al., "Strong Influence of the Nucleophile on the Rate and Selectivity of 1,2-Epoxyoctane Ring-Opening Catalyzed by Tris(pentafluorophenyl)borane, $B(C_6F_5)_3$," ACS Catalysis, 9 (12), pp. 11589-11602 (2019).
3. Bhagat, M.N., et al., "Enhancing the regioselectivity of $B(C_6F_5)_3$-catalyzed epoxide alcoholysis reactions using hydrogen bond acceptors," ACS Catalysis 9 (10), pp. 9663-9670 (2019).

Acknowledgements

The authors would like to thank the Dow Chemical Company for its financial support towards this work.

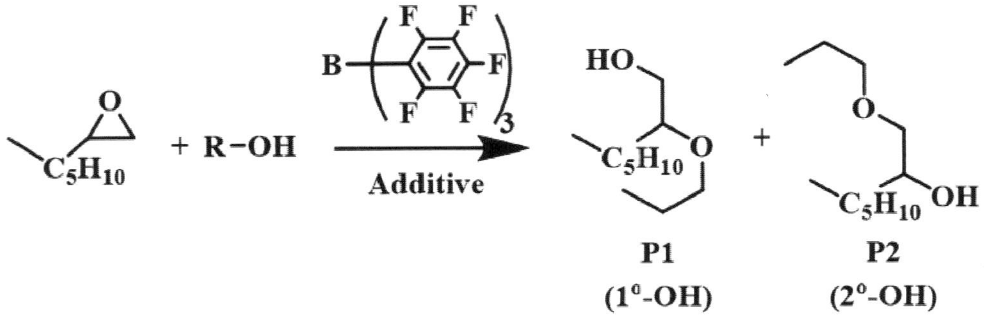

Figure 1: Ring-opening of epoxides to yield polyols P1 and P2

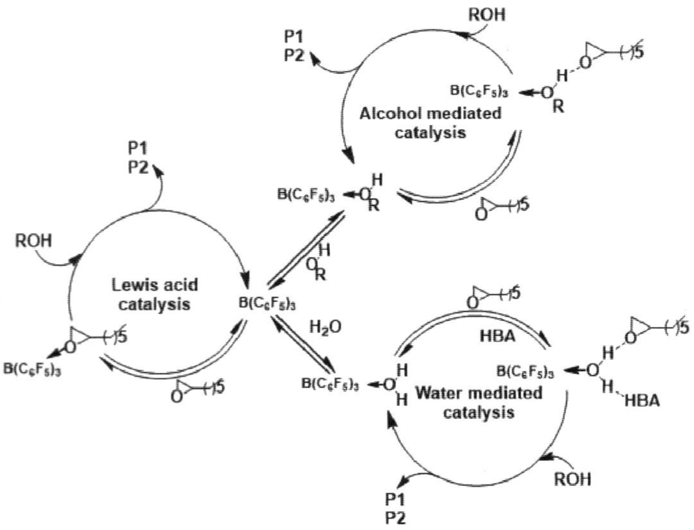

Figure 2: Proposed catalytic cycles for $B(C_6F_5)_3$-catalyzed epoxide ring opening

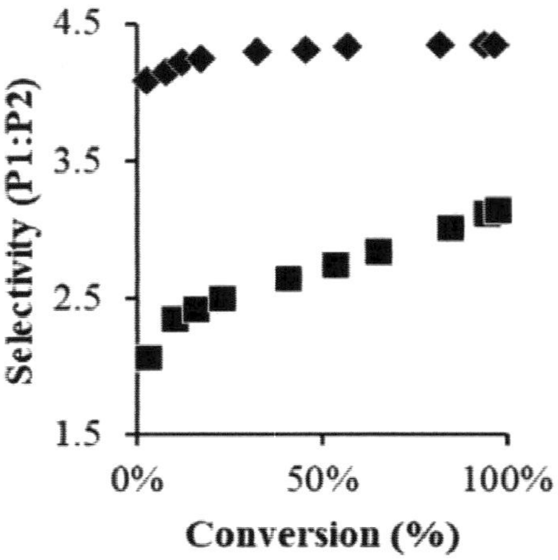

Figure 3: Enhancement in reaction selectivity with 20 eq. 1,2-propanediol (◆) in contrast to an experiment lacking additives (■)

614013 Molecular Simulations to Characterize the Assembly and Transport of Biomolecules in Solution and at Lipid Interfaces

Abstract:

Research Interests

A significant challenge in designing drug and gene delivery agents is understanding how their chemical structure and properties affect interfacial properties at the cell membrane and dictate their transport across the cell membrane. My research interest lies in developing multiscale simulations and computational techniques to understand the molecular mechanisms of how the structure of biomolecules affect: (1) their transport pathways across the cell membrane, and (2) their interaction with other biomolecules. In summary, my research focuses on utilizing the fundamentals of biophysics and chemical engineering along with computational tools to provide faster screening tools (e.g. screening of drugs for properties such as permeability across the lipid bilayer) and better designs for drug and gene delivery agents.

Research Experience

My research is focused on developing multiscale molecular dynamics (MD) simulations, state-of-the-art enhanced sampling techniques, and Python-coded analysis tools for two subareas.

(1) I study the translocation (or "flipping") of charged peptide loops across the lipid bilayer that is valuable not only to design small cationic peptides for applications in gene/drug delivery but also to understand the large-scale conformational rearrangements that involve multiple charge translocation events in naturally occurring integral membrane proteins. I developed all-atom enhanced sampling simulations (i.e. Temperature Accelerated Molecular Dynamics) protocols to predict the likelihood of loop flipping without predefining a specific translocation pathway. Further, I run the string method with swarms of trajectories to find the minimum free energy path for translocation, which gives insights into engineering materials capable of bypassing the cell membrane.

(2) I study the amphiphilic behavior of bacterial signaling molecules that will allow us to understand their behavior in different biological environments that may help to control bacterial group behaviors such as virulence factor secretion, and biofilm formation. I predict the impact of their chemical structure on their aggregate morphology using alchemical free energy calculations using all-atom MD simulations. I then investigate their interactions with other amphiphilic biomolecules, such as lipids using coarse-grained MD simulations. I further collaborate with two experimental research groups at UW-Madison to gain better understanding of the process by validating simulation results with the experimental results and developing accurate prediction models.

My research experience has led me to work as an intern at Lawrence Livermore National Laboratory, where I systematically studied the effects of different parameters for calculating the free energy barrier of drug translocation across the lipid bilayer, which led to faster screening of drugs for permeability.

614277 Enabling Selective Conversion of Biomass Derived Oxygenates to C4-C5 Dienes

Abstract:

Research Interests - Catalysis, Process Chemistry, Synthesis and characterization of materials, kinetics, process modeling

The dependence of our civilization on plastics cannot be overstated – in 2018 alone, the worldwide consumption of plastics exceeded 300 million tons *(~40 kg per person on earth)*. The carbon backbone of plastics is composed of monomeric hydrocarbons like C4 and C5 dienes. Diene market is valued at nearly $4 billion, and alongside steel and ammonia, diene production is integral to the growth of economies. These chemicals (butadiene, pentadienes, and isoprene) are currently produced by the cracking of naphtha fraction of crude oil, where ethylene is the major product. With a recent shift towards using shale-gas (primarily composed of light molecules like methane and ethane) as a cheap alternative to crude oil for ethylene production, refineries are decoupling production routes of ethylene and larger fractions including dienes. Consequently, butadiene production in the US has dropped by *ca.* 30% since 2007. As demands continue to rise and supplies fall, other on-purpose and possibly renewable routes that are not coupled to ethylene production are required.

One such chemical pathway utilizes the tandem-ring opening dehydration (hereinafter referred to as *dehydra-decyclization)* of biomass-derived saturated furanic species to produce these dienes **(Scheme 1)**. For example, four-carbon tetrahydrofuran (THF) reacts on the surface of a solid Brønsted acid catalyst to produce butadiene, while five-carbon methylated tetrahydrofuran (2-MTHF) yields linear pentadienes. Limited prior investigations of this chemistry had not rigorously tackled the three fundamental catalyst design challenges, namely assessing: rates (*how fast the reaction occurs per active site of the catalyst*), diene selectivities (*how much of fed-in carbon ends up in desirable product/s),* and lifetime *(catalyst stability under reaction conditions)*. In addition, the origin of side products (propene, formaldehyde, and alkylated aromatics), as well as the identity of key reaction intermediates were unknown. The breadth of accessible dienes would be afforded by our ability to prepare numerous substituted THF precursors, and yet it remained unclear how these substituents tune the chemistry of diene formation. Due to this lack of mechanistic knowledge, the maximum diene selectivities attainable in the chemistry had remained low (< 60%); increasing these selectivities would drive down energy-intensive downstream separation costs, and potentially lead to economic feasibility of the overall process. My dissertation has focused on gaining fundamental insights into the dehydra-decyclization of saturated furans to guide a rational search for active, selective, and stable catalysts for this chemistry.

Below, I highlight the results from three independent studies that detail my efforts in i) designing a chemical reactor capable of automated high-throughput catalytic evaluation, followed by ii) detailed kinetic evaluation of THF dehydra-decyclization to expound the catalytic mechanisms and pathways for the chemistry, and finally iii) the design and utilization of a selective and stable class of catalysts exhibiting ~89% diene selectivities at complete furan conversion.

1) *Design and implementation of a chemical reactor for automated kinetic evaluation* : Detailed mechanistic interrogations on a chemistry are preceded by high-throughput screening of a large

number of catalysts to find promising candidates and to establish rate, selectivity, and stability trends to aid this search. Typical laboratory-scale packed bed reactors (PBRs) require constant human monitoring owing to multiple independent process control elements, and separate and non-coupled reaction and separation/quantification components, making them unsuitable for high-throughput studies. Motivated by the ability to perform high fidelity kinetic studies with minimal human oversight, I constructed and implemented a method to integrate a typical PBR within a Gas Chromatograph (GC), enabling the control of reaction parameters directly from the GC circuitry **(Scheme 2)**. After carrying out reactor validation tests (e.g. residence time distributions, isothermal operation, absence of heat and mass-transfer limitations), it was found that typical reaction parameters can be obtained with reasonable accuracy with nearly no manual intervention on this low-cost setup. This automated reactor setup has since helped in the successful screening of promising catalysts for THF dehydra-decyclization (as discussed in 3)).

2) *Highlighting the mechanisms and pathways for diene formation* : We studied THF dehydra-decyclization on an aluminosilicate zeolite H-ZSM-5 by a combination of first-principle calculations, microkinetic modeling, and experimental kinetic measurements. As part of this collaborative effort, my specific contributions to this work were three-fold; *i) Experimental kinetic measurements for the chemistry:* Assessing THF conversion rates enabled the identification of two key deleterious pathways compromising diene selectivities, namely the fragmentation of THF to propene, and C-C condensation pathways to alkylated aromatics. Furthermore, nearly identical activation energies of diene and propene formation pathways guided the search for a common intermediate formed before their respective rate-limiting steps (RDS); *ii) Identification of key reaction intermediates for diene production*: Co-processing potential intermediates with THF and observing changes in the evolution of butadiene and propene enabled the unambiguous identification of the key intermediates for both desirable and undesirable pathways in the chemistry. Furthermore, a > 100x higher rate of intermediate conversion vs. THF allowed us to conclude that the dehydra-decyclization RDS preceded the dehydration of these intermediates; *iii) Experimental identification of the RDS:* C-O bond cleavage leading to the ring-opening of THF was hypothesized to be the kinetic bottleneck as it exhibited a high activation barrier of ~ 49 kcal/mol. By stabilizing the carbenium ion transition state formed during this step by methyl-substitutions on the THF ring, I was able to increase the diene formation rates by > 50x **(Figure 1a)**, experimentally demonstrating the rate-limiting nature of furan ring-opening step for the first time.

3) *Enhancing diene selectivities by weakly acidic catalysts* : Explicating the mechanisms and pathways of THF dehydra-decyclization cultivated a quantitative understanding about the origin of selectivity; dienes resulted from C-O scission, while the primary side product (propene) originated from C-C fragmentation of THF ring, meaning that *the diene selective catalysts should selectively activate C-O over C-C bonds.* I hypothesized one possible 'knob' to tune C-O/C-C bond scission rates to be the Brønsted acidic site strength, as the energetics of proton transfer from the weak acid are likely more favorable towards C-O than C-C, owing to its significantly higher proton affinity. Catalytic evaluation of weakly acidic borosilicate zeolites in different surrounding framework environments consistently exhibited *ca.* 3-30x higher C-O/C-C scission rates than strongly acidic aluminosilicate zeolites **(Figure 1b)**, evincing the inability of weak acids to cleave C-C bonds, and leading to ~10-30% increase in diene selectivities. In addition, borosilicate zeolites converted ~ 3-6x higher amounts of carbon from the feed to dienes during their lifetime than aluminosilicate zeolites **(Figure 1c)**. This remarkable stability was conferred by the inability of these sites to bind and catalyze C-C condensation pathways to alkylated aromatics, which coke up the catalyst and limit their lifetime. The suppression of competing pathways lead to an unprecedented ~ 89% combined linear C5 dienes' yield from 2-MTHF, a 30% improvement from the previous best performing catalysts.

In short, my research has provided key mechanistic insights into a sustainable catalytic route to produce commercially high-volume dienes from biomass-derived feed, which in turn, has enabled the discovery of selective and stable catalysts for this chemistry. As our society gradually moves towards renewable fuels and chemicals production, I believe that the results from this dissertation will be helpful in potential integration of this process chemistry with other hydrocarbon production routes within a biorefinery.

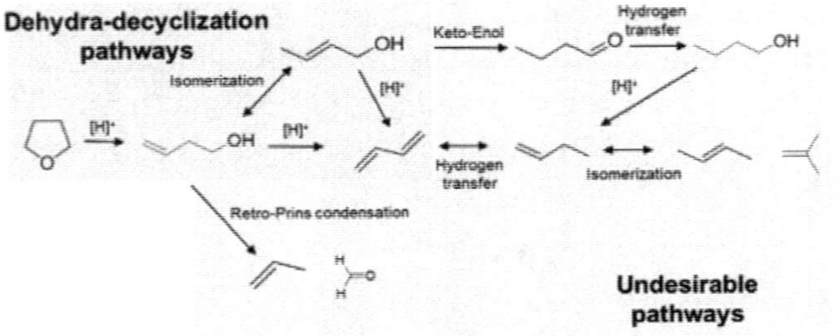

Scheme 1. Known reaction pathways for the conversion of tetrahydrofuran over solid Brønsted acids

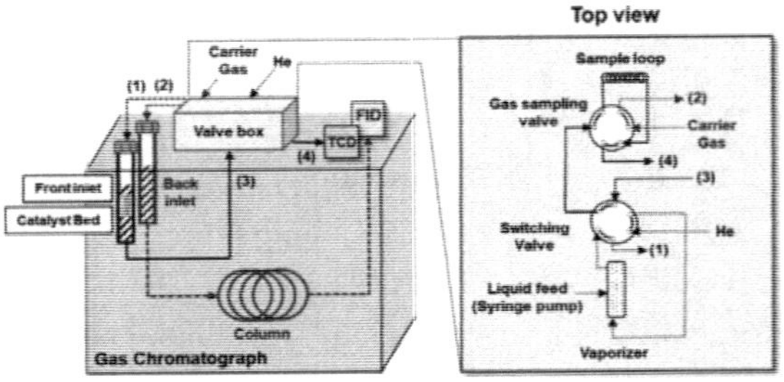

Scheme 2. Automated chemical reactor integrated within a Gas Chromatograph for kinetic evaluation of catalytic chemistries

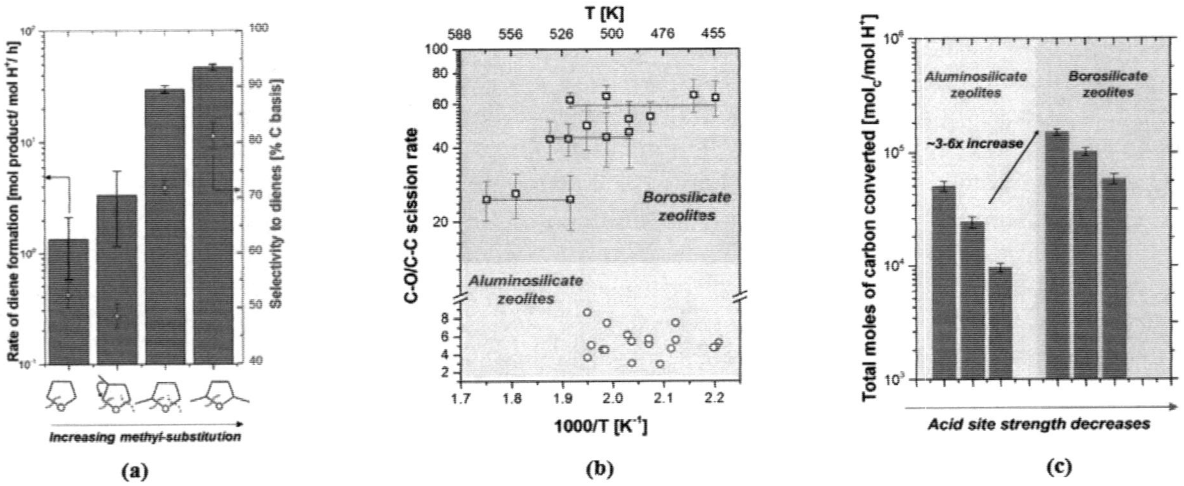

Figure 1. Plots showing (a) increasing diene formation rates observed with increasing number of methyl-substituents on tetrahydrofuran ring; (b) Increased rate of C-O over C-C bond scission in weakly acidic borosilicates compared with strongly acidic aluminosilicates observed over a range of reaction temperatures; and (c) Increased amounts of carbon converted to products over the entire lifetime on borosilicate zeolites compared with aluminosilicate zeolites during 2-methyltetrahydrofuran dehydra-decyclization

614401 Gas-Liquid Separators: Challenges and Uncertainties

Abstract:

Research Interests

The knowledge of mass transfer is a crucial parameter for any separation process. A considerable amount of research has been conducted on the absorption of gas into a liquid, while limited literature is found on the rate of desorption of gas (i.e., gas evolution) from the liquid. The rate of gas evolution is assumed to be instantaneous or within the retention time in the vessel. The rate of absorption and desorption are often assumed to be identical under the same pressure, temperature, and hydrodynamic conditions (assuming no bubbles are formed). The symmetry between absorption and desorption of gas may have been investigated using low viscous liquids. However, in the oil and gas industry, high viscous liquids are also encountered. The incomplete separation of gas can lead to plant shut down, which can cause huge revenue losses. My research interest is aimed towards better understanding the gas-liquid separation at elevated pressures and deconvoluting the factors that influence the rate of gas evolution like viscosity, presence of water in the form of emulsions, presence of solid particulates, and supersaturation ratio. The symmetry between rates of absorption and desorption would be understood as well, along with the factors mentioned.

Postdoctoral Project: Fluid Phase Kinetics – Joint Industrial Project

Under the supervision of Clint P. Aichele, School of Chemical Engineering, Oklahoma State University.

Ph.D. Dissertation: "Bio-Based Active Barrier Material and Packaging Development"

Under the supervision of Maria R. Coleman and Saleh A. Jabarin, Department of Chemical Engineering, University of Toledo/ Polymer Institute, University of Toledo

Research Experience:

Throughout my academic career, I have worked on various application-oriented research. The research areas that I have worked on are the removal of dyes (adsorption), reduction of oxygen permeability through polymers, and factors affecting the rate of gas evolution from supersaturated liquids at elevated pressures. As a result of working on these projects, I have acquired experience in many research areas: water treatment, polymers, nanomaterials, emulsions, surface tensions, and gas-liquid separations. I have become proficient as an experimentalist with an ability to develop a lab that is geared towards addressing these areas of research.

Teaching Interests:

I have been a Teaching Assistant (TA) for various courses, including Process Separation, Fluid Dynamic, Reaction Kinetics, and Unit Operation Lab. For the above course, I have been well appreciated and reviewed by the students. As an instructor, my teaching philosophy would be to embrace an active classroom environment, which encourages question/discussion and solving practical problems to better understand chemical engineering concepts.

Teaching Experience:

Throughout my doctoral program, I have undertaken various TA assignments at the University of Toledo. I have assisted in handling the Unit Operations Lab for 4 semesters and given review lectures for Process Separation and Reaction Kinetics. Having received the Best Teaching Assistant award twice (University of Toledo) reflects my ability as an instructor. I have also mentored graduate and undergraduate students in their research.

Publication:

1. **Michael Angelo Miranda**, Sayeed A Mohammad, Hariprasad J. Subramani, and Clint P Aichele "Kinetics of Gas Evolution from Super-Saturated Oils at Elevated Pressures and Temperatures" Energy & Fuels (2020)
2. **Michael Angelo Miranda**, Ashwin Yegya Raman, David M. Lavenson, Hariprasad J. Subramani, Sayeed A. Mohammad, and Clint P. Aichele, "Gas Evolution Rates in Supersaturated Water-in-Oil Emulsions at Elevated Pressures" Energy & Fuels (2019).
3. Alden B Daniel, Sayeed A Mohammad, **Michael A Miranda,** and Clint P Aichele "Absorption and Desorption Mass Transfer Rates as a Function Pressure and Mixing in a Simple Hydrocarbon System" Journal: Chemical Engineering Research and Design (2019).
4. **Michael Angelo Miranda**, Maria Coleman, Saleh. A. Jabarin "Modification of Polyethylene terephthalate (PET) using linoleic acid for oxygen barrier improvement: Impact of processing methods" Journal of Applied Polymer Science (2017).
5. **Miranda, M.A.**, M. Coleman, and S. Jabarin. Impact of Processing Method and Loading of Active scavenger (Linoleic Acid) on Properties of Polyethylene Terephthalate. in SPE ANTEC™ Indianapolis 2016.
6. **Michael Angelo Miranda**, P. Dhandapani, M. Helen Kalavathy, and Lima Rose Miranda, "Chemically activated Ipomoea carnea as an adsorbent for the copper sorption from synthetic solutions" Adsorption (2010) 16: 75–84.

614405 Rheological Characterization of Cellulose Nanomaterials for Quality Control and Processing

Abstract:

Research Interests

Rheology is the study of flow and deformation of the matter. It is a useful tool to understand the structure-property relationships of materials at various conditions. Fundamental theories and empirical equations are established to connect the rheological properties to the many material properties, which guide formulation screening and process optimization. My research interest lies in the material characterization together with model building and validation to address the challenges in material design and manufacturing.

Research Background

In my PhD study, I use my expertise in rheology to characterize various materials in many applications aiming to screen formulations and optimize processes. My thesis work focuses on rheological characterization of cellulose nanomaterials. [1] To expand the nanomaterials production capacity to industrially relevant scale and to ensure consistent production of the cellulose nanomaterial, one of the most urgent issues to be addressed is the lack of standardized, rapid and reliable characterization method for quality control during the manufacturing process. My research shows the potential of rheology as a quality control tool for these nanomaterials. Preparation and test protocols are first established to obtain reliable viscosity data at manufacturing relevant conditions. Next, a rheological model is developed, which can accurately capture the viscosity across shear rates and concentrations. The model can be used to estimate the concentration of an uncharacterized sample which is much faster than the current method to determine the concentration by drying in oven. Moreover, a flow index is developed to condense large sets of data into one single value, working as a "fingerprint" of the material. This flow index can be used to identify materials at different processing conditions, including different morphologies or surface charges. Finally, a processing method is developed to concentrate cellulose nanomaterials and obtain high loadings of well-dispersed nanomaterials in polymer composite gels. For collaborative projects, I use rheology to screen formulations for solution-based 3D printing [2], composites for fiber spinning, hydrogel fabrication and food formulations for swallowing difficulties study.

[1] Liao, J., Pham, K. A., & Breedveld, V. (2020). Rheological characterization and modeling of cellulose nanocrystal and TEMPO-oxidized cellulose nanofibril suspensions. Cellulose, 1-17.

[2] Zhang, F., Ma, Y., Liao, J., Breedveld, V., & Lively, R. P. (2018). Solution-Based 3D Printing of Polymers of Intrinsic Microporosity. Macromolecular rapid communications, 39(13), 1800274.

614425 Data-Driven Parameter Estimation of Hybrid Models

Abstract:

Data-driven modeling has proven a powerful tool for drawing useful correlations from large datasets, yet the ability to generate physical insights from purely data-driven models remains limited due to their black-box nature. Hybrid models (HMs) have the potential for increased physical interpretability by leveraging mechanistic models for known and hypothesized relationships with data-driven models for unknown relationships in a single framework. HM frameworks often assume mechanistic information is fully known, including parameter values, prior to model regression.[1] However, for many systems of interest, constructing an HM requires estimating the parameters of both data-driven and hypothetical mechanistic submodels on the same limited dataset.[2]

This submission considers novel methods for estimating parameters of dynamic models containing unknown mechanistic and data-driven parameters. Specifically considered are systems that can be formulated as a series of coupled ordinary differential equations. These methods are compared with traditional approaches to parameter estimation using forward sensitivity analysis. The comparison is further extended to the case where experimental data is low-quality and sparse. Results indicate that methods which merge data-driven models with numerical methods provide better estimates of time-evolved data and their derivatives than purely data-driven approaches. Conclusions for this presentation will identify parameter estimation approaches that accelerate the validation of interpretable models for systems when both data and mechanistic knowledge is limited.

References

1. Oliveira, R., *Combining first principles modelling and artificial neural networks: a general framework.* Computers & Chemical Engineering, 2004. **28**(5): p. 755-766.
2. Yang, A., E. Martin, and J. Morris, *Identification of semi-parametric hybrid process models.* Computers & Chemical Engineering, 2011. **35**(1): p. 63-70.

Research Interests

My research has focused on investigating methods for merging data-driven tools and mechanistic knowledge for modeling dynamic process data. The algorithms developed enable modelers to accelerate the systematic validation of hypothesized chemical-physical relationships without system-level information. I have further investigated the application of modern tools in automatic differentiation and numerical methods to support automated estimation of complex differential equations. My research interests lie primarily in developing novel solutions for modeling, control and optimization of manufacturing systems and supply chains.

614438 Process Modeling and Optimization for Monoclonal Antibody Production

Abstract:

Research Interests

In the past five years, monoclonal antibodies (mAbs) market has been increasing rapidly, and is predicted to reach $130-200 billion in year 2022[1]. There is a need to improve the overall process productivity to satisfy the increasing demands for mAb market. Two approaches to improve the process productivity include – 1) selection of high-performance process equipment, 2) application of optimum operating conditions. Continuous processing shows a significant benefit in increasing productivity, reducing the footprint, and cost-effectiveness over the conventional operations[2]. However, quantitative evaluation of the overall process performance is critical to decide the preferred mode of operation (batch or continuous) to satisfy the increasing market demands for mAb production. In addition, it is essential to procure detailed process understanding of unit operations to enhance product quality and process productivity.

In this work, we explore different simulation methods, including flowsheet modeling and mathematical modeling, to accurately predict mAb production. We implement flowsheet modeling to design and construct a fully integrated framework for continuous mAb production and compare its performance with the batch process using techno-economic analysis[3]. Scenario studies are simulated to evaluate process cost-effectiveness under varied production scales and upstream/downstream parameters. As bioreactor takes the highest percentage of the operating cost, we employ kinetic modeling method to capture the nonlinear bioprocess dynamics between operating conditions and output variables, aiming to improve the process productivity while maintaining product quality for CQAs like glycan concentration. Proposed kinetic modeling approach allows to evaluate important process variables like viable cell density, glucose, lactate, ammonia concentration, protein titers, and glycosylation during the batch culture for different temperature and pH values. Following this, we develop a surrogate model to investigate the optimum operating conditions for bioreactor. This is further integrated within the flowsheet model to capture corresponding effect on the mass balance, process scheduling, and critical quality attributes of the overall process. Thus, in conclusion, we propose applications of system engineering tools like process modeling and optimization for biopharmaceutical manufacturing, focusing on comparison of operation modes – batch and continuous, bioreactor modeling and mAb production, paving the way for cost-effective manufacturing and high quality biologics production.

Reference

1. Grilo, A. L. & Mantalaris, A. The Increasingly Human and Profitable Monoclonal Antibody Market. *Trends Biotechnol.* **37**, 9–16 (2019).
2. Yang, O., Qadan, M. & Ierapetritou, M. Economic Analysis of Batch and Continuous Biopharmaceutical Antibody Production: a Review. *J. Pharm. Innov.* **15**, 182–200 (2020).
3. Yang, O., Prabhu, S. & Ierapetritou, M. Comparison between Batch and Continuous Monoclonal Antibody Production and Economic Analysis. *Ind. Eng. Chem. Res.* **58**, 5851–5863 (2019).

614443 Single Atom Catalyst for Oxidation - Understanding Fundamentals of Synthesis and Reactivity

Abstract:

Research Interests:

1. Design of rules for synthesis of stable single-site catalysts from experiment and first principles theory

- Developing a computationally derived and experimentally validated feature-based predictive model for establishing the stability of precious group metal (PGM) precursors under solid-liquid synthesis conditions to guide the synthesis of stable PGM single-site catalysts
- Investigating active sites for low temperature CO oxidation over ceria supported single atom catalyst with validation through experiments and microkinetic analyses

2. Design of oxide@metal interface for hydrodeoxygenation chemistry of biomass derivatives

- Computational screening for predicting an optimal oxide@metal catalyst based on the fundamental properties of the oxide and the metal towards the selective C-O cleavage of oxygenated biomass derivatives, studied through Density Functional Theory (DFT)
- Investigating the electronic perturbations of the oxide through transition metal dopants and the catalytic implications for interfacial reaction chemistry

3. Tuning conductivity of high molecular weight and crystalline PEO_6 based polymer electrolytes using X-ray scattering experiments and DFT models

Poster abstract

Understanding the synthesis of single atom catalysts and the underlying metal-support interaction is crucial in improving the anti-sintering capability during chemical reactions. Strong electrostatic adsorption (SEA), which relies on an electrostatic driving force for the deposition of appropriately charged metal precursors on an oppositely charged metal oxide surface, demonstrates considerably higher weight loading threshold than for incipient wetness. Our study combines density functional theory (DFT) calculations and solution phase isothermal titration calorimetry (ITC) adsorption experiments to examine charged and solvated precursor adsorption onto ceria nano-cubes. DFT calculations bridge from an isolated gas phase metal atom and a single crystal plane of oxide surface to a ligated metal atom and bulk oxide surface (with a mixture of different crystal planes) in solution phase. We demonstrate agreement between DFT and ITC adsorption energy trends, towards developing a predictive framework for the stability of precious group metal (PGM) precursors under solid-liquid synthesis conditions. Our SEA adsorption model considers mainly electrostatics, with charge separated over the ligated complexes and the support surface. We use DFT to model the adsorption including partial deligation of the precursor, examining how the ligand chemical potential can alter adsorption thermodynamics.

We also examined CO oxidation kinetics over the single atom catalysts on ceria. DFT calculations in conjugation with experiments examine the catalytic performance and identify the synergy between individual metal atoms and the support towards CO oxidation. The emphasis is laid on specific metal-support interaction via structural relaxation or electronic structure induced by the presence of the adatom.

614448 Elucidating Structure-Function Relationships of Metal Oxides and Sulfides for Rational Catalyst Design

Abstract:

Research Interests:

From shape memory polymers to heterogeneous catalysts, my research experiences in a wide array of areas highlight my passion for materials research. Through a combination of design, synthesis, and characterization, I elucidate structure-function relationships to provide fundamental insights for rational materials design. In particular, my PhD research focuses on studying metal oxides and sulfides as catalysts for light alkane dehydrogenation, which is an important route to produce olefins —a critical feedstock for many materials and chemicals. Using a combination of catalytic testing and spectroscopic tools in my study on a series of supported molybdenum oxides, I was able to draw correlations between the catalyst surface species, reducibility, and activity. Most notably, I found that while previous studies focused primarily on bulk and high loading molybdenum oxide materials, it is sub-monolayer molybdenum oxides that warrant further study as light alkane dehydrogenation catalysts with its superior catalytic stability under the harsh reaction conditions.

Throughout my PhD, I have developed extensive expertise in many areas, from reactor engineering to materials synthesis and characterization, which readily translate to applications beyond heterogeneous catalysis. Looking forward, it is my goal to leverage the skills I have honed and continue engineering new materials to drive innovation forward.

614459 Mucin Glycopolymers Prevent Microbial Virulence While Preventing the Development of Antibiotic Resistance

Abstract:

Research Interests

I am a Ph.D. candidate and NSF Graduate Research Fellow in the Department of Biological Engineering at MIT. In Katharina Ribbeck's lab, I study how unique glycopolymers in mucus can be leveraged as antivirulence therapeutics to fight infections while circumventing antibiotic resistance. I'm interested in understanding the mechanisms of environment- and community-dependent bacterial interactions to promote human health and sustainability. I have experience in glycobiology, polymers, microfluidics, microbiology, and synthetic biology. Before coming to MIT, I earned my B.S. in Chemical Engineering at Caltech where I engineered proteins from soil bacteria with Frances Arnold. I then spent a year in Switzerland as a Fulbright Fellow, where I studied cell-free synthetic biology with Sebastian Maerkl at EPFL.

I have 4+ years of experience mentoring and teaching inside and outside the lab. I'm a trained conflict management coach through MIT REFS (since 2016). I have mentored four undergraduate students at the bench. I have been a teaching assistant for 8 courses in my career, earning me department awards for teaching excellence.

614510 Operando Study and Analysis of Crystallization for Health Care and Energy Applications

Abstract:

Research Interests

My research is focused on design and development of micro-structure platforms for (i) a superior end to end continuous process of manufacturing crystalline materials and (ii) investigation and analysis of dendrite crystallization and growth in the energy storage and conversion systems.

Research Area1: Continuous Manufacturing of Pharmaceuticals with Microstructure platforms

Description: Manufacturing of pharmaceuticals requires for multiple batch operations, long operational hours and higher costs for supply chains. The rising costs has pressurized pharmaceutical industries to employ alternative approaches and increase their efficiencies to meet the growing demands. One of the most fundamental alternative approaches is continuous flow synthesis and formulation of active pharmaceutical ingredients because of its promising lower cost of production while being more reliable and safer. As a part of pharmacy on demand (PoD) development a compact end to end manufacturing system will significantly shorten the lead times. The emergence of the continuous manufacturing of pharmaceuticals started with small molecules, yet it is still challenging to extend the continuous approach for larger molecules such as monoclonal antibodies. Moreover, overcoming these challenges becomes more substantial for some of the cases such as vaccines, cell based or gene therapies. The significance of an efficient, robust and inexpensive process for manufacturing of vaccines has become more significant specially after the most recent global pandemic of COVID-19. It is an inevitable fact that after the antibody discovery and vaccine development, the extremely high demand for these products complexes their supply chain over the world. Therefore, shifting to PoD synthesis using microstructure platforms will mitigate the expenses for production and distribution and decrease the time span of the distribution of these products around the world, especially in the underdeveloped and developing areas.

The proposed research is on the development of compact and reconfigurable platform for:

- Increasing the productivity and efficiency of manufacturing of APIs;
- Reducing the cost of R&D and further APIs;
- Faster distribution of the essential drug products and shorter the supply chain distribution time span.

In the most recent studies, the continuous manufacturing of the APIs has been perused by designing compact modules for each section. These end-to-end manufacturing process use lab scale setups and are improved over the past few years with respect to the yield of productions as well as purity of the products. Hence, a micro-structure setup which has never been studied for the continuous manufacturing. The development plan of the continuous manufacturing process with microstructure platforms for a PoD system requires deep understanding of essential synthesis, filtration, and formulation steps as well as appropriate crystallization technique and operational condition. In order to obtain the optimal design for the continuous manufacturing process, Multi-physics simulations will be employed for the fabrication and assembly of the continuous manufacturing setup.

Research Area 2: Dendrite Growth and Crystallization in Energy Storage and Conversion Systems

Description: Solid state electrocatalysis are crucial to develop electrochemical energy storage and conversion systems to sustain our future energy needs that currently depend on depleting fossil fuels. In this energy transition, lithium (Li) metal is known as a promising candidate for rechargeable battery technologies owing to its high theoretical specific capacity. Nevertheless, its practical application is limited due to high reactivity of Li metal in contact with electrolyte as well as with air components such as humidity, carbon dioxide and etc. This will lead to formation of dendrites on the surface of the Li anode upon striping/plating over the cycling process that lowers the columbic efficiency, cycling performance and more importantly, causes short circuit by punctuating the porous separator. Therefore, an overarching challenge for this technology is to preserve the Li metal anode from possible parasitic reactions, making it a safe and higher performance, rechargeable battery system. To this end, extensive studies have been carried out on the electrolyte selection, electrolyte additives, electrochemical, chemical and physical artificial solid electrolyte interphase layers, super-concentrated electrolytes, solid electrolytes, membrane modification, as well as effect of operating parameters to stabilize the surface of Li anode by suppressing the dendrite growth. However, the so-called techniques either possess poor mechanical strength, high costs, low ionic conductivities, complication in fabrication process or almost no effect on the dendrite growth and nucleation, making this technology infeasible. Therefore, deeper scientific insights are required to precisely understand the Li plating/stripping behavior and avoid dendrite formation. My research proposal is focused on

- Development microstructure devices with ability to incorporate both closed and open battery architectures for real time analysis of dendritic growth and crystallization of Li metal;
- Monitoring the dendrite crystallization can be as in-situ bright-field microscopy, confocal Raman microscopy, and X-ray diffraction;
- Screening different process parameters such as current density, the type of the electrolyte.

The information obtained by such analyses will be used to develop strategies, i.e. artificial solid electrolyte interphase layers, that can suppress the dendrite formation for higher performance and a safer battery technology.

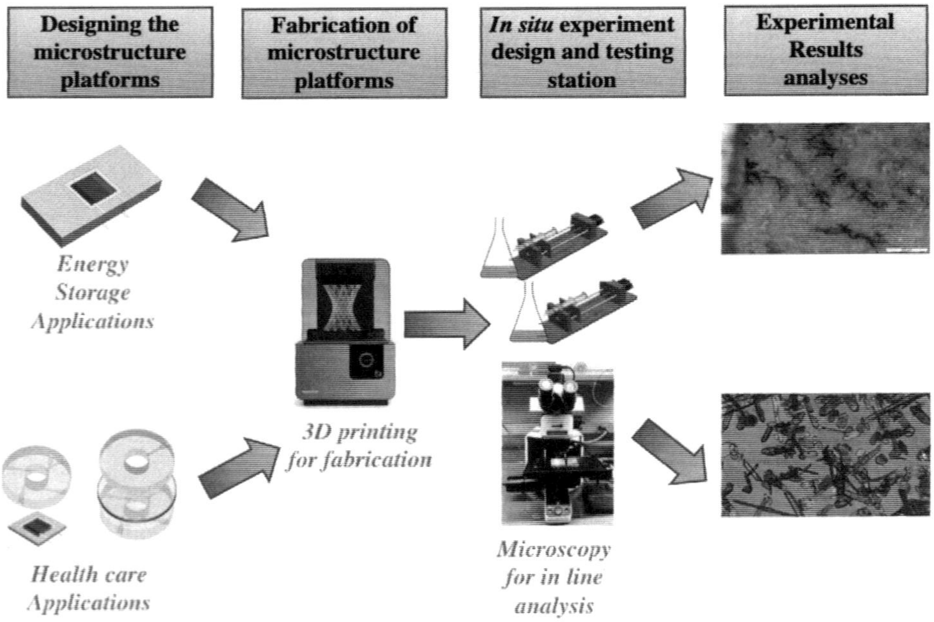

614515 Device-Level Engineering of Electrocatalytic Reactors Under Practical Operating Conditions to Enable Sustainable Small-Molecule Transformations

- Download/View Extended Abstract 3

- Download/View Extended Abstract 4

Abstract:

I am a Ph.D. student working with Prof. Karthish Manthiram in the Chemical Engineering department at MIT. I will be completing my Ph.D. in 2021. I am passionate about pushing the frontiers of sustainable chemistry swiftly and practically, so that we may have the best chance of maintaining a human-habitable planet in the coming decades; it is my hope that through my career I may contribute to this grand endeavor in some way. My doctoral research has focused on tying in chemical engineering fundamentals to practical problems in electrochemical synthesis. In the next steps of my career, I would like to move further in the direction of developing electrosynthetic reactor systems on a device level.

Research Interests

Electrochemistry is one promising route toward renewable energy storage and sustainable chemical synthesis. For example, using renewably-sourced electricity to drive water splitting is a lower-greenhouse-gas-emission alternative to steam methane reforming for the production of hydrogen; in addition, greenhouse gases such as CO_2 can themselves be transformed back into fuels and useful products using electrochemical potential. However, such processes require both fundamental study and device development in order to be economically feasible, whether or not they are accompanied by policy solutions such as carbon pricing. In modern electrocatalysis, reactions span a wide range of developmental stages. Processes such as chlorine evolution and aluminum smelting have long been staples of industry, and water splitting for the generation of hydrogen gas is nearing large-scale implementation; whereas the electrochemical reduction of carbon dioxide, while widely researched, is only beginning to be studied at high rates and moderate scales. Less developed still are reactions such as electrochemical nitrogen reduction or the direct oxidation of small organic molecules for synthetic purposes. Additionally, with some exceptions, when these reactions are studied, they are usually done so independently from their counter reactions – the reactions at the counter electrode required in order to complete the circuit.

With these points in mind, I am interested in two major approaches toward electrochemical research: (1) leveraging the state-of-the-art catalysts for small molecule transformations to study the impacts of device design upon the kinetics and transport of reactions, including the hydrogen evolution reaction, CO_2 reduction reactions, N_2 reduction reaction, and organic oxidative functionalization reactions using water as the O-atom source, under industrially-relevant conditions – for example, using realistic feedstocks – and in this way, connecting device development with fundamental understanding. (2) Examining the impacts of tuning the counter reaction on both the efficacy and the technoeconomics of the working reaction. For example: could we leverage clever flow fields to separate products made at the cathode and anode without the need for a physical separator? How do we understand the tradeoffs between the energy required to pump electrolyte, or to overcome ion diffusion resistance in physical separators, versus to perform downstream separations? Might it be possible to alter the economic prospects of reactions such as hydrogen evolution or CO_2 reduction by altering the counter reaction to perform a more lucrative partial oxidation reaction as opposed to oxygen evolution? If so, are there oxidative substrates that could be employed in such a way that electrolyte conditions enable synergistic interactions between oxidative and reductive substrates?

Ph.D. Research

Two major projects I have pursued during my doctoral research include: (1) uncovering the role of mass transport in protecting the CO_2 reduction reaction from parasitic side reactions with gas-phase contaminants, and (2) understanding solvent non-idealities in blended aqueous—nonaqueous electrolytes during electrochemical oxidative O-atom transfer reactions involving water. During this work I have extensively employed skills in electrochemical kinetic analysis, bulk electrolysis, cyclic and square-wave voltammetry, gas flow setup, gas chromatography, headspace sampling for solution thermodynamic measurements, and quantitative nuclear magnetic resonance (NMR) spectroscopy. I have additional experience in solid-state and colloidal approaches to metal oxide and nanoparticle synthesis; X-ray absorption spectroscopy (XAS); X-ray photoelectron spectroscopy (XPS); powder X-ray diffraction (XRD); UV-visible spectrophotometry; electron microscopy (SEM/TEM); elemental analysis via inductively coupled plasma (ICP); hydraulic pressing; organic synthesis; practical glassblowing; chemisorption analysis; and ultra-high vacuum (UHV) setups.

Selected Publications

Williams, K.; Corbin, N.; Zeng, J.; Lazouski, N.; Yang, D. T.; Manthiram, K. Protecting Effect of Mass Transport during Electrochemical Reduction of Oxygenated Carbon Dioxide Feedstocks. *Sustain. Energy Fuels* **2019**, *3* (5), 1225–1232.

Lazouski, N.; Chung, M.; **Williams, K**; Gala, M. L.; Manthiram, K. Non-aqueous gas diffusion electrodes for rapid ammonia synthesis from nitrogen and water-splitting-derived hydrogen. *Nature Catalysis* **2020**, *3* (5), 463-469.

Zeng, J.; Corbin, N.; **Williams, K.**; Manthiram, K. Kinetic Analysis on the Role of Bicarbonate in Carbon Dioxide Electroreduction at Immobilized Cobalt Phthalocyanine. *ACS Catalysis* **2020**, *10* (7), 4326-4336.

Lazouski, N.; Schiffer, Z. J.; **Williams, K.**; Manthiram, K. Understanding continuous lithium-mediated electrochemical nitrogen reduction. *Joule* **2019**, *3* (4), 1127-1139.

Yang, D. T.; Zhu, M.; Schiffer, Z. J.; **Williams, K.**; Song, X.; Liu, X.; Manthiram, K. Direct Electrochemical Carboxylation of Benzylic C-N Bonds with Carbon Dioxide. *ACS Catalysis* **2019**, *9* (5), 4699-4705.

Corbin, N.; Zeng, J.; **Williams, K.**; Manthiram, K. Heterogeneous molecular catalysts for electrocatalytic CO_2 reduction. *Nano Research* **2019**, 1-33.

614520 Modeling and Optimization Approaches for Process Development

- Download/View Extended Abstract

Abstract:

Research Interests:

I am a 5th-year Ph.D. candidate in the Chemical Engineering department at Massachusetts Institute of Technology. After defending in the next year, I hope to pursue a career in industry focused on modeling and optimization for process operation and design. My Ph.D. work as a member of the Process Systems Engineering Laboratory has involved both developing new numerical approaches and applying them to real systems. In particular, I have been collaborating with OCP's phosphate-processing facility in Morocco to ensure that these tools can be effectively used to direct innovation in an industrial setting. I have enjoyed the opportunity to use modeling and optimization to develop specific methods to make concrete improvements, and I hope to also have these qualities in my future career. I am particularly interested in bringing my expertise in method development to create new tools that can address traditionally challenging systems that are too complex for current approaches to handle.

Ph.D. Work:

The focus of my Ph.D. research has been using recent advances in nonsmooth equation solving to develop new methods for modeling complex chemical systems. Many processes in chemical engineering, including phase changes and batch or semi-batch manufacturing, transition between different regions, making them naturally nonsmooth. Traditionally, these processes are modeled using disjunctive or mixed-integer programs that are often nonlinear and scale poorly with system size. Instead, I have proposed simulating these systems using compact, explicitly-nonsmooth equations, which, unlike current approaches, requires only equation-solving methods, scales well with system size, and can solve for any unknown process variable. In particular, I have been collaborating with OCP to apply this approach to improve process integration and reduce resource use at their Jorf Lasfar platform. Throughout this project, I have developed and implemented process models for the major units at Jorf Lasfar using a combination of fundamental relations and empirical plant data. I have also performed analyses to identify areas with the most potential for improvement in resource use, utilized new research developments to create numerical methods tailored to simplifying this problem, and applied these methods to propose specific process designs for the sustainable expansion of Jorf Lasfar. I am currently working on developing nonsmooth optimization methods to minimize the costs of these designs by adapting global optimization algorithms. In the future, I hope to be able to continue applying these skills for process design and method development to help tackle key problems across chemical processes.

614521 Chemical Research and Development: *Process Modeling and Machine Learning*

Abstract:

Research Interests

My success as a research associate at Purdue University is the result of my skills and qualifications as a proven problem-solver and communicator who is interested in identifying interesting questions and collaborating with two members of national academy of engineering (Sangtae Kim and Doraiswami Ramkrishna) to find tangible solutions.

I believe that the pharmaceutical development and manufacturing is an excellent match for me, and the following offers a few highlights of my skills and qualifications:

- **Problem solver**: As a research associate, on a day to day basis I engage in interdisciplinary research problem using rigorous scientific methodology to analyze data, draw conclusions and communicate results. While I have a fundamental knowledge of separation, filtration, and crystallization for the pharmaceutical industry, my areas of research include transport phenomenon, reaction engineering and processing engineering. These skills have been developed over the course of my 10-year scientific career and have led me to publish papers in top ranked journals, including Nature Scientific Reports and Journal of Fluid Mechanics. These skills will provide the essential basis of my chemical engineering career.
- **Flexibility and people skills**: The range of professional environments I have worked in, including research institutions, software industry, and national research center has made me adept and comfortable interacting with anyone at any level in an organization. What I find more engaging is to understand each individual's perspective and their stakes, access their abilities and goals, and develop strategies to meet our common goal.
- **Communicator/Leader**: I am dynamic and motivated as a communicator when I have the ability to collaborate with chemist/experimentalist and to present my scientific results and ideas. In addition to my verbal presentation skills, I have proven myself an effective writer, as evidently by multiple published papers and book chapter. Lastly, I have supervised fellow workers in both scientific and non-scientific settings.

Selected publications

(out of 13 total, google scholar: https://scholar.google.com/citations?user=uEmQO5cAAAAJ&hl=en)

First-author contributions are underlined

Wang, S., Ramkrishna, D., Narsimhan, V. "Exact sampling of polymer conformations in an external field using Brownian bridges" Journal of Chemical Physics (accepted).

Wang, S., Martin, C.P. and Kim, S., 2019. Improper integrals as a puzzle for creeping flow around an ellipsoid. Physics of Fluids, 31(2), p.021101. (Invited papers on transport phenomena in celebration of Prof. Robert Byron Bird's 95th birthday)

Wang, S. and Ardekani, A.M., 2015. Biogenic mixing induced by intermediate Reynolds number swimming in stratified fluids. Scientific reports, 5, p.17448. (Highlighted on Physics.org, Futurity, Geology Page, Purdue College of Engineering News Page, Purdue Research Computing Cluster)

Wang, S. and Ardekani, A.M., 2012. Unsteady swimming of small organisms. Journal of Fluid Mechanics, 702, pp.286-297.

Author

Shiyan Wang

Purdue University

614524 Molecular Mechanisms of the AAA+ Protease Clpxp Characterized Using Single-Molecule Fluorescence Quenching

Abstract:

AAA+ proteases are enzymes responsible for removing unwanted proteins and activating cellular processes through the degradation of tagged proteins. *Escherichia coli* ClpXP is a model AAA+ protease that uses the energy of ATP binding and hydrolysis to fuel protein degradation. The hexameric ClpX ring first unfolds a specific protein before translocating the polypeptide into the ClpP degradation chamber. Various biochemical and structural studies have been used to characterize ClpXP, yet the conformational transitions and nucleotide transactions of ClpXP subunits are still unknown. Here, we engineered two novel, short-range single-molecule fluorescence quenching assays to monitor conformational switching and nucleotide-binding in a single subunit of ClpXP. We synthesized an ATP analog labeled with a dark quencher that we call ATPQ for our single-molecule nucleotide-binding assay. ATPQ supports ClpXP motor activity and permits the observation of binding events at concentrations up to 100 μM. Our conformational switching assay and DNA-based distance calibrations allow for quantitative measurements of conformational states and dynamics of individual ClpX subunits. Our results show that ClpXP subunits adopt both static and dynamic conformational states at varying nucleotide and substrate conditions. The single-molecule techniques we developed here facilitate real-time observation of molecular events in working AAA+ motors and can be applied to a wide range of molecular motors.

Research Interests

Biophysics, Biochemistry, Molecular Biology, Optical Trapping, Protein Engineering, Optics, Photonics, Fluorescence, Single-Molecule Biophysics, AAA+ Motors

614532 Paving the Way for Cybernetic Modeling of Biological Processes in Mammalian Systems

Abstract:

Abstract: Paving the way for cybernetic modeling of biological processes in mammalian systems

Our bodies are well equipped to address and respond to external injury. The body has a two-line defense system against invading pathogens: the skin which acts as a physical barrier and the immune system which triggers an inflammatory response to fight off pathogens. Significant variation in the inflammatory response can lead to complications in certain disease states (*e.g.*, cytokine storm in Covid-19 disease). Macrophages are a versatile immune cell which play a crucial role in inflammation. These cells have the ability to identify a foreign stimulus and trigger a cascade of cellular signaling events. Developing a working knowledge of these cellular events and a description of signaling pathways using mathematical models is imperative for studying immune-modulating drugs and their impact on disease outcome.

Regulation of metabolism in mammalian cells is achieved through a complex interplay between cellular signaling, metabolic reactions, and transcriptional changes. These complex interactions can be modeled through implicit accounting of regulation using a cybernetic framework, as seen in the modeling of prostaglandin metabolism in inflamed macrophage cells [L. Aboulmouna et al., Processes. 2018]. The premise of the cybernetic framework is that the regulatory processes affecting metabolism can be mathematically formulated through variables that constrain the network to achieve a specified “goal”.

Cybernetic models, developed by Ramkrishna’s group, assume that the kinetic parameters of the mechanistic fluxes are not constants and can vary with time. The kinetic parameters are represented as the product of unregulated rate constants and cybernetic control variables for induction of enzyme synthesis (u) and modulation of enzyme activity (v). The cybernetic control variables can be intuitively formulated based on the goal of the system. This model framework describes gene regulation by attributing metabolic preferences to a suitable survival goal of the organism. Previous cybernetic models use regulatory goals like maximizing growth rate [1] or carbon uptake rate [2] to describe various biological phenomena in bacterial and yeast systems, such as diauxic growth. The cybernetic goal for mammalian cells may not be based solely on survival or growth but on specific context dependent cellular responses.

This work adapts the cybernetic framework to model the eicosanoid pathway— production of prostaglandins (COX-branch) and leukotrienes (LOX-branch) from arachidonic acid— in bone-marrow derived macrophage (BMDM) cells using the *in vitro* data across four different experimental scenarios. Several models of the eicosanoid pathway precede this work [3, 4], but none consider the regulatory phenomena [5]. Using a defined cybernetic goal to maximize system inflammation, we developed a weighted formulation of cybernetic control variables by correlating the lipidomic and transcriptomic data. The model effectively describes the experimental data and provides insights into the complex regulatory patterns present in the macrophage.

Research Interests: From the cell to whole animal model—Understanding dysregulation of systems at multiple levels

My previous work has blended mathematical modeling with experimental techniques to develop an understanding of complex biological phenomena. A coupled approach to addressing biological phenomena and understanding how crucial process become dysregulated in certain disease states is required when developing therapeutics. Here, I highlight previous research initiatives.

Novel isotope tracer method for quantifying glucose metabolism in type-2 diabetes

Accurately assessing *in vivo* rates of glucose turnover, glycogen breakdown, gluconeogenesis, and futile cycling in the liver are crucial to understanding the control of whole-body glucose metabolism and how it becomes dysregulated in conditions such as diabetes and obesity. Currently, these rates are estimated using methods that do not account for the full complexity of intracellular biochemical reaction networks, and, therefore, often lead to inconsistent results. In prior work, we conducted metabolic flux analysis (MFA) studies using a unique formulation of 13C- and 2H-labeled glucose tracers and Gas Chromatography-Mass Spectrometry (GC-MS) profiling in order to understand the regulation of liver glucose fluxes in both fasted and fed Sprague-Dolly rats. By analyzing GC-MS labeling data of plasma glucose and lactate using our custom MFA software tools, we developed strategies to quantify whole-body glucose metabolism using an integrated experimental and computational approach.

Single-cell-based investigation of HL60 differentiation using a microwell array

The average cell population response to a range of stimuli inaccurately reflects individual cell response due to heterogeneity in previously assumed homogeneous cell populations. It is hypothesized that this contributes to metastatic potential in tumors, with the most invasive cells being molecularly and behaviorally distinct from the bulk. In prior work, we designed and optimized a microwell platform using standard photolithography techniques for use in high throughput single cell studies to address heterogeneity in tumor populations. The acute myeloid leukemia cell line HL60 was used as a model uncommitted precursor cell line, and we began investigating population variance by inducing differentiation along the granulocytic lineage with all-trans retinoic acid.

Future Work

I am especially interested in continuing to advance my understanding of the human system and the role inflammation plays in disrupting pathogen invasion in our bodies. As I look to pursue future research initiatives, I am excited at the opportunities available to further understand our collective understanding of what a healthy body looks like at the molecular level leading to the organ level and, ultimately, the system as a whole in order to optimize individuals' health and well-being.

614536 Lignin Depolymerization and Esterification By Carboxylic Acids to Produce Biodiesel

Abstract:

Research Interests

Past Research: As a junior student in my undergraduate program, I began my first research study on modeling and comparison of water desalination by the humidification-dehumidification and adsorption methods. My passion and interest in discovering new things, along with my hardworking, resulted in publishing a conference paper from my bachelor's final project. My master project was focused on the heat transfer of SiO_2/water nanofluid in jacket side of agitated vessels through experimental and numerical techniques. My research findings was published as two conference papers and one peer-reviewed article. I subsequently expanded my academic knowledge through collaboration with other scholars in conducting both experimental and numerical studies on heat exchangers and renewable energies. To this end, a series of studies were conducted on the hydrodynamic performance of a new wave energy converter called Searaser and evaluated its efficiency for use in the Caspian Sea.

Current Research: I receieved a PhD addimision from the Department of Mechanical Engineering at Iowa State University. My Ph.D. project included an innovative method of lignocellulosic biomass deconstruction with an emphasis on lignin. I started to work on lignin depolymerization by oxygen to produce valuable phenolic monomers. Lignin oxidation is a promising method for lignin valorization. It can produce a range of functionalized chemicals of significant economic value such as aromatic acids, aromatic aldehydes, and aliphatic carboxylic acids. I designed a small batch reactor system to perform lignin oxidation tests. This system enabled us to demonstrate the non-catalytic depolymerization of native lignin to oxygenated phenolic monomers. Oxidation tests were performed in perfluorodecalin. Perfluorodecalin is a perfluorocarbon (PFC), characterized by their chemical stability and exceptionally high solubility for O_2. High yiled of valuable phenolic monomers was achieved from lignin oxidation using perfluorodecalin. Moreover, the role of molecular oxygen was elucidated as both an oxidant to achieve oxidative depolymerization and a radical scavenger to mitigate the condensation of phenolic monomers. The lignin oxidation process was scaled up from 5 ml to 250 ml without losing yield. For scaling up the process, a stirred reactor was used. Since the reaction time is a key factor in oxidation tests, the reactor was modified.Our new batch system is a unique system with fast heating and cooling capability. My research findings on this topic have been presented in several well-known conferences, and a paper is currently under review in the Journal of Energy and Environmental Science .

As the second phase of this project, I am currently working on oxidative depolymerization of lignin in a biphasic system. In this study, technical and native lignin sources are depolymerized in the presence of oxygen inside a biphasic system composed of solvents with high lignin solubility and perfluorodecalin of high oxygen solubility. We hypothesis that organic species formed by oxidative cleavage of lignin in perfluorodecalin can be extracted and stabilized into the organic phase with much lower oxygen concentration to improve the phenolic monomer yield. I am also working on lignin depolymerization and esterification using carboxylic acids to produce phenyl ester. The net carbon emissions from biodiesel could be decreased if methanol is replaced by a biobased reactant in the production of methyl esters. We hypothesize that lignin can be used in the production of phenyl esters. To this end, a series of tests on native and technical lignin have been performed. We would like to test both one-pot and two-stage esterification of lignin.

Future Research:

Lignin-first biorefinery: To retain the intrinsic value of virgin lignin, biorefineries will have to employ either mild depolymerization processes, such as ammonia-based fractionation and low-temperature organosolv techniques employ processes that continuously stabilize phenolic products as they are released from biomass. This second approach, sometimes referred to as the lignin-first strategy for deconstructing lignocellulose, is receiving increasing attention with reductive catalytic fractionation (RCF) and formaldehyde-assisted fractionation, as prominent examples. I intend to use my knowledge in lignin chemistry and depolymerization to introduce new methods of delignification to biorefineries to produce a more reactive, less condensed lignin.

Lignin depolymerization to chemicals, fuels, and material: There are two different categories of products from lignin. The first category, which currently is the main application of lignin other than heat and power production, utilizes the polymeric structure of lignin to produce carbon nanofibers, polymer composites, adsorbents, and glues. Recently, the utilization of lignin to produce bio-based chemicals and fuels has received increasing attention. In this method, lignin is first depolymerized and then upgraded to a wide array of valuable products. There is still a long way to go to convert lignin to high-quality material and produce high yield of chemicals and fuels economically. Therefore, I intend to work on this area to improve the economic viability of biorefineries.

Fast pyrolysis and solvent liquefaction of biomass: Fast pyrolysis of biomass is a simple and robust thermochemical technology for converting biomass into bio-oil, syngas, and biochar. Solvent liquefaction has been receiving increasing attention and has the potential of becoming a robust and economical method of processing biomass . I would like to focus on both of these methods in my future research plan. Both of these methods seem promising ways to produce bio-oil from lignocellulosic biomass.

Abstract

Global energy consumption has significantly increased in the last 40 years and approximately 80% of this demand is supported by fossil fuels. Environmental and human health concerns regarding fossil fuels consumption has encouraged researchers to find sustainable alternatives for the current petrochemical industry. Within this context, biodiesel production has received increasing attention in recent years. Biodiesel is a form of diesel fuel which is composed of esters formed by chemically reacting alcohols and long chain carboxylic acids. Lignin as the only component of biomass containing aromatic structure is a great precursor for producing chemicals and fuels. Interestingly, phenols can be esterified by carboxylic acids to produce esters. We hypothesize that lignin can be used as a potential precursor for biodiesel production. Phenolic products obtained from lignin depolymerization could be esterified with a carboxylic acid to produce phenyl esters. In order to test our hypothesis, technical and native lignin sources were applied as phenolic source and carboxylic acids like acetic acid and hexanoic acid were used as acetylating agents. Both in-situ and ex-situ esterification of lignin were studied. In the in-situ esterification process, lignin was depolymerized and esterified in one pot, while in the ex-situ esterification process, lignin was first depolymerized, and then the obtained lignin-oil was esterified. Lignin-oil was further characterized by GC-FID, FTIR and 2D HSQC NMR spectroscopy.

614279 Automated Optimization and Control of Modular Chemical Systems

Abstract:

Research Interests

Modular chemical systems for advanced manufacturing have attracted increased interest in recent years due to their ability to employ process intensification, as well as their potential for increased flexibility, and productivity (e.g., see [1,2] and citations therein). This interest is particularly prominent in the chemicals, pharmaceuticals, biotechnology, and energy industries (e.g., [3-5]). This Ph.D. thesis addresses modular chemical systems from a process optimization and control perspective that enables automation. Modular chemical systems are advanced manufacturing systems with characteristics such as (1) dynamics described by combinations of algebraic, ordinary differential, partial differential, and integral equations, (2) high to infinite state dimension, (3) time delays, (4) actuator, state, and output constraints, (5) stochastic noise and disturbances, (6) parametric uncertainty. Most methodologies are developed to handle dynamical systems with a subset of these characteristics and a focus on the specifics of the process.

However, to promote efficiency in industrial practice, autonomous design of optimization and control for modular chemical systems, connected in a plug-and-play manner, should be pursued. Such control and optimization systems are mainly conducted through a computer with little to no supervision. In the work conducted in the context of this Ph.D. thesis, we describe steps taken towards such automation, through modeling, control algorithms and optimization approaches tailored to address the characteristics and challenges of such systems. Such steps include (1) automating the construction of dynamic first-principles plantwide models obtained by plug-and-play interconnection of modules that are modeled in advance; (2) automating the hierarchical plant-wide control; (3) developing methodologies to expand the range of applicability of linear input-output models in linear model predictive control, to address the high state dimension of modular systems with low on-line computational cost [6]; (4) addressing the optimization of transient periods (i.e. startup) to minimize waste and increase efficient operation times, through hybrid dynamic optimization and linear model predictive control schemes [7]; (5) creating feedback controllers with guaranteed properties for dynamic neural network models, as an alternative to first-principles model control [8]. This Ph.D. work builds upon recent publications that propose ways to modify linear model predictive control formulations to handle significant process nonlinearities [6], with applications that extend to process startup [7]. It also demonstrates progress in designing controllers with guaranteed properties for dynamic artificial neural networks [8], which can be used in the place of first-principle models, providing an alternative to nonlinear model predictive control, which usually is a nonconvex problem that requires the solution of an NLP online, with unknown feasibility and performance properties.

Interests I developed during my thesis work are related to systems modeling, simulation, and control. My current research interests include, but are not limited to, first principles modeling, and simulation, empirical modeling of dynamical systems using data, surrogate modeling, dynamic optimization, model predictive control of high dimensional systems, and autonomous design of advanced control systems. I am also interested in expanding the above methodologies to address uncertainties in systems, such as parametric uncertainties and model-plant mismatch. My research interests find applications in numerous fields, such as in the (bio)pharmaceutical industry which is rapidly moving towards continuous processing, in the chemical industry where modeling and control are always relevant, as well as in the energy industry, with the evolving subfield of battery modeling and control.

References

1. -C. Bédard, A. Adamo, K. C. Aroh, M. G. Russell, A. A. Bedermann, J. Torosian, B. Yue, K. F. Jensen, and T. F. Jamison. Reconfigurable system for automated optimization of diverse chemical reactions. Science 361(6408):1220-1225, 2018.
2. A. Paulson, E. Harinath, L. C. Foguth, and R. D. Braatz. Control and systems theory for advanced manufacturing. In Emerging Applications of Control and System Theory, edited by Roberto Tempo, Stephen Yurkovich, and Pradeep Misra, Lecture Notes in Control and Information Sciences, Springer Verlag, Chapter 5, 63-80, 2018.
3. Lakerveld, P. L. Heider, K. D. Jensen, R. D. Braatz, K. F. Jensen, A. S. Myerson, and B. L. Trout. End-to-end continuous manufacturing: Integration of unit operations. In Continuous Manufacturing of Pharmaceuticals, edited by P. Kleinebudde, J. Khinnast, and J. Rantanen, Wiley, New York, Chapter 13, pages 447-483, 2017.
4. T. Myerson, M. Krumme, M. Nasr, H. Thomas, and R. D. Braatz. Control systems engineering in continuous pharmaceutical processing. Journal of Pharmaceutical Sciences, 104(3):832-839, 2015.
5. S. Hong, K. A. Severson, M. Jiang, A. E. Lu, J. C. Love, and R. D. Braatz. Challenges and opportunities in biopharmaceutical manufacturing control. Computers & Chemical Engineering, 110:106-114, 2018.
6. Nikolakopoulou, M. von Andrian, and R. D. Braatz. Plantwide control of a compact modular reconfigurable system for continuous-flow pharmaceutical manufacturing. 2019 American Control Conference (ACC), Philadelphia, PA, USA, pp. 2158-2163.
7. Nikolakopoulou, M. von Andrian, and R. D. Braatz. Fast model predictive control of startup of a compact modular reconfigurable system for continuous-flow pharmaceutical manufacturing. 2020 American Control Conference (ACC), Denver, CO, USA, in press.
8. Nikolakopoulou, M. S. Hong, and R. D. Braatz. Output feedback control and dynamic artificial neural networks using linear matrix inequalities. Submitted.

614285 Cross-Linked Nonwoven Fibers By Room-Temperature Cure Blowing and in Situ Photopolymerization

Abstract:

Research Interests: Photopolymerization, fiber spinning, polymer processing, 3D printing

Micron-sized nonwovens serve a diverse range of applications such as medical textiles, geotextiles, filtration media, etc. Current methods for producing synthetic nonwovens rely on transforming pre-formed thermoplastic polymers by heating or adding organic solvents to facilitate drawing, which respectively results in their high energy demands and concerns about volatile solvent emissions. Furthermore, conventional synthetic fiber manufacturing is limited to thermoplastics since cross-linked thermosets do not flow like liquids. However, cross-linked (thermoset) fibers are attractive targets due to their superior thermal and chemical resistances. Herein, we describe a "cure blowing" process that addresses these issues by producing cross-linked fibers at room temperature with little or no volatile solvent. In this process, a photocurable liquid mixture of thiol and acrylate monomers was extruded through a spinning die resembling those used for commercial melt blowing and drawn into liquid filaments by high-velocity, ambient temperature air jets. The liquid filaments were cross-linked in flight into solid fibers by in-situ photopolymerization. Studies of the effects of process parameters on the fiber diameter and morphology revealed two inherent process limits: (1) a kinetic limit dictated by rate of photopolymerization relative to fiber fight time, and (2) a stability limit dictated by the surface tension and viscoelasticity of the monomer mixture. These limits manifest as fibers that are fused at their points of contact or fibers that possess surface undulations and non-uniform diameters, respectively. Several strategies to suppress or delay these limits were also identified and demonstrated experimentally to expand the cure blowing process window. Furthermore, we observed that the various parameters involved in cure blowing govern the final fiber characteristics in a convoluted and inter-dependent manner. A comprehensive analysis in terms of a few characteristic timescales corresponding to the different competing features of the process was also performed. This analysis led to the development of an universal operating diagram for cure blowing. This diagram can be used as a predictive process design tool where all the parameters of a new reactive system can be tuned a priori of fiber production. This will essentially eliminate the need to determine the appropriate process conditions through many trial-and-error experiments. These studies suggest the industrial relevance of this new cure blowing technique for producing cross-linked synthetic nonwovens for a wide range of applications.

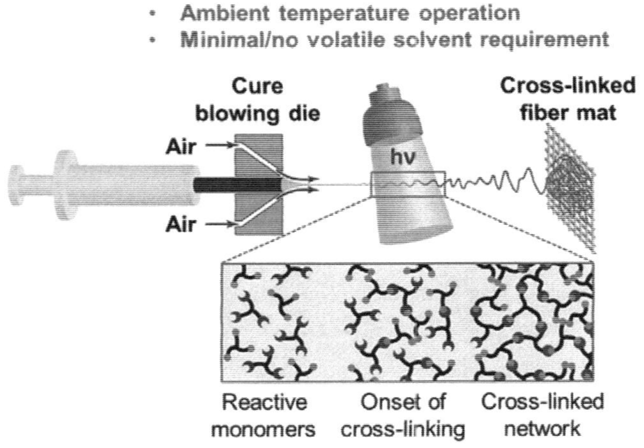

614051 Magnetic Characterization and Quantification of Magnetic Nanoparticle Endocytosis Mechanisms By Particle Tracking Velocimetry

Abstract:

Magnetic characterization is of much importance in the fields of cell labeling, cell purification, cell physiology, endocytosis, biochemical micro assays, endosome research, cell separation, drug targeting, and in vivo diagnostics. The Hyperflux^{TM} velocimeter is utilized to measure magnetophoretic mobility, size and other morphological parameters of magnetic particles and labeled cells by particle tracking velocimetry. Magnetic cytometry by velocimetry records the motion of labeled cells in an isodynamic magnetic field thereby estimating the key parameter, magnetophoretic mobility of labeled cells. The calibration capability of the instrument has been extended in order to estimate the actual particle size and thereby estimating the intrinsic magnetic properties of several commercial beads. The optical density method and chain velocity method has been explored to estimate the magnetophoretic mobility of a single bead of nanoparticle chains. The rapid estimation of magnetophoretic mobility by the instrument and collection of multiple thousand data points facilitates cellular uptake quantification and kinetic studies in less time than any other existing technique. The receptor-independent uptake by cultured CHO (Chinese Hamster Ovary) cells of 100 nm iron oxide nanoparticles with different surface coatings, namely starch, amino groups, and polyethylene glycol (PEG 2K Daltons), was studied to reveal the role of nanoparticle endocytosis mechanisms. Caveolae-mediated and clathrin-coated endocytosis are revealed by using specific mechanism-based inhibitors. The cellular survival rate, toxicity, and uptake of nanoparticles during endocytosis by different mechanisms with the different surface coatings have been investigated. This research facilitates the rapid estimation of intrinsic magnetic properties, the optimization of MNP coatings, a better understanding of cell labeling, and different cellular endocytosis mechanisms.

Research Interests: magnetic cytometry, biomedical research, cell biology, semiconductor development, process intensification

DESIGN OF A HIGH-PRESSURE CATALYTIC PHOTORREACTOR TO OPERATE AT SUPERCRITICAL CO_2 CONDITIONS.

E. Pájaro[a*], J. R. Martínez[b], V. G. Baldovino[a,c], F. Martinez[b]

[a] *Centro de Investigaciones en Catálisis (CICAT), Parque Tecnológico Guatiguará, km 2 vía El Refugio, Piedecuesta, (Santander), 681011, Colombia.*

[b] *Centro Nacional de Investigaciones para la Agroindustrialización de Especies Vegetales Aromáticas Medicinales Tropicales (CENIVAM), Universidad Industrial de Santander, Cra. 27 Cll 9, Bucaramanga, (Santander), 680002, Colombia.*

[c] *Laboratorio de Ciencia de Superficies (SurfLab), Parque Tecnológico Guatiguará, km 2 vía El Refugio, Piedecuesta, (Santander), 681011, Colombia.*

(*) estefany_pajaro@hotmail.com.

Keywords: photocatalysis, design, photoreactor, supercritical fluids.

1. Introduction.

Design a chemical reactor is a complex process since it must allow flexible operation in the ranges of temperature and pressure required by the chemical process that is to be developed. Nature of the substances involved, the phases present, the kinetics and thermodynamics of the process are considered. Especially when it comes to photocatalytic reactors since radiation plays a leading role in the reaction process these are aspects of high relevance for the selection of materials, type of operation, fluid dynamics and geometry[1,2]. Commercially, we can find reactors built in different materials such as steels, sapphire, borosilicate, PVC, some of these designed to withstand high pressures, others allow lighting inside the reactor, and another group operates in heterogeneous systems with suspended catalysts. However, having these three characteristics in a single device requires a rigorous mechanical analysis and some modifications in the commercial designs of high-pressure reactors[3]. Due to the expansion that heterogeneous photocatalysis has had in recent decades due to the variety of chemical processes that can be performed. Such as selective oxidation reactions using atmospheric oxygen[4,5], where there is a great interest in replacing organic solvents with supercritical fluids, especially CO_2. These high pressure photoreactors are required[4]. Since the reactors for heterogeneous photocatalysis using $scCO_2$ as a solvent are few and often use immobilized acid catalysts. In this work, a photoreactor was designed to allow processes of transfer of oxygen atoms to olefins using $scCO_2$ as a solvent and a dioxo-molybdenum complex supported on TiO_2 as a catalyst, activated by UV-Vis radiation[6].

2. Methodology.

In Fig. 1, the primary considerations for the selection of materials, geometry and couplings for the design of the batch photoreactor are shown. The nature of the substances, the reactions and the operating conditions to which this equipment is subjected. The design rules for high pressure vessels presented in ASME VIII code were used to establish the mechanical requirements for the construction of the high-pressure reactor. These data were simulated and adjusted using the finite element method using the CAD-type software, SolidWork, the main equations used are shown in Table 1.

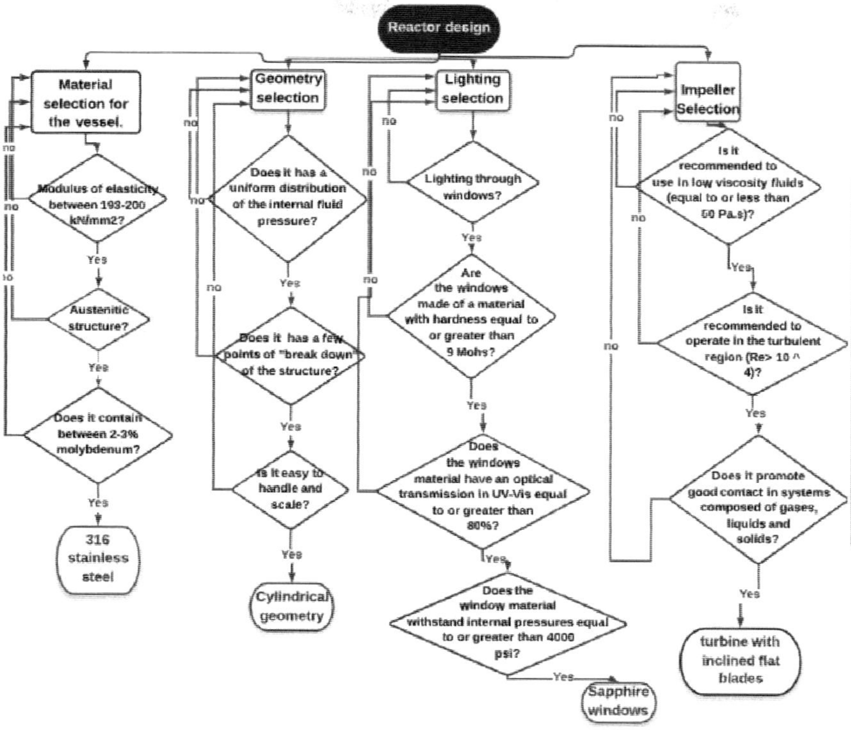

Fig 1. Design criteria for reactor.

Table 1. Main design equations.

Main design equations	
$t=\frac{PR}{SE-0.6P}$	Thickness of the reactor vessel
$H=\frac{\pi*G^2*P}{4}$	Fluid pressure inside the vessel
$H_p=\pi*G*2*m*P*b$	Packing load
$W_{m1}=H+H_p$	Bolt load in service
$W_{m2}=\pi*b*G*y$	Load of bolts in settlement
$A_m=\frac{W_{m1}}{f_a}\ o\ \frac{W_{m2}}{f_b}$	Minimum cross-sectional area of the bolt
$A_b=\frac{\pi}{4}*d_{br}^2*\#\ bolts$	Real diameter of bolt circle

3. Results.

Fig. 2 shows the scheme of the photoreactor designed to perform supercritical CO_2 processes that require lighting inside it. It was decided to make a container with a rectangular external geometry with a cylindrical cavity where the different processes would be carried out, the whole body is designed to be constructed in 316SS and the volume is 300 mL.

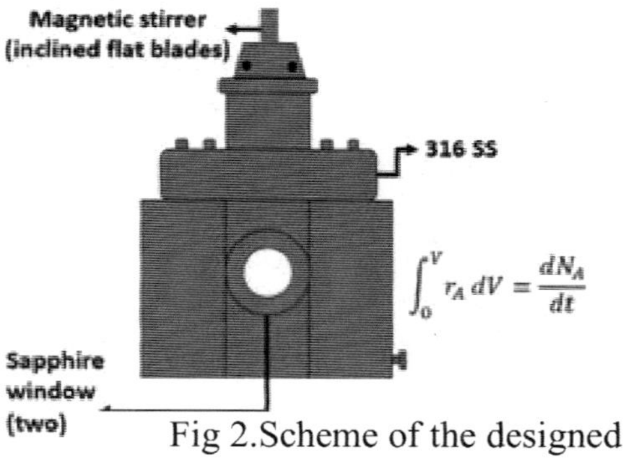

Fig 2.Scheme of the designed photoreactor.

References.

[1] A. E. Cassano, R. J. Brandi & O. M. Alfano, Trends Photochem. Photobiol, 3, 55—79 (1994)

[2] Y. Boyjoo, H. Sun, J. Liu, V.K. Pareek & S. Wang, Chemical Engineering Journal, 310, 537-559 (2017)

[3] V. Spallina, P. Chiesa, E. Martelli, F. Gallucci, M. Romano & G. Lozza, International Journal of Greenhouse Gas Control, 36, 34-50 (2015).

[4] T. Seki and A. Baiker, Chemical reviews, 109(6), 2409-2454 (2009).

[5] K. Hofstadler, R. Bauer, S. Novalic and G. Heisler, Environmental science & technology, 28(4), 670-674 (1994).

[6] H. Martínez, Á. A. Amaya, E. A. Páez-Mozo, F Martínez O., Microporous and Mesoporous Materials 265 (2018) 202–210.

614105 Tuning Catalysts for Efficient Chemical and Energy Transformation

Abstract:

The activity of platinum group metals have placed majority of these catalysts far above other types of metal catalysts, turnover frequencies have typically landed them on top of the volcano curves of most chemical reactions. However, its biggest attraction is also the reason for its lack of practicality; two of the biggest problems involving platinum group metals include lack of selectivity control, and susceptibility to poison[1]. Hydrodeoxygenation (HDO) is a key chemistry in pushing forward more bio-derived feedstocks in the industry, succinic acid -through HDO- has been highlighted as a potential alternative to maleic anhydride for the production of 1,4-butanediol, gamma-butyrolactone and tetrahydrofuran. However, upgrading these carboxylic acids proves problematic over the preferred catalysts (noble metals) because of a lack of selectivity control. These active catalysts facilitate a multitude of undesired reactions such as hydrogenolysis, decarbonylation, decarboxylation, and methanation, which all lead to a slew of low-value products[2]. It is widely reported that the addition of oxophilic "promoter metals," such as Sn can be used to tune the selectivity of noble metals during the HDO of carboxylic acids[3]; however, their mechanism of action is poorly understood. Another common problem in both upgrading and extracting energy (fuel cells) from carboxylic acids is formation of carbon monoxide, which has a tendency to inhibit reaction rates. A new approach has been brought to the forefront in modern catalyst operation, dynamic catalysis. This involves the use of an external stimuli in order to oscillate between energy state[4]; these different states will manipulate the rate determining steps (rds) in a reaction mechanism, bouncing between different rds's has the potential operate at rates magnitudes higher than operating under a static condition. In this poster, we use two separate approaches for solving the practical limitations of platinum group metals.

First, we use propionic acid HDO as a model system for tackling selectivity control over supported Pt catalysts. While succinic acid HDO might be more practical to an industrial setting, a simpler molecule is ideal for a fundamental study as it captures all the essential chemistries expected without a product stream being overcrowded. We will dissect HDO activity over monometallic Pt using experimental and computational techniques, providing macroscopic insight on reaction pathways and major inhibitors to HDO selectivity. This shall play a crucial role in understanding how the addition of an oxophilic promoter, Sn, can intrinsically tune the catalytic activity towards higher yields of HDO products.

Second, we explore the effect of dynamic catalysis on carboxylic acid oxidation using electric fields. Formic acid oxidation will be used as a model reaction for the production of carbon dioxide, this is common reaction in fuel cells. Oscillating in between different applied potentials will allow us to observe this concept experimentally and the degree to which a rate increase can be achieved using a potentiodynamic approach while also overcoming the hurdle of product inhibition.

Research Interests

Oxygenated compounds have become a major corner stone in today's industrial landscape, finding itself on the forefront of energy storage as well as chemical production. The expansion of modern catalysis is tantamount to efficiently carry out these chemical processes in order to meet social demands. As there are many aspects of a chemical process that can be approved upon, there are also many approaches one can take to solve current problems, and in some instances, a combination of approaches. Bimetallic catalysts have demonstrated it's resistance to product inhibition though the production of oxygenated products over typical undesirable ones that can be formed over highly active catalysts[3]: undesired compounds include carbon monoxide, carbon dioxide and alkanes/alkenes. Dynamic catalysis has recently been brought to the forefront as a means of increasing catalytic turn over frequencies by switching between catalytic energy states, theoretical and experimental work has exhibited reaction rates orders of magnitude higher during dynamic operation versus static operation[4-5]. It has also been reported through a theoretical investigation, that dynamic catalysis also has the capability of tuning the selectivity of pathways by making alterations to the amplitude and frequency of the stimuli promoting the dynamic operation[6]. My research interests include leveraging the favorable properties of bimetallic catalysts along with the vast potential of dynamic catalysis to tune the selectivity while increasing the rates of reactions involving heavily oxygenated compounds. There have been past studies that unknowingly demonstrated this idea to be plausible in an experimental setting[7-8], however without approaching the concept with the specific purpose of uncovering it's mystery, there will be gaps in the overall understanding that will dampen it's journey to real world applications. These research interests can find a place within many different industrial sectors such as energy extraction, biomass upgrading and pharmaceuticals; a refined solution to a commonly encountered problem.

References

1. Gurbuz, E.; Bond, J. Q.; Dumesic, J. A; Román-Leshkov, Y. 261-288. Elsevier, 2013.
2. Lugo-José, Y. K.; Monnier, J. R.; Williams, T. C. Appl. Catal. A: General 469, 410-418. 2014.
3. Vardon, D. R., Beckham, G. T., et al. ACS Catal. 7, 9. 2017.
4. M. Alexander, A.; Omar A., A.; Paul, D., Principles of Dynamic Heterogeneous Catalysis: Surface Resonance and Turnover Frequency Response. 2019.
5. Gopeesingh, Joshua; Ardagh, Matthew; Shetty, Manish; Burke, Sean; Dauenhauer, Paul; Abdelrahman, Omar (2020): Resonance-Promoted Formic Acid Oxidation via Dynamic Electrocatalytic Modulation. ChemRxiv. Preprint.
6. Ardagh, M. A.; Shetty, M.; Kuznetsov, A.; Zhang, Q.; Christopher, P.; Vlachos, D.; Abdelrahman, O.; Dauenhauer, P. J., Chem. Sci. 2020.
7. Adžić, R. R.; Popov, K. I.; Pamić, M. A., Electrochim. Acta 1978, 23 (11), 1191-1196.
8. Juan Victor Perales-Rondón; Adolfo Ferre-Vilaplana; Juan M. Feliu; Enrique Herrero. Journal of the American Chemical Society. 2014.

614106 Motility and Transport Behavior of Soft Colloids

Abstract:

Research Interests:

In this poster, I present an overview of my research on building active colloids and colloidal systems, and manipulating their transport behavior through external (imposed gradient/ microenvironment) and intrinsic (materials/ size/ shape of colloid) means. As an aspiring industry candidate, I envision working at the interface of chemical engineering, biochemistry and material science to **tackle industry challenges in formulations and formulation processes**. By applying my **colloidal and surface science expertise**, I aim to **understand the physico-chemical properties** of food, pharmaceutics, and fine chemicals formulations to **develop products that are optimized for customer needs.**

Research Experience:

My research experience of 7 years encompasses areas of active colloids, emulsions and microfluidics. Prior to my PhD work at Penn State Chemical Engineering, I worked as a research engineer at the National University of Singapore for two years, where I also received my undergraduate degree from. Here is a brief description of my research as (1) a PhD student and (2) a research engineer/ undergraduate student.

1. <u>Colloids and active matter (2016 – Present)</u>

During my time at Penn State, I worked to understand the motility and diffusion behavior of soft colloids such as enzymes, proteins and vesicles. A major part of my research comprises of characterizing movement and determining the mechanism of movement of soft colloids. Through this experience I developed expertise in areas of **colloidal synthesis and characterization, colloidal and interfacial science, and fluid dynamics.**

- Developed enzyme-powered protocell nanomotors as potential biocompatible cargo transporters.
- Characterized a new mechanism of motor movement based on solute-membrane interactions
- Applied microfluidic and microscopy techniques to study and characterize the behavior of active molecules and colloids in gradients of solutes
- Experienced in the techniques of microfluidic gradient generators, surface functionalization of diverse colloids such as polystyrene, silica, vesicles, proteins and enzymes and fabricating active colloids
- Characterization techniques like dynamic light scattering, zetasizer, confocal microscopy, fluorescence recovery after photobleaching, enzyme and protein purification, particle size and diffusion analysis

2. <u>Microfluidic emulsion based processing of colloids (2013 – 2016)</u>

During my time at the National University of Singapore, I used microfluidic emulsion generators to synthesize micro/ nanoparticles for a variety of applications. The research I did during my undergraduate thesis project and as a research engineer **culminated into a start-up spin off from my lab.** Through this experience I developed expertise in areas of **microfluidic emulsion generators, cargo encapsulation and materials formulation.**

- Developed a comprehensive research strategy to understand and evaluate the effect of polymeric materials on crystallization of pharmaceutical formulations through microfluidic emulsion based processing
- Established and optimized the study of mucoadhesive polymers for particle based ocular drug delivery and quantified the extent of mucoadhesion for various biocompatible polymers
- Developed functional food structures of sodium alginate in the form of microparticles and thin fibres using microfluidic emulsion generation techniques
- Worked on the flow based hydrogenation of organic compounds (hexene, nitrobenzene) through tube-based millifluidic reactors

614123 Bulk Property Prediction of Pharmaceutical Powders Via Bond Number Estimation

Abstract:

Research Interests

This work deals with building predictive bulk property models, intended to facilitate the use of dry powder formulations in applications such as direct compaction and continuous manufacturing, for which adequate bulk properties such as packing and flow are required before commencement of production. Such property prediction models can shorten the formulation development stage, since formulators will have an approximation of the adequacy of bulk properties before experimental testing, or level of bulk property enhancement required. These models have additional significance to continuous manufacturing, where intermediate, ad-hoc property enhancement is not possible, and adequate bulk properties are required at the start of the continuous manufacturing line. In cases where bulk property enhancement is required, dry coating via nano-silica is a proven alternative to traditional dry or wet granulation, since dry coating is a relatively quick and simple step which can save time, money and resources. During dry coating, the surfaces of host pharmaceutical powders are covered with guest nano-silica particles, decreasing the inter-particle cohesion forces between host particles, and improving bulk properties. Hence, this work aims to build predictive models, which intake particle scale characteristics to predict bulk powder properties, as this will reflect changes at the particle level, such as dry coating based processing. Bond number, which is the ratio of inter-particle cohesion force to particle weight, has shown promising bulk property prediction results in recent studies. The current work will attempt to investigate the ability of Bond number models to predict bulk properties with larger data sets comprising of commonly used pharmaceutical powders, in addition to their mixtures. Results conclude that the current version of Bond number models are able to accurately predict powder bed porosity and powder flow for many commonly used pharmaceutical powders, in agreement with previous studies. However, there are several cases where Bond number model predictions are inaccurate, such as powder bed porosity of rough surfaced particles and binary mixtures where only one component is dry coated. Attempts are made to adjust the Bond number models using particle scale characteristics, such as particle shape descriptors, in order to improve prediction where it is necessary, while maintaining adequate prediction for other powders. Finally, experimental work done for model testing produced interesting trends which may be useful for pharmaceutical formulators, since they hold true for large data sets comprising of commonly used pharmaceutical powders.

RESEARCH SUMMARY

Influence of natural catalytic environments on activation of cellulose via fast pyrolysis

Cellulose fast pyrolysis is a promising thermal deconstruction technology to breakdown large macromolecules such as cellulose to much smaller sugars and oxygenates that could be upgraded to biofuels and platform chemicals.[1] A significant challenge in realizing this technology as an economically competitive route has been an inability to gain selective control over the product diversification. Gaining such control has been closely linked with gaining a fundamental understanding of the intrinsic pyrolysis reaction chemistry.[2] However, the complex and vast pyrolytic reaction network is known to occur over very short timescales thus making it hard for experiments to distinguish between elementary pathways. Adding to this, the inherent structural complexity of cellulose obfuscates the nature of the chemical environment around reaction centers which makes it difficult to computationally model and study such systems. The research work done by the applicant therefore focuses on using molecular modeling and first principle techniques to study elementary bond breaking-making events involved in the major pathways of cellulose pyrolysis. These theoretical investigations are complemented by state-of-the-art experiments done by collaborators in the Dauenhauer research group.

Activation of 1,4-β glycosidic bonds, polymeric linkages, in cellulose has widely been considered to be the initial and most important step in pyrolysis. Transglycosylation, which involves the simultaneous breaking of glycosidic bond and formation of bicyclic intermediate levoglucosan at the chain end through an intra-molecular proton transfer, is believed to be a favorable concerted mechanism for this initial step.[3] This theoretical mechanism can potentially be validated by performing experiments that would allow one to extract kinetics for initial cellulose degradation. Pulse heated analysis of solid reactions (PHASR) is a novel experimental technique developed in the Dauenhauer group that combines thin film and pulsed heating techniques to obtain temporally resolved, intrinsic kinetics that are devoid of any transport limitations.[4,5] Using PHASR and α-Cyclodextrin, a cellulose kinetic surrogate, Zhu et al.[6] had shown that initial cellulose breakdown has two distinct kinetic regimes – 1) a high temperature (> 467 °C) regime characterized by a high activation energy (~54 kcal/mol) and high kinetic pre-factor (~ 10^{16} s^{-1}) 2) a low temperature (< 467 °C) regime characterized by a much lower activation energy (~ 23 kcal/mol) and pre-factor (~10^7 s^{-1}). While the high temperature regime appears to be in line with the concerted transglycosylation mechanism that has been proposed by previous studies, no theoretical mechanisms appear to agree with the observations made in the low temperature regime. The candidate's research has been directed towards understanding the chemistry that could potentially explain the observed experimental kinetics.

The significantly lower activation energies appeared to be closer to those reported in heterogeneous catalytic C-O bond activation rather than any reported thermal activation mechanisms. This therefore led to the idea that the cellulose matrix, rich in hydroxyl groups, could potentially create environments around the reaction center that would alter reactivity, catalyze transglycosylation and lead to these lower barriers. This hypothesis was initially tested through co-pyrolysis studies which showed that levoglucosan formation was significantly enhanced by hydroxyl containing additives such as fructose thus showing that hydroxyl groups could catalyze transglycosylation. (*) These findings were also in line with previous modeling studies that showed that small solvents such as water could enhance glucose to levoglucosan conversion.[7] Accordingly, a mechanism was proposed which involved a first step of binding of glycosidic bonds to vicinal hydroxyl in the cellulose matrix and a second, much slower step of hydroxyl catalyzed transglycosylation. Cluster models comprising of reacting oligomeric fragments in close interaction with neighboring fragments were used to probe the effects of hydroxyl groups on the proposed mechanism. DFT calculations on such models give a theoretical apparent activation energy of ~22 – 29 kcal/mol depending on the complexity of the cluster model used which are in good agreement with the earlier mentioned theoretical experiments. These claims were further validated through isotopic labeling studies done on cellobiosan using PHASR which show that at low temperatures transglycosylation indeed occurs through a low activation energy route of ~ 26.5 kcal/mol. Analysis of the transition states obtained through our theoretical studies show that these vicinal hydroxyl groups can perform cooperative catalytic roles of stabilizing the charged transition state as well as mediating proton transfer thus leading to low barriers. This work therefore suggests that at low temperatures, where it is difficult to thermally cleave bonds, the cellulose matrix can form specific hydroxyl environments that would allow for catalytic breaking of these bonds. (**)

While these studies report the electronic stabilization roles of hydroxyl groups, there is a need for studying such "3-D structural environments" through dynamic simulations techniques that can capture dynamic arrangements of the vicinal hydroxyl groups around the reaction center. Accordingly, through the use of ab initio molecular dynamics simulations, this research shows that indeed at temperatures close to experimental conditions (397 °C) multiple hydroxyl groups around the reaction center can order themselves to stabilize transition states. These simulations allow us to capture not just electronic but also entropic effects in such catalyzed reactions. These dynamic simulations are part of an upcoming publication that is in preparation. (***)

Another major focus of this research has been directed towards, understanding the role of alkaline earth metal ions that are naturally present in lignocellulosic feedstocks in catalyzing pyrolysis pathways. While some experiments exist that suggest these metal ions can significantly enhance cellulose degradation rates as well as alter product diversification, insights into the molecular roles played by these ions in the elementary reactions is still lacking.[8,9] Through a recent collaborative study (****), we have been able to show that doping cellulose surrogates with calcium salts leads to significant enhancement in degradation rates. With the aid of PHASR, kinetic parameters have been extracted that reveal an activation energy ~49 kcal/mol as well as second order rate dependence on calcium concentrations. Through comprehensive screening of favorable sites for calcium, the computational models show that these ions prefer to bind to hydroxyl groups in cellulose such that they sit between cellulosic sheets. Such binding has been shown to significantly disrupt native hydrogen bonding thus explaining the experimental observation of early and enhanced degradation. DFT derived transition states also show that besides H-bonding disruption these ions can also perform a dual role of electronically stabilize charged transition state. These two roles performed by different Ca ions potentially explains the second order rate dependence.

As part of ongoing research, the candidate is extending their scope of research to pathways that lead to other pyrolytic products, to continue to investigate the role of these natural catalytic environments and metal ions on the kinetics of these pathways. While the research performed thus far has emphasized a need to consider the structural complexity of cellulose while computationally modelling pathways it also establishes how the complexity can play a direct role in altering chemistry and kinetics of initial depolymerization reactions. Through the future upcoming research, we expect to more generally establish the importance of such complex, naturally present environments on the larger pyrolysis reaction network.

References

(1) Stamm, A. J. Thermal Degradation of Wood and Cellulose. *Ind. Eng. Chem.* **1956**, *48* (3), 413–417.

(2) Mettler, M. S.; Vlachos, D. G.; Dauenhauer, P. J. Top Ten Fundamental Challenges of Biomass Pyrolysis for Biofuels. *Energy Environ. Sci.* **2012**, *5* (7), 7797.

(3) Mayes, H. B.; Broadbelt, L. J. Unraveling the Reactions That Unravel Cellulose. *J. Phys. Chem. A* **2012**, *116* (26), 7098–7106.

(4) Paulsen, A. D.; Mettler, M. S.; Dauenhauer, P. J. The Role of Sample Dimension and Temperature in Cellulose Pyrolysis. *Energy and Fuels* **2013**, *27* (4), 2126–2134.
(5) Krumm, C.; Pfaendtner, J.; Dauenhauer, P. J. Millisecond Pulsed Films Unify the Mechanisms of Cellulose Fragmentation. *Chem. Mater.* **2016**, *28* (9), 3108–3114.
(6) Zhu, C.; Krumm, C.; Facas, G. G.; Neurock, M.; Dauenhauer, P. J. Energetics of Cellulose and Cyclodextrin Glycosidic Bond Cleavage. *React. Chem. Eng.* **2017**, *2* (2), 201–214.
(7) Seshadri, V.; Westmoreland, P. R. Roles of Hydroxyls in the Noncatalytic and Catalyzed Formation of Levoglucosan from Glucose. *Catal. Today* **2016**, *269*, 110–121.
(8) Zhu, C.; Maduskar, S.; Paulsen, A. D.; Dauenhauer, P. J. Alkaline-Earth-Metal-Catalyzed Thin-Film Pyrolysis of Cellulose. *ChemCatChem* **2016**, *8* (4), 818–829.
(9) Patwardhan, P. R.; Satrio, J. A.; Brown, R. C.; Shanks, B. H. Influence of Inorganic Salts on the Primary Pyrolysis Products of Cellulose. *Bioresour. Technol.* **2010**, *101* (12), 4646–4655.

List of applicant's research contribution

(*) Maduskar, S.; Maliekkal, V.; Neurock, M.; Dauenhauer, P. J. On the Yield of Levoglucosan from Cellulose Pyrolysis. *ACS Sustain. Chem. Eng.* **2018**, *6* (5), 7017–7025.
(**) Maliekkal, V.; Maduskar, S.; Saxon, D. J.; Nasiri, M.; Reineke, T. M.; Neurock, M.; Dauenhauer, P. Activation of Cellulose via Cooperative Hydroxyl-Catalyzed Transglycosylation of Glycosidic Bonds. *ACS Catal.* **2018**, *9*, 1943–1955.
(***) Maliekkal,V.; Dauenhauer, P.; Neurock, M. Glycosidic C-O Bond Activation in Cellulose Pyrolysis: Alpha Versus Beta and Condensed Phase Hydroxyl-Catalytic Scission *(revisions under review with ACS Catalysis)*
(****) Facas, G. G.; Maliekkal, V.; Zhu, C.; Neurock, M.; Dauenhauer, P. Cooperative Activation of cellulose with Natural Calcium. *(in preparation)*

614157 Amping up the Organic Synthesis in the Pharmaceutical Industry: Explore the Potential of Structure-Property Relationship.

Abstract:

Research Interests

Through my academic career at various universities and institutes, I have acquired a unique perspective in understanding and developing cutting-edge materials and have been exposed to various research areas, including organic chemistry, analytical chemistry, polymer chemistry chemical engineering, materials engineering and separation science. My research is highly interdisciplinary but focuses primarily on the rational design, synthesis and characterization of novel materials. The success in developing new materials, small molecules or any chemical entity in general for a certain application is usually achieved by a keen understanding of the structure-property relationships. Over more than 10 years, I have demonstrated excellence in manipulating material structure and composition to suit the requirements of the desired properties. This extensive experience could seemingly be transferred to the small molecules and new drug development by building the chemical structure-property relationship understanding needed to identify optimal structures for achieving required biological attributes and physical, chemical and pharmaceutical properties relevant to the application.

My PhD. in materials chemistry and the seven years of professional experience in the U.S. Department of Energy National Labs, enable me to acquire a great breadth of scientific knowledge and experimental, communication, technical, and leadership skills, as summarized below.

I have demonstrated excellence and extensive experience in

- Design, synthesis, characterization and modification of small organic molecules
- Porous Materials, polymers and polymer composites
- Nanoparticles synthesis
- Deep knowledge in the structure-property relationships and tuning of material properties to meet performance and reliability requirements.
- Fabrication of membranes, thin films, monolith, pellets and tablets
- Materials manufacturing and characterization
- Development of new chemical entities
- Synthetic organic chemistry
- Multistep synthesis of complex organic molecules
- Development of new synthetic routes (Organic and Inorganic)
- Analytical development and characterization using several chromatographic and biophysical characterization methods
- Separation Science
- Electrochemical techniques
- Drug delivery
- Structural analysis

Formulation chemistry
- Procedures relating to purification and separations
- Water treatment
- Heterogeneous catalysis
- Conducting work at scales spanning the range from laboratory scale to multi-kilogram scale
- Quality control and complying environmental health and safety requirements
- Delivering high-quality results
- Independent planning and execution of experiments, interpretation of data and drawing conclusions from resulting data
- Mentoring of young chemists, scientists and engineers to share best practice and knowledge toward efficiently achieved successful outcomes
- Meeting project timelines and objectives
- Leadership and Stewardship
- Working in several multi-disciplinary project teams
- Managing multiple projects and priorities
- High learning agility
- Creative problem solving
- Operating at pace and agile decision-making
- Cross-functional collaboration
- Remaining current with the latest technologies
- Writing proposals, scientific papers, patents and reports and giving presentations on technical and scientific findings.

Instrumentation skills:

Fourier-transform infrared spectroscopy (**FTIR**), Thermogravimetric Analysis (**TGA**), Ultra Performance Liquid Chromatography (**UPLC**), Gas Chromatography (**GC**), Mass Spectrometer (**MS**), High-Performance Liquid Chromatography (**HPLC**), Nuclear Magnetic Resonance (**NMR**), Single-Crystal X-ray Diffraction (**SCXRD**), Powder X-ray Diffraction (**PXRD**), Scanning Electron Microscope (**SEM**), Energy-dispersive X-ray Spectroscopy (**EDS**), Transmission Electron Microscope (**TEM**), X-ray Photoelectron Spectroscopy (**XPS**), Surface Area and Gas Analyzer, High-Pressure Gas Sorption Analyzer, Glove Box (Gas or Vacuum), Breakthrough Analyzer (**Hiden**), Inductively Coupled Plasma - Optical Emission Spectrometry (**ICP-OES**), Ultraviolet-Visible Spectroscopy (**UV-Vis**), Pycnometer, Differential Scanning Calorimetry (**DSC**), Dynamic Mechanical Analysis (**DMA**), Schlenk Line, Gas Permeation Analyzers (Pure-Gas and Mixed-Gas), Electrochemical Impedance Spectroscopy (**EIS**)

Scholarly Achievements:

- Citations: **879**, h-index: **17**, i10-index: **21,** total publications: **28**
- Peer-Reviewed Research Articles: **25** published, and **4** in preparation (**6** as a corresponding author, **19** as a first author)
- Peer-Reviewed Review Articles: **3** published and **1** in preparation (**2** as a corresponding author and **2** as a first author)
- A cover image for Cell Reports Physical Science - May issue, 2020
- Patents: **1** published and **2** in preparation
- Research Grants: **2** PI (Granted; $322K in 2020 and $120K in 2015), **2** Co-PI (Granted; $400K in 2019 and $506K in 2018), **1** fellowship, **1** conference grant, **2** participation in proposal's writing
- Invited Talks: **3** (MIT, UC-Berkeley Energy Frontier Research Center, Florida Institute of Technology)

- Conference Oral Presentations: **11**
- Conference Poster Presentations: **6**
- In the News: https://www.netl.doe.gov/node/9779; https://www.pnnl.gov/news-media/fast-and-easy-two-step-creates-new-porous-materials; https://www.pnnl.gov/news/release.aspx?id=4490; https://www.pnnl.gov/science/highlights/highlight.asp?id=4883; https://phys.org/news/2018-04-decreasing-magnetism-metallic-core-particles.html; https://www.pnnl.gov/science/highlights/highlight.asp?id=4460

Leadership:

- Leading three projects in collaboration with NETL separation team members
- Leading several projects in collaboration with other national labs and universities

Professor Ju Li (**MIT**), Dr. Praveen Thallapally (**PNNL**), Professor Brian Space (**University of South Florida**), Professor Jeffrey McCutcheon **(University of Connecticut),** Professor Moises Carreon (**Colorado School of Mines**), Professor Omar Farha (**Northwestern University**), Dr. Yoosuf Picard (**Carnegie Mellon University**) and Professor Jeffery Long (**University of California, Berkeley**), Dr. Tony Pham (**University of Tampa**), Dr. Nathalie Rosi **(University of Pittsburgh)**

- Mentored ten students (nine undergraduate students at USF and one Ph.D. student at PNNL)
- Presentations at international/national conferences (**11** talks, **3** invited talks and **6** posters)
- Leading large discussion sessions for groups of 500+ (lectures at Alexandria University) and 200+ (Inorganic Chemistry lectures at USF) students.
- Session chairing in conferences: ACS Meeting, Fall **2019**
- Conference organization: Castle Conference (USF) and International Conference on Chemistry Progress (Alexandria University, Egypt)

614161 Radiolytic Degradation of Organic Solvents in the Actinide Lanthanide Separation Process (ALSEP) for Nuclear Waste Treatment

Abstract:

Research Interests

Hydrometallurgy, liquid-liquid separation, purification, actinide science, radiochemistry

Nuclear reactors provide an attractive option for significant baseload power generation with zero carbon emissions. However, treatment of operation radioactive byproduct is an ongoing challenge. The separation of highly radiotoxic, long-lived radioactive elements from used nuclear fuel (UNF) followed by their conversion into shorter-lived nuclides is instrumental in strengthening interest and public opinion towards nuclear power. Under current industrial processes utilizing liquid-liquid separation, the highly radiotoxic long-lived minor actinide elements, americium and curium, are not efficiently removed from UNF. The selective removal of the minor actinides is difficult due to the co-extraction of lanthanides, which significantly interfere with the treatment process. The Actinide Lanthanide Separation Process (ALSEP), consisting of the extracting organic solvent containing 2-ethylhexyl phosphonic acid mono-2-ethylhexyl ester (HEH[EHP]) and N,N,N',N'-tetra(2-ethylhexyl)diglycolamide (T2EHDGA) in *n*-dodecane, was developed for the selective partitioning of minor actinides from the lanthanides. Due to the inherent radioactive nature of UNF, liquid-liquid separation systems must be resistant to ionizing radiation. Therefore, to improve our understanding under process operating conditions, the ALSEP extracting solvent was equilibrated with metal loaded acidic aqueous phase and subjected to gamma and alpha irradiation. Degradation dose constants revealed greater ligand degradation due to gamma than alpha irradiation, and equilibration with nitric acid did not significantly impact degradation kinetics. However, the presence of nitric acid appears to alter the degradation pathway by favoring the formation of higher molecular weight recombination products. Results revealed the greater radiolytic susceptibility of T2EHDGA over HEH[EHP] could disrupt the minor actinide recovery with increasing radiation exposure.

614184 A Systematic Multiscale Computational Approach for Engineering Novel Adaptive Materials for Biological Applications

Abstract:

Research Interests

My primary research interests are in the application and development of multiscale modeling frameworks for the rational design of functional and adaptive polymeric materials to address challenges related to drug screening and drug delivery. Technological innovations over the last several decades have been powered solely by our ability to uniquely understand the relationship between the chemistry of materials and their emergent macroscopic properties. In fact, these have led to the development of several novel classes of bulk materials such as thermoplastics and thermosets, which are today used widely in both pharmaceutical and manufacturing industries. Recently, to diversify the design parameter space, efforts are being focused on engineering complex materials (for ex. polymer grafted nanoparticle) that comprise two or more constituents with different chemistries. The central idea here is to leverage the intriguing behaviors arising at the interfaces formed by the constituents so that several materials with different macroscopic properties but with little variations in their chemistry can be designed. Though this approach sounds promising, each of these materials is still engineered for a specific targeted application and thus, it would be highly valuable to have one material that can be tuned systematically by suitable external perturbations such that it dynamically alters its microscopic structure and provides different functionalities.

I am particularly interested in identifying the various factors that can be used to realize such adaptive materials. Biology, especially, has umpteen number of examples where protein complexes self-assemble into different higher-order structures depending on several environmental cues such as concentrations, external forces, presence of salts, etc. For example, it is known that actin filaments inside cells form various structures like crosslinked networks, bundled filaments, branched filaments, and so on depending on their concentration and interactions with other proteins. I am interested in elucidating the molecular drivers of these delicate processes and translate them to computationally design synthetic materials with wide-ranging implications. My initial focus is on developing materials that can adaptively recapitulate cellular architectures for drug screening and targeting purposes. Such material when synthesized experimentally would enable rapid and high throughput screening of various drug candidates and identify potential drugs with increased efficiency. My approach will involve a systematic application of multiscale modeling techniques such as atomistic, coarse-graining, enhanced sampling, Monte-Carlo methods, etc. to develop computational models across varying length and time scales as required to understand these complex processes in detail.

Research Experience

My research from Ph.D. to postdoc has always focused on elucidating the fundamental molecular mechanisms connecting a material's structure with its properties. I have worked on several classes of materials ranging from bulk and interfacial polymeric systems to biological protein complexes using multiscale modeling techniques to uncover the molecular determinants of their macroscopic transport, volumetric, thermal, and mechanical properties. In one of the projects during my Ph.D., I identified the key connections between the chemistry of various polyacrylate gels and the transport of small molecules (water and ethanol) through them. In fact, I also studied how these relationships change with the introduction of an interface by supporting polyacrylate gels on another hard substrate. In these studies, I systematically characterized the effects of physical and chemical characteristics of polymers such as their flexibility, hydrophobicity, and interfacial interactions on the transport properties of penetrant molecules. These results are highly generalizable to other polymeric systems and can aid in the design of efficient membranes for separation applications. In my postdoctoral training, I worked on developing a first-of-its-kind particle-based reactive-diffusive coarse-grained model of actin filaments to unravel the various factors driving the formation of higher-order actin assemblies inside a cell. Specifically, I demonstrated how the application of external forces can alter the actin filament structure in turn affecting the overall functionalities of the cell. Although the exact mechanisms leading to the formation of these structures are not known completely, it is established that they impart cells with diverse functionalities such as their ability to move, divide, and transport cargo. A clear understanding of the mechanisms underlying these processes would enable one to engineer adaptive materials that can dynamically alter their structure and provide various functionalities. A key aspect of all my prior works is the application of multiscale computational modeling approaches such as Monte Carlo, atomistic molecular dynamics, coarse-graining, enhanced sampling, and Markov State models to seamlessly connect information across length and timescales thereby allowing to systematically investigate the emergent macroscopic behavior from microscopic mechanisms. In summary, I have more than 7+ years of experience in applying the above computational methods to study a wide variety of systems such as synthetic crosslinked polymer networks, hydrogels, supported polymers, polymer thin films, and protein complexes. I am also highly proficient in (1) implementing new theoretical methods as computational tools using C++/MPI parallel and (2) in automating simulation protocols and data analysis procedures using a combination of C++, MATLAB, Python, and Linux.

Postdoctoral work: "Multiscale Modeling of Actin Filament Networks" under the guidance of Prof. Gregory A. Voth at the University of Chicago.

Ph.D. Dissertation: "Structural and Dynamic Properties of Penetrant Molecules in Unsupported and Supported Hydrated Gels" under the guidance of Prof. Rajesh Khare at Texas Tech University

614193 Hybrid Monitoring Methods for Detection, Diagnosis, and Classification

Abstract:

Many industrial processes collect an abundance of data from different sensors. These sensors often measure a wide variety of physical properties in order to ensure that these parameters are adhering to expected values. This is essential to ensure plant and operator safety, increase economic benefits, and maintain product quality.

A number of algorithms have been developed in order to improve existing fault detection and diagnosis performance. These algorithms integrate a number of different data-driven driven tools and methods. Multiscale wavelet-based representation of data can be used in order to handle data that is autocorrelated, non-Gaussian, and noisy (Bakshi, 1998). Hypothesis testing methods such as the Generalized Likelihood Ratio (GLR) technique can be used in order to provide the best possible detection for a fixed false alarm rate (Reynolds & Lou, 2010). Moreover, certain model-based methods have been developed in order to monitor process drifts and degradations in the process model, even when a process is operating under control (Sheriff et al., 2019).

This work will discuss different hybrid monitoring algorithms that were developed, their features, and practical applications, which include fault detection, diagnosis and classification of the benchmark Tennessee Eastman Process (TEP), and online monitoring of fouling in industrial heat exchangers (Basha et al., 2020; Sheriff et al., 2017, 2018, 2019).

Research Interests: Process systems engineering with an emphasis on the development of machine learning-based methods for process modeling, estimation, fault detection, and control. The algorithms and tools developed are utilized in many applications to improve the operation of various chemical, environmental, biological, and electrical systems.

References

Bakshi, B. R. (1998). Multiscale PCA with application to multivariate statistical process monitoring. *AIChE Journal*, *44*(7), 1596–1610. https://doi.org/10.1002/aic.690440712

Basha, N., Sheriff, M. Z., Kravaris, C., Nounou, H., & Nounou, M. (2020). Multiclass data classification using fault detection-based techniques. *Computers and Chemical Engineering*, *136*, 1–11. https://doi.org/10.1016/j.compchemeng.2020.106786

Reynolds, M. R., & Lou, J. (2010). An Evaluation of a GLR Control Chart for Monitoring the Process Mean. *Journal of Quality Technology*, *42*(3), 287–310. https://doi.org/10.1080/00224065.2010.11917825

Sheriff, M. Z., Karim, M. N., Nounou, H. N., & Nounou, M. N. (2018). Process monitoring using PCA-based GLR methods: A comparative study. *Journal of Computational Science*, *27*, 227–246. https://doi.org/10.1016/j.jocs.2018.05.013

Sheriff, M. Z., Mansouri, M., Karim, M. N., Nounou, H., & Nounou, M. (2017). Fault detection using multiscale PCA-based moving window GLRT. *Journal of Process Control*, *54*, 47–64. https://doi.org/10.1016/j.jprocont.2017.03.004

Sheriff, M. Z., Nounou, H., Nounou, M., & Karim, M. N. (2019). Monitoring process degradation through operating regime based process monitoring. *AIChE Spring Meeting and Global Congress on Process Safety: Process Control Monitoring and Analytics.*

614205 Smart Biomanufacturing: A Process Systems Engineering and Artificial Intelligence Approach

Abstract:

Research Interests

My core research interests consist of data and complex systems under the general theme of decision-making. How to make better decisions is a fundamental question for individuals, organizations, and societies, and there is no easy answer. Systems research, especially process systems engineering (PSE), with the help of artificial intelligence (AI), is the approach I plan to use to tackle some of the biomanufacturing decision-making challenges.

The production and sale of monoclonal antibody (mAb) therapeutics are a hundred-billion-dollar and expanding biopharmaceutical industry to treat cancer, immune disorder, cardiovascular disease, and inflammatory diseases. However, mAbs are currently produced commercially in fed-batch cultures using media recipes and predetermined operating protocols based primarily on heuristics. Such a heuristic-based approach often leads to suboptimal production outcomes and requires costly experimentation to improve existing protocols. On-line control of mAb productivity and product quality metrics is not widely adopted for several reasons. For instance, measurement of critical quality attributes (CQAs) such as glycan distribution (the result of glycosylation) is available only after the process ends, making it impossible to adopt conventional feedback control. Cell growth, metabolism, antibody formation, and glycosylation are complex, multi-scale dynamics that are difficult to model. While currently available models can predict mAb synthesis or glycosylation dynamics accurately under specific process configurations, adapting them to new processes remains challenging.

It is tempting to contemplate adopting a purely data-driven approach; however, the unique challenges associated with modeling biopharmaceutical processes make such an approach less likely to succeed. On the one hand, the biopharmaceutical industry is not data-rich because it is expensive to generate data, and existing data are usually sparse. As a result, models trained using these data may not extrapolate well outside the training data. On the other hand, domain knowledge and mechanistic understanding of biochemical processes can compensate for the lack of data. These two considerations argue for a hybrid approach to the development of a model that is suitable for predicting and for model-based control of CQAs. Meanwhile, PSE techniques such as Kalman filtering and AI techniques such as artificial neural networks can be used to estimate CQAs based on other measurements and signals available on-line. The development of high-fidelity, hybrid, adaptable models and state estimation techniques allows us to apply existing PSE tools to design more efficient biomanufacturing processes and to control them.

614257 Dynamic Process Modeling and System Analyses for Continuous Pharmaceutical Manufacturing

Abstract:

Research Interests

Research Abstract

Given the numerous advantages including improved process efficiency and product quality, continuous manufacturing (CM) has been an active area of research in the pharmaceutical industry. This research focuses on various experimental and modeling strategies, aimed towards process understanding and development of a virtual predictive framework of the overall process. One of the major bottlenecks for successful development of such framework and subsequent commercialization of CM technology is the lack of fundamental understanding of dynamic complexities arising during powder flow. To address this bottleneck, prominent challenges involve detailed understanding and systematic integration of powder dynamics in process operation and system analyses. This is paramount for development of a predictive framework with real-time capability.

Mechanistic, first-principle based particle-scale simulations like Discrete element modeling (DEM) are increasingly employed to explore and extract particle-level information of powder flow and particle dynamics. Such tools are used to investigate powder mechanics during various process conditions and develop predictive models. During the development of such models, there is a balance that needs to be attained between the computational complexity and the model accuracy. Particle-scale simulations, when used for model development, can significantly improve model accuracy, however, at the cost of computational complexity. The computational requirement further increases when integrated with other unit operations in the process flowsheet. Alternatively, computationally cheap models might fail to capture detailed powder dynamics. Thus, there is a need to develop computationally efficient models using mechanistic particle-scale simulations, to be used for real-time prediction within integrated process flowsheets. To further improve the accuracy of these flowsheets, there is a need to develop and integrate real-time material tracking to explicitly track materials along time for lot-to-lot delineation. The process flowsheets, thus developed, can truly be dynamic in nature, while incorporating complexities of powder flow.

Following this ideology, we propose to develop computationally-efficient models of unit operations in combination with DEM, using reduced-order and approximation strategies. Unit operation models, thus developed, are integrated together, within process flowsheets. Dynamic residence time distribution models are also proposed to enable real-time material tracking. Using the improved flowsheet models, we further plan to enhance system analyses tools like sensitivity, feasibility and flexibility analysis to include process dynamics while accounting for uncertainty, due to the variable nature and material properties of pharmaceutical powders. The proposed work would, thus, play a crucial role towards the development of a predictive framework for manufacturing lines, while capturing detailed process and powder dynamics.

614266 Self-Assembly for Colloidal Crystallization and Biomimetic Structural Color

Abstract:

Research Interests: formulation of colloidal dispersions, crystallization of nanoparticles, characterization of defects.

My PhD research focuses on quantifying the relationship between crystal quality of self-assembled colloidal crystal and their optical properties, especially structural color intensity. Structural color is of interest in artificial materials because it arises from the periodicity of crystal structures, and it has better environmental stability than dyes. I fabricate colloidal crystals with structural color by controlling crystallization of colloidal particles via evaporative self-assembly. To study the connection between crystal film quality and structural color intensity, I investigate the crystal reflectivity as a function of film thickness, defect density, and impurity concentration using a combined experimental and computational approach. In the experiments, I formulated a series of colloidal dispersions of submicron-sized polystyrene spheres. By controlling the initial volume fraction of the colloidal dispersion, I varied the final thickness of crystal films after solvent evaporation and measured the thickness profiles using a profilometer. By introducing differently sized particles that function as impurities, I manipulated the defective microstructures of colloidal crystals and characterized different kinds of defects by scanning electron microscopy (SEM). I further analyzed microscope images with MATLAB to quantify crystal quality. To evaluate reflective structural color, I measured the reflection spectra of crystal films using a UV-Vis spectrophotometer and analyzed the reflection peak intensity. In the simulation, I collaborated with a colleague to model crystal structures with different quality via molecular dynamics. Then I simulated the reflection spectra of modeled crystals using the finite-difference time-domain (FDTD) method. The results show that film thickness profoundly affects structural color intensity. The peak reflection of the structural color increases approximately linearly with the film thickness, ultimately reaching reflection saturation. Self-assembled colloidal crystals cannot reach 100% reflection because they contain a variety of defects. Reduction in structural color peak reflection scales with increased defect density and impurity concentration; vacancies have a particularly detrimental effect. These quantitative relationships I have developed serves as a specific platform to guide the design of optical materials and support defect engineering in colloidal crystals.

613990 Quantifying the Effects of Hydrostatic Pressure on the Barrier Function of Mammary Epithelial Cells

Abstract:

Research Interests:

I am broadly interested in numerous areas of industrial R&D, which includes:

- Cell Biology
- Cell Culture Development
- Preclinical R&D
- Biotechnology R&D
- Agricultural R&D Related to Animals, Especially Cows
- R&D related to Skincare, Makeup, and Beauty
- Bioinformatics/Genomics/Gene Therapy
- Histology
- Oncology
- Vaccine Bioprocess R&D
- Antibodies
- Developability
- Pharmacometrics
- Reports and Regulatory Documentation
- Basic and/or Translational Research
- Upstream and Downstream R&D

Research Abstract:

The onset of pregnancy in mammals begets dramatic changes in the mammary gland to prepare for breastmilk production. Tight junctions, intercellular junctional structures that regulate paracellular transport, are crucial for regulating pregnancy-induced morphogenesis in the mammary gland. Strong tight junctions with a high "barrier function" prevent milk from leaking out of the lactating gland into the surrounding interstitium, and weak tight junctions with a reduced barrier function coincide with involution (shrinkage) of the mammary gland during weaning. Although it is known that hydrostatic pressure exerted by accumulated milk helps to weaken tight junctions during involution, the exact mechanism is unclear. To identify how pressure regulates tight junctions, we custom-built a novel cell culture system to expose mammary epithelial cells to different pressures, up to 1.5 kPa, on their apical (luminal) sides. After exposure to pressure for 6 and 24 h, we observed significant decreases in transepithelial electrical resistance (TEER), a measure of barrier function, for pressures greater than 1 kPa (**Figure**). Additionally, we found that exposure to pressure decreases the expression of E-Cadherin at intercellular boundaries. Lastly, we found that exposure to pressure has no apparent effect on either cell proliferation or death. Ultimately, these findings may have implications for better understanding mechanisms underlying complications related to breastfeeding, such as mastitis.

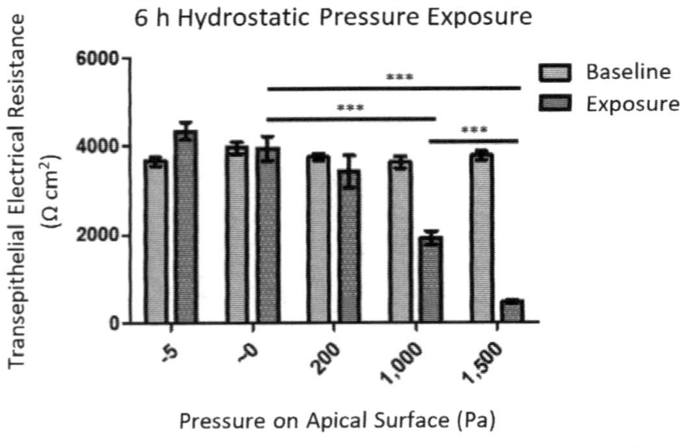

Figure: Transepithelial electrical resistance (TEER) plotted as a function of hydrostatic pressure during a 6 h exposure. The measurements for the "Baseline" group took place right before pressure exposure began. The measurements for the "Exposure" group took place right after exposure ended. Bars denote SEM. *** = $p < 0.001$, n = 4.

613992 Computational Approaches Fueling Clean Materials Chemistry - from First Principles Calculations to Data Driven Discovery and Materials Design

- Download/View Extended Abstract

Abstract:

In the search for cheaper and energetically favorable materials for sustainable applications, there is a desire to catalyze reactions at room temperature and pressure with high active surface area for a number of chemistries. Designing new catalysts in turn improve photovoltaics, chemicals, enzymes, polymers, oil and gas, and automobile industries. MXenes, a new class of 2D materials similar to graphene, are tunable through physical and chemical modifications to their surfaces and have been considered as catalysts for chemical reactions for a decade now.[1,2] For the past decade, learning methods have been used to predict and infer chemical and physical properties through analyzing datasets and building atomistic potentials.[3]

In this poster, I will discuss approaches to performing high throughput computational screening on 2D materials to generate catalytic data based on these chemical and physical alterations to their structure. Computational based discovery requires the use of combined efforts of generating data and applying advanced techniques to understand and make data useful. Previous studies demonstrate our ability to compute data using density functional theory on MXenes for a number of chemical reactions crucial to reaching a clean energy economy.[4–6] With this newly generated data, I perform data analyses of both computed and tabulated features through (a) examining correlations in the features and relationships to our target properties and descriptors of chemical reactions, based on changes to the material's stoichiometry and geometry; and (b) using data science pipelines to generate simple models for structure-property relationships and infer features that point to how our chemistries change due to materials tuning. The methods involved training machine learning models ranging from linear regression and lasso method to neural networks and decision trees through hyperparameter tuning, validation tests, and feature selection through singular value decompositions.

This work provides not only invaluable data to the 2D materials and catalysis communities but crucial analyses and fundamental understanding into how we can design materials and substances with high chemical activity through altering their electronic structure. The study also extends as a guideline to the scientific community on how to treat structured data in an appropriate manner to extract crucial insights needed to design new systems and improve upon existing ones. As a fifth year PhD candidate, the goal is to share how analytical, technical, and professional experience as a researcher, healthcare consultant, masters and undergraduate student mentor, and involved graduate student form the perfect candidate for business development and data related roles in industry.

Research Interests: materials design, catalysis, data science, electrochemistry, density functional theory

Teaching Interests: active learning, flipped classrooms, electronic note-taking, physical chemistry, reactor design, kinetics

References

1. Gogotsi, Y. & Barsoum, M. W. Two-Dimensional Transition Metal Carbides. *ACS Nano* **6**, 1322–1331 (2012).
2. Seh, Z. W. *et al.* Two-Dimensional Molybdenum Carbide (MXene) as an Efficient Electrocatalyst for Hydrogen Evolution. *ACS Energy Lett.* **1**, 589–594 (2016).
3. Kitchin, J. R. Machine learning in catalysis. *Nat. Catal.* **1**, 230–232 (2018).
4. Handoko, A. D. *et al.* Tuning the Basal Plane Functionalization of Two-Dimensional Metal Carbides (MXenes) To Control Hydrogen Evolution Activity. *ACS Appl. Energy Mater.* **1**, 173–180 (2018).
5. Johnson, L. R. *et al.* MXene Materials for the Electrochemical Nitrogen Reduction-Functionalized or Not? *ACS Catal.* **10**, 253–264 (2020).
6. Jin, D. *et al.* Computational Screening of 2D Ordered Double Transition-Metal Carbides (MXenes) as Electrocatalysts for Hydrogen Evolution Reaction. *J. Phys. Chem. C* **124**, 10584–10592 (2020).

613889 Combining Molecular Simulations and Machine Learning for Nanomaterial and Reaction Design

Abstract:

Research Interests

A significant challenge in chemical design is the large number of parameters that can be manipulated. Identifying an optimal set of parameters empirically by trial-and-error approaches is cost-prohibitive, inspiring the development of computational approaches, such as molecular simulations, to help garner insight into how these parameters relate to system properties. My research is focused on combining classical molecular dynamics (MD) simulations and machine learning techniques to systematically tune chemical properties for two relevant application areas. In both applications, I develop accurate predictive models that correlate simulation-derived descriptors to experimental data measured by collaborators, enabling the high-throughput screening of parameters using computationally efficient methods.

Application 1 is motivated by the ability of gold nanoparticles (GNP) to target biomolecules, such as receptors on cancer cells, by attaching target-specific molecules (*i.e.* ligands) on the surface of the particle. Due to the vast number of tuning parameters of the GNP (*e.g.* core size, shape, and ligand selection), it is unfeasible to experimentally screen GNPs. Therefore, I developed a generalized workflow for developing GNP systems using MD simulations for *any* arbitrary gold core shape and size, and ligand selection. I used this workflow in conjunction with unsupervised machine learning clustering algorithms to characterize GNP surface properties. These properties dictate GNP adsorption onto lipid bilayers, which I further interrogate using enhanced sampling techniques (*e.g.* umbrella sampling). These tools could be used to systematically narrow down optimal GNP parameters for drug delivery and cell therapy applications.

Application 2 focuses on converting biomass (renewable, organic material from plants) into transportation fuels or commodity chemicals. Biomass conversion is performed through acid-catalyzed reactions in aqueous solutions that are hindered by low reactivity. One way to improve reactivity is to modify the solvent composition by mixing water with organic cosolvents. Thus, I developed MD simulations to understand and predict the effects of adding an organic solvent on biomass conversion reactivity. I coupled MD simulations with convolutional neural networks, a machine learning model used to classify images, and generalized the reaction rate predictions across 84 reactant-solvent combinations; enabling fast solvent screening for enhanced reaction rates.

My research interests include developing new materials or chemical processes using computational tools, such as molecular simulations (*e.g.* molecular docking, multiscale modeling) in conjunction with machine learning techniques (*e.g.* clustering, neural networks). By joining industry, I aim to broaden my ability to generate efficient and accurate computational models that could accelerate the discovery of new, beneficial products.

613938 Materials Simulation for Manufacturing

Abstract:

Materials simulation for manufacturing

Research Interests:

Manufacturing in the pharmaceutical industry can benefit from particle-based simulations. The efficiency of the unit operations handling powders and pills depend on how those powders and pills interact and flow. Simulations often model such materials as hard spheres. Adhesion, cohesion and (sliding, rolling and twisting) friction are present however, and impact their rheology. Particle-based simulations connect those interactions with emergent behavior. On a smaller scale, molecular simulations have been effective in drug discovery, yet advances in methods and computational power make simulations well poised to address formulation challenges in the pharmaceutical industry. Simulations of nanoparticle, amphiphile and small drug undergoing directed- and self-assembly give microscopic insight and design rules for formulations. I am interested in research that adds value by increasing efficiency in energy, waste and development time. I am eager to identify and support those opportunities as customer needs and market forces changes company value areas. I am qualified for such research because of my background in soft-matter and particle-based simulations.

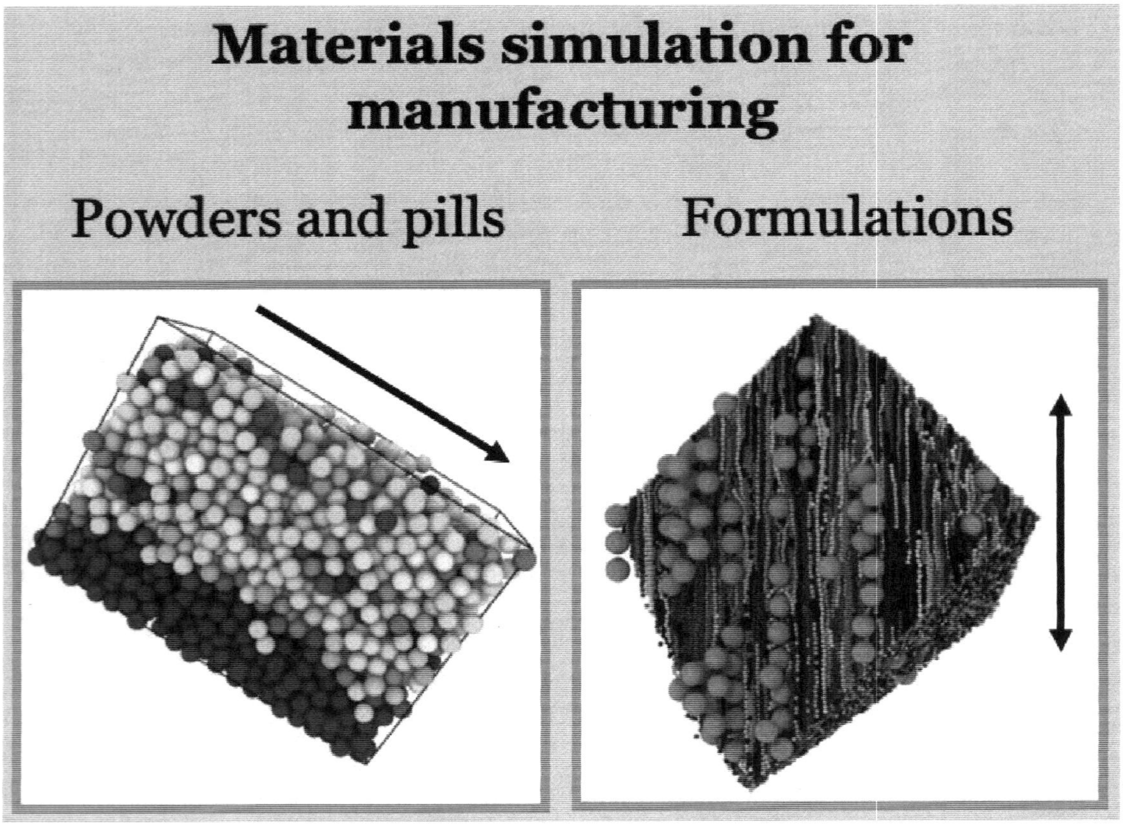

613827 Developing Experimental and Computational Single-Cell Techniques to Uncover Biological Insights

Abstract:

Single-cell genomics is a rapidly advancing field and is generating new knowledge of complex biological systems, ranging from microbial ecosystems diversity, through whole-organism clone tracing, to understanding early mammalian embryogenesis. During my graduate school, I have contributed to the field in two ways: developing a mathematical model to reconstruct cellular lineage in pre-implantation mouse embryos and creating an efficient and cost-effective mRNA enrichment method for prokaryotic cells.

Cellular lineage reconstruction: Lineage reconstruction is central to understanding tissue development and maintenance. While various tools to infer cellular relationships have been established, these methods typically involve genetic modification and have a clonal resolution. I created scPECLR, a probabilistic algorithm to endogenously infer lineages at a single cell-division resolution using 5-hydroxymethylcytosine (5hmC). When applied to 8-cell mouse embryos, scPECLR predicts the full lineage trees with greater than 95% accuracy. Furthermore, I developed a protocol to detect both 5hmC and genomic DNA from the same single cell. Information from genomic DNA, in combination with scPECLR, could allow us to identify cellular lineages more accurately and expand the reconstruction to even larger trees, where the number of possible tree topologies increase to more than 10^{26}. In addition, I showed that scPECLR can be used to map chromosome strand segregation patterns during cell division, thereby providing a strategy to test the controversial "immortal strand" hypothesis in stem cell biology.

Low-input bacterial mRNA sequencing: RNA sequencing is a powerful approach to quantify the genome-wide distribution of mRNA molecules in a population to gain understanding of cellular functions and phenotypes. However, compared to mammalian cells, mRNA sequencing of bacterial samples is more challenging due to their 100-fold lower RNA quantities and the absence of a poly-A tail that typically enables an enrichment of mRNA. To overcome these limitations, I conceived an economical and effective mRNA enrichment method called EMBR-seq. The method results in greater than 80% of the sequenced *E. coli* RNA molecules deriving from mRNA, which originally contributes to lower than 5% of total RNA in a cell. Moreover, EMBR-seq successfully quantifies mRNA from 20 picogram total RNA, a level 500-fold lower than required in existing commercial kits, without introducing technical biases and at a fraction of the cost. Lastly, due to its simplicity and efficiency, EMBR-seq could potentially be extended to a single-cell resolution to advance developments in bacterial mRNA sequencing and improve understanding of microbial systems.

Research Interests: bioengineering, molecular and computational biology, and bioinformatics

NANO LETTERS — ACS Publications — Letter

Cite This: *Nano Lett.* 2019, 19, 8191−8200 — pubs.acs.org/NanoLett

Nucleation and Early Stage Growth of Li Electrodeposits

Prayag Biswal,[†] Sanjuna Stalin,[†] Atsu Kludze,[†] Snehashis Choudhury,[§] and Lynden A. Archer[*,†,‡]

[†]Robert Frederick Smith School of Chemical and Biomolecular Engineering and [‡]Department of Materials Science and Engineering, Cornell University, Ithaca, New York 14853, United States

[§]Department of Chemical Engineering, Stanford University, Stanford, California 94305, United States

Supporting Information

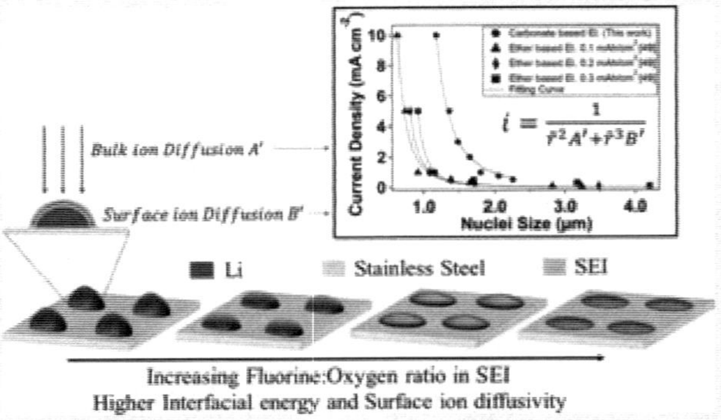

ABSTRACT: The morphologies that metal electrodeposits form during the earliest stages of electrodeposition are known to play a critical role in the recharge of electrochemical cells that use metals as anodes. Here we report results from a combined theoretical and experimental study of the early stage nucleation and growth of electrodeposited lithium at liquid−solid interfaces. The spatial characteristics of lithium electrodeposits are studied via scanning electron microscopy (SEM) in tandem with image analysis. Comparisons of Li nucleation and growth in multiple electrolytes provide a comprehensive picture of the initial nucleation and growth dynamics. We report that ion diffusion in the bulk electrolyte and through the solid electrolyte interphase (SEI) formed spontaneously on the metal play equally important roles in regulating Li nucleation and growth. We show further that the underlying physics dictating bulk and surface diffusion are similar across a range of electrolyte chemistries and measurement conditions, and that fluorinated electrolytes produce a distinct flattening of Li electrodeposits at low rates. These observations are rationalized using X-ray photoelectron spectroscopy (XPS), electrochemical impedance spectroscopy (EIS), and contact angle goniometry to probe the interfacial chemistry and dynamics. Our results show that high interfacial energy and high surface ion diffusivity are necessary for uniform Li plating.

KEYWORDS: *Lithium metal anode, nucleation and growth, surface ion diffusion, interfacial energy*

Lithium (Li) metal is among the most promising anode materials for high-energy and lightweight rechargeable batteries due to its extremely high theoretical specific capacity (3860 mAh g^{-1}), most negative electrochemical potential (−3.040 V versus standard hydrogen electrode), and low density (0.534 g cm^{-3}).[1] Rechargeable Li metal batteries (LMBs) have been extensively studied for over 40 years for prospective applications in portable electronics, electric vehicles, and grid-scale energy storage, though not yet commercialized.[1−4] These studies have revealed multiple shortcomings of Li metal anodes, including the metal's tendency to deposit during charging in low-density mossy morphologies, loosely termed dendrites; its low reversibility and poor Columbic efficiency (CE) in liquid electrolyte media; propensity to proliferate in the interelectrode space to short-circuit the battery producing thermal runaway[1−6] and to undergo large volume changes during repeated cycles of charge and discharge. The "dendritic/mossy" growth of Li exacerbates all of these problems and, additionally, imparts fragility to the metal electrodeposits that can cause them to break away from the electronic circuitry of an electrode, producing so-called "orphaned/dead" Li. The orphaned Li has recently been reported to play an important, perhaps even dominant, role in the poor reversibility and low CE of Li metal anodes.

Significant experimental[1,3,5−9,13] and theoretical[10−13] efforts have been made in recent years to understand and control mossy/dendritic Li electrodeposition. The electrodeposition is presently understood to be destabilized at current densities, i, below the classical diffusion limiting value ($i_L = 4FcD_+/\delta_D$) by at least two processes: (i) morphological instability produced by heterogeneous nucleation of Li deposits in less passivated (faster ion transport) regions of an electrode;[5,7] and (ii) metal extrusion due to heterogeneous interfacial stresses, which produces root-growth of fibrous structures.[8,12,14] For $i > i_L$ morphological instability couples to the classical hydrodynamic instability known as electroconvection to produce much faster dendrite growth.[15] A variety of technical approaches have correspondingly been reported to be effective in eliminating/slowing Li dendrite proliferation by addressing one or more of these instability modes. Among the most effective are: (i) liquid electrolytes containing additives that change the chemistry and transport properties of the solid-electrolyte interphase (SEI) formed in contact with Li;[16−22] (ii) artificial SEIs which simultaneously passivate the Li surface and enable

Received: August 28, 2019
Revised: September 19, 2019
Published: September 30, 2019

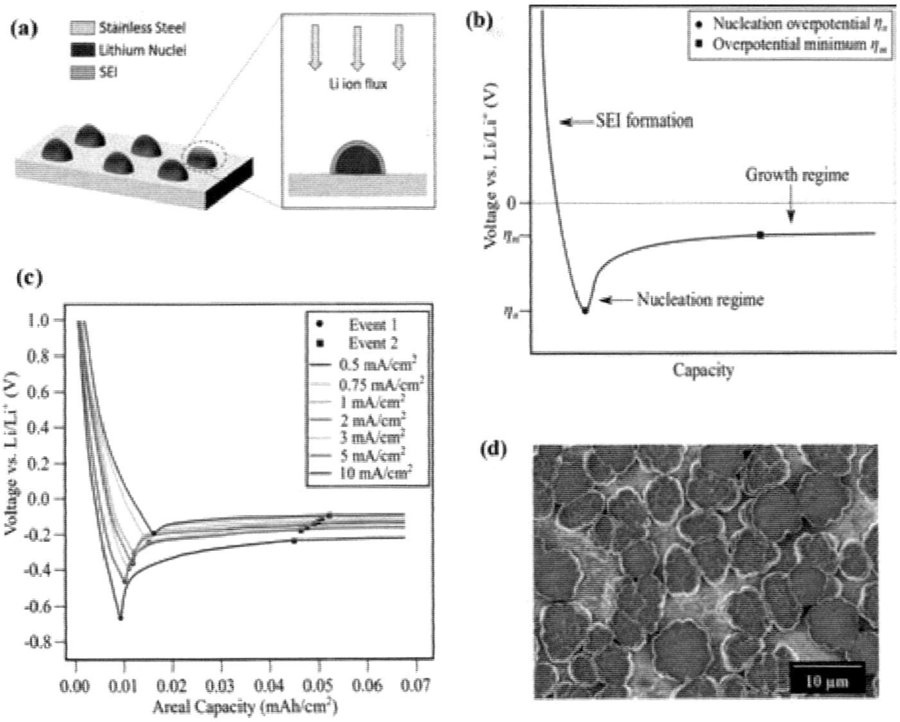

Figure 1. Fundamentals of lithium nucleation and growth. (a) Schematic showing electrodeposition of Li ion from the bulk electrolyte through the SEI on the stainless-steel substrate. (b) Schematic plot showing a typical voltage profile of galvanostatic Li deposition (black). Nucleation overpotential and overpotential minimum are marked by Event 1 and Event 2. (c) Experimental voltage profiles of galvanostatic Li deposition for a range of current densities. The electrolyte is 1 M $LiPF_6$ in EC/DMC (1:1 vol %). Nucleation overpotential and overpotential minimum are marked by Event 1 and Event 2. (d) SEM image of the electrodeposited lithium morphology captured at a current density of 0.5 mA/cm^2 and an areal capacity corresponding to that of overpotential minimum. The electrolyte is 1 M $LiPF_6$ in EC/DMC (1:1 vol %).

fast and less heterogeneous ion transport;[5,23−31] ion-transport regulators that may alter either the ion-flux to the electrode, size distribution the electrodeposit structures, or both;[32−37] (iii) high-shear modulus solid and viscoelastic liquid electrolytes, which slow localized growth of dendritic structures by interfacial mechanical and/or polymer stresses;[38−43] and (iv) advanced current collectors that enable dissipation of localized stresses developed in the metal during deposition[14,44−46] or which facilitate continuous electrical connection and, thereby, electrochemical access to orphaned metal deposits.[44] Although the majority of these studies emphasize the ability to produce anodes with nondendritic morphology after several charge/discharge cycles, the recent work by Zheng et al.[47] raises doubt about whether one is every truly able to achieve dendrite-free electrodeposition of Li on continuum length scales and in liquid electrolytes.

Here, we investigate the initial nucleation and growth dynamics of metallic lithium in liquid electrolytes with/without components believed to make the deposition nondendritic. We show that ionic diffusion in the solid−electrolyte interphase formed spontaneously on Li plays a crucial role in the development and growth of electrodeposited Li nuclei. We show further that by combining theoretical analysis of the early stage electrodeposit growth dynamics with experiments, it is possible to develop a comprehensive picture of the lithium electrodeposition process. The study builds upon the existing nucleation and growth framework proposed by Barton et al.[48] and the recent work of Pei et al.[49] to elucidate the role interphases formed on Li play in regulating ionic transport to the growing nuclei. Lithium ions are galvanostatically electrodeposited onto a heterogeneous surface composed of polished stainless-steel (rms roughness = 9.12 ± 8.78 nm) from the bulk electrolyte (Figure 1a). The electrodeposition is carried out at different current densities in carbonate liquid electrolytes with/without fluorinated carbonated additives reported to suppress the chemical and morphological instability of Li.[18,50−52] The spatial characteristics (morphology, size, number density) of Li nuclei formed in the process are analyzed ex situ using scanning electron microscopy (SEM) and the interphases formed on the deposits studied using electrochemical impedance spectroscopy (EIS), goniometry, and X-ray photoelectron spectroscopy (XPS).

To form nuclei of a particular size r at an electrode, reduced metal ions at the electrode must overcome a nucleation barrier that can be conveniently manipulated by adjusting the electrochemical supersaturation at the electrified interface. A measurable physical parameter characteristic of the electrochemical supersaturation is the overpotential.[48,49,53,54] The driving force for nucleation may therefore be divided into three components: the charge transfer overpotential, diffusion overpotential, and interfacial (surface) energy overpotential.[48,49,53,54] An additional consideration in the nucleation of reactive metals like Li concerns barriers to ionic transport through the SEI formed spontaneously at the metal/electrolyte interface (Figure 1a). These effects are captured theoretically by extending the nucleation and growth model for noninteracting hemispherical nuclei proposed by Barton et al.[48] to account for SEI diffusion

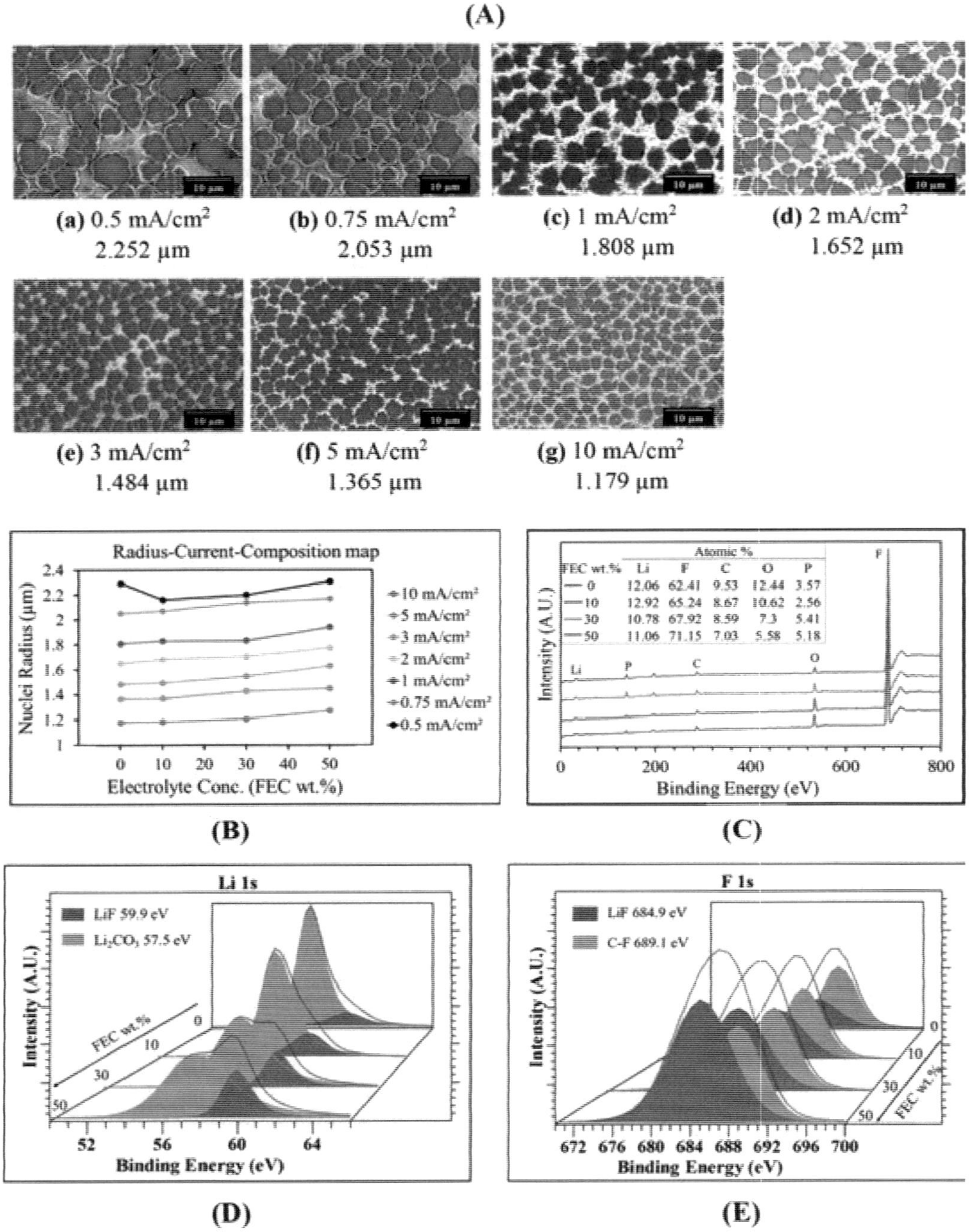

Figure 2. Effect of current density and electrolyte chemistry on the morphology and interfacial chemistry of lithium electrodeposits. (A) Ex situ SEM images of Li deposited on stainless-steel at current densities of (a) 0.5, (b) 0.75, (c) 1, (d) 2, (e) 3, (f) 5, and (g) 10 mA/cm^2, respectively. At lower current densities, the stainless-steel substrate is visible underneath the sparsely distributed Li nuclei. The electrolyte is 1 M $LiPF_6$ in EC/DMC (1:1 vol %). (B) Plot of nuclei radius for a range of current densities for varying concentration of FEC (0, 10, 30, 50 wt %) in 1 M $LiPF_6$ EC/DMC (1:1 vol %). (C) Chemical spectra of the SEI layer induced by varying concentrations of FEC (0, 10, 30, 50 wt %) in 1 M $LiPF_6$ EC: DMC (1:1 vol %). (D,E) Li 1s and F 1 spectra of the SEI layer induced by varying concentrations of FEC (0, 10, 30, 50 wt %) in 1 M $LiPF_6$ EC/DMC (1:1 vol %). The total areal capacity corresponds to that of overpotential minimum (Event 2).

$$\eta = \frac{RT}{\alpha F}\ln\frac{i}{i_0} + \frac{irRT}{D_B C_B F^2} + \frac{2\gamma V}{Fr} + \frac{ir^2 RT}{D_S C_S F^2} \quad (1)$$

Here η is the net overpotential due to all of the processes referenced above; i and i_0 are the current density and exchange current density respectively; r is the radius of the nuclei; D_B and C_B are the bulk electrolyte Li ion diffusivity and concentration respectively; D_S and C_S are the SEI Li ion diffusivity and concentration respectively, γ is the interfacial energy, and α, V, R, T, F are charge transfer coefficient, molar volume of metal, universal gas constant, temperature, and Faraday constant, respectively.

Deconvoluting each of these overpotential contributions from the voltage response during galvanostatic Li electrodeposition is difficult as all are due to electrochemical processes occurring simultaneously. At current densities $i < i_L$, the galvanostatic voltage response curve (Figure 1b) may nevertheless be divided into three distinct regimes: (1) An initial steep voltage drop, which is thought to be the result of reduction of the electrolyte components at the onset of electrode of polarization to produce an ion-conductive, solid-state electrolyte interphase (SEI) at the electrified interface. Because the SEI in typical carbonate liquids is self-limiting, for a clean heterogeneous interface with minimal intercalation or alloying the capacity corresponding to the SEI formation step is quite low.[7] (2) A sharp voltage spike corresponding to formation of Li nuclei's at the electrified interface. The voltage spike reflects the electrochemical supersaturation needed to overcome the nucleation barrier. The voltage at the spike is therefore referred to as nucleation overpotential (η_n) and is marked as Event 1 in Figure 1b. The nucleation overpotential is a consequence of simultaneous charge transfer process and Li-solution interface formation process. Hence, it can be interpreted as a combination of charge transfer overpotential and interfacial formation overpotential. (3) A final regime in which the voltage initially rises before approaching a final plateau. This regime corresponds to the postnucleation growth phase. The overpotential is mainly due to mass-transfer of lithium ions from the bulk electrolyte through the SEI to the newly formed Li nuclei. The onset of voltage plateau is associated with an overpotential minimum (η_m), marked as Event 2 in Figure 1b. The minimum has been reported to show a gradual transition of rate-limiting step from charge transfer/interface formation controlled to mass transfer controlled.[49] The plateau following the overpotential minimum event can be attributed to a combination of bulk diffusion overpotential and SEI surface diffusion overpotential. Understanding how each of these three events is influenced by electrolyte chemistry is the focus of this study. Anticipating comparisons with SEM experiments of electrodeposits, we first consider the prediction of eq 1 for η_m. Minimizing the net overpotential (η) with respect to the characteristic size (radius) of the nuclei, an inverse relationship between current density (i) and radius (r) of the nuclei is apparent (eq 2)

$$i = \frac{1}{\tilde{r}^2 A' + \tilde{r}^3 B'} \text{ where } A' = \frac{RT}{2\gamma V D_B C_B F} \text{ and } B' = \frac{RT}{\gamma V D_S C_S F} \quad (2)$$

This minimum in the net overpotential is associated with a certain characteristic nuclei size, $\tilde{r}$, and a particular Li electrodeposition capacity (Figure 1b,c). Equation 2 indicates that ionic diffusion in the bulk is associated with the term quadratic in $\tilde{r}$, while ionic transport in the SEI is produces a stronger, $\tilde{r}^3$ term in the current. Equation 2 also implies that the coefficients, A' and B', associated with the $\tilde{r}^2$ and $\tilde{r}^3$ terms can be used, respectively, to infer information about the bulk ($C_B D_B$) and surface ($C_S D_S$) ionic conductivity. In other words, from measurements of the electrodeposit size at the beginning of the growth phase, it is possible to deduce information about the bulk and interfacial ion transport in an electrolyte.

We investigated electrodeposition of Li under galvanostatic conditions at a range of current densities (Figure 1c) followed by an ex situ SEM observation of the nuclei size, morphology (Figure 1d), and distribution. A conventional carbonate electrolyte, that is, 1 M $LiPF_6$ in EC/DMC (1:1 vol %), was used in the study. At all current densities studied, the η_m was observed at a relatively small areal capacity of the electrodeposited lithium (between 0.045−0.052 mAh/cm^2, depending on the current density. If the Li electrodeposit density is assumed to be the same as the bulk metal, an equivalent of 250−300 nm of Li metal is deposited by η_m. The capacities for nucleation overpotential and overpotential minimum (Figure 1c) are higher for lower current densities. This may be due to extra capacity required owing to simultaneous formation of nascent SEI and Lithium nuclei at a lower rate of electrodeposition.[49] A lower value of voltage at overpotential minimum than nucleation overpotential suggests that it is favorable for Li ions to electrodeposit on preexisting nuclei's rather than forming new nuclei embryos (Figure 1c).

To facilitate comprehensive post-mortem studies, multiple replicates of Li electrodeposited on polished stainless-steel substrates were collected and analyzed via SEM and XPS. The areal capacity for a given current density was kept fixed for each of the replicates. The morphology of the Li electrodeposits obtained from SEM analysis is reported in Figure 2A. The bare bright stainless-steel substrate can be seen underneath the Li nuclei. The nuclei are distinct in size for different current densities with smaller and densely distributed nuclei occurring at higher current densities. Larger conjoining bloblike-nuclei are visible at lower current densities, while smaller, distinct ones can be seen at higher current densities. At the lowest current density of 0.5 mA/cm^2 nuclei twice the size (2.25 μm compared to 1.18 μm) of those formed at the highest current density of 10 mA/cm^2 are observed. Also, higher nucleation overpotentials (e.g., 0.67 V at 10 mA/cm^2 compared to 0.19 V at 0.5 mA/cm^2) and plateau overpotentials (e.g., 0.23 V at 10 mA/cm^2 compared to 0.09 V at 0.5 mA/cm^2) are observed at higher current densities. These observations can be rationalized as follows. A higher current density imparts a higher activation overpotential and mass transfer overpotential causing the overpotentials needed for the critical events to be higher. A higher nucleation overpotential η_n ensures the competing factors in nuclei formation, that is, Gibbs bulk free energy $(-\frac{4}{3}\pi r^3 F\eta_n/V)$ and the compensating surface free energy $(4\pi r^2\gamma)$ are larger. A larger compensating surface free energy is attained through an increased overpotential required for surface formation $(\frac{2\gamma V}{Fr})$, at the expense of inversely proportional critical size of the nuclei r. Hence, this leads to the formation of a greater number of smaller nuclei on the electrode surface at higher current densities.[49] A greater number of nuclei embryos ensures less capacity of lithium ions deposited per nucleus in the subsequent growth phase, hence smaller lithium nuclei are observed as compared to the low current density case. A similar argument can explain the densely distributed nuclei observed at higher current densities.

The electrolyte chemistry has been reported to play an important role in Lithium electrodeposition.[1,5,7,13] Fluorinated additives, in particular, have received intensive attention because of their reported influence on the morphology of Li electrodeposits formed in charge/discharge battery cycling experiments.[16−19,22,50−52] We studied galvanostatic electrodeposition of lithium in electrolytes containing fluoroethylene carbonate (FEC) as a fluorinated additive. Previous studies showed that this additive breaks down to form LiF and vinylene carbonate (VC), which may electropolymerize at the reducing anode potentials to form an SEI enriched in fluorinated species.[18,50−52] In order to determine the role of FEC on the morphology and interfacial properties of Li electrodeposit nuclei, 1 M $LiPF_6$-EC/DMC electrolytes containing 0 to 50 wt % of FEC were studied. The areal

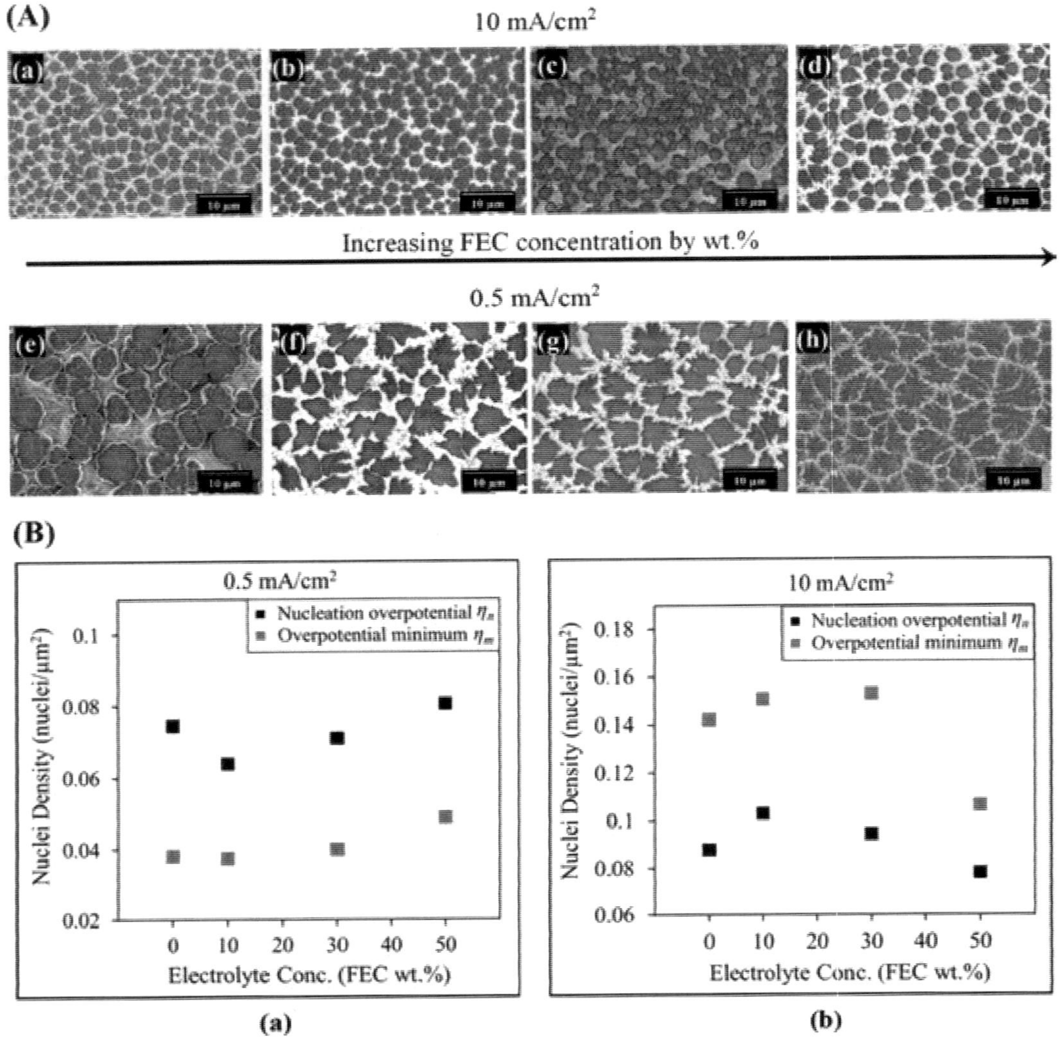

Figure 3. Effect of electrolyte chemistry on the morphology and nuclei density of lithium electrodeposits at contrasting current densities. (A) Ex situ SEM images of Li nuclei deposited on stainless-steel for current densities of 10 and 0.5 mA/cm^2 for increasing concentration of FEC (a,e) 0, (b,f) 10, (c,g) 30, and (d,h) 50 wt % in 1 M $LiPF_6$ EC/DMC (1:1 vol %). The total areal capacity corresponds to that of overpotential minimum (Event 2). (B) Plot of the nuclei density for varying concentration of FEC (0, 10, 30, 50 wt %) in 1 M $LiPF_6$ EC/DMC (1:1 vol %) at (a) 0.5 mA/cm^2 and (b) 10 mA/cm^2. The nucleation overpotential and overpotential minimum are shown in black and red, respectively.

capacity at overpotential minimum η_m was observed to vary with electrolyte composition (i.e., 0.041−0.05 mAh/cm^2 for 10 wt % FEC, 0.037−0.045 mAh/cm^2 for 30 wt % FEC, and 0.034−0.043 mAh/cm^2 for 50 wt % FEC) over the range of current densities investigated (Figure S1).

The morphology of Li-electrodeposited nuclei in the FEC-enriched electrolytes are reported in Figures S2−S4 for electrolytes containing 10, 30, and 50 wt % FEC, respectively. The results show that at a fixed current density, the nuclei are enlarged and, particularly at low current density, noticeably flatter (i.e., ink stainlike, as opposed to hemispherical) at the higher FEC concentrations. The results reported in Figure 2B show that the average radius of Li electrodeposit nuclei at η_m measured in the FEC-enriched electrolytes decrease with increasing current density, in essentially the same manner as observed in the baseline 1 M $LiPF_6$ EC:DMC case. The distribution of nuclei sizes was recovered from the images using the image analysis software ImageJ and represented as histograms in Figures S5−S8. For all electrolytes, the distribution is noticeably broader at lower current densities. For example, for the baseline electrolyte (i.e., no FEC additive) at 0.5 mA/cm^2 the relative standard deviation of nuclei sizes is about 27%, compared to 22% at 10 mA/cm^2 (Figure S5). It is also apparent that in most cases the nuclei size distributions can be crudely fitted to a normal distribution, implying that the demarcations in Figure 1b are at best crude. Specifically, although a large population of nuclei appear to form at a certain time, the breadth of the distribution suggest that smaller populations of nuclei may develop at later times and grow independently during the early stages of Li deposition.

Simultaneously growing nuclei arising from instantaneous nucleation would lead to a sharper (more peaked distribution), whereas strictly continuous nucleation would lead to a flatter distribution of sizes. The situation is further complicated by the fact that a Gaussian size dispersion can also arise from coalescence/fusing of instantaneously formed nuclei and/or heterogeneous ion transport (e.g., due to a nonuniform distribution of internucleus distances produced by the finite

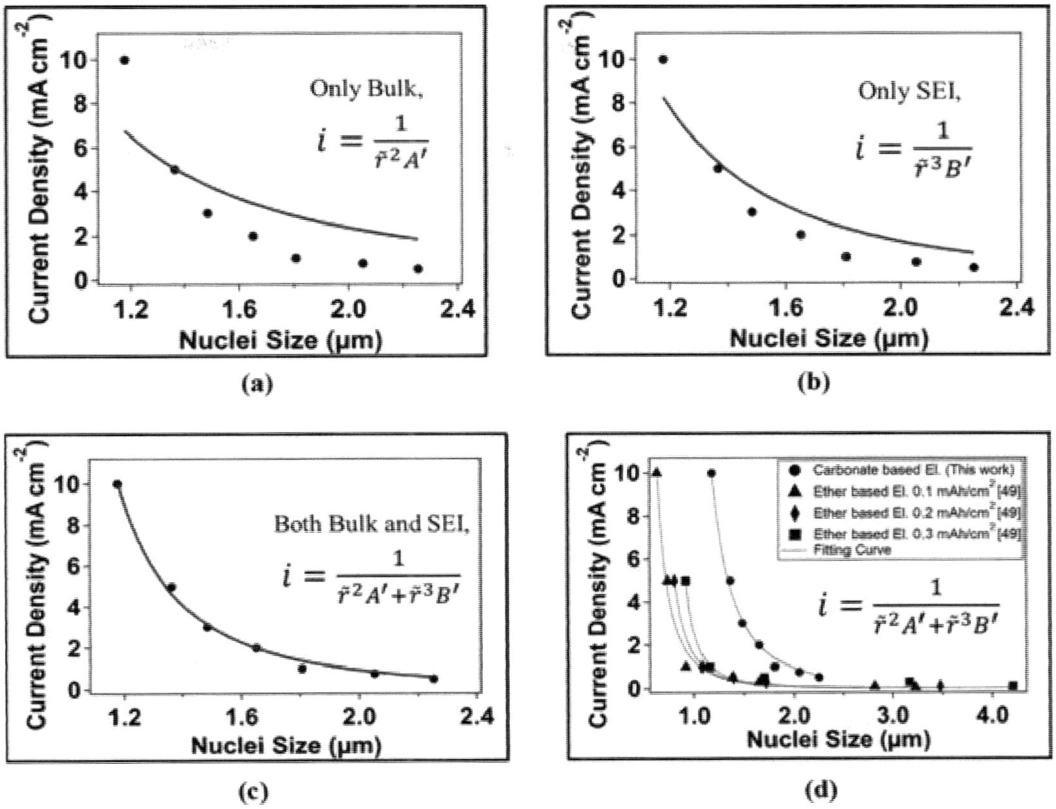

Figure 4. Proof of concept curve fitting at overpotential minimum and growth capacities. (a) The dependency of average nuclei radius on current density has been restricted to the effects of bulk diffusion overpotential only. (b) The dependency of average nuclei radius on current density has been restricted to the effects of SEI surface diffusion overpotential only. (c) The dependency of average nuclei radius on current density involving the synergistic effects of bulk diffusion overpotential and SEI diffusion overpotential. The electrolyte is 1 M $LiPF_6$ in EC: DMC (1:1 vol %) and the nuclei data is collected for areal capacities corresponding to Overpotential Minimum (Event 2). (d) The dependency of average nuclei radius on current density involving the synergistic effects of bulk diffusion and SEI diffusion overpotential for nuclei formed from carbonate-based electrolyte (this work) and ether-based electrolyte.[49]

roughness of the electrode surface) to instantaneously formed nuclei.[49,55−57] A broader distribution at lower current densities may also indicate that internuclei diffusion coupling is important at low rates of electrodeposition.[49]

X-ray photoelectron spectroscopy (XPS) was performed to investigate the surface chemistry of SEI formed in electrolytes containing FEC. The results are reported in Figure 2C−E and Table S1. Although the average nuclei sizes are indifferent to the addition of FEC, distinct changes in the chemical composition of the SEI can be observed. A single dominant peak corresponding to fluorine is observed at about 685 eV; the peak increases in height as FEC content rises in the electrolytes. Analyzing the atomic % it can be inferred that a FEC-enriched electrolyte has greater concentration of fluorine groups in the SEI, that is, 71.15 at. % F for 50 wt % FEC compared to 62.41 at. % F for 0 wt % FEC (Figure 2C). Similarly by contrasting the atomic %, it can also be inferred that the fluorine to oxygen atomic % ratio (F/O) in the SEI increases significantly with addition of FEC. For 50 wt % FEC the F/O ratio is 12.75, which is approximately 250% higher than the corresponding ratio (5.02) at 0 wt % FEC.

A closer look at the high-resolution individual element spectra of Lithium and Fluorine is necessary to further understand the nature of bonding in the FEC-induced SEI. There are two peaks observed in the Li 1s spectra: LiF at 59.9 eV and Li_2CO_3 at 57.5 eV (Figure 2. (D)). The LiF peak is enhanced for higher wt % of FEC i.e. 25.12% LiF, 74.88% Li_2CO_3 at 50 wt % FEC compared to 10.96% LiF, 89.04% Li_2CO_3 at 0 wt % FEC (Table S1). Similar inferences can be drawn for the F 1s spectra where two peaks are also observed: LiF at 684.9 eV and C−F at 689.1 eV (Figure 2E). Along with similar enhancement of the LiF peak, FEC-induced SEI shows a higher proportion of LiF than C−F i.e. 61.65% LiF, 38.35% C−F at 50 wt % FEC contrasted to 37.42% LiF, 62.56% C−F at 0 wt % FEC. The XPS results are consistent with earlier reports[18,50−52] and corroborate the fact that LiF-rich fluorinated SEI is formed via the decomposition of the FEC, even during the earliest stages of Li electrodeposition.

A closer look at the spatial evolution of these nuclei between the two critical events (nucleation overpotential and overpotential minimum) shall aid our understanding of the nucleation process and growth dynamics. We performed detailed galvanostatic electrodeposition studies at two contrasting current densities (0.5 and 10 mA/cm^2) followed by ex situ SEM observation of the nuclei size and morphology (Figure S9). The areal capacity at the η_n were recorded under these conditions: 0.009−0.016 mAh/cm^2 for 0 wt % FEC, 0.008−0.014 mAh/cm^2 for 10 wt % FEC, 0.007−0.012 mAh/cm^2 for 30 wt % FEC, and 0.006−0.011 mAh/cm^2 for 50 wt % FEC depending on 0.5 mA/cm^2 or 10 mA/cm.. Figure 3A

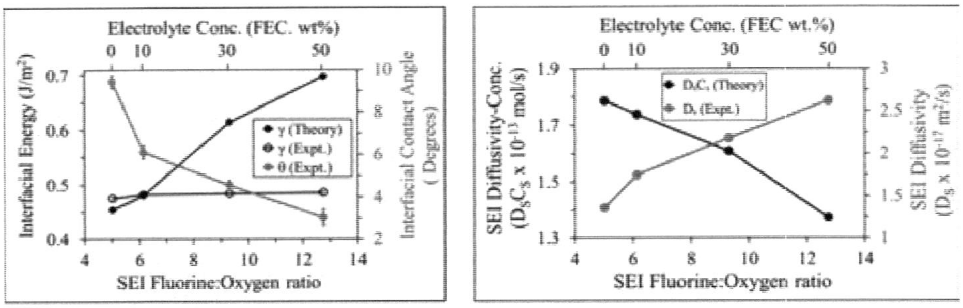

(a) (b)

Figure 5. Effect of electrolyte chemistry/surface chemistry on the interfacial energy, interfacial contact angle, surface diffusivity, and concentration and of the SEI. (a) Plot of SEI interfacial energy (theoretically/experimentally determined) and interfacial contact angle (experimentally determined) for varying concentrations of FEC (0, 10, 30, 50 wt %) in 1 M $LiPF_5$ EC/DMC (1:1 vol %) electrolyte and corresponding fluorine to oxygen ratio in the SEI layer. (b) Plot of SEI surface diffusivity-concentration (theoretically determined) and SEI surface diffusivity (experimentally determined) for varying concentration of FEC (0, 10, 30, 50 wt %) in 1 M $LiPF_6$ EC/DMC (1:1 vol %) electrolyte and corresponding fluorine to oxygen ratio in the SEI layer.

reports the analogous areal capacities recorded at η_m. The areal density of the nuclei for both events were also estimated from the SEM images and plotted in Figure 3B. Overall, densely distributed nuclei occur for the higher current density (10 mA/cm^2) at both events and at all electrolyte compositions. As mentioned earlier, a high current density gives rise to a higher nucleation overpotential at the interface, which leads to formation of a greater number of nuclei.[49]

At the lower current density of 0.5 mA/cm^2, the nuclei sizes observed at η_n are smaller than at η_m. For example, for 0 wt % FEC, the average nucleus size (radius) is about 1.81 μm compared to 2.25 μm at η_m (Figure S9(e) vs Figure 3A(e)). The areal coverage of the nuclei is also higher at η_m (e.g., for 0 wt % FEC at 0.5 mA/cm^2, the nuclei density is about 0.075 nuclei/μm^2 at η_n compared to 0.038 nuclei/μm^2 at η_m). This observation spans all electrolyte chemistries and suggests that as the formed nuclei grow, they also tend to agglomerate/fuse.[49] The decreasing overpotential after the peaking and subsequent plateauing of the overpotential curve (Figure 1c, Figure S1) in between these events is consistent with growth on preexisting nuclei. Simultaneous growth and agglomeration would lead to less dense, larger nuclei at low current density.

There is a significant change in morphology of the electrodeposits at a lower current density for the FEC-enriched electrolytes as is apparent from Figure 3A(e–h) (at η_m) and Figure S9(e–h) (at η_n). At 0.5 mA/cm^2, the nuclei size and density increase with the incorporation of FEC. The morphology of the nuclei transition from hemispherical three-dimesional (3D) spherical structures to flat two-dimensional (2D) structures. These differences are apparent even at 10 wt % FEC but are more pronounced at higher FEC content. Figure S10 shows the effect the fluorinated electrolyte chemistry at an even lower current density of 0.05 mA/cm^2. The planarizing effect is evident at even lower FEC contents and the nuclei appear to flow into each other, more fully utilizing the surface area of the stainless-steel substrate. The capacity at the critical events are also lower (see Figure S11a), implying that the 2D nuclei require lower capacity to nucleate and grow at low current densities. None of these electrolyte chemistry-dependent effects are observed at a high current density of 10 mA/cm^2 (see Figure 3A(a–d)) indicating that their origins are associated with changes in deposition dynamics induced by the FEC. Previously, XPS analysis revealed that the SEI formed on Li in a FEC-containing electrolyte is enriched in fluorine-containing species, including LiF. Also, a LiF-rich SEI has been postulated to have high surface energy and to pose lower barriers to surface diffusion.[58−61] Higher interphase mobility enhances rearrangement of Li ions prior to deposition, facilitating more uniform and compact electrodeposition. Hence, it is relevant to investigate the role of physical parameters such as interfacial energy and surface diffusion in the planarizing effect of fluorinated additives.

The theory correlating the current density with the observed nuclei size (eq 2) serves as a useful tool for unravelling the contributions made by surface energetic, bulk, and surface transport parameters. The squared/cubic relationship of the nuclei radius with the applied current density i.e. $i = \frac{1}{\tilde{r}^2 A' + \tilde{r}^3 B'}$ contains coefficients (A' and B') which have surface diffusivity (D_S), surface concentration (C_S), and interfacial energy (γ) embedded in them. A proof of concept analysis was performed to fit the current density with the observed nuclei size $\tilde{r}$ at η_m using eq 2 (see SI on curve fitting). Three distinct cases were considered to determine the relevance of each mass transfer overpotential in the growth phase as demonstrated in Figure 4a–c. In Case 1 (Figure 4a), we ignore surface diffusion (i.e., $A' \gg B'$) and in Case 2 (Figure 4b) we ignore the contribution to the transport overpotential originating from ion migration in the electrolyte bulk (i.e., $B' \gg A'$). In Case 3 (Figure 4c) we consider the case where the surface and bulk diffusion make comparable contributions to the transport overpotential. The vastly improved quality of the fit for Case 3, relative to Cases 1 and 2 confirm that a combination of mass transfer overpotentials due to bulk diffusion and SEI diffusion are important in determining η_m. A similar proof of concept analysis was carried out for all of the fluorinated/nonfluorinated carbonate-based electrolyte compositions studied as shown in Figure S12. Motivated by the ability of eq 2 to quantitatively replicate the i versus $\tilde{r}$ data for the full set of carbonate electrolytes used in the study, we also compared the model predictions to literature results for ether-based electrolyte compositions studied in the work of Pei et al.[49] The results reported in Figure 4d show that the model predictions are in quantitative accord with the observations reported by Pei et al.;[49] the corresponding parameters are provided in Table S2. We therefore conclude that in every situation, the Case 3 physics provide the best description of available $\tilde{r}$ versus i data for Li electrodeposition in liquid electrolytes.

We estimated the interfacial energy (γ) for lithium in carbonate electrolytes with different FEC contents using contact angle measurements performed in an argon-filled glovebox. The results reported in Figure 5a and Table S3 show that γ increases gradually from 475 to 486 mJ/m^2 as the FEC content is increased from 0 to 50 wt %. Electrolyte wetting of studies (Figure S13) indeed reveal better Li wettability at higher concentration of FEC. The average contact angle formed between the electrolyte and Li ranges from 9.42° to 3.03° when the FEC content ranges from 0 wt % FEC to 50 wt % (Figure 5a). The lowering of the contact angle is consistent with the higher γ. Using the best-fit values of A' and B' obtained by fitting the empirical $\tilde{r}$ versus i data using eq 2, we calculate the actual γ-values under conditions of the electrodeposition experiments. The results (Figure 5a) show that while the values of γ are comparable to the ones estimated from the contact angle experiment, those obtained by fitting the Li nucleate size data are a stronger function of FEC content (γ ranges from 454 to 697 mJ/m^2 with the addition of FEC). The mismatch between the experimentally estimated and theoretical values of γ can be 2-fold. First, Young's equation employed to calculate the interfacial energy is valid for interfaces with lower interfacial energy than the solid substrate (in this case lithium).[62] Second, the interphases formed on Li by reduction of FEC under the deposition conditions are more enriched in fluorinated species than those formed on Li in contact with the electrolyte under ambient conditions.

Electrochemical impedance spectroscopy was performed on lithium electrodeposited on polished stainless steel to determine the surface and bulk transport characteristics (Figure S15 and Figure 5b). The product of surface diffusivity and surface concentration decreases with the incorporation of FEC as shown in Figure 5b. However, the surface diffusivity of Li ions is observed to increase with increasing concentration of FEC (e.g., D_S is 2.62 × 10^{-17} m^2/s at 50 wt % FEC contrasted to 1.36 × 10^{-17} m^2/s at 0 wt % FEC), a roughly 2-fold increase in D_S. A decreasing C_SD_S value therefore means that the surface concentration of Li ions at the electrode decreases quite rapidly as FEC concentration rises (Table S4). The surface concentration C_S is 1.313 × 10^4 mol/m^2 at 0 wt % FEC compared to 0.523 × 10^4 mol/m^2 at 50 wt % FEC, that is, more than a 2-fold decrease in the concentration, which is consistent with the two-step (surface ion transport → Li-ion reduction) reaction assumed in eq 2 when the surface ion transport is rate limiting.

A higher surface diffusivity indicates relatively easier 2D migration of Li ions in the fluorinated SEI, promoting more spread-out/delocalized electrodeposition. Assuming fast reduction kinetics of Li-ions at the electrode, a higher D_S would also lower the concentration of Li ions in the SEI. This also explains the lowering of the SEI and charge transfer resistance evident in Figure S15c and is consistent with previous reports.[50] Synergistically, high surface diffusivity and high interfacial energy of the fluorinated SEI facilitates planar electrodeposits as observed earlier for fluorinated electrolytes (10−50 wt % FEC). On the other hand, the carbonated electrolyte (0 wt % FEC) lacking such interfacial advantages forms 3D, bloblike nuclei which have higher probability of succumbing to the morphological instability that drives growth of mossy/dendritic Li electrodeposit morphologies. Recent modeling studies have shown the SEI diffusivity[63−66] to be anywhere in between 10^{-16} to 10^{-26} m^2/s and SEI surface energy[12,67−69] to be in between 0.18 to 0.84 J/m^2 with higher values (10^{-16} m^2/s and 0.84 J/m^2) reported for a purely LiF interface. The range of surface diffusivity and interfacial energy values deduced by fitting eq 2 to the empirical $\tilde{r}$are in rough agreement with these values. Figure S15d shows further that the bulk diffusivity (10^{-10} m^2/s) of Li ions is about 7 orders of magnitude higher than the surface diffusivity (10^{-17} m^2/s), implying that even a highly fluorinated interphase has a throttling effect on transport of Li-ions arriving from the bulk. This mismatch in bulk and interphase diffusivities will produce a rapid piling-up of ions at the electrolyte/electrode interface, ultimately leading to nonplanar mossy/dendritic electrodeposition of Li even in fluorinated electrolytes. Similarly, accumulation of Li ion vacancies underneath the SEI during Li anode stripping has been shown to be detrimental for the adherence/passivity of the SEI layer.[70] Electrolyte and interphases that lower the bulk/surface diffusivity ratio are therefore predicted to be essential for achieving compact, planar electrodeposition of Li in liquid electrolytes.

In summary, we study the effect of electrolyte chemistry on the morphology, size, density, spatial distribution, surface chemistry of early stage Li nuclei formed at different rates of electrodeposition. The underlying physics of Lithium (reactive metal) nucleation and growth through electrodeposition is elucidated. A comprehensive understanding of the galvanostatic overpotential curve and the square-cubic inverse dependency of nuclei radius on current density is demonstrated and experimentally correlated. Such a dependency reveals both bulk ion diffusion and surface ion diffusion to play an equally important role in lithium electroplating. Theoretical formulations along with aptly designed experiments also reveal interfacial dynamics of the SEI. Designing interfaces with high surface energy and high surface diffusivity is a feasible solution to eliminate dendritic morphology of electrodeposited lithium. Additional studies dealing with understanding the nucleation and growth dynamics of reactive metals are cardinal to understand the morphological and chemical instabilities originating at early stages of electrodeposition. Understanding and elimination of the instabilities at the initiation step would enable uniform and compact plating of reactive metals.

■ ASSOCIATED CONTENT

Supporting Information

The Supporting Information is available free of charge on the ACS Publications website at DOI: 10.1021/acs.nanolett.9b03548.

Details of curve fitting, voltage response plots, nuclei distributions, SEM, XPS, impedance spectroscopy, and goniometer characterizations, coin cell fabrication, electrochemical testing methods, image analysis, and additional figures (PDF)

■ AUTHOR INFORMATION

Corresponding Author

*E-mail: laa25@cornell.edu.

ORCID

Lynden A. Archer: 0000-0001-9032-2772

Notes

The authors declare no competing financial interest.

■ ACKNOWLEDGMENTS

The research was supported as part of the Center for Mesoscale Transport Properties, an Energy Frontier Research Center supported by the U.S. Department of Energy (DOE), Office of Science, Basic Energy Sciences, under Award DE-SC0012673. The electron microscopy facilities made use of the Cornell Center for Materials Research Shared Facilities which are supported through the NSF MRSEC program (DMR-1719875). The interface characterization work made use of the facilities available in CESI, Cornell. The authors would like to thank John Grazul of CCMR, Cornell and Dr. Jiefu Yin of CESI, Cornell for their support in setting up the experiments.

■ REFERENCES

(1) Zhang, J. G.; Xu, W.; Henderson, W. A. *Lithium metal anodes and rechargeable lithium metal batteries*; Springer International Publishing: Switzerland, 2017.

(2) Whittingham, M. S. History, evolution, and future status of energy storage. *Proc. IEEE* **2012**, *100*, 1518−1534.

(3) Aurbach, D.; Cohen, Y. *J. Electrochem. Soc.* **1996**, *143*, 3525−3532.

(4) Xu, W.; Wang, J.; Ding, F.; Chen, X.; Nasybulin, E.; Zhang, Y.; Zhang, J. G. *Energy Environ. Sci.* **2014**, *7*, 513−537.

(5) Tikekar, M. D.; Choudhury, S.; Tu, Z.; Archer, L. A. *Nat. Energy* **2016**, *1*, 16114.

(6) Chianelli, R. R. *J. Cryst. Growth* **1976**, *34*, 239−244.

(7) Aurbach, D.; Zinigrad, E.; Cohen, Y.; Teller, H. *Solid State Ionics* **2002**, *148*, 405−416.

(8) Zachman, M. J.; Tu, Z.; Choudhury, S.; Archer, L. A.; Kourkoutis, L. F. *Nature* **2018**, *560*, 345.

(9) Chandrashekar, S.; Trease, N. M.; Chang, H. J.; Du, L. S.; Grey, C. P.; Jerschow, A. *Nat. Mater.* **2012**, *11*, 311−315.

(10) Monroe, C.; Newman, J. *J. Electrochem. Soc.* **2005**, *152*, A396−A404.

(11) Tang, M.; Albertus, P.; Newman, J. *J. Electrochem. Soc.* **2009**, *156*, A390−A399.

(12) Yamaki, J. I.; et al. *J. Power Sources* **1998**, *74*, 219−227.

(13) Wood, K. N.; Kazyak, E.; Chadwick, A. F.; Chen, K. H.; Zhang, J. G.; Thornton, K.; Dasgupta, N. P. *ACS Cent. Sci.* **2016**, *2*, 790−801.

(14) Wang, X.; Zeng, W.; Hong, L.; Xu, W.; Yang, H.; Wang, F.; Duan, H.; Tang, M.; Jiang, H. *Nat. Energy* **2018**, *3*, 227.

(15) Li, G.; Archer, L. A.; Koch, D. L. *Phys. Rev. Lett.* **2019**, *122*, 124501.

(16) Miao, R.; Yang, J.; Feng, X.; Jia, H.; Wang, J.; Nuli, Y. *J. Power Sources* **2014**, *271*, 291−297.

(17) Qian, J.; Henderson, W. A.; Xu, W.; Bhattacharya, P.; Engelhard, M.; Borodin, O.; Zhang, J.-G. *Nat. Commun.* **2015**, *6*, 6362.

(18) Suo, L.; Xue, W.; Gobet, M.; Greenbaum, S. G.; Wang, C.; Chen, Y.; Yang, W.; Li, Y.; Li, J. *Proc. Natl. Acad. Sci. U. S. A.* **2018**, *115*, 1156−1161.

(19) Gordin, M. L.; Dai, F.; Chen, S.; Xu, T.; Song, J.; Tang, D.; Azimi, N.; Zhang, Z.; Wang, D. *ACS Appl. Mater. Interfaces* **2014**, *6*, 8006−8010.

(20) Aurbach, D.; Pollak, E.; Elazari, R.; Salitra, G.; Kelley, C. S.; Affinito, J. *J. Electrochem. Soc.* **2009**, *156*, A694−A702.

(21) Kim, H.; Wu, F.; Lee, J. T.; Nitta, N.; Lin, H.-T.; Oschatz, M.; Cho, W. I.; Kaskel, S.; Borodin, O.; Yushin, G. *Adv. Energy Mater.* **2015**, *5*, 1401792.

(22) Choudhury, S.; Archer, L. A. *Adv. Electron. Mater.* **2016**, *2*, 1500246.

(23) Cheng, X.-B.; Zhang, R.; Zhao, C.-Z.; Wei, F.; Zhang, J.-G.; Zhang, Q. *Adv. Sci.* **2016**, *3*, 1500213.

(24) Lin, D.; Liu, Y.; Cui, Y. *Nat. Nanotechnol.* **2017**, *12*, 194−206.

(25) Zheng, G.; Lee, S. W.; Liang, Z.; Lee, H.-W.; Yan, K.; Yao, H.; Wang, H.; Li, W.; Chu, S.; Cui, Y. *Nat. Nanotechnol.* , *9*, 618−623.

(26) Zhang, J.; Yang, X.; Wang, R.; Dong, W.; Lu, W.; Wu, X.; Wang, X.; Li, H.; Chen, L. *J. Phys. Chem. C* **2014**, *118*, 20756−20762.

(27) Zhang, Y.; Wang, W.; Tang, H.; Bai, W.; Ge, X.; Wang, X.; Gu, C.; Tu, J. *J. Power Sources* **2015**, *277*, 304−311.

(28) Lee, H.; Lee, D. J.; Kim, Y.-J.; Park, J.-K.; Kim, H.-T. *J. Power Sources* **2015**, *284*, 103−108.

(29) Haregewoin, A. M.; Wotango, A. S.; Hwang, B.-J. *Energy Environ. Sci.* **2016**, *9*, 1955−1988.

(30) Umeda, G. A.; Menke, E.; Richard, M.; Stamm, K. L.; Wudl, F.; Dunn, B. *J. Mater. Chem.* **2011**, *21*, 1593−1599.

(31) Wang, X.; Hou, Y.; Zhu, Y.; Wu, Y.; Holze, R. *Sci. Rep.* **2013**, *3*, 1401.

(32) Bates, J. B.; Dudney, N. J.; Gruzalski, G. R.; Zuhr, R. A.; Choudhury, A.; Luck, C. F.; Robertson, J. D.; et al. *Solid State Ionics* **1992**, *53-56*, 647−654.

(33) Li, J.; Ma, C.; Chi, M.; Liang, C.; Dudney, N. J. *Adv. Energy Mater.* **2015**, *5* (4), 1401408.

(34) Kanno, R.; Murayama, M. *J. Electrochem. Soc.* **2001**, *148*, A742−A746.

(35) Lu, Y.; Tikekar, M.; Mohanty, R.; Hendrickson, K.; Ma, L.; Archer, L. A. *Adv. Energy Mater.* **2015**, *5*, 1402073.

(36) Song, J.; Lee, H.; Choo, M.-J.; Park, J.-K.; Kim, H.-T. *Sci. Rep.* **2015**, *5*, 14458.

(37) Cheng, X. B.; Hou, T. Z.; Zhang, R.; Peng, H. J.; Zhao, C. Z.; Huang, J. Q.; Zhang, Q. *Adv. Mater.* **2016**, *28*, 2888−2895.

(38) Stone, G. M.; Mullin, S. A.; Teran, A. A.; Hallinan, D. T.; Minor, A. M.; Hexemer, A.; Balsara, N. P. *J. Electrochem. Soc.* **2012**, *159*, A222−A227.

(39) Giles, J. R. M.; Gray, F. M.; Maccallum, J. R.; Vincent, C. A. *Polymer* **1987**, *28*, 1977−1981.

(40) Khurana, R.; Schaefer, J. L.; Archer, L. A.; Coates, G. W. *J. Am. Chem. Soc.* **2014**, *136*, 7395−7402.

(41) Pan, Q.; Smith, D. M.; Qi, H.; Wang, S.; Li, C. Y. *Adv. Mater.* **2015**, *27*, 5995−6001.

(42) Choudhury, S.; Mangal, R.; Agrawal, A.; Archer, L. A. *Nat. Commun.* **2015**, *6*, 10101.

(43) Gurevitch, I.; Buonsanti, R.; Teran, A. A.; Gludovatz, B.; Ritchie, R. O.; Cabana, J.; Balsara, N. P. *J. Electrochem. Soc.* **2013**, *160*, A1611−A1617.

(44) Yang, C. P.; Yin, Y. X.; Zhang, S. F.; Li, N. W.; Guo, Y. G. *Nat. Commun.* **2015**, *6*, 8058.

(45) Zhamu, A.; Chen, G.; Liu, C.; Neff, D.; Fang, Q.; Yu, Z.; Xiong, W.; Wang, Y.; Wang, X.; Jang, B. Z. *Energy Environ. Sci.* **2012**, *5*, 5701−5707.

(46) Ji, X.; Liu, D. Y.; Prendiville, D. G.; Zhang, Y.; Liu, X.; Stucky, G. D. *Nano Today* **2012**, *7*, 10−20.

(47) Zheng, J.; Tang, T.; Zhao, Q.; Liu, X.; Deng, Y.; Archer, L. A. *ACS Energy Lett.* **2019**, *4* (6), 1349−1355.

(48) Barton, J. L.; Bockris, J. O. M *Proc. R. Soc. London, Ser. A* **1962**, *268*, 485−505.

(49) Pei, A.; Zheng, G.; Shi, F.; Li, Y.; Cui, Y. *Nano Lett.* **2017**, *17*, 1132−1139.

(50) Zhang, X. Q.; Cheng, X. B.; Chen, X.; Yan, C.; Zhang, Q. *Adv. Funct. Mater.* **2017**, *27*, 1605989.

(51) Markevich, E.; Salitra, G.; Chesneau, F.; Schmidt, M.; Aurbach, D. *ACS Energy Lett.* **2017**, *2*, 1321−1326.

(52) Zhang, X. Q.; Chen, X.; Cheng, X. B.; Li, B. Q.; Shen, X.; Yan, C.; Huang, J. Q.; Zhang, Q. *Angew. Chem., Int. Ed.* **2018**, *57*, 5301−5305.

(53) Winand, R. *J. Appl. Electrochem.* **1991**, *21*, 377−385.

(54) Vetter, K. J. *Electrochemical Kinetics Theoretical Aspects*; Elsevier Science: Burlington, 2013.

(55) Fransaer, J. L.; Penner, R. M. *J. Phys. Chem. B* **1999**, *103*, 7643−7653.

(56) Penner, R. M. *J. Phys. Chem. B* **2002**, *106*, 3339−3353.

(57) Serruya, A.; Mostany, J.; Scharifker, B. R. *J. Chem. Soc., Faraday Trans.* **1993**, *89*, 255−261.

(58) Lu, Y.; Tu, Z.; Archer, L. A. *Nat. Mater.* **2014**, *13*, 961.

(59) Gunceler, D.; Letchworth-Weaver, K.; Sundararaman, R.; Schwarz, K. A.; Arias, T. A. *Modell. Simul. Mater. Sci. Eng.* **2013**, *21*, No. 074005.
(60) Michan, A. L.; Parimalam, B. S.; Leskes, M.; Kerber, R. N.; Yoon, T.; Grey, C. P.; Lucht, B. L. *Chem. Mater.* **2016**, *28*, 8149–8159.
(61) Sun, H. H.; Dolocan, A.; Weeks, J. A.; Rodriguez, R.; Heller, A.; Mullins, C. B. *J. Mater. Chem. A* **2019**, *7*, 17782–17789.
(62) Zhu, D.; Liao, X.; Dai, P. *Chin. Sci. Bull.* **2012**, *57*, 4517–4524.
(63) Benitez, L.; Seminario, J. M. *J. Electrochem. Soc.* **2017**, *164*, E3159–E3170.
(64) Guan, P.; Liu, L.; Lin, X. *J. Electrochem. Soc.* **2015**, *162*, A1798–A1808.
(65) Yildirim, H.; Kinaci, A.; Chan, M. K.; Greeley, J. P. *ACS Appl. Mater. Interfaces* **2015**, *7*, 18985–18996.
(66) Tasaki, K.; Goldberg, A.; Lian, J. J.; Walker, M.; Timmons, A.; Harris, S. J. *J. Electrochem. Soc.* **2009**, *156*, A1019–A1027.
(67) Liu, Z.; Qi, Y.; Lin, Y. X.; Chen, L.; Lu, P.; Chen, L. Q. *J. Electrochem. Soc.* **2016**, *163*, A592–A598.
(68) Jand, S. P.; Kaghazchi, P. *J. Phys.: Condens. Matter* **2014**, *26*, 262001.
(69) Kokko, K.; Salo, P. T.; Laihia, R.; Mansikka, K. *Surf. Sci.* **1996**, *348*, 168–174.
(70) Shi, F.; Pei, A.; Boyle, D. T.; Xie, J.; Yu, X.; Zhang, X.; Cui, Y. *Proc. Natl. Acad. Sci. U. S. A.* **2018**, *115*, 8529–8534.

613472 Jet-Mixing Reactor for Nanomaterial Synthesis and Scale-up Considerations

Abstract:

Advances in pharmaceutical discovery and increasing global healthcare demands call for innovative technologies that can produce material at high throughput while retaining quality obtained at the bench scale. Nanomaterials are being increasingly used in pharmaceutical production. The properties of nanomaterials are highly dependent on the mixing dynamics of their synthesis process. Macromixing in conventional batch syntheses increases the overall mixing time, leading to variability within batches and a wide particle size distribution. These problems can be circumvented by using microreactors that provide a small mixing time because of their associated micromixing and mesomixing. This work proposes a continuous, scalable jet-mixing reactor that is used for the synthesis of several metal and metal-oxide nanomaterials. Initially, silver nanoparticles (Ag NPs) are synthesized at ambient conditions as proof-of-concept. It is observed that Ag NPs synthesized using jet-mixing have a particle size distribution narrower by 4.5% and a 20% increase in shelf life as compared to their batch-synthesized counterpart. The jet-mixing reactor also demonstrates material economy by requiring a capping agent concentration that is four times lower than that required in batch. Next, the jet-mixing process is developed to incorporate inert conditions to synthesize pure-phase copper nanomaterials (Cu NPs) that tend to oxidize readily under ambient conditions. Cu NPs with 88% phase purity are synthesized using jet-mixing. Lastly, the jet-mixing reactor is used for the inert synthesis of multi-component core@shell Pd@TiO_2 nanomaterials. The materials obtained from batch and jet-mixing synthesis are tested as catalysts for the hydrodeoxygenation (HDO) of furfuryl alcohol. It is observed that the selectivity for the HDO product, 2-methyl furan, obtained using the jet-mixing-synthesized material is comparable to that obtained by batch synthesis. Jet-mixing is capable of a productivity of 50 g/cc min of the catalyst as compared to batch that yields 0.5 g per run, demonstrating the scalability of the synthesis. Overall, these studies demonstrate that jet-mixing is a scalable, flexible tool for the synthesis of several types of nanomaterials and is readily extendable to a variety of synthesis conditions. Currently, studies are underway to characterize the mixing dynamics of the reactor to enable process scale-up.

Research Interests: Pharmaceutical scale-up, batch-to-continuous processing, process development, product development, nanomaterial synthesis

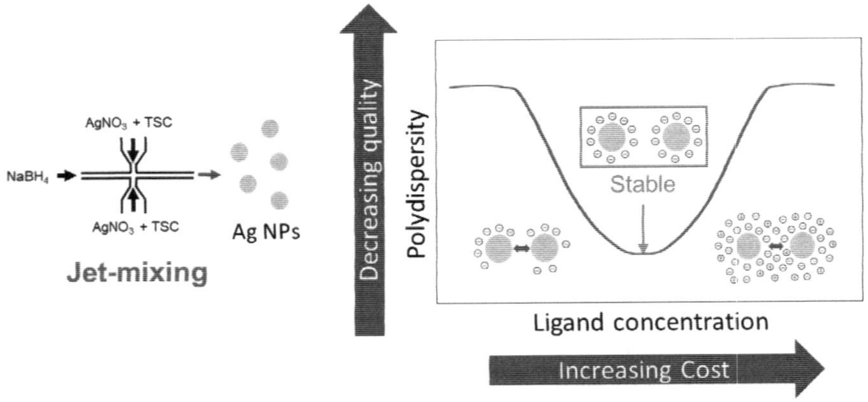

613579 Layer-By-Layer Implant Coatings for Craniomaxillofacial Bone Regeneration

Abstract:

The overall goal of my graduate work is to use layer-by-layer film assembly to answer fundamental questions about growth factor-mediated healing of craniomaxillofacial bone defects. The aim is to determine optimal release kinetics for delivery of growth factors and optimal combination of angiogenic and osteogenic growth factors from a cell-free, synthetic, biodegradable scaffold. Current work has focused on engineering layer-by-layer films that release clinically-relevant osteogenic growth factor, BMP-2, in low doses over periods of time ranging from days to weeks. In this work, we developed and optimized diffusional barrier tools to modulate release of the protein from a degradable polymer film. We completed preliminary *in vivo* pilot studies which indicate that release kinetics may influence bone growth in a critically sized rat calvarial defect. We plan to perform a follow-up large scale study in the near future including microCT tracking of bone mineral density and bone volume, histological analysis to examine maturity and quality of new bone, and mechanical testing to determine osseointegration of new bone. We are also currently working to translate these systems to 3D printed macro porous scaffolds for rabbit mandibular defect studies. Our future plans include translation of the layer-by-layer film diffusional barrier tools to dual-growth factor systems and subsequent *in vivo* analysis to determine the effect of angiogenic growth factor type and relative delivery windows of angiogenic and osteogenic growth factors on craniomaxillofacial bone healing.

Research Interests

I am very interested in the intersection of engineering and medicine, with a strong emphasis on applications-based research. Specifically, I have really enjoyed exploring how chemical engineering fundamentals can play a role in drug delivery. I am open to new experiences and would be interested in a research & development role as well as a more management-focused track.

Relevant skills

Layer-by-layer film assembly, *in vivo* surgical methods, IVIS imaging, microCT imaging, histology, ELISA and other protein assays, GPC, NMR, HPLC, tissue culture, step polymerization

613680 Fabrication of Multi-Compartment Micro-Capsules for Drug Delivery of Multiple Drugs

Abstract:

Research Interests

My research interests include the following: Micro-capsules and microfluidics, modeling and simulation, pharmaceutical drug process development and manufacturing, drug micro-encapsulation and drug delivery, data science and automation.

Abstract:

In recent years, various studies have investigated the flow dynamics of a single capsule in straight microfluidic channels such as cylindrical, square and rectangular channels, constrictions and expansions. These studies have provided a better understanding for capsules utilized in numerous engineering and biomedical applications including fabrication of single and multi-compartment micro-capsules with desirable properties, targeted drug delivery and of course its similarity to blood flow in vascular capillaries.

In the present work, our interest is concentrated on fabrication of multi-compartment micro-capsules based on interactions of single capsules. In our computational investigation, we consider spherical capsules made of a strain-hardening membrane with comparable shearing and area-dilation resistance, while we study a wide range of capsule sizes and flow rates. Under these conditions, we have identified several merging formation patterns which depend on the single capsule size. The conditions for capsules separation and the effects of lubrication physics will also be presented. Our work provides useful physical insight on the possible fabrication of multi-compartment micro-capsules via further cross-linking for drug delivery of multiple drugs as well as on erythrocytes aggregation in the micro-circulation.

613741 Tuning Aromatics Hydrogenation By Systematic Variation of Their Surface Polarity in SBA-15-Type Periodic Mesoporous Catalysts

613767 Leveraging Molecular Simulations and Data Science to Improve the Design of Biomaterials with Fine-Tuned Hydrophobicity

Abstract:

Research Interests

Hydrophobicity is an important interfacial property that determines the association of nonpolar materials in water and is highly correlated with biologically relevant properties, including immune response, protein-protein interactions, and cell uptake. Accurately predicting hydrophobicity is vital to the design of new materials with favorable properties for applications in drug delivery, biosensing, and antifouling. My research interests lie in developing and applying molecular simulations techniques and data science methods to understand how surface properties of biomaterials influence interfacial phenomena and intermolecular interactions. Then applying these newfound understandings to developing predictive models based on physical insights to rapidly screen drug and gene agents to identify candidates with desirable physicochemical properties. These models would be valuable to improving the efficiency and efficacy of new drug therapeutic design and discovery.

Research Experience

My research focuses on modeling the interface between water and self-assembled monolayer (SAM) protected gold nanoparticles using molecular dynamics (MD) simulations to understand how physical and chemical properties influence the hydrophobicity of the SAM surfaces. My research project can be divided into three fundamental questions: (1) How do physical surface properties influence hydrophobicity? (2) How do chemical surface properties influence hydrophobicity? and (3) How do physical and chemical surface properties cooperatively influence hydrophobicity? The goal of answering these questions is to better understand how surface properties alter hydrophobicity; thus, providing a design framework to engineer novel biomolecules with fine-tuned hydrophobicity. To achieve this goal, I have developed a simulation methodology that applies umbrella sampling (an enhanced sampling MD technique) to measure the hydrophobic force exerted between two SAM surfaces. I collaborated closely with experimentalists to ensure the model accurately predicts hydrophobic forces. To screen a larger range of SAM chemistries and compositions, I applied a more efficient enhanced sampling technique (Indirect Umbrella Sampling) to predict the hydrophobicity. This method required approximately ten times less simulation time than classical umbrella sampling. Next, I applied traditional analysis techniques, such as time-averaged measurements, as well as more data-centric methods, including principal component analysis and multivariate linear regression, to unbiased MD simulation data to interrogate how interfacial water structure, physical and chemical surface properties, and hydrophobicity are related. Finally, in collaboration with data-science experts, we leveraged the predictive power of machine learning algorithms to create a model that predicts hydrophobicity 400 times faster than the most efficient enhanced sampling MD simulation methods.

Microporous and Mesoporous Materials 300 (2020) 110174

Contents lists available at ScienceDirect

Microporous and Mesoporous Materials

journal homepage: http://www.elsevier.com/locate/micromeso

Seed-Assisted zeolite synthesis: The impact of seeding conditions and interzeolite transformations on crystal structure and morphology

Rishabh Jain, Jeffrey D. Rimer *

Department of Chemical and Biomolecular Engineering, University of Houston, Houston, TX, 77204, USA

ARTICLE INFO

Keywords:
Zeolite
Seeded growth
Interzeolite transformation
Crystal morphology

ABSTRACT

Seed-assisted approaches in zeolite synthesis differ from classical processes in that the seeds tend to dissolve, giving rise to an unknown memory of the parent crystal structure that facilitates the nucleation of the daughter. It has been hypothesized in literature that a shared structural feature, such as a composite building unit, between the parent and the crystals produced from a non-seeded growth solution results in identical parent-daughter framework types. In this study, our findings reveal that this scenario is true in select cases, but often the initial structure formed in seeded syntheses is metastable, resulting in interzeolite transformation(s) with prolonged hydrothermal treatment. In general, the trajectory tends to favor a final structure identical to that obtained by non-seeded growth under identical synthesis conditions. Here, we explore how seed-assisted syntheses impact zeolite properties such as size, morphology, structure, and defects. We observe that the molar composition of the growth mixture and the properties of the seed crystals play a significant role in controlling the kinetics of nucleation and the trajectory of interzeolite transformations. Furthermore, we observe that seeds offer unique routes to achieve small crystal sizes and distinct morphologies in comparison to many conventional syntheses. Advantages of seeding include shorter synthesis time and the ability to reduce or eliminate the need for organic structure-directing agents, thereby providing a facile and efficient route to design zeolites for various industrial applications. The fundamental mechanisms underlying zeolite seed-assisted crystallization are complex and elusive; however, our study provides new insight into these processes and highlights the important role of kinetics in governing parent-daughter (or seed-product) relationships.

1. Introduction

The unique properties of zeolites, including their exceptional (hydro) thermal stability, tunable acidity, and diverse range of pore networks, enable their widespread use in industrial applications spanning from adsorption and catalysis to ion exchange and separations [1–3]. There are many approaches used to tailor the physicochemical properties of zeolites, such as their crystal structure (e.g. pore sizes and dimensions), chemical composition (e.g. Si/Al ratio), and mass transport properties (e.g. crystal size and mesoporosity). A common (and often required) technique in zeolite synthesis involves the incorporation of one or more organic molecules that function as structure-directing agents (SDAs). These species typically have a size and geometry that is commensurate with the channels and cages of the targeted zeolite framework [4]. For instance, Choi et al. synthesized single unit cell ZSM-5 nanosheets using a di-quaternary ammonium-type surfactant and showed that these materials exhibit improved catalyst lifetime in the methanol to hydrocarbon reaction [5]. Recent examples have demonstrated the advantages of using computational models or machine learning to effectively screen potential SDAs [6], while rational approaches using SDAs that mimic critical transition states in chemical reactions have been used to produce optimal (reaction-designed) zeolite catalysts [7]. Cooperative effects of multiple organic SDAs as well as mixtures of organic and inorganic SDAs have also been exploited for the purposes of tailoring crystal structure [8,9], size [10,11], and/or the number and distribution of aluminum sites within the framework [12,13]. Despite recent developments, the use of organics can impose commercial limitations owing to high economic and environmental costs. One approach to either decrease or eliminate organics in zeolite synthesis is the use of seed crystals, which can reduce the total synthesis time, eliminate impurities, and alter particle size [14,15]. This approach has been successfully utilized to synthesize various zeolite frameworks: *BEA [16–18], MTW [19,20], MFI [21,22], FER [23], MEL [24], LTA [25], MOR [26], RTH [27] and FAU [28]. It is also an integral component of

* Corresponding author.
E-mail address: jrimer@central.uh.edu (J.D. Rimer).

multi-step syntheses to produce zeolite membranes for gas separations [29].

In the classical sense of seed-assisted crystallization, the seeds undergo immediate growth to bypass the induction period or can provide interfaces for heterogeneous nucleation. Conversely, seed-assisted growth in zeolite synthesis is more complex due to the presence of numerous precursors that include (alumino)silicate monomers, oligomers, and amorphous particles or gels [30]. There have been multiple mechanisms reported for seed-assisted zeolite synthesis. Xiao and coworkers proposed a core-shell growth mechanism in which the solute leads to overgrowth (shell) on the surface of seeds (core) [31]. A more commonly observed phenomenon is one reported by Sano and coworkers who proposed a complete dissolution of seeds, giving rise to locally ordered aluminosilicates (oligomers or small particles) in the growth solution that recrystallize to the final zeolite product [32]. In general, the recrystallization of the parent zeolite (seed) into a daughter zeolite can be classified into two categories. The first refers to interzeolite transformations (Scheme 1A), which occur in media containing mainly zeolite seeds (parent) and few amorphous precursors. In these processes, dissolution-reprecipiation processes lead to the conversion of one zeolite (parent) into a different zeolite structure (daughter). The second mechanism is seeded growth (Scheme 1B) where a small amount of seeds (parent) is added to a growth mixture of predominately amorphous precursors, leading to the dissolution of seeds and recrystallization into a daughter zeolite with either the same or different structure as the original seed.

There is growing interest in seed-assisted zeolite synthesis. Okubo and coworkers proposed a working hypothesis based on empirical evidence positing that common composite building units (CBUs) between the seed and crystals obtained from non-seeded growth solutions will lead to identical parent-daughter crystal structures [33]. For example, ZSM-11 (MEL) seeds introduced in a mordenite (MOR) growth solution (both having one common CBU) yielded a ZSM-11 (MEL) product. Additional examples were provided where the zeolite structure of the seed was retained when there was at least one common CBU between the seed and the zeolite obtained from the starting synthesis gel in the absence of seeds. When considering interzeolite transformations, there are also examples where parent-daughter crystals share a common structural feature. Examples include transformations of LEV into CHA [34], FAU into LEV [35], FAU into CHA, and *BEA into MFI [36] where each pair of structures share at least one common CBU.

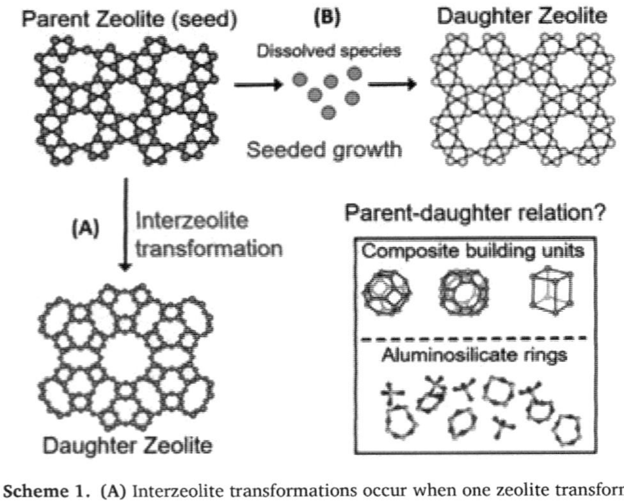

Scheme 1. (A) Interzeolite transformations occur when one zeolite transforms to a second zeolite in media containing few precursors (ca. <15 wt%). (B) Seeded growth involves the addition of a parent zeolite (seed) to media comprised primarily of amorphous precursors (ca. >85 wt%), leading to seed dissolution and recrystallization of an identical or different zeolite (daughter) structure. Comparisons of disparate parent-daughter structures can be made on the basis of shared composite building units or rings of the zeolite frameworks.

As shown in Scheme 1, structural similarities could be extended to more basic subunits, such as n-rings. For instance, Honda et al. [37] also reported transformations of FAU (*d6r, sod*) into *BEA (*mor, bea, mtw*) and MAZ (*dsc, gme*), i.e. frameworks that do not share any common CBU. In these cases, the authors attributed the transformations to common four-member rings in several CBUs (e.g. *d6r, bea* and *gme*). Goel et al. [36] reported transformation of lower density frameworks (FAU or *BEA) into high density frameworks (MFI, CHA, STF and MTW). They also posited that the presence of a common CBU permits transformations without using seeds or an organic SDA. Muraoka et al. proposed that during FAU-to-CHA transformation, a specific rearrangement of aluminosilicates occurs rather than a simple transfer of their common CBU (double-6-membered ring) wherein the final atomic configuration of CHA is dependent on the parent FAU zeolite [38]. However, examples of interzeolite transformations have been reported where both parent and daughter lack any common structural features. For instance, Qin et al. demonstrated the transformation of FAU into MFI in absence of seeds or organic SDAs [39]. Kirschhock and coworkers [40] reported the transformation of FAU into a wide range of structures (e.g. ABW, MER, and ANA) that do not share any common CBU with FAU. Indeed, Schwalbe-Koda et al. [41] used machine learning to discover that 65% of the interzeolite transformations reported in literature do not involve common CBUs. Thus, one of the prevailing questions is whether or not the criterion of shared structural features is essential for interzeolite transformations and seeded growth.

Interzeolite transformations and seed-assisted growth can provide pathways for the synthesis of zeolites with controlled elemental composition and crystal morphology [25,42,43]. Iglesia and coworkers have utilized interzeolite transformations to control the encapsulation of large metal clusters in zeolite pores, which otherwise could not be achieved by post-synthesis ion exchange [44]. Despite numerous studies of interzeolite transformations, the mechanism and driving force for these processes are not well understood. Navrotsky and coworkers established a trend for purely siliceous zeolites where the enthalpy of formation is proportional to molar volume [45], which can be used as a descriptor to rationalize many reported transformations. However, there are examples that seemingly defy this trend (e.g. MFI-to-MOR [39,46]), revealing that other factors such as elemental composition (the presence of Al or other heteroatoms) impact the enthalpy of formation [47,48] in ways that have yet to be quantified. To this end, a single (universal) descriptor is incapable of predicting interzeolite transformations. One complicating factor is that the chemical potential of the growth solution changes over the course of zeolite synthesis, reflecting the temporal changes in soluble species as precursors are consumed. This gives rise to metastable structures that can undergo a series of interzeolite transformations (akin to the Ostwald rule of stages) with prolonged hydrothermal treatment [49].

Herein, we explore the effect of parameters such as growth solution composition, seed properties, synthesis time, and temperature on the kinetics of nucleation and interzeolite transformations in seed-assisted zeolite synthesis. We examine the factors governing parent-daughter relationships by selecting zeolites with common structural features as well as those lacking structural similarity. Our findings expose several limitations to prevailing hypotheses of shared CBUs. Moreover, we explore the impact of seeds on crystal morphology and highlight the important role of kinetics in seeded growth. To this end, our findings provide insights into organic-free protocols to tailor the physicochemical properties of zeolite crystals.

2. Experimental

Materials. The following chemicals for zeolite synthesis were purchased from Sigma Aldrich: LUDOX AS-40 (40 wt% suspension in water), tetraethylorthosilicate (TEOS, $\geq$ 99.5%), sodium aluminate (57.1% Al_2O_3 and 37.2% Na_2O), 1,8-diaminooctane (DAO, 98%), tetrapropylammonium bromide (TPABr, 98%) and tetraethylammonium bromide (TEABr, 98%). Additional chemicals purchased from Alfa Aesar included sodium hydroxide (98% pellets) and tetrabutylammonium hydroxide (TBAOH, 40%). All reagents were used as received without further purification. Deionized (DI) water used in all experiments was purified with an Aqua Solutions RODI-C-12A purification system (18.2 MΩ).

Synthesis of Seed Crystals. MEL seeds with Si/Al ratios of ∞ (silicalite-2) and 33 (ZSM-11) were synthesized with organic structure-directing agents (OSDAs) TBAOH and DAO, respectively, according to previously reported protocols [24,50]. For samples with a crystal Si/Al ratio of 16, the synthesis involved the addition of $NaAlO_2$ to a NaOH solution followed by additions of TEOS and TBAOH to obtain a final molar composition of 20 TBAOH: 4 Na_2O: 100 SiO_2: 2.5 Al_2O_3: 3000 H_2O. The solution was aged for 2 h at room temperature and transferred to a 23 mL Teflon-lined stainless steel acid digestion bomb (Parr Instruments), which was placed in a Thermo Scientific gravity convection oven and heated at 160 °C for 1 day.

MFI seed crystals with Si/Al ratios of ∞ (silicalite-1) and 14 (ZSM-5) were synthesized with TPABr and without OSDA, respectively, according to previously reported protocols [51,52]. For samples with a crystal Si/Al ratio of 40, seeds were purchased from Zeolyst (CBV 8014) and were calcined in a Thermo Fisher Lindberg Blue furnace at 550 °C for 10 h (with temperature ramping rate of 1 °C min^{-1}) under the constant flow of compressed air (Matheson Tri-Gas) at 100 mL min^{-1} before application. ZSM-12 (MTW) and mordenite (MOR) seed crystals were synthesized with TEABr and without an OSDA, respectively, according to previously reported protocols [53,54]. All of the seeds synthesized using an OSDA were calcined at 550 °C for 10 h (according to the above procedure) prior to their addition to growth solutions.

Seeded Zeolite Synthesis. Seed-assisted syntheses were carried out by first adding sodium aluminate to an aqueous NaOH solution followed by the addition of LUDOX AS-40 to obtain the molar ratios listed in Table 1. To this solution was added the seeds in amounts spanning 2 to 13 wt% (calculated relative to the total silica in the growth solution). The nominal amount of seeds used in experiments was 10 wt% unless mentioned otherwise. The growth solution was aged at room temperature for ca. 24 h and subsequently transferred to a 23 mL acid digestion bomb. Hydrothermal treatment was carried out in an oven at temperatures spanning from 130 to 170 °C under autogenous pressure. Solids extracted after select time intervals were isolated from the mother liquor by two cycles of centrifugation and washing with DI water, followed by drying in an oven at 60 °C.

Characterization. Powder X-ray diffraction (XRD) patterns of dried solids were collected on a Rigaku SmartLab diffractometer with a Cu Kα source (40 kV, 30 mA). Scanning electron microscopy (SEM) images were obtained using a Zeiss Leo 1525 instrument equipped with FEG at 10 kV. All samples were carbon coated (layer thickness ≈ 20 nm) prior to imaging to reduce charging. The size of the crystals was measured from multiple SEM images of a single batch. Elemental analysis was performed by electron dispersive X-ray (EDX) spectroscopy. EDX spectra were collected using a JEOL SM-31010/METEK EDAX system at 15 kV and a 15 mm working distance.

Table 1
Molar composition of growth solutions and synthesis temperatures.

Solution[a]	Na_2O	SiO_2	Al_2O_3	H_2O	T (°C)[b]
S1	3.42	18	1	324	165
S2	30	100	1	2000	150
S3	9.6	60	1	2580	190

[a] Compositions exclude the addition of seeds; b. Seeded syntheses were performed at 150 °C.

3. Results and discussion

Parent-Daughter Structural Relationships. Here we examine the impact of growth solution composition, seed structure and properties, and synthesis time on parent-daughter relationships in seeded growth experiments. As shown in Fig. 1, we selected four commercially relevant zeolite frameworks types for analysis: MOR, MFI, MEL, and MTW. Among these zeolites, MFI and MEL share multiple common composite building units (CBUs). MOR shares a single CBU with MFI and MEL, whereas MTW only shares one CBU with MFI. The various combinations of shared structural entities provide an opportunity to test scenarios where the frameworks of the growth solution, parent crystal (seed), and daughter crystal (product) either exhibit similar structural features or lack any commonality. The molar volume of these frameworks varies (from high to low) as MOR ≈ MEL > MTW ≈ MFI. While interzeolite transformations tend to occur in order of decreasing molar volume [49], these trends are reported for purely siliceous zeolites [45]; thus, differences in the framework composition (i.e. Si/Al ratio) of parent and daughter crystals makes it difficult to ascertain the thermodynamic driving force for processes observed in this study. For example, interzeolite transformations can occur in the order of decreasing framework density (e.g. MFI-to-MOR) [39], consistent with the recent study by Schwalbe-Koda et al. [41] highlighting numerous examples of transformations occurring in sequences of increasing molar volume.

For this study we selected two growth solutions (compositions S1 and S2 in Table 1) that are commonly reported for mordenite synthesis [19, 53]. The kinetic (ternary) phase diagram in Fig. 2, which was constructed from a survey of literature (Fig. S1 and Table S1), reveals a region of overlapping synthesis compositions between MOR and pentasil zeolites MFI and MEL, while the region for MTW is distinctly separated from that of MOR. The basis for selecting these two growth solutions is that S1 lies within intersecting regions of MOR and MFI/MEL while S2 is

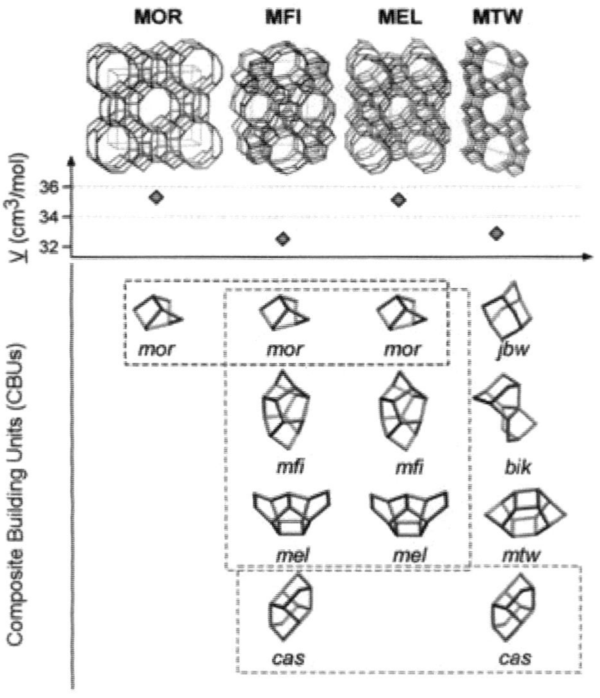

Fig. 1. (top) Comparison of four framework types (MOR, MFI, MEL, and MTW) with molar volumes $\underline{V}$ of 35.4, 32.7, 34.6, and 33.1 cm^3 mol^{-1}, respectively. (bottom) Composite building units (CBUs) with dashed boxes indicating commonalities among the four frameworks. Data for this figure was obtained from the International Zeolite Association Structure Database [68].

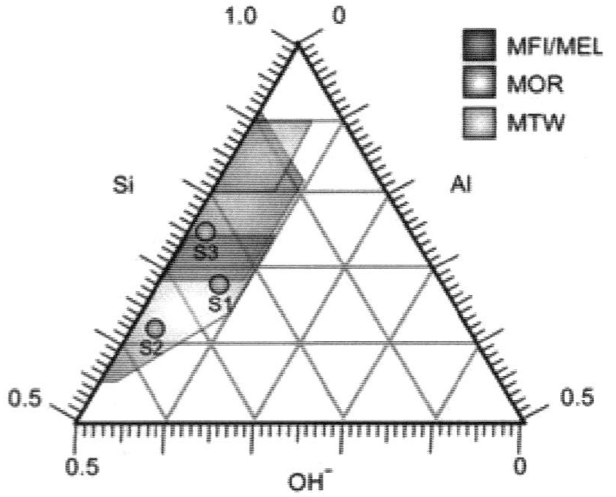

Fig. 2. Ternary (kinetic) phase diagram with growth solutions S1, S2 and S3 (grey circles). The purple, yellow, and green shaded regions depict approximate compositional regions reported in literature for syntheses of pentasil zeolites ZSM-5 (MFI) and ZSM-11 (MEL), mordenite (MOR), and ZSM-12 (MTW), respectively. The Supporting Information contains a phase diagram highlighting specific compositions reported in literature (Fig. S1) and a table of corresponding synthesis conditions (Table S1). (For interpretation of the references to colour in this figure legend, the reader is referred to the Web version of this article).

within an isolated MOR region. Results of seeded growth reported by Itabashi et al. [19] have shown that introduction of ZSM-11 (MEL) or ZSM-5 (MFI) seeds in a growth solution similar to S2 resulted in the generation of MEL and MFI products, respectively. This outcome of identical parent and daughter crystals aligned with their working hypothesis of common CBUs between the seed and the product of non-seeded growth (i.e. mordenite). It could be argued that this procedure would also be applicable to solution S1, which overlaps the MEL/MFI phase space. Therefore, we followed a similar protocol using ZSM-11 seeds (Si/Al = 16) in a S1 solution and showed that after 6 days of synthesis, the final product is mordenite, which is contrary to the expected outcome based on the prevailing hypothesis of common CBUs. To ascertain the reason for this difference, we carried out time-elapsed studies in solutions S1 and S2 to observe the trajectory of seeded growth (including potential phase transformations) using a variety of parent crystals with different Si/Al ratios and framework types.

MEL-type crystals prepared with three compositions (Si/Al = 16, 35, and ∞) were used as seeds in solutions S1 and S2 (Fig. 3A). For the synthesis employing silicalite-2 seeds (Si/Al = ∞) in growth solution S1, powder XRD patterns (Fig. 3C) of the solids extracted after 1 day of heating reveal an amorphous product, suggesting the dissolution of seeds, which is a commonly observed phenomenon in seeded growth of zeolites [20]. Alternatively, residual seed crystals (either pristine or partially disordered) may be present in quantities that are too small to detect by XRD. Additional synthesis time (e.g. 3 days) leads to the formation of pentasil zeolite. It should be noted that it is often difficult to distinguish between MEL and MFI phases in powder XRD patterns [55]; therefore, in cases where it is not evident which phase(s) is present, we denote the product(s) as a generic label of ***pentasil*** to indicate the presence of MEL and/or MFI. For experiments in S1 solutions, we never observed a pure pentasil product, but rather mixtures containing mordenite. As the silicon content of seeds increases, we observe that the time period for pentasil-to-mordenite transformation is extended to approximately 10 days (compared to 6 days for seeds with the lowest Si/Al ratio (Fig. 3C and panels A of Figs. S6–S7). Interestingly, syntheses in growth solution S2 (i.e. the composition furthest from the pentasil region in the phase diagram) yields fully crystalline pentasil (or MFI) products within the first 3 days of synthesis (panels B of Figs. S6–S8). In the case of seeds with Si/Al = 16, we did not observe the pentasil-to-mordenite transformation after 10 days of synthesis (Fig. S6B). Contrary to seeded growth in solution S1, the timeframe for pentasil-to-mordenite transformation in solution S2 increases with decreasing Si/Al ratio, such that purely siliceous seeds (silicalite-2) yield fully crystalline mordenite within 6 days of synthesis. One of the primary differences between S1

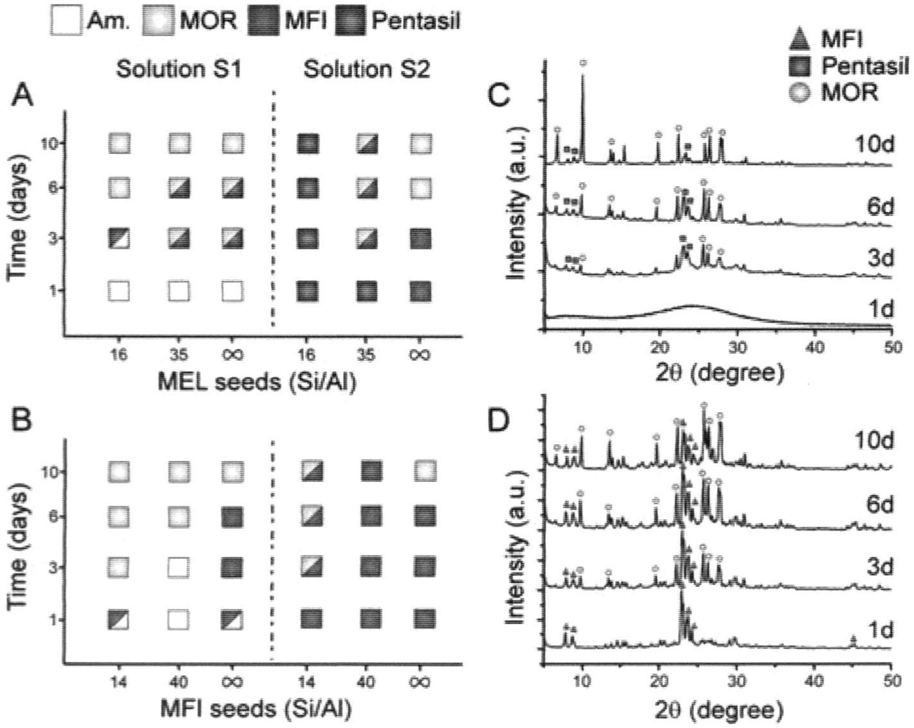

Fig. 3. Final products obtained from seed-assisted syntheses using (A) MEL and (B) MFI seeds in growth solutions S1 (left) and S2 (right) at different time intervals. Major phases were identified from powder XRD patterns where mixtures are denoted as split squares. For those samples where it could not be definitively determined if the product(s) contained ZSM-5 (MFI), ZSM-11 (MEL), or a combination thereof, the product is reported as a *pentasil*. (C and D) Representative time-elapsed powder XRD patterns for the solids extracted at various synthesis times from seeded growth experiments using (C) silicalite-2 (MEL, Si/Al = ∞) seeds in growth solution S1 and (D) ZSM-5 (Si/Al = 14) seeds in growth solution S2. Additional powder XRD patterns and SEM images are provided in the Supporting Information (Fig. S6 – S15).

and S2 is the gel Si/Al ratio for each solution (9 and 50, respectively). According to previous studies [56,57], high Si/Al ratio (>30) of the growth solution is optimal for the crystallization of pentasil zeolites while low Si/Al ratio (<30) is optimal for the MOR crystallization. This offers a possible explanation for the promotion of pentasil structures in seeded growth using solution S2, despite the fact that its location is within the MOR region of the kinetic phase diagram.

Repeating these experiments using MFI-type seeds of varying composition (Si/Al = 14, 40, and ∞) yielded less definitive trends with respect to seed Si/Al ratio (Fig. 3B); however, we generally observed for solutions S1 and S2 that higher silicon content resulted in enhanced stability of MFI phases at earlier synthesis times. In the majority of cases, ZSM-5 (MFI) was the first crystalline phase to be detected (e.g. Fig. 3D). In solution S2, all products within the first day of synthesis were pure ZSM-5 (Fig. 3D and panels B of Figs. S10–S11). The most unusual observations involved ZSM-5 seeds with Si/Al = 40 where growth solution S1 resulted in a delayed induction period followed by the generation of pure mordenite (Fig. S10A), whereas growth solution S2 produced pure ZSM-5 over the course of 10 days (Fig. S10B). The findings from these experiments reveal that under certain circumstances, the working hypothesis of common CBUs reported by Okubo and coworkers [19] is correct, in that identical parent-daughter crystal structures are achieved when the seed crystal shares at least one common CBU with the crystalline product obtained from the non-seeded growth solution (i.e. MOR in this study). It should be noted, however, that this hypothesis is highly dependent upon the selection of growth solution composition and the Si/Al ratio of the parent (seed) zeolite. It is also anticipated that synthesis temperature is an important parameter given that higher temperature expedites interzeolite transformations, which we demonstrated in a previous study [49]. Here, we observed in select cases that transformations at higher temperature (170 °C) are faster than those at lower temperature (130 °C), thus highlighting the important role of kinetics in determining parent-daughter relationships (Fig. S18).

To further examine the hypothesis of common CBUs in seeded growth we used ZSM-12 (MTW) seeds, which do not share any common structural feature with MOR and are synthesized in a different region of the kinetic phase diagram. Time-resolved studies reveal that seeded growth in both S1 and S2 solutions yields zeolite products counter to the working hypothesis. Notably, seeded growth in solution S2 results in a ZSM-12 product after 1 day of synthesis (Fig. 4A and B). This is particularly interesting given that this is an OSDA-free procedure, which is uncommon for ZSM-12 syntheses unless seeds are introduced in growth mixtures [19]. If the same synthesis is allowed to proceed for 3 days, we begin to detect mordenite in the product, while solids extracted over 10 days of synthesis continue to show a mixture of both MTW and MOR (Fig. 4C). In solution S1, the first phase detected is ZSM-5 (MFI), which shares one common CBU with MTW (Fig. S16). This is not entirely unexpected since solution S1 overlaps with the MFI region in the kinetic phase diagram (Fig. 2); however, extended synthesis time leads to a MFI-to-MOR interzeolite transformation (Fig. 4D), indicative of the general trend for seeded growth in solutions S1 and S2 to transition to mordenite with prolonged synthesis time. Therefore, the sum of observations in the seeded growth study presented here seem to indicate that shared CBUs between the seed and crystals produced from the (non--seeded) growth solution is not the dominant factor governing parent-daughter structural relationships. Indeed, the driving force for crystallization is seemingly more nuanced. The fact that interzeolite transformations are frequently observed is an indication that many of the initial products formed at short synthesis times are *kinetically trapped*. This may offer an explanation for discrepancies with Itabashi et al. [19], whose results are reported for short synthesis times (i.e. less than 24 h) that can lead to the formation of metastable products. In our study, extended periods of hydrothermal treatment lead to a progressive shift from the kinetically-trapped structure to a presumably more thermodynamically stable product that would otherwise be obtained in the absence of seeds. Collectively, these findings seem to indicate that the selection of synthesis conditions (e.g. growth medium composition, time, and temperature) are generally more influential than the structure and properties of seed crystals.

Impact of Seeds on Crystal Morphology. It is well established that the use of seeds in zeolite synthesis can lead to products with dramatically different crystal size and shape compared to the original seeds and/or the product of synthesis in the absence of seeds. Here, we examine to what extent the selection of seeds impacts the physicochemical properties of the final product. The nominal mordenite crystals obtained from solution S1 (without seeds) have an ill-defined morphology with a polydisperse size distribution (Fig. 5A). The same synthesis using silicalite-2 seeds results in a bimodal population (Fig. 5B) of rod-like mordenite crystals (>10 μm in length) and much smaller crystals (ca. 1 μm) with random (or less-defined) morphology. When using mordenite seeds (the same as those depicted in Fig. 5A), the resulting mordenite product exhibits a thin disc-like shape (Fig. 5C) and an approximately uniform distribution of crystal size. Thus, it is evident that the morphology and size of mordenite (MOR) crystals can be modified by varying the seed (parent) crystal structure, although the exact mechanism for these changes is not well understood. These findings further confirm the benefits of seed-assisted synthesis as an alternative method to tailor crystal habit compared to methods utilizing organics, such as zeolite growth modifiers (ZGMs) [58,59] or cooperative OSDAs [60,61].

To further explore the ability of seed-assisted protocols to control crystal habit, we examined a third growth solution (S3 in Table 1) corresponding to a composition that is reported for OSDA-free synthesis of ZSM-5 [53]. In the absence of seeds, this synthesis produces a relatively uniform size distribution of ZSM-5 crystals (ca. 4 μm in the *b*-direction) with the traditional coffin-like morphology (Fig. 5D). When the synthesis is repeated using silicalite-2 (MEL) seeds, the product has a similar morphology (Fig. 5E), but dramatically reduced crystal size (ca. 200 nm in the *b*-direction). For the same synthesis using mordenite seeds, there is no marked change in crystal size (Fig. 5F); however, the product contains significantly more defects (i.e. intergrowths). The latter is a commonly observed feature in syntheses of ZSM-5, but the exact origins

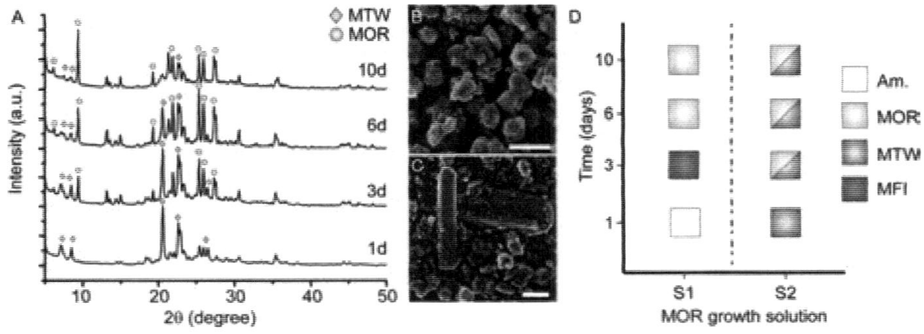

Fig. 4. Seed-assisted syntheses using ZSM-12 (MTW) seeds (Si/Al = 30). (A) Time-elapsed powder XRD patterns depicting changes in the products with increased synthesis time in growth solution S2. (B and C) Representative SEM images of the solid products obtained after (B) 1 day and (C) 10 days of synthesis (scale bars = 5 μm). The larger crystals in the latter micrograph are MOR, whereas the smaller crystals are MOR and/or MTW crystals. (D) Variations in the final product (s) obtained from syntheses using ZSM-12 seeds in solutions S1 (left) and S2 (right) at different heating times. Additional powder XRD patterns and SEM images are provided in the Supporting Information (Figs. S16 and S17).

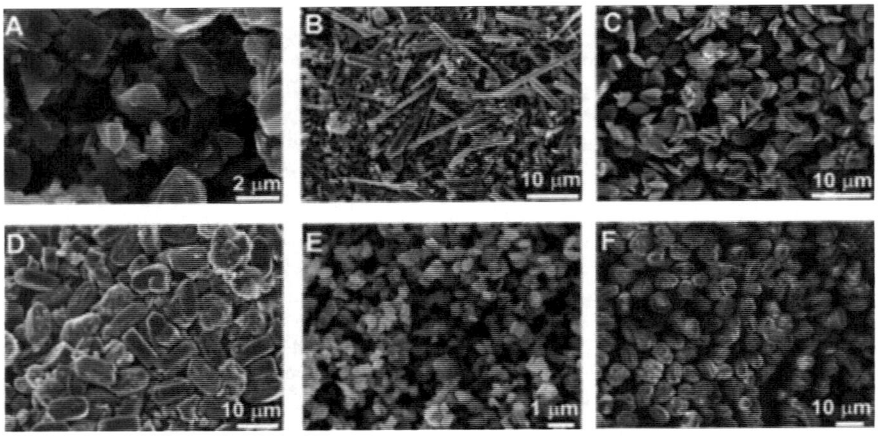

Fig. 5. Scanning electron micrographs of crystals prepared from (A–C) solution S1 and (D–F) solution S3. (A) Mordenite crystals with ill-defined morphology and polydisperse size distribution prepared without seeds. (B) Mordenite crystals prepared with the addition of silicalite-2 (MEL) seeds. (C) Mordenite crystals prepared with the addition of mordenite seeds (same as those shown in A). (D) ZSM-5 crystals prepared without the addition of seeds. (E) ZSM-5 crystals prepared with the addition of silicalite-2 (MEL) seeds. (F) ZSM-5 crystals prepared with the addition of mordenite seeds (same as those shown in A). The *b*-direction of MFI-type crystals corresponds to the thickness of the coffin-shaped platelets.

of defect formation are not well understood. One possibility is that the dissolution of mordenite seeds results in a residual "memory" of the parent structure (e.g. oligomers, rings, CBUs) [38,62]; and that these species putatively play a role in the generation of defects.

All syntheses reported to this point used a constant seed content (10 wt%); however, we observed that the final crystal size is dependent upon the quantity of seeds added to the growth mixture. An example of these findings is reported here for ZSM-5 synthesis in solution S3 (gel Si/Al = 60) using silicalite-2 seeds (Si/Al = ∞). The nominal size of ZSM-5 crystals in the absence of seeds is ca. 4 µm. In the presence of only 2 wt% silicalite-2 seeds, the crystal size is reduced by more than 100 times to an average value of 342 nm (Fig. 6A). This general trend is consistent with prior studies of MFI-type zeolite synthesis using silicalite-1 seeds [21,63]. Systematic studies of seeded ZSM-5 synthesis with increasing seed content reveals a monotonic decrease in crystal size with increasing seed content (Fig. 6C) wherein the highest seed content evaluated (13 wt%) yields 146 nm crystals (Fig. 6B). The size of the crystals was measured along the *b*-direction, which corresponds to the direction of straight channels (i.e. the less tortuous channels that more prominently impact the internal mass transport properties of ZSM-5) [64]. It has been demonstrated in literature that a reduced path length along the straight channels in MFI leads to better catalytic activity [65,66] as well as improved performance in separations [67]. It is interesting to note that varying seed content had no significant impact on the composition of zeolite products (Fig. 5C), which maintain a constant Si/Al ≈ 20. Collectively, our findings indicate that for the synthesis conditions selected in this study, only a small quantity of seeds are required to impact crystal size without altering product Si/Al ratio. The latter point is of particular importance given that it is often challenging to independently vary one property (e.g. size) without altering in some way other properties (e.g. composition). Moreover, the ability to achieve this outcome in the absence of organics increases the commercial relevance of this synthesis approach.

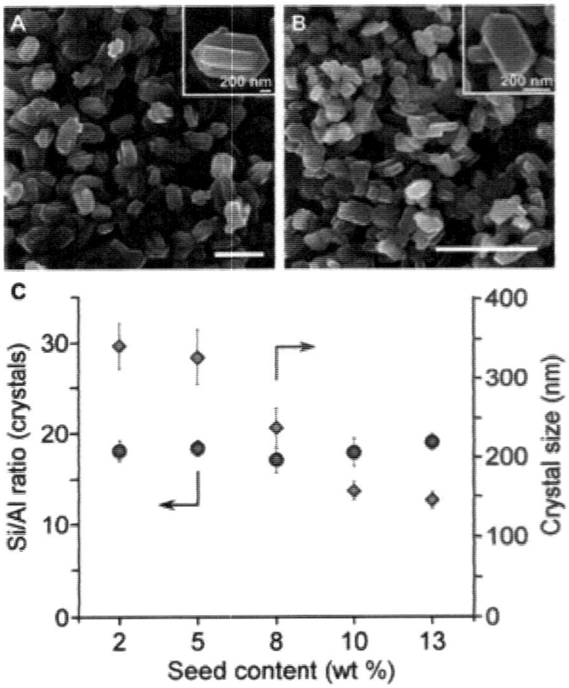

Fig. 6. Scanning electron micrographs of ZSM-5 crystals prepared with (A) 2 wt% and (B) 13 wt% silicalite-2 seeds in solution S3 with a gel Si/Al ratio of 60 (scale bars = 2 µm). Insets: high magnification images of representative crystals. (C) Average Si/Al ratio of products measured by EDX (blue circles, left y-axis) for a total of 24 points in a single batch, and the average crystal size (red diamonds, right y-axis) measured along the *b*-direction for 50 crystals in a single batch. Error bars equal two standard deviations. (For interpretation of the references to colour in this figure legend, the reader is referred to the Web version of this article).

4. Conclusion

In summary, we have examined the relationships between parent and daughter crystals in seeded growth assays using a variety of seed properties and growth solution compositions. The factors governing these relationships appear to be more multifaceted than a comparison of underlying structural features, such as the similarity of composite building units. In many instances we observe that seeded growth results in interzeolite transformations where the timescales are dependent upon the conditions selected, and the rationale for the observed trajectory is not well understood or easily predictable. In general, our observations indicate that the growth solution, which has been largely overlooked in previous studies of seeded growth, plays an important role in controlling the kinetics of interzeolite transformations. The thermodynamic driving force for these transformations cannot be easily described by a single variable, such as molar volume. The dissolution of seeds presumably leads to species that initially facilitate the nucleation of an identical structure; however, over the course of synthesis the composition of the growth solution changes, thereby altering the chemical potential (i.e. driving force), leading to the nucleation of a new phase. This highlights the importance of kinetics in seeded zeolite syntheses wherein the hypotheses of parent-daughter relationships in seed-assisted syntheses relative to non-seeded cases must take into account additional factors that include (but are not limited to) synthesis time, temperature, and the compositions of seeds and the growth medium. Although not explicitly examined in this study, it is possible that seed crystal size or the presence of defects in seeds (e.g. silanol nests, non-framework Al, etc.) can impact the kinetics of seeded growth synthesis and/or the properties of the final product.

Seed-assisted zeolite synthesis is a facile and economically viable approach to alter crystal habit. Our findings reveal that the nature of the seed has a significant impact on the final product; and that only a small quantify of seeds is needed to dramatically reduce crystal size without any unintended modification of framework composition (Si/Al ratio). This generalizable approach, while effective, is more of an art than a science given the unknown mechanisms of seeding. The proposed "memory" of dissolved seeds during the early stages of synthesis is not well understood and has not been experimentally verified; thus, the impact of residual seed species on processes of nucleation and growth is elusive. The ability of seeds to alter the pathways of zeolite crystallization, however, holds tremendous promise as a route to access properties that are otherwise unattainable by non-seeded growth. The findings presented here offer additional insight into seeded growth and the unintended interzeolite transformations that can occur during these predominantly organic-free syntheses.

Declaration of competing interest

The authors declare that they have no known competing financial interests or personal relationships that could have appeared to influence the work reported in this paper.

CRediT authorship contribution statement

Rishabh Jain: Writing - original draft, Data curation. **Jeffrey D. Rimer:** Writing - original draft, Supervision, Funding acquisition.

Acknowledgements

We wish Valentin Valtchev a happy 60th birthday and extend our thanks for his many valuable contributions to the microporous materials community, including his impact on our research over the years. This work submitted as part of his celebratory issue was sponsored by the National Science Foundation (DMREF Award 1629398). We also acknowledge additional support from the Welch Foundation (Award E-1794).

Appendix A. Supplementary data

Supplementary data to this article can be found online at https://doi.org/10.1016/j.micromeso.2020.110174.

References

[1] J.D. Rimer, A. Chawla, T.T. Le, Crystal engineering for catalysis, Ann. Rev. Chem. Biomol. Eng. 9 (1) (2018) 283–309.
[2] J. Cejka, A. Corma, S. Zones, Zeolites and catalysis: synthesis, reactions and applications 1, Wiley-VCH, Weinheim, 2010, p. 918.
[3] M.E. Davis, Ordered porous materials for emerging applications, Nature 417 (6891) (2002) 813–821.
[4] R.F. Lobo, S.I. Zones, M.E. Davis, Structure-direction in zeolite synthesis, J. Inclusion Phenom. Mol. Recognit. Chem. 21 (1–4) (1995) 47–78.
[5] M. Choi, et al., Stable single-unit-cell nanosheets of zeolite MFI as active and long-lived catalysts, Nature 461 (7261) (2009) 246–249.
[6] M. Moliner, Y. Román-Leshkov, A. Corma, Machine learning applied to zeolite synthesis: the missing link for realizing high-throughput discovery, Acc. Chem. Res. 52 (10) (2019) 2971–2980.
[7] C. Li, et al., Synthesis of reaction-adapted zeolites as methanol-to-olefins catalysts with mimics of reaction intermediates as organic structure-directing agents, Nat. Catal. 1 (7) (2018) 547–554.
[8] M. Kumar, et al., Crystallization of mordenite platelets using cooperative organic structure-directing agents, J. Am. Chem. Soc. 141 (51) (2019) 20155–20165.
[9] M. Moliner, et al., A new aluminosilicate molecular sieve with a system of pores between those of ZSM-5 and beta zeolite, J. Am. Chem. Soc. 133 (24) (2011) 9497–9505.
[10] S.H. Keoh, et al., Factors governing the formation of hierarchically and sequentially intergrown MFI zeolites by using simple diquaternary ammonium structure-directing agents, Chem. Mater. 28 (24) (2016) 8997–9007.
[11] G. Bonilla, et al., Zeolite (MFI) crystal morphology control using organic structure-directing agents, Chem. Mater. 16 (26) (2004) 5697–5705.
[12] A. Chawla, et al., Cooperative effects of inorganic and organic structure-directing agents in ZSM-5 crystallization, Mol. Syst. Des. Eng. 3 (1) (2018) 159–170.
[13] J.R. Di Iorio, R. Gounder, Controlling the isolation and pairing of aluminum in chabazite zeolites using mixtures of organic and inorganic structure-directing agents, Chem. Mater. 28 (7) (2016) 2236–2247.
[14] J. Grand, H. Awala, S. Mintova, Mechanism of zeolites crystal growth: new findings and open questions, CrystEngComm 18 (5) (2016) 650–664.
[15] K. Iyoki, K. Itabashi, T. Okubo, Progress in seed-assisted synthesis of zeolites without using organic structure-directing agents, Microporous Mesoporous Mater. 189 (2014) 22–30.
[16] Y. Kamimura, et al., Critical factors in the seed-assisted synthesis of zeolite beta and "green beta" from OSDA-free Na+–aluminosilicate gels, Chem.–Asian J. 5 (10) (2010) 2182–2191.
[17] G. Majano, et al., Al-rich zeolite beta by seeding in the absence of organic template, Chem. Mater. 21 (18) (2009) 4184–4191.
[18] B. Xie, et al., Organotemplate-free and fast route for synthesizing beta zeolite, Chem. Mater. 20 (14) (2008) 4533–4535.
[19] Y. Kamimura, K. Itabashi, T. Okubo, Seed-assisted, OSDA-free synthesis of MTW-type zeolite and "Green MTW" from sodium aluminosilicate gel systems, Microporous Mesoporous Mater. 147 (1) (2012) 149–156.
[20] K. Iyoki, et al., Synthesis of MTW-type Zeolites in the Absence of organic structure-directing agent, Chem. Lett. 39 (7) (2010) 730–731.
[21] G. Majano, et al., Seed-induced crystallization of nanosized Na-ZSM-5 crystals, Ind. Eng. Chem. Res. 48 (15) (2009) 7084–7091.
[22] J. Warzywoda, R.D. Edelman, R.W. Thompson, Crystallization of high-silica ZSM-5 in the presence of seeds, Zeolites 11 (4) (1991) 318–324.
[23] H. Zhang, et al., Organotemplate-free synthesis of high-silica ferrierite zeolite induced by CDO-structure zeolite building units, J. Mater. Chem. 21 (26) (2011) 9494–9497.
[24] Y. Shen, et al., Optimized synthesis of ZSM-11 catalysts using 1, 8-diaminooctane as a structure-directing agent, ChemPhysChem 19 (4) (2018) 529–537.
[25] Y. Bouizi, et al., Seeded synthesis of very high silica zeolite A, Chem. Mater. 19 (4) (2007) 652–654.
[26] J. Zhu, et al., Ultrafast, OSDA-free synthesis of mordenite zeolite, CrystEngComm 19 (4) (2017) 632–640.
[27] T. Yokoi, et al., Diversification of RTH-type zeolite and its catalytic application, Angew. Chem. Int. Ed. 48 (52) (2009) 9884–9887.
[28] H. Kacirek, H. Lechert, Growth of the zeolite type NaY, J. Phys. Chem. 79 (15) (1975) 1589–1593.
[29] J. Caro, et al., Zeolite membranes–state of their development and perspective, Microporous Mesoporous Mater. 38 (1) (2000) 3–24.
[30] R.D. Edelman, et al., Crystallization phenomena in seeded zeolite syntheses, Zeolites 9 (6) (1989) 496–502.
[31] B. Xie, et al., Seed-directed synthesis of zeolites with enhanced performance in the absence of organic templates, Chem. Commun. 47 (13) (2011) 3945–3947.
[32] H. Jon, et al., An insight into the process involved in hydrothermal conversion of FAU to* BEA zeolite, Chem. Mater. 20 (12) (2008) 4135–4141.
[33] K. Itabashi, et al., A working hypothesis for broadening framework types of zeolites in seed-assisted synthesis without organic structure-directing agent, J. Am. Chem. Soc. 134 (28) (2012) 11542–11549.
[34] I. Goto, et al., Transformation of LEV-type zeolite into less dense CHA-type zeolite, Microporous Mesoporous Mater. 158 (2012) 117–122.
[35] A. Yashiki, et al., Hydrothermal conversion of FAU zeolite into LEV zeolite in the presence of non-calcined seed crystals, J. Cryst. Growth 325 (1) (2011) 96–100.
[36] S. Goel, S.I. Zones, E. Iglesia, Synthesis of zeolites via interzeolite transformations without organic structure-directing agents, Chem. Mater. 27 (6) (2015) 2056–2066.
[37] K. Honda, et al., Role of structural similarity between starting zeolite and product zeolite in the interzeolite conversion process, J. Nanosci. Nanotechnol. 13 (4) (2013) 3020–3026.
[38] K. Muraoka, et al., Tracking the rearrangement of atomic configurations during the conversion of FAU zeolite to CHA zeolite, Chem. Sci. 10 (37) (2019) 8533–8540.
[39] W. Qin, et al., Organic-free interzeolite transformation in the absence of common building units, Chem.– Eur. J. 25 (23) (2019) 5893–5898.
[40] L. Van Tendeloo, et al., Alkaline cations directing the transformation of FAU zeolites into five different framework types, Chem. Commun. 49 (100) (2013) 11737–11739.
[41] D. Schwalbe-Koda, et al., Graph similarity drives zeolite diffusionless transformations and intergrowth, Nat. Mater. 18 (11) (2019) 1177–1181.
[42] R. Nedyalkova, et al., Interzeolite conversion of FAU type zeolite into CHA and its application in NH 3-SCR, Top. Catal. 56 (9–10) (2013) 550–557.
[43] T. Nishitoba, et al., Control of Al distribution in the CHA-type Aluminosilicate zeolites and its impact on the hydrothermal stability and catalytic properties, Ind. Eng. Chem. Res. 57 (11) (2018) 3914–3922.
[44] S. Goel, S.I. Zones, E. Iglesia, Encapsulation of metal clusters within MFI via interzeolite transformations and direct hydrothermal syntheses and catalytic consequences of their confinement, J. Am. Chem. Soc. 136 (43) (2014) 15280–15290.

[45] I. Petrovic, et al., Thermochemical study of the stability of frameworks in high silica zeolites, Chem. Mater. 5 (12) (1993) 1805–1813.
[46] X. Huang, R. Zhang, W. Zhengbao, Controlling crystal transformation between zeolite ZSM-5 and Mordenite without organic structure-directing agent, Chin. J. Catal. 33 (7–8) (2012) 1290–1298.
[47] K. Muraoka, W. Chaikittisilp, T. Okubo, Energy analysis of aluminosilicate zeolites with comprehensive ranges of framework topologies, chemical compositions, and aluminum distributions, J. Am. Chem. Soc. 138 (19) (2016) 6184–6193.
[48] A. Navrotsky, O. Trofymluk, A.A. Levchenko, Thermochemistry of microporous and mesoporous materials, Chem. Rev. 109 (9) (2009) 3885–3902.
[49] M. Maldonado, et al., Controlling crystal polymorphism in organic-free synthesis of Na-zeolites, J. Am. Chem. Soc. 135 (7) (2013) 2641–2652.
[50] P.M. Piccione, M.E. Davis, A new structure-directing agent for the synthesis of pure-phase ZSM-11, Microporous Mesoporous Mater. 49 (1–3) (2001) 163–169.
[51] A. Čižmek, et al., Dissolution of high-silica zeolites in alkaline solutions I. Dissolution of silicalite-1 and ZSM-5 with different aluminum content, Microporous Mater. 4 (2–3) (1995) 159–168.
[52] J.J. Keijsper, M.F.M. Post, Precursors in zeolite synthesis, in: Zeolite Synthesis, American Chemical Society, 1989, pp. 28–48.
[53] F.J. Machado, et al., Template-free synthesis and catalytic behaviour of aluminium-rich MFI-type zeolites, Appl. Catal. Gen. 181 (1) (1999) 29–38.
[54] K. Yoo, et al., TEABr directed synthesis of ZSM-12 and its NMR characterization, Microporous Mesoporous Mater. 60 (1–3) (2003) 57–68.
[55] O. Terasaki, et al., Direct observation of "pure MEL type" zeolite, Chem. Mater. 8 (2) (1996) 463–468.
[56] S.D. Kim, et al., Compositional and kinetic study on the rapid crystallization of ZSM-5 in the absence of organic template under stirring, Microporous Mesoporous Mater. 72 (1–3) (2004) 185–192.
[57] K. Muraoka, et al., Linking synthesis and structure descriptors from a large collection of synthetic records of zeolite materials, Nat. Commun. 10 (1) (2019) 1–11.
[58] A.I. Lupulescu, J.D. Rimer, Tailoring silicalite-1 crystal Morphology with molecular modifiers, Angew. Chem. Int. Ed. 51 (14) (2012) 3345–3349.
[59] M.A. Sanhoob, et al., Role of crystal growth modifiers in the synthesis of ZSM-12 zeolite, Adv. Powder Technol. 26 (1) (2015) 188–192.
[60] L. Meng, et al., A dual-templating synthesis strategy to hierarchical ZSM-5 zeolites as efficient catalysts for the methanol-to-hydrocarbons reaction, J. Catal. 361 (2018) 135–142.
[61] L. Wu, et al., Dual template synthesis of a highly mesoporous SSZ-13 zeolite with improved stability in the methanol-to-olefins reaction, Chem. Commun. 48 (76) (2012) 9492–9494.
[62] J. Zhang, et al., Interzeolite transformation from FAU to CHA and MFI zeolites monitored by UV Raman spectroscopy, Chin. J. Catal. 40 (12) (2019) 1854–1859.
[63] C.S. Cundy, P.A. Cox, The hydrothermal synthesis of zeolites: precursors, intermediates and reaction mechanism, Microporous Mesoporous Mater. 82 (1–2) (2005) 1–78.
[64] E.G. Derouane, Z.l Gabelica, Novel effect of shape selectivity molecular traffic control in zeolite ZSM-5, J. Catal. 2 (1980) 65.
[65] X. Zhang, et al., Synthesis of self-pillared zeolite nanosheets by repetitive branching, Science 336 (6089) (2012) 1684–1687.
[66] Y. Shen, et al., Deconvoluting the competing effects of framework topology and diffusion path length on methanol-to-hydrocarbons reactions, ACS Catal. 8 (2018) 11042–11053.
[67] Z. Lai, et al., Microstructural optimization of a zeolite membrane for organic vapor separation, Science 300 (5618) (2003) 456–460.
[68] International zeolite association structure database. http://www.iza-structure.org/databases.

604481 Ex Vivo Minigut Mucosal Model for Development of New Oral Vaccine

Abstract:

Research Interests

In this study, we successfully built a model for mimicking intestinal mucosal system for testing drug delivery. For this project, Intestinal stem cells was used, and Microfold cells were first differentiated by using 200 ng/ml RANKL,later on, this result were confirmed by different methods, which including qPCR, immuno-staining and western blot. Gold Nanocage was used for modeling drug transportation. In the future study, sigma 1 protein will be linked to gold Nanocage, making the Nanocage an idealistic nanomaterials for transporting antigen through M cells to tiger immune responses.

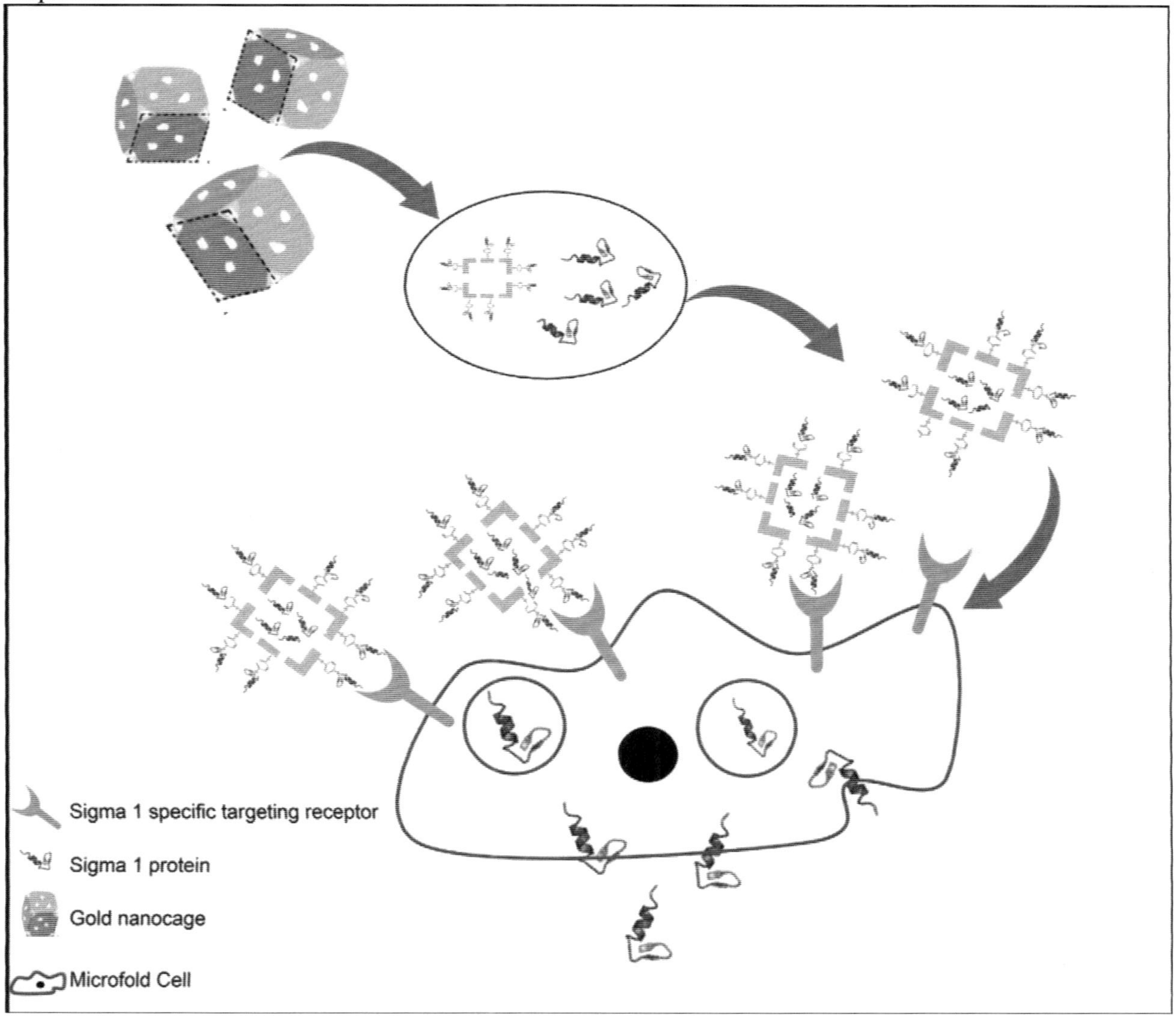

606947 Superhydrophobic Coatings Synthesized By High-Temperature Free-Radical Polymerization

Abstract:

Superhydrophobic coatings have many applications. Examples include self-cleaning and anti-ice surfaces, anti-fouling marine coatings, and anti-bacterial coatings for medical devices. To develop a superhydrophobic coating, low surface energy and high surface roughness in micro/nano scale are required. Fluoropolymers can decrease the surface energy of a coating significantly by orienting themselves in the coating-air interface. To induce surface roughness, different strategies including solvent evaporation, non-solvent induced phase separation, vapor induced phase separation, solvent/co-solvent induced phase separation, and polymerization-induced phase separation are applied. Incorporating nanoparticles into coatings is another strategy for increasing the surface roughness of coatings. Higher roughness means higher pockets of air between a liquid droplet and a solid surface, leading to a higher contact angle.

Silica is a nanoparticle widely used in high-performance coatings. Silica in a coating provides the coating with high thermal resistance, improved mechanical properties, resistance to corrosive solvents and so on. In case of covalent bonding between the silica surface and the polymeric matrix, these improvements in properties are much more significant .

Acrylate monomers with a long alkyl chain are highly hydrophobic and undergo free-radical polymerization. Such polyacrylates have already been used for the fabrication of implants thanks to their high anti-bacterial activity and biocompatibility. In this paper, we present results from in-situ polymerization of a mixture of silica and a hydrophobic acrylate monomer. Different grades of silica with different surface chemistry were used for the polymerization. After the polymerization, the nanocomposites were coated on glass substrates via spin coating. The water contact angles of the coatings were measured by the goniometry technique. Other characterization techniques including TGA, EDS, and SEM are also applied to fully characterize the products.

Research Interests:

- Polymerization science
- Polymer nanocomposites
- Mathematical modeling of polymerization reactors
- Polymer coatings and thin films
- MXene synthesis and surface modification
- Polymer melt processing

608412 Determining the Folding Landscape of α-Synuclein (35-97) Using Replica Exchange Molecular Dynamics

Abstract:

Research Interests

Synucleinopathies are a group of neurodegenerative disorders such as Parkinson's disease (PD), multiple system atrophy (MSA) and dementia with Lewy-bodies (DLB) that affects millions of people worldwide. They are characterized by abnormal aggregation or fibrillation of *α*-synuclein protein inside the nerve fibers or glial cells. These disorders are clinically characterized by memory and cognitive impairment, decline in motor or autonomic functions. α-synuclein is an intrinsically disordered protein which under pathological conditions undergoes structured transformation to oligomers containing β-pleated sheets that assists in its fibrillation process. The central amino acids from 35 to 97 (63 residues) play a major role in fibril formation and take up a unique Greek-key topology. The characterization of conformations accessible to monomeric α-synuclein is crucial in understanding the pathway to fibrillation. In this study, we use replica-exchange molecular dynamics methods to determine the folding landscape of α-synuclein fragment (35 – 97). We are able to obtain microscopic level information and thus, gain better understanding of protein folding by using an all-atom peptide model. The results provided also include the effect of temperature on these properties and the secondary structure analysis.

609841 High-Throughput and Data-Driven Strategies for the Design of Deep Eutectic Solvents

Abstract:

Research Interests:

- Energy
- Batteries
- Data Science

Within the framework of green chemistry, Deep Eutectic Solvents (DES) have been identified as promising candidates for use in many applications, including battery electrolytes. DES are characterized by two or three materials that associate with each other through hydrogen bond interactions, resulting in a eutectic mixture whose freezing point is below that of the individual materials. This design space is overwhelmingly large and poses a challenge for screening a vast and diverse set of materials. Here we present a strategic approach consisting of high throughput experimentation (HTE) coupled with data science driven analysis to identify exceptional DES candidates based on key physiochemical and electrochemical properties. Much of our HTE adopts methods that are already used frequently in the biotech and pharmaceutical industries, most notably performing parallel syntheses and analyses in 96-well-plate formats. DES samples are first synthesized using an open-sourced automated liquid handling robot. DES melting points are then determined by monitoring the melting process with an infrared camera and identifying the temperature at which the thermal conductivity of the samples changes abruptly. The solubility of battery redox-species is determined via UV-VIS well-spectrophotometers. Finally, the electrochemical stability window and cycling properties of DES electrolytes are measured in high-throughput by using screen-printed electrodes on 96-well plates adapted for use with a standard potentiostat. The ability to rapidly and efficiently collect data also creates a need for the development and use of automated processes for data analysis, which have been developed in an open-sourced format by our group. This approach to HTE also allows for the incorporation of data science techniques, such as feature extraction and machine learning, that further aid in probing a design space that is ultimately too large for experimental methods alone.

611003 Redesign and Develop Effective Liposomal Formulation for Prodrug Delivery through Synchrotron X-Ray Studies of Molecular Interactions

Abstract:

Research Interests

Capecitabine is an anticancer chemotherapy prodrug and is widely used in pharmaceutical industry to treat patients with metastatic colorectal cancer. Through two-step enzyme catalyzed degradation, capecitabine is converted to 5FU, which is an effective drug. Despite the longer half-life of the prodrug, current problems are high dose and toxicity. Many efforts had been made to increase the efficacy of the drug in terms of controlled and sustained release and thus, lower the dosage. Specifically, liposomal nanoparticle is an attractive approach to achieve those goals; however, the unique molecular structure of the prodrug makes it difficult to prepare stable nanoformulations with high drug loading and long-term stability.

In this study, we modified the prodrug molecule by attaching a cetyl chloroformate to the 5-fluorouracil head group through a simple one-pot reaction. The new prodrug (5FU-PAL) has bigger hydrophobic domain and behaves as an amphiphilic molecule at the interface. We have optimized a liposomal formulation of 5FU-PAL for prolonged blood circulation and sustained release based on the understanding of molecular interactions between 5FU-PAL and lipids, which was achieved by using synchrotron X-ray reflectivity and grazing incidence X-ray diffraction (GIXD) integrated with a Langmuir trough. Monolayer packing of 5FU-PALwith saturated neutral lipid dipalmitoyl-phosphatidylcholine (DPPC), positively charged lipid 1,2-stearoyl-3-trimethylammonium-propane (DSTAP), negatively charged lipid 1,2-dipalmitoyl-sn-glycero-3-phospho-(1'-rac-glycerol) (DPPG), unsaturated neutral lipid 1,2-dioleoyl-sn-glycero-3-phosphocholine (DOPC) and unsaturated positively charged lipid 1,2-dioleoyl-3-trimethylammonium-propane (DOTAP) were investigated. It was found that 5FU-PAL interacts strongly with positively charged lipid, especially in the head group region where the electron density of positively charged lipid increased upon adding the prodrug while the electron density of neutral and negatively charged lipid decreased. Furthermore, adding the prodrug molecule caused the packing at the air-water interface to become tighter. In addition, 5FU-PAL at the air-water interface folded into multiple layers with highly ordered crystal structures. The studied could precisely quantify the maximum drug loading in the liposomal formulation with a stable structure.

612307 Systems Engineering Approaches and Career Advancement in Bioprocess Development and Biomanufacturing

- Download/View Extended Abstract

Abstract:

Systems engineering approaches for bioprocess development and biomanufacturing can be integrated in bioprocess development and biomanufacturing during the commercialization stage of biopharmaceuticals into the engineering marketplace. Lean tools can be used as engineering techniques that can be part of management of bioprocess development and biomanufacturing projects in order to reduce different types of waste. Online or Kanban (scheduling) system allows manufacturing managers and process development engineers to create a list of requests for troubleshooting, testing and on-site inspections. Failure mode and effects analysis can be applied to reduced hazards while allowing flexibility in the layout of materials and equipment for developing a pharmaceutical production process. Life cycle analysis, techno-economic analysis, risk identification and multi-criteria decision analysis are also tools that can be applied in scenarios when single-use disposables are compared with stainless steel tanks for pilot-scale process development or large-scale biomanufacturing. With regards to the developing the workforce for this industry, mentorship and leadership are also important towards the career advancement of chemical engineers in the biopharmaceutical product development and biomanufacturing businesses. After working in the biopharmaceutical industry as a process development engineer for almost five years, I realized that building a career as a young professional is a combination of education, work experience, industry certifications, and establishing a professional network. In my poster, online resources are presented for young professionals, those with less than ten years of industrial experience, that would be useful in exploring career paths, expanding their professional network, and building relationships as they move into various positions during their careers. These resources are also helpful for researchers involved in industrial psychology, management science and human resource development.

Research Interests: Systems Engineering, Management Science, Biopharmaceutical process development, Biomanufacturing

597917 Cadherin Binding Interactions in Lipid Bilayers at the Single-Molecule Level

Abstract:

Cadherin transmembrane proteins are responsible for cell adhesion in a number of tissues and therefore, modulate a variety of biological processes, such as tissue morphogenesis, cell motility, force transduction, and macromolecular transport across tissue boundaries. Furthermore, deficiencies in cadherin-mediated adhesion are correlated with several diseases, such as a number of skin diseases and cancers. Cell-cell adhesion is facilitated via adherens junctions, consisting of adhesive, trans-interactions, and lateral, cis-interactions, but the molecular role and interplay of these interactions is a topic of discussion. Here we used single-molecule microscopy to show the drastic, double cooperativity, of cis- and trans-interactions and how this can facilitate large junction formation using a biomimetic lipid bilayer cell adhesion model. Notably, the addition of cis-binding capability resulted in a 25-fold increase in trans-binding lifetimes between epithelial-cadherin extracellular domains. The nature of this cooperativity appears to be primarily due to allostery, as opposed to avidity. It has been thought the primary role of cis-interactions is to provide molecular ordering within junctions, increasing their size, but junctions still may form in the absence of cis-interactions. Furthermore, cis-interactions have been predicted to cause an elevated cadherin surface density in the adhesive zone between cells, potentially increasing adhesive binding strength though cluster avidity. We have directly observed strong cooperativity between cis- and trans-interactions, providing vital insight into the molecular mechanism of junction formation, and therefore, cadherin-mediated cell adhesion.

Research Interests

Biomolecules at interfaces, statistical modeling, data science, single-molecule microscopy,

599438 Development of Process Models and Control Strategies to Support the Continuous Manufacturing of Pharmaceuticals

Abstract:

Research Interests

Throughout my PhD I have worked on the advancement of modeling and control for the continuous manufacturing of pharmaceuticals. I have been the primary contributor on three main projects and a secondary contributor on a few others.

The first of these projects focused on the development of a virtual plant for the manufacturing of a small molecule active pharmaceutical ingredient. The plant was used to inform process design and process control for the manufacturing system.

The second of these projects focused on the development of a model for the production of recombinant proteins in the host, *Pichia pastoris*. The model was validated with existing literature and used to understand long-term process stability and explored methods to analyze and increase this stability.

The third of these projects involves on construction of a testbed for the continuous manufacturing of monoclonal antibodies. The system is heavily instrumented so that diverse datasets can be generated for process model and control strategy development. For this, I have worked on data integration, control strategy development, and process modeling for the system.

Outside of these projects, I have a working knowledge of viral vector (specifically recombinant adeno-associated viruses) and vaccine manufacturing. I am looking for a job to leverage my experience in process modeling and control for the advancement of pharmaceutical manufacturing processes.

599981 Polyelectrolyte Surface Diffusion in a Nano-Slit Geometry

Abstract:

Research Interests

Intermittent ("hopping") surface diffusion of poly-L-lysine in a nanoslit was studied using single-molecule tracking microscopy. Three surface chemistries were employed to understand the interplay of long-range electrostatic attraction and short-range interactions: an amine-functionalized silica surface, an oligo(ethylene oxide) (OEG) modified surface, and an equally mixed surface. Diffusion increased rapidly with slit height until saturating for values <30 nm. While diffusion at a semi-infinite interface was significantly faster for OEG surfaces, the diffusion increased most rapidly with slit height for amine-functionalized surfaces, resulting in surface diffusion that was virtually independent of surface chemistry in gaps <15nm. Kinetic Monte Carlo simulations, using parameters obtained empirically from diffusion at a single interface, suggested that these trends were primarily due to strong H-bonding interactions between PLL and amine surface ligands, which led to increased rates of re-adsorption after hops and longer waiting periods between flights, and that long-range electrostatic attraction had a minor influence.

601527 Multi-Scale Modeling of Heterogeneous Catalysts Using First-Principles and Deep Learning

Abstract:

Research Interests

Although computational models have been instrumental in advancing the areas of renewable energy production and storage, traditional catalyst design primarily relies noticeably on human intuition. As such, the usual timeline for discovery, development and deployment of novel catalytic materials and chemical process is typically long-term and capital intensive. A promising solution to address this shortcoming is integration of machine learning (ML) and artificial intelligence (AI) methods, which are known to improve iteratively through experience and more data, with human intuition and creativity. *I am fascinated by this data-driven design paradigm possible through the digitization of traditional R&D pipeline using AI/ML methods; enabling teams to undertake diagnostic, predictive, and prescriptive analytics which form an integral part of Industry 4.0.*

My Ph.D. research at Purdue University is focused on developing multi-scale catalyst models that capture the essential structure-functional properties of the real-world catalyst. *These models include a combination of atomic-scale quantum mechanical simulations with subsequent reactor design and high throughput machine-learning based strategies to predict material properties.* I am using these models to investigate the active site and reaction mechanism for water-gas shift and NO_x decomposition in vehicular exhaust. The proposed theory models, as validated through our experimental collaborators, offer an improved molecular-level mechanistic understanding of the reaction. Besides this, *I am involved in developing a grand-canonical genetic algorithm to generate catalyst models which allow an improved active-site description thereby reducing the theory-experiment disparity.*

Alongside developing theoretical models, *I am investigating accelerated screening strategies for ranking catalyst candidates.* These strategies are being developed for high-entropy alloys (HEA), which typically comprise of five or more elements. HEAs have attracted considerable interest in recent years due to their superior oxidation resistance relative to that of conventional alloys catalysts. *I am using crystal-graph convolutional neural network (CGCNN) to model and rank catalyst stability with reduced computational expense compared to traditional electronic-structure optimization methods.* Appropriate featurization of the chemical moieties to encode physical-chemical properties is a crucial step in developing a ML-AI model. Crystal graphs are a promising avenue to encode these structural and chemical underpinnings. In addition to generating a surrogate model, *I am using the latent-space encodings generated through the CGCNN to discover lower dimensional manifolds that can be used for accelerated optimization of the target property.*

Besides my graduate research, *I am currently interning at Dow Chemical Company (Lake Jackson, TX) in the Chemometric and AI department.* I am part of a cross-disciplinary team comprised of process system engineers, applied statisticians, and organic chemists. *My primary projects are 1) developing process models for anomaly detection to improve unplanned event identification and 2) building a GPU-accelerated cloud-based deep-learning framework to screen large database (on the scale 10^9) of organic molecules to optimize for cost, reactivity and stability.*

606196 Dynamics of Vesicles in Strong Flows Using a Stokes Trap

Abstract:

Research Interests - My Ph.D. thesis research focuses on: (i) developing new tools for trapping and manipulating micro and nanoparticles in free solution using only fluid flow, and (ii) understanding the physics and emergent properties of vesicle suspensions using these control-based techniques. To this end, I have developed a millisecond time nonlinear model predictive control algorithm with real-time image processing (written in C++, LabVIEW and MATLAB) to control the center-of-mass and orientation of single and multiple anisotropic particle using only fluid flow. In tandem, I study the dynamics of lipid vesicles using fluidic trapping tools and advanced statistical tools. My research has direct implications to the processing of food active ingredient encapsulants, and personal care products such as shampoos and detergents. Overall, my Ph.D. research entails a distinct combination of numerical, experimental and theoretical/modeling skills involving fluid dynamics, process control techniques, and numerical methods that has provided me with the confidence to address a wide spectrum of challenges faced by modern society.

Abstract Text-

In this work, we study the shape dynamics of vesicles in precisely defined steady or time-dependent flow fields using a Stokes trap. Vesicles are membrane-bound soft containers that are often used for triggered release or reagent delivery and play an integral role in key biological processes such as molecular transport in cells. Giant unilamellar vesicles (GUVs) have been used as model systems to study the equilibrium and non-equilibrium dynamics of simplified cells that do not contain a cytoskeleton or polymerized membrane commonly found in cells. A grand challenge in the field of membrane transport lies in understanding how interfacial mechanics and fluid dynamics on the inside or outside of a soft vesicle contributes to the overall shape instabilities. Here, we study the dynamics of single floppy vesicles under large strain rates (~20 s^{-1}) using a Stokes trap, which is a new technique developed in our lab for controlling the center-of-mass position of multiple particles or single molecules in a free solution. In this way, we directly observe the vesicle shape and conformations as a function of reduced volume,which is a measure of vesicle's equilibrium shape anisotropy. We observe the formation of asymmetric dumbbell shapes, pearling, and wrinkling and buckling instabilities for vesicles depending upon the nature of flow and amount of membrane floppiness. We report the precise stability boundary of the flow-based phase diagram for vesicles in Capillary number (*Ca*)-reduced volume space, where *Ca* is the ratio of the bending time scale to that of flow time scale. We further probe the stability boundary at two different viscosity ratios to understand how the onset of dumbbell shape instability in vesicles depends on viscosity ratio. We also present results on the long-time relaxation dynamics of vesicles from high deformation back to their equilibrium spheroidal shapes after the cessation of flow. We study vesicles with shapes ranging from symmetric to asymmetric dumbbells with a long thin tether (extremely large fractional extensions with flattened thermal fluctuations), and we report on the influence of initial conditions in determining dynamic behavior. Using this approach, we study the non-equilibrium stretching dynamics of vesicles, including transient and steady state stretching dynamics in extensional flow. Our results show that vesicle stretching dynamics are a strong function of reduced volume and viscosity ratio, such that the steady-state deformation of vesicles exhibits power-law behavior as a function of reduced Capillary number. We identify two distinct relaxation processes for vesicles stretched to high deformation, revealing two characteristic time scales: a short time scale corresponding to bending relaxation and a long-time scale dictated by the membrane tension. We further discuss a model to estimate the bending modulus of lipid membranes as a function of vesicle reduced volume from the steady state stretching data. Overall, our results provide new insights into the flow-driven shape-dynamics for vesicles using new experimental methods based on the Stokes trap.

606670 Catalyst Design for Oxygen Reduction Reaction Using Molecular Modeling

- Download/View Extended Abstract

Abstract:

Fuel cell technologies are regarded as the key for clean and sustainable energy generation; however, their efficiency is limited by the cathodic oxygen reduction reaction (ORR). Pt supported on carbon black, Pt/C, has long been recognized as the standard bearer in this field; however, its high cost and scarcity really cause challenges in wide applications. Therefore, my research interest is to search, develop, and design stable and efficient catalysts for ORR using computer modeling. In this work, molecular dynamics (MD) and density functional theory (DFT) are combined to study and design complex, multifunctional catalytic materials. To understand the carbon-supported Pt electrocatalysts for both cathodic (ORR) and anodic (MOR) reactions, MD simulations were performed to show that nanoscale Pt particles can be effectively stabilized at the open edges of the vertically aligned carbon nanofibers (VACNF). DFT calculations were then carried out to determine and predict the overpotential and catalytic activities. Moreover, DFT calculations were used to reveal the electronic structures, i.e. bader charge, which can be used as a descriptor to tune ORR activity. Further, scaling relationship of electrochemical potentials versus adsorption energies on reaction intermediates or charges on the active sites can be established and employed as training set for machine learning to screen, search, and design catalysts that has higher activity for ORR.

Research Interests

My research interest is catalysts design using first-principle-based method, i.e. density functional theory, to achieve high activity for ORR. I'd like to focus on building reasonable molecular model, investigating reaction mechanism and thermodynamics, characterizing material properties, finding catalytic trend, and designing new catalyst materials.

608570 Photothermal and Immunomodulatory Nanomaterials for Rapid Tissue Repair and Wound Healing

Abstract:

Research Interests: I am a methodical and ambitious research student with interests in tissue engineering, immunology and biomedical devices. For my dissertation, I am working on developing regenerative approaches for repair of soft tissues that employ light-activated immunomodulatory biomaterials and therapeutic delivery to enhance tissue repair.

Sutures, staples and tissue glues remain the primary means of tissue approximation and vessel ligation. Despite widespread use, conventional sutures do not immediately seal approximated tissue and are susceptible to bacterial leakage, wound re-opening and infection. Laser-activated tissue sealing is an alternative approach which conventionally employs light-absorbing chromophores and nanoparticles for converting near infrared (NIR) laser to heat. The local increase in temperature engenders interdigitation of sealant and tissue biomolecules, resulting in rapid tissue sealing. We developed laser-activated nanosealants (LANS) in which, gold nanorods (GNRs) or indocyanine green (ICG) dye are embedded within a biopolymer matrix (collagen, silk or elastin-like polypeptides). We also fabricated novel laser-activated tissue-integrating sutures (LATIS) that synergize the benefits of conventional suturing and laser sealing. These laser-activated approximation devices demonstrated higher efficacies for tissue biomechanical recovery and repair in a full-thickness, dorsal surgical incision model in immunocompetent mice compared to commercial sutures and cyanoacrylate skin glue. Localized delivery of modulators of tissue repair, including histamine and copper, further improved healed skin strength following laser sealing and application. LANS films loaded with antibacterial drugs were able to significantly lower MRSA loads at the wound compared to antibacterial sutures, indicating a multifunctional strategy that synergizes rapid sealing with combating surgical site infections. In addition to incisional wounds, histamine co-delivered with silk LANS films accelerated the closure of full thickness, splinted excisional wounds in immunocompetent BALB/c mice and genetically obese and diabetic db/db mice, resulting in faster closure than Tegaderm wound dressing. Histological and immunohistochemistry analyses showed LANS-histamine treatment reduced dermal gap, promoted angiogenesis (CD31+), myofibroblast-mediated wound contraction (aSMA+), and higher TGF-β1 expression which are hallmarks of improved healing outcomes. Growth factor proteins have been investigated in tissue repair but have demonstrated limited efficacies as monotherapies. In order to address this, we first generated and characterized fusion polypeptide growth factor nanoparticles (GFNPs) that demonstrated high stabilities in wound fluid. Combination treatments of LANS-histamine and GFNPs resulted in higher efficacies of wound closure and tissue repair compared to monotherapies acting alone. An investigation into the kinetics of wound closure indicated that temporally sequential delivery of LANS-histamine followed by GFNPs, coinciding with different stages of tissue repair, demonstrated the highest efficacies of wound closure and tissue repair. Our results indicate that laser sealing and approximation, together with delivery of immunomodulatory mediators, can lead to faster healing and tissue repair, thus reducing wound dehiscence, preventing wounds moving towards chronicity and lowering incidence of surgical site infections, all of which can have significant impact in the clinic.

610141 Seseed

Abstract:

Research Interests

Metal nanocrystals are promising in a broad range of novel applications. They can bear isotropic shapes, such as nanocubes, octahedra, icosahedra, as well as anisotropic shapes such as nanorods and nanobars. Metal nanowires originate from anisotropic growth, and their aspect ratio can reach ~1000. Fivefold-twinned Cu nanowires (CuNWs) are widely used in electronic, optical, and catalytic applications. Long-chain alkylamine molecules have been employed as structure-directing agents (SDAs) in the shape-controlled synthesis of Cu nanocrystals in aqueous solutions. To understand how tetradecylamine (TDA) might function as an SDA in the growth of CuNWs, we study the adsorption of solution-phase TDA around a small and curved Cu nanoseed using a many-body metal-organic force field with molecular-dynamics (MD) simulations run with the LAMMPS code.

Given sufficient TDA, the density of TDA molecules on the Cu nanoseed surfaces is lower than that of the TDA self-assembled monolayer (SAM) on the planar Cu surfaces. Instead, TDA molecules form a distinct "bilayer" around the seed: the surface molecules (inner layer) tend to stand normal to the Cu surfaces while TDA closer to the aqueous phase (outer layer) is prone to be parallel to the surface. The bilayer act as a dense "web" to protect the Cu surfaces against the solution phase. The relatively weak TDA alky-tail interactions around the curved nanoseed allow the exchange of TDA molecules within the bilayer, while the TDA population in each layer remains dynamically stable. By fitting the exchanging probabilities into Poisson forms, we obtained the diffusion coefficient of TDA, which resembles a good agreement with NMR measurement from experimental works. Compare to the TDA SAM on the planar Cu surfaces, the TDA exchanges much more rapidly around the small nanoseed. This exchange could serve as a special mechanism for the growth of nanowire: the solution-phase $Cu2+$ complexes first attach to the TDA in the out layer, and then reach the Cu surface through the exchange with the inner layer TDA.

The significance of corners and edges, which is dominated by the size of the nanoseed, determines if TDA could form SAM or a bilayer around the Cu surfaces. To mind the gap between an infinite planar surface and a small and curved nanoseed, we create stepped Cu surfaces, which combine the features of a planar surface and a curved seed. The extent to which feature can be tuned by the dimensions of stepped surfaces. As expected, we found TDA can form SAM on the stepped surface when the effect of the planar surface overwhelms that of the edges and corners. In contrast, when the planar surface is relatively smaller, the effect of edge and corners dominates. Part of the TDA molecules escape from the SAM to form a second layer.

610616 Functional Particles for Controlled Release and Stem Cell Engineering

- Download/View Extended Abstract

Abstract:

Research Interests: controlled release, drug release, stem cell engineering, nanoparticle formulation, surface modification, polymer chemistry

Stem cells are emerging as new generation of medicine for treating various diseases and tissue defects, as they can secrete a wide range of therapeutic cytokines and growth factors. However, a major obstacle for past utilization is the limited cell secretion activities due to hostile environment or insufficient activation signal at the transplanted site. To this end, we developed a stimuli-responsive particle system, that can be tethered on stem cell surface to provide proximal stimulation and actively release stimuli to enhance the cellular secretion activity. Specifically, we devised a dual binding poly(D, L-lactic-co-glycolic) (PLGA)-based particle system through RGD and hyaluronic polymer modification. We demonstrated that when the particle-tethered stem cells were exposed to shear stress during injection, less than 10% of the particles detached from cell surface. The improved particle retention led to a better stem cell stimulation as demonstrated with a 1.5-fold increase in cellular secretion activity. Further, we designed a catalytic particle system with an enhanced molecular release profile upon reaction with an external stimuli (H_2O_2). The catalytic particle system was assembled by encapsulating MnO_2 and bioactive molecules of interest in PLGA particles. The MnO_2 catalyzed the decomposition of H_2O_2 into oxygen gas, which increased the internal pressure of particles and accelerated the release of the bioactive molecules. We demonstrated that the catalytic particles encapsulated with epigallocatechin gallate can provide cytoprotective effect against reactive oxygen species and subsequently increase the secretion activity by 2.8-fold due the enhanced molecule release effect. Our approach of functional stimuli-responsive particles therefore provides a promising pathway to engineer stem cell surface and control the release of molecules of interest.

614893 Flow behavior in complex fluids and particulates systems

- Download/View Extended Abstract

Abstract:

During my Ph.D., I focused on the flow behavior of particulate systems and specifically on flow in colloidal dispersions, polymer melts and solutions, hydrogels, and solid handling systems.

During my first major project, I investigated the mechanism controlling shear thickening in colloidal silica slurries. Shear thickening in colloidal dispersions often seriously limits formulations for coating and spraying operations, as well as flow rates for pumping concentrated dispersions. Besides, controlling shear thickening occurrence is critical in modern cement formulations, oil field extractions, and semiconductor polishing applications as well. The current study seeks ways to modify shear thickening in particulate systems and to investigate the underlying mechanisms controlling the shear thickening phenomena. Commercially available fumed silica suspensions exhibiting irreversible shear thickening at shear rates above 10,000 s^{-1} were used in this study as the base material. Using a specialized experimental protocol, the impact of spherical silica particles on shear thickening of fumed silica suspensions was investigated. It was shown that spherical particles can increase the critical shear rate when thickening occurs. In addition, rheo-Small-Angle-Light-Scattering (SALS) patterns also confirm the shift in shear thickening of fumed silica suspensions by applying spherical silica particles.

In my second major project, I studied the flow of biomass and the problems associated with biomass handling and feeding from both experimental and modeling standpoints. The goals here is to develop physics-based computational methods which will be able to capture interactions between particles and their environment as imposed by the process. To experimentally validate the physics-based models, the impact of feedstock characteristics including particles size and size distribution, moisture content, and the screw feeder operating conditions such as screw speed on the flow characteristics of terrestrial biomass feedstock was investigated. Then, the experimental results were validated by Computational Fluid Dynamics (CFD) simulations using OpenFOAM software package for a simple pipe flow case. We found that at higher screw speeds and moisture contents, the material flows easier. In addition, the viscosity of biomass in the microcompounder falls within the shear thinning region of a Cross model fit. Moreover, the impact of particle size distribution on the flow behavior of compressed biomass is insignificant. For the modeling and simulation efforts, density-dependent Bingham and Cross models were developed and validated for biomass pipe flow under pressure. Rheological values obtained from pipe flow simulations showed that the models developed are able to capture the flow behavior of compressed biomass under pressure.

Research and career interests:

Experimental and modeling of flow in particulate systems, e.g. colloidal dispersions, solid handling systems, composites and filled polymers, polymer solutions, and hydrogels.

605936 Continuous Mixing Technology: The Impact of Blend Properties and Process Parameters on Mixing Performance

Abstract:

Continuous powder mixing technology application during continuous direct compression process has emerged as a leading technology used in the development and manufacture of solid oral dosage forms [1, 2]. The critical quality attributes (CQAs) of the final product is heavily depending on the performance of the mixing step as the quality of mixing will directly influences the content uniformity of the tablet.

Therefore, the aim of this study was to investigate the impact of blend properties (bulk density, API sizes) and process parameters (process throughput, hold up mass and impeller speed) on the mixing performance. Mixing performance was characterized using the residence time distribution (RTD) of the process, which has been broadly used to characterize unit operations in pharmaceutical processes. In this work, the RTDs for a vertical continuous mixing device termed Continuous Mixing Technology (CMT) were obtained in order to evaluate mixing performance over a defined operating space. The RTDs were determined using a spike injection of tracer material into the mixer and measuring the concentration of the tracer in the outgoing material. This was achieved using a PAT interface situated at the exit of the CMT that presents the material to a near-infrared (NIR) probe. The theoretical residence time of the mixer is given by the hold-up mass divided by the throughput [1], and for each experiment, NIR spectra of the exiting material was measured for five theoretical residence times after tracer injection to obtain the full RTD.

As the CMT contains an upper de-lumping screen and a lower mixing chamber, the residence time distributions of this system can be described as two CSTRs in series with different residence times [1] where the parameter 'r' describes the ratio between these residence times. The best value of 'r' for all experiments was determined through simultaneous optimization of all measured RTDs. The results showed that the CMT operates close to a single CSTR over the whole operating space, has good ability to dampen the fluctuations from the gravimetric feeders, and that the mixing performance of the CMT is not significantly impacted by blend properties (i.e. bulk density and API particle size).

References:

[1] Blackwood, D.O., Bonnassieux, A. and Cogoni, G. Continuous Direct Compression Using Portable Continuous Miniature Modular & Manufacturing (PCM&M). In *Chemical Engineering in the Pharmaceutical Industry: Drug Product Design, Development, and Modeling*, 2nd Edition, am Ende, T., am Ende, D. J., Wiley, 2019, pp 547-560.

[2] Markarian, J. (2 April 2018). Modernizing Pharma Manufacturing. www.pharmtech.com

600731 Investigation of Critical Process Parameters of Screw Elements of a Twin Screw Extruder By an Improved Simulation Method

Abstract:

INTRODUCTION

A commonly used technique for the production of pharmaceutical products is hot melt extrusion on a co-rotating intermeshing twin screw extruder [1]. An active pharmaceutical ingredient is embedded in a molten polymeric carrier and the product is conveyed to the die [2]. The extruder screws have a modular design with, for example, conveying, kneading or mixing elements. Each element has specific properties in order to combine different unit operations in one process unit. The extrusion process is mainly influenced by the chosen screw configuration and the geometry of each screw element. [3]

Conveying elements are, depending on their pitch, responsible for the pressure build-up and the conveying of the material to the die. Kneading elements are mainly used in the melting zone because they introduce shear forces into the material according to the staggering angle of the kneading discs. Mixing elements are responsible for distributive and dispersive mixing. [4] The different types of elements can be seen in Figure 1.

Each specific screw element can be characterized by dimensionless numbers regarding the pressure and power characteristic (Figure 2). These parameters are linearly related to the dimensionless volume flow. The intersection points with the axes result in the dimensionless numbers A_1 and A_2 for the pressure build-up and B_1 and B_2 for the power input into the material. [6]

All parameters mentioned affect the residence time distribution, which indicates how long the material is exposed to the process conditions. For this reason, especially this parameter is of crucial importance for the manufacturing of pharmaceutical products with a constant product quality. Furthermore, it requires process understanding and the prediction of important process parameters. [7] Based on these challenges, the modelling has been gaining substantial interest for pharmaceutical applications within the last decade [8]. Especially the 3D simulation has moved into the focus of development due to its resolution in all special directions and its closeness to reality. Currently, it is not possible to calculate complete screw configurations of a production extruder in a reasonable time because of the high computational effort. The high amount of time is caused by the use of dynamic meshes. [9]

The aim of this work is the implementation of a 3D simulation in Ansys Fluent with the possibility to calculate a complete extruder screw configuration in a reasonable time. Furthermore, the impact of abrasion of the screw elements, which is unavoidable by the daily use in the production process, on important parameters for the extrusion process should be investigated on the basis of the explained dimensionless numbers. In addition to these correlations, the residence time distribution will be used as an important process parameter to characterize the screw elements. For this purpose, a dimensionless relationship analogous to the pressure and power characteristics will be derived.

METHODS AND MATERIALS

The simulation was performed in Ansys Fluent based on the Navier Stokes equations. A stationary, laminar and isothermal flow with an incompressible Newtonian fluid was assumed. Normally, a dynamic mesh is used for an exact representation of the screw rotation. In this case, however, an apparent movement using the method of pulse transmission was implemented. In a first step conveying elements with different pitches (20, 30, 40 mm) were constructed in Solid Works and then imported to Ansys. For the determination of the dimensionless pressure and power characteristics, the mass flow, the pressure difference across the screw element and the torque, which can be converted into a power, were evaluated. In order to investigate the effect of abrasion, screw elements currently used in the laboratory were compared with new elements and those specified by Leistritz as the lower limit. The abrasion of the barrel was not considered.

The simulation was validated by extrusion experiments performed on a twin screw extruder (ZSE27 MAXX, Leistritz Extrusionstechnik, Nuremberg, Germany) with a screw diameter of 27 mm and a customized length of five barrel elements. Only the last two elements (8D) before the die were part of the extrusion process to avoid disturbances. A silicon oil with a viscosity of 85 Pa s (Bluesil FLD 47V100000, Brenntag GmbH, Duisburg, Germany) was extruded. The pressure difference was measured with two pressure sensors (S-11, WIKA Alexander Wiegand SE & Co. KG, Klingenberg, Germany) with a distance of 30 mm. The mass flow was measured with a scale (MSE2203S-100-DR, Satorius, Göttingen, Germany) and the torque was automatically recorded by the extruder. The screw speed was set at 50 rpm and the mass flow was varied by using different die diameters (2, 4, 6 mm).

For the determination of the residence time distribution a passive scalar tracer was placed on the stationary solution of the fluid field simulation. The discretization was performed with the second order upwind method.

Experiments to validate the simulation were carried out in analogy to the measurements for the dimensionless numbers. A tracer (Eosin, Merck, Darmstadt, Germany) was added at the material inlet and the resulted signal was detected at the die with an inline UV/Vis-Spectrometer (InspectroX, ColVisTec AG, Berlin, Germany).

From the obtained residence time distribution, the Bodenstein number was calculated, which indicates the ratio of convective flows to dispersive flows and is thus a measure of the remixing within the extruder. The computations were performed with an internal python script, which fits the residence time density functions based on the axial dispersion model. This number was plotted over the screw speed for different die diameters, in order to clearly determine the resulting residence time distribution for each screw element. By a convolution of the individual residence time distributions, the resulting residence time distribution for a desired screw configuration can be determined.

RESULTS

The 3D simulation was successfully implemented in Ansys Fluent with the pulse transmission method. The dimensionless numbers regarding the pressure and power characteristics of the screw elements could be validated by extrusion experiments. Furthermore, the impact of the abrasion on the process parameters was investigated. The pressure and power characteristics were determined for conveying elements with different inner (D_{in}) and outer (D_{out}) screw diameters. Table 1 shows the results for the dimensionless pressure numbers. New, unused screw elements (new), screw elements at the lower limit (old) and actually used screw elements (used) were compared.

It can be seen that the gap between the screws and the barrel and the two screws have a great impact on the pressure build up during the extrusion process. The greater the abrasion of the elements, the less pressure build-up is possible. Next to the pressure build up, the gap width affects other process parameters, like the viscosity or the residence time distribution.

For this reason, it should be also possible to predict the residence time for individual screw elements. The residence time density function determined with Ansys Fluent was used to determine the Bodenstein number and the mean residence time and plotted against screw speed for different die diameters (Figure 3). It is obviously that the Bodenstein number increases with increasing volume flow while the mean residence time decreases.

CONCLUSION

In this work a 3D simulation in Ansys Fluent was implemented with the aim of a shorter calculation time for the representation of the extrusion process. Dimensionless numbers were used to validate the simulation and to show the impact of abrasion of the behavior of screw elements during the process. The deviations may also be significant in relation to the application of models to predict important process parameters. One of the most important parameter is the residence time, which is therefore used as a further parameter for the characterization of screw elements. By determining the Bodenstein number and the mean residence time, it is possible to predict the residence time distribution of the entire process for the specific extrusion process.

REFERENCES

1. Repka, M., et al.; Applications of hot-melt extrusion for drug delivery, Expert Opin Drug Deliv, 5 (12), 2008
2. Lang, B., McGinity, J.W., Williams, R.O.; Hot-melt extrusion - basic principles and pharmaceutical applications, Drug Dev Ind Pharm, 40 (9), 2014
3. Crowley, M. M., et al.; Pharmaceutical applications of hot-melt extrusion. Drug Dev Ind Pharm, 33(9), 2007
4. Gogos, C.G., Huiju, L., Wang, P.; Laminar Dispersive and Distributive Mixing with Dissolution and Applications to Hot-melt Extrusion, in: Douroumis, D. (Ed.), Hot-melt Extrusion: Pharmaceutical Applications, 2012
5. Sarhangi Fard, A., et al.; Tools to Simulate Distributive Mixing in Twin-Screw Extruders, Macromol Theor Simul, 21 (4), 2012
6. Eitzlmayr, A., et al.; Mechanistic modeling of modular co-rotating twin-screw extruders, Int J Pharm, 474 (1-2), 2014
7. Wesholowski, J., Berghaus, A., Thommes, M.; Investigations Concerning the Residence Time Distribution of Twin-Screw-Extrusion Processes as Indicator for Inherent Mixing, Pharmaceutics, 10 (4), 2018
8. Bochmann, E. S.; Gryczke, A.; Wagner, K. G.; Validation of Model-Based Melt Viscosity in Hot-Melt Extrusion Numerical Simulation, Pharmaceutics, 10 (3), 2018
9. Bravo, V. L., Hrymak, A. N., Wright, J. D.; Numerical simulation of pressure and velocity profiles in kneading elements of a co-rotating twin screw extruder, Polym Eng Sci, 40 (2), 2004

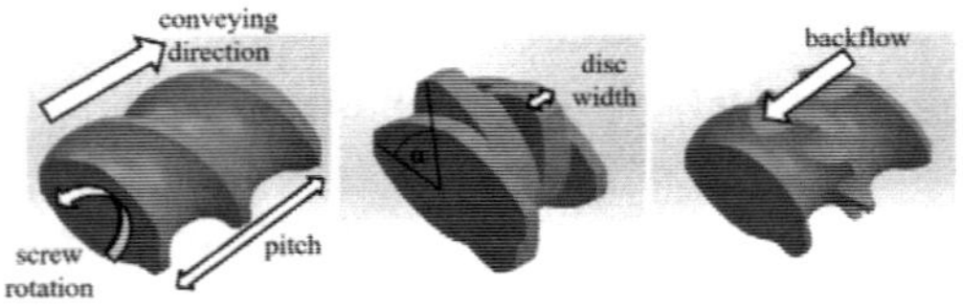

Figure 1: Conveying, kneading and mixing element, screw rotation and conveying direction are equal, according to [5]

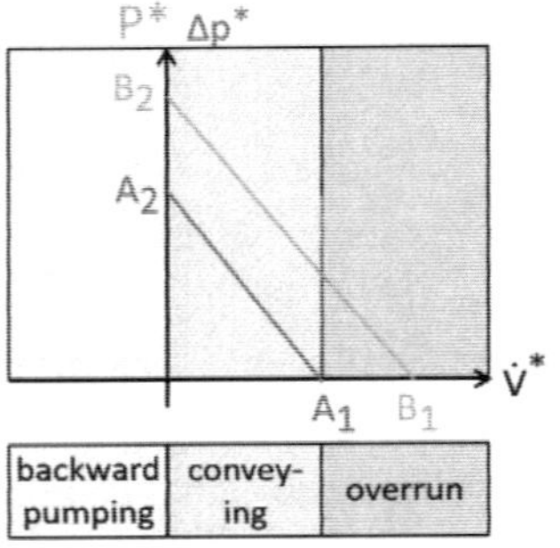

Figure 2: Dimensionless pressure and power characteristic for one specific screw element, according to [6]

Table 1: Comparison of dimensionless pressure numbers (A_1 left, A_2 right) for different geometries of GFA-2-20-30

	D_{in_used}		D_{in_new}		D_{in_old}	
D_{out_used}	0.187	3422				
D_{out_new}			0.183	3665	0.196	3372
D_{out_old}			0.198	2075	0.215	1957

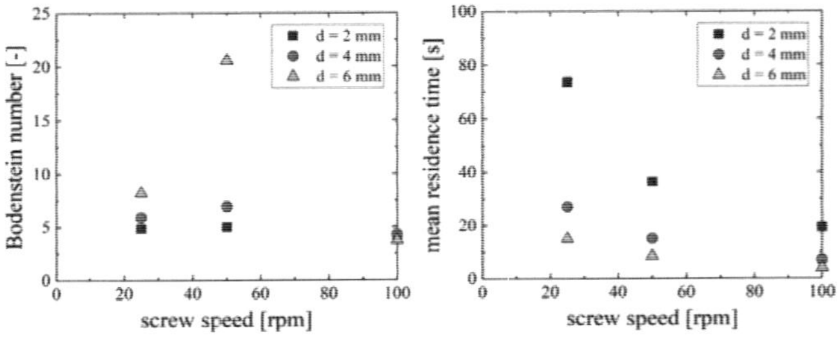

Figure 3: Bodenstein number and mean residence time plotted over the screw speed for different die diameters (new unused screw element GFA-2-20-30)

601134 Drug Product Continuous Powder Feeding for Low-Percentage Formulation Components

Abstract:

Drug product formulations typically contain low-percentage components, such as lubricants, glidants, disintegrants and sometimes drug substance. For continuous manufacturing these materials require relatively low mass feed rates. Low-end limitations of twin-screw feeders commonly used for cDC led to the investigation of a novel feeding device that uses brief, frequent pulses of compressed gas to temporarily fluidize and propel powder through a narrow, adjustable gap. For cDC relevance, the device was outfitted with an appropriately sized hopper, and additional modifications were made including vibration and agitation to disrupt powder bridging. A series of experiments were conducted with common low-percentage pharmaceutical excipients, and additionally with cohesive drug substance. An initial target of 60 grams per hour was identified which most materials could achieve, and some materials were able to be fed below 10 grams per hour. Preliminary results are promising; however, additional development of this technology is needed prior to integrating into a continuous manufacturing process train.

598164 Stochastic Analysis and Modeling of Pharmaceutical Screw Feeder Mass Flow Rates

Abstract:

Screw feeders serve as the critical first unit operation in a continuous manufacturing of drug product (CMDP) processes, influencing the mass flow rate of active pharmaceutical ingredients and excipients downstream. There is great industrial interest modeling the behavior of screw feeders accurately. Existing flow sheet models focus on deterministically simulating the powder mass flow rate [2]–[4]; however, these models only capture the average mass flow rate, and lack the stochastic behavior of the mass flow. The use of custom Discrete Element Method (DEM) models provide highly realistic simulations of particles' residence time and by extension mass flow, but prohibitively require long computation times to simulate only minutes of operation [5]. In the pursuit of designing better processes and improved controllers, there is a need for a computationally quick-to-solve flow sheet model that can simulate the stochastic nature of a real screw feeder. This is precisely the gap addressed in the present paper.

This work presents the novel characterization and modeling of the stochastic nature of a screw feeder's mass flow rate using statistical time series analysis and a deterministic flowsheet model. First, a battery of experiments was performed using different powders and a variety of operating speeds on a Coperion K-tron KT20 screw feeder. Then, for each experiment, the parameters of a hybrid mechanistic-empirical screw feeder model, based upon the work of [4], were estimated. After, the stochastic element of the mass flow is isolated by subtracting the flowsheet model's deterministic mass flow rate from the feeder-reported instantaneous mass flow rate. Next, each experiment had an autoregressive moving average model (ARMA) [6] fit to its stochastic remainder, characterizing its mass flow variation. Finally, the set of ARMA models was used to develop a predictive model, relating the flow rate stochasticity to powder properties and operating conditions. This predictive model was integrated into the current deterministic feeder model, yielding a novel hybrid mechanistic-empirical-stochastic flow sheet model that simulates a realistic, high variance mass flow rate and is suitable for the development of CMDP processes and controllers.

[1] S. García-Muñoz, A. Butterbaugh, I. Leavesley, L. F. Manley, D. Slade, and S. Bermingham, "A flowsheet model for the development of a continuous process for pharmaceutical tablets: An industrial perspective," *AIChE Journal*, vol. 64, pp. 511–525, 2017.

[2] Y. Yu, "Theoretical modelling and experimental investigation of the performance of screw feeders," PhD thesis, 1997.

[3] F. Boukouvala, V. Niotis, R. Ramachandran, F. J. Muzzio, and M. G. Ierapetritou, "An integrated approach for dynamic flowsheet modeling and sensitivity analysis of a continuous tablet manufacturing process," *Computers & Chemical Engineering*, vol. 42, pp. 30–47, 2012.

[4] D. Bascone, F. Galvanin, N. Shah, and S. Garcìa-Muñoz, "A hybrid mechanistic-empirical approach to the modelling of twin screw feeders for continuous tablet manufacturing," *Industrial & Engineering Chemistry Research*, 2020.

[5] P. Toson and J. G. Khinast, "Particle-level residence time data in a twin-screw feeder," *Data in brief*, vol. 27, p. 104672, 2019.

[6] G. E. P. Box, G. M. Jenkins, G. C. Reinsel, and G. M. Ljung, *Time series analysis: Forecasting and control*. John Wiley & Sons, 2015.

595586 Tablet Pan Coating Performance during Scale up and between Batch and Continuous Processes

Abstract:

Abstract

Dimensionless analytical models for cylindrical pans under fixed bed conditions were developed that provided mechanistic relationships between material, equipment, and process parameters; and coating uniformity for development- and production-scale processes in batch and continuous operation. The batch coating model was verified with the exact analytical solution derived using the binomial distribution and was found to agree well with literature data. The models were then used to provide insights into formulation changes that resulted in higher coating uniformity for batch and continuous coating processes. A stochastic model was used to check the validity of the non-cylindrical pans and dynamic bed conditions. Both analytical and practical solutions for bridging coating performances between batch and continuous processes, and different formulations are presented.

Extended Abstract

The main goal of a tablet coating process is to arrive at processing conditions that result in high-quality, evenly coated tablets. It follows that the control strategy for coating processes should be the coating uniformity endpoint but, apart from the drug-layering process, this is rarely done in the industry largely due to a lack of knowledge and/or the high cost of coating uniformity characterization. This work was undertaken to provide an analytical comparison for easier visualization of performance differences between batch and continuous coating processes, as well as between development-scale and production-scale processes for seamless technology transfers. This is achieved through derivations of dimensionless mechanistic models by applying simplification to a few process/equipment parameters provided in a coating uniformity simulation patent (Choi, 2009).

Model Derivation

Statistical and analytical coating uniformity models were derived based on the coating schematics shown in Figures 1 and 2 for the batch and continuous mode of operation, respectively. The models assume the same quantity of coating is received per pass through the coating zone and the probability of entering the coating zone per cycle is x_c. This assumption proved to be a good approximation regardless of the shape of the coating distribution since all approach normal distribution with an increasing number of cycles according to the central limit theorem.

The exact analytical models were derived by summing all probabilities of tablets passing through the coating zone. The resulting binomial distribution models for the batch and continuous (see Choi and Meisen, 1997) coating processes, respectively, were:

Equation (see attachment) (1)

And Equation (see attachment) (2)

The statistical model for the batch coating schematic shown in Figure 1 was derived using the central limit theorem as follows:

Equation (see attachment) (3)

Substituting the geometric relationship for a cylindrical pan shown in Figure 3; assuming a static bed; and using the tablet wall velocity relationship given by Chen et al. (2010); Equation 3 could be expressed as

Equation (see attachment) (4)

where Equation (see attachment) (5)

and Equation (see attachment) (6)

Results

The exact analytical model (Equation 1) results were indistinguishable from those of the statistical model (Equation 3); therefore, Equations 1 and 3 are considered the same. The statistical model (Equation 4) agreed well with the Chen et al. data: the bed loading had the index of 0.58 compared with 0.5 in both Pandey et al. (2006) and Chen et al. (2010). The model predicted 3.85% compared to the experimental result of 3.8%[ii] when the pan speed was changed. The effect of tablet diameter could be matched exactly to the experimental data (Chen et al., 2010). The statistical model, by using a formulation coefficient for x_c, also predicted the higher coating uniformity results seen in a high-solids-load coating formulation (Karan et al., 2019) for batch and continuous coating processes.

To test the validity of the assumptions using static bed and cylindrical pan, stochastic simulations for a non-cylindrical pan and dynamic bed were conducted. Stochastic simulations of a 60" pan with 6" cone extensions on either side pan showed no discernable difference in coating uniformity from that of a cylindrical pan with the same diameter. The stochastic simulations showed 5% increase in coating uniformity for a 22% weight gain while the statistical model gave a 4% difference between the initial and the final bed size with 22% weight gain.

Despite the simplicity of assumptions used, the models derived in this work agreed well with the literature data and stochastic simulations.

Implications

The implications of this work is that the analytical models can be used to select and monitor coating uniformity, typically expressed in relative standard deviation or Acceptance Value, as the critical quality attribute to ensure the desired performance in a manufacturing control strategy for both functional and non-functional tablet coating processes.

For process bridging, the models can be used to calculate the relationship between different scales of equipment as well as between batch and continuous coating processes. Figure 5 shows an example of how the coating time needs to be adjusted with increasing pan diameter for the same relative bed loading and spray coverage. Figure 6 shows the characteristics of continuous coating design that is needed to match the coating uniformity of batch coating process.

Symbols

d_p tablet diameter, Sauter or surface-volume diameter used

D pan diameter

F_B coating distribution for the batch process

F_C coating distribution for the continuous process

h bed height

i denotes the position in the number of cycles or the number of passes through the spray zone

k coefficient

n number of cycles

P_i probability of i

RSD relative standard deviation

x_c probability of tablet going through the spray zone

x_e exit probability of tablets in continuous processes

Δ depth of the spray zone

φ pan rotational speed

σ population standard deviation

θ angle shown in Figure 1

θ_Δ angle accounting for the thickness of the coating zone depth

τ total coating time

References

Chen, W., Chang, S-Y, Kiang, S., Marchut, A., Lyngberg, O., Wang, J., Rao, V., Desai, D., Stamato, H., and Early, W., "Modeling of Pan Coating Processes: Prediction of Tablet Content Uniformity and Determination of Critical Process Parameters", J Pharm. Sci., 99 (7), July, 2010.

Choi, M., "Pan Coating Simulation for Determining Tablet Coating Uniformity", *US Patent #10/981,812*, December, 2009.

Choi, M. and Meisen, A., "Sulfur Coating of Urea in Shallow Spouted Beds", Chemical Engineering Science, v52, p1073 (1997).

Karan, K., Hach, R., Lenick, L., During, T., Marjeram, J., Stevenson, G., Kuczynski, C., and Botnick, M. "Evaluation of Tablets Coated in a High Throughput Continuous Coater Using an Ultra-high Solids Film Coating System", Poster T1330-08-56, presented at the Annual AAPS Meeting & Convention in San Antonio (2019).

Pandey, P., Katakdaunde, M., and Turton, R., "Modeling Weight Variability in a Pan Coating Process Using Monte Carlo Simulations", *AAPS PharmSciTech*, 7 (4), 2006.

Equations

$$F_B = \begin{cases} P_0 = (1 - x_c)^n \\ P_1 = \binom{n}{1} x_c \cdot (1 - x_c)^{n-1} \\ \vdots \\ P_i = \binom{n}{i} x_c^i \cdot (1 - x_c)^{n-i} \\ \vdots \\ P_n = x_c^n \end{cases} \quad (1)$$

$$F_C = \begin{cases} P_0 = \frac{x_e \cdot (1 - x_c)}{1 - (1 - x_c)(1 - x_e)} \\ P_1 = \frac{x_e \cdot x_c}{1 - ((1 - x_c)(1 - x_e))^2} \\ \vdots \\ P_i = \frac{x_e \cdot x_c^i \cdot (1 - x_e)^{i-1}}{1 - ((1 - x_c)(1 - x_e))^{i+1}} \\ \vdots \end{cases} \quad (2)$$

$$RSD = \sqrt{\frac{1}{n} \cdot \left(\frac{1}{x_c} - 1\right)} \quad (3)$$

$$RSD = k \sqrt{\frac{\theta + \sin\theta}{\varphi^{2/3} \cdot d_p^{1/6} \cdot \tau} \cdot \frac{2\theta_\Delta - \sin(2\theta_\Delta)}{(2\theta - \sin(2\theta)) - (2\theta_\Delta - \sin(2\theta_\Delta))}} \quad (4)$$

where $\theta = \cos^{-1}\left(1 - \frac{h}{D/2}\right)$ (5)

and $\theta_\Delta = \cos^{-1}\left(1 - \frac{h - k_\Delta d_p}{D/2}\right)$ (6)

Figures

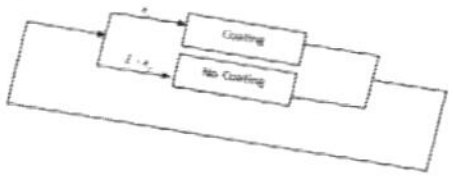

Figure 1: Schematic for the batch coating model.

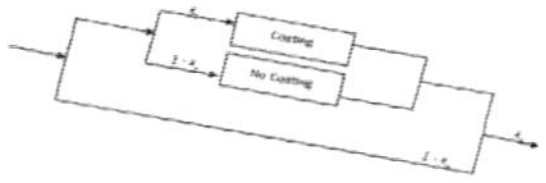

Figure 2. Schematic for the continuous coating model.

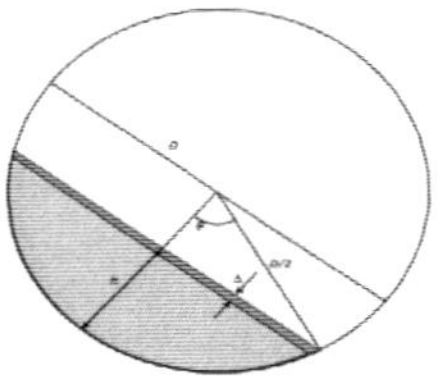

Figure 3: Geometry of pan and tablet bed (coating zone is depicted by the dark green shade)

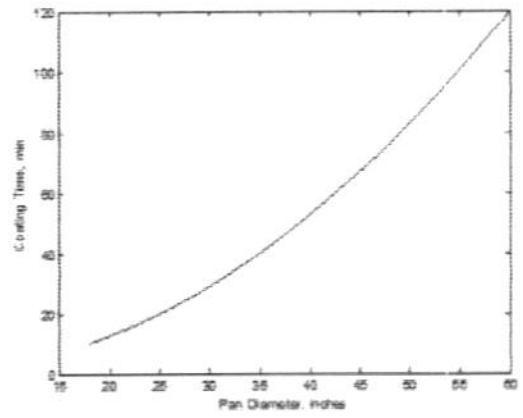

Figure 4: Coating times needed to achieve the same RSD with increasing pan size.

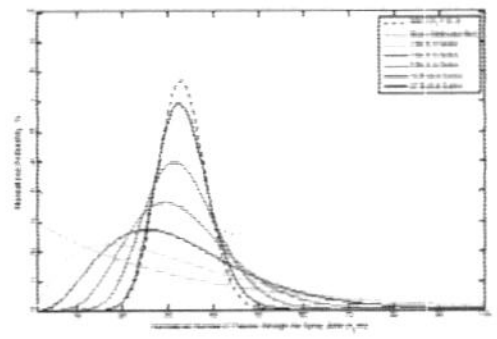

Figure 5: Comparison of coating distribution between continuous and batch coating processes.

608174 Design Method for Injectable Manufacturing Processes Considering Continuous and Batch Compounding

Abstract:

We present a process design method considering continuous and batch modes in the compounding unit for injectable manufacturing. The method can economically optimize the design variables such as the compounding mode, the lot size, and the number of filling needles. The method integrates two types of models. One is "process-oriented models" which captures the continuous nature of the process by residence time distribution (RTD), among others. The other is "discrete-oriented models" which describes the aspects such as failure events, operating shifts, and inventory by discrete and stochastic modeling. Know-how at the manufacturing shopfloors are fully used to estimate important model parameters such as the trend of failure events happening in the process. Net present value (NPV) is used as the objective function, considering pharma-specific costs such as change control and quality control. This economic evaluation model is an extension of our previously presented model for lot-sizing in the injectable manufacturing [1].

We applied the this method in an industrial case study. Using the parameters obtained in an actual injectable manufacturing process, the design variable, especially the choice between continuous and batch compounding, could be optimized. Furthermore, a generalized map was also created by varying the specified parameters in the possible ranges of products, process, and the market characteristics. The map provided quantitative insights on the cases where continuos compounding could be economically advantageous than batch compounding. The result could be useful for both retrofitting and grassroot design for injectable manufacturing processes.

[1] M. Yamada, S. Badr, S. Fukuda, M. Nakaya, Y. Yoshioka, H. Sugiyama, Journal of Pharmaceutical Innovation, accepted

600815 Efficient Production of Multi-Component Amorphous Solid Dispersions By Spray Drying

Abstract:

One of the major challenges in drug development nowadays is low water solubility (BCS classes II and IV). A widely used approach that had been proven successful in improving drug solubility is the formulation of amorphous solid dispersions (ASDs). Typically an ASD is composed of the poorly soluble drug and the hydrophilic polymer. If the combination is correct, the polymer propels the concentration of the poorly soluble drug into a supersaturated state during dissolution (spring effect). However, often the drug precipitates back into the solution, causing the concentration and thus bioavailability to drop. This issue can be circumvented by using a second polymer, which acts as a precipitation inhibitor (parachute effect). Yet, finding out the correct ratios in this tri-component system (drug: polymer 1: polymer 2) can often prove to be difficult.

In this work, we produced multi-component solid dispersions by spray drying in an efficient manner. As a widely recognized model system, we chose cyan, magenta, and yellow (CMY) printer dyes and lactose as a carrier. We can conclude that the ratio between the components in certain collected batch (experimental values) can be different from the ratio between the components in the inlet feed (theoretical values) as the residence time of particles in the system can be greater than the collection period. We evaluated the cross-contamination of samples by UV-Vis spectroscopy. Across all 15 samples, which formed a ternary diagram, the average deviation from the theoretical compositions was approximately 1.7 %. The first part of the figure below shows all collected samples arranged in the ternary diagram, forming visually confirmable color gradients. In the second part, the results from UV-Vis spectroscopy are plotted into the ternary diagram. These results show that we can produce multicomponent solid dispersions efficiently without significant cross-contamination between samples.

Acknowledgements:

Authors acknowledge financial support by the Technology Agency of the Czech Republic, project number TJ02000095 (Zeta Program).

a) Samples produced

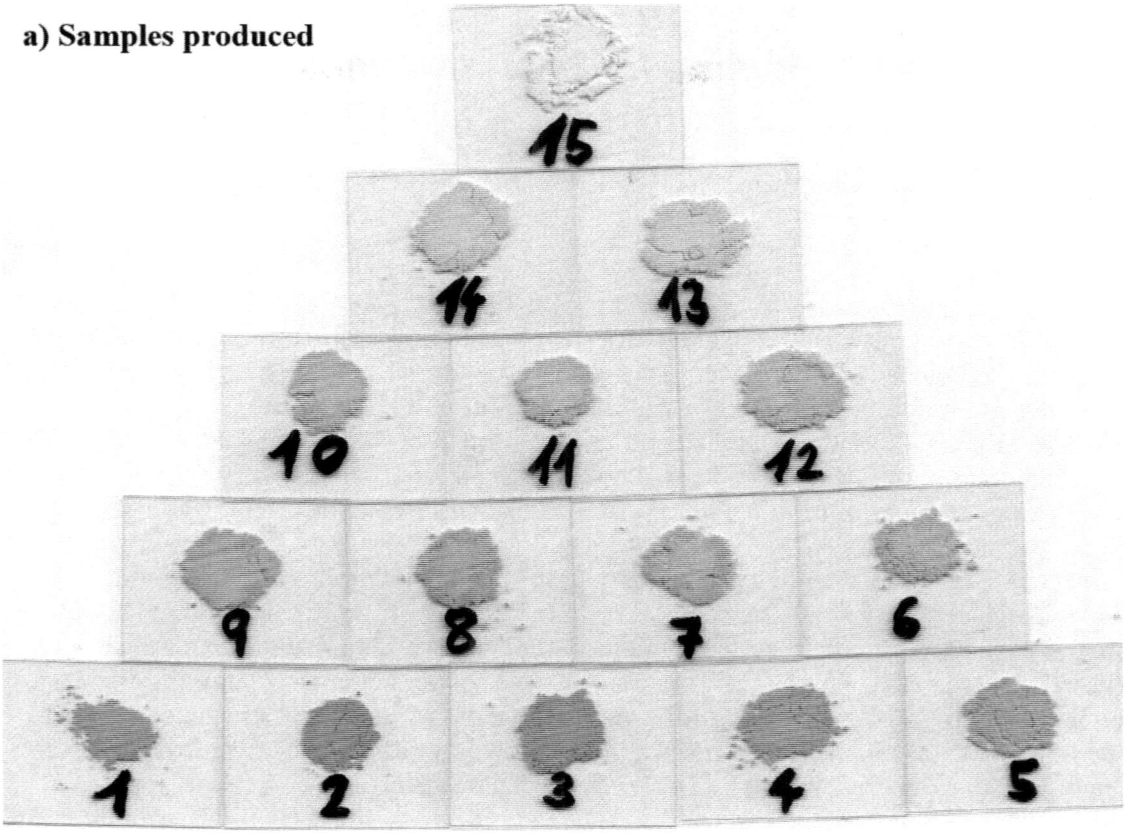

b) Results in ternary diagram

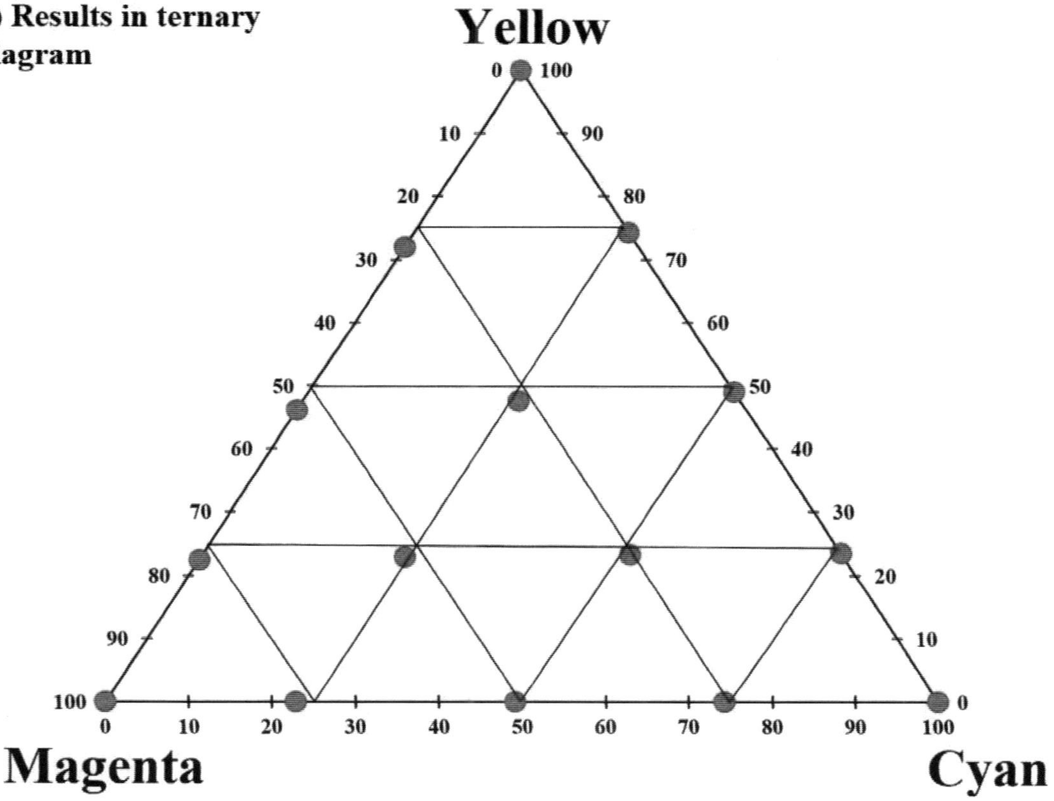

603414 Leveraging of Modeling and Automation in Development of a Control Strategy for a cGMP Buchwald-Hartwig Cross-Coupling

Abstract:

The Buchwald-Hartwig cross-coupling has become one of the most important aryl-heteroatom bond connection strategies available to the pharmaceutical industry. Owing to high selectivity, mild processing conditions, and broad availability of palladium sources and ligands, Buchwald-Hartwig reactions are conceptually well-suited for late introduction into the GMP synthesis. Despite these advantages, poor predictability upon scale-up as well as complex and/or expensive options for palladium removal can combine to limit the overall use of such transformations in the synthetic route.

This contribution focuses on control strategy development for a Buchwald-Hartwig cross-coupling introduced late in a drug substance synthetic route. Key to the control strategy was the control of the residual aryl halide process intermediate, process induced impurities, and palladium to single digit ppm levels in the drug substance. The talk outlines how automation and data-rich sampling were leveraged to enable reaction characterization. Informed by the data, a kinetic model was developed that incorporated both the catalytic cycle and mass transfer of the coordinated catalyst to the surface of the heterogeneous base. Once verified, the model was applied to determine critical process parameters and tested to predict potential failure modes of the process Critical Quality Attributes (CQAs). In addition, development of a homogeneous coordination and liquid-liquid extraction strategy for palladium removal will be discussed. Optimization of a liquid-phase approach afforded a palladium removal process with considerable cost and operational advantages relative to adsorption on functionalized silica. The overall control strategy for both the reaction and subsequent Pd removal was successfully transferred from the lab to full cGMP production scale to deliver more than 600 kg of drug substance.

601407 Moving to a Chromatography-Free Synthesis: Establishing Product-Related and Polymeric Impurity Control Strategy during Process Development

Abstract:

The process to make an active pharmaceutical ingredient (API) evolves as a project advances through clinical stages, requiring project teams to implement innovative control strategies throughout drug development to maintain API quality. An enabling process is typically used to generate small quantities of API quickly for early clinical trials, yet these enabling processes typically are not optimized and in some cases rely on costly chromatographic purifications to ensure the API meets quality requirements. As processes evolve and maturate, a common goal among project teams is to improve the overall process efficiency and scalability (e.g. obviate the need for chromatography) while producing high-quality API.

In this presentation, we highlight the evolution of the control strategy employed for a complex small molecule API as it transitioned from an early-stage enabling process to a more commercially viable process. As part of this process development, a new synthetic route was implemented; we chose to use the penultimate step as the synthetic quality gatekeeper to mitigate any risk to the final API quality. There were substantial challenges to achieve comparable purity of the penultimate between the enabling and desired process. While the penultimate step of the enabling process consisted of simple nucleophilic deprotections using chromatographed input, the penultimate step of the intended long-term process consisted of a complex telescoped reaction (intramolecular phosphoramidate cyclization + 2 nucleophilic deprotections) with substantially less pure inputs. In an attempt to move away from chromatographic purifications, we designed and implemented a two-drop crystallization procedure to remove product-related impurities, obtaining solids with > 99 LCAP. Despite high purity, the potency was low due to impurities generated from the polymerization of a by-product of the cyclization reaction. Controlling these oligomers and polymers required two mitigations: extractive work-up to remove the precursor monomer and recrystallization to remove the polymer. Through this process development, we have realized the targets of a commercially viable process by increasing the overall yield 10-fold, reducing our process mass intensity (PMI) by 98%, removing all chromatographic purifications, and, critically, maintaining API quality requirements.

605432 Process Control Strategy Development for an Integrated Continuous Platform for Monoclonal Antibody Manufacturing

Abstract:

Process and product understanding is at the root of manufacturers' efforts to produce safe and effective medicines. The 2009 ICH and FDA Guidance for Industry Q8(R2) Pharmaceutical Development describes a quality by design (QbD) approach that "emphasizes product and process understanding and process control, based on sound science and quality risk management" (1). During the product development stage, the critical quality attributes required to ensure product quality are defined and their relationship to the critical process parameters which impact them are determined. Manufacturers can use a risk-based approach to define a design space within which the combination of process inputs and process parameters have been demonstrated to provide an assurance of product quality (2,3). Consistent product quality is ensured by development of a control strategy that maintains the process within the bounds of the design space.

Process models are an efficient way to define a design space, from the level of individual unit operations or of connected unit operations up to the level of the entire manufacturing process. Process models can be used by manufacturers to ensure product quality, to make predictions, to modify procedures, and for control. Insights gained from process models can lead to improved troubleshooting capabilities during manufacturing and improved control of the critical quality attributes through long-lasting changes in process operating protocols.

Here we present a platform for the cultivation of this understanding via a fully instrumented, integrated continuous manufacturing testbed for monoclonal antibodies (mAbs). The testbed consists of 4 parallel upstream bioreactor setups including 4 perfusion devices, with one reactor assembly integrated with a fully continuous downstream purification system including Protein A chromatography, in-house designed viral inactivation, and ion exchange chromatography. This presentation describes the design and implementation of integrated control strategies for the system. The control strategy discussion begins with tuning of lower level controls (i.e. pH, DO, VCD, media addition, etc.) during process development and for extended cultivations. Following experimental demonstration of the control strategies for integrated operation, some first-principles and data-driven models (4) for the prediction of product-related quantities (i.e. N-linked glycoform distribution) are constructed and validated experimentally. These models are then used to inform a control strategy for the stable, continuous manufacturing of a monoclonal antibody that explicitly controls critical product quality attributes rather than only process quality attributes and process operations.

602472 Modulating Antibody Glycosylation Using Traditional and PAT Methods

Abstract:

Control of critical quality attributes (CQAs) is essential for the manufacture of high-quality biologics. Control over glycosylation of antibodies within specified ranges can be challenging but is crucial for product efficacy, immunogenicity and pharmacokinetics. Particularly for biosimilar products, the glycosylation ranges for demonstrating similarity to the innovator product can be very small. Changes in bioreactor process parameters at the manufacturing scale could impact glycosylation outside the specified ranges.

Several bench-scale experiments were conducted to understand the relationship between TAF and early culture pH. It was demonstrated that the TAF of the product harvested was affected by pH early in the culture. Given this information, pH control was improved using traditional univariate setpoint control in the early phase of bioreactor culture. This resulted in lots with acceptable TAF.

More sophisticated bench-scale studies were carried out to understand the effect of initial pH in combination with other levers with the aim of constructing a predictive model. The pH was intentionally manipulated in early culture to create a defined disturbance in TAF. A Process Analytical Technology (PAT) strategy was used to monitor and control TAF at large-scale.

609532 Model Predictive in Vitro Dissolution Testing for Real-Time Release in Pharmaceutical End-to-End Integrated Continuous Manufacturing: An Equivalence Study

Abstract:

The real-time release testing (RTRT) is one of the important cornerstones to achieve end-to-end continuous manufacturing in the pharmaceutical industry. Beside many innovative chemometric model-based process analytical technologies (PAT) to infer the critical quality attributes (CQAs) of drug product, mechanistic model-based predictive testing for target product profiles (TPPs) is also emerging rapidly in the past decade, e.g., tablet dissolution profile [1]. In this study, a predictive mathematical model for tablet dissolution was developed and implemented in an end-to-end integrated continuous manufacturing pilot plant at CONTINUUS to produce immediate release tablets with the specific Extrusion-Molding-Coating (EMC) unit operation. Within the EMC, the active pharmaceutical ingredient (API, a hydrochloride salt) particles are dispersed in the melted polymeric excipients [2]. The model development considered the unique dissolution mechanism of the EMC tablets, which was mainly controlled by the swelling and eroding of the polymeric matrix due to liquid penetration. The major governing equations of the mechanical model consisted of the dissolution, diffusion, and population balance of API particles in the swollen polymeric matrix, as well as the mass balance of API solute in the buffer solution. During in-line implementation, PAT tools, such as laser diffraction for API particle size distribution and near infrared spectroscopy for tablet dosage strength, were OPC connected to the computational server as model inputs. The model output of dissolution time of 70% total API in the tablet was sent to the control system for process monitoring and facilitating the decision-making of real-time release. More importantly, an equivalence study of the model-based predictive dissolution testing was conducted by comparing the model prediction to the experimental dissolution profiles conducted according to *USP42-NF37*General Chapter <711> *Dissolution*. Similarity factor f_2 values of >50 were obtained for two levels of drug dosing. A f_2 value of >50 indicates the similarity between the predicted and experimental dissolution profiles. Sensitivity of the model parameters and risk analysis of the model-based dissolution testing were also investigated in this study. Last but not the least, the developed predictive dissolution testing strategy is not limited to the specific API and can be generalized to other APIs for RTRT implementation using the EMC technology.

References:

[1] Zaborenko N, Shi Z, Corredor CC, Smith-Goettler BM, Zhang L, Hermans A, Neu CM, Alam MA, Cohen MJ, Lu X, Xiong L, Zacour BM. First-principles and empirical approaches to predicting In Vitro dissolution for pharmaceutical formulation and process development and for product release testing. *The AAPS Journal*. 2019; 21:32.

[2] Hu C, Testa CJ, Wu W, Shvedova K, Shen DE, Sayin R, Halkude BS, Casati F, Hermant P, Ramnath A, Born SC, Takizawa B, O'Connor TF, Yang X, Ramanujam S, Mascia S. An automated modular assembly line for drugs in a miniaturized plant. ChemComm. 2020; 56: 1026-1029.

599843 Or-Gym: A Reinforcement Learning Library for Operations Research Problems

Abstract:

OR-Gym: A Reinforcement Learning Library for Operations Research Problems

Christian Hubbs[1,2], Owais Sarwar[1], Hector Perez[1], Nikolaos Sahinidis[1], and Ignacio Grossmann[1]

[1]Department of Chemical Engineering, Carnegie Mellon University, Pittsburgh, PA 15213

[2]Digital Fulfillment Center, Dow Chemical, Midland, MI 48642

Keywords: Machine Learning, Reinforcement Learning, Optimization, Scheduling, Stochastic Program- ming

1 Abstract

We introduce OR-Gym, an open-source benchmark in the form of OpenAI Gym for developing reinforcement learning algorithms to address operations research problems [Brockman et al., 2016]. Reinforcement learning has been widely applied to game-playing and surpassed the best human-level performance in many domains, yet there are few use-cases in industrial or commercial settings.

We seek to provide a standardized library for the research community who wishes to extend RL into commercial applications by building on top of the preceding work and releasing OR-Gym, a single library that relies on the familiar OpenAI interface for RL, but contains problems relevant for industrial use. To this end, we have incorporated the benchmarks in Balaji et al. [2019], while extending the library to address the traveling salesman problem, knapsack problem [Kellerer et al., 2004], multi-dimensional bin packing [Coffman et al., 2013], multi-echelon news vendor problem, portfolio optimization [Dantzig and Infanger,

1993], and vehicle routing problem [Pillac et al., 2013]. These problems cover logistics, finance, engineering, and are common in many business operation settings. In each case, we select a prototypical version from the literature to benchmark reinforcement learning and other optimal approaches against.

RL problems are formulated as Markov Decision Processes (MDP), meaning they are probabilistic, se- quential decision making problems. MDP's are defined by the state, which informs an agent to select an action, which then receives a reward during the transition to the subsequent state. This framework is not widely used in the optimization community, so we make explicit our thought process as we reformulate many optimization problems to fit into the MDP mold without loss of generality.

Many current RL libraries such as the canonical OpanAI Gym, have many interesting problems, but problems that are not directly relevant to industrial use. Moreover, many of these problems (e.g. the Atari suite) lack the same type of structure as classic optimization problems, and thus are primarily amenable to model-free RL techniques, that is RL algorithms that learn with little to no prior knowledge of the dynamics of the environment they are operating in. Bringing well-studied optimization problems to the RL community may encourage more integration of model-based and model-free methods to reduce sample complexity and provide better overall performance.

It is our goal that this work encourages further development and integration of RL into optimization and the OR community while also opening the RL community to many of the problems and challenges that the OR community has been wrestling with for decades.

We explicitly present the design decisions we made to translate these problems to MDP's, as well as the results of the benchmark algorithms. All code is open-source and available at www.github.com/hubbs5/or- gym.

References

B. Balaji, J. Bell-Masterson, E. Bilgin, A. Damianou, P. M. Garcia, A. Jain, R. Luo, A. Maggiar, B. Narayanaswamy, and C. Ye. ORL: Reinforcement Learning Benchmarks for Online Stochastic Op- timization Problems. 2019. URL http://arxiv.org/abs/1911.10641.

G. Brockman, V. Cheung, L. Pettersson, J. Schneider, J. Schulman, J. Tang, and W. Zaremba. OpenAI Gym. pages 1–4, 2016. ISSN 00217298. doi: 10.1241/johokanri.44.113. URL http://arxiv.org/abs/1606.01540.

E. G. Coffman, J. Csirik, G. Galambos, S. Martello, and D. Vigo. Bin Packing Approximation Algorithms: Survey and Classification. In Handbook of Combinatorial Optimization, pages 455–531. Springer, Heidel- berg, 2013.

G. B. Dantzig and G. Infanger. Multi-stage stochastic linear programs for portfolio optimization. Annals of Operations Research, 45:59–76, 1993.

H. Kellerer, U. Pferschy, and D. Pisinger. Knapsack Problems, volume 53. Springer Verlag, Heidelberg, 2004. ISBN 9788578110796. doi: 10.1017/CBO9781107415324.004.

V. Pillac, M. Gendreau, C. Gueret, and A. L. Medaglia. A review of dynamic vehicle routing problems. European Journal of Operational Research, (225):1–11, 2013.

600671 Updated LEAPS2 for Surrogate Recommendation

Abstract:

The growing need for accurately modeling complex physical systems is increasing the complexity of high-fidelity models. Simpler, often analytical, computationally inexpensive surrogates or meta-models offer an attractive alternative. Surrogates are data-driven models that mimic input and output patterns in the data by means of response surfaces. Selecting a surrogate model that approximates a complex system most accurately is critical. A straightforward approach for this selection problem is to "try" several surrogates and select the best. Several other approaches exist in the literature. Genetic Programming (GP) has been used (Koza 1994; Streeter and Becker 2003; Lessmann, Stahlbock, and Crone 2006) to derive an optimum combination of operators and simple basis functions defining a surrogate model. Cozad, Sahinidis, and Miller (2014) developed a low-complexity, accurate model (ALAMO) that uses MILP optimization to identify the best mix of basis functions. Apart from these, MINLP formulations (Cozad and Sahinidis 2018) and extended GP (Kaizen Programming) (Rad, Feng, and Iba 2018) have also been used to good effect. All these works must be exhaustively applied to every new data set to determine the best surrogate. However, a smarter, faster, and learning-based approach is to first unearth patterns that match the meta-features or attributes of a data set with surrogate model performance. Such meta-learning can help select the surrogate for a future data set. Cui et al. 2016 and Garud, Karimi, and Kraft (2018) developed this basic idea into CRS and LEAPS2 frameworks respectively. Later, Davis, Cremaschi, and Eden 2018 studied the performance of surrogates with respect to sample sizes, input dimensions, and shapes of input functions. Although LEAPS2 addressed the limitations of CRS with respect to sample sizes, dimensionality, and surrogates, LEAPS2 has its shortcomings. It was trained only on noise-free synthetic data, so it may recommend an over-fitting model for real-world, noisy data. Moreover, LEAPS2 used an error-based metric that requires splitting the data into train/test sets. Furthermore, LEAPS2 includes both data-distribution based attributes such as local and global fluctuations, dimensionality, etc. along with statistical data-based attributes like mean, standard deviation, gradient, etc. Intuitively, it appears that only distribution-based attributes should determine surrogate performance, as it is the underlying trends or features in data that determine surrogate performance.

In this work, we modified and broadened the scope of LEAPS2 in several significant ways. First, we incorporated noisy and real-world data sets to address a key challenge in surrogate modeling. Second, we added one more metric for surrogate selection, namely a complexity-based metric called AIC weight. This metric provides an alternative for surrogate selection when splitting the dataset into train/test sets is not feasible. Third, we essentially revamped the attribute set of LEAPS2 to use only those attributes that quantify the underlying features of "data-distribution", rather than the data itself. Some of these new attributes quantify the degree and variations of non-linearity in the data, asymmetry, and flatness of response with respect to standard distribution. Thus, we now have fewer (11 vs 14) but intuitively more appealing attributes in LEAPS2. Fourth, we have improved the surrogate recommendation strategy by developing simple heuristics. Finally, we have updated our surrogate pool by adding 10 new surrogates in LEAPS2. Our improved LEAPS2 framework was evaluated with respect to the two metrics (Garud et al. 2018), namely "Total Degree of Success" (TDoS) that quantifies the success in recommending the best surrogates, and "Total Coefficient of Reward" (TCoR) that combines the success and computational savings in a single score. The new framework gives a TDoS = 91% and a TCoR = 42% for the error-based metric, and TDoS = 83% but a much higher TCoR = 63% for AIC weight on test data. However, they improved during the learning process. We tested the new framework on two case studies with real data, one on a compressor, and the other on COVID-19 data. In both cases, our improved LEAPS2 achieved a TDoS of 100%. This framework acts as a smart tool for surrogate selection to model complex physical systems.

References:

Cozad, Alison, and Nikolaos V. Sahinidis. 2018. "A Global MINLP Approach to Symbolic Regression." *Mathematical Programming* 170 (1): 97–119. https://doi.org/10.1007/s10107-018-1289-x.

Cozad, Alison, Nikolaos V. Sahinidis, and David C. Miller. 2014. "Learning Surrogate Models for Simulation-Based Optimization." *AIChE Journal* 60 (6): 2211–27. https://doi.org/10.1002/aic.14418.

Cui, Can, Mengqi Hu, Jeffery D. Weir, and Teresa Wu. 2016. "A Recommendation System for Meta-Modeling: A Meta-Learning Based Approach." *Expert Systems with Applications* 46 (March): 33–44. https://doi.org/10.1016/j.eswa.2015.10.021.

Davis, Sarah E., Selen Cremaschi, and Mario R. Eden. 2018. "Efficient Surrogate Model Development: Impact of Sample Size and Underlying Model Dimensions." In *Computer Aided Chemical Engineering*, 44:979–84. Elsevier. https://doi.org/10.1016/B978-0-444-64241-7.50158-0.

Garud, Sushant S., Iftekhar A. Karimi, and Markus Kraft. 2018. "LEAPS2: Learning Based Evolutionary Assistive Paradigm for Surrogate Selection." *Computers & Chemical Engineering* 119 (November): 352–70. https://doi.org/10.1016/j.compchemeng.2018.09.008.

Koza, JohnR. 1994. "Genetic Programming as a Means for Programming Computers by Natural Selection." *Statistics and Computing* 4 (2). https://doi.org/10.1007/BF00175355.

Lessmann, Stefan, Robert Stahlbock, and Sven F Crone. 2006. "Genetic Algorithms for Support Vector Machine Model Selection," 7.

Rad, Hossein Izadi, Ji Feng, and Hitoshi Iba. 2018. "GP-RVM: Genetic Programing-Based Symbolic Regression Using Relevance Vector Machine." *ArXiv:1806.02502 [Cs]*, August. http://arxiv.org/abs/1806.02502.

Streeter, Matthew, and Lee A Becker. 2003. "Automated Discovery of Numerical Approximation Formulae via Genetic Programming," 32.

602522 When Deep Learning Meets Sparse Model Identification: Online Adaptive Sparse Identification of Systems (OASIS)

Abstract:

Over the past few decades, nonlinear model-based control is receiving significant attention as linear models are inadequate in describing inherently nonlinear and complex industrial processes. This requires developing methods for nonlinear model identification. Modeling using first-principles is desirable only when there is sufficient knowledge of the process available. For the processes that are complex and poorly understood, it is difficult to derive such models. In view of this, there has been an increasing interest in data-driven system identification for prediction and control purposes. Specifically, subspace identification methods such as multi-variable output-error-state-space (MOESP) [1], numerical algorithms for subspace state-space identification (N4SID) [2], and canonical variate analysis [3] are popular in the process control domain. Although these methods are successful in identifying state-space models for various industrial applications using input-output data, they do not provide a physical understanding of the process. More importantly, for adaptive modeling applications, it is advantageous to have a model that can provide interpretability of the changing dynamics, which will help guide in evaluating the process operating conditions to take appropriate actions. Lately, sparse identification of nonlinear dynamics (SINDy) has delivered promising results for various nonlinear processes [4]. The SINDy algorithm is based on the techniques of sparse regression and compressive sensing. It fits the input-output data to a library of candidate functions such that only the functions describing the original dynamics are identified. Thereby, using this approach, a parsimonious and interpretable model is obtained. Additionally, any prior knowledge about the process from, for example, physics and thermodynamic laws can be included in the candidate library to quicken the model identification process. Due to these reasons, there has been a growing interest in applying SINDy to a variety of applications, which includes identifying rational nonlinearities [5], sparse learning of reaction kinetics [6], reduced-order modeling of a complex process [7], model recovery from abrupt system changes [8], and for control [9].

Despite the simplicity of SINDy algorithm, it is challenging to use SINDy for model-based control, especially at any instance of plant-model mismatch or process upset. This is because re-training the model using SINDy is computationally expensive and cannot guarantee to catch up with rapidly changing dynamics. Hence, we propose online adaptive sparse identification of systems (OASIS) framework that extends the capabilities of SINDy for accurate, automatic, and adaptive approximation of process models. The key novelty is to combine the usefulness of SINDy in discovering an interpretable model with a deep neural network (DNN) to adaptively model and control the process dynamics in real-time. The proposed method is implemented in two steps: system identification and controller design. For the system identification step, we utilize several sets of process historical data that are available for various input settings and identify their corresponding models using SINDy. Next, we train a DNN using the previously collected historical data sets and their respective SINDy models such that the DNN approximates the relationship between process data and SINDy models. We use this trained DNN to design a controller wherein the DNN updates the SINDy model by utilizing a new set of measurements at every sampling time to accurately predict the future process behavior. In this way, the OASIS method supports the application of SINDy for real-time model identification and control. We demonstrate the OASIS methodology on the model identification and control of a continuous stirred tank reactor. The closed-loop results showed that the proposed OASIS framework can be effectively used for adaptive modeling and control of nonlinear processes.

Literature cited:

[1] Verhaegen M, Dewilde P. Subspace model identification part 2. Analysis of the elementary output-error state-space model identification algorithm. International Journal of Control. 1992; 56(5):1211-1241.

[2] Van Overschee P, De Moor B. N4SID: Subspace algorithms for the identification of combined deterministic-stochastic systems. Automatica. 1994; 30(1):75-93.

[3] Larimore WE. Canonical variate analysis in identification, filtering, and adaptive control. In: 29th IEEE Conference on Decision and Control, vol. 2. Piscataway, NJ: IEEE. 1990; pp. 596-604.

[4] Brunton SL, Proctor JL, Kutz JN. Discovering governing equations from data by sparse identification of nonlinear dynamical systems. Proceedings of the National Academy of Sciences. 2016; 113(15):3932-3937.

[5] Mangan NM, Brunton SL, Proctor JL, Kutz JN. Inferring biological networks by sparse identification of nonlinear dynamics. IEEE Transactions on Molecular, Biological and Multi-Scale Communications. 2016; 2(1):52-63.

[6] Hoffmann M, Frhner C, No F. Reactive SINDy: Discovering governing reactions from concentration data. The Journal of Chemical Physics. 2019; 150(2):025101.

[7] Narasingam A, Kwon JSI. Data-driven identification of interpretable reduced-order models using sparse regression. Computers & Chemical Engineering. 2018; 119:101-111.

[8] Quade M, Abel M, Nathan Kutz J, Brunton SL. Sparse identification of nonlinear dynamics for rapid model recovery. Chaos: An Interdisciplinary Journal of Nonlinear Science. 2018; 28(6):063116.

[9] Brunton SL, Brunton BW, Proctor JL, Kutz JN. Koopman invariant subspaces and finite linear representations of nonlinear dynamical systems for control. PloS one. 2016; 11(2):e0150171.

599332 Applying Reinforcement Learning to Control Batch Processes

Abstract:

Reinforcement Learning (RL) is one of the three basic machine learning paradigms, alongside supervised and unsupervised learning. RL focuses on how a smart agent can learn the optimal policy from the maximization of cumulative rewards from its environment. The recent development of model-free RL has achieved remarkable success in various control tasks, where multiple applications have been reported in literature including parameter tuning for existent single PID control loops [1], supply chain management [2] and robotic control [3].

In the process control domain, there is an extensive state of art that applies RL algorithms to different types of chemical processes [4,5]. Nevertheless, most of these studies focus on the implementation and application of a specific RL method. These studies usually do not consider comparison with other RL algorithms or with traditional methods in the process control area. Additionally, there is no guideline regarding the RL algorithm selection.

In this work, we will provide a comprehensive investigation of RL applications in the process control area. The above practical issues would be addressed through the case study of a batch bio-reactor. Different types of RL methods including value-based, policy-based, and actor-critic algorithms will be compared in this case study. Specifically, their training stability, convergence, reproducibility and sample efficiency would be evaluated. Beyond the algorithm level evaluation, RL performance on common control scenarios, including control for optimization, control to set point, and control with state constraints would also be performed. Furthermore, the performance of the RL algorithms would be compared with a model predictive controller. Based on these comparisons, we will summarize some guidelines and suggestions on current model-free RL approaches in the control area, as well as some future perspectives regarding RL development and its application in the chemical industry.

Reference

[1] Badgwell, Thomas A., et al. "Adaptive PID controller tuning via deep reinforcement learning." U.S. Patent Application No. 16/218,650.

[2] Oroojlooyjadid, Afshin, et al. "A Deep Q-Network for the Beer Game: A Deep Reinforcement Learning algorithm to Solve Inventory Optimization Problems." arXiv preprint arXiv:1708.05924 (2017).

[3] Peng, Xue Bin, et al. "Sim-to-real transfer of robotic control with dynamics randomization." 2018 IEEE international conference on robotics and automation (ICRA). IEEE, 2018.

[4] Spielberg, Steven, et al. "Toward self-driving processes: A deep reinforcement learning approach to control." AIChE Journal 65.10 (2019): e16689.

[5] Petsagkourakis, Panagiotis, et al. "Reinforcement learning for batch bioprocess optimization." Computers & Chemical Engineering 133 (2020): 106649.

601603 Inverse Linear Optimization: Conceptual Framework, Two-Phase Algorithm, and Adaptive Sampling

Abstract:

Real-world decision making generally involves selecting the best among multiple possible options. It can be conjectured that these decisions are solutions to an underlying optimization problem which, in many cases, is unknown even to the decision maker. Uncovering these hidden optimization problems can act as a medium to generate insights into the decision-making process and predict future decisions. Examples where this can be useful are as diverse as learning the strategy of an expert decision maker operating a complex process, understanding the preferences of customers, or analyzing decision-making strategies in biological systems known to have evolved to behave optimally for their survival. Several of such decision-making problems can be formulated or approximated as linear programs (Burgard & Maranas, 2003; Saez-Gallego et al., 2016).

In this work, we develop a data-driven framework for uncovering the unknown cost vector when the decision-making process can be represented as a linear optimization problem. This is referred to as inverse optimization (Ahuja & Orlin, 2001). Research on inverse optimization started with a single, deterministic observation and has more recently progressed to the more realistic setting of multiple noisy observations coming from different sets of input parameters (Aswani et al., 2018). The latter is often referred to as data-driven inverse optimization and is the focus of this work.

The two main theoretical contributions of this work are as follows: (1) We propose a novel two-phase algorithm that performs denoising of observations and cost vector estimation separately. Our approach derives from a polyhedral understanding of the problem and is shown to be effective in alleviating the limitations of existing methods found to lead to severely restricted feasible sets of cost vector estimates. (2) We develop a rigorous adaptive sampling strategy that systematically reduces the set of feasible estimates and therefore increases the confidence in the resulting point estimate. Our extensive computational experiments show that the proposed approach can find the true cost vector (or the best possible estimate) with significantly less samples compared to naïve random sampling.

References

Ahuja, R. K., & Orlin, J. B. (2001). Inverse optimization. *Operations Research*, *49*(5), 771–783.

Aswani, A., Shen, Z. J. M., & Siddiq, A. (2018). Inverse optimization with noisy data. *Operations Research*, *66*(3), 870–892.

Burgard, A. P., & Maranas, C. D. (2003). Optimization-based framework for inferring and testing hypothesized metabolic objective functions. *Biotechnology and Bioengineering*, *82*(6), 670–677.

Saez-Gallego, J., Morales, J. M., Zugno, M., & Madsen, H. (2016). A Data-Driven Bidding Model for a Cluster of Price-Responsive Consumers of Electricity. *IEEE Transactions on Power Systems*, *31*(6), 5001–5011.

610309 Connections between Residual Networks and Explicit Numerical Integrators, with Applications to Identification of Noninvertible Dynamical Systems

Abstract:

Noninvertibility, an intrinsic property of certain discrete-time dynamical systems, is defined as non-uniqueness of the backward-time dynamics. Moreover, this noninvertibility concept could be also be adapted to neural network predictions that approximate the ground-truth fpr some unknown system [1]. In many cases, the governing equations for discrete-time systems could be treated as a discretized form of a continuous time, ordinary differential equation, which could be solved numerically by, say, explicit or implicit Euler integration methods. These methods, to some extent, are similar in architecture to residual neural networks (ResNets), which have recently seen wide use in machine learning [2].

We begin our work with numerical simulations of chemical oscillatory dynamics. Extending previous work [1], we implemented an algorithm to determine the range of noninvertible regions for the Euler discretization map. Then, training a neural network to discrete time data, we show how noninvertibility arises naturally in neural networks, and how to quantify its occurrence.
In addition to this study of the noninvertibility itself, we investigate its dynamical consequences: dynamic instabilities and pathological phenomena that occur when this feature of the discrete map interacts with the dynamics and bifurcations of the underlying continuous-time system. This study yields useful insights into the long-term prediction power of network architectures. Additionally, our work establishes connections between a network's invertibility and the estimation of its Lipschitz constants, a connection which has value beyond dynamical systems applications [3].

[1] R. Rico-Martinez, I. G. Kevrekidis and R. A. Adomaitis, "Noninvertibility in neural networks," IEEE International Conference on Neural Networks, San Francisco, CA, USA, 1993, pp. 382-386 vol.1.

[2] K. He, X. Zhang, S. Ren, and J. Sun. Deep residual learning for image recognition. arXiv preprint arXiv:1512.03385, 2015.

[3] J. Behrmann, W. Grathwohl, R.T.Q. Chen, D. Duvenaud, and J. Jacobsen. Invertible Residual Networks. arXiv preprint arXiv: 1811.00995, 2019.

602599 Crystal Morphology Modification in a Nucleation Dominant Crystallization: A Tool from Traditional Chemistry

Abstract:

Crystal morphology engineering is an important component of process development in the pharmaceutical industry, since the crystal shape or habit directly influences downstream processing, formulation, and the physical properties of the API. Most commonly the desired habits of crystals are prepared through the modification of crystallization process parameters such as temperature, the crystallization solvent, and/or the addition of a co-solvent or an additive. However, in some cases, these methods do not successfully deliver the desired morphology. Here we present the use of Soxhlet extraction conditions as a valuable tool for morphology modification in a nucleation dominant crystallization process. In this case study, the process to produce a cocrystal between Vertex compound A and coformer X resulted in a fibrous morphology in every attempt under various conditions, leading to mixing and filtration issues and low yields. The use of the Soxhlet extraction method provided very slow mixing of the two compounds coupled with temperature cycling close to the solvent's boiling point to generate a needle morphology, which provided better filtration times, processability and yields, enabling a path forward for further API development.

599684 Process Development to Avoid Meta-Stable Liquid-Liquid Phase Separation during Crystallization of API

Abstract:

Crystallization is a critical unit operation in the production of small molecule active pharmaceutical ingredients (APIs), as it often determines the form and powder properties of the product. Therefore, crystallization process development focuses on identifying conditions that are controlled and scalable. In general, liquid-liquid phase separation or oiling out of product are undesirable as they may lead to uncontrolled nucleation and growth of particles that impacts critical quality attributes, such as particle size and level of impurities. Therefore, understanding liquid-liquid phase equilibria and solubility of the compound is critical in designing crystallization processes.

This presentation will describe how we experimentally investigated the liquid-liquid phase separation that occurred during crystallization of a small molecule API in a ternary solvent system of tetrahydrofuran, acetone and water. Although these solvents are typically miscible, the presence of the API resulted in liquid-liquid phase separation at the seed point under some conditions. In-line process analytical tools such as ParticleViewer and FBRM were incorporated to detect and understand the phenomena. A detailed phase diagram was generated for the crystallization design space as a function of solvent composition, concentration of API in solution and temperature. Along with solubility data, the phase diagram was used to design a process that avoids liquid-liquid phase separation throughout the entire crystallization. In addition, we will describe the impact of wet-milling during seed age that was implemented to ensure that full desupersaturation was achieved, which was critical in avoiding phase separation.

596447 Decoupling Nucleation and Growth for Continuous Crystallization in Microfluidics

Abstract:

1. Introduction

Microfluidics provide several advantages such as high efficiency, easy control and safety over conventional, and mostly batch reaction systems[1]. Therefore it can be seen as a replacement for batch crystallization as one of the most important separation and purification process especially in pharmaceutical industries[2,3].

The aim of this project is the design of a modular and decoupled microfluidic system consisting of separate nucleation and growth sections. The strategy persued in this project is using the two- phase interface in microreactors as a heterogeneous nucleation site. In this regard micro- bubbles are introduced in the system to control the location of the nucleation site and particle aggregates and to limit wall interactions. The velocity of the bubbles determines the residence time and using a gas phase enables a straight forward separation, see Figure 1.

2. Experimental study

2.1. Nucleation section

For designing the nucleation section, in the first step, the generation of micro-bubbles (i.e. nitrogen as an inert gas) in capillaries is studied. The total bubble surface area per unit volume of the reactor is related to the bubble diameter and the number of bubbles present in the reactor volume. Bubble diameter, bubble velocity and bubble generation frequency were extracted from the images captured by the high-speed camera using an inhouse-generated Matlab script. Using this information, the bubble surface area available as a heterogeneous nucleation site is further calculated and the effect of gas and liquid flowrate is investigated, see Figure 2.

This is followed by a systematic study of nucleation of an organic compound (i.e. cooling crystallization of paracetamol) using the developed system employing microbubbles. The particle size distribution is measured at the outlet of the channel subjected to the microbubble residence time and the nucleation and growth kinetics are extracted.

The continuous crystallization experiments provided proof that the micro-bubbles act as a heterogeneous nucleation site. An enhanced nucleation rate compared to the operation without bubbles was observed, and the presence of micro-bubbles results in the formation of more crystals, which indicates that nucleation is faster than without bubbles, i.e. the metastable zone width is reduced, see Figure 3. Furthermore, quantifying the crystal yield showed an increase by more than a factor of 2 in the presence of micro-bubbles.

2.2. Growth section

Further on a growth section is added to the nucleation section which is kept at the constant temperature equal to the outlet of nucleation section to allow formed nuclei to grow. Sample is collected at the outlet of the growth section for further analysis of particle size distribution, see Figure 4.

4. **Conclusions**

A decoupled crystallization set-up consisting of different nucleation and growth sections is designed using microreactors. Nucleation is enhanced using microbubbles acting as heterogeneous nucleation sites providing lower surface energy interface for the nucleation to start. Further a growth part is added to allow the formed nuclei grow within the constant temperature growth section.

In case of the continuous crystallization of paracetamol the results allow the conclusion that the low-energy heterogeneous surface introduced by the bubbles promotes crystal nucleation. In addition, adding micro-bubbles enables the continuous crystallization for long operating times, as the bubbles act as carriers for the formed crystals and effectively prevent wall deposition and thus clogging.

References

[1] Gunther, K. F. Jensen, Multiphase microfluidics: from flow characteristics to chemical and materials synthesis, Lab Chip , 2006, 6, 1487–1503

[2] S. Flowers, R.L. Hartman, Particle Handling Techniques in Microchemical Processes, Challenges 2012, 3, 194-211

[3] Wu, S. Kuhn, Strategies for solids handling in microreactors. Chim Oggi. 2014; 32(3):62-66.

[4] Fatemi, Z. Dong, T. Van Gerven, S. Kuhn, Microbubbles as Heterogeneous Nucleation Sites for Crystallization in Continuous Microfluidic Devices. Langmuir (2018). doi:10.1021/acs.langmuir.8b03183

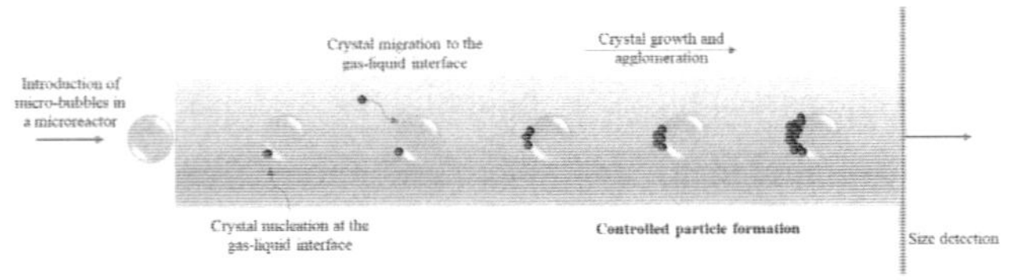

Figure 1. Microbubbles used as a heterogeneous nucleation site [4]

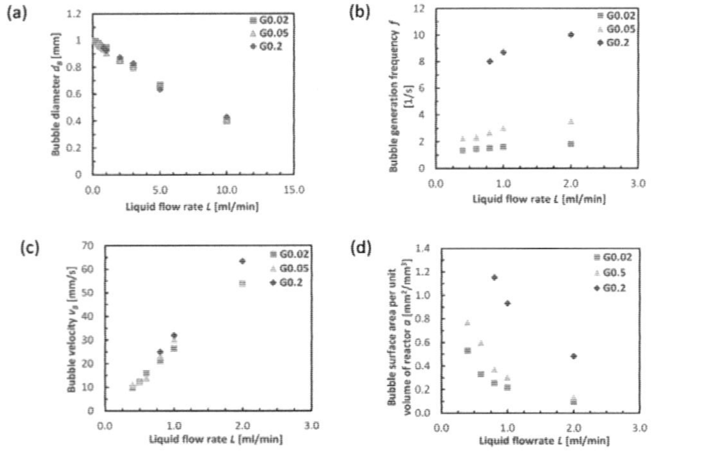

Figure 2. Effect of gas and liquid flow rates on (a) size of the microbubbles; (b) frequency of microbubble generation; (c) microbubble velocity; and (d) microbubble surface area per unit volume of the reactor for [4]

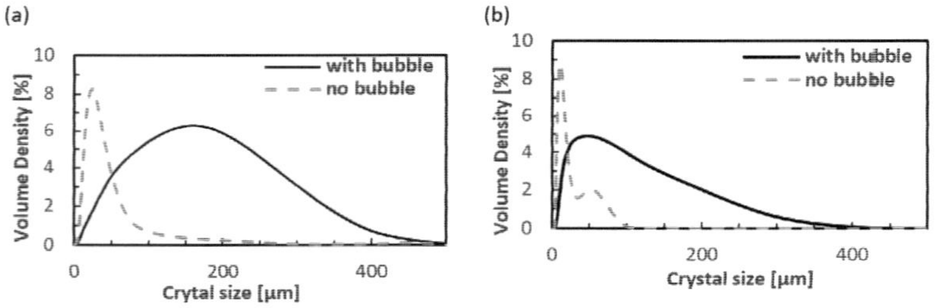

Figure 3- Effect of micro-bubbles on the crystal size distribution for a growth time of 1 minute and a supersaturation of S=1.81 and the gas flow rate G=0.05 ml/min and liquid flow rate of: (a) L=0.5 ml/min, (b) L=1 ml/min

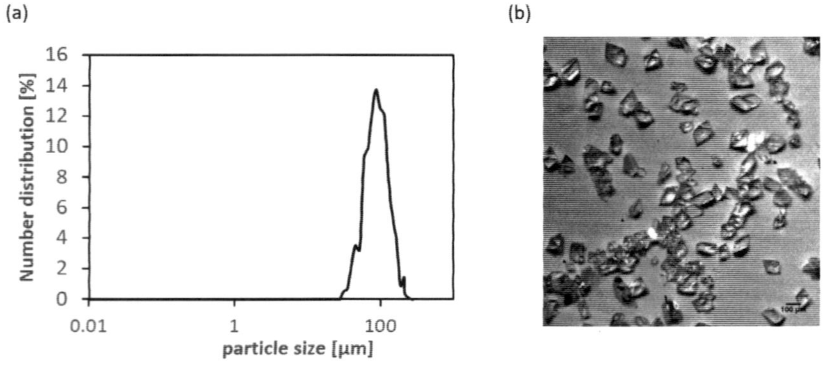

Figure 4- (a) particle size distribution, (b) microscope images of paracetamol crystals for growth section with the length of 1.4m at T=25°C with a supersaturation of S=1.81 and the gas and liquid flowrates 0.05 and 1 ml/min, respectively

597227 Effects of Volume Reduction on Glycine Nucleation

Abstract:

Crystallization in confinement is a technique used both as an approach for nanocrystal development and as way to gain access to unstable crystal polymorphs. To elucidate this, the effect of decreasing volumes in the nucleation of glycine in water was explored. Using a supersaturated aqueous solution of glycine at decreasing volumes from 1000 μL to 25 μL, induction times for primary nucleation were obtained experimentally. Analysis of probability distribution of β-glycine induction times showed that as sample volume decreases, induction time increases. Image analysis revealed that β-glycine formation serves as the primary nucleation event; however, primary nucleation shifts toward γ-glycine nucleation at volume sizes below 100 μL. These results demonstrate that feasibility of accessing metastable and unstable crystal polymorphs with the aid of confinement. However, confinement also increases induction times which relates to very low probabilities of a nucleation event.

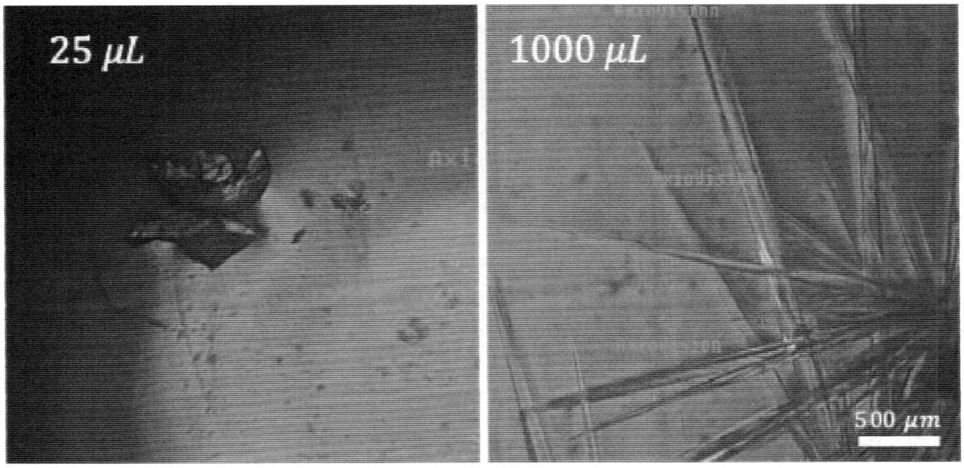

597873 Formation of Solid Solutions As a Key Impurity Retention Mechanism in Solution Crystallization in the Presence of Structurally Similar Impurities

Abstract:

The main purpose of crystallization is to purify the product by removing undesired impurities. In pharmaceutical industry, crystallizations at different steps in the synthesis are relied upon to control the final quality of the active pharmaceutical ingredient (API). During these operations, the main challenge often stems from dealing with structurally similar impurities, which tend to be difficult to remove. Presented herein is an investigation into how this type of challenging impurities are purged mechanistically in the crystallization, using pharmaceutical case studies. Through these examples, it is shown that formation of substitutional solid solutions is the underlying mechanism for impurity retention. Impurities molecularly replace the product molecules in the crystal lattice during the crystallization. This lattice incorporation entrapment occurs unevenly during the crystallization resulting in a gradient in impurity levels throughout the material. A mathematic framework combined with a previously reported SLIP test is used to elucidate this varying impurity entrapment through a so-called *material impurity distribution* (MID). The impurity levels are seen in many cases to change by an order of 20 across the material. The effect of the entrapment can also be observed in the physical properties of the material, e.g. by a significant reduction in crystallinity, increase in solubility and earlier onset melting event.

609881 Design of a Continuous Crystallization for a Drug Substance Using a Dynamically Mixed Reactor

Abstract:

The pharmaceutical industry has predominantly relied on batch processes for manufacturing of drug substance as well as drug products. However, longer campaign times, labor-intensive nature of batch manufacturing and batch to batch variations in the product quality pose significant drawbacks. In view of this, the Pharma industry is turning towards continuous manufacturing, going beyond flow chemistry to continuous crystallization and downstream operations for the synthesis of drug substances. Continuous crystallization offers many advantages such as better control over the solid form and particle size, and the potential for an end-to-end continuous manufacturing process.

The current work describes the design of a continuous crystallization process for a drug substance in order to integrate with upstream and downstream continuous processes. It was not only important to achieve the targeted solid state form (trihydrate, Form I) but also ensure that the filtration rate could be increased to enable integration with the downstream continuous filter and minimize the cleaning frequency. The presented case study discusses in detail the approach, starting from lab-scale batch experimentation to development of a continuous crystallization process at lab followed by scale-up design for a capacity of multi-tons per month.

As a first step, fundamental solubility data was collected followed by identification of critical parameters impacting nucleation and transformation between the various forms. It was observed that the solvent (alcohol) to antisolvent (water) ratio, seed loading and temperature had a significant impact on the solid form. A batch crystallization process was first established by combining antisolvent and cooling crystallization to consistently generate the desired Form I, while ensuring the absence of other solid forms.

The translation of the batch process to continuous was initially attempted in MSMPRs in series but proved unsuccessful as an undesired form started to appear over time despite seeding. It was hypothesized that several more MSMPRs would be required to achieve the desired form but this would occupy a huge footprint and increase hold-up at scale. While plug flow reactors would be suitable, the risk of fouling is high, particularly when flow rates are reduced to achieve sufficient residence time. Scouting for an appropriate technology lead to evaluation of dynamically mixed reactors from AM Technologies®, which were able to handle a slurry and provide a mixing profile close to plug flow decoupled from the residence time. At lab scale, the Coflore® Agitated Cell Reactor (ACR), which is a bench-top mechanical agitated reactor comprising ten small cells, was able to aid in controlled crystallization of the desired Form 1 with well defined crystals and a consistent particle size distribution. Since additional reactors and residence time were required to gradually cool the slurry from 65°C to 25°C, the ACR was combined with three CSTRs in series (forming a mixed suspension mixed process set up). The trials were executed by combining a stream of alcohol containing the dissolved solute (at 65°C) and another stream of hot water containing seeds (at 65°C) in the ACR. The slurry exiting from the reactor was continuously transferred into the CSTRs, where it was subjected to stage-wise cooling in each CSTR. The residence times were designed to achieve maximum crystallization in the ACR and minimize secondary nucleation in the subsequent CSTRs.

After establishing the success and robustness of the continuous crystallization process at lab-scale, the process was scaled up in a 100L, production-scale version of the dynamic mixing reactor - Coflore®

Rotating Tubular Reactor (RTR). Optimization of the process parameters revealed that maintenance of higher temperatures (~65 °C), , enhanced mass transfer at the point of nucleation along with the presence of Form 1 seeds, and higher ratios of anti-solvent were critical for achieving the desired polymorph while low mixing shear aided in crystal growth. The growth of these fine, needle-shaped crystals was essential to achieving a sufficiently high filtration rate for the downstream continuous separation of solids. The RTR, which operates as a ten-stage, actively mixed, self-baffled reactor, provided high wall shear with radial mixing while maintaining a low mixing shear, thus enabling crystal growth and improving the filtration rate while minimizing fouling.

In summary, the work presents key aspects of design and scale-up of a continuous crystallization for a drug substance with regard to both process and technology selection. One of the main advantages of this continuous process was consistency in the solid form, particle size distribution resulting in improved filtration of the slurry, and ability to integrate with upstream and downstream continuous processes.

Internal communication number: IPDO IPM-00637

604490 Digital Design Via Mechanistic Modelling of a Continuous Crystallization Process with Periodic Wet Milling

Abstract:

Introduction

The attainable regions for critical quality attributes, such as Particle Size Distribution (PSD) and impurity levels, in the manufacture of Solid Oral Dosage Forms (SODF) can be highly dependent upon the type and operation of the crystallization process. In this work, we outline a step-wise workflow consisting of model validation, model-based technology transfer and process optimisation employed for the digital design of a continuous cascade cooling crystallization and wet milling process for manufacturing an Active Pharmaceutical Ingredient (API). Following this workflow, the mechanistic model was first validated using batch crystallization data and subsequently applied to describe the continuous crystallization process and with continuous and periodic wet milling, and explore the impact of varying process parameters, including duration of milling, rotor speed and frequency of milling operation during the crystallization process.

The step-wise approach taken for model validation and its subsequent application was as follows:

1. Model validation of crystallization kinetics for the pure API solid phase from batch de-supersaturation experimental runs.
2. Application of the validated crystallization kinetics to describe the continuous crystallization of pure API in a three-stage cascade process, as shown in Figure 1 below (wet milling was not considered at this stage).
3. Model validation of breakage kinetic parameters using batch data from a wet milling unit operated in a recycle with a batch crystallizer. A rotor-stator wet mill was utilised for this process, with a range of rotor frequencies and milling head generator configurations probed to assess the impact on the PSD over time and to fit the breakage kinetic parameters. Subsequent refinement of the breakage kinetic parameters utilizing continuous crystallization experiments, where intermittent/periodic wet milling was employed using an immersion wet mill in the first stage crystallizer.
4. Scenario analysis of the continuous crystallization and periodic wet milling process for this API, looking at the impact of process parameters on the Critical Quality Attributes (CQAs) of the product, including product PSD and dimer phase formation (impurity).

Conclusions

The main conclusions of the work include the following:

- The continuous crystallization process was unable to achieve the desired product PSD without wet milling, due to very low levels of secondary nucleation in the system. Optimal PSD quantiles (d10, d50 and d90) predicted for the product were significantly higher than the target PSD quantiles required to achieve the desired product performance for this compound.
- Addition of a wet milling step in a recycle with the continuous crystallization significantly increased the lower end of the attainable region in terms of PSD quantiles and impurity level,

allowing the process to comfortably achieve the desired range for product PSD quantiles.

- Mechanistic modelling approaches can be utilised to significantly reduce the development timelines, material consumption and efficiency of continuous API crystallization production processes, in particular cases where periodic operation of some units, in this case the wet mill, are employed.

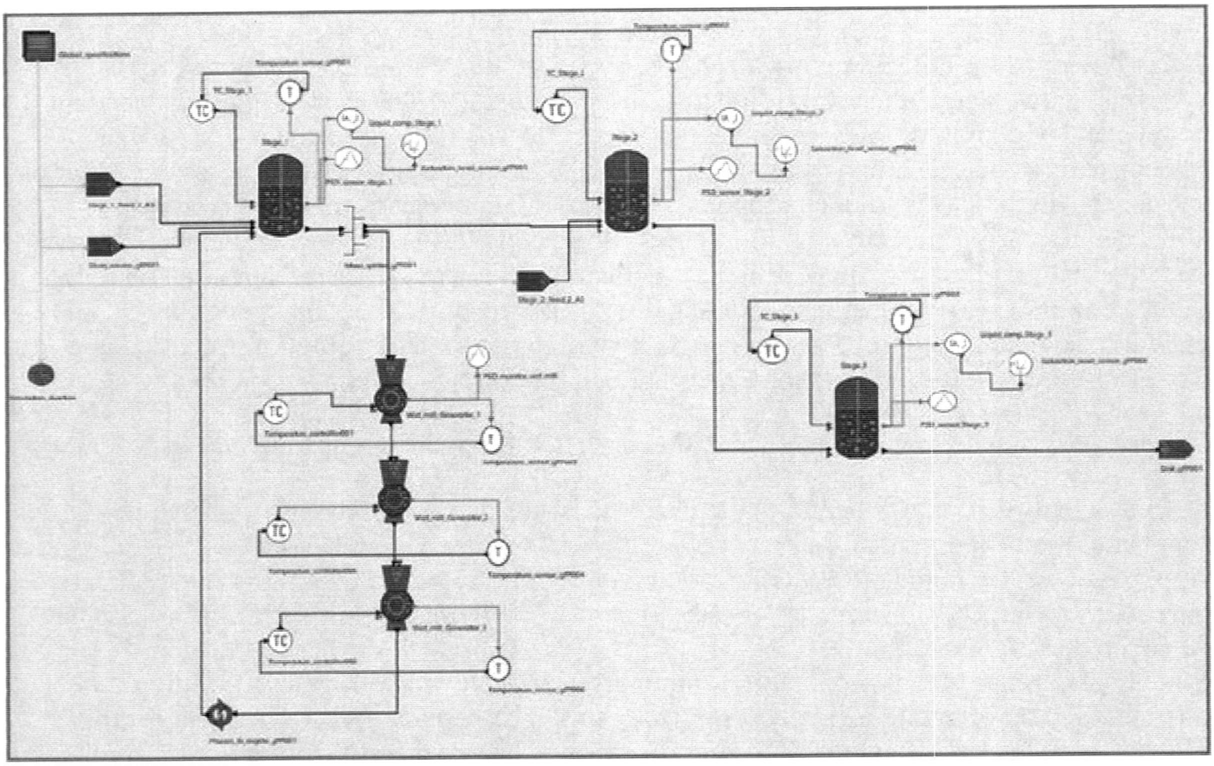

Figure 1: Simulation flowsheet used model for optimisation of continuous crystallization and wet milling process.

606179 Continuous Generation of Glycine Crystals with Pure Form and Narrow Size Distribution in Slug Flow

Abstract:

In pharmaceutical industries, continuous manufacturing of medicine has the potential to reduce capital, time and increase efficiency[1]. Continuous crystallization as the bridge between synthesis and formulation[2], the process and quality control is vital such as aggregation of crystals can affect the following blending and coating. Segmented/slug flow crystallization is one of the effective continuous process demonstrated to allow enhanced control of organic crystal properties such as size and shape[3–6].

This presentation will elaborate in the following topics[7]: (1) generate pure-form uniform size crystals in slug flow rapidly; (2) explore the operation region (meaning no secondary nucleation nor aggregation) in slug flow crystallizer. The crystal forms at different conditions generated by both slug flow crystallizer and batch crystallizer are discussed and compared. Experimental results validated the proposed crystallizer reduced production time and equipment cost while increased the operational region significantly compared to batch crystallization.

Reference

(1) Schaber, S. D.; Gerogiorgis, D. I.; Ramachandran, R.; Evans, J. M. B.; Barton, P. I.; Trout, B. L. Economic Analysis of Integrated Continuous and Batch Pharmaceutical Manufacturing: A Case Study. *Ind. Eng. Chem. Res.* 2011, *50* (17), 10083–10092.

(2) Lee, S. L.; O'Connor, T. F.; Yang, X.; Cruz, C. N.; Chatterjee, S.; Madurawe, R. D.; Moore, C. M. V.; Yu, L. X.; Woodcock, J. Modernizing Pharmaceutical Manufacturing: From Batch to Continuous Production. *J. Pharm. Innov.* 2015, *10* (3), 191–199.

(3) Jiang, M.; Zhu, Z.; Jimenez, E.; Papageorgiou, C. D.; Waetzig, J.; Hardy, A.; Langston, M.; Braatz, R. D. Continuous-Flow Tubular Crystallization in Slugs Spontaneously Induced by Hydrodynamics. *Cryst. Growth Des.* 2014, *14* (2), 851–860.

(4) Eder, R. J. P.; Schrank, S.; Besenhard, M. O.; Roblegg, E.; Gruber-Woelfler, H.; Khinast, J. G. Continuous Sonocrystallization of Acetylsalicylic Acid (ASA): Control of Crystal Size. *Cryst. Growth Des.* 2012, *12* (10), 4733–4738.

(5) Jiang, M.; Papageorgiou, C. D.; Waetzig, J.; Hardy, A.; Langston, M.; Braatz, R. D. Indirect Ultrasonication in Continuous Slug-Flow Crystallization. *Cryst. Growth Des.* 2015, *15* (5), 2486–2492.

(6) Lu, J.; Litster, J. D.; Nagy, Z. K. Nucleation Studies of Active Pharmaceutical Ingredients in an Air-Segmented Microfluidic Drop-Based Crystallizer. *Cryst. Growth Des.* 2015, *15* (8), 3645–3651.

(7) Mou, M.; Jiang, M. Fast Continuous Non-Seeded Cooling Crystallization of Glycine in Slug Flow: Pure α-Form Crystals with Narrow Size Distribution. *J. Pharm. Innov.* 2020, 22–24.

609118 Continuous Crystallisation Oscillatory Flow Platform for Protein Purification and Bio-Separation

Abstract:

Crystallization of proteins could be a potentially cost effective and robust method to purify protein molecules in the pursuit of the next generation of protein based biopharmaceutic therapies. However, the step from traditional small-scale high throughput screening methods, to crystallizers which can produce high yields of pure crystalline protein material, is still in its infancy.

Here, we present a platform for continuous oscillatory flow crystallization (COFC) (Figure 1) and its application for the crystallization of lysozyme[1,2]. A workflow is developed from μl screening experiments, to scaled up batch oscillatory flow crystallization (BOFC) and COFC experiments.

The induction time for lysozyme is found to be inversely proportional to the initial protein concentration, with the same relationship found between the induction time and the frequency and amplitude of the oscillation. However, it is also observed that the decrease in induction time comes at a cost to the crystal size and quality. In turn, the high shear environment from increased amplitude and frequency of the COFC can also damage the fragile protein crystals, resulting in the formation of more fine particles and an undesirable size distribution. This led to further investigation of slug flow mixing in the continuous crystalliser, where the crystal size and shapes are compared to the particles produced from the osciallatory flow reactor. Further work around optimization of the crystallization parameters and crystallizer geometry using the gCrystal platform to find the optimum scale up conditions are also discussed. The use of seeding and potential templating materials is also covered with a view to expanding the use of this crystallizer to less well characterized protein materials.

It is demonstrated that purification of proteins can be achieved by crystallization in a continuous mode. This route offers the possibility to revolutionise downstream separations of biopharmaceuticals

<u>References</u>

[1] H.Y. Yang, W.Q. Chen, P. Peczulis, J.Y.Y. Heng (2019): **Development and Workflow of a Continuous Protein Crystallization Process: A Case of Lysozyme**, *Cryst. Grow. Des*; 19, 983-991.
[2] H.Y. Yang, P. Peczulis, P. Inguva, X.Y. Li, J.Y.Y. Heng (2018): **Continous Protein Crystallisation Platform and Process: Case of Lysozyme**, *Chem. Eng. Res. Des.* 136, 529-53

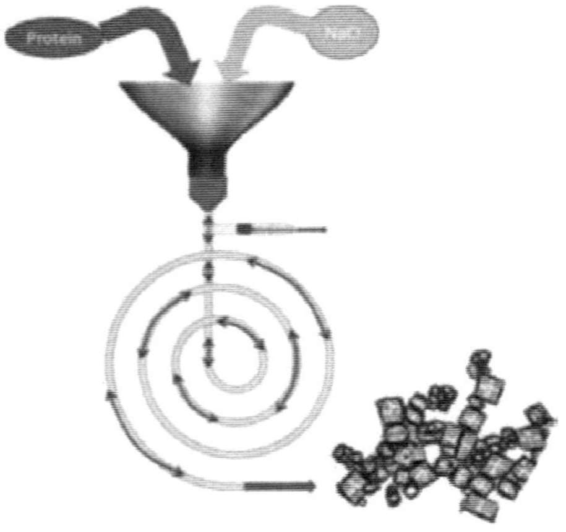

Figure 1: Schematic of the continuous oscillatory flow crystallisation platform

609250 Temperature Control Approaches for Cooling Crystallization Using a Continuous Microfluidic Mixer

Abstract:

Cooling crystallization is a popular crystallization technique in pharmaceutical and fine chemical industries. The driving force for this process is the temperature gradient that consequently results in increasing the supersaturation ratio. However, it is very challenging to minimize the gradient range and maintain a constant supersaturation without sacrificing the mixing performance. In this study we have we have developed two techniques to enable temperature gradient within the crystallization zone using the previously developed continuous microfluidic mixer. For the first method, we have designed a cooling bath around the circular zone which has separated inlet and outlet. The cold stream is circulated around the zone using a peristaltic pump while the hot streams enter from the four main inlets of the device. Due to the temperature difference between the hot and cold streams, the solution within the crystallization zone will attain an equilibrium temperature lower than the hot streams. In the second method, saturated hot and cold streams will enter the circular zone and mixed to impose supersaturation inside the mixing zone. Both systems are modeled using COMSOL Multi physics and equilibrium temperature and imposed supersaturation are calculated respectively. These microfluidic devices are then 3D-printed using the Formlabs 3D-printer for experimental verification. Using these two systems we have performed cooling crystallization of L-Glutamic acid and L-Histidine using water as solvent. Here we have compared eight different supersaturation for each compound and measured the face specific growth rate followed by morphology and size distribution of the crystals. Comparative analysis has also been performed using a batch microtiter plate.

602048 “Filling the Gap” in Continuous Pharmaceutical Technologies: A Novel Continuous Filtration Drying System

Abstract:

In recent years, the world’s top pharmaceutical companies have started major research programs to switch from classical batch to continuous manufacturing of their active drug substances. There is a significant gap in technology for the continuous isolation. Alconbury Weston Ltd (AWL), a leading supplier of continuous crystallization and filtration equipment, has designed an innovative continuous carousel filter dryer (CCFD) which filters, washes and dries solids in small aliquots simultaneously in a continuous manner.

The core technology is based on well-established Nutsche filtration, processing thin cakes in a fully automated system. A state-of-the-art vision system can be used to track the cake and liquid heights during filtration and washing, improving product purity, yields and quality attributes. The CCFD is a complete plant which replaces a traditional batch production facility and fits into a walk-in fume cupboard.

The presentation will describe the challenges and issues observed in batch filtration systems and demonstrate how these challenges can be addressed using continuous filtration. The talk will demonstrate the collaborative efforts required for the development of a novel continuous processing technology through to a commercial product. AWL has worked with Strathclyde University and CMAC (Glasgow, UK) and Purdue University (Indiana, US) to investigate the advantages of continuous filtration. The presentation will include case studies carried out at both academic centres of excellence.

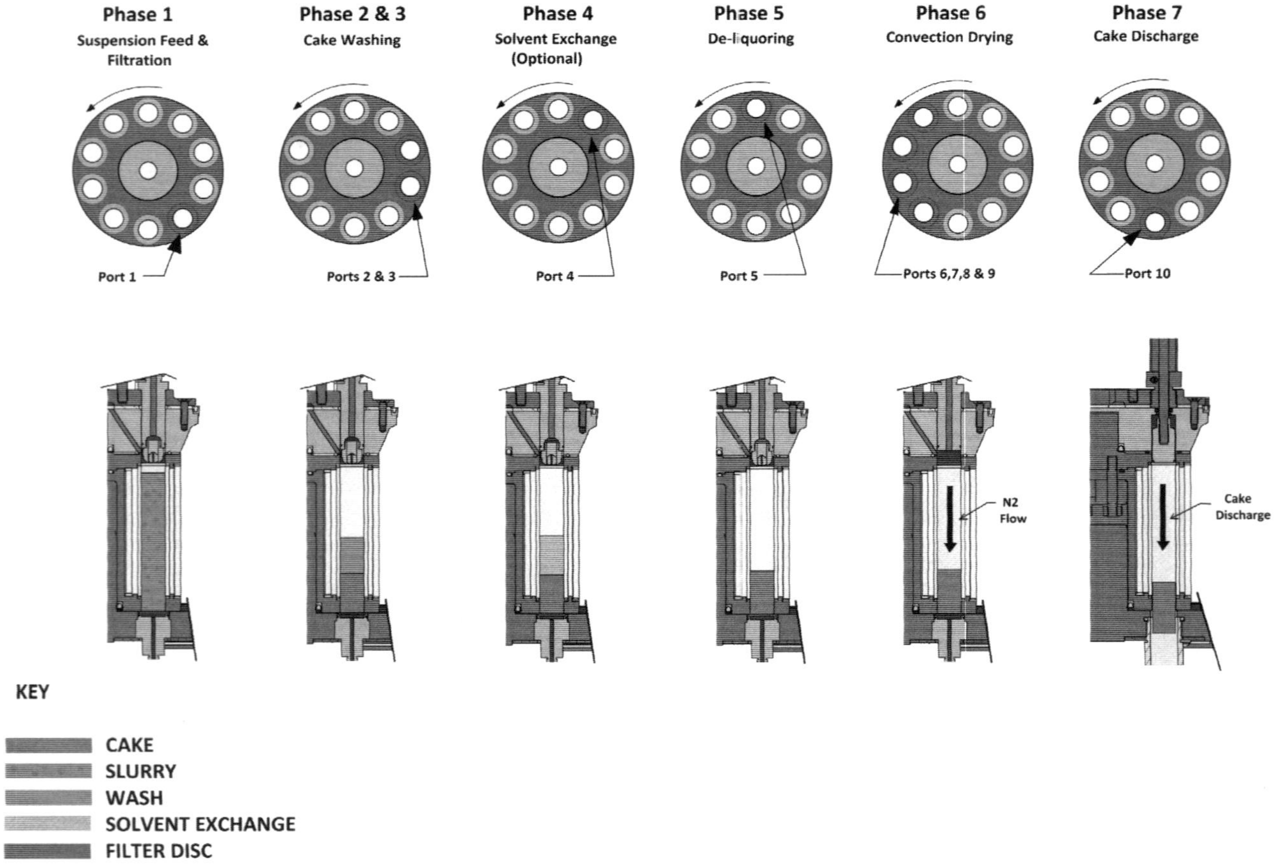

608687 Integrated Continuous Crystallization and Spray Drying of Insulin

- Download/View Extended Abstract

Abstract:

Insulin injection is commonly used for the treatment of diabetes. Pulmonary delivery through inhalation of dry insulin powders is commercially also available and offers a more convenient route compared to injection.[1] The manufacture of such powders is challenged by the need to produce crystalline particles with an aerodynamic diameter in the narrow range of 1 – 5 μm. Particles outside the optimal size range will not deposit well into the lungs. Amorphous particles produced from spray drying an insulin solution are limited by their fast dissolution,[2] which leads to a therapeutic effect of only several hours. Furthermore, in general, crystalline particles have a better stability and aerosolization performance and different drug release characteristics compared to amorphous particles.[3,4] Therefore, the manufacture of a dry powder with insulin crystals is of interest to potentially improve pulmonary delivery of insulin. The crystallization of insulin has been studied in various crystallizers.[5–7] However, those processes generally do not deliver crystals in the optimal size range for pulmonary delivery. Therefore, there is a need for novel crystallization processes that can produce insulin crystals with tailored size for the application of dry powder inhalation more efficiently.

The objective of this work is to develop and characterize a novel continuous crystallization process for insulin that is tailored for dry powder inhalation by integrating a segmented-flow crystallizer with spray drying for rapid solvent removal. Continuous crystallization has received growing interest in the fine-chemical and pharmaceutical industries due to typical advantages such as easier process control and a more consistent product quality compared to batch crystallization.[8] The benefits of integrating continuous crystallization with spray drying for the manufacture of dry powders for inhalation has been demonstrated by earlier work from our group for different small-molecule active pharmaceutical ingredients.[9] The present work is, to the best of our knowledge, the first demonstration of such process for a therapeutic protein.

The process consisted of a tubular crystallizer and a spray dryer. A calibrated two-channel peristaltic pump was used to introduce the feed and the precipitant solution into the crystallizer at the same flow rate through a T-mixer. A gas flow controller was used to introduce nitrogen with controlled flow rate via another T-mixer to create a segmented flow, which offers the benefit of a narrow residence time distribution.[10,11] The supersaturation was generated by both cooling and a change in pH. The tubular crystallizer, which consisted of two connected pieces of tubing of 14m long each, was immersed in a thermostatic bath to allow temperature control. At the outlet of the crystallizer, a buffer vessel was used for separation of nitrogen and to provide a steady flow to the spray dryer using another peristaltic pump. Finally, the dry powder was collected in a powder collector located below a cyclone. Initially, the crystallizer was operated stand-alone to identify optimal crystallization conditions. Subsequently, optimal crystallization conditions were applied for the integrated process. The morphology of the particles and their approximate size were characterized with optical microscopy and scanning electron microscopy (SEM). The mass median aerodynamic diameter (MMAD) and fine particle fraction (FPF) fraction was measured with a Next Generation Impactor (NGI).

The recovery of insulin from stand-alone crystallization experiments was in excess of 90% when using a combined flow rate of 1.2 ml/min and an insulin concentration of 2.4 mg/ml. A lower recovery was obtained at higher combined flow rates, and thus shorter residence times, in the range of 1.8 ml/min to 2.4 ml/min. A constant recovery could be approached in a short period of about 10 to 15 minutes when running at a high initial concentration in the range of 2.0 mg/ml to 2.4 mg/ml. Such fast start-up time is consistent with plug-flow conditions, which is an inherent benefit of a segmented-flow crystallizer compared to a mixed-suspension mixed-product-removal crystallizer. The produced particles appeared crystalline from SEM and were close to the desired size range for pulmonary drug delivery. The particles produced with the integrated process under optimal conditions were characterized by a MMAD in the range of 2 μm to 6 μm and an FPF of 20 - 40%, which indicated a good potential for pulmonary drug delivery. In conclusion, the demonstrated benefits of the novel process include a short total residence time, high recovery from crystallization, and suitable product quality attributes for pulmonary drug delivery of insulin without any added excipients.

1. References

1. Easa, N., Alany, R. G., Carew, M. & Vangala, A. A review of non-invasive insulin delivery systems for diabetes therapy in clinical trials over the past decade. *Drug Discovery Today* vol. 24 440–451 (2019).
2. Bailey, M. M., Gorman, E. M., Munson, E. J. & Berkland, C. Pure insulin nanoparticle agglomerates for pulmonary delivery. *Langmuir* **24**, 13614–13620 (2008).
3. Shenoy, B., Wang, Y., Shan, W. & Margolin, A. L. Stability of crystalline proteins. *Biotechnol. Bioeng.* **73**, 358–369 (2001).
4. Basu, S. K., Govardhan, C. P., Jung, C. W. & Margolin, A. L. Protein crystals for the delivery of biopharmaceuticals. *Expert Opin. Biol. Ther.* **4**, 301–317 (2004).
5. Hirata, G. A. M., Bernardo, A. & Miranda, E. A. Crystallization of porcine insulin with carbon dioxide as acidifying agent. *Powder Technol.* **197**, 54–57 (2010).
6. Parambil, J. V., Schaepertoens, M., Williams, D. R. & Heng, J. Y. Y. Effects of oscillatory flow on the nucleation and crystallization of insulin. *Cryst. Growth Des.* **11**, 4353–4359 (2011).
7. Chen, F. *et al.* Crystallization of bovine insulin on a flow-free droplet-based platform. in *AIP Conference Proceedings* vol. 1820 030003 (American Institute of Physics Inc., 2017).
8. Wood, B., Girard, K. P., Polster, C. S. & Croker, D. M. Progress to Date in the Design and Operation of Continuous Crystallization Processes for Pharmaceutical Applications. *Org. Process Res. Dev.* **23**, 122–144 (2019).
9. Hadiwinoto, G. D. *et al.* Integrated Continuous Plug-Flow Crystallization and Spray Drying of Pharmaceuticals for Dry Powder Inhalation. *Ind. Eng. Chem. Res.* **58**, 16843–16857 (2019).
10. Neugebauer, P. & Khinast, J. G. Continuous Crystallization of Proteins in a Tubular Plug-Flow Crystallizer. *Cryst. Growth Des.* **15**, 1089–1095 (2015).
11. Jiang, M. *et al.* Continuous-flow tubular crystallization in slugs spontaneously induced by hydrodynamics. *Cryst. Growth Des.* **14**, 851–860 (2014).

607564 Real-Time Modeling of Crystal Nucleation, Growth, and Breakage in an Agitated Bioreactor

Abstract:

Real-time modeling of Crystal Nucleation, Growth, and Breakage in a Agitated Bioreactor

We use GPU-based CFD/DEM simulations to model crystal nucleation, growth and break-up in an agitated tank in real-time. This approach, which pairs lattice-Boltzmann-based direct numerical simulations with fully resolved discrete element modeling (DEM), enables mechanistic and time-accurate modeling of crystal growth via solvent reactions and crystal growth via reactions and particle interactions. In addition to particle growth and agglomeration, particle break-up is modeled at the level of individual particles, providing insights into the competition between grown and break-up. Additionally, since the particle trajectories are modeled using Newtons second law, the effects of particle growth and break-up on settling on are also predicted. Since the particle field is solved in tandem with the fluid field, incorporate the effects of particle size and concentration into the local fluid viscosity when modeling fluid flow. Using this model, we can predict the spatiotemporal particle size distribution, and show how this distribution is informed by reactor operating conditions, solvent chemistry, particle interactions, particle settling, and particle break-up.

In fluid mechanical systems, momentum transport can be described by the Navier-Stokes equations. The Navier-Stokes equations link fluid acceleration and the applied body and surface forces, which typically include gravity, buoyancy, and shear stress gradients. Although only a few exact solutions to the Navier-Stokes equations exist, many numerical solution approaches are available. One of the most general and computationally efficient approaches for solving the transient Naiver-Stokes equation is Lattice-Boltzmann. The Lattice-Boltzmann method scales exceptionally well on GPU-based architecture, enabling high-resolution large eddy simulation of turbulent flows at industrial scale. Following this approach, CFD simulations running lattice points at lattice updates per second are routine on a multi-GPU workstation.

In solid granular systems, particle trajectories are governed by Newton's second law. Newton's second law links the acceleration of a particle to the sum of the forces acting on it, including gravity, buoyancy, fluid-interaction, and particle contact. Exact solutions are not available for generalized granular systems containing more the two particles. Computational approaches must be applied. One of the most general and computationally efficient approaches for solving this N-body problem is to use a velocity Verlet integration scheme for particle trajectories with a GPU-based BVH tree to account for neighbor-neighbor interactions. Following this approach, DEM simulations tracking interacting particles at particle updates per second are routine on a multi-GPU workstation.

From a modeling perspective, it is practical to combine fluid mechanical systems and solid granular systems into a unified multi-phase system, wherein both transport processes are solved simultaneously. In this approach, the local and instantaneous properties of the fluid inform the local and instantaneous forces on the particles. Likewise, the properties of the particles inform the local properties and forces applied to the fluid. Since these processes are solved in real time and directly on the GPU, it is straightforward to model time-varying physics, including reactions, phase change, and particle growth/breakage within this unified model.

607567 CFD Predictions of Power Dissipation and Related Hydrodynamic Effects for a Retreat-Bladed Impeller in a Stirred Vessel Under Different Baffling Configuration

Abstract:

The manufacture of active pharmaceutical ingredient (API) in the pharmaceutical industry typically involves a number of fluid mixing and reaction operations, many of which are conducted in mechanically stirred, glass-lined tanks and reactors provided with a torispherical bottom and equipped with a single baffle and a retreat-blade impeller (RBI). The use of glass lining is critical to provide corrosion resistance, allow ease of cleanliness, and reduce product contamination. Still it requires coating with glass not just the internal wall of the tank, but also all components that may come in contact with the liquid contents, including the impeller and the baffles. This often results in the use of partial baffling, typically a beavertail baffle, which is mounted from the top of the reactor and inserted between the shaft and the tank wall. However, in other industrially relevant systems the more conventional four vertical baffles are mounted on the vessel wall, although in some pharmaceutical product manufacturing and applications (e.g., crystallization, biopharmaceutical processes, solid suspensions, etc.) unbaffled stirred vessels can also be found.

In previous work by our group, the power dissipated by the impeller, *P,* and the impeller Power Number, Po, which arecritical for scale-up purposes, especially when mass transfer operations are involved, were experimentally determined for systems commonly found in the pharmaceutical industry. This was achieved by measuring the power dissipated by an RBI under fully baffled, partially baffled, and unbaffled conditions, with different liquids, and under different hydrodynamic regimes (1<Re< 300,000) using a 61-Liter vessel that was built as an actual scaled-down replica of the glass-lined vessels typically used for API manufacturing in the pharmaceutical industry.

In our present work, Computational Fluid Dynamics (CFD) was utilized to conduct detailed hydrodynamic simulations for the same systems, not only to determine the flow features of the different systems computationally but also to obtain relevant critical quantities such as the power dissipation and Power Number. Simulations were conducted for a variety of baffling configurations and operating conditions using different CFD approaches, turbulence models, and simulation software. The results indicate that the computational predictions typically compared well with the experimental data in all cases, thus validating the CFD approach. The information provided by this approach to understanding the different fluid-dynamics features of vessel-baffling configurations and the relevant effect of operating parameters is clearly beneficial for the selection, design and use of suitable vessels and reactors for industrial applications, and it is expected to be of value to the industrial practice.

595684 Predicting Filtration Performance By Studying the Pore Network of the Filter Cake: The Case of Needle-like Crystals

Abstract:

Introduction

Filtration is extensively used in pharmaceutical manufacturing to recover the crystals of active ingredients and intermediates. This operation is often a slow processing step, accounting for a substantial quota of the process time. Furthermore, its design requires extensive experimental investigations at different scales as the impact of most driving factors is not sufficiently well understood to allow quantitative modeling of the process [1]. The structure of the filter cake is a result of the packing of its building blocks, i.e., the crystals and the way they arrange within the structure. The resulting network of pores ultimately affects the filtrate flow, hence the processing time. Here, we propose a combined modelling and experimental strategy to quantitatively understand the packing of powders based on their particle size and shape distributions. Needle-like crystals are here chosen as case study due to their extensive presence in pharmaceutical manufacturing [2,3]. Our results show how the structure of the cake can be linked to the performance of the filtration unit and therefore used to predict filtration times.

Methods

In this work we present a computational methodology to simulate the cake formation of needle-like crystals and to determine the filtration performance of the cake. Monte Carlo sampling is used to simulate the pressure driven packing of crystals. The spherocylinder, i.e. a cylinder capped by hemispheres at each end, is used as model shape [4]. Systems with statistically relevant number of particles and with polydispersities that resemble experimental distributions are used [5]. A graphical example of a filter cake produced with such methodology is presented in Fig.1(a).

We also provide an experimental counterpart to analyze filter cake properties and to validate the simulation protocol. For this purpose, X-ray tomography is used in this work to determine the structure of the pore space and how this develops along the cake [6]. Furthermore, an in-house segmentation algorithm is used to identify the shape, positions and orientation of individual needle-like crystals forming the experimental cake structure. Fig. 1 b. presents a small section of a real filter cake and shows the segmented single particles with different colors.

The structure of the pore space that forms inside filter cakes, both experimental and simulated, is then represented through a pore-network model [7,8]. Within pore network modeling, the connectivity of the pores is approximated via cylindrical channels, namely the throats, whose flow properties can be easily estimated using the Hagen-Poiseuille equation. An example of how the pores are distributed and interconnected inside the filter cake is presented in Fig. 2. This approach allows us to determine the permeability of the structure and therefore the filtration time.

Results

The modeling and experimental methodology described above has been used to evaluate the filtration dependency on particle morphology for a series of compounds, some of which taken from current development projects of AstraZeneca. The results collected confirm the strong dependency of filtration performance on particle morphology. The ability of these simulation routines to account for diverse particle size and shape distributions enables us to explore the effect of polydispersity on cake structure. Polydispersity appears to have a substantial effect on cake porosity, comparable to the one of the particle aspect ratio. This highlights that size control strategies employed during crystallization can have a tremendous impact on filtration time and that crystallization control is paramount.

The simulation strategy proposed in this work has been applied to investigate the packing of crystals of β L-Glutamic Acid with known particle size and shape distributions (determined in a previous experimental work [1]). The outcome of this investigation is reported in Fig. 3 [5]. The porosity of the simulated cake structures is compared to the one of the experimental cakes. It can be observed in Fig. 3(a) how both datasets follow the same trend, but that there is a constant offset between the two. The linear relation between the two (R^2=0.93) is further highlighted by the parity plot in Fig. 3(b). This confirms that the model is able to accurately predict the porosity trend using exclusively information about the particle size and shape distributions. The constant shift between the datasets is likely caused by the fact that we are simulating smooth spherocylinders that have a significantly different morphology and surface roughness compared to real faceted crystals. Furthermore, our simulations do not cover substance and system-specific properties (particle-particle and particle-solvent interactions).

Comparisons between simulation and experimental results also highlight that the porosity and permeability of the filter cakes are not directly related to each other. Our efforts using pore network modeling show that the properties of the filter cake structure (i.e., the throats sizes, their distribution and interconnectivity) play a vital role in determining the permeability as well. Using our combined approach of simulating packings and then processing them with the help of a pore network model, however, allows us to predict the permeability trend in dependence of the particle size and shape distributions of the underlying powders.

Conclusions and outlook

The modeling protocol presented here sheds light on the crucial importance of the polydispersity of crystal populations on filtration performance. It also shows its ability to successfully predict the porosity of real cakes of crystals. Quantitatively knowing a priori the filterability of a population of crystals will ultimately allow pinpointing the optimal distribution of particle size and shape to minimize filtration time. This could then be used as a decision criterion in the process development phase where the gain in filtration time should be counter-weighed with the effort required in the crystallization process to obtain a tailored particle size distribution. Our contribution therefore opens up new avenues to integrate different unit operations (crystallization and filtration) into an overall optimized process with high productivity.

References

[1] Perini G., Salvatori F., Ochsenbein D.R., Mazzotti M., Vetter T., *Sep. Purif. Technol.*, **211**, 768–781 (2019)

[2] Wilson D. et al., *Powder Technol.,* **339**, 641–650 (2018).

[3] MacLeod C. S. and Muller F. L., *Org. Process Res. Dev.*, **16**, 3, 425–434 (2012).

[4] Meng L., Lu P., Li S., Zhao J., Li T., *Powder Technol.*, **228**, 284–294 (2012).

[5] Perini G., Avendaño C., Hicks W., Parsons A. R., Vetter T., "Predicting filtration of needle-like crystals: a Monte Carlo simulation study of polydisperse packings of spherocylinders", *in preparation.*

[6] Löwer E., Pham T.H., Leißner T., Peuker U.A., *Powder Technol.*, **363**, 286–299 (2020).

[7] Lin C.L., Miller J.D., *Chem. Eng. J.*, **77**, 79–86 (2000).

[8] Rabbani A., Jamshidi S., Salehi S., *J. Pet. Sci. Eng.*, **123**, 164–171 (2014).

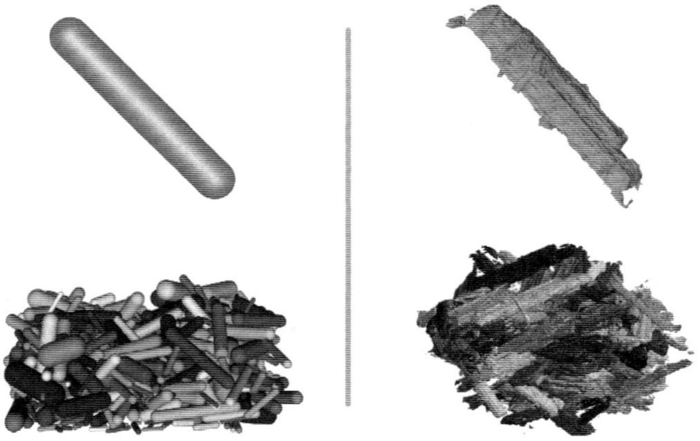

(a) *(b)*

Fig. 1 – Structure of a filter cake of needle-like crystals. *(a)* A simulated structure generated through Monte Carlo sampling of spherocylinders. *(b)* An experimental cake characterized through X-ray computed tomography. The color scheme is used only to distinguish the particles.

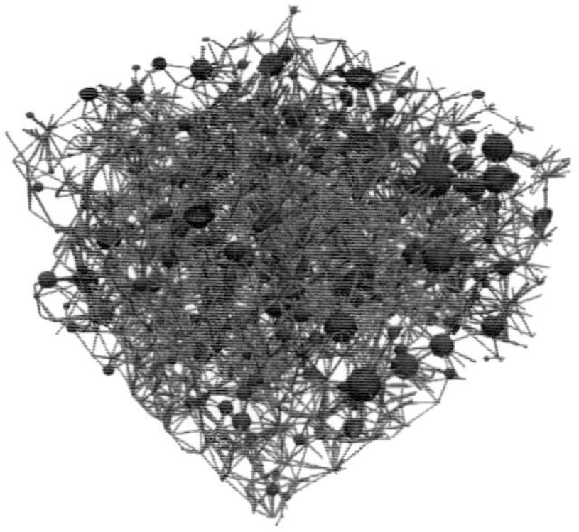

Fig. 2 – Illustration of the internal porous structure of a section of a filter cake. This was obtained through pore network modeling. The pores are shown in red, while the throats are represented by the gray tubes.

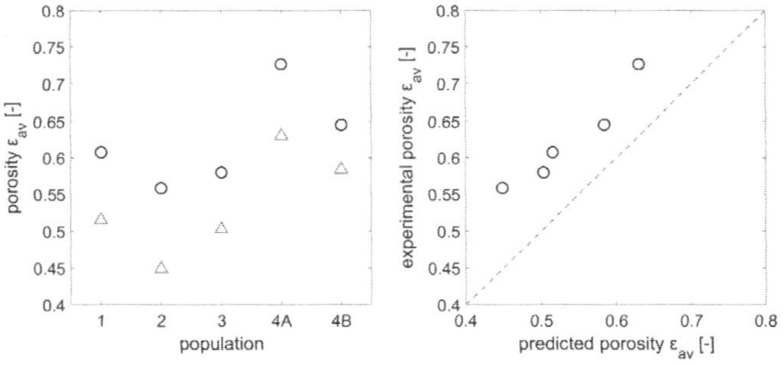

(a) *(b)*

Fig. 3 – *(a)* Comparison of porosity of filter cakes predicted via Monte Carlo sampling (triangles) and experimentally assessed by pressure filtration of β L-Glutamic Acid crystals at 0.75 barg (circles) [1]. *(b)* The same data are displayed as parity plot.

607531 Modeling CHO Cell Glycosylation Process Using Dynamic Kriging

Abstract:

Monoclonal Antibodies (mAbs) are highly demanded therapeutic products that are used for the treatment of cancers, autoimmune diseases, infectious and microbial diseases[1]. It has been reported that mAb amounts for almost 37% of biologic drugs market[2]. For most mAbs, glycosylation process is critical to protein folding, secretion, signal induction, and it impacts the product quality and efficacy [3]. To understand and improve the glycoform distribution, first-principle models have been built to quantify the effects of operating conditions, including process feed-variables, temperatures, pH to protein production, and glycosylation[4-6]. First-principle models capture the complex biochemical pathways of cell metabolism and protein glycosylation and correspond to nonlinear dynamic set of equations that are usually computational expensive. To reduce the computational complexity, a dynamic surrogate model can be used to capture the input and output relations describing the bioreactor operation. Kriging is a nonparametric modeling technique that predicts the target point based on the weighted sum of the observed function values at sampling points around the predicted point [7, 8]. Among different data-driven model techniques, kriging in most cases, requires a limited number of samples while achieving better efficiency and accuracy. It also provides error estimation for its predicted value[8, 9].

In this study, we build a dynamic kriging model to replace the first-principle model, which is used to capture the dynamic behavior of cell growth and simulate protein glycosylation with high computational efficiency. First, an unstructured kinetic model is built to capture the CHO cell culture process in fed-batch bioreactor and predict viable cell, glucose, lactate, and protein concentration. This model is coupled to a structured single-cell glycosylation model to determine the secreted glycoprotein fractions. Then a series of simulations are performed under different operating conditions based on a full factorial design, which provides inputs and outputs under all the time points to build the dynamic kriging surrogate model. By varying pH and the concentration of media supplements ($MnCl_2$), the dynamic kriging model is able to capture time dependent responds of the system, including viable cell density, glucose, lactate concentrations and glycan fractions by solving the kriging iteratively at each time step [8, 10]. This model is used to find the optimal operating conditions for specific product specifications and integrate to a bioreactor model to achieve on-line product prediction and control.

Reference

1. Newswire, P.R., *Monoclonal Antibodies (mAbs) Market Analysis by Source (Chimeric, Murine, Humanized, Human), by Type of Production, by Indication (Cancer, Autoimmune, Inflammatory, Infectious, Microbial, Viral Diseases), by End-use (Hospitals, Research, Academic Institutes, Clinics, Diagnostic Laboratories) and Segment Forecasts, 2018-2024.* 2017: Market Research Report No. GVR-1-68038-280-8.
2. Mullin, R., *Contract services for biologics evolve.* Chemical & Engineering News, 2017. **95**(33): p. 34-35.
3. Bieberich, E., *Synthesis, Processing, and Function of N-glycans in N-glycoproteins.* Advances in neurobiology, 2014. **9**: p. 47-70.
4. Sou, S.N., et al., *Model-based investigation of intracellular processes determining antibody Fc-glycosylation under mild hypothermia.* Biotechnol Bioeng, 2017. **114**(7): p. 1570-1582.

5. Kotidis, P., et al., *Model-based optimization of antibody galactosylation in CHO cell culture.* 2019. **116**(7): p. 1612-1626.
6. Villiger, T.K., et al., *Controlling the time evolution of mAb N-linked glycosylation - Part II: Model-based predictions.* 2016. **32**(5): p. 1135-1148.
7. Bhosekar, A. and M. Ierapetritou, *Advances in surrogate based modeling, feasibility analysis, and optimization: A review.* Computers & Chemical Engineering, 2018. **108**: p. 250-267.
8. Boukouvala, F., F.J. Muzzio, and M.G. Ierapetritou, *Dynamic Data-Driven Modeling of Pharmaceutical Processes.* Industrial & Engineering Chemistry Research, 2011. **50**(11): p. 6743-6754.
9. Lucifredi, A., C. Mazzieri, and M. Rossi, *APPLICATION OF MULTIREGRESSIVE LINEAR MODELS, DYNAMIC KRIGING MODELS AND NEURAL NETWORK MODELS TO PREDICTIVE MAINTENANCE OF HYDROELECTRIC POWER SYSTEMS.* Mechanical Systems and Signal Processing, 2000. **14**(3): p. 471-494.
10. Shokry, A. and A. Espuña, *Sequential Dynamic Optimization of Complex Nonlinear Processes based on Kriging Surrogate Models.* Procedia Technology, 2014. **15**: p. 376-387.

618102 Sponsored Technology Workshop - Microstructure Characterization for Drug Development: Novel AI Analytics and Release Prediction - DigiM Solution

Abstract:

From microspheres to tablets, microstructures are at the core of drug performance, formulation design, and process development. In this open discussion workshop, the DigiM team will showcase a suite of high resolution image-based tools for quantifying and correlating microstructures with drug performance and behavior. Novel techniques including AI microstructure quantification, image-based release prediction, and digital drug formulation will be presented. Applications of these techniques for complex dosage forms and solid dosage process development will be discussed.

595965 High-Throughput Approach to Studying Adsorption on Zeolite Surfaces

Abstract:

Zeolites are nowadays abundantly used as adsorbents and catalysts for many industrial applications. Computational studies have recently complemented the widely available experimental data and observations on zeolites by shedding light on reaction mechanisms and the active sites. The computational studies of zeolites, however, are challenging due to the numerous possible active sites. A zeolite structure can have many unique tetrahedral sites (T). Substitution of a trivalent aluminum atom to a tetravalent silicon atom in the zeolite framework leads to a net negative charge that can be compensated by a proton or a cation. However, there are numerous possibilities for the placement of Al and the compensating proton or cation. The number of possibilities grow exponentially for higher Al/Si ratios in the framework. Due to this complexity, most theoretical studies have focused on a limited number of T sites based on physical intuitions. For example, many studies of the MFI zeolite, which has 12 T-sites, have focused only on one or two T sites based on the accessibility of guest molecule to be the active site. These choices are justifiable given the high computational cost of accessing the full descriptor space, however, they probably miss promising active sites. In this work, we attempt to thoroughly explore the possible active sites using a high-throughput approach. We apply our approach to the problem of NO adsorption on metal-exchanged zeolites. It is known that Pd-exchange CHA, for example, is an effective adsorbent to limit NOx emissions. Full exploration of this problem requires enumerating the possible Al-Al pairs on the starting zeolites, identifying the favorable sites for the compensating protons and cations and searching for a global minimum in each case, which means thousands of calculations even for simple zeolites. The problem is further compounded when the cation is a metal that might have multiple oxidation states. We show how we systematically explore the possible sites, starting with a lower level of theory on the initial calculations, and then narrowing the potential candidates and using a higher level of theory afterwards. The automation of this calculation reduces the amount of manual labor and highly improve the efficiency of the workflow. It presents a new approach to tackle the complicated problems related to zeolite catalysis.

607612 Machine Learning-Guided Flow Synthesis of Inorganic Metal Halide Perovskite Quantum Dots

Abstract:

The controlled synthesis of advanced colloidal nanomaterials is limited by the currently available reactors utilized for their synthesis and the disjointed workflow between experimentation and data processing. Manual flask-based experimentation strategies, although applied across most research laboratories, are unlikely to sufficiently capture and control the complex and highly sensitive reaction systems of many colloidal nanomaterial syntheses. These batch reactors are slow, labor-intensive, and subject to batch-to-batch variations in experimental data sets. In response, over the last decade different microfluidic synthesis strategies have been developed and effectively applied towards the controlled synthesis of different classes of colloidal nanomaterials. Such flow synthesis strategies notably enhance the efficiency of colloidal synthesis studies, while simultaneously unlocking previously unattainable precision in synthesis parameter control. Despite the effectiveness of flow synthesis microreactors in accelerating reaction condition screening of colloidal nanomaterials, they still rely on user-directed experiment selection. Broad reaction space screening using design of experiments is materially inefficient, and expertly guided experiments are often too slow to effectively cover a synthesis space within the lifetime of unstable chemical precursors. These bottlenecks can be addressed through development of self-optimizing fluidic reactors and integrating fully automated experimentation, materials diagnostics, and data analysis with artificial intelligence-guided experiment-selection algorithms. Such autonomous materials synthesis platforms will be able to isolate optimal reaction parameters entirely without user intervention.

Colloidal semiconductor nanocrystals (known as quantum dots) are a particularly suitable testbed for autonomous materials development strategies. Quantum dot syntheses possess a large and highly sensitive reaction parameter space.[1] Proper control of their reaction conditions is difficult to attain both within a single set of precursors and across research labs (batch to batch variation). As a result, despite extensive efforts and constantly growing applications, many synthesis paths are either unexplored or not fully optimized. One of the recently emerging classes of quantum dots is colloidal metal halide perovskites, which have highly favorable properties for applications in optoelectronic devices. Due to the combinatorially increasing synthesis space of metal halide perovskite quantum dots, many of their possible synthesis routes are novel and unexplored. Therefore, these reaction systems present an exciting opportunity for exploration within an autonomous flow synthesis platform.

In this work, we present the first machine learning guided-fluidic reactor for the controlled synthesis of high-quality colloidal quantum dots. While the strategies and techniques detailed here may be easily applied to a large number of solution-processed materials, as a case study we analyzed the halide exchange of cesium-lead-bromide perovskite quantum dots with zinc halide salts and two different surface ligands.[2] We developed a novel *in situ* material diagnostic module to attain accurate real-time access to peak emission energy, emission linewidth, and photoluminescence quantum yield of the in-flow synthesized perovskite quantum dots. These output parameters, corresponding to the physicochemical and optoelectronic properties of quantum dots, were then simultaneously optimized through an objective function integrated within a custom neural network ensemble model and various decision-making policies (used to select conditions from the model). Utilizing the developed autonomous materials development platform, we evaluated the performance of eight different experiment selection algorithms (spanning machine learning to evolutionary algorithms) over the course of more than 1400 experiments. The controlled syntheses of perovskite quantum dots were performed without user intervention and required less than 400 μL of starting quantum dot solution per experimental condition. The methods detailed in this work offer autonomous reaction optimization across multiple vital performance parameters. Further use of similar self-optimizing systems in the field of colloidal nanoscience would reduce the cost and expedite the process of materials development and discovery.

References

[1] R. W. Epps, K. C. Felton, C. W. Coley, M. Abolhasani, *Lab Chip* 2017, *17*, 4040.

[2] K. Abdel-Latif, R. W. Epps, C. B. Kerr, C. M. Papa, F. N. Castellano, M. Abolhasani, *Adv. Funct. Mater.* 2019, *29*, 1900712.

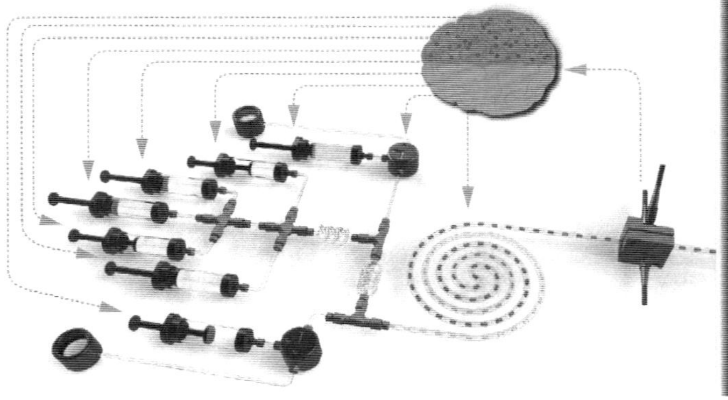

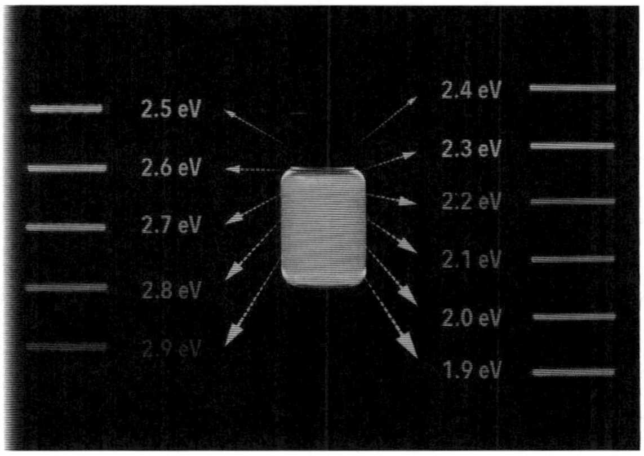

609449 High-Throughput and Data-Driven Strategies for the Design of Deep Eutectic Solvent Electrolytes

Abstract:

Within the framework of green chemistry, Deep Eutectic Solvents (DES) have been identified as promising candidates for use in many applications, including battery electrolytes. DES are characterized by two or three materials that associate with each other through hydrogen bond interactions, resulting in a eutectic mixture whose freezing point is below that of the individual materials. This design space is overwhelmingly large and poses a challenge for screening a vast and diverse set of materials. Here we present a strategic approach consisting of high throughput experimentation (HTE) coupled with data science driven analysis to identify exceptional DES candidates based on key physiochemical and electrochemical properties. Much of our HTE adopts methods that are already used frequently in the biotech and pharmaceutical industries, most notably performing parallel syntheses and analyses in 96-well-plate formats. DES samples are first synthesized using an open-sourced automated liquid handling robot. DES melting points are then determined by monitoring the melting process with an infrared camera and identifying the temperature at which the thermal conductivity of the samples changes abruptly. The solubility of battery redox-species is determined via UV-VIS well-spectrophotometers. Finally, the electrochemical stability window and cycling properties of DES electrolytes are measured in high-throughput by using screen-printed electrodes on 96-well plates adapted for use with a standard potentiostat. The ability to rapidly and efficiently collect data also creates a need for the development and use of automated processes for data analysis, which have been developed in an open-sourced format by our group. This approach to HTE also allows for the incorporation of data science techniques, such as feature extraction and machine learning, that further aid in probing a design space that is ultimately too large for experimental methods alone.

602248 Development of a Hybrid Model to Describe the Dynamics of a Partially Known System

Abstract:

Mathematical modeling plays a critical role in the design, optimization, and control of chemical processes [1]. In order to apply models to a wide range of process conditions, it is often desirable to describe the system using a first principles model. However, the formulation of such models can be challenging if the underlying mechanisms are only partially understood [2, 3, 4]. Under such circumstances, a data-driven model can be built from measured data, and the resultant model can describe the system adequately, even when mechanistic understanding is limited [5, 6]. However, a data-driven model has narrow applicability as it is tailored to describe the input-output relationships contained in the training datasets [7]. As an alternative, a hybrid modeling approach that combines first-principles and data-driven modeling has been proposed to describe a partially known process [7, 8]. Hybrid models have better prediction capabilities than first-principle models, and thus will have better generalizability and interpretability than data-driven models.

In this work, we proposed a new numerical scheme to construct a hybrid model for a system that has partial mechanistic information. A hybrid model was constructed by introducing correction terms for estimating process-model mismatches between mechanistic model predictions and experimental measurements [9]. Here, a nominal model refers to a mechanistic model based on our prior knowledge of the system of interest, and the correction terms are assigned to differential equations of the nominal model so that the resultant hybrid model has improved prediction capabilities. The proposed methodology consists of three steps. First, observability analysis was performed for the nominal model to determine a subset of states that are observable from the available measurements. This step is critical to reduce the computational load of the estimation and avoid potential overfitting issues. Second, the available measurements were used to estimate the correction terms associated with the observable states through solving a L2-regularized least-squares problem [10]. Lastly, once the correction terms were estimated, the trajectories of the states were simulated based on the hybrid model. Then, an artificial neural network (ANN) model was developed for predicting the correction terms for a given set of state values, and the developed ANN was used within the hybrid model. Through incorporating the ANN, the hybrid model was able to predict dynamic responses to an input, which was not in the measurement data used for estimating the correction terms.

The validity of the proposed method was demonstrated by constructing a hybrid model for the NFκB signaling pathway induced by both lipopolysaccharide (LPS) and brefeldin A (BFA). The signaling dynamics induced by LPS have been well studied, and hence, a previously published model was used as the nominal (mechanistic) model [11]. Since the dynamics initiated by BFA are less characterized, the nominal model could not capture the signaling dynamics induced by both BFA and LPS well. By implementing the proposed methodology, the developed hybrid model was able to predict the NFκB signaling dynamics induced by both LPS and BFA accurately.

References

[1] Hangos, K. M.; Cameron, I.T. Process modelling and model analysis. Academic Press: London, 2001.

[2] Cho, K.-H.; Choo, S.-M.; Jung, S.H.; Kim, J.-R.; Choi, H.-S.; Kim, J. Reverse engineering of gene

regulatory networks, IET Systems Biology 2007, 1, 149 - 163.

[3] Willis, M.J.; von Stosch, M. Inference of chemical reaction networks using mixed integer linear programming, Computers and Chemical Engineering 2016, 90, 31-43.

[4] Lee, D.; Jayaraman, A.; Kwon, J.S. Identification of a time-varying intracellular signalling model through data clustering and parameter selection: application to NF-κB signalling pathway induced by LPS in the presence of BFA, IET Systems Biology 2019, 13, 169-179.

[5] Thompson, M.L.; Kramer, M.A. Modeling chemical processes using prior knowledge and neural networks, AIChE Journal 1994, 40, 1328-1340

[6] Narasingam,A.; Kwon, J.S. Development of local dynamic mode decomposition with control: Application to model predictive control of hydraulic fracturing, Computers & Chemical Engineering 2017, 106, 501-511.

[7] von Stosch, M.; Oliveira, R.; Joana Peres, J.; de Azevedo, S.F. Hybrid semi-parametric modeling in process systems engineering: Past, present and future. *Computers & Chemical Engineering* **2014**, *60*, 86-101.

[8] Faizan Bangi, M.S.; Kwon, J.S. Deep hybrid modeling of chemical process: Application to hydraulic fracturing, Computers & Chemical Engineering 2020, 134, 106696.

[9] Engelhardt, B.; Fröhlich H.; Kschischo, M. Learning (from) the errors of a systems biology model, Scientific Reports 2016, 6, 20772.

[10] Howsmon, D.P.; Hahn, J. Regularization Techniques to Overcome Overparameterization of Complex Biochemical Reaction Networks, IEEE Life Sciences Letters 2016, 2, 31 – 34.

[11] Lee, D.; Ding, Y.; Jayaraman, A.; Kwon, J.S. Mathematical modeling and parameter estimation of intracellular signaling pathway: application to LPS-induced NFκB activation and TNFα production in macrophages, Processes 2018, 6, 21.

607279 Ffiber: Framework for Fluorescent Neuroimaging Based Experimental Routines

Abstract:

Background: The modern lab produces more data than it can analyze using traditional experimental techniques. For neuroimaging labs, terabytes of imaging data can easily be generated in the course of a single experiment. However, only a small portion of this data is accessible for extracting results based on existing methods. To enhance the portion of data that can be analyzed, many labs are turning to data science for high throughput processing and analysis. However, to synthesize highly skilled domain scientists with an effective data science strategy, a robust method for developing well-designed and data informed experimental pipelines must be followed. We developed a framework for fluorescent neuroimaging based experimental routines (FFIBER) that implements key practices of data science, software engineering, chemical engineering, and neuroimaging. The goal of this pipeline is to produce experimental workflows for high throughput neuroimage acquisition and analysis. For each experiment we follow a six-step process: (1) develop a data awareness, (2) design a data management plan, (3) determine an optimal experimental pipeline, (4) build out supporting data science infrastructure, (5) perform primary and supplemental imaging, and (6) produce interpretable visualizations of results. The *objective* of this project is to apply our FFIBER method during initial experimental stages and retroactively to generate, process, and analyze high volumes of neuroimaging data without an overall increase in labor-hours.

Methods: To develop a data awareness, we performed a data assessment of relevant metadata, data storage, and data relationships. When designing a data management plan, we identified the optimal data storage locations, file structure for raw data and results, and personnel responsibilities. To determine an optimal experimental pipeline, we performed software reviews, addressed labor-hour bottlenecks, and identified necessary programs, packages, and scripts. Then we built out the supporting data science infrastructure utilizing Python and Jupyter notebooks as the base script for pipelines, electronic laboratory notebook integration, and version control with the team through GitHub and Google Drive. After the pipeline was built, we completed the primary set of experimental imaging using confocal microscopy of brain slices and fed the data through the pipeline for analysis. We then completed supplemental imaging to experimentally support a representative set of results obtained through FFIBER. Finally, we wrote Python scripts to output consistent interpretable visualizations of our analyses and align these with other experimental results.

Results: We applied FFIBER to a data science experiment analyzing the phenotype of microglial cells in *ex vivo* brain slices to assess extent of glial cell activation. We were able to create and modify a pipeline during the initial stages of (1) an oxygen-glucose deprivation experiment and retroactively for (2) an inflammation-sensitized model using *E. coli derived* lipopolysaccharide, (3) and a neonatal hypoxia-ischemia (HI) model. A preliminary data assessment revealed a storage necessity of 5 GB per experiment with five individual file types from five separate personnel. We completed the data assessment and built a data management plan with university-supported Google Drive as our storage location, *.tiff*, *.csv,* and *.ipynb* as our file types, and established a slice-based, double-blind file structure. We wrote Python scripts automating image upload, cleaning and automatic feed to our analysis pipeline that reduced labor-hours from five hours per image set to less than ten minutes. Our software review led us to integrate a published cell morphometrics package, VAMPIRE, with our diff_register package for skeletonized morphometrics of microglia. Our image analysis pipeline developed with FFIBER split uploaded confocal images into fourths, segmented, skeletonized and identified the microglia in the images, and uploaded the images into our integrated morphometric analysis. FFIBER structured the development of our work pipeline that decreased work hours from a full day to less than an hour to analyze an entire experiment. Finally, our FFIBER structured pipeline enabled us to detect and quantify shape differences between non-treated and injured *ex vivo* slices – providing additional information for our initial oxygen-glucose deprivation experiment and retroactive insight to our inflammation sensitized and neonatal HI models. Our first application of FFIBER decreased our human labor time from > 24 hours per 5 GB image set to less than an hour, decreased our storage budget to $0, and produced information about cell morphometrics previously unavailable to our lab due to lack of analysis software and time constraints.

Conclusions: By applying FFIBER, we were able to create workflow pipelines both during initial stages of experimental design and retroactively in neuroimage sets already collected. We decreased labor hours for data sets from 24 hours to less than an hour through using Python scripts rather than manual work and a well-designed pipeline. Our framework creates a high throughput approach for experimental design in research dependent on neuroimaging by integrating traditional experimental methods with modern data science applications. We can further iterate the methodology to include metadatabases for easier data access and data connections, budget vs storage matrices for determining best data storage location, and templates for electronic lab notebooks, file structures, data relationships, and generalized processing pipelines. Additionally, while FFIBER was developed for fluorescent neuroimaging datasets, the current workflow is robust, and flexible enough for application to other image analysis techniques, other organs, and other fluorescent imaging methods.

597491 Data-Driven Discovery of Polymeric Vehicles for Gene Editing:Serendipity-Inspired Design Directions.

Abstract:

Synthetic materials such as cationic polymers are poised to replace engineered viruses as vehicles for therapeutic nucleic acids. Polymeric materials offer a vast design space where desired material properties can be engineered by deft optimization of chemical composition and polymer architecture. However, rational polymer design is impeded by the "curse of dimensionality" since systematic examination of numerous variables such as composition, architecture, length and formulation parameters is paramount. To accelerate polymeric vector discovery, we need to develop models that predict the impact of key design attributes such as phase behavior, basicity, surface charge, nucleic acid binding, and hydrophobicity on the ultimate biological responses observed. In this work, we eschew low-throughput hypothesis-driven design of polymers in favor of a data-driven approach that exploits combinatorial polymer libraries and high-throughput experimentation. Thereafter, we employ statistical learning to extract the molecular attributes of the hit polymers that promoted gene editing efficiency and apply this knowledge to provide experimental guidance on designing subsequent polymer libraries.

Drawing inspiration from high-throughput experimental approaches used in the pharmaceutical industry to screen drug libraries for therapeutic activity, we employed parallel polymer synthesis, formulation and well plate based biological assays for rapid screening of gene editing efficiency We synthesized a chemically diverse library of copolymers combining different ratios of 1) cationic monomers bearing amines spanning a broad range of basicity and 2) neutral monomers of varying hydrophilicity. Our combinatorially designed library allows for systematic investigation of the effect of amine basicity by studying variations in polymer pKa resulting from the use of 4 cationic monomers, in ratios of 100, 75, 50 and 25 %. Subsequent to parallel polymer synthesis, extensive physicochemical characterization was completed using automated tools -: composition and molecular weight analysis, pK_a, polyplex size distribution, binding assays, and ζ-potential measurements were acquired in high-throughput modes to generate a rich dataset. Polymers were complexed with ribonucleoprotein (RNP) payloads and gene editing was quantified by estimating the proportion of mCherry positive cells. To resolve the trade-off between sensitivity and experimental throughput, we employed image cytometry and developed an image processing algorithm to quantify mCherry expression in a robust and automated fashion from a bank of images arising from 200 unique formulations. At the end of the high-throughput screening campaign, we obtained a high-performing hit polymer that outperformed state-of-the-art synthetic transfection reagents.

Having identified a hit formulation from our library, the challenge was to: 1) unravel the relationship between polymer attributes and gene editing efficiency. 2) build on these structure-activity relationships to guide the design of future polymer libraries that will yield a higher "hit rate". In order to derive these predictive relationships, we sought to understand how 10 polymer descriptors influenced biological performance. We turned to principal component analysis (PCA), to deal with the dimensionality challenge created by our dataset. Through PCA, we concluded that a single polymer descriptor cannot be used in isolation to guide the design of future libraries. Rather, complex non-linear relationships between several molecular attributes were responsible for editing performance. Discarding preconceived notions of how various chemical functional groups will influence delivery, we screened a large chemically diverse polymer library to discover design guidelines that do not conform to traditional heuristics as well as a promising hit polymer, which may not have been accessible through hypothesis-testing. If statistical learning and automated experimental workflows are applied in tandem, we can overcome challenges originating from the complexity of structure-function relationships governing polymeric gene delivery.

605093 Artificial Intelligence-Assisted Full-Map Understanding of Strain Sensing Devices and Designated Predictions for Soft Robotic Systems

Abstract:

Compliant strain sensing devices are widely applied in real-time mechanical feedback for soft robotic systems. However, it remains challenging to reach customized sensing devices with high sensitivity coordinated with designated working windows due to the complex and unclear structure-composition-property relationships which are hard to predict and pre-design. In this work, through active learning and developed algorithms, we propose machine assisted full-map structure-composition-property understandings of strain sensors consist of classical nanocomponents of $Ti_3C_2T_x$ MXene, carbon nanotube, and poly(vinyl alcohol). Important variables of nanocomponent weight loading, nanocoating thickness, and nanocoating surface morphology are systematically investigated, and an informative data base (5000+ experimental resistance points, 500+ nanocoatings, and 200+ sensing devices) is constructed for training of machine learning models. Besides, comprehensive algorithm optimizations are performed for better machine performance. Ultimately, the machine learning models show low average scattered numerical error of 24% and give customized predictions in 1) sensing performance in a given material recipe; 2) suitable material recipes of sensing devices with high sensitivity and designated working windows in soft robotic systems. The reported strategies here open new perspectives towards design-free high-performance strain sensing devices and customized applications in soft robotic systems.

More specifically, in our material systems, at least 33,700 non-repeating recipes are available for strain sensing devices fabrication although we limit the step length in each variables of material recipes. With such a complex system and very limited understanding, there are no available *in silico* simulation methods to interpret or predict the sensing devices. In addition, it is nearly impossible to screen the whole design space *via* Edison approach because the sensing needs are customized every time. Herein, we have proposed a data-driven framework to exploratively learn the full-map of the strain sensing devices. The data is collected with thirteen active learning rounds which investigate both the mean minimal distance (L_2) between suggestion points and existing data points, and the numerical uncertainty of well-trained models during each round. After a representative dataset is constructed, optimization strategies including GA combination, customized loss function, and data augmentation are used for improving the model performance. This best model has achieved a low average scattered numerical error ((Predicted value - Real value)/Real maximum value in device) of 24%.

610072 Calibration-Free Approach for End-Point Determination of a Challenging Powder Blending Process: Scientific Case Study

Abstract:

Background: Ensure adequacy of mixing to assure uniformity and homogeneity and assure in-process testing and finished product conforming to specifications has been essential. As such, innovation on in-process monitoring and control of powder blending represents one exciting area for oral solid dosage form development and manufacturing. Over the past 18 years, quantitative calibration modeling approach using Process Analytical Technology (PAT) tools has been well documented in public literatures and regulatory submissions. In contrast, there are much less documents in the public domain regarding development of qualitative calibration-free approach for pharmaceutical applications. In this presentation, we will discuss a case study of integration of Design of Experiments (DOE) and PAT real time process monitoring for pharmaceutical blending end-point determination by calibration-free approach.

Method: Following a Quality-by-Design (QbD) and risk-based approach to illustrate key steps of in-process control strategy establishment for powder blending case study: (a) identification of most influential formulation and process variables by applying multivariate data analysis techniques (such as principal component regression (PCR) to the powder blend characterization data; (b) further narrowing down to the most critical process parameters for in-process monitoring and control development by comparing the multivariate approach and univariate approach for powder segregation tendency modeling; (c) Integration of Design of Experiments (DOE) and Near Infrared (NIR) real time process monitoring for a narrow therapeutic index (NTI) drug powder blending process for process endpoint determination via calibration-free approach, recognizing that all blend quality attributes and process performance matrix are functions of time and will approach steady state values gradually. Two calibration-free approaches, i.e., moving block standard deviation method across the entire wavelength range, and examining the dynamics of specific API characteristic peak over time, are discussed.

Results and discussions: Results from the chemometric modeling in this work suggests that moving window standard deviation (MWSD) method in conjunction with appropriate spectral data preprocessing algorithms, can be an effective means to evaluate the overall progression of the powder blending and blending process dynamics. Significant blending dynamics was observed for batches with large particle size difference between excipients. By monitoring specific NIR peak intensity of the API, it can further improve the method specificity.

Conclusions: This work illustrated the feasibility of using integrated real time PAT monitoring to establish a calibration-free approach for end-point determination of a challenging powder blending process. It covers critical parameter identification via multivariate data analysis techniques, implementation of a real time PAT monitoring for a narrow therapeutic index (NTI) drug powder blending with particle size difference as the main factor, chemo-metrical data analysis and modeling with initial MWSD method and enhanced specific NIR peak method.

606472 Prediction of Entire Tablet Compositions from Pure Powder Spectra Using a Nonlinear Iterative Optimization Method

Abstract:

Process analytical technology in the pharmaceutical industry often requires monitoring of critical quality attributes (CQA) through calibrated models. The development, implementation, and maintenance of these quantitative models are both resource and time-intensive[1].

The methodology proposed in this study is based on Iterative Optimization Technology[2], and it requires minimum calibration and allows simultaneous prediction of the entire formulation instead of only the API with just one standard and pure component spectral data. Unlike Partial Least Squares (PLS), which needs tedious development of standards to incorporate variations in the process, this non-destructive methodology minimizes significant calibration effort by developing a mathematical model that uses only one standard and spectral information of pure powders present in the tablet. In continuous manufacturing, this approach will enhance real-time monitoring of CQA by predicting entire formulation using single calibrated model. The spectra of pure components do not require additional preparation as physical information from tablets can be extracted from the standard. It is a two-step calibration-minimum approach of which the first step is to compute the model parameter (λ^n) for each wavelength using a single standard tablet spectrum (X_{mix}) and spectra of pure powders ($X_{pure,i}$) as model development followed by the second step as predicting formulations (r_i). In terms of model maintenance, it is flexible in incorporating factors such as change of spectroscopic instruments, variations in raw materials, environmental conditions, and methods of tablet preparation. Robustness of the proposed approach for variation in compaction (physical changes) and variation in composition (chemical changes) was evaluated. A PLS method was also constructed for reference.

[1] Shi Z., et al., Applied Spectroscopy, 2012. **66**(9): p. 1075-1081.

[2] Muteki K., et al., Industrial & Engineering Chemistry Research, 2013. **52**(35): p. 12258-12268.

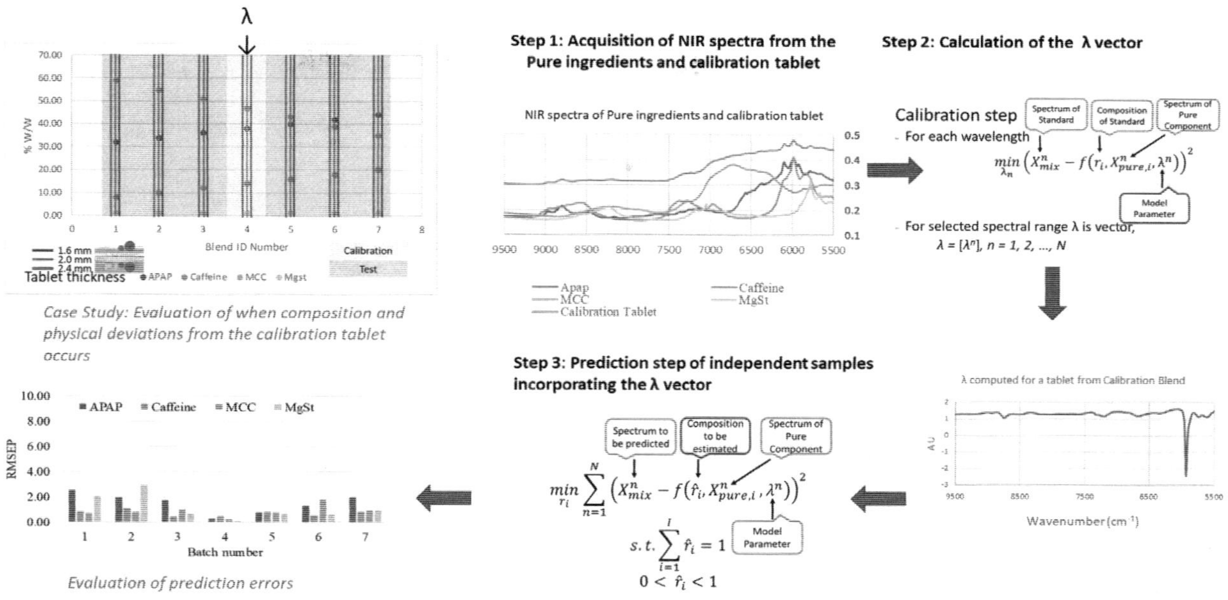
Step 1: Acquisition of NIR spectra from the Pure ingredients and calibration tablet
Step 2: Calculation of the λ vector
Calibration step
For each wavelength
For selected spectral range λ is vector, λ = [λ^n], n = 1, 2, ..., N
Step 3: Prediction step of independent samples incorporating the λ vector
Case Study: Evaluation of when composition and physical deviations from the calibration tablet occurs
Evaluation of prediction errors

603982 Real-Time Monitoring of Powder Mass Flowrates for MPC/PID Control of a Continuous Direct Compaction Tablet Manufacturing Process

Abstract:

To continue the shift from batch operations to continuous operations for a wider range of products, advances in real-time process management (RTPM) are necessary. The key requirements for effective RTPM are to have reliable real-time data of the critical process parameters (CPP) and critical quality attributes (CQA) of the materials being processed, and to have robust control strategies for the rejection of disturbances and setpoint tracking.

Real-time measurements are necessary for capturing process dynamics and implement feedback control approaches. The mass flow rate is an additional important CPP in continuous manufacturing compared to batch processing. The mass flow rate can be used to control the composition and content uniformity of drug products as well as an indicator of whether the process is in a state of control. This is the rationale for investigating real-time measurement of mass flow of particulate streams. Process analytical tools (PAT) are required to measure particulate flows of downstream unit operations, while loss-in-weight (LIW) feeders only provide initial upstream flow rates. The X-ray based sensor investigated by Ganesh et al.[1] and a novel capacitance-based sensor offered by Tech4Imaging LLC, (Columbus, OH), which have recently been investigated in our group both demonstrated the ability to effectively measure powder mass flow rates as well as to capture solid flow dynamics in the downstream equipment.

Robust control strategies can be utilized to respond to variations and disturbances in input material properties and process parameters, so CQAs of materials/products can be maintained and the amount of off-spec production can be reduced. The hierarchical control system (Level 0 equipment built-in control, Level 1 PAT based PID control and Level 2 optimization-based model predictive control) was applied in the pilot plant at Purdue University and it was demonstrated that the use of active process control allow more robust continuous process operation under different risk scenarios compared to a more rigid open-loop process operation within predefined design space.[2-3] With the aid of mass flow sensing, the control framework becomes more robust in mitigating the effects of upstream disturbances on product qualities. For example, excursions in the mass flow from an upstream unit operation, which could force a shutdown of the tablet press and/or produce off-spec tablets,[4] can be prevented by proper control and monitoring of the powder flow rate entering the tablet press hopper.

In the first part of this study, the impact of mass flow sensing on the control performance of a direct compaction line is investigated by using flowsheet modeling implemented in MATLAB/Simulink to examine the effects of sampling time, measurement precision and time delay. In the second part of this study, pilot plant studies are reported in which the mass flow sensor is integrated into the tableting line at the exit of the feeding-and-blending system and system performance data is collected to verify the effect of mass flow sensing on the performance of the overall plant-wide supervisory control.

References

[1] Ganesh, S., Troscinski, R., Schmall, N., Lim, J., Nagy, Z., & Reklaitis, G. (2017). Application of X-ray sensors for in-line and noninvasive monitoring of mass flow rate in continuous tablet manufacturing. Journal of pharmaceutical sciences, 106(12), 3591-3603.

[2] Su, Q., Moreno, M., Giridhar, A., Reklaitis, G. V., & Nagy, Z. K. (2017). A systematic framework for process control design and risk analysis in continuous pharmaceutical solid-dosage manufacturing. Journal of Pharmaceutical Innovation, 12(4), 327-346.

[3] Su, Q., Ganesh, S., Moreno, M., Bommireddy, Y., Gonzalez, M., Reklaitis, G. V., & Nagy, Z. K. (2019). A perspective on Quality-by-Control (QbC) in pharmaceutical continuous manufacturing. Computers & Chemical Engineering, 125, 216-231.

[4] Martinetz, M. C., Rehrl, J., Aigner, I., Sacher, S., & Khinast, J. (2017). A continuous operation concept for a rotary tablet press using mass flow operating points. Chemie Ingenieur Technik, 89(8), 1006-1016.

595414 Regulatory Case Studies on in-Process Controls and Scale up of Pharmaceutical Powder Blending

Abstract:

Powder blending is a critical unit operation in the manufacture of solid dosage forms. The uniformity of final blend has a direct impact on the unit dose uniformity of the finished drug product. Since blending is a complex operation that involves a large parametric space including processing parameters (i.e., blender RPM, fill level, loading order, and blending duration), equipment design, and material attributes, assessing powder blending uniformity (BU) is critical to ensure delivering the intended dose of the drug to the patient consistently. In-process testing of powder blends to demonstrate "adequacy of mixing" is a CGMP requirement (21 CFR 211.110). Since the withdrawal of FDA's draft guidance document for industry "Powder Blends and Finished Dosage Units—Stratified In-Process Dosage Unit Sampling and Assessment" in 2013, various approaches have been developed to assess blend and/or dosage unit uniformity and submitted to the FDA in NDAs and ANDAs by applicants.

In this presentation, we will provide case studies on in-process controls (IPCs) of blending uniformity and content uniformity (BU/CU) as well as FDA's assessment on the IPCs as an integral part of the overall control strategy of the drug product manufacturing process. Our focus will be on how our assessment the proposed IPCs is linked to the critical quality attributes (CQAs) of the finished drug product. Additionally, considering the recent publication of ICH Q12, we will also discuss how ICH Q12 will affect our assessment of process parameters and IPC testing, especially with regards to applicant's overall control strategies on scale-up for commercial production. By analyzing these cases, we wish to shed some light on FDA assessors' current thinking on these important topics.

609052 Use of Process Data for Granule Moisture Estimation in the Consigma 25^{TM} Semi-Continuous Fluid Bed Dryer

Abstract:

Introduction

Continuous pharmaceutical manufacturing lines containing a wet granulation unit operation - like twin screw granulation - require the implementation of a subsequent drying step in order to achieve the desired granule moisture. Several continuously and semi-continuously operated dryers are available to the pharmaceutical industry, e.g., [1–3]. The semi-continuously operated ones [1,3] have in common, that the continuous material stream is split into separated mini-portions of material, each of them being dried in an individual cell of the dryer. In this study, the ConsiGma 25^{TM} dryer will be investigated in detail. It consists of six drying cells that are consecutively filled and – after drying has finished – emptied.

The most important material property during the drying process is the moisture content of the granules. If this quantity would be available for all the dryer cells, sophisticated feedback control strategies could be developed and implemented, ensuring that the granule moisture after drying is the same for all the processed granules. However, the systems that are commonly installed in industry only allow for the measurement of granule moisture in one cell and not in all of the cells. The moisture is mainly affected by the granules initial moisture content, the inlet air temperature, the air volume flow and the drying time. Therefore, a mathematical model that uses this data for predicting the moisture content in the individual cells is of interest. Further, the model should not be computationally demanding in order to allow its real-time execution in parallel to the process. This work presents such a model.

Method

A mathematical model has been developed, which uses available process data in a ConsiGma 25^{TM} drier for estimating the granule moisture content. The proposed model is based on mass- and energy balances. Some parameters of the model are known (e.g., material properties like specific heat capacity of water), whereas others need to be identified (e.g., heat and mass transfer coefficients). Since up to six cells need to be simulated simultaneously, the modeling approach was kept as simple as possible by proposing appropriate simplifications (e.g., lumped parameter model, perfect mixing). A set of experiments was planned and executed and the generated process data was used to identify the model parameters. A second set of experiments - at different process settings - was conducted for validating the proposed model.

Results

The proposed model could predict the granule moisture of the validation dataset by an error of approximately 1wt.% (wet basis) of moisture, which is considered to be sufficiently accurate for the intended purpose. Further, it was demonstrated that the simulation could be executed approximately 400 times faster than real-time on an Intel XEON E3-1245V2, 3.4Ghz, desktop computer, suggesting that its execution is also feasible in real-time on a less powerful PLC.

Conclusion

A model for granule moisture prediction in a ConsiGma25^{TM} drier has been successfully developed and implemented in simulation. By means of that model, in a next step advanced process control concepts can be realized in order to keep the granule moisture close to the desired value.

References

1. Glatt MODCOS line Available online: https://www.glatt.com/en/products/continuous-technologies-pharma/ (accessed on Apr 30, 2019).
2. Bohle Qbcon 1 Available online: https://www.continuous-production.com/continuous-granulation-continuous-drying (accessed on Apr 29, 2020).
3. GEA ConsiGma line Available online: https://www.gea.com/en/products/consigma-ctl.jsp (accessed on Apr 30, 2019).

606043 Monitoring Loss-in-Weight Pharmaceutical Feeders through State Estimation

Abstract:

In recent years the pharmaceutical industry has been putting a lot of effort and resources for operating in a more continuous fashion [1]. Continuous manufacturing of solid oral dosage forms presents many benefits with respect to traditional batch production, and has been encouraged by the U.S. Food and Drug Administration [2]. One important bottleneck for this shift of paradigm is the implementation of an efficient feeding system, as any disturbance coming from the feeders will critically affect the downstream units and the quality of the product, even irreversibly [3,4].

Loss-in-weight feeders are usually resorted to in drug product manufacturing. They are made up of a hopper, where the powder is collected and eventually refilled, an agitator, connected to a motor, and a screws system. Realtime monitoring of the feeder operation is inherently challenging, as the only available process measurement is the net weight of powder in the hopper. The mass flowrate of powder delivered by the feeder can only be obtained indirectly, as a discrete time derivative of the mass measurements in the hopper. The propagation of noise from the measured mass to the inferred mass flowrate can compromise the control of the powder flow and the monitoring of the whole manufacturing line. Commercial feeders tackle the problem by providing the user with a smoothed version of the indirect measurement of the mass flowrate, but this introduces a time delay with respect to the actual mass flowrate fed downstream. In presence of gross errors in the measurement of the mass in the hopper, the inaccuracy in the calculation of the mass flowrate is even greater.

In this study we improve the accuracy of estimation of the mass flowrate coming out of a feeding unit by the implementation of a state estimator [5]. State estimators are widely employed in the process industry, but few contributions in the literature consider their application to continuous pharmaceutical manufacturing [6]. The implemented state estimator is based on a recently developed hybrid model [7] and is tested on experimental data coming from a feeding system. The estimation framework can accurately estimate the states of the system, namely the mass of powder in the hopper (deprived of the measurement noise), the actual mass flowrate provided by the feeder and the effective density in the hopper (an additional variable that cannot be measured online but that is important for process monitoring). The state estimator demonstrates to be robust to the presence of gross measurement errors and faults. The estimated states can be used for detecting faults in the feeding unit and to monitor the downstream operation.

References

[1] Ierapetritou, M., F. Muzzio and G. Reklaitis (2016). Perspectives on the continuous manufacturing of powder-based pharmaceutical processes. *AIChE J.*, **62**, 1846–1862.

[2] Lee, S. L., T. F. O'Connor, X. Yang, C. N. Cruz, S. Chatterjee, R. D. Madurawe, C. M. V. Moore, L. X. Yu and J. Woodcock (2015). Modernizing Pharmaceutical Manufacturing: from Batch to Continuous Production. *J. Pharm. Innov.*, **10,** 191–199.

[3] García-Muñoz, S., A. Butterbaugh, I. Leavesley, L. F. Manley, D. Slade and S. Bermingham (2017). A flowsheet model for the development of a continuous process for pharmaceutical tablets: an industrial perspective, *AIChE J.*, **64**, 511–525.

[4] Blackshields, C. A. and A. M. Crean (2018). Continuous powder feeding for pharmaceutical solid dosage form manufacture: a short review. *Pharm. Dev. Technol.,* **23**, 554–560.

[5] W. H. Ray (1981), *Advanced process control*, McGraw-Hill, New York (U.S.A.).

[6] Liu, J., Q. Su, M. Moreno, C. Laird, Z. Nagy, G. Reklaitis (2018). Robust state estimation of feeding–blending systems in continuous pharmaceutical manufacturing. *Chem. Eng. Res. Des.*, **134**, 140–153.

[7] Bascone, D., F. Galvanin, N. Shah and S. García-Muñoz (2020). A hybrid mechanistic-empirical approach to the modelling of twin screw feeders for continuous tablet manufacturing. *Ind. Eng. Chem. Res.*, *in press*

604543 Towards an Understanding of Fluid and Particle History in a Turbulent Mixing Tank Using Large Eddy Simulation: Examining Effect of Different Impeller Types.

Abstract:

Prior work has shown the differences in energy dissipation zones in different portions of the vessel, e.g. in the trailing vortex, far from the impeller, etc. The path that the fluid or particle takes between these zones is difficult to measure experimentally. Computational Fluid Dynamics is an attractive approach to directly predict the path of the fluid or particles in the system and measure the history of exposure to the differing energy dissipation regimes. In particular, the large eddy simulation model is particularly suited for describing the random transient behaviors inside a fully turbulent tank.

Following previous work [1], we will present the time history of energy-dissipation in mixing tanks across multiple scales using the LES model. We will extend the prior work by extending the range of systems studied including changing pumping direction and impeller types.

First, the time-averaged energy-dissipation predictions will be compared to literature reported values, and the accuracy and limitations of the LES modelling approach will be discussed. The energy dissipation history can be filtered through signal processing operations that leads to an understanding of power being applied to the fluid at different scales. We will show that there are fundamental relationships that collapse the time history information onto a universal curve. We will show the effect of changing the system beyond a geometrical scale up, including by varying impeller type.

[1] Flamm, Raudenbush, Sirota, Cote. AICHE Annual Meeting 2019.

607638 Building the Predictive Process Engineering Toolbox to Define Mixing Parameters for Early-Stage Drug Product Manufacturing of Large Molecules

Abstract:

Mixing is an integral operation used throughout the manufacturing process train of pooling, formulation, and primary container closure filling for large molecule sterile drug products. As such, evaluating appropriate mixing parameters and associated ranges is necessary to understand their impact on product uniformity and physical stability. For this work, a broad toolbox of at-scale, computational, and scale-down approaches were used to demonstrate successful mixing operating ranges. Saline and water were used as a model system to evaluate mixing at-scale (65 L) with agitation rates ranging from 37 to 100 RPM. In parallel, computational fluid dynamics (CFD) was used to predict at-scale homogeneity with further refinement and validation from the at-scale mixing results. It was observed that an agitation rate of 80 RPM yielded a uniform mixture within 30 minutes. At these mixing conditions, the computational model predicted maximum shear rates in the impeller region to be approximately 1,000 s^{-1}. A scale-down approach was developed using a rheometer to subject active large molecules to uniform shear rates based on the predicted range. The stressed molecules were then analyzed for particle aggregates as an indication of physical stability at shear rates similar to those shown in the CFD models. This scale-down approach is particularly useful during early development to evaluate the shear sensitivity of large molecules under anticipated mixing conditions when material quantities are limited. Setting the operating parameters for the scale-down model required the predictions from the CFD models, which were optimized with data collected from at-scale mixing. When used together, the toolbox of at-scale, computational, and scale-down approaches implemented within this work are broadly applicable to evaluate and define successful mixing parameters for the manufacture of large molecule sterile drug products.

605968 Hold Tank Mixing Profile Optimization for Potency Control in Filling a Shear Thinning Suspension Formulation

Abstract:

This study discusses the complex mixing considerations arising throughout the time course of filling a shear-thinning suspension drug product formulation. As feed tank working volume is drawn down, rotational speed of the off-set, bottom mounted mixer must balance air entrainment, regions of stagnation, and apparent active particle buoyancy and creaming in the transitional regime without impacting potency control of the suspension drug product being filled. M-Star Simulation's® Digital Mixing Tank (DMT) is employed to characterize the complex phenomena to inform mixing profile optimization.

610514 Efficient Process Development in the Micromixing-Mesomixing Space: The Bourne Protocol and Beyond

Abstract:

Reaction and crystallization processes often need to balance competing rates in order to achieve the desired results. Mixing can affect certain rate processes, so understanding the role of mixing is a critical part of process development and scaling. Process developers in the chemical and pharmaceutical industries need efficient ways to understand how mixing matters to their processes.

Since it was introduced in 2003, the Bourne Protocol has been used to efficiently understand the effect of impeller speed, feed time, and feed location on semi-batch or gradual-addition processes. Bourne's simple experimental plan creates a framework to understand if micromixing (mixing at the molecular or particle length scale) or mesomixing (mixing at feed plume length scale) or macromixing (mixing at the length scale of the vessel) matters in a process. Applications and enhancements to the Bourne Protocol have been discussed in a variety of venues over the past decade (see References).

Most recently, we obtained more appropriately visualized regions of acceptable process performance by plotting Bourne Protocol results in a micromixing-mesomixing space (Sarafinas, 2019). That study looked at examples from the literature and showed how regions of different process failures could be characterized in the micromixing-mesomixing space, enabling improved and efficient understanding of the acceptable process operating window for scale-up or scale-down. There are many potential undesirable process responses that can be affected by mixing: low yield or conversion, poor selectivity, unacceptable purity, unmanageable particle size, poor crystal morphology, inappropriate polymorph, etc. Understanding how a process responds to different levels of micromixing and mesomixing can enable more successful scaling, suggest potential process improvements, and speed troubleshooting of industrial processes.

This paper continues the exploration of the micromixing-mesomixing space to characterize our processes. Beginning with how to apply the Bourne Protocol to efficiently show how mixing matters in a process, we explore how process practitioners can expand the Protocol to better understand their process for scaling (up or down) in development and troubleshooting. This approach results in achieving more robust processes and more optimal equipment designs, more quickly. Further literature examples and some new examples will be shown.

References:

Bourne J.R., "Mixing and the Selectivity of Chemical Reactions", **Org. Process Res. Des. Dev.**; 2003; 7(4) pp. 471-508.

Sarafinas, A. "Efficiently Characterize Process Scalability in the Micromixing-Mesomixing Space Using the Bourne Protocol," **AIChE 2019 Annual Meeting**, Orlando, FL, November 2019.

Sarafinas, A. "Test Process Mixing Sensitivities Using the Bourne Protocol," **Scientific Update Webinar**, 13 November 2018.

Sarafinas, A. "Using the Bourne Protocol to test mixing sensitivities in the lab," **DynoChem Guest Webinar**, 8 August 2018.

Sarafinas, A. "A further look at the Bourne Protocol for efficient investigation of mixing sensitivities during process development," **AIChE 2017 Process Development Symposium**, Toronto, Canada, June 2017.

Sarafinas, A. and C.I. Teich in Kresta, S.M., A.W. Etchels III, D.S. Dickey, V.A. Atiemo-Obeng, **Advances in Industrial Mixing: A Companion to the Handbook of Industrial Mixing**, John Wiley and Sons, 2016.

Teich, C.I., A. Sarafinas, P.M. Morton, **AIChE 2010 Annual Meeting**, Salt Lake City, UT, November 2010 (2 papers):

- "Can this process be saved? A search for understanding using the Bourne Protocol and advanced process development tools"
- "Taking the min to the max: a case study in small scale process development using on-line reaction calorimetry and in-situ particle characterization"

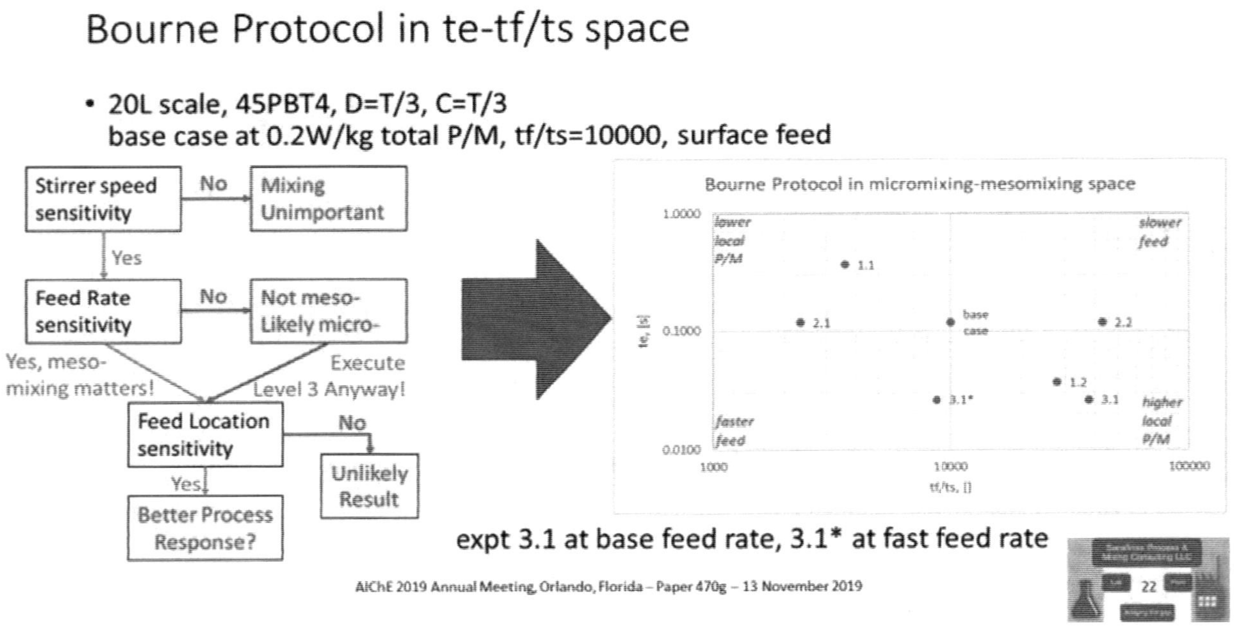

598802 Development of a Mixing Index Provides a Means to Solve Many Industrial Mixing Problems

Abstract:

Industrial mixing success is measured by practical and economical product production, not by extensive description of mixing behavior. Most often the cost and time required for a scientific research project to find the best mixing option is not feasible. The majority of industrial mixer applications are for formulation processes that combine and blend components, as opposed to complicated chemical reactions with multiple possible products. The ability to conveniently quantify mixing intensity can be an effective path to improving or solving many process problems.

A further limitation to most industrial mixing processes is the requirement to use existing mixing equipment. A new mixer is rarely an option for process improvement or even new product production. The successful use of existing equipment often depends on the ability to predict the capabilities of the mixer for the planned product or modified process. A 1 to 10 Scale of Agitation for liquid mixing intensity was introduced in a series of articles published in Chemical Engineering in 1975 and 1976. That intensity scale was linked to a "bulk fluid velocity" and applied to the selection of mixing equipment. Because the concept focused on new mixing equipment, it was presented in a way that was difficult to apply to existing equipment. A convenient and consistent method for predicting mixing intensity for turbine style mixers in liquid applications has been developed to facilitate the evaluation of mixing equipment.

The concept of a 1 to 10 scale of mixing intensity has been reduced to the straightforward calculation of a Mixing Index. This index is based on otherwise obvious variables that influence mixing success, including liquid volume, impeller diameter, rotational speed, and impeller power number. The Mixing Index (MI) can be calculated using the following formula for turbulent conditions: (equation - see uploaded file)

The expression for MI can be rearranged to solve for a rational speed based on an impeller diameter and volume or solved for an impeller diameter at a speed and volume. The simple expression can be used for either evaluation of existing mixing conditions or mixer design based on the selection of other variables. The expression needs a correction factor for reduced impeller pumping at higher viscosity. The correction factor can be used over a typical range of Reynolds numbers for turbine mixers.

The calculation of a MI with a value between 1 and 10 can be compared with CFD images or used for scale-up calculations. A non geometric scale-up example and associated CFD plots will be used to demonstrate the ability of the MI to calculate or compare mixer capabilities. The basic equation can be used to adjust for different volumes and with different impeller types. Scale-up can even keep multiple mixing variables constant by solving multiple equations with multiple unknowns.

While not sophisticated or precise, the MI gives a sufficient estimate of mixing intensity for many industrial applications. It successfully parallels other methods of mixer evaluation and can be applied using spreadsheet calculations and basic engineering skills. Like any other method for solving mixing problems, the MI has its limitations, but it does provide a flexible method for getting estimates of mixing capacities quickly and cost effectively.

$$\mathrm{MI} = 4.05\times10^{-4}\,\frac{\mathrm{N_P}^{1/3}\,N[\mathrm{rpm}]\,(D[\mathrm{in}])^{5/2}}{(V[\mathrm{gal}])^{1/2}}$$

598109 Implementation of a Vapor-Liquid Equilibrium and Vapor Purge Model for Scale-up of a Transfer Hydrogenation

Abstract:

Scale-up of hydrogenation reactions can uncover mixing sensitivities due to heterogeneous catalyst use or catalyst deactivation by impurity adsorption. This presentation will describe the control of carbon dioxide (CO2) formed as a by-product during a palladium-on-carbon catalyzed transfer hydrogenation reaction. At small scale, CO2 was adequately purged via nitrogen sparging to ensure sufficient mass transfer of reagents to the catalyst. This resulted in a first-order kinetic model with respect to the limiting reactant. However, during an initial 5x scale-up, zeroth-order reaction kinetics were obtained and were attributed to CO2 inhibition of the catalyst. The reaction had been scaled to maintain nitrogen mass transfer rate coefficient (kLa) through adjustment of agitation and reactor configuration. While this achieved the desired kLa, the headspace turnover time (i.e. headspace volume over nitrogen flow rate) was much longer on-scale. This resulted in increased CO2 concentrations in the headspace and, therefore, the batch. Utilizing Process Analytical Technology and data from scale-down experiments, a coupled kinetic and mass transfer model was developed to account for CO2 generation, vapor-liquid equilibrium, and headspace purge. The kLa and headspace turnover requirements predicted by the model were then successfully demonstrated at scale. This model allows accurate prediction of reaction times and should be considered for other reactions where catalyst inhibition is a risk.

607809 Coupling Reaction Kinetics, Calorimetry and Mixing Studies for Successful Scale-up of Exothermic Reactions

Abstract:

In this work we showcase a procedure to troubleshoot a reaction that had very good outcome in the lab (an over-reaction key impurity was maintained at ≤1.2% area, which allows to meet the dry solid specification of 0.15% in area) while in the first large scale batch (30 kg batch size), the level of this impurity reached 30% area at reaction end-point, which did not allow its purge during downstream, failing to meet the impurity specification in the dry solid. The outline course of action was to build a kinetic model for the formation of product and the key impurity and use historical data to follow as close as possible the Bourne protocol (Bourne 2003) to understand the impact of mixing on the reaction outcome.

According to the reaction mechanism (**A** + **B** --> **P**, **P** + **B** --> **Imp**), the only way for very low levels of **Imp** impurity is for its formation reaction rate to be much lower than the product formation reaction rate, since they are competing-consecutive reactions. Even though this seems obvious, we wanted to confirm this hypothesis with a kinetic model.

A set of experiments was carried out to better understand the formation of the impurity, namely adding to a solution of product **P**, a solution of 1 eq. of reagent **B**, at two temperatures; Additionally, four other experiments were carried out namely adding a solution of **B** (1eq, 3 vol) to a 3 vol solution of starting material **A** at the same two temperatures and two feeding durations (2h and 1h, respectively), and the reverse addition (adding a solution of **A** to a solution of **B**, with the same temperature and feeding durations described previously). Furthermore, we also added data from a DoE (which explored reaction temperature, agitation speed and feeding duration) to give more robustness to the kinetic model. Data obtained from an RC1 experiment was used as well to assess the reaction exothermicity and, thus, determine the ΔHr, which is a critical aspect for scaling-up this reaction.

To assess the impact of the agitation on the reaction, we used some of the experiments from the previously described DoE, to apply the Bourne protocol (Bourne 2003), and thus help provide the needed insights to understand which of the micro-, meso- or macro-mixing mechanism would be the controlling mechanism, and thus have a science based approach to scale-up the mixing conditions of this reaction.

It was proven that by resorting to a mechanistic modelling approach and a reduced number of experiments we were able to successfully investigate the root cause of the failed batch and scale-up the reaction to large scale conditions and avoid further failed batches in the future.

References

Bourne, John R. 2003. "Mixing and the Selectivity of Chemical Reactions." *Organic Process Research and Development* 7 (4): 471–508. https://doi.org/10.1021/op020074q.

610202 Molecular Recognition and *In Vivo* Detection of Temozolomide and 5-Aminoimidazole-4-Carboxamide for Glioblastoma Using Fluorescent Nanosensors

Abstract:

The efficacy of chemotherapeutics such as temozolomide (TMZ) and its metabolic product 5-aminoimidazole-4-carboxamide (AIC) is often affected by the timing, quantity and frequency of dosages. There is strong interest in facilitating the ability to monitor efficacy in individual patients for specific subtypes of cancer. Real-time, dynamic measures of potency may also supplement or in some cases replace reliance on bio-imaging. Towards this end, in this work we develop new, synthetic molecular recognition sites for TMZ, and its decomposition product AIC, grafting them onto near infrared fluorescent nanoparticles capable of forming optode or other biosensor interfaces to monitor drug efficacy in real-time. Infrared fluorescent single-walled carbon nanotubes, wrapped using specific DNA oligonucleotides, are encapsulated within biocompatible, poly(ethylene glycol) diacrylate hydrogels and enable the selective recognition of an anti-cancer drug, TMZ, on U-87 MG human glioblastoma cells. In both solution phase and hydrogel form, the sensors exhibit a fluorescence response to TMZ with a detection limit of 30 M. Furthermore, U-87 cells exhibit no changes in viability for 7 days in contact with the hydrogel. The sensors were used to track the progression of glioblastoma death following TMZ administration. In addition, the sensors were able to monitor *in vivo* AIC and TMZ dynamics in SKH-1E mice. These results enable real-time detection of chemotherapeutic concentration and metabolism at the cellular and subcellular levels, with potential to increase the efficacy of cancer treatments.

606459 An Enzymatic Electrochemical Biosensor for Real-Time Detection of Physiologically Relevant Nicotine Concentrations

Abstract:

The glucose biosensor, built upon the redox enzyme glucose oxidase, is the most commercially successful and studied enzymatic biosensor. However, the lack of available and functionally validated enzymes is prohibiting the development of redox-based sensors for other important analytes. Herein, we present the development and assessment of an electrochemical nicotine biosensor, using genomic screening to identify the gene for a known nicotine catabolizing redox enzyme. The resulting nicotine biosensor demonstrated a specific, sensitive, and stable operational profile with a limit of detection of 27 μM over the range of 0-200 μM. This range is well within the physiological concentrations of nicotine present in smoker urine. Specificity and cross-reactivity were measured against structurally similar compounds to nicotine as well as to known physiological by-products. Our results highlight that this novel enzymatic electrochemical nicotine biosensor possesses operational capabilities for monitoring of nicotine in physiologically relevant conditions. The screening methodology can be generalized for the discovery of enzymes for novel sensor development.

608267 Simulations Guide Optimization of Electroenzymatic Biosensors for Neurotransmitters and Enable Proper Interpretation of Sensor Response *In Vivo*

Abstract:

Electroenzymatic biosensors for neurotransmitters are uniquely compact, sensitive, and selective devices that provide the neuroscience community with a method to monitor fluctuating chemical concentrations in the brain. Many studies primarily consider glutamate, the most common excitatory neurotransmitter, and study the relation between sub-second neurotransmitter dynamics and various stimuli, behaviors, or diseases. Challenges, however, have remained in the optimization of sensor design and in clarifying the significance of assumptions made during *in vivo* use. Accordingly, a number of simulations were carried out to guide the development of glutamate sensors and to test the limitations of the most prominent assumption, that sensor response can be linked to the immediate extracellular neurotransmitter concentration using a predetermined calibration factor. Recommended modifications to sensor design were developed and successfully implemented, improving sensitivity six-fold (up to 320 nA/μM/cm^2) with an accompanying order of magnitude decrease in response time (down to ~80 ms). Since even this improved response time is an order of magnitude slower than the timescale of synaptic neurotransmitter release, it is clearly wrong in many cases to assume that the use of a calibration factor to translate sensor response to neurotransmitter concentration without any signal deconvolution is accurate. To investigate the significance of this assumption and the conditions for which this practice remains reliable, the model was expanded with a new set of partial differential equations to describe the extracellular space of the brain. Considering the effects of variations in the local biological environment bordering a sensor, a range of representative parameters for some biological processes were selected. Comparisons were made between the theoretical responses expected from *in vitro* calibrations, tonic steady-state glutamate presence, and bolus-type glutamate releases that mimic synaptic release. Surprisingly, in steady-state cases, the sensor response could indicate a concentration over twice the actual value, while in cases where sensors were 10 μm or more from the source of glutamate being detected, sensor response was < 50% of the expected value. In some cases, transient concentrations due to a single synaptic-type release of neurotransmitters at a moderate distance from the sensor were not detected by the sensor at all, and when they were, the time course of signal decay did not match the rates of reuptake expected in the brain without significant signal deconvolution. Ultimately, results show that additional characterization of sensor performance *in vivo* may be necessary to enable proper interpretation of sensor data.

611068 Theoretical and Experimental Studies on an Electrochemical Enzyme Immunosorbent Biosensor

Abstract:

Electrochemical immunosensors integrate biological recognition molecules (e.g., antibodies) with redox enzymes (e.g., horseradish peroxidase (HRP) to combine the advantages of immunoassays (extremely high sensitivity and selectivity) with those of electrochemical biosensors (reproducible, quantitative electrical output). Mechanistic mathematical models that describe the multiple mass-transfer and chemical reaction steps that give rise to the electrical output are needed to help design, optimize, and validate electrochemical immunosensors for medical applications. In this study, we developed such a model of an electrochemical immunosensor that used HRP as a reporter enzyme on the secondary and validated it experimentally.

The model consists of a system of non-linear mass-balance equations that describe simultaneous mass transfer and chemical reactions of the reactants (catechol and hydrogen peroxide) that HRP oxidizes, as well as the HRP product, O-quinone, which is reduced at the working electrode to give the amperometric output. A bi-substrate, ping-pong reaction mechanism was used to describe the enzymatic reaction, and the *Butler-Volmer* kinetic model was used to describe the reduction of o-quinone back to catechol at the working electrode's surface. The system of equations was solved numerically using the BVP4C function in MATLAB.

Experimental data for model validation was obtained using electrochemical immunosensors for a surrogate protein antigen (mouse IgG). The immunogens interface was fabricated on the gold working electrode of DropSens screen printed electrode arrays. The steady-state amperometric output of the resulting immunosensors was measured over a range of four independent variables: the concentrations of the two reactants, the voltage of the working electrode, and pH. The experimental results were used to determine optimal operating conditions and to validate the model. The validated model was then used to calculate Damkohler numbers and flux-control coefficients to identify mass-transfer and reaction steps that most strongly affected the magnitude of the immunosensors' output and the output's sensitivity to analyte concentration.

610485 HTL Derived Biochar and Graphene Nanoplatelets for Biosensor Applications

Abstract:

HTL Derived Biochar and Graphene Nanoplatelets for Biosensor Application

Bharath K Maddipudi, Vinod Amar, Hope Dosch, Anuradha Shende and Rajesh V Shende*Department of Chemical and Biological Engineering, South Dakota School of Mines and Technology, Rapid City, South Dakota 57701, United States of America

***Corresponding Author:** Rajesh Shende, Department of Chemical and Biological Engineering, South Dakota School of Mines and Technology, Rapid City, South Dakota 57701, United States of America (E-mail: *Rajesh.Shende@sdsmt.edu*)

Abstract

Biosensors with Au electrode layer immobilized with peptides are currently being developed for the detection of microbial contamination. It is anticipated that the use of a low-*k* dielectric layer modified with graphene will improve the sensor sensitivity. The low-*k* dielectric layer can be modified with the biochar derived from hydrothermal liquefaction of different biomass substrates. In this investigation, HTL of lignocellulosic substrates was performed under non-catalytic and catalytic conditions and by-products were analyzed. The biochar was thermally treated in a tubular reactor under inert environment from 400-1100°C without any pretreatment. In other set of experiments, HTL derived biochar was acid treated and subsequently, thermal activation was performed. After the acid pretreatment and thermal treatment steps, the porosity, the specific surface area (SSA) and the morphology of the char was analyzed using BET and SEM/EDX. The char obtained after thermal treatment was combined with low-*k* dielectric layer and biosensor was fabricated. The sensor output was compared with the biosensor fabricated with low-*k* and graphene nanoplatelets. In addition, the sensor output was correlated with the char materials obtained under different thermal conditions. The results obtained on hydrothermal liquefaction of biomass substrates, characteristics of biochar before and after thermal treatment, biosensor fabrication and sensor output will be presented.

604780 Construction of a Red Emission Bodipy-Based Probe for Tracing Lysosomalviscosity Changes in Culture Cells

Abstract:

A red emission BODIPY-based lysosome-targeted viscosity probe (Lys-VBOD) was synthesized for monitoring lysosomal viscosity in live cells. The increase of viscosity can restrict the twisting of double bond between indole
and BODIPY fluorophore, also restrict the rotating of single bond between quinoline and BODIPY, which lead to the enhancement of fluorescence intensity in high viscosity environment. The changes of fluorescence intensity
exhibit a good linear relationship with viscosity values. Besides, Lys-VBOD is quite stable in various pH solvents with given viscosity, suggesting that Lys-VBOD is suitable for applying in a wide pH range. Moreover, Lys-VBOD
can visualize the changes of lysosomal viscosity in live cells under the stimulus of dexamethasone. BSA titration experiment demonstrated that Lys-VBOD is stable in the presence of macromolecules and proteins. Notably, the
fluorescence lifetime of Lys-VBOD in large viscosity solvents is extremely high. These results indicate that LysVBOD is an excellent tool for detecing viscosity changes in vitro and live cells.

610593 Super Sensitive Cerium Oxide-Based Composite Sensor for the Detection of Hydroxyl Radicals

Abstract:

Free radicals are recognized as essential molecules for upholding the function of normal cells. Immune systems, such as bacteria elimination and virus inactivation, are examples of the crucial role that free radicals play in human cells. Even though free radicals have benefits, an optimum production is key to maintain their advantages, as an excess can cause the development of oxidative stress conditions, such as cancer, Alzheimer's, and Parkinson's diseases. Among all free radical species, hydroxyl radicals (•OH) are the most reactive and dangerous, and they can be used as biomarkers for the detection of those diseases at their initial stages. Thus, a very sensitive analytical sensor is needed to detect and monitor low concentrations of •OH. The integration of an electrochemical technique with a sensing element is regarded as a promising method for •OH detection due to a rapid and direct measurement without the pretreatment of samples. Thus, in this study, a glassy carbon electrode was modified with a sensing element comprised of ultrasmall cerium oxide nanoclusters (<2 nm) grafted onto a superconductive carbon (Figure 1a). XPS results showed 46.3% Ce^{3+} sites, which are those able to react with •OH. Cyclic voltammetry (CV) was implemented to characterize the interaction of the composite sensor with •OH, and they showed that the composite sensor was able to differentiate between solutions with and without •OH by an increased redox response in presence of •OH (Figure 1b). Furthermore, we observed a linear relationship between the redox response and the concentration of •OH in the range of 0.1 mM to 10 mM, with a limit of detection as low as 0.1 μM when using only 0.5 wt% CeO_x loading.[1]

(1) A.C. Alba-Rubio, D.S. Kim, S. Duanghathaipornsuk; Super sensitive sensor for the detection of hydroxyl free radicals with scavenging properties. Application number 62/977,945 (2020).

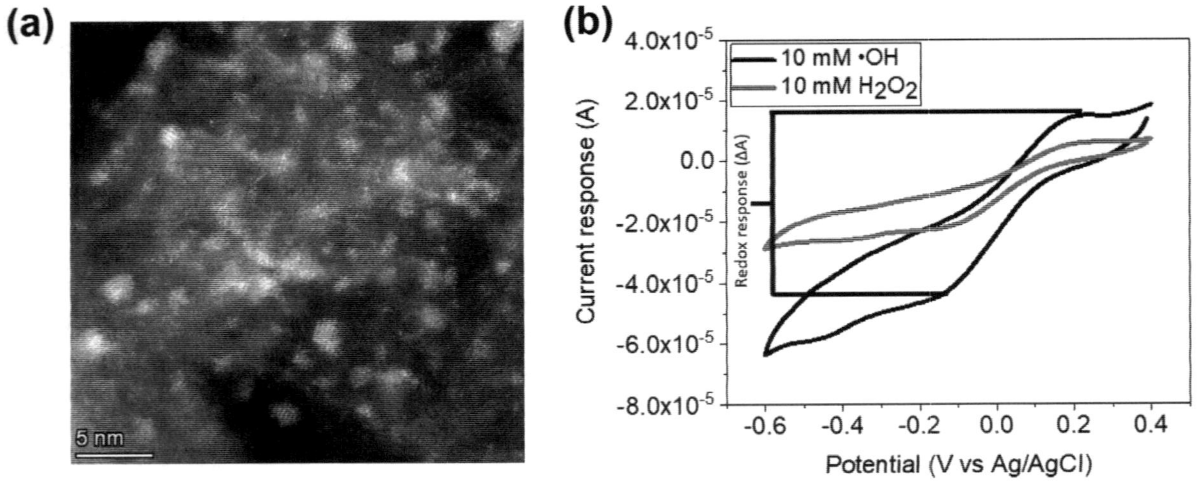

Figure 1. (a) STEM image of 0.5 wt% CeO_x/Carbon and (b) Current response to hydroxyl radicals.

611380 High Throughput Antibiotic Susceptibility Testing with Optical Nanosensors

Abstract:

Infections in a clinical setting can result from biofilm-forming species, which presents two key challenges: increased antimicrobial resistance, and poor diagnostic approaches to categorize antimicrobial susceptibility. Traditional antimicrobial susceptibility assay approaches are built upon planktonic bacteria (bacteria that are growing in solution). Biofilms are bacteria that grow in a community on a surface, and this community structure changes resistance to antimicrobials that is not reflected in these planktonic assays. While current biofilm-based assays exist, they are both time and material intensive, and can still provide antibiotic susceptibility results that may not reflect the actual biofilm response. Here we present our nanosensor based approach to monitor biofilm metabolic response during the administration of antibiotics. We grow biofilms with oxygen-monitoring nanosensors embedded throughout the film, and these nanosensors continuously report out the metabolic state of the biofilm. As we add antibiotics, we can measure what concentration can stop metabolism, and infer the minimum biofilm inhibitory concentration (MBIC) of the specific strain. Our diagnostic approach functions in 96 well plates with minimal materials and time (important for high throughput screening), and works with both lab and clinically relevant strains (demonstrated with Pseudomonas aeruginosa).

599605 Data-Driven Process Development and Scale-up of an Acetal Protection Reaction and Crystallization

Abstract:

Integrated process design provides many benefits such as supplying fundamental process knowledge, identifying important scaling parameters to ensure quality performance upon scale-up, and detecting safety and quality risks within the process. Herein, the principles of integrated process design were employed in the development and scale-up of an acetal protection reaction and subsequent crystallization.

The challenges associated with the acetal protection reaction were instability of the product at end-of-reaction, solvent related impurity formation, and side reaction between reagents. Kinetic, thermodynamic, and multivariate statistical models, constructed through data-rich experimentation using parallelized reactors equipped with automated sampling, elucidated key sensitivities controlling reaction rates and selectivity. This enabled optimization of the reagent loadings and conditions to achieve reasonable reaction rates and stability at the end of reaction. In the subsequent crystallization, particle agglomeration and attrition caused slow filtration. Process analytical technology in scale-down studies of shear rates revealed shear sensitivity and particle attrition of agglomerates as well as the effect of the particle size distribution on filtration rate. Mixing models guided manufacturing operating conditions to minimize particle nucleation and attrition to achieve fast filtration. These process conditions were successfully scaled-up.

599444 Model Development for Enzymatic Reactive Crystallization of β-Lactam Antibiotics: A Reaction-Diffusion-Crystallization Approach.

Abstract:

The β-lactam antibiotics amoxicillin and cephalexin are among the most widely used antibiotic Active Pharmaceutical Ingredients (API) [1]. Penicillin G Acylase (PGA) can catalyze the reaction between an activated acyl donor (e.g. 4-hydroxyphenylglycine methyl ester) and a nucleophilic molecule containing the β-lactam core (e.g. 6-aminopenicillanic acid) to produce the final antibiotic (e.g. amoxicillin). Similar to many other industrial enzymatic processes, working with immobilized enzyme is essential for downstream separation of the catalyst from the product and retaining the valuable enzyme in continuous manufacturing processes. Working with immobilized enzyme necessitates understanding the effect of transport limitations on the enzyme activity and reaction rate. Moreover, this process meets difficulties such as potentially low conversion and yield due to hydrolysis of the API by the same enzyme such that the desired product is an intermediate in the overall reaction network. One way to improve this process and to protect the antibiotic from being hydrolyzed by the enzyme is isolating it through crystallization. The reactive crystallization also provides process intensification, as a separate downstream crystallization process for purification is no longer required [2].

The complex interplay of three phenomena determine the process attributes: enzyme-catalyzed chemical reaction, species transport into and out of the biocatalyst, and crystallization of the API molecule. In this talk, we will describe our efforts to develop a mathematical model for the enzymatic synthesis of β-lactam antibiotics using immobilized enzyme. The effective diffusion coefficients for two types of commercially available supports (agarose and methacrylic-based beads De=1-5×10^-10 m2/s) were measured experimentally and the results were used to describe the effect of pore diffusion on the apparent enzyme activity and reaction rate. Furthermore, coupling the reaction-diffusion module with a crystallization population balance, a mathematical model that encompasses all three phenomena is developed. Application of the formulated model for model-based design and optimization will be discussed. In particular, this approach can aid in design of a biocatalyst to optimize the process attributes of conversion, selectivity, and productivity. To optimize the biocatalyst design into a systematic procedure, three directly controlled variables for biocatalyst are defined: (1) enzyme loading, (2) support type, and (3) support size. The interaction among these factors can be probed using the model for selecting the optimal immobilization condition. As one might expect, using an immobilization support with a radius smaller than a critical value Rc = F(enzyme kinetic, support porosity), selectivity and conversion comparable to that of the soluble enzyme can be achieved (Fig.1). On the other hand, increasing the biocatalyst size above an API-dependent value leads to significant mass transfer limitations and deviation from inherent kinetics. Furthermore, our analysis illustrates that two often neglected nonidealities in the enzyme immobilization procedure, nonuniform enzyme distribution and variance in bead size distribution, may affect the apparent reaction rate and the catalyst performance and cause deviations from targeted operating point (Fig.2).

[1] Giordano, R. C., Ribeiro, M. P., & Giordano, R. L. (2006). Kinetics of β-lactam antibiotics synthesis by penicillin G acylase (PGA) from the viewpoint of the industrial enzymatic reactor

optimization. Biotechnology advances, 24(1), 27-41.
[2] McDonald, M. A., Bommarius, A. S., Rousseau, R. W., & Grover, M. A. (2019). Continuous reactive crystallization of β-lactam antibiotics catalyzed by penicillin G acylase. Part I: Model development. Computers & Chemical Engineering, 123, 331-343.

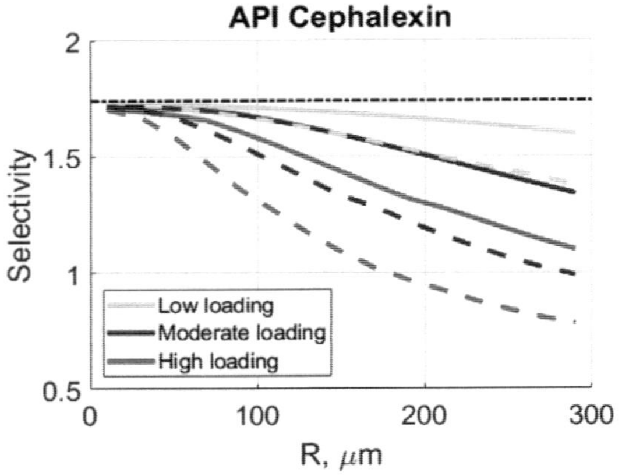

Fig. 1. Selectivity of the API to API hydrolysis product as a function of support size and enzyme loading in biocatalyst. Larger support size results in significant mass transport limitations, enzyme hydrolyzing the API before it leaves the catalyst environment. Horizontal line shows soluble enzyme values, solid curves correspond to $D_e = 5\times10^{-10}$ m^2/s and dashed curves to $D_e = 1.3\times10^{-10}$ m^2/s.

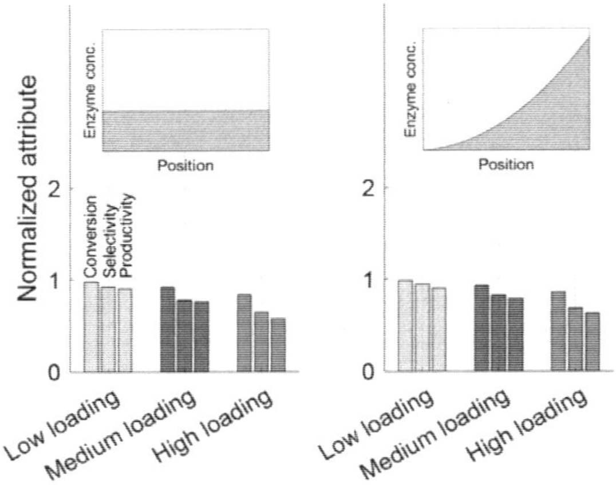

Fig. 2. Effect of nonideal enzyme distribution profile on process attributes at different enzyme loadings for a support of fixed radius.

600860 Optimization of Process Risk, Cost, and Environmental Sustainability during Late Stage Drug Substance Control Strategy Development

Abstract:

During late stage pharmaceutical process development, objectives such as cost, environmental sustainability, and risk to product quality, are important to consider. Optimization at this stage balances these objectives against the complexity of evaluating and incorporating process adjustments. Here, we present a series of process improvements made during late stage development of a small-molecule drug substance. These improvements were supported by empirical and mechanistic process understanding incorporating measures of risk, cost, and sustainability.

To direct the investment of limited development resources, a quantitative analysis was done to calculate key process metrics. A key insight from this analysis was that the process to produce one specific intermediate was estimated to account for three quarters of the drug substance's manufacturing cost. This analysis directed investment into improving the process yield across the eight chemical transformations, involving a wide variety of unit operations to make this intermediate. Specifically, significant cost reduction was achieved through multivariate optimization of an oxidation reaction, identification and mitigation of mass loss during extraction and distillation, and multivariate optimization of yield vs. impurity purge during crystallization.

Another improvement later in the synthesis involved the reduction of significant process risk present in a three-reaction telescope through the development of an additional isolation after the first reaction in this sequence. The reduced process risk was quantitated as a failure rate, using multivariate Bayesian modeling to ensure process robustness despite potential parameter variations. Finally, changes were made to the penultimate process step that improved sustainability and reduced cost without increasing process risk. These changes were enabled through generation of mechanistic models that enhanced kinetic and thermodynamic understanding of the reaction and the subsequent water uptake of the product. Developing fundamental engineering understanding and enhancing it with quantitative metrics both informs where to focus development resources and provides a fully comprehensive data package summarizing knowledge, residual risks, and documenting improvements in a format that is easy to communicate to stakeholders and enables quality by design.

607951 Digital Design of an Intensified Filtration-Drying Unit for Continuous Manufacturing

Abstract:

In recent years, pharmaceutical companies have been making a transition from traditional batch manufacturing to continuous production mode, due to the well-known advantages of continuous operation [1]. Even though continuous implementations already exist for many pharmaceutical unit operations [2-3], major limitations leave continuous end-to-end manufacturing lagging behind, especially regarding the isolation of the drug substance [4]. In this work we address the state-of-the-art challenges in continuous filtration and drying for upstream manufacturing through the modeling, design and optimization of a novel intensified filtration-drying carousel. The unit is based on the prototype manufactured by Alconbury Weston Ltd [5], which has already been successfully tested in an integrated continuous crystallization-filtration experimental framework [6-7]. The carousel consists of five cylindrical ports, where slurry loading, filtration, washing, deliquoring and drying are carried out simultaneously. After a fixed time interval, the carousel rotates, moving each port to the next processing position. In position 1 the slurry from the crystallizer is introduced into the port, while in positions 2-4 the filtration, washing and deliquoring steps are carried out, with the order designed by the operator. A final gas drying step is carried out in position 5, before the product is discharged. Strict conditions on the solvent and impurity content of the discharged dry cake must be satisfied. The introduction of the gas drying equipment is a novel feature of the carousel, which was not present in previous applications [5-7]. Operating the carousel requires deciding on the value of multiple decision variables, namely the pressure drop for filtration and deliquoring, the type and flowrate of the washing solvent and the flowrate and temperature of the gas flow for the drying step. Above all, due to the intrinsic carousel mechanism, the rotation time has an important impact on the final product quality. Hence, the total cycle time has to be regulated considering a trade-off between the residence time at the processing positions. All this considered, the traditional design, based on empirical knowledge, of the filtration-drying operating conditions is very challenging and time-consuming. For this reason, in this work we develop a comprehensive mathematical model for the unit, to investigate an optimal design for the operating conditions of the carousel. The employed equations are standard macroscopic and microscopic mass, energy and momentum balances. We consider the effect of the crystal size distribution of the solid phase coming from the crystallizer on the cake properties, such as porosity [8] and specific resistance [9]. Finally, upon validation with experimental data, we use the model for determining the optimal values of the decision variables of the carousel for different API and solvent systems. Future work will involve model-based online control of the carousel.

References

[1] Lee, S. L., T. F. O'Connor, X. Yang, C. N. Cruz, S. Chatterjee, R. D. Madurawe, C. M. V. Moore, L. X. Yu and J. Woodcock (2015). Modernizing Pharmaceutical Manufacturing: from Batch to Continuous Production. *J. Pharm. Innov.,* **10,** 191–199.

[2] Teoh, S. K., C. Rathi, and P. Sharratt (2016). Practical Assessment Methodology for Converting Fine Chemicals Processes from Batch to Continuous. *Org. Process Res. Dev.*, **20**(2), 414–431.

[3] Su, Q., M. Moreno, C. Laird, Z. Nagy and G. Reklaitis (2019). A perspective on Quality-by-Control (QbC) in pharmaceutical continuous manufacturing, *Comput. Chem. Eng.*, **125**, 216–231.

[4] Simon, L. L., A. A. Kiss, J. Cornevin, and R. Gani (2019). Process engineering advances in pharmaceutical and chemical industries: digital process design, advanced rectification, and continuous filtration. *Curr. Opin. Chem. Eng.*, **25**, 114–121.

[5] Ottoboni, S., C.J. Price, C. Steven, E. Meehan, A. Barton, P. Firth, A. Mitchell and F. Tahir (2019). Development of a Novel Continuous Filtration Unit for Pharmaceutical Process Development and Manufacturing. *J. Pharm. Sci.*, **108**(1), 372–381.

[6] D. Acevedo, R. Peña, Y. Yang, A. Barton, P. Firth, and Z. K. Nagy (2016). Evaluation of mixed suspension mixed product removal crystallization processes coupled with a continuous filtration system. *Chem. Eng. Process. Process Intensif.*, **108**, 212–219.

[7] Liu, Y. C. , A. Domokos, S. Coleman, P. Firth, and Z. K. Nagy (2019). Development of Continuous Filtration in a Novel Continuous Filtration Carousel Integrated with Continuous Crystallization- *Org. Process Res. Dev.*, **23**(12), 2655–2665.

[8] Ouchiyama, N., and T. Tanaka (1986). Porosity estimation from particle size distribution. *Ind. Eng. Chem. Fundam.*, **25**(1), 125–129.

[9] Bourcier, D., J.P. Féraud, D. Colson, K. Mandrick, D. Ode, E. Brackx and F. Puel (2016). Influence of particle size and shape properties on cake resistance and compressibility during pressure filtration. *Chem. Eng. Sci.*, **144**, 176–187.

600793 Integrated Filtration and Washing Modelling to Predict Continuous Isolation Performance

Abstract:

The pharmaceutical industry is starting to adopt continuous active pharmaceutical ingredient (API) manufacturing in order to reduce production costs, improve manufacturing flexibility, reduce infrastructure costs, reduce manufacturing lead time (from typically 6 months to 10 days) and to improve sustainability. A further driver is reduction of variance in API quality critical attributes. To facilitate the complete transition from batch to continuous manufacturing it is necessary to "smartly" integrate single continuous unit operations to achieve a continuous material flow from synthesis to formulation. To achieve this smart integration of unit operations, a combination of modelling, online measurement and advanced control techniques are vital to predict product property outcomes, to monitor and control processes and to reduce the risk of non-conforming products.

Another challenge the pharmaceutical industry faces is to reduce the quantity of material consumed during process development. A stretch goal is to consume just 100g of API (and the corresponding precursors) and to complete development in 100 days. Digital design of continuous API manufacturing offers a path to achieving this goal. This includes modelling and predicting process performance as a function of the operating conditions for both individual continuous unit operations and for the integrated processes with the aim of optimizing process design and reducing the laboratory time and cost needed to develop new products. Whilst a few examples of modelling integrated continuous unit operations using flowsheet models, have been published, these are mainly focused on secondary drug product manufacture rather than API synthesis and isolation.

The team in the Continuous Manufacturing and Advanced Crystallization (CMAC) Future Manufacturing Research Hub have addressed this gap by modelling primary drug manufacturing using an integrated unit operations approach. A digital tool capable of transferring material property information between operations to predict the product attributes in integrated synthesis and purification processes has been developed. The focus of the work reported here combines filtration and washing operations used in API purification and isolation by combining predicted and experimental data generated during upstream crystallization process. The model approach described is subdivided into two levels of increasing system complexity where the filtration model approach is held constant, while different washing modelling approaches are used for the two levels in order to describe different washing scenarios. The filtration model approach considers the case where the different components of the input suspension are separated between the retained filter cake with residual impure mother liquor and the impure liquid filtrate removed during filtration. The assumptions considered for the washing modelling described in level 1 are that the impure mother liquor is fully miscible with the wash solvent selected and no changes in solid phase are considered (no particle dissolution or growth). Level 1 thus can describe three different washing mechanisms: pure displacement (level 1a), dilution with perfect liquid mixing (level 1b) and dilution with axial dispersion (level 1c). The more complex level 2 allows, additionally, the possibility of cake and impurity dissolution or precipitation during washing. The level 2 washing model considers cake dissolution due to solubility variation during washing, the possibility of API or impurity precipitation from solution due to a drastic solubility drop (the antisolvent effect occurring during washing) as an instantaneous process. This level presumes that kinetic aspects can be neglected and equilibrium is reached instantaneously. In detail; the level 2a model describes a washing process where the system is assumed to be homogeneous and the same solvent composition is assumed for the entire cake. In level

2b, a more realistic washing mechanism is modelled in which the system shows vertical heterogeneity and there is a composition gradient along the cake height, which is described in the model

The integrated modelling tool uses information on the product crystal suspension characteristics predicted using gPROMS/Matlab to predict filtration time, filtrate flow rate and the composition of the filter cake and filtrate generated during filtration. The washing of the wet filtered cake is then simulated to predict; washing efficiency and to generate washing curves, cake and filtrate composition, indicate the probability of particle size variation caused by cake dissolution, and residual cake moisture content and composition. To validate the scenarios described using the integrated models a design of experiments (DoE) approach was used and two experimental case studies were investigated using mefenamic acid and paracetamol as representative isolation processes. In each case the objective of the work was to meet a product purity specification and minimize changes to the crystalline particle attributes occurring during the isolation process. The validation is also useful to identify which model level best reflects the experimental data, with the aim of identifying the minimum criteria a filtration and washing model needs to have to simulate experimentally verified isolation outcomes.

604953 Comparison of Surrogate Modeling Techniques for Design Space Approximation and Surrogate-Based Optimization: Effect of Sampling Technique and Sample Size

Abstract:

Comparison of Surrogate Modeling Techniques for Design Space Approximation and Surrogate-Based Optimization: Effect of Sampling Technique and Sample Size

Bianca Williams, Selen Cremaschi

Session: Advances in Machine Learning and Intelligent Systems

Surrogate models, also known as response surfaces, black-box models, metamodels, or emulators, are simplified approximations of more complex, higher order models. These models are used to map input data to output data when the actual relationship between the two is unknown or computationally expensive to evaluate (Han & Zhang, 2012). Surrogate models can also be constructed for use in surrogate-based optimization when a closed analytical form of the relationship between input data and output data does not exist or is not conducive for use in traditional gradient-based optimization methods. Surrogate modeling techniques are of particular interest where high-fidelity, thus expensive, simulations are used (Han & Zhang, 2012) or when the fundamental relationship between the design variables and output variables is not well understood, such as in the design of cell or tissue manufacturing processes (Du et al., 2016).

With all the surrogate modeling techniques currently available, there is a need for a systematic procedure for selecting the appropriate technique for a given application. Current common practices for selecting the appropriate surrogate model form rely on process-specific expertise. Numerous studies have been conducted comparing the performance of surrogate modeling techniques for approximation purposes (Bhosekar & Ierapetritou, 2018; Davis et al., 2017). The majority of these only compare a few models on a limited number of data sets or for specific applications (Ju et al., 2016; Luo & Lu, 2014). Progress has been made in recent works in generalizing the process for selecting a surrogate model to approximate a design space by using meta-learning approaches to build selection frameworks (Cui et al., 2016; Garud et al., 2018), avoiding expensive trial-and-error methods. However, few of the developed meta-learning frameworks take model complexity into account, which can lead to overfitting, or consider that multiple models might perform similarly to the one identified as best in terms of their accuracies. The selection of surrogate models for surrogate-based optimization remains an open challenge.

Our goal is to comprehensively investigate and compare the performance of several different surrogate modeling techniques for both approximating functional relationships and for surrogate-based optimization to link that performance to the characteristics of the data involved in the application. Previous work on this topic has shown that the performance for approximation is dependent on data characteristics such as the input dimension and the underlying function shape (Davis et al., 2017; Williams & Cremaschi, 2019). The specific data characteristics being investigated in this study are the shape of the underlying function being modeled, the number of input dimensions, the sampling method used to generate the data, and the number of sample points in the dataset. The surrogate-modeling techniques considered include Artificial Neural Networks, Automated Learning of Algebraic Models using Optimization (ALAMO), Radial Basis Networks, Extreme Learning

Machines, Gaussian Progress Regression, Random Forests, Support Vector Regression, and Multivariate Adaptive Regression Splines (MARS). These techniques are used to construct surrogate models for data generated using the 47 optimization challenge functions from the Virtual Library of Simulation Experiments (Surjanovic & Bingham, 2013). The sampling methods studied are Sobol sequence sampling, Halton sequence sampling, and Latin Hypercube sampling. Four performance measures are used to evaluate the accuracy of the surrogate models: root mean squared error (RMSE), maximum percent error (MPE), the R-squared value, and the adjusted R-squared value. The surrogate models' ability to locate the extrema of the functions are evaluated by calculating the distance between the extreme point(s) estimated by the model and the actual function extrema. The results provide guidance on which surrogate models generate the best predictions and give general "rules of thumb."

Using information extracted from the surrogate modeling comparison experiments and building upon previous meta-learning approaches, we constructed a tool to provide recommendations for the appropriate modeling techniques for the datasets based only on the characteristics of the data being modeled. Characteristics, i.e., attributes, were calculated for each dataset with the goal of representing its overall behavior. Attributes were calculated based only on input and output values in the dataset. The attributes that have the strongest relationships with the performance metrics are determined using feature reduction methods, including the ReliefF algorithm (Kira & Rendell, 1992) and principal component analysis (Hotelling, 1933). These attributes were used as inputs, with designated performance metrics as outputs, to train models to make predictions on the performance of the surrogate modeling techniques. The performance metrics used as outputs for training the recommendation tool are the adjusted R-squared value for the design space approximation application and the normalized distance between the extreme point(s) estimated by the models and the actual extrema of the true model for surrogate-based optimization. The adjusted R-squared value takes into account both the surrogate model accuracy and its complexity (Miles, 2005). The tool identifies which surrogate modeling techniques are recommended for use for either approximating a design space or for surrogate-based optimization given a set of data.

References:

Bhosekar, A., & Ierapetritou, M. (2018). Advances in surrogate based modeling, feasibility analysis, and optimization: A review. Computers & Chemical Engineering, 108, 250-267.

Cui, C., et al. (2016). A recommendation system for meta-modeling: A meta-learning based approach. Expert Systems with Applications, 46, 33-44.

Davis, S., et al. (2017). Efficient Surrogate Model Development: Optimum Model Form Based on Input Function Characteristics. In A. Espuna, M. Graells & L. Puigjaner (Eds.), 27th European Symposium on Computer Aided Process Engineering (ESCAPE 27) (Vol. 40, pp. 457-462). Barcelona, Spain: Elsevier.

Du, D., et al. (2016). Statistical Metamodeling and Sequential Design of Computer Experiments to Model Glyco-Altered Gating of Sodium Channels in Cardiac Myocytes. IEEE J Biomed Health Inform, 20, 1439-1452.

Garud, S. S., et al. (2018). LEAPS2: Learning based Evolutionary Assistive Paradigm for Surrogate Selection. Computers & Chemical Engineering, 119, 352-370.

Han, Z., & Zhang, K. (2012). Surrogate-Based Optimization. In O. Roeva (Ed.), Real-World Applications of Genetic Algorithms (pp. 343-362). Rijeka, Croatia: InTech Open.

Hotelling, H. (1933). Analysis of a complex of statistical variables into principal components. Journal of Educational Psychology, 24, 498-520.

Ju, Y. P., et al. (2016). Artificial intelligence metamodel comparison and application to wind turbine airfoil uncertainty analysis. Advances in Mechanical Engineering, 8.

Kira, K., & Rendell, L. A. (1992). A Practical Approach to Feature-Selection. Machine Learning /, 249-256.

Luo, J. N., & Lu, W. X. (2014). Comparison of surrogate models with different methods in groundwater remediation process. Journal of Earth System Science, 123, 1579-1589.

Miles, J. (2005). R Squared, Adjusted R Squared. In Encyclopedia of Statistics in Behavioral Science: John Wiley & Sons Ltd.

Surjanovic, S., & Bingham, D. (2013). Virtual Library of Simulation Experiments. In (Vol. 2018). Simon Fraser University.

Williams, B., & Cremaschi, S. (2019). Surrogate Model Selection for Design-Space Approximation and Surrogate-Based Optimization. In S. Munoz, C. Laird & M. Realff (Eds.), Ninth International Conference on Foundations of Computer-Aided Process Design (FOCAPD-19) (pp. 353-358). Copper Mountain, CO, USA: Elsevier B.V.

609659 Investigating Performance Metrics and Machine Learning Models Towards Automating the Diagnosis of Membranous Nepthropathy

Abstract:

Membranous nephropathy (MN) is a glomerular disease that causes nephrotic syndrome among adult patients. Records indicate that approximately one third of the patients that possess MN further progress into end-stage renal diseases in 5 to 15 years; among patients that are above 60 years old, 20-30% of the cases are found to be associated with malignancy such as lung cancer and prostate cancer. [2][3]

Diagnosis of MN is exclusively achieved by renal biopsy with a focus on the glomerular basement membrane area, which involves examination to the images of histochemically stained tissue samples captured by either light microscopy (LM) or electron microscopy (EM) [1]. For precise identification of glomerular features, the pathoanatomical examination often involves using whole-slide images (WSIs) of tissues stained in multiple histochemical approaches, which is a labor-intensive process.

Motivated by this, we aim to develop an automated examination method based on artificial intelligence and investigate its efficacy towards reducing the workload of clinicians and improving the consistency of diagnostic results. The basis for this effort is artificial intelligence algorithms to design an efficacious data-driven classification model.

In this work, a Convolutional Neural Network (CNN)[4] based model was structured and trained to determine the existence of MN within the images of glomeruli. To further address the issue of model accuracy not satisfying the requirement of the application, a method called high confidence acceptance (HCA) was designed to boost the precision of the model to an application specified target of 95\%. With HCA, we are aiming to use only strongly confident predictions as automated diagnostic results, leaving the relatively lower confidence instances to manual inspections performed by renal pathologists. In such scenario, a partial replacement of the manual examination by renal pathologists at the demanded accuracy can be accomplished. A metric called Acceptance Rate (AR) was accordingly defined and used in our scheme to measure HCA performance; AR represents the percentage of workload being automated and replaced by the CNN model.

As a result, the model trained in this work which initially had an overall accuracy of approximately 86% demonstrated an AR of approximately 34%, with a precision that met target accuracy as high as 95%. This result implied that one third of the workload will be fully automated at target accuracy, and the rest of the less confident instances (declared inconclusive) will be sent to renal pathologists for manual inspection. Furthermore, the model developed in this work also presented an acceptance rate of 23% when applied to a dataset solely comprised of HE (Hematoxylin-Eosin) stained biopsy images, which indicated a potential reduction to the staining effort which conventionally requires up to 4 different staining methods and usage of excessive specialist labor. A further test on images of larger field of view was also carried out; a performance consistent with tests on glomeruli-level images was obtained and the practical value of the model and HCA method is thus illustrated.

The result illustrated the feasibility of automatically diagnosing membranous nephropathy with CNN, which has considerable value to the relevant medical profession. Meanwhile, the HCA method is an universal technique that can be applied to improve and measure the applicability of any classification model that is not satisfying the required accuracy of the target application, and may thus further broaden the horizon of artificial intelligence in practical implementation.

References:

[1] Agarwal, S. K., Sethi, S., & Dinda, A. K. (2013). Basics of kidney biopsy: A nephrologist'sperspective.Indian Journal of Nephrology,23(4), 243–252. doi: 10.4103/0971-4065.114462

[2] Couser, W. G. (2017). Primary membranous nephropathy.Clinical Journal of the AmericanSociety of Nephrology,12(6), 983–997. doi: 10.2215/CJN.11761116

[3] Lai, W. L., Yeh, T. H., Chen, P. M., Chan, C. K., Chiang, W. C., Chen, Y. M., . . . Tsai,T. J. (2015, 2).Membranous nephropathy: A review on the pathogenesis, diagnosis,and treatment(Vol. 114) (No. 2). Elsevier. doi: 10.1016/j.jfma.2014.11.002

[4] LeCun, Y., Haffner, P., Bottou, L., & Bengio, Y. (1999). Object Recognition with Gradient-Based Learning. In (pp. 319–345). Retrieved fromhttp://link.springer.com/10.1007/3-540-46805-619doi: 10.1007/3-540-46805-6{\}19

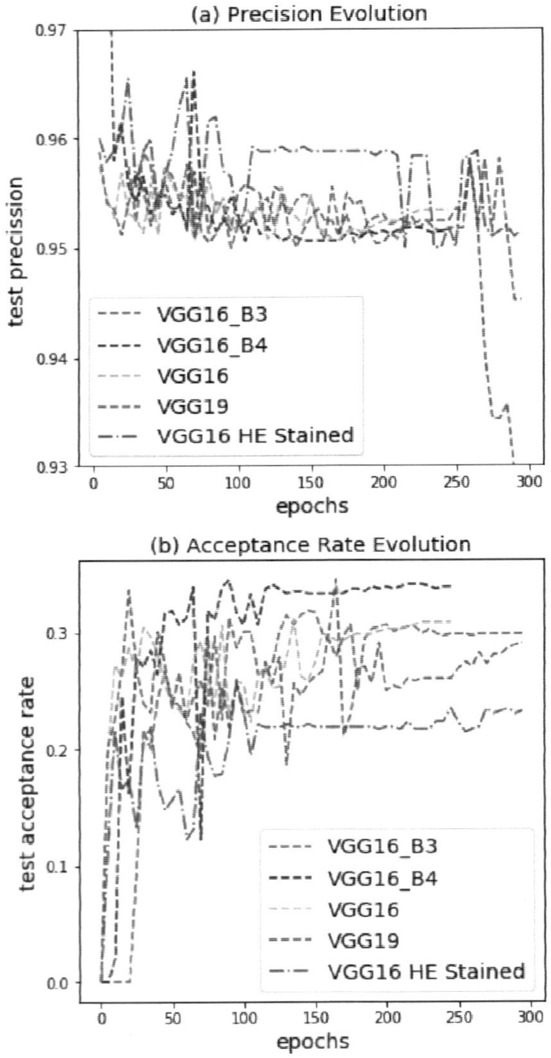

611386 High-Throughput Adsorbent Screening for Gas Separation Via Artificial Neural Networks

Abstract:

Pressure swing adsorption (PSA) is an advanced technology widely used for gas separation and storage applications. Several types of adsorbents including zeolites, metal organic frameworks (MOFs) and activated carbon have been characterized and synthesized with a wide array of applications in PSA operations [1–4]. Numerous hypothetical zeolites and MOFs can be also conceptualized [5,6]. The available feedstock specifications and the desired product specifications determine the choice of adsorbent and process design. A majority of the previous material-screening studies either use computationally expensive methods, and are able to screen only a few prospective adsorbents, or use significant modeling assumptions thereby leading to uncertainties in the reported results. This incentivizes the search of more predictive methods and metrics that can correctly estimate material performance in a real processing operation. We have developed a high-throughput material screening framework that is capable of quickly screening millions of potential adsorbent materials. The framework integrates a high-fidelity adsorption process model with an artificial neural network (ANN) model to describe breakthrough dynamics in adsorber columns. Detailed Grand Canonical Monte Carlo (GCMC) simulations are performed for obtaining equilibrium adsorption capacity data for different adsorbent-adsorbent pairs. We introduce ANN-based surrogate models for predicting breakthrough dynamics in an adsorber column. The resulting models have much less complexity compared to the detailed first-principles model, and can be leveraged for rapid screening of a large adsorbent database. The developed framework can be used to evaluate material separation performance for different process designs, operating conditions and adsorbent materials. For example, it can predict the so-called "breakthrough time" that can be used as a single metric to capture the effects of both adsorption capacity and selectivity [7]. Breakthrough time also influences the column size and regeneration frequency. This metric is shown to facilitate a more efficient and realistic screening of adsorbent materials as it combines the effects of adsorption capacity and selectivity to indicate PSA performance. To demonstrate the utility of the framework, we screen 196 pure-silica zeolites in the Structure Commission of the International Zeolite Association (IZA-SC) database for post-combustion carbon capture and natural gas purification. Using the framework, we have identified zeolites WEI and JBW as the top two candidate adsorbents, as they are consistently ranked among the top zeolites for both applications.

References

[1] L.J. Murray, M. Dincua, J.R. Long, Hydrogen storage in metal--organic frameworks, Chem. Soc. Rev. 38 (2009) 1294–1314.

[2] D. Farrusseng, Metal-organic frameworks: applications from catalysis to gas storage, John Wiley & Sons, 2011.

[3] A.U. Czaja, N. Trukhan, U. Müller, Industrial applications of metal--organic frameworks, Chem. Soc. Rev. 38 (2009) 1284–1293.

[4] J.-R. Li, R.J. Kuppler, H.-C. Zhou, Selective gas adsorption and separation in metal--organic frameworks, Chem. Soc. Rev. 38 (2009) 1477–1504.

[5] A.N. Dickey, A.Ö. Yazayd\in, R.R. Willis, R.Q. Snurr, Screening \ce{CO}2/\ce{N2} selectivity in metal-organic frameworks using Monte Carlo simulations and ideal adsorbed solution theory, Can. J. Chem. Eng. 90 (2012) 825–832.

[6] K. Sumida, D.L. Rogow, J.A. Mason, T.M. McDonald, E.D. Bloch, Z.R. Herm, T.-H. Bae, J.R. Long, Carbon dioxide capture in metal--organic frameworks, Chem. Rev. 112 (2012) 724–781.

[7] R. Krishna, J.R. Long, Screening metal-organic frameworks by analysis of transient breakthrough of gas mixtures in a fixed bed adsorber, J. Phys. Chem. C. 115 (2011) 12941–12950.

606814 A Hybrid Modelling Framework for on-Line Data-Driven Modelling of Particle Processes

Abstract:

Particles play a key role in many industrial productions, where particle processes are frequently used for removal of insolubles, product isolation, purification and polishing. Because particle processes are quite complex and fundamental understanding of underlying phenomena is lacking, control and monitoring these processes are often challenging tasks. To overcome these challenges, it has previously been examined whether hybrid modelling could capture particle dynamics and be used to optimize these processes. Shallow neural networks have here been used to estimate particle phenomena kinetics and combined with low-order population balance models to produce predictions of particle size evolution. During the last two decades, hybrid models have been employed in various particle processes, including crystallization operations [1,2,3] and a pharmaceutical milling process [4]. While these models have shown good predictive capabilities, they have been rather case-specific and been trained either using indirect process variables or using fairly small data sets of particle size distributions due to historical limitations of particle analysis methods. With recent developments in particle monitoring tools, such as dynamic image analysis and focused beam reflectance measurements, it is now possible to measure particle properties with a much higher frequency, allowing for better capture of the process dynamics.

In this work, a hybrid modelling framework is presented for particle processes that include all the common particle phenomena, including nucleation, growth, shrinkage, agglomeration and breakage. The hybrid model is trained using time-series data from an on-line/at-line particle analysis sensor, along with other measured process states that may have an impact on the phenomena kinetics. The particle phenomena kinetics are estimated using a deep neural network, where all the available measured and calculated process states are used as inputs. The kinetic model is combined with a medium-resolution discretized population balance model and mass balances, ensuring that physicochemical constraints are not violated. With the high flexibility and the automatic feature selection capabilities of the deep neural network, it becomes possible to model an arbitrary particle process without relying on excessive prior process understanding. The only requirement is that the process dynamics related to the particle phenomena and the particle size distribution can be measured during the process.

The presented modelling approach utilizes automatic differentiation for training of the hybrid model, which significantly speeds up the model training from an order of O(2n) using a finite difference method to O(1), where n is the number of weights in the neural network. This opens up for training the hybrid model in real-time, where new sensor data can be continuously incorporated into the hybrid model during process operation. It furthermore allows for easy scaling of the neural networks to higher complexity without significant additional computational costs.

The versatility of the hybrid modelling framework is demonstrated through three case studies of varying nature and scale, where on-line/at-line dynamic image analysis has been used to measure particle properties during process operation. The case studies include a lab-scale crystallization of lactose, an industrial scale crystallization of an active pharmaceutical ingredient and a lab-scale flocculation and breakage of silica particles. We demonstrate how data quality and quantity affects the model performance and compare with traditional first principles modelling.

It is furthermore illustrated how computational chemistry methods can bring the framework an additional understanding of nanoscale particle interactions [5] by including first-principles model-based soft-sensors as inputs to the deep neural network. This includes estimations of solid-liquid interfacial tension (IFT), which can predict attractive or repulsive forces between particles in a robust way, much faster than experimental measurement of surface forces.

Lastly, a couple of future perspectives are presented, including the use of the hybrid model in model predictive control (MPC) where the hybrid model can be trained in parallel with solving the MPC problem.

[1] P. Lauret, H. Boyer, J. Gatina, Hybrid modelling of the sucrose crystal growth rate, Int. J. Model. Simul. 21 (2001) 23-29. doi:10.1080/02286203.2001.11442183.

[2] V. Galvanauskas, P. Georgieva, S. F. D. Azevedo, Dynamic optimisation of industrial sugar crystallizationprocess based on a hybrid (mechanistic+ann) model (2006). doi:10.1.1.124.7855.

[3] Y. Meng, S. Yu, J. Zhang, J. Qin, Z. Dong, G. Lu, H. Pang, Hybrid modeling based on mechanistic and data-driven approaches for cane sugar crystallization, J. Food Eng. 257 (2019) 44-55. doi:10.1016/j.jfoodeng.2019.03.026.

[4] P. K. Akkisetty, U. Lee, G. V. Reklaitis, V. Venkatasubramanian, Population balance model-based hybrid neural network for a pharmaceutical milling process, J. Pharm. Innov. 5 (2010) 161-168. doi:10.1007/s12247-010-9090-2.

[5] N. Nazemzadeh, L. W. Sillesen, R. F. Nielsen, M. N. Jones, K. V. Gernaey, M. P. Andersson, S. S. Mansouri, Integration of Computational Chemistry and Artificial Intelligence for Multi-scale Modeling of Bioprocesses, Accepted in 30th European Symposium on Computer Aided Process Engineering, 2020.

609201 Integration of Control with Process Operation through Reinforcement Learning

Abstract:

Increased market volatility, sustainability concerns and process flexibility are a few of the main challenges contemporary process industries are facing. At the same time major manufacturing areas, like the pharmaceutical industry, have started to investigate the continuous processing paradigm as more profitable and efficient. Moving in these directions, integration of control with operations has received increasing attention over the last decade [1,2]. Integrated planning, scheduling and control (iPSC) aims to exploit the inherent reciprocal information flow among the three problems so as to enhance feasibility, robustness and profitability of the operations in the process industries. The resulting problem tends to be computationally intensive and so far has been studied mostly under deterministic assumptions [3,4] with few exceptions who can be employed under restrictive assumptions [5]. As far as the incorporation of control considerations within these problems, the use of model predictive control [2,3], multi-parametric programming [6,7] as well as surrogate-based and data-driven approaches have been proposed [8,9].

Despite the research effort in this area the development of systematic, computationally efficient and data-driven methods remains an open challenge. On top of that, the underlying dynamic optimization of the control problem suffers from three conditions: (i) there is no precise known model for most industrial-scale processes (plant model mismatch), leading to inaccurate predictions and convergence to suboptimal solutions; (ii) the process is affected by endogenous uncertainty (i.e. the system is stochastic) & (iii) state constraints must be satisfied due to operational and safety concerns. Therefore constraint violation can be detrimental. To solve the above problems, Reinforcement Learning (RL) Policy Gradient method is proposed, which satisfies chance constraints with probabilistic guarantees [10]. The resulting optimal policy is a neural network designed to satisfy the optimality conditions, and the optimal control actions can be evaluated fast as the policy requires only function evaluations. In this work, a novel framework for closed-loop integration iPSC under dynamic disturbances and uncertainty is proposed. The key elements are the integration of novel RL-based optimal policy for the control task of an uncertain dynamic physical system with the optimization-based algorithm for the efficient rescheduling that mitigates the impact of exogenous disturbances on the real-time implementation. Finally, the proposed framework is tested on the iPSC of a nonlinear industrial process illustrating its merits and providing key insights about the interdependence of the iPSC decisions and the importance of RL for its real-time implementation.

1. Grossmann, I. E. (2012). Advances in mathematical programming models for enterprise-wide optimization. Comput. Chem. Eng., 47, 2-18.
2. Chu, Y., and F. You. (2015) Model based integration of control and operations: Overview, challenges, advances, and opportunities. Comput. Chem. Eng. , 83, 2-20.
3. Dias, L. S., & Ierapetritou, M. G. (2016). Integration of scheduling and control under uncertainties: Review and challenges. Chem. Eng. Res. Des., 116, 98-113.
4. Georgiadis, G.P., Elekidis, A.P. and Georgiadis, M.C. (2019). Optimization-Based Scheduling for the Process Industries: From Theory to Real-Life Industrial Applications. Processes, 7, 438.
5. Charitopoulos, V. M., Aguirre, A. M., Papageorgiou, L. G., & Dua, V. (2018). Uncertainty aware integration of planning, scheduling and multi-parametric control. Comput. Aid. Chem. Eng. 44, 1171-1176.
6. Charitopoulos, V. M., Papageorgiou, L. G., & Dua, V. (2019). Closed-loop integration of

planning, scheduling and multi-parametric nonlinear control. *Comput. Chem. Eng.*, *122*, 172-192.

7. Burnak, B., Katz, J., Diangelakis, N. A., & Pistikopoulos, E. N. (2018). Simultaneous process scheduling and control: a multiparametric programming-based approach. *Ind. Eng. Chem. Res.*, *57*(11), 3963-3976.

8. Du, J., Park, J., Harjunkoski, I., & Baldea, M. (2015). A time scale-bridging approach for integrating production scheduling and process control. *Comput. Chem. Eng.*, *79*, 59-69.

9. Dias, L. S., & Ierapetritou, M. G. (2019). Data-driven feasibility analysis for the integration of planning and scheduling problems. *Optim. Eng.*, *20*(4), 1029-1066.

10. Petsagkourakis, P., Sandoval, I. O., Bradford E., Zhang, D. & del Rio-Chanona, E. A. (2020). Constrained Reinforcement Learning for Dynamic Optimization under Uncertainty. Accepted for publication in 21st IFAC World Congress

596264 Control Lyapunov-Barrier Function-Based Predictive Control of Nonlinear Processes Using Real-Time Machine Learning Modeling

Abstract:

Safety considerations have been integrated into feedback control system design to yield cooperative actions to ensure both operational safety and economic performance of system operation via control Lyapunov-barrier functions (CLBF) [1,2]. Specifically, in [2] a CLBF-based model predictive control scheme was proposed to optimize process performance accounting for process stability and safety requirements. While MPC systems implemented in an industrial setting often utilize linear data-based empirical models to compute control actions to maintain optimal process operation while accounting for process and control actuator constraints, chemical processes are inherently nonlinear and often require nonlinear models in order to be controlled efficiently. Modern machine learning modeling tools provide an efficient way to nonlinear process modeling, taking advantage of advances in training, efficient coding and computational power [3, 4, 5].

In this work, we develop a machine-learning-based CLBF-MPC based on an ensemble of recurrent neural network (RNN) models that are widely-used to model nonlinear dynamic systems for prediction to control an input-constrained nonlinear process accounting for stability and safety considerations. Specifically, RNN models are first developed to model a general class of nonlinear systems using process operating data, and sufficient conditions that account for bounded modeling error between the RNN model and the actual nonlinear process are provided to achieve closed-loop stability and safety for the nonlinear process under CLBF-MPC. Additionally, following the design of machine-learning-based CLBF-MPC, the CLBF-based economic MPC using RNN models is proposed to optimize process economic benefits as well. Moreover, to handle model uncertainty issue in real-time implementation of controllers, on-line learning of RNN models is also employed within CLBF-MPC and CLBF-EMPC to update process models in the presence of time-varying disturbances.

[1] Romdlony, M. Z., and Jayawardhana, B. Stabilization with guaranteed safety using control Lyapunov–barrier function. Automatica, 66, 39-47, 2016.

[2] Wu, Z., F. Albalawi, Z. Zhang, J. Zhang, H. Durand and P. D. Christofides, "Control Lyapunov-Barrier Function-Based Model Predictive Control of Nonlinear Systems," Automatica, 109, 108508, 2019.

[3] Kosmatopoulos, E. B., Polycarpou, M. M., Christodoulou, M. A., and Ioannou, P. A. High-order neural network structures for identification of dynamical systems. IEEE transactions on Neural Networks, 6, 422-431, 1995

[4] Mohanty, S. Artificial neural network based system identification and model predictive control of a flotation column. J. Process Control, 2009, 19, 991−999.

[5] Wu, Z., A. Tran, D. Rincon and P. D. Christofides, "Machine Learning-Based Predictive Control of Nonlinear Processes. Part I: Theory," AIChE J., 65, e16729, 2019.

608985 Constrained Reinforcement Learning for Process Optimization and Control

Abstract:

The optimization of chemical processes presents distinctive challenges to the stochastic systems community given that they suffer from three conditions:

- There is no precise known model for most industrial scale processes (plant model mismatch), leading to inaccurate predictions and convergence to suboptimal solutions.
- The process is affected by disturbances (i.e. it is stochastic).
- State constraints must be satisfied due to operational and safety concerns, therefore constraint violation can be detrimental or even dangerous.

To solve the above problems, we propose a Reinforcement Learning (RL) Policy Gradient method, which satisfies chance constraints with probabilistic guarantees.

Machine learning is helping to address complex problems in the chemical and process industries, such as process optimal control [1,2], estimation and online monitoring [3,4]. However, less studies have been conducted to investigate the applicability and efficiency of RL in process engineering, and none include the efficient handling of constraints.

RL would be a natural choice to address nonlinear, uncertain and stochastic process control problems as it effectively addresses stochastic environments [5]. Unfortunately, present RL algorithms fail to reliably satisfy state constraints even when initialized with feasible initial policies [6]. Various approaches have been proposed in the litera-ture, where usually penalties are applied for the constraints.Such approaches can be very problematic, easily losing op-timality or feasibility [7] especially in the case of a fixed penalty. The main approaches to incor-porate constraints in this way make use of trust-region and fixed penalties [9],as well as cross entropy [7]. As it is observed in [8], when penalty methods are ap-plied in policy optimization, depending on the value of the penalty parameter the behaviour of the policy may change.

We propose a constrained RL algorithm which guarantees the satisfaction of joint chance constraints. To accomplish this, we propose the introduction of backoffs, which are computed simultaneously with the feedback policy. Backoffs are adjusted with Bayesian optimization using the empirical cumulative distribution function, which can, therefore, guarantee the satisfaction of joint chance constraints.

[1] Bradford, E.; Schweidtmann, A. M.; Zhang, D.; Jing, K. and del Rio-Chanona, E. A., Dynamic modeling and optimization of sustainable algal production with uncertainty using multivariate Gaussian processes, 118, 143-158, 2018

[2] del Rio-Chanona, E. A.; Fiorelli, F.; Zhang, D.; rashid Ahmed, N.; Jing, K.; and Shah, N, An efficient model construction strategy to simulate microalgal lutein photo-production dynamic process, 114(11), 2518-2527, 2017

[3] do Carmo Nicoletti, M. and Jain, L. C., Computational Intelligence Techniques for Bioprocess Modelling, Supervision and Control, Volume 218 of Studies in Computational Intelligence, Springer Science & Business Media, 29 Jun 2009

[4] Xiong, Z. and Zhang, J., Modelling and optimal control of fed-batch processes using a novel control affine feedforward neural network, Neurocomputing, 61, 317-337, 2004

[5] Petsagkourakis, P.; Sandoval, I. O.; Bradford E.; Zhang, D. and del Rio-Chanona, E. A., Reinforcement Learning for Batch Bioprocess Optimization, 133, 2020

[6] Wen, M., Constrained Cross-Entropy Method for Safe Reinforcement Learning, Neural Information Processing Systems (NeurIPS), 2018

[7] Ray, A.; Achiam, J. and Amodei, D., Benchmarking Safe Exploration in Deep Reinforcement Learning, Deep RL Workshop NeurIPS 2019, arXiv:1910.01708, 2019

[8] Achiam, J; Held, D.; Tamar, A. and Abbeel, P., Constrained Policy Optimization, International Conference on Machine Learning (ICML) 2017

[9] Tessler, C.; Mankowitz, D. J.; and Mannor, S.; Reward Constrained Policy Optimization, International Conference on Learning Representations (ICLR) 2019

609526 Finite-Size Effects on the Chemical Potentials of Molecular Solids

Abstract:

Numerous important properties of solid oral therapeutics are associated with the chemical potential of the active pharmaceutical ingredient (API) in the solid form. These properties include aqueous solubility, dissolution rate, and physical stability in its crystalline form among others. Molecular modeling of the crystal structures of APIs to estimate the chemical potentials has become an increasingly popular approach in recent years for confirming the suitability of API solid forms for further development *in silico*. However, due to the computational expense of these approaches, the models are limited to time and length scales far smaller than macroscopic materials, and rely on simplifications such as microscopic cells with periodic boundary conditions, as well as cut-offs for long-range interactions. These approximations can have a significant effect on the chemical potential of organic small molecule crystals, particularly at small system sizes.

Here we systematically examine the effect of supercell size and long-range cutoff distance on the computed chemical potentials of pharmaceutically relevant molecular crystals using classical molecular dynamics simulations. The Gibbs free energies of the crystal polymorphs and liquid phases of carbamazepine and sulfamerazine were computed from 0 K to 600 K for various system sizes and long-range cutoff radii. We compare the estimated chemical potentials with small systems and short cutoffs to those of the corresponding larger systems and show how these differences change with temperature. Furthermore, we show the effects that these small perturbations to the chemical potentials have on the macroscopic observables of enantiotropic transition temperature and solid-liquid melting temperature. Finally, we make recommendations for future molecular simulations of API crystal structures to balance the trade-off of chemical accuracy and computational efficiency.

609123 Computer-Aided Design of Solvent Blends for the Crystallisation of Ibuprofen – Regulating Crystal Morphology By the Spiral Growth Model

Abstract:

Solid dose pharmaceuticals continue to be the most common means of drug delivery worldwide; because of this, pharmaceutical production is dependent on effective crystallisation systems. The majority of industrial crystallisers are solvent based – hence, decisions surrounding solvent choice can drastically affect the outcome of the crystallisation process. Although attaining a high crystal yield of the active pharmaceutical ingredient (API) is typically the dominant objective during the solvent selection process, solvent and crystal properties can also influence health, safety, environmental, and operational performance. Indeed, the shape of the crystalline product can greatly impact downstream processes[1]; filtration can be hampered by needle- or plate-like particles that form dense cakes, high aspect ratio crystalline material may not readily flow during transportation, and tabletting via direct compression can fail if the solid product fractures during compaction. Moreover, crystal morphology can affect the *in-vivo* efficacy of the final pharmaceutical product.

Despite the many material, time and cost constraints present during initial design stages, solvent selection for the crystallisation of novel API molecules is currently performed via time-consuming and expensive experiments, heavily reliant on past experience and rules-of-thumb[2]. Over the last decades, computer-aided molecular design (CAMD) methods have been developed to overcome such difficulties[3], with the aim of guiding lab-based experiments towards optimal candidate molecules for crystallisation[4,5]. Whilst it is generally agreed that crystal shape has a significant role when designing optimal solvents, the integration of shape considerations within CAMD methods has principally focused on empirical approaches[4] combining an understanding of the underlying chemistry with prior experimentation. Direct, quantitative methods, where mechanistic interpretations of crystal growth are considered[6], have not yet received significant attention in formulating the solvent design problem. This is likely due to the absence of applicable group contribution approaches to these models, as well as the additional complexity of formulating and resolving a mixed-integer optimisation problem to represent these design choices.

We extend the recent work of Watson et al.[7], in which the SAFT-γ Mie equation of state[8] was employed in a CAMD framework to identify optimal solvent blends and process conditions for integrated cooling and antisolvent crystallisation, to include the target morphology of the pharmaceutical product. By assuming a low supersaturation growth environment – a standard condition for industrial crystallisation – it is possible to calculate the relative growth rates of corresponding crystal faces[6], thus providing a quantitative measure of aspect ratio of the crystalline API. The design formulation is implemented in gPROMS and is applied to the design of solvent mixtures for the crystallisation of ibuprofen, whereby the impact of crystal shape on the optimal solvent molecules and process conditions is investigated.

1. Dandekar, P., Kuvadia, Z.B., Doherty, M.F., 2013. Annual Review of Materials Research 43, 359-386.
2. Brown, C.J., McGlone, T., Yerdelen, S., et al., 2018. Molecular Systems Design & Engineering 3(3), 518-549.

3. Gani, R., 2004. Computers & Chemical Engineering 28(12), 2441-2457.
4. Karunanithi, A.T., Achenie, L.E., Gani, R., 2006. Chemical Engineering Science 61(4), 1247-1260.
5. Jonuzaj, S., Gupta, A., Adjiman, C.S., 2018. Computers and Chemical Engineering 116, 401-421.
6. Li, J., Tilbury, C.J., Kim, S.H., Doherty, M.F., 2016. Progress in Materials Science 82, 1-38.
7. Watson, O.L., Galindo, A., Jackson, G., Adjiman, C.S., 2019. Computer Aided Chemical Engineering 46, 949-954.
8. Papaioannou, V., Lafitte, T., Avendano, C., et al., 2014. The Journal of chemical physics 140(5), 054107.

596083 Machine Learning for Thermodynamic Properties of Materials: Towards the Prediction of Solubility

- Download/View Extended Abstract

Abstract:

Introduction

Solubility is one of the most important properties in the development and design of drugs. It plays a big role in pharmacokinetics and pharmacodynamics, where it relates to dosing, toxicity and the effect of drugs on their targets. In the field of process design, knowledge on the solubility of drugs and other intermediate products are crucial in the design of separation techniques and for the transitioning from batch to flow processes. Computer-aided screening of the solubility of API's can dramatically shorten the drug discovery cycle and the costs related to the development and manufacturing of new drugs.

State of the art solubility models focus on the prediction of octanol-water partitioning coefficients (logP) and solid aqueous solubility (logS) because of their role in the evaluation of the drug-likeness of API's in an initial screening process. Although models for octanol-water partitioning coefficients have improved a lot the last couple of years, the prediction of aqueous solubility for a broad range of drug-like compounds remains challenging. In view of downstream manufacturing and process intensification, the interest in solubility prediction is not limited to octanol or water as a solvent. Computer-aided screening of potential solvents can contribute significantly to a reduction of waste streams and ease the upscaling towards continuous processing. However, models for solubility predictions in other organic solvents are limited by the scarce availability of experimental data.

This works aims at using machine learning techniques and physical relationships to predict the solubility of gases, liquids and solids in a variety of solvents. Theoretical quantum calculations are used to compensate for scarce and biased experimental databases. Furthermore, the physical relation between gas-liquid solvation energies, liquid-liquid partitioning coefficients and solid solubility is used to aid solubility predictions for drug-like molecules in a broad range of solvents.

Methodology

Graph convolutional neural networks are used to encode the molecular structure of both the solute and solvent. A molecular embedding is learned separately for the solute and solvent with the directed message passing neural network, as implemented in Chemprop. The solute and solvent embeddings are concatenated and fed through another dense layer for the property prediction.

Quantum chemical calculations are done to determine the gas-liquid solvation free energies for various solute and solvent combinations. These calculations are used to pretrain the message passing neural network, such that physical interactions between solvents and solutes can be learned. Contrary to the experimental gas-liquid solvation free energies, this dataset is not biased towards certain solvents (eg. water). Using theoretical data to pretrain the model also reduces the effect of aleatoric errors, caused by data noise, on the final model predictions. The dataset can be extended towards classes of solutes and solvents for which no or limited experimental data is available, improving the generalizability of the model.

For the prediction of liquid-liquid partitioning coefficients and solid solubilities, the physical relationship with the well-established gas-liquid solvation free energies is enforced. Liquid-liquid partitioning coefficients are directly related to the ratio of the gas-liquid partitioning coefficients of the same solute in both solvents. Some error is introduced by the mutual solubility of the solvents, which will be accounted for in the loss function. The solid solubility is related to the gas-liquid partitioning coefficients and the fusion free energy of the crystal structure. To account for the fusion free energy, an additional model is trained on an experimental dataset of melting temperatures and fusion enthalpies.

Results

A preliminary theoretical dataset is constructed with COSMO*therm*. It consists of >20k data points and includes 78 solvents and 244 solutes. Solvation free energies of the theoretical dataset are predicted by the model with a RMSE of 0.08 kcal/mol and a MAE of 0.04 kcal/mol. Analysis of the solute and solvent molecular embeddings demonstrate the learned physical interactions. A principal component analysis of the solvent molecular embeddings clusters solvents according to their physical properties relevant to solvation. Analysis of the atom messages for some solute molecules highlights atoms with an important hydrogen bonding/donating character.

The model trained on the theoretical dataset is further refined using experimental data for solvation free energies. The experimental dataset is gathered from different sources including the Minnesota solvation database, the Compsol database and different publications related to the regression of Abrahams solute and solvent parameters. The total experimental dataset has more than 8900 entries for solute-solvent combinations with over 300 different solvents. The machine learning model uses molecular embeddings for the solute and solvent molecules that are learned from the theoretical dataset. The dense layer for property prediction is initialized by the learned weights from the theoretical dataset and optimized with the experimental solvation free energies. Solvation free energies in a randomly selected test set can be predicted with a RMSE of 0.60 (±0.04) kcal/mol and MAE of 0.27 (±0.02) kcal/mol. These results are similar compared to those of a model without pretraining. However, if no pretraining is done on a theoretical dataset, analysis of the molecular embeddings do not directly demonstrate learning of any physical interactions. Moreover, the predictions on "unseen" solvents, i.e. those not included in the training set, improve significantly if pretraining is applied.

For the prediction of partitioning coefficients, a dataset available in OChem is used. This dataset includes 3600 entries for partitioning coefficients between water and various solvents including chloroform, cyclohexane, toluene and hexadecane. The model trained on theoretical gas-liquid solvation energies is used only for prediction of the experimental logP data with a RMSE 1.18 and MAE 0.66. Further efforts on improving predictions of partitioning coefficients will include (1) collection of a broader dataset (2) refine a pretrained model on experimental data enforcing the physical relation with gas-liquid solvation energies.

Solid solubility can be related to gas-liquid solvation energies of the solute in the respective solvent and the solute fusion free energy. The fusion free energy of a crystal structure can be approximated with a known melting temperature and fusion enthalpy. Experimental databases on latter two properties are gathered from different sources, including OChem and NIST. The database is used to train a machine learning model with a directed message passing neural network to construct the molecular embeddings, as implemented in Chemprop. Fusion enthalpies are predicted with a RMSE of 1.69 (±0.14) kcal/mol and melting temperatures with a RMSE of 33.7 (±1.6) K. These predictions will be combined with the model for gas-liquid solvation energies to create a model for solid solubilities in a broad range of solvents.

Conclusions

A machine learning model is developed for the prediction of gas-liquid solvation free energies. The molecular embeddings for solvent and solute are done with a directed message passing neural network. Experimental solvation free energies are predicted using the new model, pretrained on theoretical data, with a RMSE of 0.60 kcal/mol. The pretraining on a theoretical dataset ensures that the model captures the physical interactions between solvent and solute molecules and improves the performance of the model on unseen solvents.

The physical relationship between gas-liquid solvation energies, partitioning coefficients and solid solubility will be used to enable the prediction of those properties in a broad range of solvents. The solute fusion enthalpy and melting temperature are required for the solid solubility predictions. Those are predicted by another model with a RMSE of 1.69 kcal/mol and 33.7 K respectively.

Figure

Figure 1. Parity plots for the predictions of gas-liquid solvation free energies ΔG_{solv}, fusion enthalpies ΔH_{fus} and melting temperatures T_m.

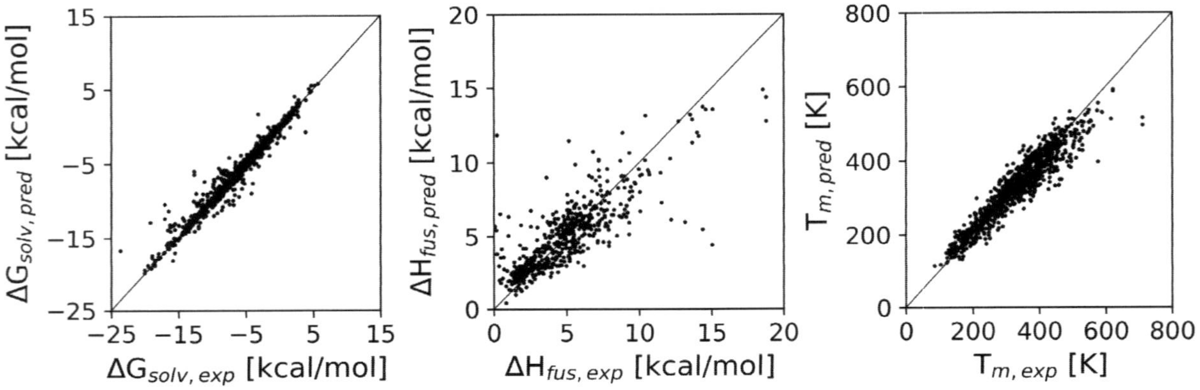

604581 Approaching Coupled Cluster Accuracy with a General-Purpose Neural Network Potential

Abstract:

In this talk, we will present a fully transferable deep learning potential that is applicable to complex and diverse molecular systems well beyond the training dataset. Recently we introduced ANAKIN-ME (Accurate NeurAl networK engINe for Molecular Energies) or ANI in short. ANI is a new *method and sampling procedure* for training neural network potentials that utilizes a special kind of symmetry functions to build single-atom atomic environment vectors (AEV) as a molecular representation.

The AI methods that focuses on the use of large and diverse data sets in training new potentials, has consistently proven to be universally applicable to systems containing the atomic species in the training set. Focusing on parametrization for organic molecules (with CHNOSFCl atoms so far), we have developed a universal neural network potential which is *highly accurate compared to reference QM calculations at speeds* 10^7 *faster*. The potential is shown to accurately represent the underlying physical chemistry of molecules through various test cases including: *chemical reactions (both thermodynamics and kinetics), thermochemistry, structural optimization, and molecular dynamics simulations*. The results presented in this talk will provide evidence of the universal applicability of deep learning to various chemistry problems involving organic molecules.

604074 Kinetic Modelling and Risk Assessment of an Exothermic Fed-Batch Reaction Using Calorimetry

Abstract:

The digital design of manufacturing processes using mechanistic models has been playing an increasingly important role in drug substance process development. It enables the study of unknown kinetics using experimental data, risk assessment at extreme process conditions where experimentation may be infeasible, and efficient exploration of decision space for critical process parameters.

Calorimetry studies are an integral part of process development of synthesis steps in the pharmaceutical industry. Experimental kits, such as the Mettler-Toledo RC1 are used for this purpose with the typical aim of understanding the heat of the reaction, heat evolution, adiabatic temperature rise and therefore, safety of the chemical reactions under scrutiny. They can also see usage in reactive systems to regress kinetics of reactions, which are otherwise difficult to detect through offline composition analysis. In order to assess the thermal risks associated with a chemical reaction, both the desired and undesired (i.e., decomposition) reactions must be studied. Synthesis steps in pharmaceutical processes can be accompanied by significant heat release and must be understood well to manage them successfully at pilot and plant scales. In this study, we use mathematical modeling to predict the behavior of a reaction under different heat transfer profiles and simulate what would happen in the event of process disturbances such as a cooling jacket failure, or loss of control of fed batch (e.g. reactant added too quickly).

The reaction was carried out in a 1L stirred tank reactor and the reactor temperature was maintained by a cooling jacket. When the reactor temperature is well regulated, only the primary reaction occurs, however in the event of a cooling failure, an undesired decomposition reaction occurs. Experimental data from RC1 heat flow measurements were obtained using experiments executed at different process temperatures. A fed-batch reactor model was developed and simulated in gPROMS FormulatedProducts. Reaction kinetics for the primary reaction were extracted using parameter estimation by fitting the RC1 data (Figure 1).

In order to study the effect of temperature control failure, a pre-defined decomposition reaction was added to the model. Global system analysis (GSA) is an effective tool to explore design space, determine safe operating regimes for exothermic reactions, quantify effects of uncertainty, manage risk and identify opportunities for optimization. This tool was used to perform a sensitivity analysis by varying the heat transfer coefficients and initial temperatures within a specific range. The results of GSA give a clear picture of: 1) what the maximum temperature in the reactor would be if temperature control fails and hence the operation conditions that can trigger synthesis of undesirable products and impurities; 2) when the cooling failure occurs during the process, i.e. during the reactant addition or following reaction. The output from this may help to inform the control strategy for this reaction process, such as the setting of maximum reactant addition flow rates to ensure safe operation.

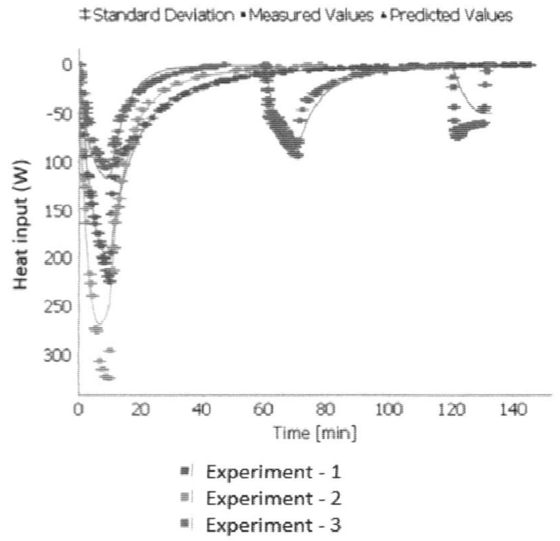

603492 Kinetic Modeling of a Non-Homogeneous Metal-Catalyzed Reaction System for Drug Synthesis

Abstract:

Mathematical modelling of chemical reaction kinetics has proven to be a very effective tool for pharmaceutical drug synthesis development, where complex organic reaction networks are often encountered. The chemical and molecular complexity of modern drug molecules has been rapidly increasing, and this requires design and optimization of extensive synthesis routes which presents a challenge to process scientists. First-principles modeling can play an important role in streamlining this process. It helps with analyzing, interpreting and discerning the relationship between various process parameters and reaction profiles with minimal experimentation requirements.

In this presentation, a modelling workflow to understand the Buchwald-Hartwig cross-coupling reactions is described. These reactions are ubiquitous in process chemistry. This is a non-homogeneous, metal-catalyzed reaction system encompassing multiple partially soluble reactants, a biphasic liquid mixture, and homogenous catalytic cycle in the organic phase. Challenges arise from subtle case-to-case variability, with different influential factors like mass transfer limitations, catalyst activation and water sensitivity.

Using this case, the presentation will illustrate the following steps in a typical mechanistic modeling lifecycle: 1) Design and perform kinetic experiments; 2) Recreate experiments *in silico*; 3) Estimate reaction kinetic parameters; 4) Evaluate model predictions; 5) Postulate new mechanisms, kinetic rate expressions and/or perform additional targeted experimentation until predictions are satisfactory

HPLC measurements were available for key reactants and products from experiments conducted for a range of operating conditions. A mathematical model was developed in gPROMS FormulatedProducts starting from preliminary fitting of primary reaction pathway(s) assuming simple reaction mechanisms and solubility correlations.

Using a systematic parameter estimation approach this was further expanded to a more rigorous model that includes effect of particle size distribution, mass transfer effects and relative concentration effects. The regression of kinetic parameters takes into account uncertainty in measurements and process parameters, which provides a robust model that can be used to make process predictions, and aid scale-up/technical transfer decisions.

The model can be further used to understand the behavior of the system over a wide range of input conditions. This becomes particularly important to make scale up decisions and identify critical operating regimes early in the development of a pharmaceutical compound. Global system analysis was used to explore the design space and identify key factors that influence reaction selectivity and impurity generation. This allows for better understanding of the process and thus provides useful insights to support control strategy, optimization goals, and risk assessment.

Keywords: reaction kinetics, parameter estimation, mechanistic modelling, digital design, catalyst

605922 Continuous Mixing Technology: Development of Digital Twin

Abstract:

Continuous manufacturing has many advantages over batch processing, including smaller footprints, faster turnaround times, and better control of process parameters and product quality. Direct compression is a minimal setup for a continuous tableting line and consists only of three unit operations: feeding of the individual components, mixing and tableting. The main goal of the mixing step is to achieve content uniformity in the final product, which is a critical quality attribute (CQAs).

This work focuses on a vertical continuous mixing device termed Continuous Mixing Technology (CMT). The main of goal CMT goal is to mix incoming powder feed so that the content uniformity of the exiting blend stays in specification even if there are variations in the mass flows of the individual components from the feeding process. To enhance the process understanding of this mixing process and expedite the development time with minimal material consumption, we have been developing computational model of CMT device using Discrete Element Method. Our previous work focused on (DEM) simulations to gain a very detailed process insight using the commercial software package XPS (Extended Particle System) at low mass throughput of 10kg/h [1].

Recently, DEM model has been extended to simulate production-scale processes with a hold-up mass of up to 1000g and mass throughput of 30kg/h with 5.7million particles [2]. These simulations yield particle-level residence times, making it possible to create RTDs as function of process parameters (e.g. hold-up mass, throughput, impeller speed) and material parameters (e.g. different material lots with different flowability). Extensive validation of these simulations is carried out by comparing against RTD data collected through tracer experiments over a wide operating space. This validation study show that simulation results are in good agreement with the experimental data establishing DEM model as a Digital Twin of the physical process.

References:

[1] Toson et al., Int. J. Pharm. 552, 1–2 (2018), 288–300. doi:10.1016 IJP pharm 2018.09.032.

[2] Toson et al., AICHE meeting 2019.

602070 Integrated Continuous API Purification and Direct Formulation of Drug Products Via Confined Suspension Agitated Bed Crystallization and Fluidized Bed Coating

Abstract:

Continuous manufacturing processes have seen increasing uptake in the pharmaceutical industry with a number of notable demonstrations of their utility within the industry resulting in increased process intensity, simplification in scale-up, in both drug substance and drug product production. Applications of new CM platforms within industry to date have maintained the traditional break between drug substance and drug product mirroring batch operations where API crystallization and isolation almost uniformly demarks a transition from Drug Substance to Drug Product manufacturing, occurring in separate facilities. This study investigates new architectures that forgo traditional API isolation and instead utilize isolation free crystallization for purification to deliver a purified liquid stream to an integrated spray coater to directly produce highly engineered drug product intermediates.

To this end a novel confined suspension agitated bed crystallization process that produces a purified API solution effluent has been applied to purify a model API with high purity and yield. The purified API solution from this process was then deposited onto microcrystalline cellulose (MCC) beads via fluidized bed spray coating. This process is shown schematically in Figure 1. Fluidized bed spray coating operations onto excipient carriers are capable creating final bulk powders with excellent flow properties in addition to incorporating, disintegrants, controlled release polymers and coatings enabling direct isolation of final highly engineered drug product intermediates ready for example capsule filing or direct compression in a DS facility.

As spray coating is a complex process with a large number of interconnected process parameters, a DOE based optimization approach was used to develop a robust process with regard to loading efficiency and the degree of drug crystallinity on the coated beads. A five-factor two-level fractional factorial study was produced to screen for the parameters with the most significant influence on the spray coating process. The mass of API sprayed was kept constant for each treatment and factors investigated were process temperature (50 to 60 °C), solvent volume (50 to 100 ml), binder mass (1 to 3 g), binder type (PVP or HPMC) and solvent type (methanol or ethanol). From this study, spraying conditions were identified that maximised both the amount of API loaded onto the beads and the degree of crystallinity and were then used to spray purified crystallization product directly onto the MCC beads (Figure 2). The product from the spray coater allows for both direct formulation of the product and for further processing and coating in order to mediate drug release. As both processes are easily scalable and the spray coating transferable to an analogous continuous operation, this work demonstrates a promising method for isolation free purification and direct formulation of Drug Products.

Acknowledgements:

This work is supported by the Irish Research Council Government of Ireland Postgraduate Ireland Postgraduate Research Scholarship under grant number (GOPI/2017/1980)

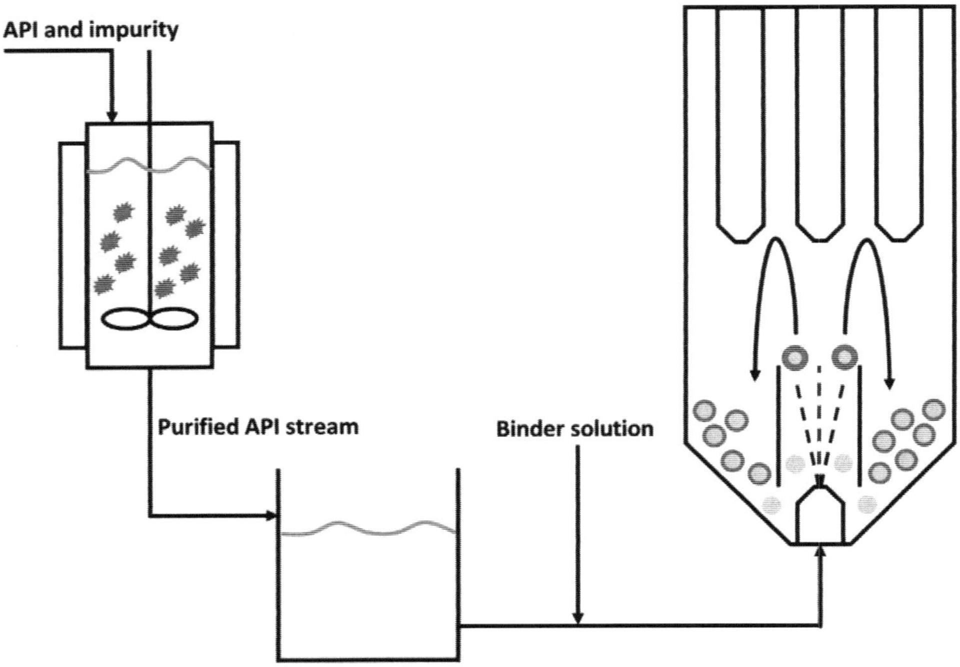

Figure 1: Schematic of proposed telescoped purification and direct formulation approach

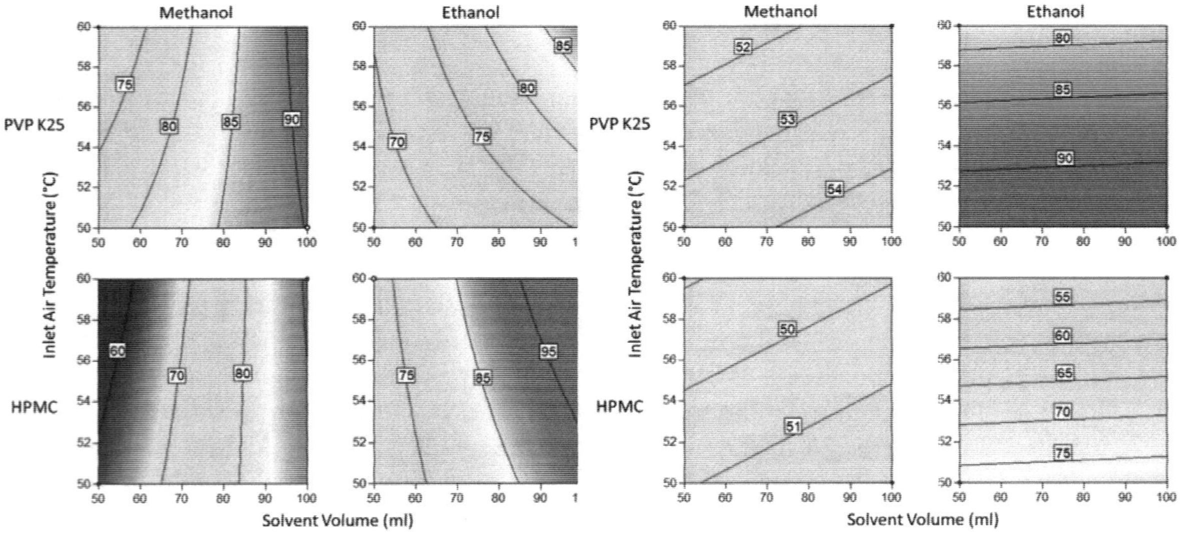

Figure 2a: Example set of DOE results of the effect of process parameters on drug loading used for optimization of the spray coating process

Figure 2b: Example set of DOE results of the effect of process parameters on crystallinity used for optimization of the spray coating process

610008 Developing HME-Based Drug Products Using QbD and Emerging Science

Abstract:

Introduction

Hot melt extrusion (HME) is a continuous manufacturing process primarily carried out using co-rotating intermeshing twin-screw extruders (TSE). The process is mostly used in order to produce amorphous solid dispersions of poorly soluble active pharmaceutical ingredients (APIs) by dispersing them in polymer carriers. In addition, it can also facilitate the development and production of products with crystalline API embedded into a polymer matrix and nanopharmaceuticals. Formulation screening for HME based drug product presents a significant challenge in the product development cycle. On the one hand this can be attributed to the complex formulation, process and final product relationship where it is not entirely clear how the chosen formulation is transformed into the final dosage form under selected HME process conditions. On the other hand, being a continuous manufacturing technology, HME traditionally requires a large amount of API (kilograms of premix vs grams available in the early development stages) for successful formulation screening and early process setup. The process setup itself is modular, allowing the process to be tailor-made for any formulation. This might however be problematic in the context of choosing the appropriate and efficient setup for an unknown formulation and it is an additional reason for relatively high quantities of material usually needed for the screening. Effectively solving the multidisciplinary challenge of formulation development, early process screening, effective scale-up and transfer to GMP production represents one the key challenges of the pharmaceutical industry and is especially challenging for novel and imported production technologies like HME.

Methodology

As a response to these challenges, our group has focused on developing scientific tools that allow for a fast and minimum-risk development of HME-based formulations using several advanced tools. These include advanced material screening, small-scale formulation test beds, the design of small-scale processes and the scale-up to GMP production of clinical batches. During the formulation development and screening phase traditional approaches (e.g., assessment of biopharmaceutical and physicochemical properties of the formulation) or modern in-silico approaches (e.g., molecular dynamics, machine learning). Since to this date most of the process setup and scale-up activities are performed experimentally and empirically, one of the goals of our group was to create *in silico* tools for a rational, science-based process setup and scale-up, while addressing other important aspects, such as an API degradation and overall product quality[1], [2]. The fundamental idea behind process modelling is the break down and the detailed analysis of the key process aspects, among which the most prominent might be the analysis of flow patterns developed as a result of the rotation and geometry of the individual screw element pairs[3]–[5]. A further step is the development of numerical models aimed at understanding and predicting the connection between the independent process variables like the screw speed, throughput, barrel temperature, and screw configuration to the dependent process variables like the resulting melt temperature distribution, SMEC distribution and local and overall residence time distribution[1], [6]–[8]. Once an accurate prediction of the internal process state is possible, a push towards connecting the internal process state, the formulation characteristics and resulting product quality can be made.

Results

Following this approach formulation and process development will be discussed and analyzed. A detailed analysis of the different extruder elements and equipment scales used, based on the results of detailed smoothed particle hydrodynamics (SPH) simulations will be presented and discussed. Moreover, a detailed process analysis based on mechanistic 1D simulations will be performed, comparing various configurations and equipment scales, in the context of scalability and resulting process quality. The overall goal is to present a systematic and holistic framework for rapid, reliable and waste free product development of HME based formulations. Following the Quality by Design (QbD) framework, the pharmaceutical product development should be based on sound scientific principles and quality risk management, with an emphasis on predefined objectives, scientific product and process understanding and process control. Hence, our approach aims at connecting the different formulation properties, desired product quality attributes and the critical process parameters with the help of mechanistic process simulations and advanced formulation design, laying the foundation for automated process setup and scale-up with a prescribed product quality in mind.

Literature

[1] J. Matić, A. Witschnigg, M. Zagler, S. Eder, and J. G. Khinast, "A novel in silico scale-up approach for hot melt extrusion processes," *Chem. Eng. Sci.*, vol. 204, pp. 257–269, Aug. 2019.

[2] J. Matić, A. Paudel, H. Bauer, R. A. L. Garcia, K. Biedrzycka, and J. G. Khinast, "Developing HME-Based Drug Products Using Emerging Science: A Fast-Track Roadmap from Concept to Clinical Batch," *AAPS PharmSciTech.*

[3] A. Eitzlmayr and J. G. Khinast, "Co-rotating twin-screw extruders: Detailed analysis of conveying elements based on smoothed particle hydrodynamics. Part 1: Hydrodynamics," *Chem. Eng. Sci.*, vol. 134, pp. 861–879, Sep. 2015.

[4] A. Eitzlmayr and J. G. Khinast, "Co-rotating twin-screw extruders: Detailed analysis of conveying elements based on smoothed particle hydrodynamics. Part 1: Hydrodynamics," *Chem. Eng. Sci.*, vol. 134, pp. 861–879, Sep. 2015.

[5] A. Eitzlmayr, J. Matić, and J. G. Khinast, "Analysis of flow and mixing in screw elements of corotating twin-screw extruders via SPH," *AIChE J.*, vol. 63, no. 6, pp. 2451–2463, Jun. 2017.

[6] A. Eitzlmayr *et al.*, "Experimental characterization and modeling of twin-screw extruder elements for pharmaceutical hot melt extrusion," *AIChE J.*, vol. 59, no. 11, pp. 4440–4450, Nov. 2013.

[7] A. Eitzlmayr *et al.*, "Mechanistic modeling of modular co-rotating twin-screw extruders," *Int. J. Pharm.*, vol. 474, no. 1–2, pp. 157–176, Oct. 2014.

[8] R. Baumgartner, J. Matić, S. Schrank, S. Laske, J. G. Khinast, and E. Roblegg, "NANEX: Process design and optimization," *Int. J. Pharm.*, vol. 506, no. 1–2, pp. 35–45, Jun. 2016.

605470 First-Principles Dynamic Simulation of an Integrated Continuous Biomanufacturing Platform

Abstract:

Biopharmaceuticals, which are also widely known as biologics or biologic drugs, are products derived from biological organisms for treating or preventing diseases. The global sales of biopharmaceuticals have continually increased for many years, with hundreds of approved products on the market and over 7000 medicines in development. Monoclonal antibodies (mAbs) are the highest selling class of biopharmaceuticals due to their specific action and reduced immunogenicity. With continued growth in sales of existing mAb products and a growing pipeline of mAb product candidates being developed, the total sales of mAb products and all biopharmaceuticals will continue to increase in the coming years. The development of mAbs is expected to grow further as more diseases are understood at molecular and cellular levels [1].

Most biomanufacturing unit operations use empirical (black box) or semi-empirical (grey box) models based on experimental data because biomanufacturing processes are challenging to model using first principles. However, data-based models are not suitable for process prediction and optimization because they perform poorly on extrapolation from their original data set and are not based on any process understanding [2]. First-principles models, on the other hand, are based on understanding of fundamental physical phenomena such as conservation laws, thermodynamics, chemical kinetics, and transport phenomena. These models are useful for process development because less experimental data is needed. Integrated first-principles models for the entire biomanufacturing plant are needed to consider that individual unit operations do not operate in isolation, and changes in a unit operation can affect a process further downstream [3,4].

This presentation describes a software tool for carrying out high-fidelity dynamic simulations of integrated continuous mAb manufacturing plants. The simulations include first-principles models of individual biomanufacturing unit operations, including bioreactors, chromatography columns, crystallizers, viral inactivation units, and filtration units. Thoroughly validated models were not available for the quantitative prediction of some relationships. The lack of validated models was addressed by using the best models available in the literature and then validating them using experimental data collected for the individual unit operations from an automated integrated continuous manufacturing platform at MIT. These validated individual units are then combined into a plant-wide dynamic model that includes the effects of model parameter uncertainties and disturbances, which is then used to validate the integration of the individual models and to map the raw materials and operations to the critical quality attributes and other variables of interest anywhere in the system.

Plant-wide first-principles predictive models can be used to design, compare, and evaluate various control and real-time release testing strategies [5]. This approach enables evaluation of the impact of choosing the underlying manufacturing model and control strategy on process performance. The plant-wide simulation software is designed to make the replacement of individual models and unit operations seamless. This plug-and-play ability enables the simulation-based evaluation of multiple process options before equipment is swapped in and out of the real physical manufacturing plant, including for less established technology such as protein capture and purification via crystallization [1,6].

References:

[1] M. S. Hong, K. A. Severson, M. Jiang, A. E. Lu, J. C. Love, and R. D. Braatz. Challenges and opportunities in biopharmaceutical manufacturing control. *Computers and Chemical Engineering*, 110:106–114, 2018.

[2] D. B. Boedeker and D. J. Magnus. Opportunities and limitations of continuous processing and use of disposables. *American Pharmaceutical Review*, 20(1):66+, 2017.

[3] Z. Xing, B. M. Kenty, Z. J. Li, and S. S. Lee. Scale-up analysis for a CHO cell culture process in large-scale bioreactor. *Biotechnology and Bioengineering*, 103(4):733–746, 2009.

[4] R. P. Harrison, S. Ruck, N. Medcalf, and Q. A. Rafiq. Decentralized manufacturing of cell and gene therapies: Overcoming challenges and identifying opportunities. *Cytotherapy*, 19(10):1140–1151, 2017.

[5] M. Jiang, K. A. Severson, J. C. Love, H. Madden, P. Swann, L. Zang, and R. D. Braatz. Opportunities and challenges of real-time release testing in biopharmaceutical manufacturing. *Biotechnology and Bioengineering*, 114(11):2445-2456, 2017.

[6] B. Smejkal, N. J. Agrawal, B. Helk, H. Schulz, M. Giffard, M. Mechelke, F. Ortner, P. Heckmeier, B. L. Trout, and D. Hekmat. Fast and scalable purification of a therapeutic full-length antibody based on process crystallization. *Biotechnology and Bioengineering*, 110(9):2452-2461, 2013.

609427 A Comprehensive Investigation of Process Design and Parameters of Spray Dried Proteins

Abstract:

Biopharmaceuticals, such as monoclonal antibodies, represent an important field in the pharmaceutical industry. Development and production of these valuable and complex substances is costly and the resulting price on the market for a single dose can reach more than thousand euros. This highlights the demand of processing these biopharmaceuticals in a sensitive way to preserve their activity and stability. The majority of biopharmaceutics is formulated as liquid solutions. Proteins are more susceptible to physico-chemical degradation in liquid state, thus various stabilizers are needed to prevent self-aggregation and denaturation. The conversion of these biologics from the liquid to the solid state by removing their natural environment does not appear to be intentional at first glance. However, removing water decreases the proteins mobility and stabilizes the protein. Additionally, dried protein powders convince in eased handling, storage and transportation.

Freeze drying is currently the gold standard, when protein formulations are transferred into dry solid powders. Briefly, freeze drying of a protein formulation is based on converting the liquid protein formulation to a solid state by freezing and eventually removing the water by applying vacuum (i.e. sublimation). However, freeze drying is time intensive during processing and handling. A promising alternative is spray drying of proteins. The protein formulation is atomized by a spray nozzle. The finely dispersed droplets are dried by a heated gas flow (e.g. air, nitrogen) and subsequently collected. Spray drying allows for a vast decrease in processing time, processing at atmospheric pressure and mild drying temperatures, the control of powder particle properties, and remaining water content.

We developed a stable drying process for different model proteins by optimizing the yield and residual moisture. The structural stability, activity and properties in solid state of the protein were assessed by a standardized set of methods. The talk will highlight the lessons learned and provide further steps towards understanding the impact of spray drying. An outline to support industrialization of the process design and development for spray drying as alternative technology will be presented to fully exploit its benefits.

607069 H_2O_2 Decontamination for Sterile Manufactruing: Impact of Sorption By Polymers on the Duration of Aeration

Abstract:

Hydrogen peroxide decontamination is the key technology to sterilize isolators or clean rooms to produce injectables. Typically, injectables are manufactured batch-wise, where decontamination serves as the change-over operation between batches. Decontamination is known to be time-intensive, especially the aeration part to reduce the residual H_2O_2 to a target concentration (e.g., 1 ppm, or even lower). Decontamination is becoming relevant, along with the market growth of biopharmaceuticals/vaccines that are usually provided as injectables, and also with the trend of small lot-size production and frequent change-overs.

We analyzed the impact of H_2O_2 sorption by polymers on the duration of aeration. Five polymers, which are typically used as materials/parts in sterile isolators, were investigated: chlorosulfonated polyethylene, silicone, polyethylene, polyoxymethylene, and polyvinyl chloride. For assessing different polymers, we defined the storage capacity and diffusion coefficients of H_2O_2 in the polymer as the indicators. Experiments were conducted to assess the abovementioned five polymers, and also to develop models to estimate the required duration of aeration, given the target H_2O_2 concentration. We extended the model calculation to map-out the duration, given the properties of the polymers. In the simulated setup, polyoxymethylene and silicone showed the tendency to require long duration for aeration. The result suggests that the application of these polymers in the isolator requires attention. Also, the target H_2O_2 concentration affected the superiority of the polymers largely, which is another practical finding from the analysis. This result suggests the benefit of integrating product information into the design of isolators, in order to achieve a rapid cycle of decontamination and aeration.

611767 Simulation of Drying of Pharmaceutical Wet Granules in a Fluidized Bed Dryer Using Coupled CFD-DEM Approach

Abstract:

Introduction/Purpose:

Fluidized bed drying is a topic of interest for scientists and engineers because of its complicated physics and its presence in a wide range of industrial processes e.g. pharmaceutical product manufacturing, food processing, wood processing, reduction of iron ore, flue gas cleaning, the roasting of sulfide ores, drying of coal, catalyst industry and many more. Here, we propose a model based on coupled Computational Fluid Dynamics (CFD) and Discrete Element Method (DEM) approach to simulate the drying of granular particles in fluidized bed dryer associated to the drug manufacturing process. The drying process of wet granular particles in a fluidized bed dryer involves momentum, heat and mass transfer between the particles and the drying medium which is inherently a multiphase-multicomponent flow problem.

Methods:

The model was implemented using an opensource software CFDEM coupling where the fluid phase was solved by OpenFOAM codes and the motion of particles were calculated by LIGGGHTS codes. The information between the fluid phase and particles were exchanged at certain time intervals. The software CFDEM coupling was used because of its flexibility to include new models and its ability to handle large scale systems.

Results:

The model was validated with the results available in existing literatures. The CFDEM coupling was capable of capturing the physical fluidization phenomena and predicting the fluidization velocity correctly for a packed particle bed. Our model was capable of capturing the heat and momentum transfer between the particles and the drying medium. Present work is going on the modification of the model to include moisture transfer between the particles and drying medium. It was observed that the fluidization of particles significantly improved the performance of the fluidized bed dryer.

Conclusions:

This validated coupled CFD-DEM model provides a better qualitative and quantitative understanding of the effects of different process parameters on the drying process which can be beneficial for the industry. This model can further help set up reliable scale up, troubleshooting or optimization schemes by greatly replacing the burden of the cost of design of experiments.

610041 Continuum Simulations of Heat and Mass Transfer in Wet Granular Mixtures

Abstract:

The rheology and hydrodynamics of wet granular mixtures are important in many industrial processes such as pharmaceutical and agricultural production, and petrochemical refining. Heat and mass transfer considerations are also important in key industrial processes involving solvent removal by thermal drying of powders and the product of wet granulation. Continuum simulations involve solving the averaged equations for conservation of mass, momentum, and energy by treating the granular medium as a continuum. In this study, we develop an analytical model for verifying Computational Fluid Dynamics (CFD) simulation results in a cylinder filled with stationary wet granular material. The analytical model solves for the unsteady axisymmetric temperature field during transient heating of the wet granular material to the saturation temperature and the steady axisymmetric temperature field during drying of the wet granular mixture. In order to obtain tractable analytical solutions, it is assumed in the analytical model that there is no mass transfer during the transient heating phase and that all mass transfer occurs at the steady temperature distribution corresponding to the drying stage through a specified sink term in the heat equation. Continuum simulations of heat transfer in the stationary granular bed are performed and the results are verified with those of the analytical model. Further realities of particle drying due to solvent boiling and wet bulb evaporation mechanisms are incorporated into a coupled heat and mass transfer CFD simulation to obtain more realistic results. The assumptions made in the analytical model are also quantitatively examined using the continuum simulations.

608701 New Insights into Local Residence Time Distributions of Twin Screw Elements Used in Hot Melt Extrusion Via Smoothed Particle Hydrodynamics Simulation

Abstract:

Motivation

Hot Melt Extrusion (HME) is a continuous manufacturing process used in the pharmaceutical industry to produce drug products with a controlled release profile or with increased bioavailability. Due to the complex flow in Twin Screw Elements (TSE), some parts of the polymer melt reside longer in it then others. This effect is commonly described with Residence Time Distributions (RTD). The RTD can be investigated experimentally for entire Hot Melt Extrusion (HME) processes. However, when kinetic effects (e.g., product formation or degradation) play a role local RTDs are required and these are difficult to obtain experimentally.

Method

Here we show that the numerical simulation method Smoothed Particle Hydrodynamics (SPH) is suitable for determining local RTDs in silico. SPH is a lagrangian meshless fluid dynamics method and offers computational advantages for flow simulations with complex moving boundaries.[1–3]

Results

We compared our computed local RTDs with experimental results. Furthermore, we investigated a variety of geometrically different screw elements and how their local RTDs varied with screw speed, throughput and screw length. Computational tracer experiments were used to visualize the underlying flow mechanism leading to differences in the local RTD.

References

[1] Eitzlmayr A, Khinast J. Co-rotating twin-screw extruders: Detailed analysis of conveying elements based on smoothed particle hydrodynamics. Part 1: Hydrodynamics. Chem Eng Sci 2015. https://doi.org/10.1016/j.ces.2015.04.055.

[2] Eitzlmayr A, Khinast J. Co-rotating twin-screw extruders: Detailed analysis of conveying elements based on smoothed particle hydrodynamics. Part 2: Mixing. Chem Eng Sci 2015;134:880–6. https://doi.org/10.1016/j.ces.2015.05.035.

[3] Eitzlmayr A, Matic J, Khinast J. Analysis of Flow and Mixing in Screw Elements of Co-Rotating Twin-Screw Extruders via SPH. vol. 00. 2016. https://doi.org/10.1002/aic.

599138 Towards a Generic Model for Twin-Screw Wet Granulation: Calibration and Validation of an Improved PBM for Multiple APIs.

Abstract:

Recently the pharmaceutical industry has undergone changes in the way of producing oral solid dosages (tablets) from traditional inefficient and expensive batch production to continuous. Recent advances in the pharmaceutical industry include increased use of twin-screw wet granulation (TSWG) in the manufacturing of solid dosage and application of advanced modeling tools such as Population Balance Models (PBM). The twin-screw wet granulation unit is a unit operation of the ConsiGmaTM-25 continuous powder-to-tablet process line from GEA Pharma Systems. However, improved understanding of the physical process properties within the TSWG, expanding knowledge of the effect of the active pharmaceutical ingredient (API)/formulation on the granulation mechanisms, and the improvement of current PBM models are necessary to successfully operate this continuous production process.

As an initial effort to look into process mechanisms present in the twin-screw wet granulator, a unique dataset was collected at different locations inside the granulator and reported in (Verstraeten et al., 2017). Here, the main factor that affects the granule size distribution (GSD) is the liquid-to-solid ratio (L/S). At low L/S ratio, the GSD exhibits bimodality, whereas at high L/S ratio the GSD shows unimodal behavior. From this work, a 1D-PBM compartmental model was developed for predicting the GSD at the outlet of the granulator starting from the pre-blend using aggregation and breakage as the main phenomena that take place in the granulation process. The full model is composed of three PBM models in series (Van Hauwermeiren et al., 2018). One for each zone for which data was gathered: the wetting zone (pre-blend to after the addition of the liquid binder), kneading zone 1 (from after the wetting zone to after the first kneading elements), kneading zone 2 (just before the second kneading elements up until the end of the granulator barrel).

Since none of the traditional kernels from the literature could model the bimodal behavior observed from the measurements in the wetting zone, a new aggregation kernel was proposed that was capable to capture different types of behavior at different L/S ratio conditions with one single kernel. The breakage kernel (only used in the kneading zones) is a combination of attrition and uniform breakage (Van Hauwermeiren et al., 2018). This work was conducted for a single formulation, therefore, for general applicability, this needs to be extended by adding the effect of formulation properties. In that way, a true generic PBM can be constructed.

In the present work the 1D-PBM compartmental model was calibrated with a new dataset collected for two hydrophobic and two hydrophilic formulations, with different process conditions, at 4 different locations: the wetting zone, kneading zone 1, kneading zone 2, and at the end of the granulator.

To attain the desired behavior in the wetting zone, some mathematical modifications were made to the aggregation kernel to accurately predict the location and the size of the two peaks for the bimodal behavior and simultaneously incorporate the influence of the L/S ratio. Further, the model parameters that affect the bimodal behavior were linked to specific formulation properties.

The aforementioned changes in the aggregation kernel, constitute improvements to the model reducing the number of parameters to be calibrated from five to two. This parameter reduction makes the model more clear and interpretable. Further, it removes any parameter identifiability issues so that a clear link between model parameters and both process settings as well as formulation properties can be determined.

The PBM was calibrated and validated for multiple formulations (with different APIs content) and process settings. These developments allow us to assess the effect of the formulation properties on granulation behavior. For a new formulation, its properties can now be used to make a simulation of the possible GSDs so that with only a limited number of experiments, the optimal range can be determined. This general PBM for TSWG will greatly reduce the amount of material needed for experimentation and it can reduce the time-to-market for a new product.

607202 Developing a Digital Twin for the L.B. Bohle Tablet Coating Process

Abstract:

Tablet coating is a common unit operation in pharmaceutical manufacturing. Process engineers in the pharmaceutical industry are interested in optimizing the perforated pan coater operation to get low inter-tablet coating variability. In a tablet coater, tablet mixing, spraying and drying occur simultaneously. Tablets are mixed as a result of drum rotation and the action of helical baffles mounted inside the drum. The coating solution is sprayed on the tablet bed using spray nozzles. Hot air is directly supplied to ensure efficient drying of the coated tablets. Performing physical experiments with coaters to determine the optimal set of process parameters is, more often than not, impractical. Computer simulations incorporating the detailed physics of tablet coating offer a viable alternative to experiments.

In the current study, a coupled CFD-DEM model is developed to understand the impacts of coater hardware and process parameters on the inter-tablet coating variability in an L.B. Bohle tablet coater (Figure-1). Impact of operational parameters like drum rotation speed (Figure-2), fill level, tablet shape & size, spray rate, spray area (determined by spray gun-to-tablet distance and spray pattern air flow rates), is studied. The model was also used to compare the performance of the coaters from laboratory to industrial scales. The results from the simulation were used to understand which parameter plays a crucial role in determining the coating variability. The model results were also used to evolve guidelines for scale-up.

Figure-1: L.B. Bohle Tablet Coater (Model)

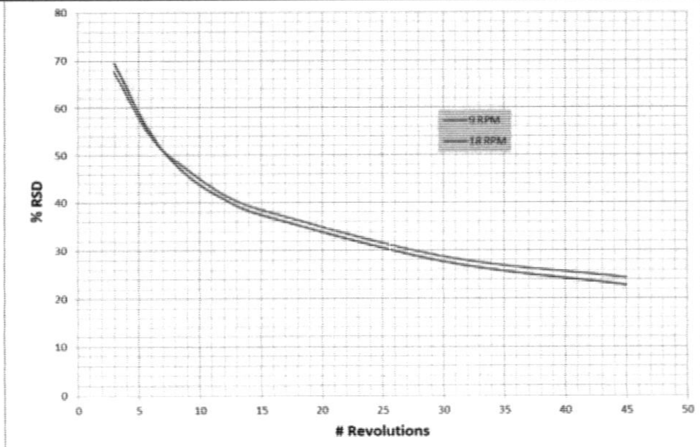

Figure-2: Impact of Drum speed on Inter-Tablet Coating Variability (Model Output)

594628 Implementation of a Dynamic Mechanistic Model Towards Optimal Design and Scale-up of an Atypical Freeze-Drying Process

Abstract:

Freeze-drying is widely used by pharmaceutical companies for manufacturing materials that are otherwise sensitive to moisture or high temperatures. A freeze-drying cycle is energy intensive since it happens over several days with the primary drying step being the longest. The energy and time required to maintain the desired shelf temperature and support the vacuum for several days at a time contributes towards a significant cost factor. In an optimal freeze-drying cycle, the drying time must be designed to be energy efficient as well as meet the product critical quality attributes (CQAs).

This work illustrates process optimization of a previously designed sub-optimal cycle, following a model-based design of experiment approach to minimize the cycle time while meeting the product CQAs. This cycle optimization was unique in that the formulation contained acetic acid and no bulking agent. These atypical formulation components led to a rare freeze-drying cycle optimization opportunity.

A mechanistic, non-steady state process model with a moving boundary (sublimation interface) was used in this work [1]. The cycle time was reduced by 50% (~2.5 days). The process map obtained from the model was used to establish the proven acceptable ranges (PARs) on the critical process parameters and identify the design space. The process was then scaled from lab to pilot plant based on the identified design space. This entire process optimization workflow required less than ~20 grams of the active pharmaceutical ingredient (API) in experimentation.

References

[1] Liapis and Bruttini, 1994, A theory for the primary and secondary drying stages of the freeze-drying of pharmaceutical crystalline and amorphous solutes: comparison between experimental data and theory. Sep. Technol., 4, 144-155

605034 Numerical Modeling of Deagglomeration of Cohesive Particles By Turbulence

Abstract:

The transport and deposition of fine particulates in turbulent flows play important roles in many engineering and medical systems. Examples include dry powder inhalers for drug delivery and fluidized bed reactors. Micron-sized particles tend to form aggregates due to inter-particle cohesion. The dynamical evolution and morphology of these aggregates involve a complex interplay between turbulent stresses and inter-particle cohesive forces. As a result, particle clumping can arise under various circumstances, which is known to compromise the performance of the aforementioned systems. Of particular interest to the present study is turbulence-induced breakup of fine particulate aggregates. Solid particles with diameters smaller than the Kolmogorov length scale ($d_p < \eta$) are initially aggregated into a spherical 'clump' of diameter $D > \eta$ and placed in homogeneous isotropic turbulence. Parameters are chosen relevant to powder suspended in air such that cohesion due to van der Waals is important. Simulations are performed using a CFD-DEM framework that models two-way coupling between the fluid and solid phases and resolves particle-particle interactions. Aggregate breakup is investigated for different Adhesion numbers Ad, Taylor micro-scale Reynolds numbers Re_λ and nondimensional clump sizes D/d_p. The intermittency of turbulence is found to play a key role on the early-stage breakup process, which can be characterized by a turbulent Adhesion number Ad_η that relates the potential energy of the van der Waals force to turbulent shear stresses. A scaling analysis shows that the time rate of breakup for each case collapses when scaled by Ad_η and an aggregate Reynolds number proportional to D. A phenomenological model of the breakup process is proposed that acts as a granular counterpart to the Taylor Analogy Breakup (TAB) model commonly used for droplet breakup. Such a model is useful for predicting particle breakup in coarse-grained simulation frameworks, such as Reynolds-averaged Navier–Stokes, where relevant spatial- and temporal-scales are not resolved.

605325 Regimes of Impact-Induced Erosion of Particles of Varying Strength

Abstract:

Particle breakage is relevant to many applications, from pharmaceutical processing to jet printing to defense technologies. Fundamental understanding of particle behavior during and after impact as a function of particle strength and impact velocity is needed for process optimization. In addition, detailed information about the number, shape, size and velocity of resulting fragments is difficult to obtain, particularly at conditions of high impact velocity. Discrete element method (DEM) simulations provides the capability to simulate the physics of particle impact and breakup at the micro-scale. This work explores varying particle breakage behavior upon impact with a flat surface. Impacting particles are modeled as spherical agglomerates consisted of a number of smaller constituent particles held together via a constant adhesive force characterized by a surface energy. Under the influence of a wide range of impact velocities and particle surface energies, five distinct behavioral regimes – rebounding, resting, fragmentation, pancaking, and shattering – are identified. In the rebounding regime, the coefficient of restitution decreases linearly as impact velocity increases. In the fragmentation regime, the rebound velocity generally decreases with increasing fragment size. These trends are consistent with the experimental results obtained in the study of Hassani-Gangaraj *et al.* (2018). Some regimes are only present at certain velocities while the extent of these regimes varies with particle strength. The type of regime depends largely on how the impact force is transmitted and dissipated through particle-particle contacts within the agglomerate.

Hassani-Gangaraj, M.; Veysset, D.; Nelson, K.A.; Schuh, C.A. "Melt-Driven Erosion in Microparticle Impact," *Nature Communications* **2018**, *9*, 1-7

606339 Breakage of Structured Particles: A Comparison between Stress-Strain and Impact Events

Abstract:

The breakage of structured particles is important in the food, automotive, electronic, and ceramic industries. Many of the structured particles created consist of relatively uniform or well-graded primary powder particles designed to have minimal segregation that are then made into larger particles that have intricate structures and designs. The tendency for structured particles to break depends on the strength of the bonds formed between the primary particles as well as on the structure of the larger particles. Breakage also depends on the set of forces that act on a structured particle during a breakage event. A population balance model is a reasonable means to describe breakage events and can be used to examine how breakage might occur. However, current formulations of the population balance model do not include the effects of global structure on breakage. They also do not include the effect of particle scale bond strengths, nor do they include any description of the influence of the different types of stress-strain or impact-induced stresses that are present with the material during processing. There is a difference between the breakage mode when the primary influence of breakage events consists of impact events as compared to stress-strain events.

The goal of this paper is to look at particle scale strength as a function of particle size distribution and then examine the differences in breakage of structured particles during stress-strain events versus impact events. A population balance model is used to determine the breakage selectivity coefficients. These coefficients are then correlated with the mode of breakage (impact versus stress-strain) using FEM models of stress induced by each type of breakage. However, the breakage rate constants are correlated to the strength as a function of the particle size distribution of the primary particles. Combining the breakage pattern computed by FEM analysis of particles with particle scale strength models and the population balance model provides a rich toolset to understand and predict breakage. A set of structured particles (balls, rods, crosses, and grid-shaped) were created from a consistent set of cohesive particles. These particles were then subjected to stress-strain events and impact events. The breakage studied by the population balance model and FEM analysis of structured particles were used to determine the breakage selectivity behavior. A polydisperse particle scale strength law was used to determine strength as a function of the primary particle size which was correlated to the breakage rate constants.

608742 Abradable Particles in DEM through Spherical Harmonics

Abstract:

Abrasion is a particular form of particle attrition in which a particle changes in shape by becoming smoother over time through the preferential removal of protrusions on the particle's surface. Fines are produced by abrasion which are much smaller than the particle's size. Abrasion plays a significant role in industrial applications, as does attrition more broadly. Fines can reduce flowability which can impair processing operations such as conveying, blending or tableting [1]. Attrition affects the bulk density, specific surface area, segregation behaviour and dissolution rate which has major implications for the quality of pharmaceutical products [2, 3]. The rehydration characteristics of infant formula are significantly affected by attrition [4]. Abrasion of fluid cracking catalysts is a significant cost factor in these processes as it necessitates the periodic replacement of the catalysts [5].
Because of its importance to a wide range of industrial particulate processes, quantifying abrasion, and ultimately understanding abrasion in order to better control it, are of fundamental importance. DEM simulations have contributed greatly to our understanding of granular materials behaviour in recent years [6], as they give access to particle-scale information such as interparticle contact forces which are difficult, if not impossible, to acquire directly from experiments. Unfortunately, most DEM simulations are based on spherical particles, while our everyday life experience shows that abrasion depends heavily on the particle shape: angular particles are more susceptible to surface breakage than rounded ones. Additionally, simulations of abrasion in DEM are very few and are mostly application-based [7].
In view of these considerations, a novel approach is presented for modelling non-spherical, abradable particles in DEM. The particle's shape is described by an expansion in spherical harmonic functions [8], with the high-degree terms mainly responsible for microscopic details of the shape, e.g., the surface texture. Mathematical considerations [9], together with recent experimental results [10], show that higher harmonics, corresponding to the surface texture, are the first to be eroded, while ellipsoidal shapes, found in abundance in nature, take much longer to become spherical, i.e., the sphere is the equilibrium shape [9]. Therefore the abrasion process can be represented through the sequential removal of the highest spherical harmonics (see Fig. 1). During abrasion, mass is lost in the form of fines, and an additional set of scaling factors ensures that the high-degree spherical harmonic expansion, representing the original shape, bounds the expansion at any lower degree: the abraded shape. The sequence of spherical harmonic expansions with decreasing order allows prediction of a particle's shape evolution during the abrasion process. Finally, this abrasion is related to microscopic wearing laws, based on energy considerations. A comparison between the abraded shapes obtained through the microscopic wearing laws and the predicted shapes is discussed. The implementation of this novel method in the open-source LAMMPS code is presented along with possible sources of performance improvement.

[1] A. Lekhal, K. P. Girard, M. A. Brown, S. Kiang, B. J. Glasser, J. G. Khinast (2003), Impact of agitated drying on crystal morphology: KCl-water system, Powder Technol. 132(2-3): 119–130.
[2] C. Hare, M. Ghadiri, R. Dennehy (2011), Prediction of attrition in agitated particle beds, Chem. Eng. Sci. 66(20): 4757–4770.
[3] X. Hua, J. Curtis, Y. Guo, B. Hancock, W. Ketterhagen, C. Wassgren (2015), The internal loads, moments, and stresses in rod-like particles in a low speed, vertical axis mixer, Chem. Eng. Sci. 134: 581–598.
[4] K. J. Hanley, E. P. Byrne, K. Cronin, J. C. Oliveira, J. A. O'Mahony, M. A. Fenelon (2011), Effect

of pneumatic conveying parameters on physical quality characteristics of infant formula, J. Food Eng. 106(3): 566 236–244.
[5] J. Reppenhagen, J. Werther (2000), Catalyst attrition in cyclones, Powder Technol. 113(1-2): 55–69.
[6] G. Lu, J. R. Third, C. R. Müller (2015), Discrete element models for non-spherical particle systems: From theoretical developments to applications, Chem. Eng. Sci. 127: 425–465.
[7] P. W. Cleary, R. D. Morrison (2016), Comminution mechanisms, particle shape evolution and collision energy partitioning in tumbling mills, Minerals Eng. 86: 75–95.
[8] E. J. Garboczi (2002), Three-dimensional mathematical analysis of particle shape using X-ray tomography and spherical harmonics: Application to aggregates used in concrete, Cem. Concr. Res. 32(10): 1621–1638.
[9] J. F. Bloore (1977), The shape of pebbles, Math. Geol. 9: 113–122.
[10] I. Deiros Quintanilla, G. Combe, F. Emeriault, C. Voivret, J. Ferellec (2019), X-ray CT analysis of the evolution of ballast grain morphology along a Micro-Deval test: key role of the asperity scale, Granul. Matter 21: 30-1–30-12.

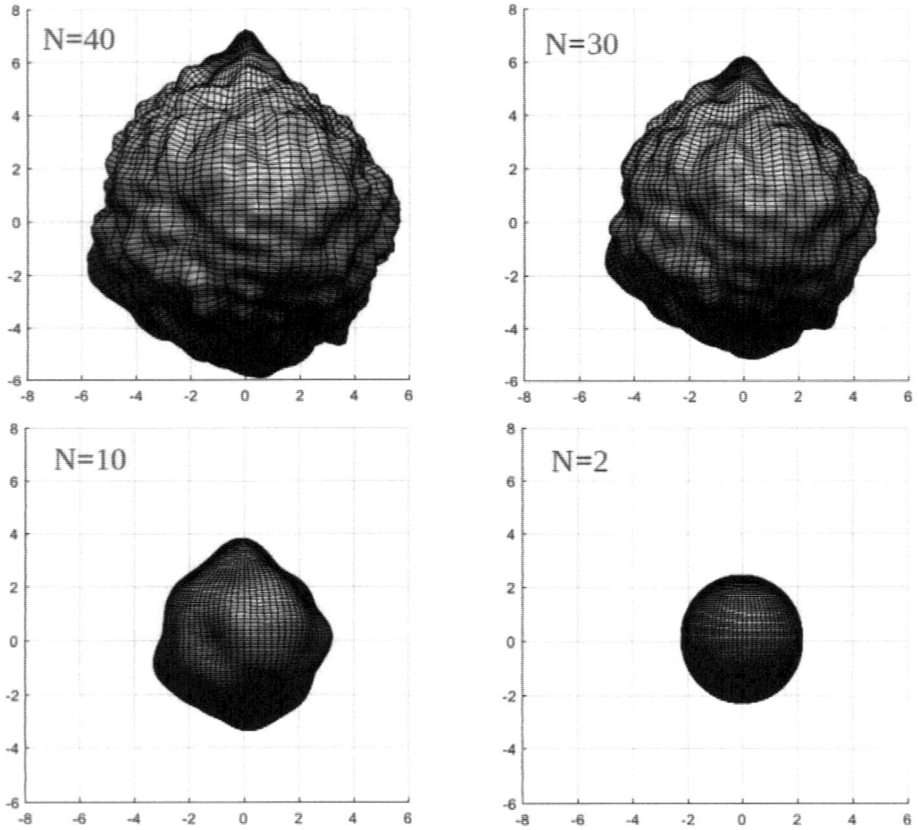

Figure 1. Four expansions for N=40, 30, 10 and 2 obtained by eliminating higher harmonic terms from the expansion and multiplying by suitable the scaling factors.

610419 Investigation into the Role of Holdup on the Gas and Particle Flow Pattern in a Spiral Jet Mill

Abstract:

Both breakage and classification occur within the main chamber of a spiral jet mill. The result is a complex fluid flow field that is difficult to analyse using traditional experimental techniques, whilst underlying mechanisms of particle size reduction remain poorly understood [1, 2]. Using Discrete Element Method (DEM) modelling, coupled with Computational Fluid Dynamics (CFD) the effect of hold-up has on both the fluid and particle phase within the mill. The design of the mill domain is based on the Hosokawa Micron AS-50 spiral jet mill (Runcorn, UK) and five mass loadings were used: 0.4 g, 0.8 g, 1.2 g, 1.6 g & 2.0 g. It was found that as the mass of material held within the bed increases, the average particle velocity decreased, along with the span of particle velocity distribution. As a result, the energy associated with an average collision decreased. It was also found that the bed has its own velocity gradient, and the solid particles travel with a greater velocity towards the surface. The result is that greatest amount of energy transfer takes place where the largest presence of shearing is found. This is located along the bed surface and on the back-face of each jet, as shown in Figure (1). These results are consistent with the experimental work of Kurten and Rumpf [3] and Luczak *et al.* [4]. Finally, it was found that the fluid velocity surrounding the classifier decreases with increasing particle load. The decrease in fluid velocity is due to damping of the fluid field, as momentum is transferred to stabilise the particle bed.

1. MacDonald, R., et al., *The spiral jet mill cut size equation.* Powder Technology, 2016. **299**: p. 26-40.
2. Bnà, S., et al., *Investigation of particle dynamics and classification mechanism in a spiral jet mill through computational fluid dynamics and discrete element methods.* Powder Technology, 2020. **364**: p. 746-773.
3. Kürten, H. and H. Rumpf, *Strömungsverlauf und Zerkleinerungsbedingungen in der Spiralstrahlmuhle.* Chemie Ingenieur Technik, 1966. **38**(11): p. 1187-1192.
4. Luczak, B., et al., *Visualization of flow conditions inside spiral jet mills with different nozzle numbers– Analysis of unloaded and loaded mills and correlation with grinding performance.* Powder Technology, 2019. **342**: p. 108-117.

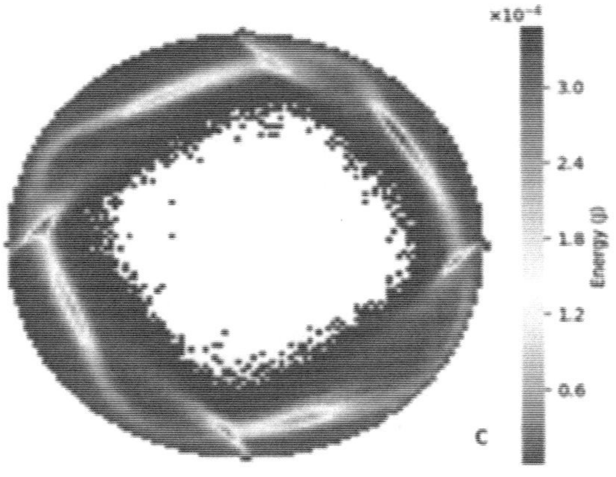

1: Particle contour plot highlighting total energy dissipated (J) from 1.6 g loading case recorded over 0.03 s.

611549 Breakage Functions for High Aspect Ratio Particles

Abstract:

Filtration of particles is influenced by the size-shape distribution of the particles. Therefore, it is necessary to determine how upstream processes affect the size-shape distributions in upstream processes before entering the filter. One such upstream process is particle breakage in stirred vessels such as crystallizers . Experiments on high aspect ratio particles such as paracetamol and urea demonstrate how the size-shape distribution changes with time. This work presents recent developments in modeling fragmentation of high aspect ratio crystals based on the experimental results in stirred suspensions.

602062 A Nonlinear Elasto-Plastic Bonded-Sphere Model for the Discrete Element Simulation of Biomass Particle Breakage

Abstract:

To enhance the current understanding of biomass fracturing mechanism so as to optimize the mechanical preprocessing techniques for biomass size reduction, a discrete element method (DEM) model has been proposed and developed with customized bonding laws designed specifically for biomass materials. The customized bonding laws are able to capture the elasto-plastic deformation within a single biomass particle under compression, bending, and twisting. To better match the experimental observations, two additional nonlinear force-displacement form for normal force contact freedom are proposed as well, i.e., quadratic form and square root form. The fracturing of biomass particles is simulated by allowing the bond to break based on the normal and shear strength criteria. Simulation cases of fracturing tests of loblolly pine cube have been developed and further validated by comparing to the physical experiments. It is shown the elasto-plastic DEM model with a square root force-displacement form can replicate the mechanical failure of pine cubes under shearing with a higher fidelity comparing to the linear elastic bonded-sphere model. By assigning different bonding strengths across the growth ring, the DEM model is able to simulate the anisotropic behavior of pine cubes with different fiber orientations, which is a key factor that should be considered in the biomass processing modeling. Future research will focus on understanding the grinding process of biomass particles for equipment optimization and energy consumption minimization.

594139 Mathematical Modeling of Continuous Milling Processes Using a Cell-Based PBM Approach: A Parametric Study

Abstract:

Particles with a high specific surface area usually exhibit desirable functional responses, such as high reactivity, fast dissolution, and good content uniformity. For this reason, various industries usually increase the specific surface area of particles, like pharmaceuticals, pigments, and minerals, using milling processes that decrease the size of particles for downstream processing or end use. However, current milling technologies are both energy-intensive and energy-inefficient [1]. Therefore, designing a milling process that can operate at or near an optimal condition would benefit the overall performance, but finding optimal milling operational conditions by trial and error is costly. For this reason, computer simulations of the process could provide significant operational insights while enhancing understanding of the milling process, thereby aiding in rational process design and optimization at a minimum cost.

Several modeling approaches, such as the discrete element method and the population balance model (PBM), have attempted to develop a fundamental understanding of the milling processes. Among these approaches, only the PBM provides a quantitative analysis of the spatio-temporal evolution of the particle size distribution (PSD) during milling. In this theoretical study, we developed a cell-based PBM that takes into account the complex nonlinear breakage kinetics and the degree of mixing in an open-circuit continuous dry mill with a discharge screen, unlike previous studies [2–6]. In the context of dry milling, the proposed mathematical model is the first in literature that accounts for (i) the elusive nonlinear breakage kinetics that originate from the multi-particle interactions, (ii) a non-ideal degree of mixedness, and (iii) particle separation via an ideal discharge screen. Various back-mixing ratios (R) and number of cells (n) were used to emulate different extents of particle mixing in the mill including the idealized cases of perfect mixing and no back-mixing (plug flow). Other simulations were performed to examine the impact of cushioning action of fine particles and the screen size. The simulation results were reported in terms of cumulative PSD at steady-state and during the transient period. Apart from the PSD of the product at the mill outlet, spatial variations of the PSD inside the mill were also examined.

For continuous mills without a screen, the cell-based PBM with $n = 1$ exactly emulated the PBM solution of an idealized perfect mixing scenario, whereas the cell model with $n = 60$ nearly represented the ideal plug-flow (no back-mixing) scenario, with an increase in n converging to the PBM solution for the plug-flow obtained previously in the literature. These findings demonstrated the consistency of the cell-based PBM. The simulation results also suggest that an increase in the number of cells and/or a decrease in the back-mixing ratio, emulating lower extent of back-mixing, led to finer product. While the cushioning action of finer particles led to coarser product PSD in the absence of a screen, this nonlinear impact becomes less important when a screen is present and especially when the screen opening size is small. This study has demonstrated for the first time in literature that the use of a discharge screen with a fine opening not only allows for passage of the desired fine particles and retention of the relatively coarse particles, but also it reduces the retardation impact of the fines on breakage kinetics, thus ultimately resulting in a finer product PSD.

References

[1] B.A. Wills, An introduction to the Practical Aspects of Ore Treatment and Mineral, Butterworth-Heinemann, Oxford, 2016.

[2] L.G. Austin, P.T. Luckie, D. Wightman, Steady-state simulation of a cement-milling circuit, Int. J. Miner. Process. 2 (1975) 127–150.

[3] H. Benzer, L. Ergun, M. Oner, A.J. Lynch, Simulation of open circuit clinker grinding, Miner. Eng. 14 (2001) 701–710.

[4] E. Bilgili, B. Scarlett, Numerical simulation of open-circuit continuous mills using a non-linear population balance framework: incorporation of non-first-order effects, Chem. Eng. Technol. 28 (2005) 153–159.

[5] O. Genc, S.L. Ergun, A.H. Benzer, The dependence of specific discharge and breakage rate functions on feed size distributions, operational and design parameters of industrial scale multi-compartment cement ball mills, Powder Technol. 239 (2013) 137–146.

[6] C. Mihalyko, T. Blickle, B.G. Lakatos, A simulation model for analysis and design of continuous grinding mills, Powder Technol. 97 (1998) 51–58.

598350 Spherical Agglomeration: Process Modelling and Validation for Kinetic Estimation

Abstract:

Background

Manufacturing particulate products with optimal performance requires tightly controlled processes across several industries such as pharmaceuticals. A key unit operation at the drug substance to drug product interface is crystallisation which is a critical separation and purification technique for the recovery of active pharmaceutical ingredients. It is therefore essential to understand and control crystal properties as they have shown to strongly impact downstream processes and the manufacture of pharmaceutical tablets [1].

Spherical agglomeration (post-crystallisation) is an emerging particle formation method which has the capability to improve problematic particle shapes (needles, rods, flakes) and increase particle size [2]. This involves the controlled addition of a partially-miscible solvent termed the bridging liquid to pre-suspended crystals through a single intensified unit operation. The bridging liquid should have a high affinity with the drug to agglomerate the crystals and under optimised conditions, compact and dense spherical agglomerates are produced. As a result, the produced agglomerates display high bulk densities, better flow properties, easier filtration and downstream processing in addition to suitability for direct compression of active pharmaceutical ingredients [3].

Process Modelling

Analogous to wet granulation mechanisms, spherical agglomeration exhibits different rate processes: wetting/nucleation of primary crystals by the bridging liquid, consolidation/growth of agglomerate nuclei, and attrition/breakage of the agglomerates through an immersion and distribution mechanism (Figure 1). A recent novel mathematical model was developed to predict kinetics and formation of agglomerate nuclei occurring in an immersion spherical agglomeration process [4]. Importantly, three different regimes for agglomeration nucleation were found; *immersion rate limited regime*, *collision rate limited regime* and *intermediate regime*. The mechanistic model was then successfully integrated along with key material properties and process parameters into a gPROMS FormulatedProducts® (PSE) flowsheet to develop a population balance model framework.

The aim of this work is to identify and assess the impact of material attributes, equipment and process parameters on product attributes through population balance modelling. The influence of key input model parameters such as the bridging liquid mean particle, liquid flow rate and addition time on the particle size, size distribution and porosity is investigated. Furthermore, characteristic process trends such as initiation and duration of nucleation are studied. As such, a comprehensive sensitivity analysis will be evaluated for risk mitigation approaches and greater process understanding of the predictive model. For kinetic estimation and model validation, carefully designed small-scale decoupled experiments will be performed to account for the different rate processes in spherical agglomeration.

References:

1. Variankaval, N., A.S. Cote, and M.F.J.A.J. Doherty, *From form to function: Crystallization of active pharmaceutical ingredients.* 2008. **54**(7): p. 1682-1688.

2. Pitt, K., et al., *Particle design via spherical agglomeration: A critical review of controlling parameters, rate processes and modelling.* 2018. **326**: p. 327-343.
3. Amaro-González, D. and B.J.P.t. Biscans, *Spherical agglomeration during crystallization of an active pharmaceutical ingredient.* 2002. **128**(2-3): p. 188-194.
4. Arjmandi-Tash, O., et al., *A new mathematical model for nucleation of spherical agglomerates by the immersion mechanism.* 2019. **4**: p. 100048.

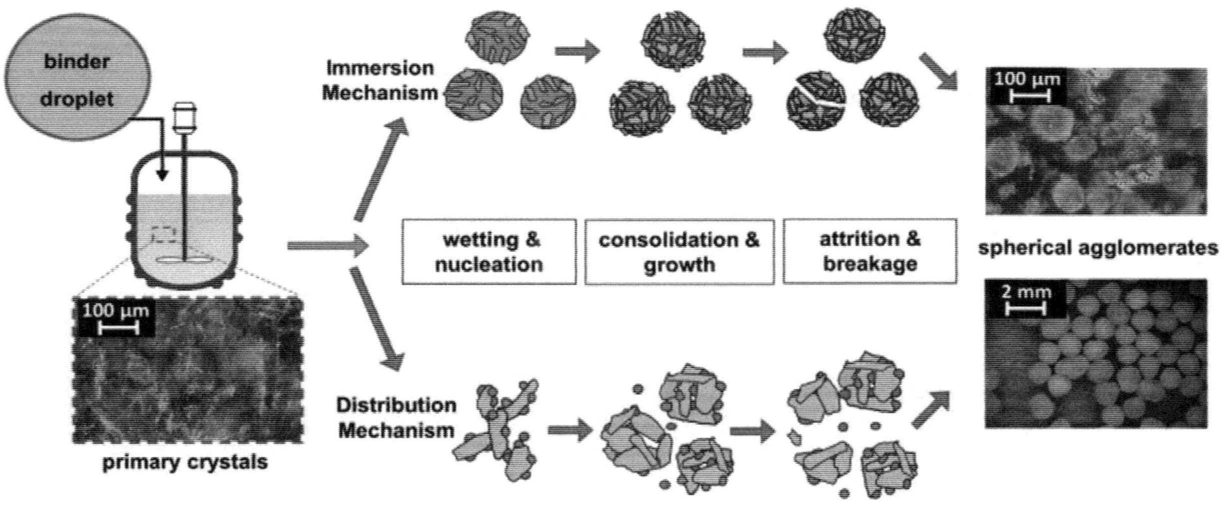

Figure 1: Schematic of rate processes and possible mechanisms occurring during spherical agglomeration [4].

599488 Understanding AUTO-Agglomeration of Dry ACTIVE Pharmaceutical Ingredients

Abstract:

Introduction:

Dry Active Pharmaceutical Ingredient (API) auto-agglomeration is a type of unwanted powder transformation that occurs without any additives. If uncontrolled, formation of hard, i.e. capable to withstand drug manufacturing process, API agglomerates may lead to an inadequate API distribution within the final formulation exhibiting the real risk to safety, efficacy and processability of drug products. Both, the fact that API auto-agglomeration may occur at, almost, any stage of drug manufacturing and/or handling processes as well as that a variety of general powder transformation mechanisms have been previously reported (for instance, mechanical, chemical, and triboelectric modes of bulk powder caking) make API auto-agglomeration phenomenon highly complex and challenging to anticipate and to prevent. However, despite the key role it has in the drug product performance and manufacturing, the number of studies addressing auto-agglomeration is currently limited and its underlying root-causes remain largely unknown. As a result, pharmaceutical industry relies mainly on the cost-ineffective trial and error approach when facing issues related to the unwanted API agglomeration.

The aim of this paper is to explore the correlation between particle properties, its processing/handling conditions and auto-agglomeration phenomenon as well as to establish a method/tool that would allow anticipation and, hence, better control of API auto-agglomeration leading to enhanced product quality, sustainability and cost-effectiveness. The experimental work was carried out to test the following hypothesis: API surface chemistry impacts its auto-agglomeration tendency. Ibuprofen recrystallised in different solvents was used as the main model API. Its auto-agglomeration tendency was tested using mechanical vibration that corresponds to drug handling conditions. Size distribution and particle shape of ibuprofen were analysed using SEM and G3 morphology, FTIR was used to confirm the identification of ibuprofen structure, and Instron to study the strength of the ibuprofen particles before and after the mechanical vibration.

Results and Discussion:

Four batches of ibuprofen of distinct morphologies were recrystallised in order to test the effect of crystal surface chemistry on its auto-agglomeration tendency. Collected FTIR data confirmed the ibuprofen structure in each case. SEM and morphology G3 analyses revealed that the shape regularity of recrystallised ibuprofen decreases with decreasing solvent polarity as follows: ibuprofen recrystalised from methanol (IbuMeth) > ibuprofen recrystalised from ethanol (IbuEth) > ibuprofen recrystalised from acetonitrile (IbuAce) > ibuprofen recrystalised from hexane (IbuHex), in line with previously reported data in the literature. Mechanical vibration experiments resulted in formation of visually detectable, relatively soft agglomerates that differ in size and shape across the batches. In addition, morphology G3 results indicated formation of agglomerates at microscale in the case of IbuEth and IbuHex being vibrated: d10, d50, and d90 of recrystallised IbuEth approximately doubled after mechanical vibration, whereas in the case of IbuHex this shift was less considerable. The strength of both IbuEth d90 as recrystallised and after mechanical vibration was tested using Instron revealing that the latter is approximately ¾ of the former.

Conclusion:

Detection of ibuprofen agglomerates upon mechanical vibration suggests there is a correlation between ibuprofen particle morphology, hence, particle surface chemistry and ibuprofen auto-agglomeration tendency. Further investigations of ibuprofen agglomerates as well as other model APIs agglomerates will include analysis of their strength (Instron, Nanoidenter), internal structure (X-ray micro computed tomography), and their surface chemistry (X-ray photoelectron spectroscopy).

595680 Modeling of Granule Size Distribution in Pharmaceutical Roller Compaction Process

Abstract:

Pharmaceutical roller compaction (RC) is a particle size enlargement process in which granules are produced by making ribbons from an active or placebo compound and milling them into smaller agglomerates. Determination of the granule size distribution (GSD) as the major product attribute requires extensive experimental efforts which are time consuming and material demanding. Modeling the impacts of process parameters and ribbon properties on granule attributes provides a quantitative insight into the RC granulation process. In this talk, we present a computational approach to simulate the size distribution of granules produced in the milling step of pharmaceutical roller compactors. RC process was emulated by making slugs and milling them with a lab hand-mill apparatus. The model was structured based on the GSD data obtained from different active compounds of Aspirin and Ibuprofen with various drug loads, including two commonly used excipients of Microcrystalline Cellulose PH102 and Lactose FastFlo 316. Experimental data were simulated using a Guassian distribution function, and the determined model parameters were correlated to ribbon Young's Modulus. Model was validated by data which were excluded from the training set. Results demonstrated the success of the model in predicting the GSD obtained from experimental measurements. This model holds only two fitting parameters which can be determined from regular material profiling techniques. This approach can be integrated with an RC model that can predict the ribbon attributes, and be utilized as a powerful computational tool for RC process space design.

608093 Infleunce of Natural and Synthetic Binders on Paracetamol Granules Properties

Abstract:

The role of binders in the formulation of tablets widely varies based on the intended application. Binders, or excipients, account for most drug formulations and provide enhanced properties that aid in drug delivery and bioavailability. Excipients are distinguished by their functionality, efficacy, safety, and qualify. Currently, scientists are seeking more naturally occurring sources (plant-based) to replace common synthetic binders such as polysorbates, povidone, polyvinylpyrrolidone (PVP) and polyethylene glycol (PEG). Natural excipients are considered non-toxic and more biocompatible. Overall, the tablet or granule cohesiveness, stability, manufacturability, and release characteristics will provide an indicator of the feasibility for using a given excipient in industrial-scale manufacturing applications. The purpose of this research is to investigate the influence various natural and semi-synthetic excipients on granule properties.

Paracetamol granules were prepared using the previously proven method of binder dropping. All excipients were dissolved into deionized water prior to placement on the powder bed. Analysis of the granules included weight and size analysis using Image-Pro Premier software, friability testing, compression testing, disintegration testing, and dissolution testing. Size measurements included diameter, aspect ratio, and projected area. Dissolution and disintegration tests were performed at a dip rate of 10 DPM in phosphate buffer solution at a pH of 7.0. The concentration of paracetamol in the solution was determined using UV-Vis spectrometry. Friability testing occurred for 100, 150, and 200 rotations. With the exception of PEG, the concentration of the excipient influenced the behavior of the paracetamol granules during dissolution and friability testing. However, the extent of the influence varied between the excipients. For most excipients, the increase in concentration produced larger granules that were more susceptible to breakage during friability testing. The bonding of the excipients determined whether the formed paracetamol granules immediately disintegrated or dissolved over time.

602835 Identifying Optimum Liquid-to-Solid Ratio Range for Microcrystalline Cellulose Based Formulations in Twin Screw Wet Granulation

Abstract:

Twin screw wet granulation (TSWG) is gaining increasing interest in the pharmaceutical industry as an effective continuous manufacturing process for solid oral dosage forms. Although multiple variables have been shown to impact various characteristics of granules produced by this technique, liquid-to-solid ratio is known to be one of the primary factors affecting granule attributes. Hence, identifying an optimum range of liquid-to-solid ratio can help streamline formulation development and minimize material requirements for TSWG process development. Here we present the results of TSWG studies conducted with microcrystalline cellulose-based (MCC) formulations of varying drug loads, using micronized acetaminophen as a model active compound. The blends were granulated using a GEA ConSigma-1 25 mm twin screw wet granulator with a standard kneading element screw configuration at a wide range of liquid-to-solid ratios until pasting was visually determined. It was observed that the paste points for different granulations were dependent on the level of MCC in the formulation. We couple this process-focused analysis with a lab-scale approach to establish viable ranges of liquid-to-solid ratios in advance of actual process development. Comparisons of lab-scale mechanical properties of wet granular material, various process parameters, and material and granule characteristics will be discussed, with guidelines for optimizing liquid addition levels proposed.

601969 Parameter Sensitivity Analysis of a High-Shear Wet Granulation Model for Experimental Design and Parameter Estimation.

Abstract:

A typical workflow in developing a new product via high shear wet granulation (HSWG) involves experimentally testing every combination of input variables, across 3 to 5 scale of operation. When done using traditional DoE approaches, the experimental effort increases as 2^n where n is the number of process and formulation parameters. This makes HSWG a costly process to formulate a design of experiments in time, money, and materials. Therefore, this work aims to apply a model-driven design workflow to identify the most critical process parameters and thus reduce and better target experiments to be performed.

The model-driven design focuses on developing a predictive and well-calibrated process model. The process model is based on a population balance modelling framework. Mechanistic understanding of the rate processes is incorporated through appropriate kernels. The most impactful modelling parameters of these kernels must be identified in addition to the operating parameters for the process. To do this, we adopt a Gaussian Process (GP) surrogate modelling approach to directly interrogate the wet granulation computational model. The GP is trained from an input space of the wet granulation computer model produced from randomly sampling the model parameters, normally distributed about the mean. Predicting the model output using a GP enables a reduction in the considerable computational effort required to analytically calculate the Sobol' indices for a Global Sensitivity Analysis (GSA).

Cross-validation ensures the GP surrogate model is capable of predicting the model outputs given all twenty modelling parameters. Subsequently, calculations of the Sobol' indices demonstrate that only four of the twenty parameters have sufficient impact to influence the DoE. The most critical process parameter is the liquid spray rate, dominating all four models outputs. The collision coefficient and the critical pore saturation are together the most important modelling parameters with respect to the fine fraction and D50, with each variable ascribing to 20% of the output variance alone and above 60% when including cross-effects. Whereas, for fine fraction the breakage coefficient is the most significant modelling parameter, resulting in a first-order Sobol' indices of 25% and negligible interactions. Interestingly, the total Sobol' index of the nuclei-to-drop diameter ratio is 50% making it the most dominant variable for granule porosity but it has little relevance towards the other outputs. Overall, the GSA has identified the critical process parameter and the impactful modelling parameters, and has enabled a proposal of an appropriate experimental design and model calibration workflow. By applying this workflow within model-driven design, industrial development is improved by determining more beneficial production conditions and reducing experimental effort by 30-60% compared to conventional approaches.

604611 Analysis and Evaluating Design Space of Continuous High Shear Granulation and Drying Process Using Gproms

Abstract:

Continuous wet granulation (high shear granulation) process is widely used in the chemical and pharmaceutical industries. This work aims to investigate the product properties and production efficiency of the continuous high shear granulation with experimental studies and simulation using gPROMS. In the continuous high shear granulation process, powder and liquid binder contact in a short resident time, where particles experience nucleation, growth and breakage. This work is also the first time to examine the effects of varying process parameters in a Schugi mixer (continuous high shear granulator). At the meantime, gPROMS is used to simulate the processing efficiency (yield) and the product properties (PSD, porosity, etc.) with the different formulation and process parameters in the granulation process.

600975 Robotic Platform for Execution, Analysis, and Optimization of Multistep Reactions in Continuous Flow

Abstract:

Automated systems for chemical synthesis integrated with computer-aided synthesis planning and algorithm-guided optimization of reaction conditions have helped relieve chemists from routine tasks and accelerated process development [1,2,3]. The synthesis of organic compounds during process development often involves multiple steps with both discrete (e.g., catalyst, solvent) and continuous (e.g., temperature, time) process variables. In continuous flow synthesis where multiple steps can be telescoped, an important question is whether for a given optimization objective (e.g., yield, productivity), the combination of optimal conditions for each step is the same as the globally optimal conditions for the entire sequence. The ability to analyze the process inline between two steps in addition to at the end of a multistep synthesis can provide valuable insight into the effect of process variables on individual steps.

Here, we present a robotically reconfigurable flow chemistry platform capable of executing, analyzing, and optimizing multistep reactions. The platform contains a library of process modules that can be placed in any order onto a process stack by the robot for performing reactions, separations, and inline analysis (liquid chromatography–mass spectrometry (LC-MS) and Fourier transform infrared (FT-IR) spectroscopy). Reagents are delivered to the process stack via reconfigurable fluidic connections which the robot can place onto a reagent "switchboard", enabling switching between multiple reagent candidates. The hardware was coupled to an optimization algorithm that employs optimal design of experiments (DoE) to intelligently navigate the design space with as few experiments as possible, and branch and bound (B&B) to handle discrete variables in addition to continuous variables [4]. The LC-MS and FT-IR modules were utilized simultaneously in multistep syntheses to analyze the process at multiple locations, providing reaction-specific information. Several case studies highlight the platform's capabilities for chemical synthesis and explore the question of how to find globally optimal reaction conditions for multistep flow synthesis.

Continuous production of a chiral amine in a packed bed reactor with co-immobilized amine dehydrogenase and formate dehydrogenase

Robert D. Franklin, Joshua A. Whitley, Adam A. Caparco, Bettina R. Bommarius, Julie A. Champion, Andreas S. Bommarius*

Abstract

A continuous flow packed bed reactor system was developed for the asymmetric reductive amination of 5-methyl-2-hexanone with a chimeric amine dehydrogenase (AmDH) and the formate dehydrogenase (FDH) from *Candida boidinii*. Both enzymes were co-immobilized onto the Nuvia® IMAC resin from Bio-Rad which employs the well-established binding chemistry between 6xHis-tagged proteins and immobilized Ni^{2+}. Binding was robust under reaction conditions and continuous operation in a packed-bed flow column for more than five days was achieved. At 40 °C, conversion of 20 mM 5-methyl-2-hexanone ranged between 19 and 48%, with volumetric productivities between 443 and 107 g/L·day. The apparent biocatalyst half-life was 159 hours and the absence of both internal and external mass transfer limitations was confirmed. Conversion was limited by modest competitive product inhibition of (*R*)-amine on ketone and an excess of enzyme over cofactor concentration, which precludes the use of Michaelis-Menten kinetics and results in a rate decrease of 80% compared to conditions where [S] >> [E]. The relationship between flow rate, conversion, and productivity was examined, with important lessons for reaction engineering. While there is still more optimization to be done, the reported work represents the first stable application of AmDHs in a continuous flow reactor for the production of chiral amines.

1. INTRODUCTION

1.1 Background

Amine dehydrogenases (AmDHs), first engineered in 2012,[1] are a family of enzymes which catalyze the reductive amination of prochiral ketones to form chiral amines. Chiral amine functional groups can be found in many of the current top selling small molecule drugs such as sitagliptin (Januvia®), oseltamivir phosphate (Tamiflu®), and dolutegravir (Tivicay®)[2] and their efficient production is an important challenge in the pharmaceutical industry. The chiral purity of many active pharmaceutical ingredients (APIs) must be greater than 99% to ensure efficacy or avoid adverse side effects.[3] The naturally high enantioselectivity of enzyme catalysts, among other benefits, have made them increasingly interesting to both industry and academics[4-7] for the production of chiral amine-containing APIs or their precursors.

The family of AmDHs has been under continuous development by groups around the world since their first publication eight years ago. The first examples were the leucine amine dehydrogenase (L-AmDH) engineered from the natural leucine dehydrogenase (LeuDH) from *Geobacillus stearothermophilus*[1] and the phenylalanine amine dehydrogenase (F-AmDH) engineered from the natural phenylalanine dehydrogenase (PheDH) from *Bacillus badius.*[8] For both L-AmDH and F-AmDH, homologous residues in the active site responsible for binding the carboxylic acid group on the natural amino acid substrate were mutated to instead promote binding to methyl ketones. A similar strategy has been employed by other groups to generate AmDHs based on other PheDH and LeuDH scaffolds.[9-11] Recently, an AmDH was engineered from the ε-deaminating L-lysine dehydrogenase from *Geobacillus stearothermophilus* which showed activity toward substrates not as easily accessible with the previous engineered AmDHs.[12]

Through high-throughput screening and metagenomic mining several naturally occurring AmDHs were also recently identified.[13, 14]

The application of continuous flow chemistry to biocatalysis has also been increasingly attractive in the pharmaceutical and fine chemical industries.[15] The current paradigm for biocatalysis in most of the pharmaceutical industry calls for large stirred batch reactors into which enzyme, either soluble or immobilized, is added as a reagent. After the reaction is complete, processing steps are required to remove the enzyme. Continuous flow reactions with immobilized enzymes combine the reaction and biocatalyst separation steps into a single unit operation. The dimensionless operational lifespan of the enzyme, known as the total turnover number (TTN), can also be significantly higher in flow reactors due to the stabilizing effects of immobilization[16] and the ease of enzyme reuse compared to batch reactors.

There is a growing interest in the pharmaceutical industry to replace the nearly ubiquitous batch reactors with continuous flow chemistry where possible.[17] Continuous processing can

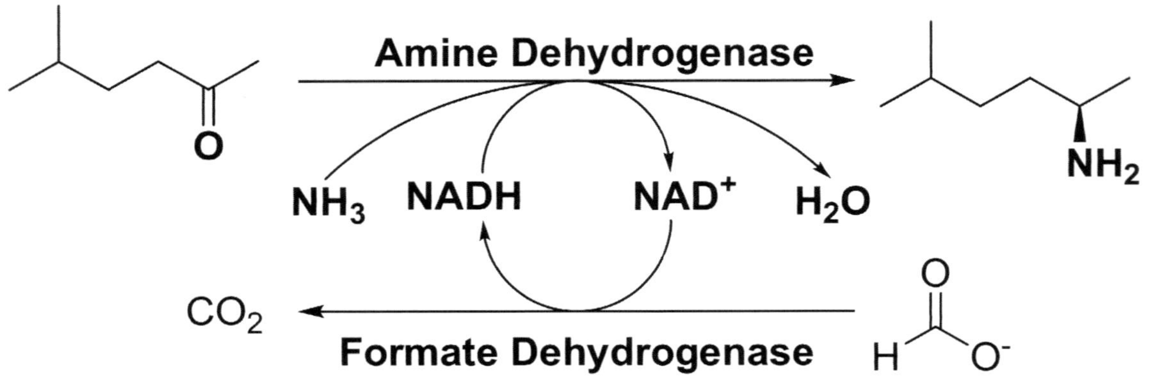

Figure 1 Reductive amination of 5-methyl-2-hexanone with cofactor regeneration

offer key advantages over batch reactors in terms of reducing manufacturing cycle time, plant footprint, waste, and thus both reduced capital and operating costs. Since 2013, the U.S. Food and Drug administration has been encouraging pharmaceutical companies to pursue continuous processes.[18] Continuous processing is especially attractive for heterogeneous catalysis, as product separation from the solid phase occurs simultaneously with reaction. Continuous flow processes are also easier to monitor and control, as they tend to operate at a single controlled point rather than in a transient mode. Ease of control over reaction conditions tends to result in tighter control over product purity. For enzymes specifically, continuous processing can alleviate product inhibitions due to the *in-situ* removal of products.

The primary goal of the reported work was to demonstrate a stable application of continuous flow chemistry to the production of chiral amines with an AmDH. The chimeric amine dehydrogenase cFL1-AmDH,[19] which will subsequently be referred to here as AmDH, was chosen for its thermostability and high expression levels. AmDH, along with the formate dehydrogenase (FDH) from *Candida boidinii*, was used to convert 5-methyl-2-hexanone (5M2H) to (*R*)-5-methyl-2-aminohexane with ammonium formate and NADH as co-substrates (see Figure 1). Initially, the biocatalysts in solution were fed into a continuously stirred enzyme membrane reactor such as the one employed by Rogers *et al.* to study the deactivation of TEM-1 beta-lactamase.[20] Unfortunately, the tendency of AmDH to aggregate in solution under reaction conditions rendered continuous soluble operation impractical. Our focus then shifted toward co-immobilizing AmDH and FDH onto a solid support for use in a packed bed reactor (PBR), which showed much greater promise.

1.2 Enzyme immobilization

The immobilization support for AmDH and FDH needed to be robust under reaction conditions, have tight binding without significant leaching, be amenable to packing into a flow column with good flow properties, and be readily available. Several groups in the past five years have explored co-immobilization of other amine dehydrogenases with FDH and with other enzymes such as ω-transaminases and alcohol dehydrogenases.[21-24] While the reported immobilization strategies met some of the listed criteria, none seemed to meet all of them. In this work, all twelve of the Immobead® enzyme carriers offered commercially by ChiralVision, which included ten different binding chemistries, were screened for their effectiveness. Despite promising binding efficiency on a few of the beads, the physical properties of the Immobeads® yielded significant channeling effects when tightly packed into a column which rendered them unsuitable.

Ultimately the immobilization strategy chosen for this study was to use a commercial immobilized metal affinity chromatography (IMAC) resin from Bio-Rad called Nuvia IMAC. This resin is charged with divalent nickel ions which have been immobilized to the 49 μm acrylamido polymer beads with covalently bound nitriloacetic acid (NTA). A polyhistidine (6xHis) tag is recombinantly expressed at either the C- or N- terminus of a target enzyme, and under the right conditions this 6xHis tag tightly chelates the immobilized nickel ion.[25] Once chelated to the nickel, the 6xHis-tagged proteins are tightly bound but still retain their folded structure and activity. While IMAC resins have traditionally been used for protein purification rather than enzyme immobilization, Merck and Co. recently demonstrated the use of Nuvia IMAC for the production-scale immobilization of multiple enzymes in the production of their Phase 3 HIV drug islatravir (MK-8591).[26] EnginZyme in Sweden has also developed a range of IMAC resins for enzyme immobilization with slightly different metal ion immobilization chemistry. The Turner group demonstrated the use of EziG Amber, which employs immobilized Fe^{3+} rather than Ni^{2+}, for use with co-immobilized AmDH and FDH in a packed bed flow reactor.[27] The authors in that study reported 68% conversion of a 10 mM inlet stream of 4-fluorophenylacetone to form 4-fluoroamphetamine and a volumetric productivity over 300 g/L·day. However, this level of productivity was only stable for a reported 3 hours before rapid deactivation of the catalyst. In the present study, we report the successful application of the Nuvia IMAC resin to immobilize AmDH and FDH in a flow reactor with productivity values slightly superior to those found for the EziG immobilization, but with an apparent half-life exceeding five days instead of three hours.

2. MATERIALS AND METHODS

2.1 Sources of Raw Materials

Materials used for enzyme expression and purification, as well as product quantification with benzoyl chloride were published previously.[28] Additional materials are listed below.

Ni-charged Nuvia® IMAC resin (Bio-Rad Laboratories, Hercules, CA, USA), 5-methyl-2-hexanone 99% (Alfa Aesar, Ward Hill, MA, USA), 5-methyl-2-aminohexane 98% (Santa Cruz Biotechnology, Dallas, TX, USA)

2.2 Buffers Used for Expression, Immobilization and Reaction

Binding Buffer: 1X phosphate-buffered saline (PBS), 300 mM additional NaCl, 10 mM imidazole; **Wash Buffer 1:** 1X PBS, 300 mM additional NaCl, 20 mM imidazole; **Wash Buffer 2:** 1X PBS, 300 mM additional NaCl, 40 mM imidazole; **Elution Buffer:** 1X PBS, 300 mM additional NaCl, 250 mM imidazole; **Reaction Buffer:** 2M ammonium formate (NH_4COOH) adjusted to pH 8.5 with 2M NH_4OH, 20 mM 5-methyl-2-hexanone, variable concentration of NAD^+ between 0.1 and 4 mM; **Autoinduction Media:** 1X PBS, 5 g/L NaCl, 20 g/L tryptone broth, 5 g/L yeast extract. This solution is autoclaved in 500 mL batches in 2.8 L baffled shake flasks. After it has cooled, a 50 mL mixture containing 12.5 mg kanamycin, 6 mL glycerol, 0.5 g glucose, and 2 g lactose is filtered through a 0.22 μm syringe filter into the baffled flask.

2.3 Enzyme Expression

Engineered cFL1-AmDH[19] and wild-type *cb*-FDH are expressed separately, but with identical procedures. The gene is stored in the pet28a vector with NdeI and XhoI restriction sites, giving the enzyme an N-terminal 6xHis-tag. After transformation into *E. coli* BL21(DE3) cells, frozen cell stocks are stored at -80 °C in 10% DMSO. 5 mL overnight cultures (standard LB media with 25 mg/L kanamycin) are seeded from the frozen stock and allowed to grow for 16 hours shaking at 37 °C and 250 rpm. The 5 mL starter culture is used to seed 500 mL of autoinduction media. After incubation for 24 hours at room temperature in a 125 rpm orbital shaker, cells are spun down at 2,800 g for 35 minutes. The supernatant is removed, and cell pellets are stored at -20 °C or -80 °C for short-term or long-term storage, respectively.

2.4 Enzyme Purification

Frozen cell pellets are resuspended in an amount of binding buffer based on the size of the expression culture which produced the pellet. A pellet from a 500 mL expression is resuspended in 40 mL of binding buffer. After resuspension, the mixture is split into 20 mL aliquots and sonicated to break open the cells. After cell lysis, the crude lysate is centrifuged at 12,000 g for 40 minutes at 4 °C. Each 20 mL aliquot of clarified lysate is incubated in a separate gravity filtration column for 1 hour at room temperature with 2 mL of Bio-Rad Nuvia IMAC resin which had been previously washed with 30 mL of DI water and 40 mL of binding buffer. After incubation, the column is drained, and subsequent washes are completed first with 10 mL of wash buffer 1 followed by 10 mL of wash buffer 2. After washing, elution buffer is added in 2.5 mL increments and is collected after incubation at room temperature for 5 minutes. Subsequent fractions are collected until the protein concentration in the outlet is less than 1 mg/mL as measured by the Bradford assay.[29] Elution fractions are buffer exchanged back into the binding buffer using a PD10 desalting column (GE Lifesciences, Chicago, USA) with standard procedures. After desalting, pure protein solutions are stored at 4 °C until needed.

2.5 Enzyme Immobilization

Empty Nuvia IMAC Ni-NTA resin is stored as a 66% slurry in 20% ethanol. Per documentation by Bio-Rad, the resin has a particle size which ranges from 38 to 53 µm.For immobilization, the desired volume of resin was first washed in a gravity filtration column with 10 bed volumes of deionized water, followed by 10 column volumes of binding buffer. With the column equilibrated in binding buffer, enzyme can be introduced to the resin in form either of purified protein or clarified lysate. After incubation overnight on a rotary mixer (Roto-Mini™ Rotator Series R2020, Benchmark Scientific Inc, Sayreville, NJ, USA), the resin is washed with 10 bed volumes of wash buffer 1, followed by 10 bed volumes of 2.0 M NH_4COOH/NH_4OH at pH 8.5. The resin is now ready to be packed into a column or transferred to a batch reaction. The procedures followed for packing the resin in the column were provided in the instruction manual for the IMAC resin provided by Bio-Rad.

2.6 Binding Capacity Determination

The binding capacity of the Nuvia IMAC resin was determined for both AmDH and FDH separately in a continuous flow binding experiment. The reactor system was set up as shown in Figure 2. Stock solutions of AmDH (100 mL, 0.72 mg/mL) and FDH (100 mL, 0.468 mg/mL) in Binding Buffer were prepared and kept at 4 °C or on ice for the duration of the experiment. For each enzyme, 0.666 mL (settled bed) of fresh resin were packed into the reactor column (Omnifit™ EZ SolventPlus™ 10 mm x 100 mm chromatography column with one adjustable endpiece, Diba Industries, Danbury, CT, USA) and the column was attached into the flow system. After equilibration of the column with at least 20 mL of binding buffer, the inlet was switched to the enzyme solution, which flowed slowly at 0.168 mL/min through the column. 384 fractions of 230 μL were collected into four UV-transparent 96-well plates using a programmable fraction collector. The change in absorbance at 280 nm between the enzyme stock solution and the reactor outlet could then be used to determine the amount of enzyme which had bound to the enzyme out of each mL that flowed through the resin. For both enzymes, the amount of stock solution was not large enough to fully saturate the resin, so the remaining portion of the binding isotherm was extrapolated. Integration of the binding curves (see Figure S1) yields the specific binding capacities of the resin for each enzyme.

2.7 AmDH to FDH Binding Ratio Optimization

Based on the binding capacities obtained from Section 2.6, and the assumption that the two enzymes have the same rate of binding, the amount of AmDH and FDH needed to saturate 0.2 mL of Nuvia IMAC resin plus 25% was calculated for five mass ratios. The ratios and masses of each enzyme required for each ratio are listed in Table S1 in the Supplement Information~~**Error! Reference source not found.**~~. For each ratio, 0.2 mL of resin are prepared as described in Section 2.5, and purified enzyme is prepared as described in Section 2.4. The required amounts of each enzyme (in binding buffer) are combined in a 15 mL conical tube and diluted up to 15 mL with binding buffer. The enzyme mixtures are then used to slurry transfer the resin from the gravity columns to the conical tubes. After incubation overnight, rotating at 20 rpm at room temperature, the five immobilized biocatalyst samples are washed with 20 bed volumes of 2.0 M NH_4COOH/NH_4OH at pH 8.5 and then transferred to 50 mL conical tubes, each containing 50 mL of reaction buffer with 1 mM NAD^+. The batch reactions proceed for 21 hours at room temperature rotating at 20 rpm, with samples taken at selected time points by stopping the rotisserie mixer and allowing the resin to settle for 5 minutes before collecting 100 μL of supernatant, which is stored at 4 °C until all time points have been collected.

2.8 Continuous Flow Reactor System Setup

A representative schematic of the continuous flow reactor used in the present study can be found in Figure 2.

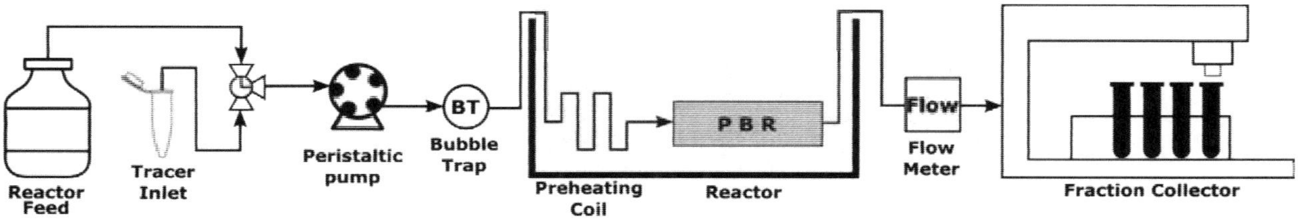

Figure 2 Block flow diagram of the continuous flow packed bed reactor system

Flow into the system comes either from the substrate feed bottle or tracer inlet tube. A selection valve (V-100L, Idex Health & Science, Oak Harbor, WA, USA) allows for rapid switching between inlets. A peristaltic pump (Ismatec™ IPC-4, Cole-Parmer, Vernon Hills, IL, USA) with 1.30 mm ID 2-stop santoprene tubing (Precision Glassblowing, Centennial, CO, USA) pumps fluid first through a bubble trap (Omnifit™ 006BT, Diba Industries, Danbury, CT, USA). The fluid then flows through 1.6 meters of 1/16 inch O.D. stainless steel tubing followed by the reactor (Omnifit™ EZ SolventPlus™ 10 mm x 100 mm chromatography column with one adjustable endpiece, Diba Industries, Danbury, CT, USA). The steel tubing and reactor are both submerged in a water bath with the temperature controlled by a thermal immersion circulator (Precision™ Cooker, Anova Culinary, San Francisco, CA, USA). Once through the reactor, fluid flows through a digital flow meter (SLI-1000, Sensirion AG, Stäfa, Switzerland) and finally into an automated fraction collector (BioFrac™ Fraction Collector, Bio-Rad Laboratories, Hercules, CA, USA).

2.9 Flow Meter Calibration

The flow meter used in this study must be calibrated for each solution. At various pump speed settings, the desired fluid is pumped through the flow meter, and is collected for precise amounts of time into pre-tared tubes with the fraction collector. The mass of fluid in each tube is converted to a volumetric flow rate by dividing by the collection time and by the density as measured by an electric density meter (DMA35, Anton Paar Instruments, Graz, Austria). Comparison of the real flow rate to averaged flow rate as measured by the flow meter is used to produce a pump curve and calibration curve (not shown).

2.10 Residence Time Distributions

Flow through the reactor was characterized through the measurement of the residence time distribution at various flow rates. The reaction system is set up as described in 2.8 with 1X PBS in the substrate feed bottle and 20 mM NADH in the tracer inlet tube. The reactor column was packed with 2 mL (settled bed) of empty Nuvia IMAC resin. The fraction collector was configured to collect with the required sampling time collect 200 μL samples into the wells of two 96-well flat bottom polystyrene plates. After purging of the tracer inlet line and subsequent equilibration with 1X PBS, an RTD experiment was commenced by simultaneously starting the fraction collector sequence and rapidly switching the selection valve to allow for the flow of the tracer solution into the reactor for one second, then rapidly switching back to PBS. Once 186 fractions were collected into the two plates, the absorbance of each well at 340 nm was measured using a BioTek Synergy H4 Hybrid Reader (BioTek Instruments Inc., Winooski, VT). Absorbance values were normalized by dividing by the average absorbance of the process fluid at steady state.

For each experiment, the residence time distribution, $E(\mathrm{t})$, was obtained from the normalized absorbance values, $C(\mathrm{t})$, with (1).

$$E(t) = \frac{C(t)}{\int_0^\infty C(t)\, dt} \tag{1}$$

The mean residence time between the tracer inlet valve and the fraction collector outlet, τ_{sys}, is obtained from the *E(t)* by (2).

$$\tau_{sys} = \int_0^\infty t\, E(t) dt \tag{2}$$

Once the mean residence time has been determined, the residence time distribution can be nondimensionalized to obtain *E(θ)* by dividing the function by τ_{sys} as shown in (3).

$$E(\theta) = \frac{E(t)}{\tau_{sys}} \tag{3}$$

The dimensionless residence time, θ, is defined as t/τ_{sys}. Finally, the variance, σ^2 of the residence time distribution is given by (4).

$$\sigma^2 = \int_0^\infty (t - \tau_{sys}) E(t) dt \tag{4}$$

The one-dimensional dispersion model with Danckwerts boundary conditions was utilized to estimate the Peclet number (Pe) for each of the collected RTDs by solving (5) numerically for Pe in Excel or Mathematica.[30]

$$\frac{\sigma^2}{\tau_{sys}^2} = \frac{2}{Pe} - \frac{2}{Pe^2}(1 - e^{-Pe}) \tag{5}$$

2.11 Conversion Determination With HPLC

Conversion in the reactor was determined by direct measurement of 5-methyl-2-aminohexane concentration after derivatization with benzoyl chloride. The procedures for derivatization and quantification of primary amines with reverse-phase HPLC have been described previously.[28]

3. RESULTS

3.1 Enzyme Leaching Under Reaction Conditions

It had been previously speculated that high concentrations of ammonia/ammonium in solution could lead to elution of bound proteins from Ni-NTA resins. Instruction manuals from multiple manufacturers for purification of 6xHis-tagged proteins with Ni-NTA functionalized beads list ammonium chloride as an alternative to imidazole for protein elution. To test if the ammonium formate in the reaction buffer would elute the enzymes used in this study from the Nuvia IMAC, AmDH (11.3 mg) and FDH (5.1 mg) were each immobilized separately onto 0.25 mL of resin. After incubation of the samples at room temperature in 10 mL of a 2.5 M NH_4COOH solution at pH 8.5 for two hours, no protein was detectable by the Bradford assay in the bulk solution for either enzyme. The previously reported elution of 6xHis-tagged proteins from IMAC resins in the presence of NH_4Cl is likely due to a pH effect of un-buffered NH_4Cl solutions, rather than competitive binding of the ammonium ion. The absence of detectable enzyme leaching in the short term indicated the resin was suitable for further testing.

3.2 Enzymatic Activity and Stability in Reaction Conditions

Samples of separately immobilized AmDH and FDH from Section 3.1 were combined and employed in a 10 mL batch conversion of 20 mM 5M2H (5-methyl-hexan-2-one) for 24 hours at room temperature. The immobilized enzyme was then left in the solution from the completed batch experiment for 30 days on the bench. After 30 days, there was no detectable protein concentration in the supernatant, which confirms the lack of significant leeching of either enzyme from the resin

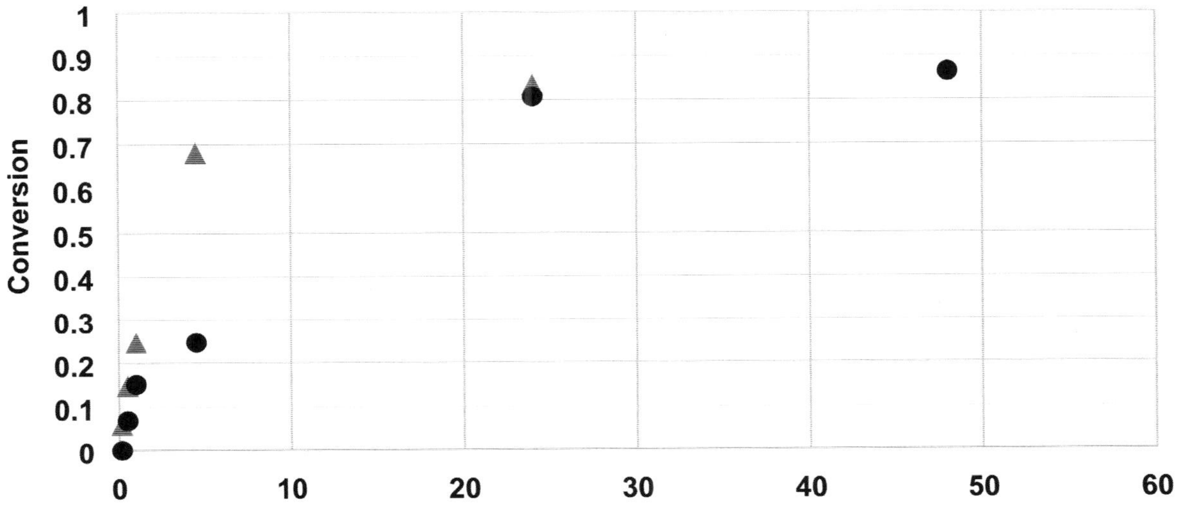

Figure 3 Comparison of conversion time courses of the conversion of 5-methyl-2-hexanone with immobilized AmDH and FDH before (black circles) and after (red triangles) 30 days of storage in NH_4COOH. Reaction buffer contained 20 mM 5-methyl-hexan-2-one, 1 mM NAD^+, and 2 M NH_4COOH adjusted to pH 8.5 with 2 M NH_4OH. Reactions were carried out at room temperature, approximately 21 °C, with a 10 mL reaction volume in a 15 mL conical centrifuge tube and constant end-over-end rotation at 20 rpm.

due to the reaction buffer. After the resin was washed with NH_4COOH buffer at pH 8.5 to remove any remaining amine product, the 10 mL batch conversion was repeated to assess the long-term stability of the immobilized enzymes. In both cases, roughly 85% conversion was reached. Importantly, the measured time course for day 30 was virtually indistinguishable from the one measured on day 0, as can be seen in Figure 3.

3.3 Binding Capacity

The binding capacities of the Nuvia IMAC resin for both AmDH and FDH were determined using a continuous flow experiment through a packed column. A known concentration of enzyme was pumped through a packed column with a known quantity of empty Nuvia IMAC resin. The outlet from the reactor was collected into 96-well plates, and the change in enzyme concentration from the stock solution of each fraction was determined by measuring the absorbance at 280 nm. All the enzyme which is missing in the outlet solution was assumed to have bound to the resin. For both experiments, the enzyme stock solution ran out before the resin was completely saturated, so the binding curve was extrapolated to infinite time by assuming symmetry of the binding isotherm about the inflection point. Integration of the binding curves shown in Figure S1 yields the total binding capacity of the resin for each of the two enzymes. The resin can bind 78.3 mg of AmDH per mL of settled bed, or 440 mg per gram of dry resin. However, the binding capacity for FDH was just 42.1 mg per mL of settled bed, or 236 mg per gram of dry resin. Dry resin weight was calculated based on 64% random packing efficiency for spheres, the 80% reported pore volume for Nuvia IMAC resin,[31] and 1.11 g/mL density of polyacrylamide. The difference in binding capacity cannot be explained by differences in molecular weight, as the monomer masses of AmDH and FDH are 43.3 and 42.5 kDa, respectively. Instead, the inconsistency in binding capacity is likely due to differences in quaternary structure. FDH is known to form dimers in solution,[32] while the AmDH likely shows a different dominant oligomeric state, likely a tetramer.[unpublished results]

3.4 AmDH to FDH Binding Ratio

While the binding capacities of the two enzymes on Nuvia IMAC resin are significantly different, the rate of binding was assumed to be the same. Mixtures containing saturating enzyme concentrations at different mass ratios of AmDH to FDH (enzyme ratio ranged from 1:5 to 5:1) were incubated with 0.2 mL of resin for 24 hours. Then, the immobilized biocatalyst beads were employed on 50 mL scale for conversions of 20 mM 5M2H for 21 hours. Reaction conversion increased as a function of AmDH to FDH ratio, with a maximum conversion of 51.5% observed for the 5:1 case. The lowest conversion, 38.8%, was observed for the 1:5 case. Importantly, there was nearly 8 times as much AmDH in the 5:1 reaction compared to the 1:5 case but the conversion only increased by 12.7%. The benefits of increased AmDH concentration were likely offset by slower cofactor regeneration. An optimum AmDH to FDH ratio must exist somewhere above 5:1, but it was not investigated further in the reported work. Saturated Nuvia IMAC resin with an AmDH to FDH ratio of 5:1 was used for the final reactor setup.

3.5 Flow Characteristics of the Reactor System

After the complete flow reactor setup described in Section 2.8 was assembled, the physical parameters and flow characteristics were characterized first using empty resin absent of bound enzyme. After calibration of the electronic flow meters and generation of pump curves, the flow properties of the packed bed could be examined. When the Nuvia IMAC resin is properly packed, 2 mL of settled bed compresses to 1.58 mL in the column based on a measured bed height of 2.01 cm and the 1 cm diameter of the bed. The small particle diameter (38-53 μm) and dense packing of the resin leads to significant pressure drop across the column which limits the total bed length. Using the Ergun equation,[33] the pressure drop across the length of the 2.01 cm packed bed used in the present study was estimated across the range of flow rates of interest. The predicted pressure drop increased from 0.43 bar at a 100 μL/min flow rate up to 3.46 bar at an 800 μL/min flow rate. To avoid increased pressure drop as the process is scaled to larger bed volumes, the bed diameter should be increased rather than the bed height.

The flow properties of the system were examined using standard pulsed tracer experiments with NADH as the tracer. By following the absorbance at 340 nm in the outlet over time, the residence time distribution (RTD) of the system at varying flow rates could be determined. A representative example of a collected RTD is shown in Figure S2. For an inlet flow rate of 0.574 mL/min, the mean residence time of the system, τ_{sys} was determined to be 9.78 minutes. Importantly, this residence time includes not just the reactor, but all the tubing between the tracer injection point and the outlet of the fraction collector. The shapes of the collected RTDs indicate a flow pattern in the system similar to a plug-flow reactor with a small dispersion effect, as evidenced by an estimated axial Péclet number of 128. The RTD of the reactor alone in the absence of the rest of the system could not be measured experimentally, so it is unknown how much of the observed dispersion occurs in the actual packed bed. Also, the single clean peak in the RTD indicates proper packing of the reactor without significant channelling or dead volume. After collection of residence time distributions at varied flow rates and at two different packing volumes, the specific effective volume of the packed bed, and the relationship between flow rate and mean residence time could be determined (see Supplementary Information). When properly packed in the described reactor system, 2 mL (settled bed) of Nuvia IMAC resin has an effective volume of 1.27 mL.

3.6 Continuous Amine Production: Initial Tests

For an initial proof of concept, AmDH and FDH were coimmobilized directly from clarified lysate onto 2 mL (settled bed) of Nuvia IMAC resin. While the exact amount of protein immobilized onto the resin for this test is not known, it is estimated that roughly 25 mg of FDH and 60 mg of AmDH were bound to the resin. This estimate comes from the measured yields of protein purified from a fraction of the same fermentation batch as was used for immobilization.

The total amount of enzyme on the beads could be measured by elution of the bound protein from the resin with either a high concentration of imidazole or EDTA followed by quantification with the Bradford Assay. However, because the two enzymes were co-immobilized, it is not possible to determine the relative concentration of the two enzymes after elution with benchtop assays.

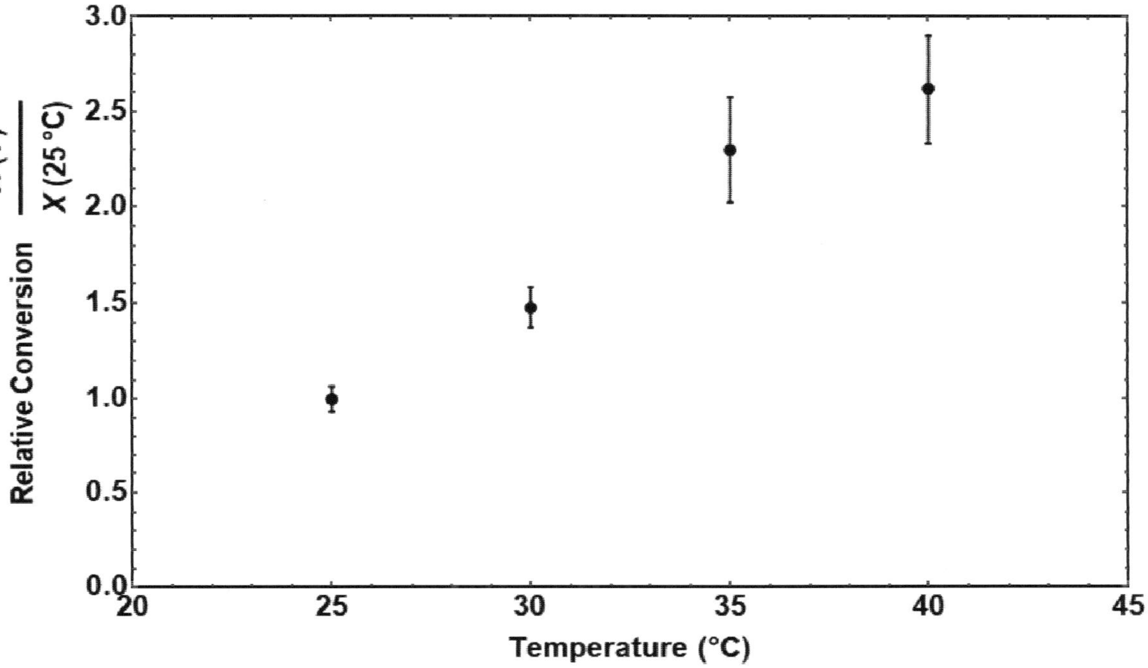

Figure 4 Effect of temperature on reaction conversion of 5-methyl-2-hexanone in the packed bed reactor. Feed to the reactor consisted of 20 mM 5M2H, 1 mM NAD^+, and 2 M NH_4COOH/NH_4OH adjusted to pH 8.5. Inlet flow rate was fixed at 0.5 mL/min.

Further development of the reactor platform was conducted with purified enzyme rather than lysate to give better control over relative enzyme concentrations on the beads. After packing the resin into a column, the effect of temperature on conversion was determined. As shown in Figure 4, increasing the temperature from 25 °C to 40 °C increased conversion by roughly 2.5-fold. When the natural log of the relative conversion is plotted versus the inverse of absolute temperature to produce an Arrhenius plot, a linear relationship is observed as expected (See Figure S4).

The relative conversion has a linear relationship to the apparent rate constant, so its use in an Arrhenius plot is valid. Based on (6), the apparent activation energy, E_a^{app}, of the AmDH/FDH system is given by the slope of the Arrhenius plot multiplied by the gas constant, R, which has a value of 8.314×10^{-3} kJ mol^{-1} K^{-1}. From (6), the apparent activation energy of the combined AmDH/FDH reactions was calculated to be 51.9 kJ/mol.

$$\ln\left(\frac{X(T)}{X(25\,°C)}\right) = \frac{-E_a^{app}}{R}\left(\frac{1}{T}\right) + \ln\left(A_0\right) \tag{6}$$

Because increasing temperatures often leads to decreased enzyme stability, the apparent deactivation rates of the immobilized biocatalysts were determined at two temperatures. For both 30 °C and 40 °C, the change in conversion over five days due to deactivation appeared to follow first-order kinetics. Deactivation profiles over 120 hours at the two temperatures are shown in Figure 5. There was no significant decrease in stability at 40 °C compared to 30 °C. The apparent half-life at 30 °C was 124 hours, while the apparent half-life at 40 °C was even longer, at 159 hours.

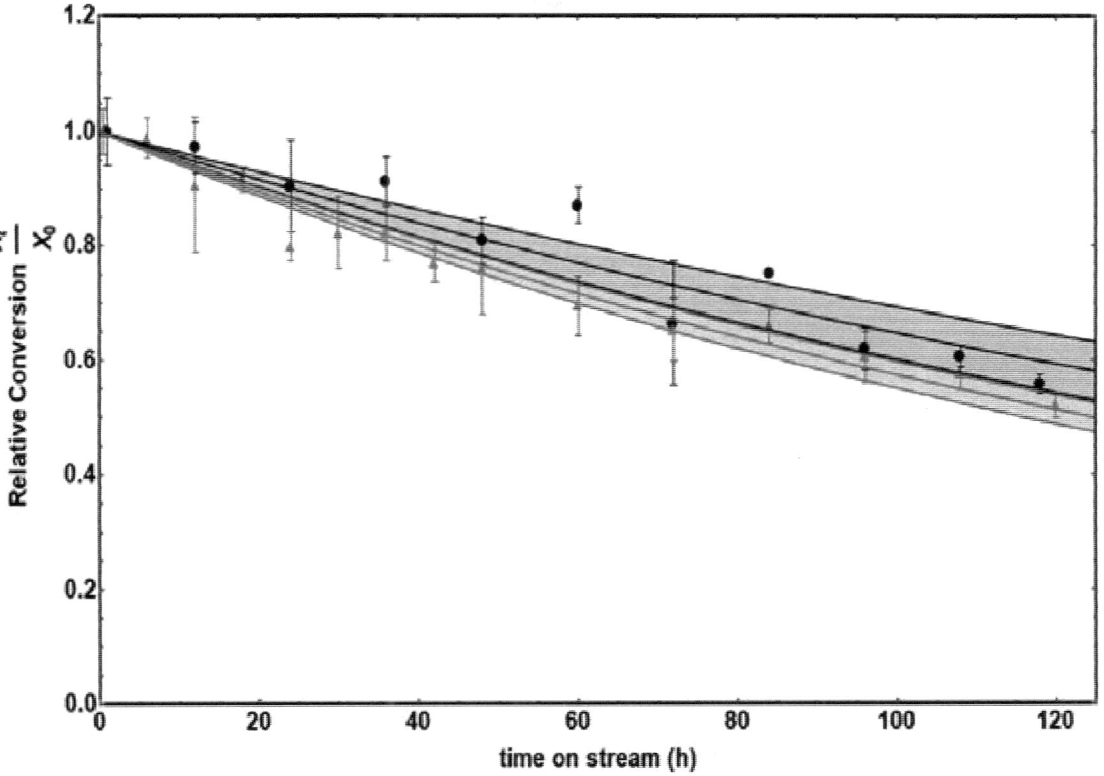

Figure 5 Apparent deactivation kinetics of immobilized FDH and AmDH under continuous flow conditions. Red points and curves represent reaction at 30 °C, while black points and curves and points represent reaction at 40 °C. Shaded regions represent the 95% confidence region on the first order deactivation fit. Feed to the reactor consisted of 20 mM 5M2H, 0.5 mM NAD^+, and 2 M NH_4COOH/NH_4OH adjusted to pH 8.5. Inlet flow rate was fixed at 0.5 mL/min.

3.7 Improved Reaction Conditions

Using the lessons learned from previous experiments, a final batch of immobilized enzyme was prepared and characterized. A 2 mL sample of Nuvia IMAC resin was saturated with AmDH and FDH in a 5:1 ratio by mass. With the reactor held at 40 °C, inlet flow rate was varied to determine the effect of residence time on conversion and productivity. The results of this experiment are displayed in Figure 6.

Figure 6A shows the relationship between flow rate and reactor performance, while Figure 6B displays data from the same experiment, but in terms of residence time in the reactor rather than flow rate. The two different views of the data are useful for different ways of thinking about the experiment. The shape of the residence time versus conversion plot indicates an asymptotic conversion somewhat less than 100% at very long residence times. The flow rate plot, Figure 6A is more useful as a tool for operating the reactor, as flow rate is a readably controllable parameter. The maximum observed conversion, 48%, was at the largest residence time of 11.9 minutes and a flow rate of 106.8 μL/min, as expected. Conversion steadily decreased as flow rate was increased, with a minimum of 18.8% at 816 μL/min and a residence time of 1.55 minutes. Productivity was reasonably high at all flow rates, with a minimum value of 166 g/L·day at 106 μL/min and a maximum of 433 g/L·day at the second highest flow rate, 715 μL/min.

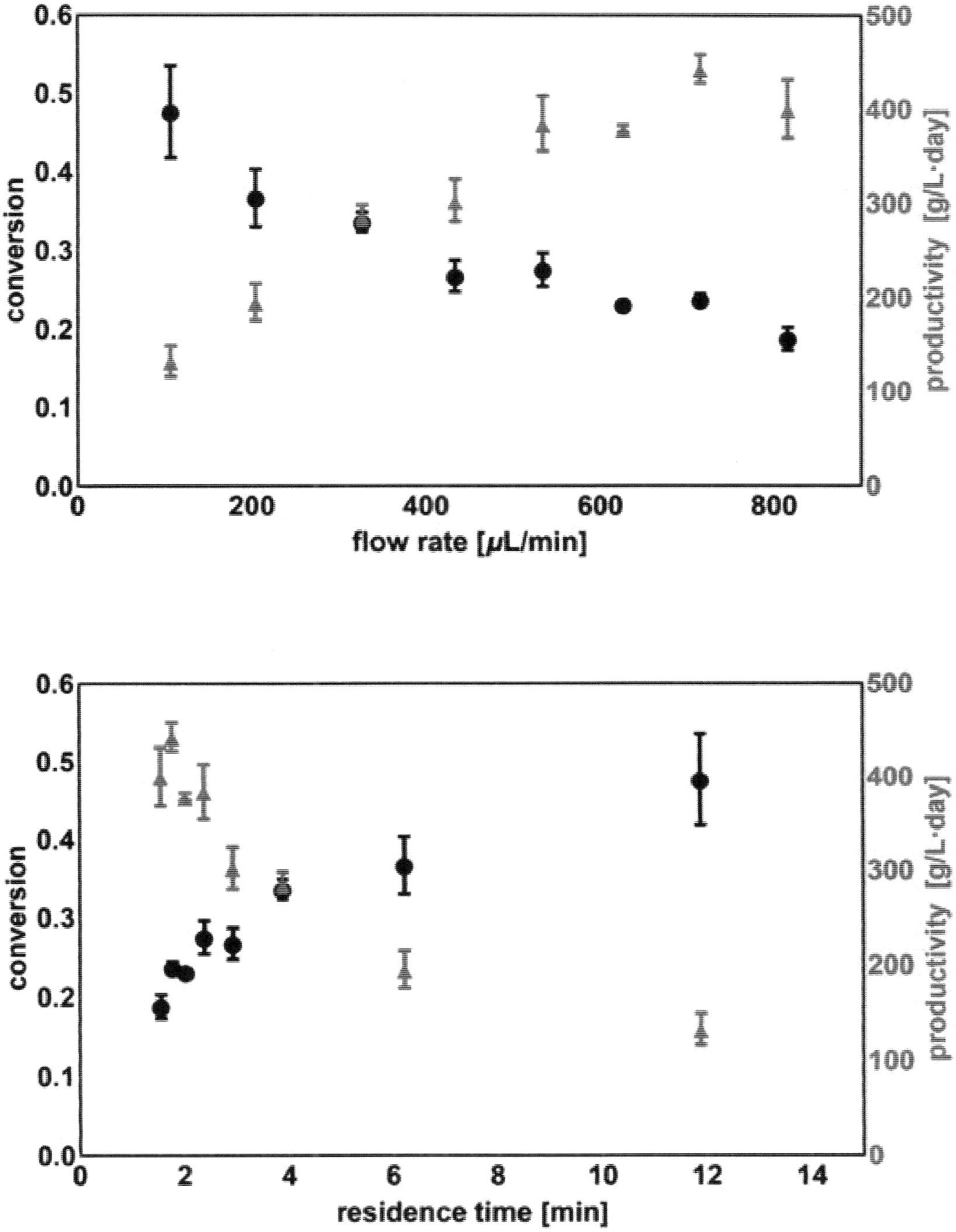

Figure 6 Impact of flow rate on conversion (black circles) and productivity (red triangles) in the AmDH/FDH packed bed reactor. Feed to the reactor consisted of 20 mM 5M2H, 0.5 mM NAD^+, and 2 M NH_4COOH/NH_4OH adjusted to pH 8.5. Temperature was fixed at 40 °C.

4. DISCUSSION

Despite high volumetric productivities reported in Section 3.6, overall conversion was limited to just under 50% for the longest included residence time. Batch reactions with immobilized AmDH and FDH showed that conversions of 5-methyl-2-hexanone exceeding 85% are possible. However, the shape of the residence time vs. conversion plot in Figure 6B indicates it would likely require a residence time in excess of one hour to reach ≥ 85% conversion at the given reaction conditions, which would yield much decreased volumetric productivity. An exploration of the possible causes of limited conversion is important for future reactor development.

4.1 Enzyme Kinetics

4.1.1 Assumptions

The most basic analyses of biocatalytic processes assume that the expression for the rate of reaction takes the form of the Michaelis Menten equation.[34]

$$-r_A = \frac{V_{max}\,[S]}{K_M + [S]} = \frac{k_{cat}[E][S]}{K_M + [S]} \tag{7}$$

While the real reaction kinetics of the reactor system presented in this study are much more complicated than (7),[28] the simplifying assumptions needed for (7) to be potentially useful are reasonable and are listed below.

1) Full conversion of NAD^+ by FDH is fast enough that the concentration of NADH is constant over the length of the reactor and is close to the inlet concentration of NAD^+, 0.5 mM.
2) The concentration of NH_3 in the reaction is constant over time.
3) There is no significant product inhibition.
4) The expression $V_{max} = k_{cat}[E]$ is true for the relevant enzyme concentration [E] in the reactor of 2.5 mM

Based on reported literature values for FDH, the k_{cat} value is estimated to be 8.06 s^{-1}, which is significantly higher than estimated value for AmDH of 1.75 s^{-1}. Thus, Assumption 1 is likely valid to a reasonable approximation. Since the concentration of NH_4COOH in the feed is 2.0 M and the maximum possible change in ammonia concentration is 20 mM, Assumption 2 is also reasonable.

In the literature on continuous reaction engineering with immobilized enzymes, it is common to further assume (Assumption 5) the rate of the reaction in the reactor can be accurately described as a first-order reaction with a rate constant equal to either V_{max}/K_M or $V_{max}/C_{A,0}$ where $C_{A,0}$ is the inlet concentration of the substrate.[35-45] This results in a very simple expression for $Da_{(I)}$ shown in (8) and equally simple expressions for $Da_{(II)}$ and the Thiele modulus ϕ_m.

$$Da_{(I)} = \frac{k_{cat}[E]}{K_M}\tau \qquad (8)$$

For the experimental results shown in Figure 6, the enzyme concentration [E] was 2.5 mM, and initial rate experiments yielded k_{cat} and K_M values of 1.75 s^{-1} and 4 mM, respectively. Applying (8) with these values at a residence time of 3.9 minutes yields an expected value for $Da_{(I)}$ of 254.6, which would imply a conversion value of 100%. Experimental results, instead, showed a conversion of 33.6% when the residence time was set at 3.9 minutes. In fact, conversion would be expected to be 99.5% with a residence time of 5.0 seconds if (8) were valid for the reactor system studied here.

4.1.2 High Enzyme Concentration

One of the fundamental assumptions of the Michaelis Menten equation is that the concentration of enzyme in solution is much less than the concentrations and K_M values of the substrates involved in the reaction. In the experiments which produced Figure 6, the enzyme concentration was 2.5 mM, while the concentration on NADH in the reactor was at most 0.5 mM. Additionally, the $K_{M,NADH}$ for AmDH is approximately 0.0125 mM. The reaction conditions are not consistent with the low enzyme concentration assumption when NADH is concerned. A more general rate law, (9), can be used to describe single-substrate kinetics without inhibition at any ratio of enzyme to substrate.[46, 47]

$$-r_A = \frac{2\, k_{cat}[E]\, [S]}{K_M + [S] + [E] + \sqrt{(K_M + [S] + [E])^2 - 4[S][E]}} \quad (9)$$

Based on assumption 1 from Section 4.1.1, the concentration of NADH is expected to be relatively constant at around 0.5 mM, which is much higher than the K_M value of 0.0125 mM and much lower than the 2.5 mM AmDH concentration. As a result, the substrate is completely saturated with enzyme and adding additional enzyme will not increase the reaction rate. This the opposite situation of the classic Michaelis Menten case, where enzyme is saturated when [S] >> K_M. Under the chosen reaction conditions, Assumption 4 from Section 4.1.1 is clearly invalid. As a consequence, the reaction rate which would be predicted by (7) is roughly five-fold larger than the rate predicted by (9) at the relevant concentrations of enzyme and cofactor. The apparent maximum velocity $V_{max,app}$ is given by (10), where $V_{\max}$ is multiplied by the ratio of (9) and (7) and then simplified for the case that [NADH] >> K_{M} and [E] >> K_{M}, as is the case in the reactor.

$$V_{max,app} = \frac{2([S] + K_M)}{K_M + [S] + [E] + \sqrt{(K_M + [S] + [E])^2 - 4[S][E]}} k_{cat}[E]$$

$$V_{max,app} = \frac{2[S]}{[S] + [E] + \sqrt{([S] - [E])^2}} k_{cat}[E] \qquad (10)$$

Equation (10) was evaluated for [NADH] = 0.5 mM and [E] = 2.5 mM to give the final expression for $V_{max,app}$ shown in (11).

$$V_{max,app} = 0.203 k_{cat}[E] \qquad (11)$$

4.1.3 Product Inhibition

The extent of product inhibition of AmDH by 5-methyl-2-aminohexane was examined *via* initial rate experiments with soluble enzyme. The reaction rate was determined at various simulated levels of conversion by mixing various combinations of concentrations amine and ketone, each of which added to a total of 20 mM. AmDH is competitively inhibited by the amine product,[48] and the collected data fits to (12) reasonably well (see Figure 7).[49]

$$-r_A = \frac{V_{max,app} C_{A,0}(1 - X)}{K_M\left(1 + \frac{C_{A,0}X}{K_I}\right) + C_{A,0}(1 - X)} \tag{12}$$

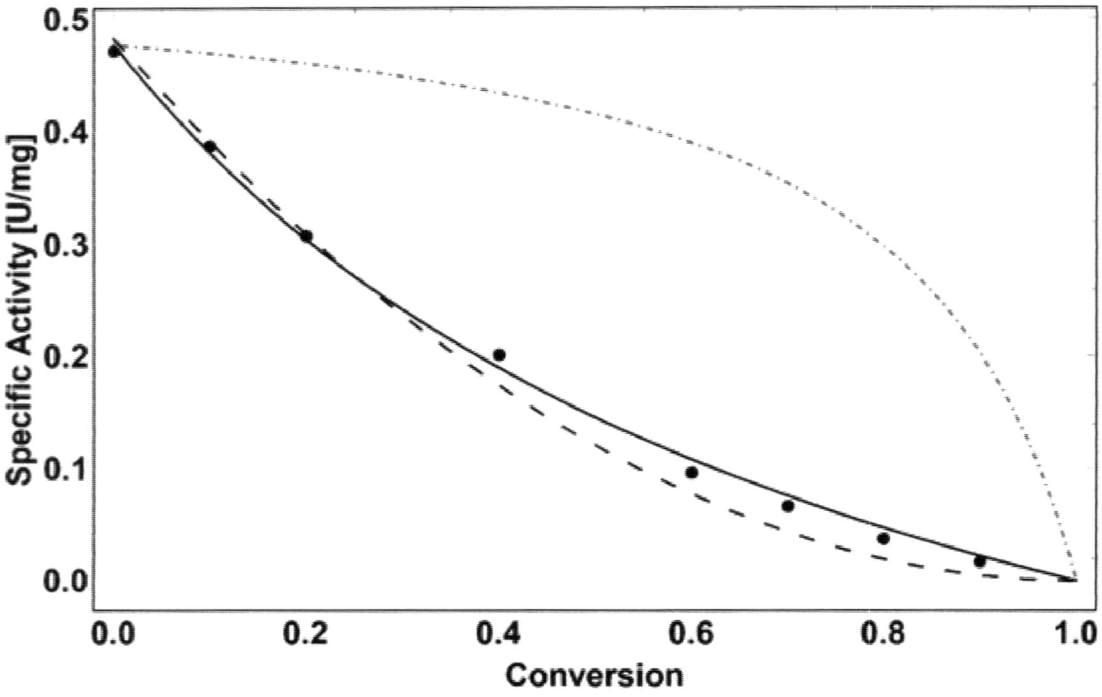

Figure 7 Relationship between conversion and reaction rate for AmDH amination of 5-methyl-2-hexanone. Circular data points represent the results of individual initial rate assays with simulated levels of conversion of 20 mM 5-methly-2-hexanone between 0.0 and 0.9. The solid black curve represents a nonlinear fit of the initial rate data to (12), the rate law for competitive inhibition, the dash-dotted red curve represents the expected activity vs. conversion curve for Michaelis Menten kinetics (7), and the dashed black curve is a fit of the initial rate data to a standard second order reaction kinetic, $-r_A = k\ C_{A,0}(1\text{-}X)$ with a fit rate constant of 0.02 U/(mg mM).

In this rate law, $C_{A,0}$ is the starting ketone concentration, X is conversion, K_M is the Michaelis constant for the ketone, and K_I is the inhibition constant for the amine product. The fit values for K_M and K_I were 3.54 mM and 1.54 mM, respectively, resulting in an inhibition ratio *K*-

$_M/K_I$ of 2.3. Given these results, both Assumption 3 and Assumption 5 from 4.1.1 are invalid for the current AmDH reaction.

4.1.4 The Improved Rate Law and an estimation for $Da_{(I)}$

Given the results from Sections 4.1.2 and 4.1.3, an improved form of the rate law as a function of conversion was developed which will be used in reactor characterization calculations in subsequent sections.

$$-r_A = \frac{V_{max,app}C_{A,0}(1-X)}{K_M\left(1+\frac{C_{A,0}X}{K_I}\right)+C_{A,0}(1-X)} = \frac{0.203\ k_{cat}[E]C_{A,0}(1-X)}{K_M\left(1+\frac{C_{A,0}X}{K_I}\right)+C_{A,0}(1-X)} \quad (13)$$

As shown in Figure 7, the rate of reaction changes constantly with conversion, and is neither in the first-order nor the zero-order regime. Thus, there is not a single value of the first Damköhler number, $Da_{(I)}$, which is valid over the entire length of the reactor. To arrive at a single average value for $Da_{(I)}$, one can take the integral of $Da_{(I)}$ with respect to conversion from 0% to 100% at a given residence time τ, as shown in (14).

$$Da_{(I)} = \int_0^1 Da_{(I)}[X]dX = \int_0^1 \frac{-r_A[X]}{C_{A,0}(1-X)}\tau dX \quad (14)$$

For a residence time of 3.9 minutes, (14) predicts a $Da_{(I)}$ value of 4.6. Because the AmDH kinetics roughly approximate a second order power law kinetic (see Figure 7), (15) can be used to give an estimate of expected conversion for a 3.9 minute residence time.[30]

$$X \approx \frac{Da_{(I)}}{1 + Da_{(I)}} = \frac{4.6}{1 + 4.6} = 0.82 \qquad (15)$$

While 82% conversion is still much higher than the observed conversion of 33.6%, it is not unreasonably high, and is much more accurate than the conversion value estimated in Section 4.1.1.

4.2 Internal Mass Transfer

If the rate of internal diffusion through the pores of the beads in a packed bed is slower than the rate of intrinsic kinetics, then the overall conversion in the reactor will be lower than would be observed if all the enzyme were at the bead surface and thus internal diffusion were unimportant. The Thiele modulus, ϕ_m, is often used to describe the relationship between a reaction rate at the catalyst surface to a rate of diffusion within the catalyst particle.[30] The Thiele modulus can be defined in terms of a form of a Damköhler number for internal pore diffusion, $Da_{(II),i}$ where the characteristic length l is defined as $0.5d_p/6$, the ratio of the volume of a sphere with diameter d_p to its surface area.[50]

$$Da_{(II),i} = \phi_m^2 = \left(\frac{0.5d_p}{3}\right)^2 \frac{-r_{As}}{D_{A,eff}C_{As}} \qquad (16)$$

In (16), d_p is the bead diameter (49×10^{-6} m) and $-r_{As}$ is the reaction rate per unit volume at surface concentration C_{As} of 5-methyl-2-hexanone as defined in Section 4.1.4. The effective diffusivity of ketone inside the pores, $D_{A,eff}$, is the bulk diffusivity multiplied by a factor of $\varepsilon_{bead}/\tau_{bead}$ which is defined as the porosity ($\varepsilon_{bead} = 0.8$)[31] divided by the tortuosity (estimated to be 3.0) of the Nuvia IMAC beads. The bulk diffusivity of 5M2H at 40 °C was assumed to be roughly equal to the infinite dilution diffusivity at that temperature, 1.06 x 10^{-9} m^2/s.[51] This results in an effective diffusivity inside the pores of 2.83 x 10^{-10} m^2/s. While the assumed value of the diffusivity is likely inaccurate due to the high ionic strength present in the reaction buffer, recalculation of the transport parameters in the following sections with a 100-fold lower estimation of the diffusivity did not change the conclusions. At very low values of the ϕ_m, pore diffusion is much faster than the reaction, so diffusion limitations can be neglected. At the reactor inlet, where the reaction rate is the highest and the effects of mass transfer limitations would be greatest, ϕ_m was estimated to be 0.091. The very low value of ϕ_m yields an internal effectiveness factor, η_I, greater than 0.99 which indicates no impact of internal mass transfer limitations in the reactor.[30]

The volumetric productivities (in g/L·day) from the continuous reactor measured in Section 3.6, after unit conversion to $mol/m^3 \cdot s$, are equivalent to the average observed rate of reaction over the entire reactor, $-r_A(obs)$, and were used to further verify the absence of limitations due to pore diffusion using the Weisz-Prater criterion.[52]

$$C_{W-P} = \frac{-r_A(obs)(0.5\,d_p)^2}{9D_{A,eff}C_{A,f}} \ll 1 \quad (17)$$

$C_{A,f}$ in (17) is the outlet concentration of ketone after the system had reached steady state. For all inlet flow rates recorded C_{W-P} was less than 0.01, which satisfies the criterion. The very small 49 μm particle diameter of the Nuvia IMAC resin is very effective at avoiding internal diffusion resistance for small molecules.

4.3 External Mass Transfer

Even though internal mass transfer limitations can be neglected, the impact of external mass transfer from the bulk fluid to the resin beads needed to be considered. Predictions of the impact of external mass transfer were made by computing the external mass transfer Damköhler number, $Da_{(II),ex}$, which is the ratio of the reaction rate to the rate of diffusive mass transfer from the bulk fluid to the bead surface. The value for $Da_{(II),ex}(X)$ at any point along the reactor is a function of the conversion, because the reaction rate and diffusion rate are both strongly dependent on the concentration of substrate. To give an average value of $Da_{(II),ex}$ across all concentration ranges, the instantaneous Damköhler number is integrated from X=0 to X=1.

$$Da_{(II),ex} = \int_0^1 Da_{(II),ex}[X]dX = \int_0^1 \frac{\phi_{bed}}{1-\phi_{bed}} \frac{-r_A[X]}{k_L a\, C_{A,0}(1-X)} dX \qquad (18)$$

In (18), X is conversion, $C_{A,0}$ is the inlet ketone concentration (20 mol/m^3), and ϕ_{bed} is the packed bed void fraction which was assumed to be 0.3. The values of k_L, the external mass transfer coefficient, and a, the volume averaged interfacial area, were estimated using (19) and (20).

$$k_L = \frac{D_A}{d_p} Sh \qquad (19)$$

$$a = \frac{6(1-\phi_{bed})}{d_p} \qquad (20)$$

The Sherwood number, *Sh* was estimated based on (21), a correlation incorporates the Reynolds (*Re*) and Schmidt (*Sc*) numbers and is valid under conditions of creeping flow.[53] For all flow rates in the reported work, *Re* is less than $1x10^{-4}$, which is consistent with creeping flow. Sc, which is independent of flow rate, is estimated to be $8.04x10^5$.

$$Sh = 2 + 0.991(Re\,Sc)^{1/3} \tag{21}$$

Where,

$$Sc = \frac{\nu}{D_A}, \quad Re = \frac{U_{SF}\,d_p}{\nu}, \text{and} \quad U_{SF} = \frac{v_f}{3.141(0.5\,d_r)^2}$$

The kinematic viscosity ν for 2M NH_4COOH was measured using a capillary viscometer at 22 °C and was found to be 30% higher than that of pure water at the same temperature and it was assumed the relationship holds true at 40 °C.[54] In the equation above for superficial fluid velocity, U_{SF}, the reactor diameter d_r and volumetric flow rate v_f were measured directly. Typical values of Sh for the flow rates relevant to this study are predicted to be between 3.0 and 4.2. For all flow rates and conversion values relevant to the reported reaction system, the predicted value of $Da_{(II),ex}$ is less than 0.003. From the Damköhler number, the external effectiveness factor, η_E was determined using (22) and was greater than 0.99 for all relevant flow rates and conversions.[36] While (22) is only strictly valid for a second order kinetics, Figure 7 shows that a second order rate law is a reasonable approximation for the real kinetics in the system.

$$\eta_E = \frac{\sqrt{1 + 4Da_{(II),ex}} - 1}{2Da_{(II),ex}} \tag{22}$$

As was the case with internal mass transfer, no limitations on conversion are expected due to external mass transfer. This is due to the very small diameter of the Nuvia IMAC beads. A 10-fold increase in the bead diameter to 490 μm would yield a 100-fold increase in $Da_{(II),ex}$, and η_E values as low as 0.8. Further confirmation of the lack of external mass transfer limitations was accomplished with a version of the Mears criterion which has been modified for enzyme kinetics.[55, 56] External mass transfer limitations can be ignored if (23) is satisfied.

$$\frac{-r_A(obs)}{k_L\,C_{A,f}}\left(\frac{d_p}{6}\right) < 0.05\frac{1+\hat{K}}{\hat{K}} \tag{23}$$

Where,

$$\hat{K} = \frac{K_{M,eff}}{C_{A,f}}, \quad and \quad K_{M,eff} = K_M\left(1 + \frac{C_{A,0} - C_{A,f}}{K_I}\right)$$

Over the range of flow rates relevant to this study, values of the left-hand side of (23) ranged from $1.59\text{x}10^{-4}$ to $2.34\text{x}10^{-4}$ while the values of the right-hand side of (23) ranged from 0.06 to 0.08. Thus, for all flow rates examined in the reported work, the modified Mears criterion is satisfied, and external mass transfer limitations can be neglected.

5. CONCLUSION

We have demonstrated the first continuous flow reaction system that creates enantiomerically pure (*R*)-amines with co-immobilized AmDH and FDH which is stable for more than 24 hours of operation. The system remained operational, with reductions in activity, for more than a week and the system is as stable at 40 °C as it is at 30 °C: the half-lives of apparent enzyme activity were 159 and 124 hours, respectively. Conversion of 20 mM 5-methyl-2-hexanone increased with decreasing flow rate and ranged between 19% and 48%. However, decreased conversion resulted in increased volumetric productivity, which ranged from 107 g/L·day at 48% conversion up to a maximum of 443 g/L·day at 24% conversion.

Binding of AmDH and FDH to the Nuvia IMAC resin was shown to be tight and stable under reaction conditions. The specific binding capacities of the resin for AmDH and FDH were determined to be 78.3 mg/mL and 42.1 mg/mL respectively. The very small diameter of the resin beads of 49 μm resulted in excellent mass transfer properties, and thus no limitations due to internal or external diffusion rates. After batch experiments at varying enzyme ratios, a 5:1 ratio of AmDH to FDH was chosen to maximize conversion. However, when saturated resin was packed into the reactor, the AmDH concentration was much higher than the cofactor concentration, which caused about 80% of the enzyme to go un-utilized in the packed-bed reactor. Correspondingly, Michaelis-Menten kinetics, which assumes that [S] >> [E], is not valid and modified equations have to be used to calculate kinetic parameters. The underutilization of enzyme can be alleviated by either increasing the cofactor concentration or by utilizing the same amount of enzyme but spread over more resin. Modest competitive product inhibition by (*R*)-amine on ketone with an inhibition ratio K_M/K_I of 2.3 contributed to decreased observed conversion compared to an ideal reactor.

The relationship between conversion and productivity is a key design constraint to consider when engineering a continuous chemical process. The choice of whether to favor conversion at the expense of productivity in the reactor or *vice versa*, i.e. the choice of optimum operating point of the reactor, should be based on the requirements and costs of the unit operations surrounding it and cannot be made based solely on the results in the present study. High costs for the reactor and inexpensive separation favour lower conversion, possibly with a post-separation recycle stream. However, expensive ketone from amine separation favour higher conversion per-pass, at the expense of higher costs for enzyme or resin.

While more can be done to optimize the system, the reported work represents an encouraging and promising first step toward continuous manufacturing of chiral amines with amine dehydrogenases.

6. ACKNOWLEDGEMENTS

Funding for the reported work by the United States National Science Foundation through NSF CBET grant 1512848 and NSF I/UCRC grant 1540017 to the Center for Pharmaceutical Development is gratefully acknowledged. We also wish to acknowledge the Biopolymer Characterization (BPC) core facility at the Parker H. Petit Institute for Bioengineering and Bioscience at the Georgia Institute of Technology for the use of its shared equipment, services, and expertise.

REFERENCES

[1] M.J. Abrahamson, E. Vazquez-Figueroa, N.B. Woodall, J.C. Moore, A.S. Bommarius, Development of an amine dehydrogenase for synthesis of chiral amines, Angew Chem Int Ed Engl 51 (2012) 3969-3972.

[2] N.A. McGrath, M. Brichacek, J.T. Njardarson, A Graphical Journey of Innovative Organic Architectures That Have Improved Our Lives, Journal of Chemical Education 87 (2010) 1348-1349.

[3] L.A. Nguyen, H. He, C. Pham-Huy, Chiral drugs: an overview, Int J Biomed Sci 2 (2006) 85-100.

[4] G. Grogan, Synthesis of chiral amines using redox biocatalysis, Curr Opin Chem Biol 43 (2018) 15-22.

[5] D. Ghislieri, N.J. Turner, Biocatalytic Approaches to the Synthesis of Enantiomerically Pure Chiral Amines, Topics in Catalysis 57 (2014) 284-300.

[6] H. Kohls, F. Steffen-Munsberg, M. Hohne, Recent achievements in developing the biocatalytic toolbox for chiral amine synthesis, Curr Opin Chem Biol 19 (2014) 180-192.

[7] M.D. Truppo, Biocatalysis in the Pharmaceutical Industry: The Need for Speed, ACS Medicinal Chemistry Letters 8 (2017) 476-480.

[8] M.J. Abrahamson, J.W. Wong, A.S. Bommarius, The Evolution of an Amine Dehydrogenase Biocatalyst for the Asymmetric Production of Chiral Amines, Advanced Synthesis & Catalysis 355 (2013) 1780-1786.

[9] F.F. Chen, Y.Y. Liu, G.W. Zheng, J.H. Xu, Asymmetric Amination of Secondary Alcohols by using a Redox-Neutral Two-Enzyme Cascade, Chemcatchem 7 (2015) 3838-3841.

[10] L.J. Ye, H.H. Toh, Y. Yang, J.P. Adams, R. Snajdrova, Z. Li, Engineering of Amine Dehydrogenase for Asymmetric Reductive Amination of Ketone by Evolving Rhodococcus Phenylalanine Dehydrogenase, Acs Catalysis 5 (2015) 1119-1122.

[11] A. Pushpanath, E. Siirola, A. Bornadel, D. Woodlock, U. Schell, Understanding and Overcoming the Limitations of Bacillus badius and Caldalkalibacillus thermarum Amine Dehydrogenases for Biocatalytic Reductive Amination, Acs Catalysis 7 (2017) 3204-3209.

[12] V. Tseliou, T. Knaus, M.F. Masman, M.L. Corrado, F.G. Mutti, Generation of amine dehydrogenases with increased catalytic performance and substrate scope from epsilon-deaminating L-Lysine dehydrogenase, Nat Commun 10 (2019) 3717.

[13] O. Mayol, S. David, E. Darii, A. Debard, A. Mariage, V. Pellouin, J.L. Petit, M. Salanoubat, V. de Berardinis, A. Zaparucha, C. Vergne-Vaxelaire, Asymmetric reductive amination by a wild-type amine dehydrogenase from the thermophilic bacteria Petrotoga mobilis, Catal. Sci. Technol. 6 (2016) 7421-7428.

[14] A. Caparco, E. Pelletier, J.-L. Petit, A. Fossey, B. Bommarius, V. de Berardinis, A. Zaparucha, J. Champion, A. Bommarius, C. Vergne-Vaxelaire, Metagenomic Mining for Amine Dehydrogenase Discovery, Advanced Synthesis & Catalysis DOI: 10.1002/adsc.202000094 (2020).

[15] M.P. Thompson, I. Penafiel, S.C. Cosgrove, N.J. Turner, Biocatalysis Using Immobilized Enzymes in Continuous Flow for the Synthesis of Fine Chemicals, Organic Process Research & Development 23 (2019) 9-18.

[16] B. Brena, P. González-Pombo, F. Batista-Viera, Immobilization of Enzymes: A Literature Survey, in: J.M. Guisan (Ed.) Immobilization of Enzymes and Cells: Third Edition, Humana Press, Totowa, NJ (2013), pp. 15-31.

[17] P. Poechlauer, J. Colberg, E. Fisher, M. Jansen, M.D. Johnson, S.G. Koenig, M. Lawler, T. Laporte, J. Manley, B. Martin, A. O'Kearney-McMullan, Pharmaceutical Roundtable Study Demonstrates the Value of Continuous Manufacturing in the Design of Greener Processes, Organic Process Research & Development 17 (2013) 1472-1478.

[18] S.L. Lee, T.F. O'Connor, X.C. Yang, C.N. Cruz, S. Chatterjee, R.D. Madurawe, C.M.V. Moore, L.X. Yu, J. Woodcock, Modernizing Pharmaceutical Manufacturing: from Batch to Continuous Production, Journal of Pharmaceutical Innovation 10 (2015) 191-199.

[19] B.R. Bommarius, M. Schurmann, A.S. Bommarius, A novel chimeric amine dehydrogenase shows altered substrate specificity compared to its parent enzymes, Chem Commun (Camb) 50 (2014) 14953-14955.

[20] T.A. Rogers, R.M. Daniel, A.S. Bommarius, Deactivation of TEM-1 beta-Lactamase Investigated by Isothermal Batch and Non-Isothermal Continuous Enzyme Membrane Reactor Methods, ChemCatChem 1 (2009) 131-137.

[21] A.A. Caparco, B.R. Bommarius, A.S. Bommarius, J.A. Champion, Protein-Inorganic Calcium-Phosphate Supraparticles as a Robust Platform for Enzyme Co-Immobilization, Biotechnology and Bioengineering DOI: 10.1002/bit.27348 (2020).

[22] W. Bohmer, T. Knaus, F.G. Mutti, Hydrogen-Borrowing Alcohol Bioamination with Coimmobilized Dehydrogenases, ChemCatChem 10 (2018) 731-735.

[23] J. Liu, B.Q.W. Pang, J.P. Adams, R. Snajdrova, Z. Li, Coupled Immobilized Amine Dehydrogenase and Glucose Dehydrogenase for Asymmetric Synthesis of Amines by Reductive Amination with Cofactor Recycling, Chemcatchem 9 (2017) 425-431.

[24] H. Ren, Y. Zhang, J. Su, P. Lin, B. Wang, B. Fang, S. Wang, Encapsulation of amine dehydrogenase in hybrid titania nanoparticles by polyethylenimine coating and templated biomineralization, J Biotechnol 241 (2017) 33-41.

[25] V. Gaberc-Porekar, V. Menart, Perspectives of immobilized-metal affinity chromatography, J Biochem Biophys Methods 49 (2001) 335-360.

[26] M.A. Huffman, A. Fryszkowska, O. Alvizo, M. Borra-Garske, K.R. Campos, K.A. Canada, P.N. Devine, D. Duan, J.H. Forstater, S.T. Grosser, H.M. Halsey, G.J. Hughes, J. Jo, L.A. Joyce, J.N. Kolev, J. Liang, K.M. Maloney, B.F. Mann, N.M. Marshall, M. McLaughlin, J.C. Moore, G.S. Murphy, C.C. Nawrat, J. Nazor, S. Novick, N.R. Patel, A. Rodriguez-Granillo, S.A. Robaire, E.C. Sherer, M.D. Truppo, A.M. Whittaker, D. Verma, L. Xiao, Y. Xu, H. Yang, Design of an in vitro biocatalytic cascade for the manufacture of islatravir, Science 366 (2019) 1255-1259.

[27] M.P. Thompson, S.R. Derrington, R.S. Heath, J.L. Porter, J. Mangas-Sanchez, P.N. Devine, M.D. Truppo, N.J. Turner, A generic platform for the immobilisation of engineered biocatalysts, Tetrahedron 75 (2019) 327-334.

[28] R.D. Franklin, C.J. Mount, B.R. Bommarius, A.S. Bommarius, Separate sets of mutations enhance activity and substrate scope of amine dehydrogenase, ChemCatChem 12 (2020) 2436-2439.

[29] M.M. Bradford, A rapid and sensitive method for the quantitation of microgram quantities of protein utilizing the principle of protein-dye binding, Anal Biochem 72 (1976) 248-254.

[30] H.S. Fogler, Elements of Chemical Reaction Engineering, fifth ed., Prentice Hall Upper Saddle River, New York (2016).

[31] M. Zhu, G. Carta, Protein adsorption equilibrium and kinetics in multimodal cation exchange resins, Adsorption 22 (2016) 165-179.

[32] H. Slusarczyk, S. Felber, M.-R. Kula, M. Pohl, Stabilization of NAD-dependent formate dehydrogenase from Candida boidinii by site-directed mutagenesis of cysteine residues, European Journal of Biochemistry 267 (2000) 1280-1289.

[33] S. Ergun, Fluid flow through packed columns, Chem. Eng. Prog. 48 (1952) 89-94.

[34] L. Michaelis, M.L. Menten, Die kinetik der invertinwirkung, Biochemische Zeitschrift 49 (1913) 333-369.

[35] R. Lortie, D. Pelletier, Comparison between dispersion and plug-flow models for fixed-bed enzyme reactors, AIChE Journal 38 (1992) 1477-1480.

[36] P. dos Santos, M.A.A. Meireles, J. Martínez, Production of isoamyl acetate by enzymatic reactions in batch and packed bed reactors with supercritical CO2, The Journal of Supercritical Fluids 127 (2017) 71-80.

[37] R. Lortie, Evaluation of the performance of immobilized enzyme reactors with michaelis—menten kinetics, Journal of Chemical Technology & Biotechnology 60 (1994) 189-193.

[38] A. Sadana, Enzyme Deactivation in Reactors, Biocatalysis 2 (1989) 175-216.

[39] J.W. Parker, C.S. Schwartz, Modeling the kinetics of immobilized glucose oxidase, Biotechnol Bioeng 30 (1987) 724-735.

[40] A.E. Al-Muftah, I.M. Abu-Reesh, Effects of internal mass transfer and product inhibition on a simulated immobilized enzyme-catalyzed reactor for lactose hydrolysis, Biochem. Eng. J. 23 (2005) 139-153.

[41] I.M. Abu-Reesh, B.F. Abu-Sharkh, Comparison of Axial Dispersion and Tanks-in-Series Models for Simulating the Performance of Enzyme Reactors, Industrial & Engineering Chemistry Research 42 (2003) 5495-5505.

[42] C. Horvath, J.-M. Engasser, External and internal diffusion in heterogeneous enzymes systems, Biotechnology and Bioengineering 16 (1974) 909-923.

[43] B. Romas, K. Juozas, P. Linas, Modelling the enzyme catalysed substrate conversion in a microbioreactor acting in continuous flow mode, Nonlinear Analysis: Modelling and Control 23 (2018).

[44] R. Khalilpour, R. Roostaazad, Development and verification of a model to describe an immobilized glucose isomerase packed bed bioreactor, Biochem. Eng. J. 40 (2008) 328-336.

[45] V.I

Effects on the Performance of an Immobilized Lipase Packed Bed Reactor During the Hydrolysis of Rice Bran Oil, The Canadian Journal of Chemical Engineering 83 (2005) 780-783.

[46] S. Cha, Kinetic Behavior at High Enzyme Concentrations: MAGNITUDE OF ERRORS OF MICHAELIS-MENTEN AND OTHER APPROXIMATIONS, J. Biol. Chem. 245 (1970) 4814-4818.

[47] A.R. Tzafriri, Michaelis-Menten kinetics at high enzyme concentrations, Bulletin of Mathematical Biology 65 (2003) 1111-1129.

[48] R.D. Franklin, J.A. Whitley, J.M. Robbins, A.S. Bommarius, Engineered amine dehydrogenase exhibits altered kinetic mechanism compared to parent with implications for industrial application, Chemical Engineering Journal 369 (2019) 634-640.

[49] L.G. Lee, G.M. Whitesides, Preparation of optically active 1,2-diols and .alpha.-hydroxy ketones using glycerol dehydrogenase as catalyst. Limits to enzyme-catalyzed synthesis due to noncompetitive and mixed inhibition by product, The Journal of Organic Chemistry 51 (1986) 25-36.

[50] E.W. Thiele, Relation between Catalytic Activity and Size of Particle, Industrial & Engineering Chemistry 31 (1939) 916-920.

[51] C.L. Yaws, Diffusion Coefficient at infinite dilution in Water – Organic Compounds, in: C.L. Yaws (Ed.) Transport Properties of Chemicals and Hydrocarbons, Gulf Publishing Company, Oxford (2014), pp. 614-703.

[52] P.B. Weisz, C.D. Prater, Interpretation of Measurements in Experimental Catalysis, in: W.G. Frankenburg, V.I. Komarewsky, E.K. Rideal (Eds.) Advances in Catalysis, Academic Press, Cambridge, Massachusetts (1954), pp. 143-196.

[53] R.B. Bird, E.N. Lightfoot, W.E. Stewart, Interphase Transport in Nonisothermal Mixtures, Transport Phenomena, J. Wiley, Hoboken, New Jersey (2002).

[54] L. Korson, W. Drost-Hansen, F.J. Millero, Viscosity of water at various temperatures, The Journal of Physical Chemistry 73 (1969) 34-39.

[55] D.E. Mears, Tests for Transport Limitations in Experimental Catalytic Reactors, Industrial & Engineering Chemistry Process Design and Development 10 (1971) 541-547.

[56] G.K. Lee, R.A. Lesch, P.J. Reilly, Estimation of intrinsic kinetic constants for pore diffusion-limited immobilized enzyme reactions, Biotechnology and Bioengineering 23 (1981) 487-497.

Highlights

- IMAC resin immobilization yielded 6.5 d half life and ≤ 443 g/(L·d) productivity.
- Overall reaction rate is limited by excess enzyme over cofactor but not mass transfer.
- As [S] < [E], the general rate law applies but not standard Michaelis-Menten kinetics.
- Conversion or volumetric productivity can be optimized but not both simultaneously.

September 3, 2020

613619 Scale-up of Co- and Counter-Current Irradiated Photochemical Reactors Using Common LEDs for Industrial Applications

Abstract:

Common cross-currently illuminated photochemical reactors, i.e. radially irradiated reactors, lack the energy efficiency to be competitive in industry. The use of co- and counter-current illumination was previously proven to increase reactor performance, but using a less efficient collimated LED.[1,2] By studying the use of non-collimated LEDs (illuminates the reactor along its axial axis) for the use in co- and counter-currently illuminated reactors efficiency can potentially be improved (Figure-1).

For this study, 4 types of LEDs were used ranging in total viewing angle from 20° to 120°, the reflectivity of the reactor material was varied, the absorbance (specific absorbance and path length) was altered and the tube diameter was investigated. The model results indicate that the optimal light source for achieving efficient light absorption in the reactor was the most collimated LED possible, in this case with a total viewing angle of 20°. The optimal reactor set-up uses the most reflective material preferably aluminum or silver to recuperate diverging light rays for the used wavelength. Furthermore, the wall thickness of the glass reactor should not be excessively large. In this case we saw no performance increase below 1.5 mm reactor thickness. Regarding reagents and absorbance, it is shown to best use a higher concentration and reduce the reactor length as this increases reactor performance under the condition that quantum yield is stable as function of concentration. Via the use of non-collimated light, it was determined that the entrance efficiency can be increased compared to a fully collimated light source, at the cost of reflection losses that increase with path length. Thus, the optimal parameters were determined to utilize co- and countercurrent illumination in the industrial context while keeping efficiency high.

References

1. Meir G, et al.,. *J Adv Manuf Process*. doi:10.1002/amp2.10044
2. Meir G, et al., *J Adv Manuf Process*. 2020;(February):1-16. doi:10.1002/amp2.10045

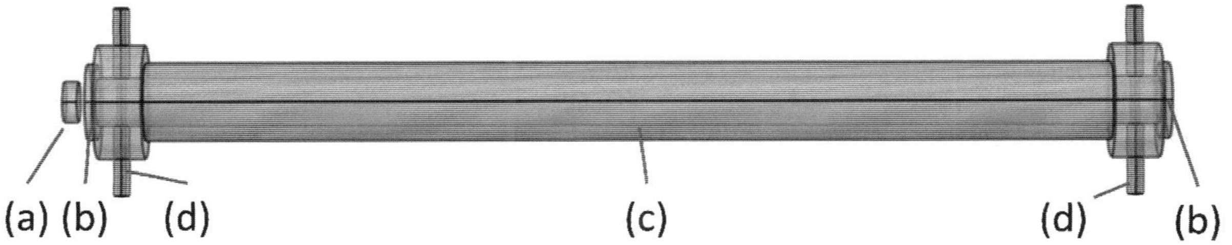

613614 Design of Multichannel Photocatalytic Reactors

Abstract:

Photocatalysis is a promising technology to perform oxidation reactions without the use of harmful chemicals based on heavy metals. Despite the interest in the field, no practical example of a photocatalytic reactor in the industry exists. This is mainly due to the lack of synergy between photon and mass transfer in photocatalytic reactor design. A widely investigated design in the past decade is the microreactor. Despite the high volume productivity, microreactors are difficult to illuminate and are not scalable. The illumination and the scalability problem is solved by incorporating multiple microreactors in translucent structures. These structures contain multiple channels which can be coated with catalyst. The large surface area of structured reactors is easy to illuminate and allows to apply thin catalyst coatings while maintaining a sufficiently high catalyst loading.

The design flexibility of translucent structured reactors causes a more extensive design procedure than single channel reactors. The number of incorporated channels and the catalyst layer thickness are the important design parameters. Plenty of work in this field is either of experimental nature or based on numerical simulations. Researchers interested in designing their reactors to test particular catalysts or specific chemical systems want design methods or tools which can be easily used without the need for multiple experiments or complex numerical simulations. In this work, a graphical tool and an analytical design equation is proposed based on the diffusion equation in the catalyst layer. The presented procedure allows to predict the optimal catalyst layer thickness and number of structural layers based on initial calibration experiments. This reduces the number of experiments researchers have to perform to obtain an optimal design.

602564 Visible Light Mediated Switchable Selectivity in C-O, C-C and C-S Bond Formation Induced By Disulfide

Abstract:

Chemists are beginning to turn their attention to the development of catalysts whose activity in a given chemical processes can be switched by an external stimulus. To achieve a mild, selectivity-controllable and general C-O, C-C and C-S bond formation, the facile selectivity-switchable functionalization of 1,3-dicarbonyl compounds represents an ideal protocol, yet a significant challenge. We thus focused on developing a bioinspired strategy. In this work, a visible light-mediated a-functionalization of 1,3- dicarbonyl compounds with switchable selectivity induced by disulfide is disclosed for the first time. Upon irradiation with visible light, the metal- and base-free α-hydroxylation and α-hydroxymethylation reactions proceeded smoothly through a disulfide-catalyzed oxidation with air under mild conditions and generated the desired products. The highly tunable selectivity between the hydroxylation (21 examples, up to 98% yield) and hydroxymethylation (20 examples, up to 97% yield) is controlled by simple changes in the stoichiometry of the styrene additive. In contrast, conducting the reaction in darkness prevented the S-S bond from being photoactivated, and the disulfides served as a sulfurization reagent and promoted the α-sulfenylation (15 examples, up to 95% yield). The reaction efficiencies of the hydroxylation and hydroxymethylation could be further improved by using a continuous-flow reactor. The combination of a continuous-flow strategy and enzyme-like switchable catalysts allowed the facile preparation of synthetically useful intermediates and products meanwhile providing a digital control and manipulation of chemical reaction paths.

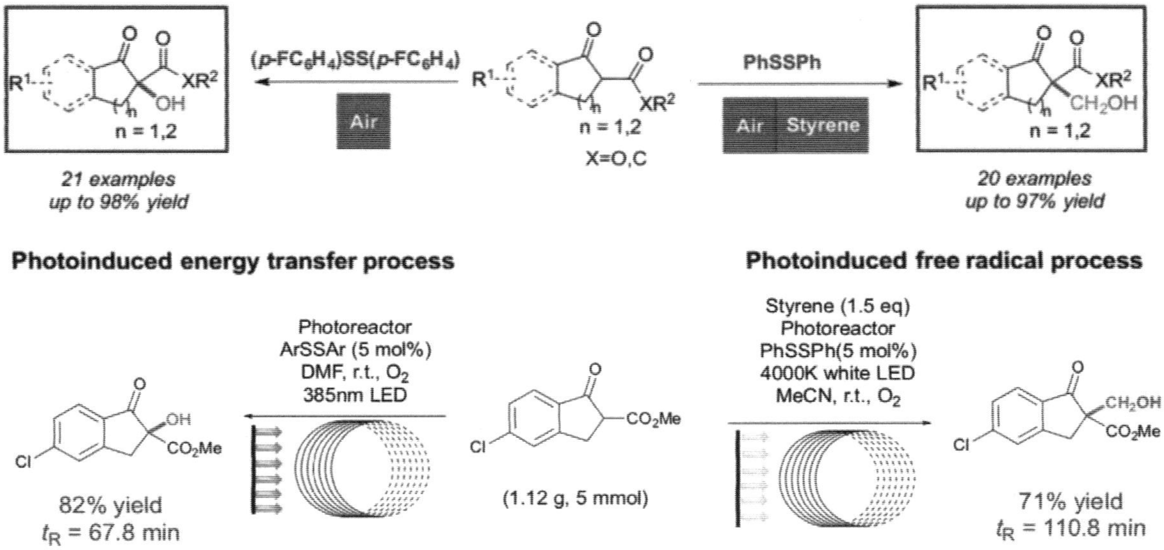

611834 Elucidating Molecular Origins of Drug Selectivity for Rational Design of Selective Drugs

Abstract:

Proteins families that evolve to bind particular ligand also contain diverse members that exhibit a wide range of affinities for that particular ligand. However, this variance in affinity resulting from evolution of the family members is usually at odds with the high sequence and structure similarity shared by its members. In this talk, we discuss three such protein families i.e. human protein kinases, Cannabinoid receptors and Strigolactone receptors in plants. These proteins show several orders of magnitude difference in their ligand affinity but share high sequence and structural similarity. The characterization of the free energy landscapes for ligand binding starting could shed light on the molecular origin of their ligand selectivity mechanisms. In order to achieve this goal, we have develop computational methods for evaluation of their binding mechanism using molecular simulations. These examples served as model system for obtaining generalizable insights about the free energy landscapes of ligand binding that could help design of drugs with desired selectivity profiles.

600211 Computer Aided Generation of Diverse and Synthesizable Molecular Library

Abstract:

Modern drug discovery is extending into chemical spaces with increased complexity. It is desirable to explore these spaces more efficiently. In order to maximize the information gained from testing new molecules, the concept of molecular library has been widely adopted in medicinal chemistry.

Molecular libraries are collections of compounds that have, or expected to have, similar structures or properties. Accessing all molecules in a molecule library can enable the rapid generation of structure-activity relationships during the hit-to-lead or lead optimization phases, which can greatly enhance the efficiency of drug discovery. Small molecule libraries can be constructed in many different ways. A library may be a collection of compounds extracted from the literature with similar functionality;[1,2] or it may be designed by enumerating possible side groups as decorations of a common core scaffold;[3–5] it may also be designed using any one of the increasing number of *in silico* tools for the generation of drug-like compound libraries.[6–8]

Despite the significant progress in molecular library generation, there is a usually a tradeoff between accessibility and diversity – libraries that are constructed by simply substituting functional groups on the same scaffold are usually limited in diversity, and libraries with larger structural diversity tend not to be all accessible through similar reactions. In this work, we leverage informatics tools to maximize the diversity of molecular libraries as well as their synthetic accessibility. Starting from a given molecule, using retrosynthesis analysis and reaction prediction models, we can generate libraries that can be accessed through alternative late-stage functionalization. The diversity of the molecular libraries is assessed by average pairwise similarity, so that the most diverse molecular library can be identified. This work provides insight on expanding the exploration of surrounding chemical space of a molecule with minimal additional effort.

1 U. F. Röhrig, S. R. Majjigapu, P. Vogel, V. Zoete and O. Michielin, *J. Med. Chem.*, 2015, **58**, 9421–9437.

2 B. T. Xin, E. M. Huber, G. De Bruin, W. Heinemeyer, E. Maurits, C. Espinal, Y. Du, M. Janssens, E. S. Weyburne, A. F. Kisselev, B. I. Florea, C. Driessen, G. A. Van Der Marel, M. Groll and H. S. Overkleeft, *J. Med. Chem.*, 2019, **62**, 1626–1642.

3 K. Kannan Sivaraman, A. Paiardini, M. Sieńczyk, C. Ruggeri, C. A. Oellig, J. P. Dalton, P. J. Scammells, M. Drag and S. McGowan, *J. Med. Chem.*, 2013, **56**, 5213–5217.

4 R. Fleeman, T. M. Lavoi, R. G. Santos, A. Morales, A. Nefzi, G. S. Welmaker, J. L. Medina-Franco, M. A. Giulianotti, R. A. Houghten and L. N. Shaw, *J. Med. Chem.*, 2015, **58**, 3340–3355.

5 G. M. Keseru, D. A. Erlanson, G. G. Ferenczy, M. M. Hann, C. W. Murray and S. D. Pickett, *J. Med. Chem.*, 2016, **59**, 8189–8206.

6 R. Gómez-Bombarelli, J. N. Wei, D. Duvenaud, J. M. Hernández-Lobato, B. Sánchez-Lengeling, D. Sheberla, J. Aguilera-Iparraguirre, T. D. Hirzel, R. P. Adams and A. Aspuru-Guzik, *ACS Cent. Sci.*, 2018, **4**, 268–276.

7 M. H. S. Segler, T. Kogej, C. Tyrchan and M. P. Waller, *ACS Cent. Sci.*, 2018, **4**, 120–131.

8 A. Gupta, A. T. Müller, B. J. H. Huisman, J. A. Fuchs, P. Schneider and G. Schneider, *Mol. Inform.*, 2018, **37**, 1700111.

605512 *De Novo* Design of Engineered Cytokine Mimetics As Targeted Immunotherapeutics.

Abstract:

Cytokines are soluble factors that signal through stimulation of their cognate transmembrane receptors on target cells to perform critical biological functions, particularly those related to immune homeostasis. As a result of their essential immune activities, cytokines have great potential as immunotherapeutics, both to activate the immune response to fight diseases such as cancer and infectious diseases, as well as to suppress the immune response to treat autoimmune disorders or for transplantation medicine. However, the pleiotropic activities and unfavorable pharmaceutical properties of natural cytokines have limited their clinical performance.

Recent efforts to re-engineer native cytokines to improve their selectivity and stability have significantly advanced progress in cytokine therapeutic development. However, there are inherent challenges to developing therapeutics from natural proteins. First, most natural cytokines are only marginally stable, hence mutations aimed at increasing efficacy can decrease expression or cause aggregation, making manufacturing and storage difficult. More substantial changes, such as the deletion or fusion of functional or targeting domains, can also dramatically alter pharmacokinetic properties and tissue penetration. In addition, immune responses targeting a variant of a natural protein may cross-react with the endogenous molecule with potentially catastrophic consequences. Finally, the high potency and multifarious functions of cytokines hinder their efficacy and can result in harmful off-target effects and systemic toxicities.

Here, we combine a best-in-class computational protein design software with directed evolution technologies to generate *de novo* cytokine mimetics that circumvent limitations for naturally-derived cytokine drugs. Rather than modifying existing molecules, we adopt a bold new approach to create exceptionally stable proteins with customized receptor interaction properties as robust and efficacious targeted therapeutics. We applied this strategy to develop a mimetic of the interleukin-2 (IL-2) cytokine, which plays a pivotal role in T and natural killer (NK) cell function. We created a hyper-stable *de novo* protein that is biased toward activation of IL-2 signaling pathways on immune effector cells, and showed that this molecule inhibits tumor growth and prolongs survival in mouse models of colorectal cancer and melanoma, while also mitigating the toxic side effects typically associated with IL-2 therapy. We built upon this exciting advance to design *de novo* cytokines that mimic IL-4, which promotes pro-regenerative immune programs following tissue damage. We demonstrated the extreme thermal stability and biased receptor activity of our engineered IL-4 mimetic, and showed that it elicited pro-regenerative immune responses in both cellular and animal models. Collectively, our work pioneers a novel cytokine engineering platform that integrates computational and experimental approaches to establish a general paradigm for advancing the clinical translation of cytokine therapeutics.

606100 Integrated Prediction of Chemical Effects on Important Pathways of Embryo Development Based on in Vitro Embryonic Stem Cell Test and Machine Learning

Abstract:

Currently, there are thousands of chemicals in common use; however, only a small number of them have sufficient toxicity evaluation, and even less information for the specific embryo-related toxicity field: embryotoxicity and developmental toxicity. Embryonic stem cell test (EST) is the only accepted *in vitro*method for assessing embryotoxicity without animal sacrifice. However, EST for regulatory embryotoxicity screening are impeded by its complexity, time-consuming, limited differentiation lineage endpoints, and poor availability for use to develop a reliable prediction model (PM) for embryotoxicity assessments.

A key step in the understanding of embryo development is to construct a systematic and comprehensive model of network with the development-related pathways involved. In our research, we mainly focus on the essential markers in the pathways related to ESCs apoptosis and proliferation, pluripotency, and specific-lineage differentiation. Quantitative structure-activity relationship (QSAR) models which predict the biological activities from chemical structures information is an efficient and less time-consuming strategy compared to experimental-based studies. Therefore, an integrated prediction pipeline for *in vitro*high throughput screening (HTS) based on engineered murine embryonic stem cells (mESCs) expressing a reporter, enhanced green fluorescent protein (EGFP), driven by human promoters for key genes in the pathways associated with apoptosis, pluripotency, and tissue-specific differentiation cultured in novel multi-well plates is developed and used to obtain new embryotoxicity and developmental toxicity data needed for the training of machine learning models such as support vector machines (SVM) and deep neural network (DNN). The application of advanced machine learning algorithms can significantly improve the accuracy and efficiency of the PM for embryotoxicity assessment of compounds with unknown developmental toxicity. This study will provide the technology needed for fast and accurate embryotoxicity assessments. The data obtained from our integrated analysis can provide the information needed for the regulation of industrial and emerging chemicals.

607548 A Computational Approach to Identifying Synergistic Compounds for Treating Antimicrobial-Resistant Pathogens

Abstract:

Antimicrobial resistance (AMR) is increasingly a threat to global public health [1]. AMR-related infections limit the efficacy of lifesaving antibiotic therapy necessary for the treatment of infectious disease and limit the success of advanced surgical procedures like organ transplants [2, 3]. To combat AMR pathogens, either new antibiotics or the cocktails of existing antibiotics with inhibitors of antimicrobial resistance proteins should be explored. Since the development of new antibiotic has been slowed down significantly recently, identifying compounds/antimicrobials with synergistic effects becomes the major approach to addressing the AMR crisis. In this work, we demonstrate a computational approach to identifying plant compounds that have synergistic effect with Fosfomycin to inhibit *Listeria monocytogenes*, a food-borne pathogen that could cause listeriosis disease, especially on immune-compromised people [4, 5]. Experimental results further validate the our finding.

L. *monocytogenes* was found to infect and adversely affect patients' liver and spleen. In addition, *L. monocytogenes* can penetrate blood-brain barrier and blood-placenta barriers to harm the central neural system of pregnant woman and infant [4]. Since *L. monocytogenes* is an intracellular pathogen, antimicrobials used to treat listeriosis should be able to be transported into host cells. Penicillin, ampicillin and amoxicillin were commonly used antibiotics in the treatment of listeriosis [6]. However, antibiotic resistance genes have been continuously found in *Listeria* strains. For example, the strain that has resistance to Penicillin G was isolated from vegetables in 2016 [4, 7]. As a natural product, Fosfomycin was found effective against clinical isolates of *L. monocytogenes* and used as a novel therapeutic antibiotic for listeriosis clinical treatment [8]. In addition, Fosfomycin is able to penetrate the blood-brain barrier and reach clinically relevant concentrations. Thus, it has the potential to eliminate *L. monocytogenes* which would cause neuron damage[9]. However, stronger Fosfomycin resistance was found in the *L. monocytogenes* isolates with Fosfomycin resistance proteins detected [10]. In particular, a resistant gene FosX (LMO1702,402bp) was identified and expressed in *L. monocytogenes* EGDe (strain ATCC BAA-679), a typical well-studied strain [11]. The FosX enzyme catalyzed the hydrolysis of Fosfomycin and resulted in the Fosfomycin resistance in *L. monocytogenes* EGDe [5, 10]. Therefore, there is an urgent need to identify compounds that can inhibit FosX enzyme to revive the efficacy of Fosfomycin to treat *L. monocytogenes.*

Automated molecular docking is the most commonly used computational approach that evaluates the binding of small-molecule ligands like compounds to a target receptor with a known crystal 3D structure [12]. Molecular docking provides an avenue for a high-throughput virtual screening of ligands, and it has been widely implemented in drug discovery research for hit identification[13]. Docking programs have been improved recently to provide more accurate prediction on ligand-target binding by optimizing docking algorithms and scoring functions [14]. Among those existing docking programs, Molsoft ICM was evaluated with 93% accuracy in flexible docking and 90% successful rate in covalent docking. This was significantly better than the performance of Autodock, DOCK, FlexX, Gold, FITTED and MOE [13-15]. Since structures and activities of the FosX protein in *L. monocytogenes* have been well studied, we used an integrated ICM-docking and experimental approach to identify FosX inhibitors that are of synergistic effect with Fosfomycin in treating resistant *L. monocytogenes*. Specifically, automated ligand docking was implemented to perform virtual screening of Indofine natural-product database and FDA-approved drugs to identified potential inhibitors. *In vitro* bacterial growth inhibition test was then utilized to verify the effectiveness of identified compounds combined with Fosfomycin on inhibiting the resistant *L. monocytogenes*. We demonstrated that two phenolic acids, i.e., caffeic acid and chlorogenic acid, were predicted as high-affinity FosX inhibitors from the ligand-docking platform. Experiments with these compounds indicated that: the cocktail of either caffeic acid (1.5mg/mL) or chlorogenic acid (3mg/mL) with Fosfomycin (50mg/L) was able to significantly inhibit the growth of the pathogen [16]. The finding of this work implies that the combination of Fosfomycin with either caffeic acid or chlorogenic acid is of potential to be used in the clinical treatment of *Listeria* infections.

References

1. Velez, R. and E. Sloand, *Combating antibiotic resistance, mitigating future threats and ongoing initiatives.* J Clin Nurs, 2016. **25**(13-14): p. 1886-9.
2. Michael, G.B., et al., *Emerging issues in antimicrobial resistance of bacteria from food-producing animals.* Future Microbiol, 2015. **10**(3): p. 427-43.
3. Davies, J. and D. Davies, *Origins and evolution of antibiotic resistance.* Microbiol Mol Biol Rev, 2010. **74**(3): p. 417-33.
4. Radoshevich, L. and P. Cossart, *Listeria monocytogenes: towards a complete picture of its physiology and pathogenesis.* Nat Rev Microbiol, 2018. **16**(1): p. 32-46.
5. Scortti, M., et al., *Epistatic control of intrinsic resistance by virulence genes in Listeria.* PLoS Genet, 2018. **14**(9): p. e1007525.
6. Temple, M.E. and M.C. Nahata, *Treatment of Listeriosis.* Annals of Pharmacotherapy, 2000. **34**(5): p. 656-661.
7. de Vasconcelos Byrne, V., et al., *Occurrence and antimicrobial resistance patterns of Listeria monocytogenes isolated from vegetables.* Brazilian Journal of Microbiology, 2016. **47**(2): p. 438-443.
8. Lepe, J.A., et al., *In vitro and intracellular activities of fosfomycin against clinical strains of Listeria monocytogenes.* International Journal of Antimicrobial Agents, 2014. **43**(2): p. 135-139.
9. Falagas, M.E., et al., *Fosfomycin.* Clin Microbiol Rev, 2016. **29**(2): p. 321-47.
10. Fillgrove, K.L., et al., *Mechanistic diversity of fosfomycin resistance in pathogenic microorganisms.* J Am Chem Soc, 2003. **125**(51): p. 15730-1.
11. Fillgrove, K.L., et al., *Structure and mechanism of the genomically encoded fosfomycin resistance protein, FosX, from Listeria monocytogenes.* Biochemistry, 2007. **46**(27): p. 8110-20.
12. Meng, X.Y., et al., *Molecular docking: a powerful approach for structure-based drug discovery.* Curr Comput Aided Drug Des, 2011. **7**(2): p. 146-57.
13. Kitchen, D.B., et al., *Docking and scoring in virtual screening for drug discovery: methods and applications.* Nat Rev Drug Discov, 2004. **3**(11): p. 935-49.
14. Lionta, E., et al., *Structure-Based Virtual Screening for Drug Discovery: Principles, Applications and Recent Advances.* Current Topics in Medicinal Chemistry, 2014. **14**(16): p. 1923-1938.
15. Scarpino, A., G.G. Ferenczy, and G.M. Keserű, *Comparative Evaluation of Covalent Docking Tools.* Journal of Chemical Information and Modeling, 2018. **58**(7): p. 1441-1458.
16. Zhang, F., et al., *Synergistic Effect of Chlorogenic Acid and Caffeic Acid with Fosfomycin in Growth Inhibition of a Resistant Listeria monocytogenes Strain.* ACS Omega, 2020. **https://dx.doi.org/10.1021/acsomega.0c00352**.

594532 Liquid Phase Membrane Supported Cyclical Flow Synthesis of Oligonucleotides for Therapeutic Applications

Abstract:

There now exists a consensus that RNA based gene inference techniques such as Anti-Sense Oligonucleotides and siRNA represent a significant therapeutic modality in the pharmaceutical industry. While the applications and understanding of the underlying biology of these systems for use as therapeutic platforms have advanced rapidly, little change in the commercial synthesis of these compounds has occurred. Metric tonnes per annum of oligonucleotides are projected to be required due to the high number of oligonucleotide-containing drugs in late phases of clinical trials and also projected for individual ASO therapies as larger patient populations are targeted in clinical trials following a number of clinical successes in small orphan patient populations.[1][2] The existing state-of-the-art synthesis of oligonucleotides is carried out with the use of a solid phase support and batch sizes currently limited to approximately 5 kg due to the constraints of solid phase synthesis.[3] In contrast, liquid phase supports reduce many of the limitations that are associated with solid phase supports.[4] Liquid phase synthesis of oligonucleotides was established by using linear and branched polyethylene glycol (PEG) as the soluble support[5] Membrane enhanced synthesis of oligonucleotides and peptides using PEG as a soluble support have been shown to be a potential solution for the problems associated with solid phase synthesis.[6][7]

The current work demonstrates proof of concept for a novel liquid phase manufacturing process utilizing cyclical flow synthesis combined with a novel membrane separation (**Figure 1**). This should enable fully scalable cyclical flow synthesis of oligonucleotides with increased productivity within a compact and scalable reactor footprint. This novel process intends to meet the growing demands of the industry and promises reduced capital costs and increased product quality control.

Modifications to soluble supports were utilized to both optimize the nucleotide loading process but also to optimize membrane flux and rejection factor, and hence yield and purity with commercially available membranes. Both liquid batch and phase cyclical flow synthesis of short oligonucleotides were optimized for coupling yield, phosphoramidite consumption, solvent consumption and overall productivity. The ability to use optimum reaction/separation timescales which are inherent to coupling reactions in oligonucleotide synthesis enabled improved system performance and in turn the ability to rapidly supply oligonucleotides on demand.

Acknowledgements:

This publication is supported by Science Foundation Ireland (SFI) through SSPC, The SFI Research Centre for Pharmaceuticals (Grant number 12/RC/2275_P2).

[1] Y. S. Sanghvi, *Curr. Protoc. Nucleic Acid Chem.* **2011**, 1–22.

[2] A. G. Molina, Y. S. Sanghvi, *Curr. Protoc. Nucleic Acid Chem.* **2019**, *77*, 1–17.

[3] A. R. Lajmi, L. Schwartz, Y. S. Sanghvi, *Org. Process Res. Dev.* **2004**, *8*, 651–657.

[4] P. R. J. Gaffney, J. F. Kim, I. B. Valtcheva, G. D. Williams, M. S. Anson, A. M. Buswell, A. G. Livingston, *Chem. - A Eur. J.* **2015**, *21*, 9535–9543.

[5] B. Burcovich, F. M. Veronese, V. Zarytova, G. M. Bonora, in *Nucleosides and Nucleotides*, Marcel Dekker Inc., **1998**, pp. 1567–1570.

[6] M. Mutter, H. Hagenmaier, E. Bayer, *Angew. Chemie* **1971**, *83*, 883–884.

[7] F. Brandstetter, H. Schott, E. Bayer, *Tetrahedron Lett.* **1973**, *14*, 2997–3000.

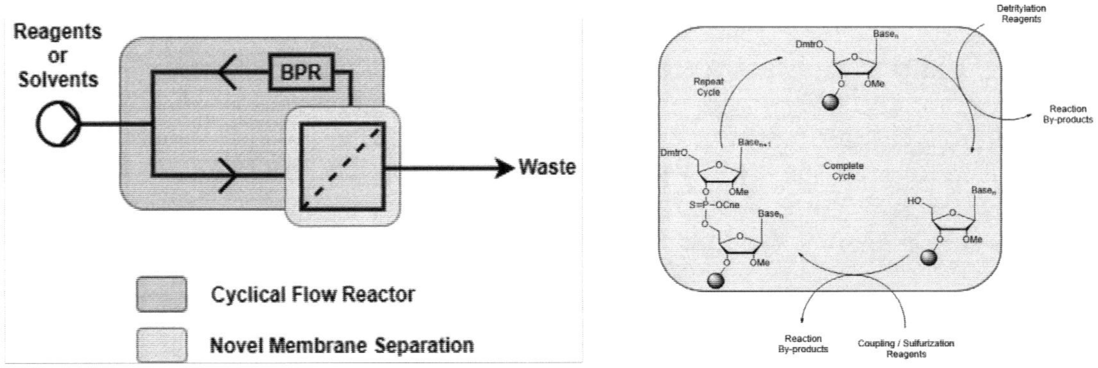

Figure 1: Schematic of system configuration (left) and Overview of cyclical synthetic route (right)

605370 Predict New Cocrystals Via Mechanochemistry

Abstract:

Within the pharmaceutical industry around 60% of recently developed drugs have solubility challenges while 80% of in-production line drugs are classified as being poorly soluble. Solubility is a critical property for drug performance (e.g. bioavailability) [1,2]. Cocrystal formation is one possible solution to overcome solubility problems. Pharmaceutical cocrystals, formed by combining an active pharmaceutical ingredient (API) with a cocrystal former (coformer), are often manufactured in a variety of methods that rely on the utilization of solvents. Pharmaceutical companies use between 25-100 kg solvent per kg of product, causing major water and air pollution. Hence, state-of-the-art solvent-based methods need to be replaced with more sustainable alternatives. Additionally, the formation of a cocrystal requires its components to be soluble in the same solvent. These limitations hinder the processability and development of novel pharmaceutical products.

Mechanochemistry utilizes mechanical forces to induce reactions. The application of mechanochemistry to form cocrystals without solvents is a novel, promising and sustainable method [1,3]. Scientists have exploited the features of mechanochemistry, but only in a random, non-systematic manner for the production of new cocrystals. Indeed, with the enormous number of potential API-coformer combinations, it is currently impossible to predict if two crystalline materials will cocrystallize under mechanochemical forces.

In this work, a high throughput screening approach combined with data analysis (machine learning) has been developed to predict if two crystals will cocrystallize under mechanochemical conditions. Strategically distinct pairs of APIs and crystalline coformers were screened by neat grinding (solventless) in a Retsch MM 400 vibration-mill. Powder X-ray diffraction has been applied to characterize all the samples and to determine the formation of new crystal structures. With the help of these results and molecular descriptors a machine learning algorithm (random forest model) has been successfully developed, effectively predicting the propensity of two molecules to cocrystallize under mechanochemical conditions. Additional cocrystallization events are predicted and experimentally validated. The technical feasibility of further API-coformer combinations can be estimated to create novel cocrystals without any solvent limitations, expanding the scope of mechanochemistry as a technique for drug discovery.

[1] C. B. Aakeröy and A. S. Sinha. Co-crystals. Monographs in Supramolecular Chemistry. The Royal Society of Chemistry, 2018.
[2] M. A. E. Yousef and V. R. Vangala. Pharmaceutical cocrystals: Molecules, crystals,formulations, medicines. Crystal Growth & Design, 19(12):7420–7438, 2019.
[3] S. Lou, Y. Mao, D. Xu, J. He, Q. Chen, and Z. Xu . Fast and Selective Dehydrogenative C–H/C–H Arylation Using Mechanochemistry. ACS Catalysis, 6 (6) , 3890-3894, 2016.

597737 Investigating Indirect Ultrasound on Particle Size, Shape and Nucleation of APIs through Process Intensification

Abstract:

Background

Particle engineering of active pharmaceutical ingredients (API) is an increasingly studied area of research & development. During drug substance manufacturing, particle formation typically originates from crystallisation which is a critical multiphase unit operation for the purification and separation of APIs. It is imperative that product properties such as particle size, shape and polymorphic form are well defined and well controlled as they affect a drug's bioavailability, shelf-life and to effectively formulate the API into a finished drug product. [1]

However, achieving control over these properties is not necessarily met at the point of API crystallisation and isolation. Process intensified strategies through incorporation of ultrasound devices has shown wide interest in the last decade for manipulating particle properties during or immediately after the crystallisation step. For instance, the integration of ultrasound with crystallization allows for multipurpose functionality such as particle size & shape modification, [2] de-agglomeration, [3] scale-up, [4] accelerating the rate of crystallization kinetics and presents an overall more flexible approach to particle engineering. [5]

Ultrasound technology is commonly applied across several industries through insertion of a probe horn in direct contact with a liquid medium. For pharmaceutical crystallisation, this can be advantageous in process development especially when small seed particles <15 μm are preferred for controlling the crystal growth and thus, final product properties. [6] Whilst still a popular method, direct contact of the probe horn with particles has well reported limitations such as surface erosion, [7] substantial temperature deviations, [8] excessive noise in the vicinity of health and safety, rapid dissipation of energy transferred per unit volume and continuous operability. Therefore, it is worthwhile exploring alternative particle engineering strategies which can avoid the aforementioned limitations whilst ensuring the manufacturability and quality of the drug substance.

Case-Study

In this work, indirect ultrasound is explored as a viable and robust means for particle engineering of APIs. A semi-continuous platform which applies indirect ultrasound for inducing nucleation, particle size reduction and shape modification is experimentally investigated via process intensification. The process consists of an inline flow-cell which generates intense ultrasonic vibrations via electrical stimulation (Figure 1). These mechanical vibrations are then transferred to a glass tube which in turn sonicate the incoming process stream without direct contact.

Two APIs are selected for this study both of which have been found hard to nucleate with standard operating conditions under current industrial manufacturing processes. To establish key relationships between process parameters and product attributes, a series of isothermal desupersaturation experiments were conducted which monitored the impact of power amplitude (0, 20%, 40% and 80%) on inline count profiles, chord length distribution (CLD), particle size distributions (PSD) and solution concentrations. Nucleation rates affected from ultrasound impact were then extracted through population balance modelling as a proof-of-concept demonstration. To our knowledge, this is the first study to test and implement this indirect ultrasound flow-cell strategy on APIs for nucleation acceleration and control of particle attributes. The overall platform integration and ease of operation were additionally assessed for potential application in continuous manufacturing of pharmaceuticals.

References:

1. Variankaval, N., A.S. Cote, and M.F.J.A.J. Doherty, *From form to function: Crystallization of active pharmaceutical ingredients.* 2008. **54**(7): p. 1682-1688.
2. Zeiger, B.W. and K.S.J.J.o.t.A.C.S. Suslick, *Sonofragmentation of molecular crystals.* 2011. **133**(37): p. 14530-14533.
3. Cote, A. and E.J.A.P.R. Sirota, *CRYSTALLIZATION: the pursuit of a robust approach for growing crystals directly to target size.* 2010. **13**(7): p. 46.
4. Kim, S., et al., *Crystallization process development of an active pharmaceutical ingredient and particle engineering via the use of ultrasonics and temperature cycling.* 2003. **7**(6): p. 997-1001.
5. Ruecroft, G., et al., *Sonocrystallization: the use of ultrasound for improved industrial crystallization.* 2005. **9**(6): p. 923-932.
6. Jiang, M., et al., *Indirect ultrasonication in continuous slug-flow crystallization.* 2015. **15**(5): p. 2486-2492.
7. Price, C.J., *Application of Ultrasound in Crystallization (Sonocrystallization)*, in *Engineering Crystallography: From Molecule to Crystal to Functional Form.* 2017, Springer. p. 301-313.
8. Zhang, Z., et al., *Enhancement of crystallization processes by power ultrasound: current state-of-the-art and research advances.* 2015. **14**(4): p. 303-316.

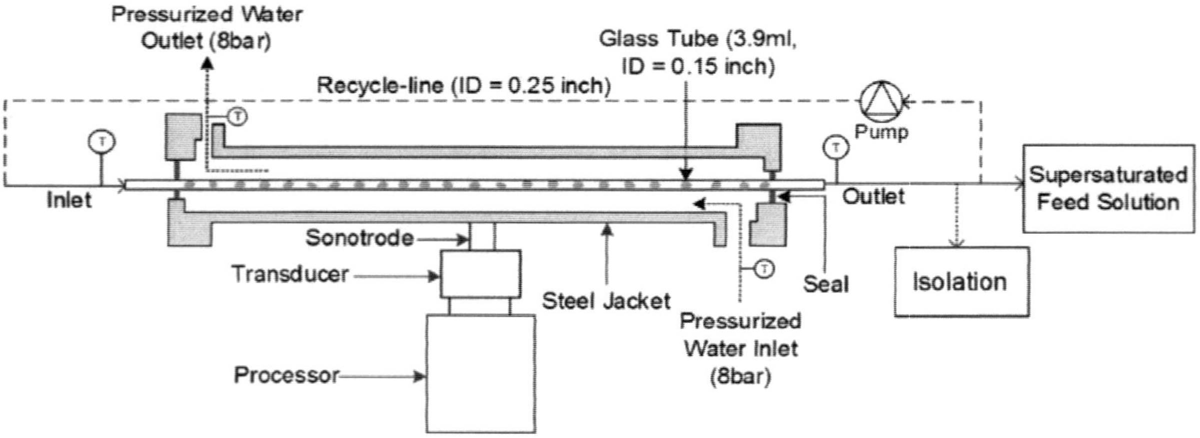

Figure 1: Sketch of the implemented indirect ultrasound flow-cell used in this study

607549 Experimental Investigation of Wet-Mill Integrated Crystallization with Temperature Cycles for Size and Aspect Ratio Control of a Commercial Pharmaceutical Ingredient

Abstract:

The control of the crystal properties has been in the focus of pharmaceutical industries and it represents a challenging aspect of the active pharmaceutical ingredient (API) production processes. To ensure that the desired critical quality attributes (CQAs) of the product are met, such as size, shape or polymorphic form, as well as to ease the downstream operations such as filtration or tableting, a careful design of the crystallization processes is critical. In this work we demonstrate the systematic design of a crystallization system for a commercial API, (compound A) from Takeda Pharmaceuticals International. The challenges related to the crystallization of the API is that the process is nucleation dominated and crystals are difficult to grow to the desired particle size (150-250 μm) by a simple cooling process. Moreover, the compound A tends to form high aspect ratio (AR) crystals, which can yield manufacturability problems. The aim of the crystallization design is to produce low AR and sufficiently large crystals with narrow distribution. Two methods were applied to reach these goals: (1) implementing temperature cycles to internally remove the fines, and (2) application of immersion milling to further control the shape and size of crystals.[1,2,3] These approaches were suggested in the literature based on the results of comprehensive model-based optimization studies.[4]

The work shows the implementation of a Quality-by-Control (QbC) guided wet-mill integrated crystallization design to produce large crystals with low AR. In this work, nine batch cooling crystallization experiments were performed. The first experiment was a seeded crystallization with very slow linear cooling rate. The remaining eight experiments were divided into two sets: (i) temperature cycling using simultaneous internal wet-mill; and (ii) crystallizations using temperature cycling but without milling. All experiments were performed with similar batch times but increasing the temperature cycle number from zero to four. Then, by using the same temperature profiles and operating procedure, simultaneous milling was applied for 75 % of the duration of the first temperature cycle as suggested in the literature, by using an immersion mill.[3]

A batch seeded cooling crystallization with very slow cooling (-0.02 °C/min) was performed, which showed that even this very slow cooling was insufficient to obtain the desired large crystals (> 150 μm) because of strong secondary nucleation in the process. Instead, applying internal fines removal by temperature cycles led to considerable increase of the product crystal size. Increasing the number of temperature cycles had a significant effect on producing large crystals (with mean size of 150-250 μm) by eliminating the fines produced by nucleation. These conclusions were also evidenced by using FBRM and in-line particle images. The results suggested that at least 2 temperature cycles were required to reach the desired product size for compound A. In the second part of the study, experiments with the same temperature profiles with wet-milling applied during the first cycle were performed. In these experiments, large crystals with low AR (>150 μm with mean AR around 2.5) could be successfully produced. The reason for using the wet-mill only at the beginning of the process is that it helps create large amount of small, low AR particles, which are partly removed by the first heat-up stage. In addition, it was shown with the 4 cycle-experiment that the milling rate does not have a significant effect on the CSD within the range applied (4000 and 8000 RPM) due to the achievable smallest size limit that characterizes the milling equipment. Therefore, the milling intensity was set to 4,000 RPM for all experiments. The experiments demonstrated that wet-mill contributed significantly to reaching the desired product crystal size and shape. This was quantified by 2D size distribution measurements based on optical microscopy images. Additionally, unseeded experiments were also performed, in which in situ seed generation was achieved by (1) cooling or (2) application of the immersion-mill as a high share nucleator. Both of these experiments demonstrated that applying wet-milling during the first temperature cycle not only significantly improves the CSD and AR of the product, but also considerably decreases the uncertainty caused by variation in the seed quality, and in combination with internal seeding via cooling or using the immersion mill as a nucleator can increase the overall efficiency and robustness of the process.

References:

[1] D. Ramkrishna and M. R. Singh, "Population Balance Modeling: Current Status and Future Prospects," Annu. Rev. Chem. Biomol. Eng., vol. 5, no. 1, pp. 123–146, 2014.

[2] D. Acevedo, V. K. Kamaraju, B. Glennon, and Z. K. Nagy, "Modeling and Characterization of an in Situ Wet Mill Operation," Org. Process Res. Dev., vol. 21, no. 7, pp. 1069–1079, 2017.

[3] Salvatori, F., Binel, P., & Mazzotti, M. (2019). Efficient assessment of combined crystallization, milling, and dissolution cycles for crystal size and shape manipulation. Chemical Engineering Science: X, 1, 100004.

[4] B. Szilagyi and Z. K. Nagy, "Model-based analysis and quality-by-design framework for high aspect ratio crystals in crystallizer-wet mill systems using GPU acceleration enabled optimization," Comput. Chem. Eng., vol. 126, pp. 421–433, 2019.

608902 Beyond Structure Determination: Crystallisation As a Purification Unit Process for Proteins and Peptides

Abstract:

The purification of high molecular weight proteins and peptides, for their use in biopharmaceutical drug therapeutics, can still account for a significant proportion of the drug development cost, which averages at between $1-6 billion. Crystallisation of these materials has mainly been used only in the field of structural determination, using small scale methods such as hanging drop vapour diffusion. However, advances in the optimisation of crystallisation conditions has seen significant increase in the scalability of protein crystallisation, including the development of continuous crystallisation approaches. This has seen crystallisation emerge as a more economic and greener alternative to the existing chromatographic methods typically utilised in peptide and protein purification.

The Made Smarter review of 2017 identified that manufacturing industries should embrace transformative digital technologies to streamline their unit processes. Mechanistic process modelling, based on population balance equations to represent the underpinning physical phenomena, has proved a computationally efficient method of predicting optimal crystallisation conditions of the purification of small molecule active pharmaceutical ingredients. However, this has yet to be expanded into the crystallisation of biopharmaceutical proteins.

Here, we utilise a combined process modelling and experimental approach to scale up the crystallisation of lysozyme and insulin from the µl to the tens of ml scale. We successfully optimise conditions, such as stirring rate and operation, pH, concentration and templating additives to not only scale up, but also increase the achieved crystallisation yield.

Insulin, being more susceptible to denaturing, utilised shaking to induce mixing of the protein rich buffer and salt precipitant. Small scale hanging drop screening experiments found that the addition of amino acids could reproducibly encourage the crystallisation of insulin, offering the promise of control over the nucleation. Scale up experiments to 15ml allowed the investigation of the impact of shaker mixing from 50-150 rpm. The insulin case study also highlights how to circumvent common issues with protein crystallisation, such as working with extremely low solubility materials and identifying the nucleation and growth points during the crystallisation process.

The lysozyme protein is more resistant to denature than insulin, whereby it can be subjected to mechanical stirring. Here we show successful batch and continuous crystallisations, utilising both shaking and mechanical stirring, up to the 30ml scale. Induction times were measured and classical nucleation kinetics parameters were derived from this data, with the pre-exponential factor and the surface energies being consistent with previous studies. Further optimisation of stirring methods and initial protein concentration was investigated to optimise the size and shape of the produced lysozyme crystals, whereby their purity was compared to the as purchased lipolyzed powder.

Finally, process modelling techniques were applied to predict the crystallisation of lysozyme and insulin in the batch reactor conditions. Such models are well parameterised to describe the crystallisation of small molecule active pharmaceutical materials. However, the classical nucleation and growth equations, as well as the models to describe the liquid state, have been less commonly applied to biological materials. Kinetic parameters to describe the processes that were extracted from the literature were used as a starting point for the model, where careful optimisation of these parameters were used to identify problems and where discrepancies between modelled and experimental particle size distributions and concentration vs time data originated. The kinetic parameters that were optimised against reliable experimental data were confronted with further experimental data to examine the wider applicability of protein crystallisation process model.

It was found that the classical Mullin nucleation and power law growth kinetic equations could be successfully regressed against experimental concentration vs time data, even for long nucleation times (Figure 1).

Figure 1: Concentration vs time for a 30ml batch crystallisation of lysozyme, using a sodium acetate buffer and sodium chloride precipitant. Blue dots are experimental data with approximate 10% error from nanodrop UV vis measurements and the red line in the regressed fit using Mullin classical nucleation kinetics and power law growth kinetics

Figure 1 shows that a good fit of the process model to the experimental concentrations vs time data was achieved. It is interesting to observe that the classical two step growth kinetics failed to find a fit, where we speculate that the diffusion term associated with the large lysozyme molecule limits the growth of this material in the model. Further investigation of the exact parameters in the equations will be presented.

This study demonstrates a novel experimental and process modelling workflow for optimising the crystallisation of large biological molecules, paving a path for this to be a viable unit process in biopharmaceutical drug development. This study gives guidance and insights into the challenges of protein crystallisation and modelling of this process, along with what are the next steps to make both processes more robust for further scale-up.

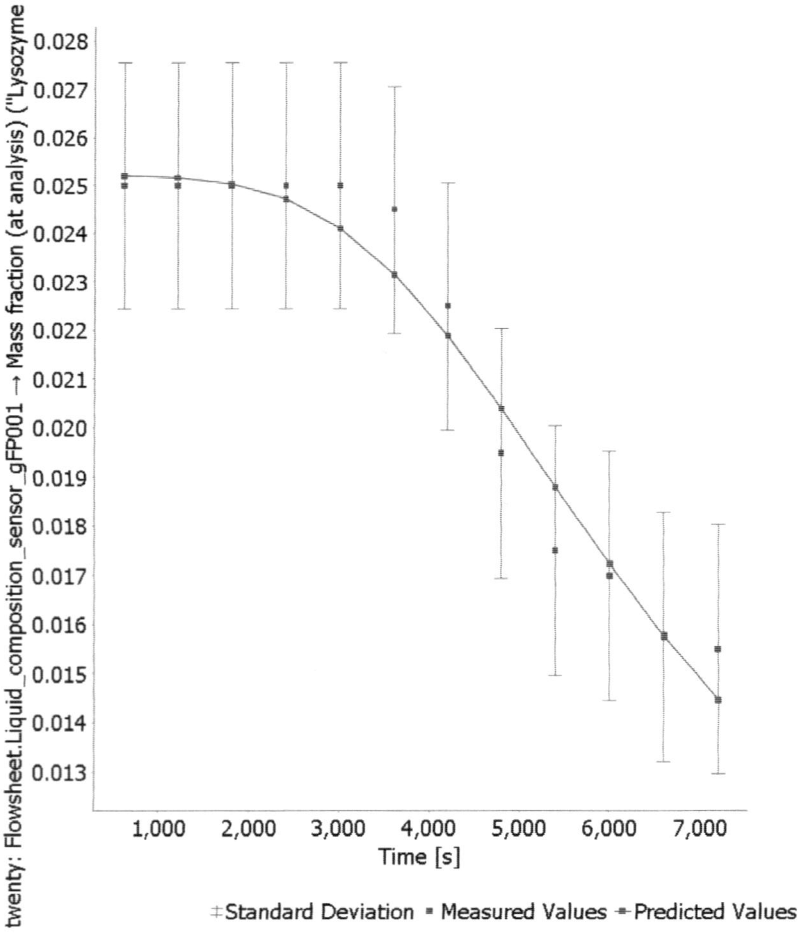

605706 Maneuvering Around Liquid-Liquid Phase Separation (LLPS) to Enable Development of Robust Crystallization Processes.

Abstract:

Anti-solvent and/or cooling crystallizations are commonly employed in purification of pharmaceutically relevant molecules. In addition, the crystallizations are also typically tasked with isolating solids with controlled physical properties. During development of multi-solvent crystallizations, it is not unusual to observe Liquid-Liquid Phase Separation (LLPS), commonly referred to as "oiling out". LLPS is generally undesirable due to its detrimental impact on purity and physical properties as well as due to the challenges it brings in during scale-up. LLPS is impacted by multiple parameters such as solute concentration, solvent composition, and temperature. Understanding of such a dynamic phase diagram, in addition to the kinetic parameters, is paramount to achieve a predictable process scale-up.

In this presentation, we detail a methodology to develop a hybrid anti-solvent / cooling crystallization process maneuvering around the LLPS boundaries. Use of automated Lab reactors (ALR) equipped with Process Analytical Technologies (PAT) is highlighted for performing data-rich experiments quickly. The generated data is further modelled with thermodynamic equations for enhanced process understanding. Finally, we will present a case study to showcase how the above-referenced methodology was utilized for enhanced process understanding and for development of a robust and scalable crystallization process.

All authors are employees of AbbVie and may own AbbVie stock. AbbVie sponsored and funded the study; contributed to the design; participated in the collection, analysis, and interpretation of data, and in writing, reviewing, and approval of the final publication.

603539 Utilizing Mechanistic Modelling to Support Crystallization Seeding Design

Abstract:

Introduction

Digital design of manufacturing processes using mechanistic models is fast becoming an essential tool during Active Pharmaceutical Ingredient (API) early process development activities. It enables rapid and effective exploration of the decision space for Critical Process Parameters (CPP), helping to reduce risk and product time-to-market, and aiding in the effective and safe production of high quality pharmaceutical products. In this work, we demonstrate how mechanistic modelling can be utilised to support the early stage design of seeding strategy for a batch cooling crystallization process. We outline a step-wise workflow consisting of model validation from which primary nucleation and growth kinetics are estimated, as a minimum list of crystallization mechanisms. The data used during the bespoken parameter estimation was experimentally measured in an unseeded small-scale crystallization process.

As part of this work, the model predictions for the seeded crystallization process was subsequently verified experimentally. This model-based approach along with rich experimental data yielded a more efficient workflow for seeding design, with reduced experimental effort compared to a more traditional purely empirical approach. A range of process data was utilized for the purposes of qualitative mechanism discrimination and quantitative model validation as follows:

- Qualitative mechanism & model discrimination
  - FBRM data to inform active crystallization mechanisms to consider
  - SEM and/or optical images of product crystals to inform crystallization mechanisms to consider within the mechanistic model
- Quantitative model validation
  - FBRM data to provide indication of onset of nucleation and access to the rate of nucleation in the unseeded crystallization runs.
  - Solute concentration data over time both online (IR/Raman) &/or offline (HPLC)
  - Particle size measurements of the product crystals from laser diffraction

The validated mechanistic model was subsequently used for model-based design of seeding policy, to assist in the seeding design (seed mass, seed addition point, and/or seed particle size distribution) required to achieve the target PSD specifications (defined by the volume-based quantiles d10, d50 and d90 of the product PSD). Uncertainty and sensitivity analysis were performed to understand the attainable region for the product PSD in terms of quantiles for the crystallization process and to identify the relative importance of process parameters, including seed mass and seed PSD.

Conclusions

The main conclusions of the work include the following:

- It has been shown that a model-based approach can be utilised to support the early stage development of crystallization processes to achieve the desired quality attributes of the product particles.
- A model-based approach has been shown to help to reduce the degree of experimental effort

required compared to a purely empirical approach.

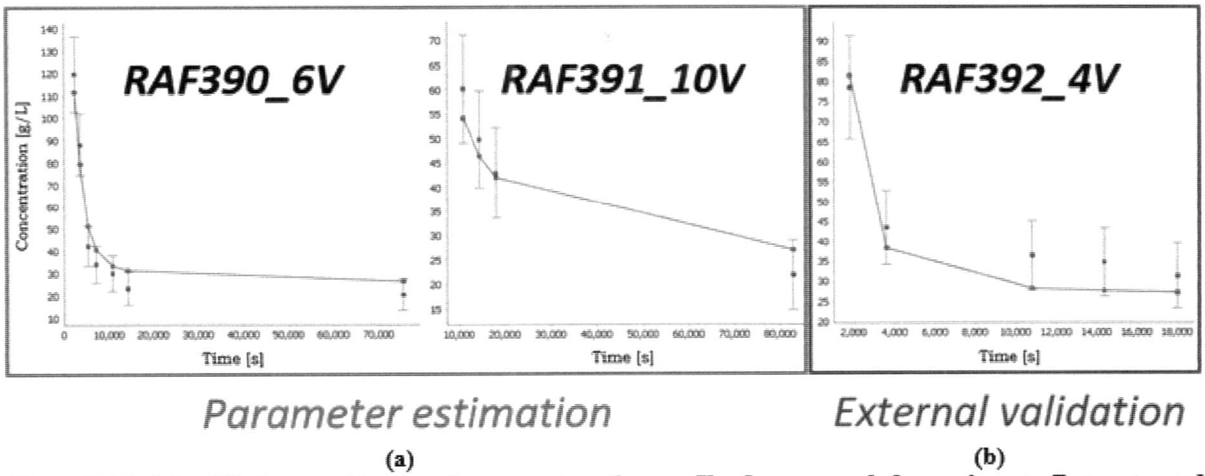

Figure 1: Model validation results using de-supersaturation profiles from unseeded experiments. Data sets used for (a) parameter estimation and (b) external model validation.

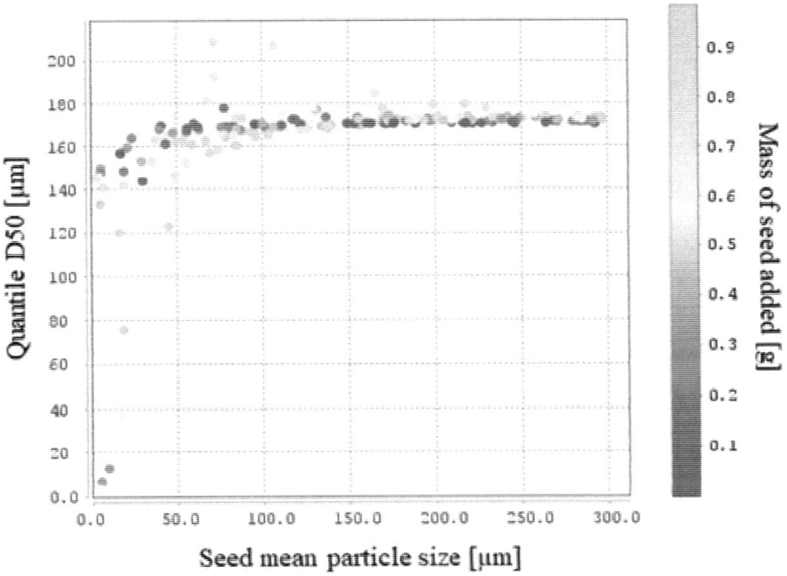

Figure 2: Sensitivity analysis results showcasing the impact of mass of seed added and seed mean particle size on the D50 quantile of the crystallized material.

611464 Process Optimization for Controlling the Morphology of Cof-5 Crystals

Abstract:

Covalent Organic Frameworks (COFs) have gained significant attention in the field of material science over last two decades. 2D boronate-ester linked frameworks like COF-5, are porous, crystalline, organic polymers with low density, that can be used for numerous applications such as gas storage and as membranes for various industrial purification applications. COF-5 exhibits high thermal stability, high surface area and permanent porosity widening the scope of its applicability. To maximize its applicability, it is essential to optimize and control the morphology of the crystallites by tweaking appropriate processing conditions. COF-5 can be synthesized by evaporative crystallization using 1,4-diboronic acid (PBBA) and Hexahydroxy Triphenylene (HHTP), in an equimolar solvent mixture of Dioxane and Mesitylene. This process is heterogenous, uncatalyzed and slow, needing three days to reach a considerable and acceptable yield. Researchers have previously made efforts to reduce the reaction barrier and reaction time to 20 hrs. However, a systematic understanding of the effects of alterations in processing conditions on crystallite yield and morphology remain unanswered. Effective and accelerated crystallization of COF-5 while controlling the structure of these crystals is the prime focus of our research. This study has addressed the optimization of processing conditions, catalyst content, and solvent environment to synthesize and control the morphology of COF-5 at a higher rate. Additionally, these conditions were utilized to grow COF-5 films on substrates, while ensuring that their crystallite properties remained intact. Experimentally derived kinetic models using the optimized conditions, have been coupled with population balance equations to get a deeper insight on the crystal growth mechanisms of COF-5.

608089 Crystallization Model Discrimination Using in Situ Chord Length Monitoring and a Size Conversion Model (CLD2PSD model).

Abstract:

A new framework has been recently introduced to correlate in situ chord length distribution (CLD) measurements and offline particle size distribution (PSD) data (Irizarry et al, 2017) and applied to different particle morphologies (Schoell et al, 2019). Subsequently the model was expanded to consider bimodal distributions and inferences of modality and morphology changes (Irizarry et al, 2020). This work applies the proposed approach to estimate the kinetics parameters for seeded batch crystallization processes of different APIs in combination with population balance modelling. First, a data-driven CLD2PSD model was built for the studied pharmaceutical compounds since the differences in optical properties and particle morphology require a new model for every single substance. Then, time-resolved PSDs have been extracted from in situ CLD data of seeded batch processes and the CLD2PSD model. Finally, the online PSDs have been combined with online solute concentration data to allow for the estimation of crystal growth, secondary nucleation and agglomeration kinetics, depending on the supersaturation level encountered during the process. The validity of this approach has been confirmed by comparison to kinetic data determined using a traditional approach relying on offline data. The main advantage of the proposed approach is that all data is generated online and in situ, thus avoiding sampling errors.

References

- Irizarry, R. et al., 2017. Data-driven model and model paradigm to predict 1D and 2D particle size distribution from measured chord-length distribution. Chemical Engineering Science, 164, pp. 202-218.
- Schoell, J. et al., 2019. Determining particle-size distributions from chord length measurements for different particle morphologies. AIChE Journal, 65, e16560

606939 Continuous Solid-State Deracemization Via Temperature Cycles – a Model-Based Process Development

Abstract:

Introduction

Although preferential crystallization is a standard industrial process to obtain enantiopure crystals of conglomerate forming compounds, its maximum theoretical yield is 50%, unless one adds to it recycle and racemization steps [1]. On the contrary, one can produce an enantiopure powder from an initially racemic one with 100% yield through solid-state deracemization by exploiting several techniques [1,2]. Among such techniques, one based on temperature cycling is a promising candidate for industrial applications due its simplicity of operation [1,3,4]. However, the process itself is complex and hence it warrants the need for a mathematical model to understand the effect of initial conditions and of operating parameters on the process outcome, i.e. the production of the desired enantiomer, the process time, etc.

Continuous processes have been shown to outperform batch operation. In addition, they can recover from process disturbances, because they run at stable steady state [5]. Continuous crystallization processes have been adapted for the separation of conglomerate forming enantiomers, in order to exploit the advantage of continuous crystallization techniques. In case racemization of enantiomers in the solution proves difficult, continuous preferential crystallization without a racemization reaction is a widely studied choice [6]. If the racemization reaction is available, continuous preferential crystallization with racemization reaction or continuous Viedma ripening can be performed. Alternatively, deracemization via temperature cycles in mixed suspension mixed product removal crystallizers (MSMPRC) can also be performed, but this is an unexplored method of choice.

Mathematical model

A population balance equation (PBE) -based mathematical model has been developed to provide a clear explanation of the phenomena involved in the deracemization process. The model describes the interplay of several phenomena involved in a temperature cycling process (size-dependent solubility, crystal growth and dissolution, agglomeration, attrition/secondary nucleation and racemization) accounting for the dependence of their thermodynamic and kinetic parameters on the crystal size and on the temperature [5].

Results and discussion- batch

Using the PBE-model that we have introduced, first, we have simulated temperature cycles in batch crystallizers. A mathematical model of deracemization is a useful tool to perform process analysis: it provides information on all the relevant physical quantities during the whole process and enables decoupling of the individual phenomena, and investigating the effect of variations in e.g., the breakage rate intensity, the agglomeration intensity, etc. After identifying the necessary phenomena to deracemize using temperature cycles, process characterization was performed using the developed model, in order to identify an optimal process. Parametric studies varying operating parameters like the temperature and cooling rate were performed and based on the outcome it is shown that the simulation outcome qualitatively describes experimental results [1,4].

Moreover, by investigating the effect of slight variation in the initial conditions, an explanation to the large variations in deracemization time and in process outcome — that are typically observed in experiments — is provided using the mathematical model.

Results and discussion- continuous

The inherent disadvantage of the batch crystallization process is that it suffers from batch-to-batch product variability and requires control techniques to obtain the product with the specific quality of interest. As tuning the many degrees of freedom blindly while performing the actual experiments is ill-advised [7], this contribution aims at providing a mathematical tool that can be used to design robust continuous processes.

Therefore, we have extended our batch non-isothermal population balance-based model of solid-state deracemization via temperature cycles to simulate the proposed continuous configuration. To avoid undesired nucleation of the counter enantiomer, different process variants are considered, e.g., utilization of recycle loop, wet mill, fines dissolution loop. We have investigated the steady state operation of temperature cycles carried out in MSMPRC. Periodic operation of the continuous crystallizer via temperature cycling affects the transient behaviour of the process and thereby the attainable steady state. We have run simulations to study the interplay of the cycle time and the residence time with respect to the physical properties of the system, i.e., how the process characteristic times compare to the characteristic times of growth, dissolution and racemization.

The aim of this analysis was to verify whether one of the continuous configurations envisaged performs better than the batch alternative. The effect of the initial conditions (variations in the initial enantiomeric excess, in the suspension density, etc.) and of the operating parameters on the periodically perturbed process and its steady state in terms of productivity and attainable enantiomeric excess are identified. Together with this assessment, we aim to provide strategies for the optimal implementation of the process.

Acknowledgments

This research has received funding as part of the CORE project (October 2016 – September 2020) from the European Union's Horizon 2020 research and innovation programme under the Marie Sklodowska-Curie grant agreement No 722456 CORE ITN.

References

[1] Breveglieri, G. M. Maggioni, M. Mazzotti, Cryst. Growth Des. 18 (2018) 1873-1881.
[2] L. Noorduin, W. J. P van Enckevort, H. Meekes, B. Kaptein, R. M. Kellogg, J. C. Tully, J. M. McBride, E. Vlieg, Angew. Chem., Int. Ed. 49 (2010) 8435-8438.
[3] Iggland and M. Mazzotti, Cryst. Growth Des. 18 (2011) 4611-4622.
[4] Suwannasang, A. E. Flood, C. Rougeot, G. Coquerel, Org. Process Res. Dev. 21 (2017) 623-630.
[5] Köllges and T. Vetter, Cryst. Growth Des. 17 (2017) 233-247.
[6] Vetter, C. L. Burcham, M. F. Doherty, AICHE J. 61 (2015) 2810-2823.
[7] Bodák, G.M. Maggioni, M. Mazzotti, Cryst. Growth Des. 18 (2018) 7122-7131.

$$\frac{\partial f_i}{\partial t} + \frac{\partial(G_i(S_{\infty,i}, T, L)\, f_i)}{\partial L} = B_i(L, c_i, T) - D_i(L, c_i, T) + \frac{\dot{m}_{in} f_{in,i} - \dot{m}_{out} f_i}{m}$$

$$\frac{dc_i}{dt} = -k_v \rho \frac{d\phi_{3,i}}{dt} + R_i(c_i, c_j, T) + \frac{\dot{m}_{in} c_{in,i} - \dot{m}_{out} c_i + \dot{m}_{rec} c_i}{m} \qquad (i = D, L)$$

- $\phi_{n,i} = \int_0^{+\infty} L^n f_i(L,t)\,dL$
- $ee = \frac{\phi_{3,D} - \phi_{3,L}}{\phi_{3,D} + \phi_{3,L}}$

Initial and boundary conditions:

$$f_D(L,0) = f_D^0(L)$$
$$f_L(L,0) = f_L^0(L)$$
$$f_i(0,t) = 0$$
$$c_D(0) = c_L(0) = c_\infty(0)$$

Figure 1. Summary of the model equations that are the basis for the model of our continuous configuration. L: size of the crystal, T: temperature, S_∞: supersaturation, f_i: particle size distributions, G_i : growth rates, B_i and D_i: birth and death terms, $\dot{m}_{in}$, $\dot{m}_{out}$, $\dot{m}_{rec}$: mass flow rates, c_i: concentrations, R_i: racemization term

609443 Real-Time Process Monitoring in Continuous Manufacturing of Oral Solid Dosages Using Raman Process Analytical Technology

Abstract:

The growing recognition of the limitations of batch manufacturing and recent advances in Process Analytical Technology (PAT) have led to the shift towards continuous manufacturing (CM). However, the implementation of continuous manufacturing has its challenges and continues to be an area of intense research interest. Near-infrared (NIR) spectroscopy has been the choice of PAT tool in the pharmaceutical industry for online measurements of powder composition in drug product manufacturing, but it has its limitations in measuring low concentrations. Raman spectroscopy is an alternative vibrational spectroscopy method that has also received attention to monitor the drug concentration in powders and is studied in this work.

Important components in implementing the CM strategy are: (1) real-time measurement and monitoring, (2) process control, (3) fault detection, (4) diagnosis of exceptional events, and (5) tracking and isolation of non-compliant product[1]. This research focuses on the first foundational component – real-time measurement and monitoring – using the spectroscopy-based PAT tools. The measurement of the active pharmaceutical ingredient (API) concentration in low drug load blends has been a known challenge[2]. Raman spectroscopy can be utilized to predict the API concentration in such low drug load products. Also, it is important to have the right spectroscopic sensor at the right sensor location to achieve the true benefits of the in-situ measurement. The feed frame of the tablet press has been selected as a promising location as it provides various benefits[3] and has only been explored in to a limited extent in the literature. The blend composition measured multiple unit operations upstream of tableting may not be truly representative of the API concentration in the final tablets. Locations close to tablet compression provide better API concentration estimates in the final product. Also, real-time measurement of API at the tablet press location can aid in immediate rejection of the out of specification tablets.

The present study aims to develop a reliable PAT method using Raman spectroscopy to determine the API concentration in real-time in continuous tablet manufacturing of Low Drug Load Product. This includes checking the model robustness against process conditions, understanding the difference in Raman spectra between static powder sample and moving powder stream inside feed frame, and understanding the issue of fluorescence with some of the ingredient excipients. Previous study has shown the use of Raman spectroscopy in measuring API concentration with 8% API (w/w)[4]. The second objective is to integrate the Raman sensor in the existing pilot plant facility at Purdue to send the data and results of concentrations to the Distributed Control System (DCS) using proper communication protocols to enable effective statistical process control of the critical quality attributes (CQA) and pave way to develop feedback control. This study investigates the application of Raman spectroscopy for real-time measurement of a continuous tableting process and the feasibility of Raman-based feedback control in the manufacturing of pharmaceutical solid dosage forms.

Reference:

1. Moreno M, Liu J, Su Q, et al. Steady-State Data Reconciliation Framework for a Direct Continuous Tableting Line. *J Pharm Innov*. 2018. doi:10.1007/s12247-018-9354-9

2. Nagy B, Farkas A, Borbás E, Vass P, Nagy ZK, Marosi G. Raman Spectroscopy for Process Analytical Technologies of Pharmaceutical Secondary Manufacturing. *AAPS PharmSciTech*. 2019;20(1). doi:10.1208/s12249-018-1201-2
3. De Beer T, Burggraeve A, Fonteyne M, Saerens L, Remon JP, Vervaet C. Near infrared and Raman spectroscopy for the in-process monitoring of pharmaceutical production processes. *Int J Pharm*. 2011;417(1-2):32-47. doi:10.1016/j.ijpharm.2010.12.012
4. Li Y, Anderson CA, Drennen JK, Airiau C, Igne B. Method Development and Validation of an Inline Process Analytical Technology Method for Blend Monitoring in the Tablet Feed Frame Using Raman Spectroscopy. *Anal Chem*. 2018;90(14):8436-8444. doi:10.1021/acs.analchem.8b01009

596387 Multipurpose Automated Optimization Platform for Reactions Involving Solids

Abstract:

Continuous manufacturing of pharmaceuticals and fine chemicals is attractive due to its small footprint, consistent product quality, and demonstrated benefits from safety, economic, and environmental perspectives.[1-4] However, handling solids in research-scale flow reactors creates hurdles, as the solids often lead to reactor channel clogging. To tackle this problem, we present a continuous stirred-tank reactor (CSTR) cascade that can handle slurries/solids during a chemical transformation in flow. Moreover, we implement a mixed-integer nonlinear program (MINLP) algorithm for multi-objective optimization by simultaneously modulating discrete variables (catalyst types) and continuous variables (residence time, temperature, and catalyst loading). The optimization strategy involves a sequential adaptive response surface methodology along with optimal design of experiments and the global search strategy branch and bound.[5]

We demonstrated the autonomous optimization of a Suzuki-Miyaura cross-coupling reaction involving solid substrates and catalysts in an automated flow platform comprising a CSTR cascade, slurries feeding pumps, and an inline HPLC. The hardware control and automation were achieved with an integration of LabVIEW™,[6] MATLAB®,[7] and online analysis. This research-scale fully automated flow platform for reaction self-optimization with solids/slurries feeding and handling while consuming a reduced amount of raw materials facilitate identification of optimal reaction conditions for manufacturing process development.

References:

1. Kevin P. Cole, Brandon J. Reizman et al. (2019) *Org. Process Res. Dev.* 23, 5, 858-869

2. Kevin P. Cole, Brandon J. Reizman et al. (2019) *Org. Process Res. Dev.* 23, 5, 870-881

3. Christopher L. Burcham, Alastair J. Florence, and Martin D. Johnson (2018) *Annu. Rev. Chem. Biomol. Eng.* 9:253–81

4. Kevin P. Cole and Martin D. Johnson (2018) Continuous flow technology vs. the batch-by-batch approach to produce pharmaceutical compounds, *Expert Review of Clinical Pharmacology*, 11:1, 5-13

5. Lorenz M. Baumgartner, Conner W. Coley, Brandon J. Reizman, Kevin W. Gao, and Klavs F. Jensen (2018) *React. Chem. Eng.* 3, 301-3011.

6. LabVIEW™ is a trademark of National Instruments. This publication is independent of National Instruments, which is not affiliated with the publisher or the author, and does not authorize, sponsor, endorse or otherwise approve this publication.

7. MATLAB is a registered trademark of The MathWorks, Inc. See mathworks.com/trademarks for a list of additional trademarks.

595407 Implementation of Model-Predictive Control in Continuous Crystallization

Abstract:

Automation and process control are essential when operating a continuous (flow) process to minimize waste and start up time. Typical operating protocols involve implementing the steady state inputs such as feed concentration and reactor jacket temperature immediately at start-up, then divert the out-of-spec material until the process achieves steady-state. From that point the product is collected. In this study, we implement a model-predictive controller to start up a continuous crystallizer optimally, achieve and maintain steady-state conditions faster than open-loop recipe driven operations. Reactor temperature and solution concentration are the measured variables. The solution concentration is measured using a ReactIR probe calibrated using absorbance as a function of solution concentration. The feed concentration of the inlet stream and the reactor jacket temperature are the manipulated variables. This presentation will compare open-loop and closed-loop controlled concentration profiles in a continuous crystallization for a small molecule API and for a chemical intermediate at Biogen. The closed-loop controller results in steady-state achieved in approximately three residence times, whereas it takes five residence times for the open-loop case to reach steady state. Crystal habit and size are also monitored and show size trends as a function of supersaturation.

607903 Model-Based Dynamic Optimization of Feeder Refill Strategy Used in Continuous Pharmaceutical Manufacturing Process

Abstract:

The process model is in the heart of 'digital twin initiative' of the continuous pharmaceutical manufacturing process [1-3]. The process model can be used to optimize the process and design and tune the control system. The feeder is the first and one of the most crucial unit operations of the continuous manufacturing of pharmaceutical products. It directly influences the drug content of the tablets and thereby, patient safety. The feeder needs to be refilled very frequently for continuous manufacturing, and its performance depends significantly on the refill strategy followed. The feeder refill strategy involves three operational parameters that need to be identified before using the feeder for continuous manufacturing. These parameters are 'feeder mass at which the refill needs to be started,' 'refill size,' and 'refill rate.' If the refill strategy is not correctly optimized, then the overall quality of the product will be compromised. Presently, all the settings involved in refill strategies are being experimentally determined using a heuristic approach, which is time and resource-intensive and often leads to suboptimal solution. Therefore, a systematic dynamic optimization framework is needed through which the feeder refill strategy can be optimized.

In this work, a systematic framework including the methods and tools has been developed for dynamic optimization of the feeder refill strategies. The feeder unit operation has been modeled using 'gPROMS' FormulatedProducts software (PSE),' and the optimization method has been developed in MATLAB (Mathworks). Both tools are communicating through the gO:MATLAB toolbox (PSE). The deviation of the outlet mass flow of the feeder from the targeted flow rate has been minimized to obtain the optimum value of the feeder refill parameters. Once optimal refill strategy has been obtained then it has been transferred to distributed control system (DCS) to implement it into the continuous pharmaceutical manufacturing pilot-plant [4]. The material properties also affect the refill strategy meaning that the feeder refill strategy need to be frequently optimized if there are any changes in the materials and plant. Therefore, the developed feeder model and dynamic optimization tool can save the time and recourses significantly. The heuristic approach currently used in industry could also lead to suboptimal refill strategy and thereby poor feeder performance and poor product quality. The proposed method can save the time and resources needed for continuous manufacturing as well as can improve the product quality significantly.

The objective of this presentation is two-fold; first to highlight the developed systematic framework for model-based dynamic optimization of the feeder refill strategy and then demonstrate its application for continuous pharmaceutical manufacturing process.

References

1. Metta, N., Ghijs, M., Schafer, E., Kumar, A., Cappuyns, P., Assche, I. V., Singh, R., Ramachandran, R. De Beer, T., Ierapetritou, M., Nopens, I. (2019). Dynamic flowsheet model development and sensitivity analysis of a continuous pharmaceutical tablet manufacturing process using the wet granulation route. Processes Journal, 7(4), 234.
2. Singh, R. (2018). System engineering for a novel continuous pharmaceutical manufacturing process. Pharma. Issue 30. https://www.pharmafocusasia.com/manufacturing/system-engineering-pharmaceutical-manufacturing-process.
3. Barros, F. N., Bhaskar, A., Singh, R. (2017). A validated model for design and evaluatic control architectures for continuous tablet compaction process. Processes Journal, 5(4), 76. doi:10.3390/pr5040076
4. Singh, R., Sahay, A., Fernando Muzzio, Ierapetritou, M., Ramachandran, R. (2014). A systematic framework for onsite design and implementation of the control system in continuous tablet manufacturing process. Computers & Chemical Engineering Journal, 66, 186-200. http://dx.doi.org/10.1016/j.compchemeng.2014.02.029

609039 Real Time, Dynamic (Non-Steady State) Process Models for Osd Continuous Manufacturing Control

Abstract:

The creation of a dynamic, real-time, mass balanced based process model for a continuous, direct blend encapsulation line will be discussed. The ability to control the response of a continuous process to deviations in feeder rate is a critical aspect of all oral solid dosage form continuous manufacturing systems. Real time modeling is a robust and flexible approach for this type of control strategy. This type of dynamic modeling can predict concentration changes in the system simultaneously with changes in throughput and hold up mass with as few as two fit parameters.

612158 Big Data-Based Fault Detection with Advanced Analytics in the Pharma Industry

Abstract:

Motivated by market pressure to reduce product costs, regulatory incentive to modernise manufacturing and a growing environmental concern to reduce waste has led the pharmaceutical industry to improve its manufacturing process. The impetus to adapt and modernise the manufacturing process can largely be attributed to the promotion of the Quality by Design paradigm (QbD) and guidance on Process Analytical Technology (PAT) by the FDA in 2004. The promotion of these methods coincided with a time where the cost of computing and data storage had declined, but also with developments in non-invasive online sensor technology, such as Raman and Near-Infrared (NIR) spectroscopy. This culminated into an increase intro the application of multivariate analysis (MVA) and specifically Multivariate Statistical Process Control (MSPC).

MSPC exploits the Big Data generated from a complex chemical process such as pharmaceutical production and can be used for process understanding, troubleshooting and on-line fault detection. MSPC tools include multivariant process charts and soft sensors and can be combined with APC methods to improve overall plant efficiency. MSPC aims to use big data to better understand a process, to develop a definition of normal operation and then use that to detect faults on-line. There are three main categories to classify as SPC methods (1) model-based (2) knowledge-based (3) data-based. Data-based methods are often the preferred method for pharmaceutical manufacturing. The advantages of data-driven methods are that they do not require first principals understanding of the process or a unit as they construct a model using input and output data. The most popular data-driven methods are multivariate latent/MVA methods (which include PCA, PLS and their adapted versions). These methods project data into the latent space and work as a feature extraction method and reduce the dataset into a smaller or manageable set that can be used in further applications of fault detection.

This paper presents an industrial case study which focused on fault detection towards operational improvements for a key API production, pursued in collaboration with a major multinational UK pharmaceutical company (GSK). The aim of this project is to exploit data availability towards constructing and validating a data-driven model of a specific API production unit of the GSK plant (Montrose, Scotland, UK). The project involves using large data sets of process data (temperatures, partial pressures, flow rates, and stream compositions, from an on-site GCMS analysis station), and implementing advanced MVA methods in MATLAB, in order to diagnose operational patterns and recommend feasible improvements towards reliable API production intensification.

LITERATURE REFERENCES

[1] E. Tomba et al., "General procedure to aid the development of continuous pharmaceutical processes using multivariate statistical modeling-An industrial case study," Int. J. Pharm., vol. 444, no. 1–2, pp. 25–39, 2013.

[2] T. Kourti, "Application of latent variable methods to process control and multivariate statistical process control in industry," Int. J. Adapt. Control Signal Process., vol. 19, no. 4, pp. 213–246, 2005.

[3] J. F. MacGregor and T. Kourti, "Statistical process control of multivariate processes," Control Eng. Pract., vol. 3, no. 3, pp. 403–414, 1995.

[4] T. Kourti, "Multivariate dynamic data modeling for analysis and statistical process control of batch processes, start-ups and grade transitions," J. Chemom., vol. 17, no. 1, pp. 93–109, 2003.

[5] J. F. MacGregor, P. Nomikos, and T. Kourti, "Multivariate Statistical Process Control of Batch Processes Using PCA and PLS," IFAC Proc. Vol., vol. 27, no. 2, pp. 523–528, 1994.

[6] S. Garcia-Munoz and D. Settell, "Application of multivariate latent variable modeling to pilot-scale spray drying monitoring and fault detection: Monitoring with fundamental knowledge," Comput. Chem. Eng., vol. 33, no. 12, pp. 2106–2110, 2009.

[7] M. Boiret, D. N. Rutledge, N. Gorretta, Y. M. Ginot, and J. M. Roger, "Application of independent component analysis on Raman images of a pharmaceutical drug product: Pure spectra determination and spatial distribution of constituents," J. Pharm. Biomed. Anal., vol. 90, pp. 78–84, 2014.

598331 Recent Advances in Kinetic Parameter Estimation Toolkit (KIPET) with Spectra

Abstract:

Developed over the past four years, the Kinetic Parameter Estimation Toolkit (KIPET), is an open-source toolbox for the determination of kinetic parameters from a variety of experimental datasets including spectra and concentrations. *KIPET* seeks to overcome limitations of standard parameter estimation packages by applying a unified optimization framework based on maximum likelihood principles and large-scale nonlinear programming strategies for solving estimation problems with nonlinear differential algebraic equations (DAEs). The software package includes tools for data preprocessing, determination of parameter confidence levels for a variety of problem types, and informative wavelength selection to improve the lack of fit (Schenk et al., 2020). All these features have been implemented in Python with the algebraic modeling package *Pyomo*. *KIPET* exploits the flexibility of *Pyomo to* formulate and discretize the dynamic optimization problems that arise in the parameter estimation algorithms. Solutions of these optimization problems are obtained with the nonlinear solver *IPOPT* and confidence intervals are obtained through the use of either *sIPOPT* or a newly developed tool, *k_aug*.

In this talk, we discuss a recently enhanced *KIPET* package (Short et al., 2020) that considers multiple experiments with potentially different reactants and kinetic models, different dataset sizes with shared or unshared individual species' spectra, leading to fast parameter estimation and confidence intervals based on the NLP sensitivities. In addition, we present a new variance estimation technique based on maximum likelihood derivations for unknown covariances from two sample populations. This approach leads to a straightforward deconvolution of variances between noise in model variables and in spectral measurements. Moreover, we discuss a new estimability analysis approach that systematically determines a ranked subset of kinetic parameters with well-defined confidence intervals (Chen and Biegler, 2020). With nonlinear kinetic models and limited measurements, it is often difficult to correctly estimate all the parameters, due to linear dependence and low correlation among the parameters. A common approach is to estimate a subset of the parameters by fixing the others at reasonable (often literature) values. However, it is often challenging to determine which parameters can be properly estimated. In this talk we present an efficient approach that ranks the estimable parameters, and discards those that cannot be estimated accurately. Based on reduced the reduced Hessian information with simple Gauss-Jordan elimination, the proposed approach leads to fast parameter selection and estimation within a simultaneous collocation framework. This approach becomes much more efficient for large problems than competing approaches based on multiple eigenvalue decompositions (Quaiser and Mönnigmann, 2009). Several case studies with increasing complexity are presented to demonstrate the performance of this proposed approach.

1. W. Chen, L. Biegler, "Reduced Hessian Based Parameter Selection and Estimation with Simultaneous Collocation Approach," submitted for publication (2020)
2. Schenk, M. Short, J. S. Rodriguez, D. Thierry, L. T. Biegler, S. Garcia-Munoz, W. Chen, "Introducing KIPET: A novel open-source software package for kinetic parameter estimation from experimental datasets including spectra," Computers and Chemical Engineering, 134(4), 106716 (2020)
3. Short, L. T. Biegler, S. Garcia-Munoz, W. Chen, "Estimating Variances and Kinetic Parameters from Spectra Across Multiple Datasets Using KIPET," Chemometrics and Intelligent Laboratory Systems, to appear (2020)
4. Quaiser, T., Mönnigmann, M. "Systematic identifiability testing for unambiguous mechanistic modeling-application to JAK-STAT, MAP kinase, and NF-κB signaling pathway models," BMC Systems Biology, 3,50 (2009).

601039 Teaching Computers to Interpret Mdsc Thermograms

Abstract:

We present a novel algorithm for the automatic data analysis of modulated Differential Scanning Calorimetry of thermograms designed to minimize subject matter expert input reduce the man reliability.

Differential Scanning Calorimetry (DSC) and mainly modulated DSC (mDSC) analysis are widely used to support the production of amorphous solid dispersions for the identification of kinetic and thermodynamic events such as glass transition, crystallization and melting. However, interpretation of thermograms require careful analysis of subject matter experts (SMEs) which can be time-consuming and thus can be a bottleneck to, e.g., formulation development where large numbers of samples are generated are require prompt analysis for reiteration. Furthermore, often there are elusive phenomena that can sometimes go unnoticed or cause dubious interpretation from different SMEs.

The presented Python-based software tool's computation of the derivatives and the maximums, minimums and inflections of each curve, is the base of the software. It then uses this information to search for events, based on their magnitude. One essential advantage of using an automated software is the ability to use the derivatives of the heat flow curves to improve the precision of the results. The derivatives are essential not only to the detect the exact temperature at which events occur but also to help define their onsets. In order to tackle the inherent noise of the DSC analysis and data acquisition, the proposed tool employs a low pass, Fourier transform-based filter, which enables the use of the thermogram derivatives.

We show case studies where the software analysis increases the detection rate of events that can easily be unnoticed by visual assessment due to their magnitude or shape, as they only cause very subtle variations. We also present case studies where smaller events are distorted or confounded with a nearby dominant events. However, following the digital detection of these events a more detailed visual assessment confirms their presence and often indicative of the onset of undesirable events such as phase separation or crystallization. Furthermore, the presented digital analysis tool promotes a faster and objective analysis of generated DSC thermogram data.

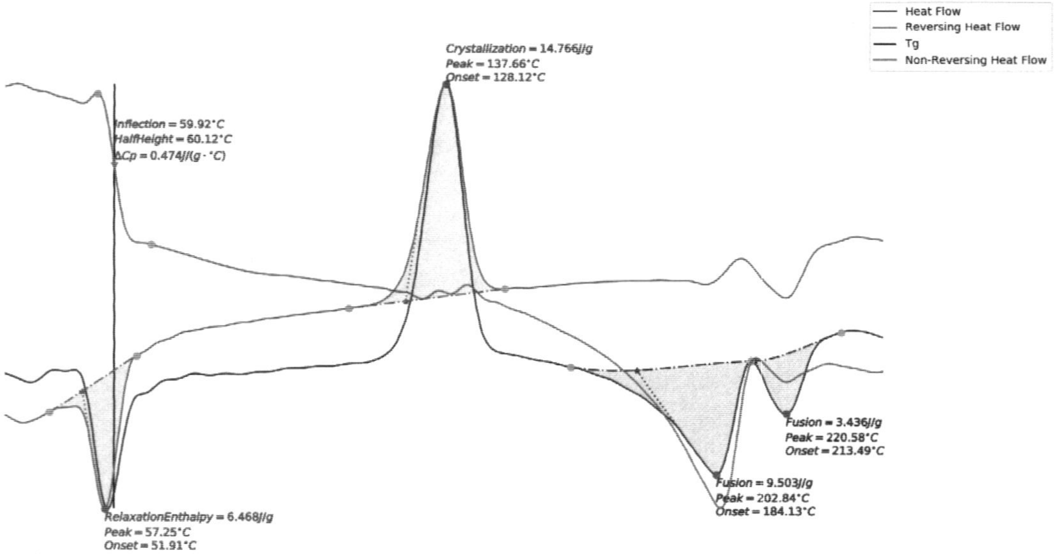

595673 Data-Smart Machine Learning Methods for Predicting Youngs Modulus of Directly Compressed Blends of Pharmaceutical Powders

Abstract:

The ability to predict the mechanical properties of compacted powder blends of Active Pharmaceutical Ingredients (API) and excipients solely from those of the individual components can reduce the amount of "trial-and-error" involved in formulation design. Machine Learning (ML) models can reduce model development time and effort with the imperative of adequate historical data. This work evaluates linear and non-linear ML for predicting the Youngs Modulus of directly compressed arbitrary blends of known excipients and API from readily available information about the API. The training data obtained from three BCS Class I APIs and four excipients demonstrate "data-smart" strategies to train ML models efficiently. The results indicate that even simple linear ML model provide reasonable prediction accuracy. The practical benefits of this method and how it compares with other mechanistic models are discussed. Finally, we demonstrate an application of the model to enable Quality-by-Design in drug product development.

593356 Advances and New Developments on Eiot for PAT Applications in Pharmaceutical Development and Manufacturing

Abstract:

The EIOT method[1] was proposed as an extension to the IOT[2] method as an alternative approach to extracting information from spectroscopic data into interpretable mass fractions of the resolved chemical species in the powder mixture.

The method in essence seeks to use Beer-Lambert's law as the deterministic functional relationship between spectra and mass fractions. The main difference with IOT is that EIOT does not assume the spectrum of the pure components is a measurable quantity; but a quantity that can only be estimated from measurements of known composition. The EIOT method proposes to decompose the measured spectra as the summation of the chemical signal plus the effect of the non-chemical interferences. The chemical signal is given by the summation of the apparent pure spectrum weighted by the mass fraction of each specie; the non-chemical interferences as the summation of the non-chemical interference signals weighed by their respective strength. The method relies on optimization techniques to estimate the apparent pure spectra and the signal of the non-chemical interferences; and when the method is used in real-time, to estimate the mass fractions and the strengths of the non-chemical interferences in each new sample.

Since the method was published, a number of enhancements have been done to the method, this talk presents such enhancements that i) the consideration of supervised signatures and ii) the consideration of exclusivity in a set of signatures (i.e. only one in a set is allowed to contribute). Details around the maintenance and long-term ownership of the EIOT method are discussed for its potential inclusion as a central component for a control strategy in drug product manufacture.

EIOT (as well as its predecessor IOT) bring a disruptive change in the practice of chemometrics; retiring the use of fixed regression models, to the use of real-time optimization and computer decision-making instead. As such, the requirements and expectations for use and long-term maintenance might have to be revised and adapted.

[1] Z. Shi, J. Hermiller, S.G.J.A.J. Muñoz, Estimation of mass-based composition in powder mixtures using Extended Iterative Optimization Technology (EIOT), (2018).

[2] K. Muteki, D.O. Blackwood, B. Maranzano, Y. Zhou, Y.A. Liu, K.R. Leeman, G.L. Reid, Mixture component prediction using iterative optimization technology (calibration-free/minimum approach), Industrial & Engineering Chemistry Research, 52 (2013) 12258-12268.

608805 Solvent Mixture Design for the Integrated Purification in Pharmaceutical Manufacturing.

Abstract:

Solvents play a dominant role in the manufacturing of pharmaceutical products, as they are extensively used to facilitate synthetic reactions, enable separation/purification and take part in the final drug formulation. More than 80% of waste and byproducts generated in a typical active pharmaceutical ingredient (API) manufacturing process (i.e., 25-100 kg of byproducts/kg API [1]) is related to solvents used in purification processes [1,2]. Hence, selecting suitable solvents in different purification stages can affect the overall process efficiency and final product quality, as well as cost, environmental, health and safety metrics [3,4]. However, choosing optimal solvents (or solvent mixtures) and the best operating conditions for a given process can be challenging due to the discrete nature of the problem (that can lead to combinatorial explosion) and the trade-offs between competing objectives (e.g. enhance product quality attributes can lead to increased cost and/or environmental impact). In current practice, most pharmaceutical companies employ solvent selection guides to (i) consider the physicochemical properties and/or environmental profile of solvents, and (ii) focus on reducing solvent consumption in order to minimise cost and environmental footprint [4,5]. Despite these advances, most of the methods used are based on heuristic approaches or time-consuming experimental investigations. Such trial-and-error practices can often be expensive and are liable to lead to sub-optimal designs that fail to consider the integrated nature of all process decisions. Thus, the development of efficient methodologies and more sophisticated tools that explore all possible options and consider integrated process decisions would be an indispensable tool in the design of new pharmaceutical processes.

Several solvent mixtures design approaches [6-9] are based on the computer-aided mixture/blend design (CAM^{b}D) framework [10] and are applied to the crystallisation of active pharmaceutical ingredients. To date, most of the proposed methods consider optimising a single purification process unit (e.g., crystallisation) with fixed operating conditions (e.g., fixed temperature). Recently, this has been broadened to the consideration of temperature as a variable, to investigate hybrid anti-solvent and cooling crystallisation [9]. In this work, the scope of design is extended further and a comprehensive solvent mixture design methodology for the integrated crystallisation and isolation processes of pharmaceutical compounds is presented. The proposed integrated approach considers a combined cooling and antisolvent crystallisation process, where optimal solvent and antisolvent mixtures, their proportions in solution, and optimal process temperatures are determined simultaneously, while maximising crystal yield and reducing solvent consumption [9]. In addition, critical interlinked design decisions, such as API solubility across both crystallisation and isolation stages, as well as the miscibility of crystallisation and wash solvents, are taken into account. Furthermore, environmental, health and safety performance measures are considered in the comprehensive model, so that only safe and environmentally friendly solvents are identified.

The design method is applied to identifying high-performance solvent blends for the crystallisation and isolation of mefenamic acid, while removing an impurity, chlorobenzoic acid, from the system. This work demonstrates that optimising simultaneously several design decisions and different performance criteria can lead to significant improvements in overall process efficiency over conventional designs.

1. W. Cue, J. Zhang, 2009. Green Chemistry Letters and Reviews 2, 193-211.

2. D. Curzons, C. Jiménez-González, A.L. Duncan, D.J.C. Constable, et al., 2007. The International Journal of Life Cycle Assessment 12, 272-280.
3. J. Brown, T. McGlone, S. Yerdele, V. Srirambhatla, et al., 2018. Molecular Systems Design & Engineering 3, 518-549.
4. Jiménez-González, P. Poechlauer, Q.B. Broxterman, B-S. Yang, et al., 2011. Organic Process Research & Development 15, 900-911.
5. Prat, J. Hayler, A. Wells, 2014. Green Chemistry 16, 4546-4551.
6. T. Karunanithi, L.E.K. Achenie, R. Gani, 2006. Chemical Engineering Science 61, 1247-60.
7. Jonuzaj, P.T. Akula, P. Kleniati, C.S. Adjiman, 2016. AIChE Journal 62, 1616-33.
8. Jonuzaj, A. Gupta, C.S. Adjiman, 2018. Computers & Chemical Engineering 116, 401-421.
9. L. Watson, A. Galindo, G. Jackson, C.S. Adjiman, 2019. Computer Aided Chemical Engineering 46, 949-954.
10. Gani, R., 2004. Computers & Chemical Engineering 28, 2441-2457.

602099 In-Line Monitoring of Low Drug Concentration of Flowing Powders in a New Sampler Device

Abstract:

A novel sampler device for flowing powders was tested to quantify drug concentrations as low as 0.76% w/w in pharmaceutical powder blends. The sampler device was developed based on the powder flow behavior within a tablet press feed frame, following the principles laid down in the Theory of Sampling. Two Near-Infrared (NIR) spectroscopic calibration models were developed with powder blends that varied from 0.52 to 2.52% w/w and 1.51 to 4.52% w/w. The calibration models were able to determine caffeine concentration in test set blends with root mean square error of predictions and bias below 0.1% w/w. Samples were collected from the sampler device and analyzed by ultraviolet-visible (UV-Vis) to determine the caffeine concentration. A high agreement between the in-line NIR predictions and the sampled UV-Vis results was found. The paddle wheel speed in the sampler can be varied up to ±10% without affecting NIR predictions; however, the models did not respond adequately to a 25% increase in this speed. Variographic analysis showed that the sampler device may quantify low drug concentrations with nugget effects below 0.0050 $(\%w/w)^2$. This study demonstrate that the sampler device may handle throughputs up to 45 kg/h, without significantly affecting the physical properties of powder blends.

605905 Fabrication of Curcumin-Loaded Microparticles from Oil-in-Water Emulsion By Supercritical Assisted Atomization

Abstract:

For oral administration of hydrophobic drugs, a novel oral drug delivery system based on emulsion has been widely implemented, where hydrophobic drug is encapsulated in the hydrophobic core. However, emulsion in liquid state is lack of long-term stability under the action of Ostwald ripening, coagulation, creaming and cracking. This can be avoided by the transformation from liquid emulsion to solid particles and an effective technology is necessary.

In this work, liquid emulsion was first prepared with ultrasound where ethylcellulose was as oil phase matrix and gum arabic (GA) as emulsifier and wall material. The oil-in-water emulsion formulation was then processed by an improved supercritical assisted atomization technique (SAA-HCM) and curcumin-loaded microparticles were successfully fabricated. Influences of operation conditions including mixer pressure, temperature, gas/liquid mass flow ratio on diameters and size distributions of microparticles were investigated in detail, and particle formation mechanism was discussed. It is verified that fast extraction of organic solvent from oil phase by supercritical CO_2 in mixer is favorable to entrap curcumin in the matrix of oil phase preventing drug leaching at stage of atomization, and high drug loading efficiency (53.91±1.6%) was obtained when the GA concentration was up to 20 g/L. CLSM indicated that the shell of curcumin-loaded microparticles was composed of GA and nanoparticles dried from emulsion droplets were enclosed in the inner. The chemical stability of curcumin loaded in microparticles was identified by FI-IR, and the amorphous form of curcumin attributed to fast formation of drug-loaded particles was confirmed by DSC and X-ray. *In vitro* drug release behavior of microparticles exhibited controlled release and enzyme trigger properties.

SAA-HCM is a green micronization technique, and this work will expand the application of SAA-HCM technique to emulsion, which will be of great progress in the design of drug-loaded microparticles with elaborate structures.

606780 Development of Targeted Exosomes for Drug Delivery

Abstract:

Exosomes have great potential to deliver drugs due to its low antigenicity. However, low yield and lack of targeting capability have hindered its application. This study aimed at creating a biomanufacturing platform of cancer-targeted exosomes. First, we created a scalable, high-productivity, cell line-based monoclonal antibody tagged exosomes (mAb-Exo) production process, and characterized the product with transmission electron microscopy, NanoSight and Western blotting. Second, we confirmed the cancer-specific targeting using flow cytometry and confocal laser scanning microscopy and using live animal imaging system. Finally, we loaded Romidepsin, a histone deacetylase inhibitor, in mAb-Exo and validated the targeted drug delivery capability in tumor xenograft animal model by testing the cancer targeting specificity and anti-cancer efficacy. We developed new technique enabling targeted drug delivery.

603201 Linking Process & Performance of Drug Products with Micro-Image Analytics

Abstract:

Microstructures play an increasingly important role in drug formulation design and product development. The physical properties of API domains, excipients, and pores, dictate both manufacturability and performance of a final drug product. Furthermore, to achieve specific drug release patterns and rates, functional coatings and performance-enabling excipients have become increasingly common.

This poster will introduce a novel platform for evaluating the roles of microstructures from micro-images. The micro-scale nature of components and their interplay in a drug product demand a high-resolution technique to evaluate their role in performance. Three-dimensional micro-imaging, including X-Ray Microscopy (XRM) and Focused Ion Beam Scanning Electron Microscope (FIB-SEM), can be used separately and/or correlatively to qualitatively visualize these microstructures.

What's more, the spatial and chemical distributions, as well as image-based modeling of release behaviors, can be assessed quantitatively and semi-quantitatively. The performance impact of microstructures on a PLGA long-acting formulation and spray-dried particles, including performance relation to process and manufacturing conditions, will be used to illustrate this micro-image based platform.

Image-based characterization can be utilized in drug product development for a fundamental understanding of the process-property-performance interplay, optimizing formulation process and design, and increasing manufacturability. The combination of micro-imaging, AI-based image analytics, and image-based modeling presents a new platform for advancing drug design and evaluation, with significantly reduced evaluation time, improved drug performance, and lowered costs of *in vitro* and *in vivo* experiments.

608085 Surrogate Modeling of Dissolution Behavior Toward Efficient Design of Tablet Manufacturing Processes

Abstract:

Speed is critical in the research and development of drug products and processes. The challenge here is the existence of various process alternatives and product characteristics in tablet manufacturing processes, which makes the decision-makings complicated. As for the process alternatives, the choice of dry or wet granulation is known to affect product quality significantly, while this decision is usually conducted without rigorous comparison. As for the product characteristics, solubility of active pharmaceutical ingredients (APIs) depends on the product, and influences the dissolution behavior of tablets largely. In the literature, first-principle as well as statistical modeling approaches has been researched for tablet manufacturing (e.g., [1] and [2]) for assisting more efficient process design. The power of the first-principle modeling is recognized so well that commercial process simulators, e.g, gPROMS, are available for tablet manufacturing processes. However, there is still a hurdle of computational time to use first-principle models in the actual decision-makings.

This work presents surrogate modeling of dissolution behavior for enhancing the use of first-principle modeling in the design of tablet manufacturing processes. Our research approach consists of the following three steps: (i) collect process input and output data using gPROMS, (ii) perform surrogate modeling, and (iii) test the developed model. In the first step, 200 data are collected per each manufacturing method by simulating flowsheet models. Secondly, random forests are used for developing surrogate models from collected data sets after fitting Weibull models for dissolution behaviors. Finally, in the third step, sensitivity analysis is performed using the developed surrogate model to generate a proposal for the efficient process design.

Case studies of dry granulation and wet granulation methods were performed, where paracetamol was used as an API of the tablets. The simulation results showed the differences in tablet properties between manufacturing methods, e.g., hardness, porosity, and dissolution. After the execution of surrogate modeling, the sensitivity of input parameters was assessed using rank correlation coefficients, which determined critical parameters in the process, and product design phases. In ongoing work, the generality and the effectiveness of the developed models are discussed by performing case studies using a different API.

[1] D. Van Hauwermeiren, M. Verstraeten, P. Doshi, M. T. am Ende, N. Turnbull, K. Lee, T. De Beer, I. Nopens, Powder Technol, 341, 116–125 (2019)

[2] K. Matsunami, T. Nagato, K. Hasegawa, H. Sugiyama, Int J Pharm, 579, 119160 (2020)

607115 Optimization of Hydride Vapor Phase Epitaxy (HVPE) Deposition Reactor Manufacturing III-V Materials Using Multiple Large-Eddy Simulations (LES)

Abstract:

Compared to traditional silicon-based solar cells, III-V materials possess several advantageous properties for photovoltaic applications. First, their large light-absorption coefficient allows for thin and lightweight solar cells, thereby enabling portable applications. Second, their conversion efficiency from light to electricity is superior to that of silicon-based materials. Metalorganic vapor-phase epitaxy (MOVPE) [1] is the preferred manufacturing process of III-V materials, as it has been successful at producing highly efficient solar cells [2]. However, this technique induces large manufacturing costs which make III-V materials viable only for specific applications. The large cost is due to the cost of the reactants used, and the rate at which the solar cells grow, which prevents industrial scaling of the technology.

An alternative process is the hydride vapor phase epitaxy (HVPE) [3] which has been shown to provide a growth rate two orders of magnitude higher than that of MOVPE, while using cheaper reactants [4]. The process has been recently refined into a dynamic-HVPE (D-HVPE) and has been shown to provide satisfying efficiency [5, 6]. Although promising, the faster growth rate of the solar cells and the cheaper materials used come with new manufacturing challenges that need to be addressed. The overarching objective of this work is to propose a design that will address these challenges and make D-HVPE viable in an industrial context. The specific challenges are described below:

- Thin film solar cells consist several layers, each one made of a different III-V compound. When the growth rate of the solar cell is low enough, the substrate can be kept idle in the same chamber while different reactants are injected over time. With a higher growth rate - such as in HVPE deposition reactors - this strategy can lead to a non-uniform composition in each layer, which is detrimental for the efficiency. To avoid this issue, here, the substrate is moved to a different chamber each time a new layer needs to be grown (see figure enclosed). Ideally, the chambers operate at the same time on a different substrate to maximize the throughput of the reactor. While efficient, this configuration poses a fluid dynamic problem: if the reactants of one chamber escape to a neighbor chamber, it could contaminate the layer grown in the neighbor chamber, thereby hindering the solar cell efficiency. Additionally, the deposition of reactants between chambers can create maintenance issues as the reactants are typically poisonous. The first challenge is to reduce the leak between chambers.

- In HVPEs, the gaseous halides used for the deposition is obtained by mixing HCl with the group-III metal (here, Ga or In) at high temperature. Since the products of the reaction need to be maintained at a high temperature, it is advantageous to have the chemical reaction occur inside the deposition chamber (see figure enclosed), while the chamber is externally heated. This step creates a complex fluid flow in the chamber due to the buoyancy effects. At the same time, reactants should be deposited fast enough to avoid spurious reactions, and should mix fast enough to ensure uniformity of the layer grown on the substrate. The second challenge is to find an injection scheme that will satisfy both conditions.

In this work, multiple large-eddy simulations are conducted to better understand the manufacturing challenges posed by D-HVPE deposition reactors, and improve their design. The simulations are run with a variable density low-Mach solver based on OpenFOAM [7]. The results are processed with different sensitivity analysis techniques including active subspaces [8] and Sobol indices [9] to identify the features that affect the most out-of-chamber leak and reactants uniformity on the substrate. The results of the sensitivity analysis are used to propose new designs and optimize the geometry of the deposition reactor. In the end, this work participates in the improvement of D-HVPE deposition reactors and will enable the emergence of portable photovoltaic applications in a variety of sectors.

References

[1] Manasevit, H., "The use of metalorganics in the preparation of semiconductor materials: Growth on insulating substrates," Journal of Crystal Growth, Vol. 13, 1972, pp. 306–314.

[2] Chen, H.-L., Cattoni, A., De Lépinau, R., Walker, A. W., Höhn, O., Lackner, D., Siefer, G., Faustini, M., Vandamme, N., Goffard, J., et al., "A 19.9%-efficient ultrathin solar cell based on a 205-nm-thick GaAs absorber and a silver nanostructured back mirror," Nature Energy, Vol. 4, No. 9, 2019, pp. 761–767

[3] Lourdudoss, S. and Kjebon, O., "Hydride vapor phase epitaxy revisited," IEEE Journal of Selected Topics in Quantum Electronics, Vol. 3, No. 3, 1997, pp. 749–767.

[4] Horowitz, K. A., Remo, T. W., Smith, B., and Ptak, A. J., "A techno-economic analysis and cost reduction roadmap for III-V solar cells," Tech. rep., National Renewable Energy Lab.(NREL), Golden, CO (United States), 2018.

[5] Metaferia, W., Schulte, K. L., Simon, J., Johnston, S., and Ptak, A. J., "Gallium arsenide solar cells grown at rates exceeding 300 μm/h by hydride vapor phase epitaxy," Nature communications, Vol. 10, No. 1, 2019, pp. 1–8.

[6] Schulte, K. L., Simon, J., and Ptak, A. J., "Multijunction Ga0.5 In0.5 P/GaAs solar cells grown by dynamic hydride vapor phase epitaxy," Progress in Photovoltaics: Research and Applications, Vol. 26, No. 11, 2018, pp. 887–893.

[7] Hassanaly, M., Koo, H., Lietz, C. F., Chong, S. T., and Raman, V., "A minimally-dissipative low-Mach number solver for complex reacting flows in OpenFOAM," Computer and Fluids, Vol. 162, 2018, pp. 11–25.

[8] Constantine, P. G., Dow, E., and Wang, Q., "Active subspace methods in theory and practice: applications to kriging surfaces," SIAM Journal on Scientific Computing, Vol. 36, No. 4, 2014, pp. A1500–A1524.

[9] Sobol, I. M., "Global sensitivity indices for nonlinear mathematical models and their Monte Carlo estimates," Mathematics and computers in simulation, Vol. 55, No. 1-3, 2001, pp. 271–280.

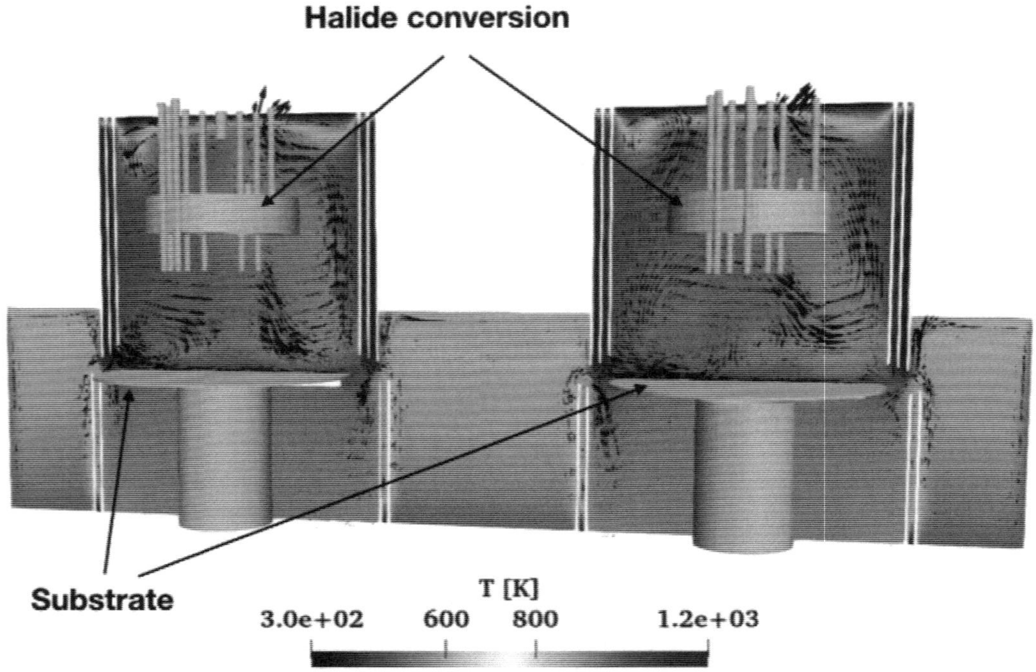

Figure 1: Contour of temperature in the HVPE deposition reactor. White elements denote the substrate, its support, and the reactant injection scheme. Arrows indicate the instantaneous orientation of velocity. Note that multiple substrates are operated at once in neighboring chambers, and that the gaseous halide is produced inside each chamber.

603903 Process Intensification and Integration Potential in Manufacturing Pollution Prevention

Abstract:

Process intensification (PI) aims to dramatically improve manufacturing processes through the application of novel process systems and equipment. The novel approaches can be used to overcome bottlenecks, such as those imposed by thermodynamics, or to combine processing phenomena into fewer processing units with a concurrent reduction of capital and operation and maintenance costs and energy, water and materials intensity. PI approach goes beyond the incremental improvements achieved through optimizing existing equipment and process systems and achieves step changes in energy and materials efficiency, total life-cycle cost reduction, and environmental impact by minimizing wastes at the sources via various hierarchy pollution prevention techniques.

Our goal is to modify industrial processes so that services and manufactured goods can be produced without waste. But it is important to understand that some manufacturing processes inherently produce wastes, even after all reasonable efforts at pollution prevention. Thus in some cases the use of a conversion technology may be more appropriate than a program of pollution prevention: many industrial wastes can be processed to render them viable as material inputs to another industry or to part of an industrial cluster of several connected industries - as part of the movement of "industrial ecology". Two separate case studies are presented that highlight a profitable industrial "by-product-to-new products (Das, 2020).

Reference: Das, T.K. (2020). Industrial Environmental Management: Engineering, Science and Policy, Wiley, Hoboken, NJ.

601217 Process Intensification for Ceramic Part Production – Field Assisted Sintering of Core/Shell Powders Fabricated By Particle Atomic Layer Deposition

Abstract:

Yttria-stabilized cubic zirconia (YSZ) is a common electrolyte material for solid oxide fuel cells (SOFC) due to its moderate oxygen-ion conductivity and chemical stability at high temperatures. To maximize ionic conductivity, YSZ ceramics must be near theoretical density which conventionally requires placing the sample in a furnace and slowly increasing the furnace temperature to ~1450°C over a few hours. This requires significant energy use and can adversely affect other layers of the SOFC during manufacturing. Alternatively, flash sintering has been demonstrated to densify YSZ at furnace temperatures as low as 750°C in a few seconds, resulting in cost savings, higher ionic conductivities, and smaller grain sizes as compared to conventional processing. During flash sintering, an electric field is applied across the sample and at a minimum threshold furnace temperature, the sample conductivity and power dissipation rapidly increase causing densification. Typically, small quantities of aluminum oxide (Al_2O_3) are added to YSZ ceramics by mechanical mixing to alter grain growth behavior and increase the final density. In this work, particle atomic layer deposition (ALD) was used to uniformly coat YSZ particles in thin films of amorphous Al_2O_3, homogeneously dispersing the Al_2O_3 prior to flash sintering. Analysis of the flash sintering behavior revealed that the Al_2O_3 thin film reduced the electrical conductivity of the starting powder more significantly than an equivalent Al_2O_3 concentration added as particles, enabling adjustment of the onset flash temperature with minimal secondary phase addition. Small quantities of Al_2O_3 increased the sample relative density at a constant sample temperature, suggesting that the Al_2O_3 acts as a sintering aid and enhances the densification mechanism. This is supported by in situ XRD measurements, where the dissolution of Al_2O_3 into the YSZ lattice is observed and suggests that Al_2O_3 changes the chemical environment of the YSZ lattice and grain boundaries to enhance densification. The flash sintering of core/shell powders fabricated by particle ALD results in dense parts with fine grain sizes and has a significant cost/performance advantage over conventional processing.

611704 Sinker Electric Discharge Machining As an Inexpensive Method to Prototype Microscale-Based Lamina-Plates with High Aspect Ratio Features

Abstract:

Microscale architectures – sought for their enhancement of heat and mass transfer in microscale-based technologies – can be difficult and/or expensive to manufacture with conventional means. Photochemical machining is a standard microscale fabrication method that is limited by low feature aspect ratios and low maximum etch depth. On the other hand, metal additive manufacturing provides a viable route to obtaining high aspect ratio features, but it is not an ideal manufacturing process for device prototyping due to post processing requirements, expensive materials and tools, and developmental infancy. Sinker Electric Discharge Machining (SEDM) is explored here as an inexpensive method for prototyping high aspect ratio features of lamina-plates that comprise a microscale-based device. Two SEDM lamina-plate prototypes are studied here. First, plates of an integrated reactor/heat-exchanger for the direct synthesis of dimethyl-ether (DME) are examined. Second, the flow plate of a lab scale multiphase microchannel separator (MMS) is discussed. In these designs, SEDM enables both scale-up and numbering-up, as well as consistent microscale-feature sizes with high aspect ratios. Physical characterization of the lamina-plates is presented. The limit in feature spacing attenable with SEDM is tested and found to be approximately 400μm.

608189 Process Design and Development of a Small Scale Hybrid Manufacturing System for the Cancer Drug Lomustine

Abstract:

Each year, various low volume, high value medications are reported on drug shortage lists with no generic alternative. Here at Purdue University, a small-scale integrated, modular, reconfigurable platform was developed in order to meet the market demand for these drugs. The compound of interest selected for the case study was Lomustine; first produced in 1976, Lomustine is an anti-cancer drug for treating brain tumors and Hodgkin's lymphoma[1]. In 2014, manufacturing rights changed ownership and Lomustine began to be produced under the brand name of Gleostine. Production of generics is not typically newsworthy; however, over the last few years, the price of the medication increased from $50 to $768 per capsule[2]. The current cost of the drug may prevent patient access to the medication even if it could positively impact their health, highlighting the need of an alternative process for production to lower overall costs.

This presentation details the framework for the design and development of a small-scale end-to-end process for continuous manufacturing of Lomustine at a fraction of the current dosage costs. In order to achieve the successful design and development of the Lomustine process, researchers implemented a novel flow synthesis pathway for Lomustine, which was first developed by researchers Jaman *et al.* at Purdue University[3] within the process. The synthesis consists of two continuous reaction steps, which eliminates the need to handle the Lomustine intermediate, 1-(2-chloroethyl)-3-cyclohexylurea; inherently improving process safety by decreasing intermediate handling and exposure. Additionally, solvent screening, solubility studies, and impurity mapping for the process are reported along with individual unit operation design and construction. Finally, flow synthesis integration with continuous liquid-liquid extraction and crystallization modules showcase an end-to-end continuous synthesis and purification platform capable of producing the anti-cancer drug Lomustine.

[1] Lee, F. Y.; Workman, P.; Roberts, J. T.; Bleehen, N. M., Clinical pharmacokinetics of oral CCNU (lomustine). *Cancer chemotherapy and pharmacology*. **1985**, 14 (2), 125-131.

[2] Loftus, P. Cancer Drug Price Rises 1,400% With No Generic to Challenge It. The Wall Street Journal [Online], 2017 https://www.wsj.com/articles/cancer-drug-price-rises-1400-with-no-generic-to-challenge-it-1514203201 (accessed Mar 15, 2019).

[3] Jaman, Z.; Sobreira, T. J. P.; Mufti, A.; Ferreira, C. R.; Cooks, R. G.; Thompson, D. H., Rapid On-Demand Synthesis of Lomustine under Continuous Flow Conditions. *Organic Process Research & Development*. **2019**, 23 (3), 334-341.

Selection of functional solvent factory for automatic aldehyde and ketone separation in aldol reaction: A combined COSMO-RS and experimental approach

Chenyue Zhang1,2, Zhen Song3, Can Jin4, Job Nijhuis1, Teng Zhou3, Timothy Noël1, Harald Gröger5, Kai Sundmacher3, Jan van Hest2, Volker Hessel1,6

1 Department of Chemical Engineering and Chemistry, Eindhoven University of Technology, Netherlands

2 Department of Biomedical Engineering, Eindhoven University of Technology, Netherlands

3 Process Systems Engineering, Max Planck Institute for Dynamics of Complex Technical Systems, Germany

4 College of Pharmaceutical Sciences, Zhejiang University of Technology, China

5 Organic Chemistry I, Faculty of Chemistry, Bielefeld ,Bielefeld University, Germany

6 School of Chemical Engineering and Advanced Materials, The University of Adelaide, Australia

KEY WORDS

Computing and Systems Engineering, Process Design & Development, Pharmaceutical Development.

ABSTRACT

We present a new microfluidic concept based on complex multiphase systems, which may be switched by temperature to a higher or lower number of fluidic phases. This would allow to fluidically open and close interim reaction compartments - close when reaction is required (= bringing all reaction components together) and open when separation is required (= giving each reaction component at best an own phase). We termed this 'Green-Spaciant Solvent Factory' (create spaces = 'spaciant') which provides 'horizontal hierarchy' – as opposed to the 'vertical hierarchy' of common multi-step flow syntheses (or batches) with their many consecutive reactors-separators. This is inspired by nature, as the cell with its internal organelles is a horizontally compartmentalized soft-matter reactor. The vision of the FET-OPEN project 'ONE-FLOW' (www.one-flow.org) is that such flow processing suits multi-step cascade reactions, since it ideally just needs one reactor passage for a complex synthesis task.
The 2-step cascade from 3-chlorobenzaldehyde to (1R,3S)-1-(3-chloro-phenyl)butane-1,3-diol, well-established by Groeger's group, [1] is studied as a model system. This is part of a chemoenzymatic cascade combining organo- and biocatalysis, which represents in general an attractive process concept for producing pharmaceutical medicines. In this work, we are exploring to achieve the functionalities of product separation and reactant/catalyst recovery solely by multi-phases with designer solvents. This would allow to reduce complex reaction and separation equipment as well as catalyst modification to one single tube reactor using a standard catalyst (one-flow). Firstly, a large database of conventional organic solvent and/or ionic liquids (ILs) is generated from the COSMOthermX software. Then, the reactivity potential of different families of solvent with the reactants is considered to narrow the screening space. Furthermore, relevant physical properties and the environmental effect of solvents (LCA assessment) is taken into account to screen out the suitable solvents. Afterwards, the top solvent candidates were carefully

evaluated by experiments.[2, 3] The reaction equation and whole 'One-Flow' solvent screening methodology for aldol reaction are given in Scheme 1 [2].

As key result, dodecane as best aldol reaction solvent and two 1-butyl-3-methylimidazolium based ionic liquids as candidate solvents in enzymatic reduction are identified. Up to 92mol% conversion and with (R)-aldol product yield (63 mol%) for aldol reaction is obtained, shown in Figure 1. Meanwhile, we propose the compatibility of nanoreactor into the micro-flow reactor to protect and separate catalysts. Our final goal is to achieve multi-step reactions in one run (ONE-FLOW) in multi-phasic systems using this methodology, i.e. to mimic the metabolic cascades of nature. In Scheme 2, [2] we have envisioned how such Spaciant Solvent Factory for a two-step catalytic reaction may look alike.

ACKNOWLEDGEMENTS

The research work is supported by FET-Open EU project ONE-FLOW (grant no. 737266). The authors acknowledge the "Deutsche Forschungsgemeinschaft" SFB/TRR 63 project InPROMPT and the grant (15120868/218) of the University of Adelaide.

REFERENCES AND CITATIONS

[1] Rulli G, et al. Angewandte Chemie International Edition, 2011, 50(34): 7944-7947.
[2] Zhang C, Song Z, Jin C, et al. Chemical Engineering Journal, 2020, 385: 123399.
[3] Morales-Gonzalez O M, Zhang C, Li S, et al. Chemical Engineering Science: X, 2019, 3: 100024.

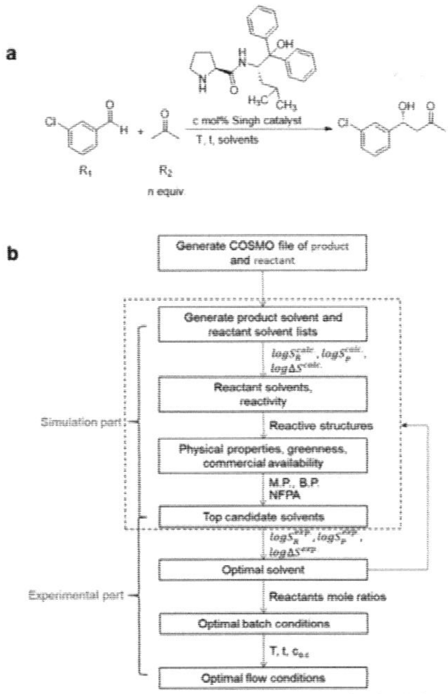

Scheme 1. (a) Aldol reaction equation. (b) Methodology for reaction solvent screening in aldol reaction.

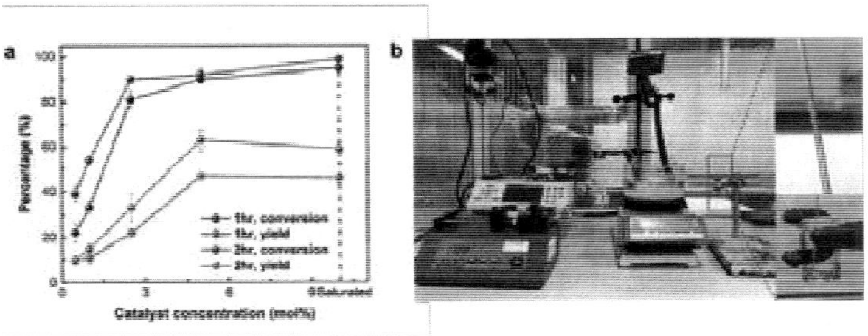

Figure 1. (a) The effect of Singh catalyst concentration on the reaction conversion and yield. (b) Biphasic system and segmented flow formed during the reaction with $c_{cat} \geq 2.5$ mol%, $t \geq 1$ hr. [2]

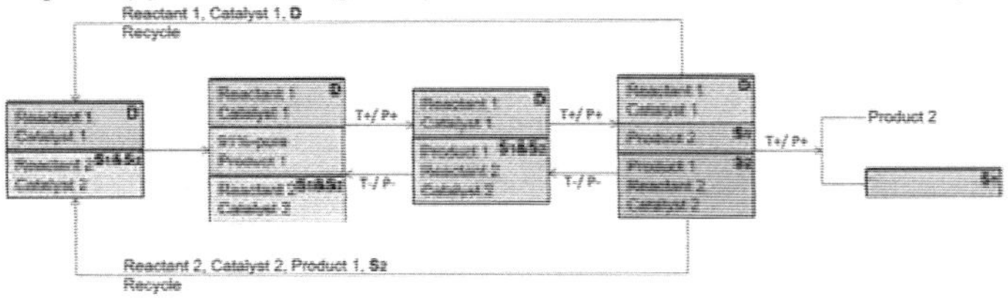

Scheme 2. Outline of the ONE-FLOW Spaciant Solvent Factory.[2]

601142 Next Generation Lipid-Based Excipients: Overcoming the Challenges of Spray Drying Applications on the Development of Inhalable Lipid-Microparticles

Abstract:

Lipid-based excipients (LBE) are natural and biodegradable materials that have been extensively used in pharmaceutical dosage forms. The characteristic solid-state transitions of LBE, viz. polymorphism and phase separation, often make their application challenging. These transitions are strongly linked to the limited processability of LBE through certain pharmaceutical manufacturing processes such as spray drying. During spray drying, the crystallization of LBE into metastable polymorphs or phase separation into low-melting-point fractions impair the formation of solid particles. In a previous work of our group novel LBE, polyglycerol esters of fatty acids (PGFAs), were introduced (brand name: Witepsol PMF). The molecular structure and composition of PGFAs hinders polymorphism and provide crystallization into a monophasic system. In this work, the processability of PGFAs via spray drying, in order to produce inhalable lipid microparticles for inhalation, in the final dosage form of dry powder (DPI) is presented. Applications in organic solvent and aqueous-based spray drying processes were evaluated. Distinctive particle attributes of readily inhalable lipid-microparticles were envisaged to provide systemic or local effect after pulmonary administration.

Lipid-microparticles loaded with ibuprofen intended for systemic delivery of analgesics were manufactured under an organic solvent-based spray drying process. For effective systemic delivery, large particles (VMD>3 μm) of low density (<0.4 g/cm^3) are intended to provide median mass aerodynamic diameter in the range of 1 - 5 μm for reaching lung deposition and diminish lung clearance. PG3-C22 partial ester (WITEPSOL® PMF 123) was used as the particle matrix in a lipid:drug ratio of 70:30. A solution of 1.5% w/w solid content in tetrahydrofuran was sprayed through a 0.2 mm-nozzle at a feed rate of 3 g/min into a closed-loop system at 71°C. The process showed no impairments and resulted in high yield of solid particles (70.1%). Unimodal particle size distribution, volume mean diameter (VMD) of 6.6 ± 1.1 μm and density of 0.389 ±0.007 g/cm^3 were achieved. Residual solvent < 50ppm was determined, which is below the limit for pharmaceutical products (<720ppm). The application of the particles as DPI was tested in Aerolizer dry powder inhaler coupled to a next generation impactor. The MMAD of 3.57 ± 0.11 μm evidenced the inhalability of the particles. *In vitro* lung deposition with high fine particle fraction of 45.2 ± 1.3% was reached. Crystalline dispersion of ibuprofen and PG3-C22 partial ester with eutectic composition was observed in the particles. Dissolution of the deposited particles in simulated lung fluid showed 38.9 ± 6.9% drug released in the first hour and 60.1 ± 0.1% after 6 hours.

Lipid-microparticles for local delivery of rifampicin were produced through an aqueous-based spray drying process. Particle attributes were designed for a median particle size < 4 μm (optimal 2.1 μm) to facilitate phagocytosis from alveolar macrophages in the treatment of tuberculosis. An aqueous suspension of drug-loaded PG2-C22 full ester (WITEPSOL® PMF 222) was prepared by melt-emulsification and stabilized with the lipid-based emulsifier IMWITOR® 372P (glyceryl stearate citrate). Lipid-microparticles loaded with 2.5% w/w rifampicin, encapsulation efficiency of 95.2 ± 0.0% and median particle size of 2.11 ± 0.02 μm were achieved. The suspension with 0.5% w/w solid content was spray dried through a 0.4 mm-nozzle at a feed rate of 2 g/min into an open-loop system at 75°C. The process yielded solid particles (83%) without impairments. The produced particles showed a unimodal and sharp particle size distribution (span = 1.8) with a median particle size of 2.53 μm.

Crystalline state of the lipids within the particles was determined. Neither polymorphism nor phase separation was observed.

The use of PGFAs on organic solvent and aqueous-based spray drying processes led to high yield of drug-loaded solid particles. The systemic and local application of the produced particles is currently under investigation on *in vitro* cell culture studies. The processability of PGFAs is attributed to the absence of solid-state transitions, especially polymorphism. Conventional LBE crystallize in metastable polymorphs, which increases the degree of supercooling, namely the difference between melting and recrystallization temperature. Although crystallization via spray drying is mostly driven by solvent evaporation, melt-crystallization also occurs in LBE. The high degree of supercooling of conventional LBE delays the formation of solid particles. The absence of polymorphism in PGFAs led to a low degree of supercooling, thus achieving high processability via spray drying. In conclusion, the use of PGFAs is an attractive approach for developing lipid-based engineered particles for diverse applications in the pulmonary field.

Acknowledgment: IOI Oleo GmbH, the Austrian Funding Agency (FFG) and the Doctoral Academy NanoGraz, University of Graz.

610506 Comparative Evaluation of Dissolution Behavior and Bioequivalence of Various Capsules

Abstract:

Two-piece hard capsules continuously serve as widely used oral drug and dietary supplement dosage forms. Hard gelatin capsules are used in 20% of all oral dosage forms but have numerous shortcomings especially with potential crosslinking with the active ingredient, the growing consumer preference for non-animal products and concerns about bovine spongiform encephalopathy (BSE or mad cow disease). The most common polymer used to replace gelatin in hard capsules is hydroxypropyl methylcellulose (HPMC) or hypromellose. In this study, Propranolol HCl was used to evaluate dissolution behavior and bioequivalence of various capsules especially between gelatin and various types of HPMC capsules.

DuPont's FloVitro ^{TM} technology is a flow through approach utilizing a series of solid transfer cells which incorporates drug pharmacokinetic characteristics to simulate a biomimetic system. It is a biorelevant *in vitro* test that reflects physiological environment in the test conditions with a purpose of correlating *in vitro* drug release with *in vivo* drug absorption. FloVitro ^{TM} has the unique ability to generate dissolution curves that match plasma profiles directly without the need for mathematical modeling. This study will use Modified USPII and FloVitro ^{TM} technology to study dissolution behavior of Propranolol HCl in various commercial gelatin and HPMC capsules.

Using USP II method, there was approximately 5 to 6 min delayed release of Propranolol HCl from HPMC capsules without gelling agent, and approximately 2 to 23 min delayed release of Propranolol HCl from HPMC capsules with gelling agent. Using FloVitro ^{TM} technology, it was found that the key pharmacokinetic parameters, C_{max} and T_{max}, of HPMC capsules without gelling agent were not statistically different to that of gelatin capsules. Additionally, HPMC capsules without gelling agent showed more consistent dissolution performance (less coefficient of variation) than that of HPMC capsules with gelling agents (see Table 1).

In summary, dissolution of an active in the dosage form is critical for absorption and bioavailability, and for product quality and consistency assessments during drug product life cycle. It was found that the key pharmacokinetic parameters, C_{max} and T_{max}, of HPMC capsules without gelling agent were not statistically different to that of gelatin capsules using FloVitro ^{TM} technology. Using DuPont's FloVitro ^{TM} technology, pharmaceutical scientists can relatively quickly assess biorelevant information on specific APIs and dosage forms. The prediction of *in vivo* performance of Propranolol HCl in various commercial gelatin and HPMC capsules will help formulation scientists gaining insights on key factors that impact the drug efficacy and safety in the patients.

Acknowledgement

We would like to thank Alejandro Carbo, Jonathan Gilinski of CapsCanada for their discussion and support of this report.

References

1. AAPS 2008 Poster Presentation: Exploring Suitability and Feasibility of a Novel *In Vitro*

Dissolution System A. Selen, L. F. Buhse, E. G. Chikhale, W. H. Doub, S. K. De, Z. Gao, A. S. Gehris, L. Hughes, R. Lu, H. Mahayni, P. K. Maturu, T. D. Mehta

2. Eddington ND, Rekhi GS, Lesko LJ, Augsburger LL. Scale-Up Effects on Dissolution and Bioavailability of Propranolol Hydrochloride and Metoprolol Tartrate Tablet Formulations. AAPS PharmSciTech. 2000; 1(2): article

3. AAPS 2010 Poster Presentation: Evaluation of Suitability, Transferability, and Bio-relevance of a Novel *In-Vitro* Dissolution Technique (Rohm & Haas Dissolution System, i.e. FloVitro™ Technology*) A. Selen, L. F. Buhse, E. G. Chikhale, W. H. Doub, S. K. De, Z. Gao, A. S. Gehris, L. Hughes, R. Lu, H. Mahayni, P. K. Maturu, T. D. Mehta

4. AAPS 2011 Poster Presentation: Application of FloVitro™ Technology to Evaluate Dissolution of Furosemide and Danazol in Simulated Media at Fed and Fasted Conditions A. Selen, W.J. Rodriguez, W. H. Doub, L. F. Buhse, E. G. Chikhale, S. K. De, Z. Gao, A. S. Gehris, L. Hughes, R. Lu, H. Mahayni, P.K. Maturu, T. D. Mehta 13

5. Hughes, Lyn, Dissolution Test Equipment and Method. US Patent No. 2004/US6799123B2, Sept 28th, 2004

Table 1. Comparison of key characteristics of biorelavent dissolution performance.

Sample Description	% Recovery	%RSD	C_{max} (mg/L)	%RSD	T_{max} (min)	%RSD
IR (Inderal LA IR capsules average)[2]	103	0.5	34.66	0.08	110.5	2.3
in vivo (human data based on Inderal IR capsules)[2]	-	-	34.64	-	120	-
Gelatin Capsules	94.9	2.2	33.57	2.68	127.3	4.9
HPMC Capsules without gelling agents	91.0	1.7	33.34	4.08	135.3	5.1
HPMC capsules with gelling agents	94	3.5	33.25	2.3	139.5	11.5

592713 Development of High-Solids, Quick-Set Pharmaceutical Tablet Coatings.

Abstract:

Pharmaceutical tablet coatings are commonly based on hydroxypropyl methyl cellulose (HPMC), poly(vinyl alcohol) (PVOH), and acrylic polymers. Currently available commercial pharmaceutical tablet coating technology includes spraying of a dilute aqueous polymer over the tablets, leading to a time-consuming and high cost drying process. Development of tablet coating technology that minimizes this drying step would be commercially advantageous since it would reduce coating manufacturing time and energy by up to 40%, leading to significant cost savings. The proposed amphoteric latex coating materials contain weak acid and strong base stabilizing moieties on the latex particle surfaces, and exist as low-viscosity dispersions at high solids at mildly acidic pH where the weak acid is not ionized. However, an increase in pH, triggered by the loss of CO_2 or addition of base, deprotonates the carboxylic acid group and leads to its ionic linkage with the quaternary ammonium groups also present.[1] This ionic coacervation between and among the latex particles enables the coating to set quickly (Figure 1(a)).

This study[2] focused on the evaluation of amphoteric polyacrylate latex-based film forming materials for tablet coatings. A series of amphoteric latexes was prepared from monomers already used in the synthesis of FDA-approved polyacrylate polymers. Polymer composition was systematically varied in order to understand the effects on polymer film formation temperature, setting efficiency as a function of pH, and film properties such as tack, puncture, gloss, optical clarity, water permeability, etc.

Latex films were evaluated using pharmaceutical test methods, and showed equal or superior performance compared to commercially-used cellulosic and poly(vinyl alcohol) coatings. The quick-set efficiency of these latexes was evaluated based on their aggregation behaviors in a series of buffers with pH ranging from 5 to 10. The effect of composition variables on the setting efficiency of latexes was studied. Several performance characteristics for tablet coatings have been measured, including high solids (~40 wt. %), good optical properties (clarity > 90%, transmission > 93%, and haze < 6%), no tackiness, and low permeability (water vapor permeability $< 2 \times 10^{-7}$ g / Pa s m). Mechanical properties of films made from amphoteric latexes are similar to those of PVOH films, but they are softer and more elastic than HPMC films.

We have demonstrated that latexes with a balance of quick-setting efficiency and tunable coating properties can be prepared. Amphoteric latex-based tablet coatings can be performance- and cost-competitive to HPMC- and PVOH-based coatings, easier to apply from a high solids formulation, and their composition can be easily adjusted to provide coatings with a desirable balance of properties. The polyacrylate latex films exhibited excellent mechanical and optical properties, high transmission, high clarity, and low haze. The water vapor permeability of latex films was lower than that of current commercial HPMC and PVOH films. These results suggest that amphoteric latexes can be used to prepare tablet coatings with good appearance and desirable performance. Typical coated tablets are shown in Figure 1(b).

References: 1. (a) Schmidt, D.; Mussell, R.; Rose, G. *Coat. Technol.* **2003**, *75*, 59-64. (b) Rose, G. *et al. Langmuir* **2005**, *21*, 1192-1200. 2. Ladika, M.; Kalantar, T.; Shao, H.; Dean, S.; Harris, K.; Sheskey, P.; Coppens, K.; Balwinski, K.; Holbrook, D. *Appl. Polym. Sci.* **2014**, *131(7)*, 539-550.

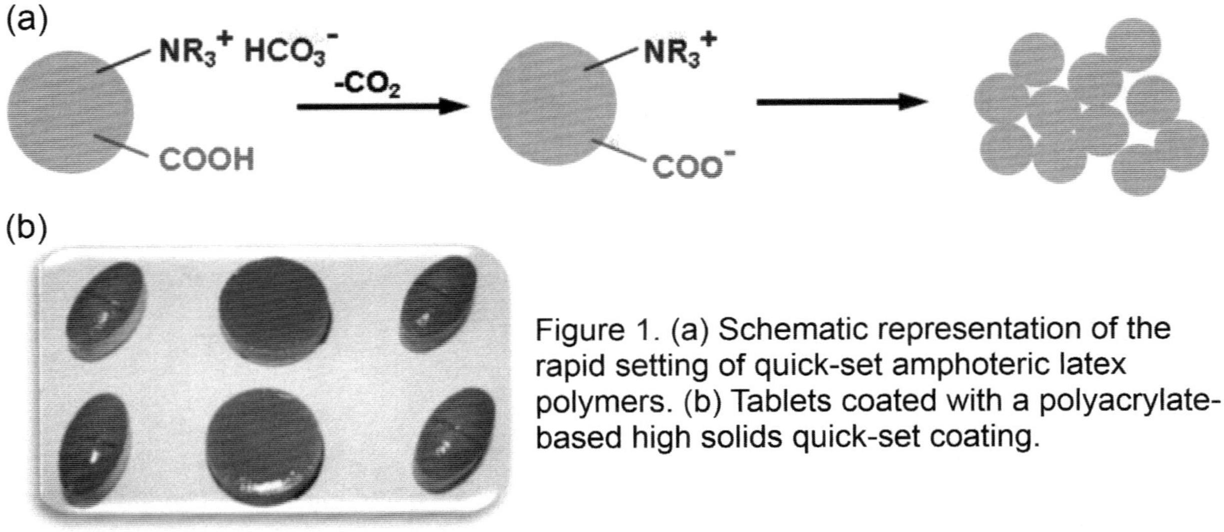

Figure 1. (a) Schematic representation of the rapid setting of quick-set amphoteric latex polymers. (b) Tablets coated with a polyacrylate-based high solids quick-set coating.

602051 Influence of External Lubricants and the Environmental Conditions on the Puncturing Properties of Gelatin and HPMC Capsules for Inhalation

Abstract:

Introduction

Hard gelatin capsules have been applied in single dose dry powder inhalers for several decades. Recently, hydroxyl-propyl methylcellulose (HPMC) capsules are increasingly being used in DPI devices due to their favourable properties, i.e. lower moisture content and mechanical characteristics less sensitive to variations in humidity. The capsules are externally lubricated during production to prevent sticking to each other in capsule filling operations and subsequent packaging [1], [2]. In order to emit the powder from the device, capsules are pierced by one or more needles equipped in inhaler device during dose administration manoeuvres [3], [4]. Different lubricants could alter adhesion to the capsule material surface and thus mechanical properties of capsules. This could in turn influence the puncturing properties and subsequently shape and size of the opening and the flap and subsequent powder release. This study attempts to evaluate the puncturing properties of the RS01 Plastiape® device of gelatin and HPMC capsules treated with different external lubricants. Moreover, the goal is to investigate the relationship between the mechanical properties of differently lubricated capsules and their puncture characteristics and to indirectly compare those to the aerodynamic performance.

Material and Methods

Size 3 capsules of gelatin and HPMC were obtained from Qualicaps (Qualicaps Europe, S.A.U., Spain). These capsules were externally lubricated with magnesium stearate (MgSt), sodium lauryl sulphate (SLS) and carnauba wax (CW) and were compared to unlubricated capsules (w/o). After a storage period of one week in dry (22% RH) and humid (51% RH) conditions, the capsules (n=6) were punctured using a compression and tensile measuring device (Instron 5943, Instron® GmbH, Germany). The needle from an inhaler (RS01®, Plastitape, Italy) was disassembled from the device and attached to the force transducer. The applied forces (N) were recorded, as previously described by Torrisi et al. [3]. The needle moved with 1mm/sec towards the centre of the capsule lid and the maximum force applied was referred to as the breakthrough force of the capsule. Subsequently, macroscopic pictures of the created openings were taken using a DSLR camera with a reversed 18mm objective (Canon 60D, 0.6sec, ISO500) and the contours were traced using ImageJ software. The number of the pixels in the opening area was calculated and compared to an area of the external diameter of the needle (1mm, 100% of the opening area), resulting in a calculated opening value. Mechanical properties of the conditioned capsules were measured at ambient conditions (22°C ± 2°C, 35% RH ± 3%) using a MCR compact Rheometer (MCR 300, Anton Paar, Austria). The resulting normal force (N) and displacement (mm) values were recorded every 0.06 mm, resulting in the stress-strain curves. The Young modulus was determined by linear regression from the slope of the elastic region of the stress-strain curves. Subsequently, these results were indirectly compared to the release of a powder mixture from the capsules after a filling process. For this purpose, the unconditioned capsules were filled with 25mg of a model blend at the previously mentioned room conditions (22%

and 51% RH) with a capsule filling machine (FlexaLab, MG2, Pianoro, Italy) [5]. The model blend consisted of 1% (w/w) Budesonide (Laboratorios Liconsa, Spain) and 99% (w/w) Inhalac 230 (MEGGLE, Germany). The machine settings were maintained for all capsule types and aerodynamic performance was measured via the fast screening impactor (FSI).

Results and Discussion

Puncture profiles for gelatin and HPMC capsules were very different and were influenced largely by the storage conditions, which has also been previously described [3], [4]. Gelatin capsules required overall higher piercing forces (average of all capsules: 6.86N ± 1.14N at 22% RH to 5.90N ± 1.20N at 51% RH) than HPMC capsules (3.89N ± 0.65N at 22% RH to 2.99N ± 0.41N at 51% RH). The piercing forces decreased with increased humidity for both types of capsules. Even though there were slight differences in puncture properties between the differently lubricated capsules of the same material, this differences were not significant (data not shown). Our results showed that the small amount of lubricant does not significantly impact puncturing properties (shown in Table 1). This could be explained by the very low amount of lubricant used in the after-treatment of the production (2g lubricants on 10kg of capsules), which gives results in a very small layer in the nanometer scale.

Further, taking the size and shape of the openings into account, visible differences between gelatin and HPMC could be noticed. While gelatin capsules have relatively large openings under all conditions (90.8% opening ± 6.2% (average of all capsules)), the position of the flap in HPMC capsules was closing the puncture to a greater extent, resulting in a lower value for the openings (39.5% ± 18.9% (average of all capsules)). Provided that the flap might return towards the capsule material between the time point of the puncturing measurements (Instron) and the time point when the pictures were taken, the size of the opening was calculated without the area covered by the returning flap. HPMC capsules still showed substantially lower opening values than gelatin capsules (77.3% ± 8.5% (average of all capsules/conditions). It is assumed that, when capsules are opened within an inhaler, the flap has substantially less time to return. Also, these experiments were performed in absence of airflow, which is actually the driver of *in vitro* or *in vivo* aerosolisation. Therefore, we assume that the opening of HPMC capsules in the actual use within inhaler could be larger as well. Observing further the shape of the opening, it was evident that the edges of the punctures become smoother with increasing storage humidity (Figure 1).

The percentage values on Figure 1 show larger openings at lower humidity for all gelatin capsules and most of the HPMC capsules (exception were HPMC capsules lubricated with CW). However, looking into differently lubricated capsules, only few differences in opening sizes may be noticed for both materials. While at 22%RH, HPMC w/o (80.72% ± 8.98%), SLS (85.4% ± 10.41%) and MgSt (78.1% ± 15.38%) showed similarly large openings, HPMC capsules lubricated with CW (61.8% ± 11.75%) resulted in much smaller openings. Differences in the size and shape of the openings among the differently lubricated gelatin capsules were less pronounced. All gelatin capsules stored at 22% RH exhibited similarly large openings, (SLS 95.87% ± 5.09%, CW 95.76% ± 8.42%, w/o 92.26% ± 4.85% and MgSt 88.55% ± 2.99%). At 51% RH storage, opening area seemed to slightly decrease (w/o 91.76% ± 5.17%, SLS 90.17 ± 7.31%, CW 83.99% ± 7.16% and MgSt 87.94% ± 8.40%). The force needed for puncturing correlated well with the elasticity properties of the capsules as observed by Young modulus described previously by authors [6]. Elasticity of gelatin capsules increased with higher storage humidity and is probably the consequence of the increase of the amount of water molecules bound to the helical fragments of proteins in gelatin. The elasticity of HPMC capsules was found to increase at 51% RH, indicating the potential plasticization of the material [6]. Lubricants did not show large effect on the elasticity of both capsule types.

The aerodynamic performance of all filled capsules was tested using FSI measurements and was related to the capsule openings Figure 2. It was found that the main difference in API release from capsules was arising from the capsule material, although, there were some differences observed as a consequence of storage and lubrication. In general, gelatin capsules resulted in smaller fine particle dose (FPD) compared to HPMC capsules, suggesting that complete opening of the capsules is not necessarily indicator of the higher FPD. It is possible that due to the modified opening structure on both materials, different air flows are created in the inhaler, resulting in different shear forces directly at the exit point, which influence the detachment of an API. On the other hand structural differences of the different capsule materials on the inner surface could potentially influence the release of finer particles. Interestingly, opening size for HPMC and gelatin capsules did not seem to have that much impact on the FPD (Figure 2). HPMC capsules without lubricants filled at 51% RH and capsules with MgSt and SLS filled at 22% RH achieved a FPD of almost 60µg, whereas the average for gelatin was only about 25µg.

Conclusion

In summary, the use of different external lubricants on the surface of capsules could not be directly related to the force required to pierce capsules. However, it showed slight variations in the puncture sizes and shapes. The piercing forces were higher for gelatin than for HPMC capsules and have decreased with increased RH for both materials. The aerodynamic performance of both capsules showed that the larger size of the opening is not necessarily decisive for the higher dose delivery, however, both the capsule material and the storage humidity had an important role on the effective dose delivery from the device.

Table 1: Force needed to pierce a capsule lid measured by Instron, (n=6).

	Gelatin				HPMC			
	22% RH		51% RH		22% RH		51% RH	
Lubricant	Maximum Load [N]	stdv	Maximum Load [N]	stdv	Maximum Load [N]	stdv	Maximum Load [N]	stdv
wo	7.27	1.03	5.81	1.49	3.53	0.60	2.74	0.36
SLS	6.24	0.85	6.59	1.26	4.17	0.66	3.14	0.49
CW	7.40	1.93	5.90	0.99	3.73	0.66	3.24	0.48
MgSt	6.53	0.76	5.28	1.05	4.11	0.67	2.85	0.30

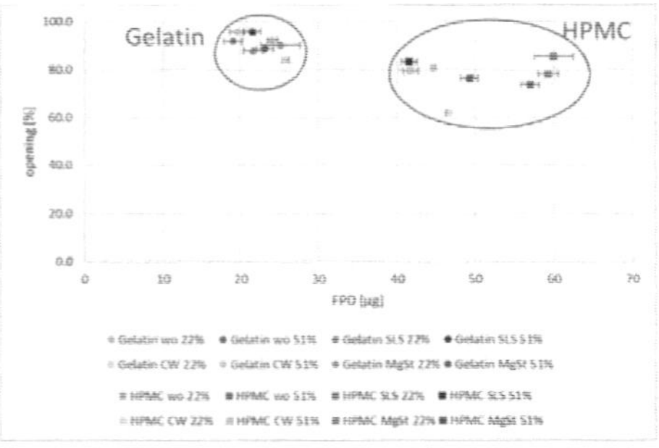

Figure 2: Capsule opening vs. fine particle dose (FPD) (n=3)

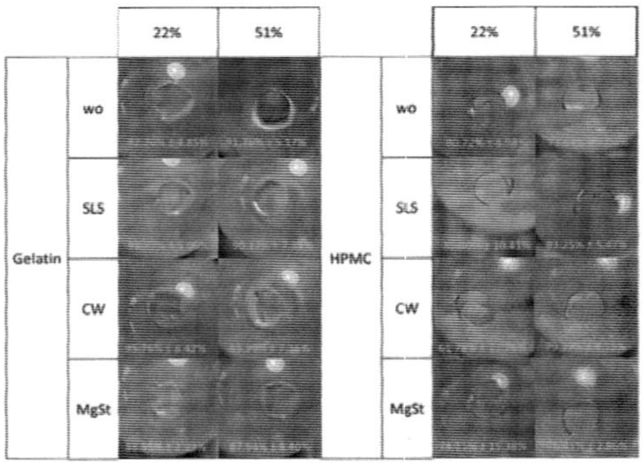

Figure 1: Opened capsule lids after piercing experiments in Instron, (n=3).

607008 Formulation, Stability, and Scalability of Fast-Releasing Lumefantrine Nanoparticles for the Treatment of Malaria

Abstract:

Lumefantrine, an anti-malaria drug, suffers from low bioavailability due to its hydrophobic character. Via Flash Nanoprecipitation (FNP), we kinetically trap the drug as a nanoparticle in an amorphous state, which increases solubility and therefore bioavailability. 200-400 nm lumefantrine nanoparticles are produced via FNP using safe and inexpensive stabilizers: hydroxypropylmethylcellulose acetate succinate (HPMCAS), zein protein, and lecithin phospholipid. FNP and spray drying are combined to generate solid powders. These powders are stable under hot and humid conditions. Here we demonstrate the use of FNP and spray drying as a continuous and scalable platform to generate and recover nanoparticles without compromising the dissolution kinetics of the drug. Our process for continuous nanoparticle synthesis and recovery is also inexpensive, which makes it viable for low-income countries. Three scales of mixers were utilized, allowing nanoparticle production rates ranging from a few milligrams up to around 1 kg/day, all with similar nanoparticle size and polydispersity.

We compare lyophilized and spray dried lumefantrine NP powders and confirm that using spray drying as our solidification method does not compromise the dissolution kinetics and can therefore be used as a cost effective and scalable drying method. The dissolution kinetics for the spray dried NP powders remain constant under fasted and fed conditions for over a month in accelerated stability testing conditions (50°C, 75% RH, open vial). Via Powder X-ray diffraction, differential scanning calorimetry, and solid-state nuclear magnetic resonance, we confirm that the lumefantrine in the core of the nanoparticle is amorphous. The combination of FNP and spray-drying offers a low-cost, scalable, and continuous nanofabrication platform to produce amorphous nanoparticles in a solid dosage form.

602060 Monitoring and Detection of Byproduct Crystallization Using in-Situ Microscopy

Abstract:

In-situ monitoring of solid-phase composition during crystallization is often accomplished by Raman spectroscopy, but not all compounds in the system of interest are strongly Raman active. We are developing an image-based analysis to detect a second undesired solid phase with in-situ microscopy before other PAT tools detect the appearance of the solid. Our system involves the reactive crystallization of beta-lactam antibiotics, wherein a byproduct of the enzyme-catalyzed reaction can crystallize within the antibiotic slurry.

The current target of our research is cephalexin, a critical antibiotic, with a byproduct of phenylglycine; however, several other beta-lactam antibiotics exhibit the same behavior, including ampicillin, cefaclor, and cefazolin. The byproduct phenylglycine is present in solution at concentrations on the order of tens of millimoles per liter with a supersaturation less than 1.5; detecting a concentration decrease of about 10 millimolar (indicating initiation of phenylglycine crystallization) within a mixture of several components at >100 millimolar concentration proved difficult with in-situ liquid-phase tools like ATR-FTIR. Phenylglycine shares many functional groups with the desired product and reactants, making its spectroscopic fingerprint similar to the product and reactants. With in-situ Raman spectroscopy, we were unable to detect low loadings of solid phenylglycine in a cephalexin slurry, either because of overlap with the complex solution of similar functional groups or because of low Raman activity compared to cephalexin. Changes in focused beam reflectance measurement (FBRM) counts and chord length distributions did not show up at low phenylglycine solids loadings, nor did changes in turbidity. The solution pH exhibited the strongest change upon formation of phenylglycine solids, but pH is not a robust measure of the appearance of phenylglycine as many (often benign) system perturbations give similar pH responses. As crystals of beta-lactam antibiotics are (predominately) needle-shaped and phenylglycine crystals are plate- or column-shaped, the difference in shape, as well as backscatter intensity, can be used to differentiate the two classes of particles. Additionally, phenylglycine crystals are known to partition to air-water interfaces, a fact that can be exploited as the crystals tend to clump on bubble surfaces making them easier to identify in images (Hoeben et al. 2009).

Image analysis enables features and statistics of in-situ images are extracted and recorded in real time, and we have shown that phenylglycine can be detected from in-situ images. First, objects within the image are identified by edge detection. Then properties of the objects such as size, aspect ratio, pixel intensity, and more, as well as distributions of these properties for every object in each image are used to detect phenylglycine. By tracking these features and statistics, a classifier can be constructed to identify phenylglycine from image data and can be implemented in real-time for monitoring and detection. The statistics can also be recorded as a historical record, which is more feasible than storing the experiment as complete images.

Hoeben, M. A., et al. (2009). "Design of a Counter-Current Interfacial Partitioning Process for the Separation of Ampicillin and Phenylglycine." Industrial & Engineering Chemistry Research **48**(16): 7753-7766.

598054 Integration of In-Silico Tools and High-Throughput Automation for Enhanced Crystallization Development

Abstract:

The presentation will discuss a case study, where multiple in-silico solubility prediction tools and high throughput automation were utilized to accelerate solubility and morphology screening in binary and ternary solvent systems. Regression-based prediction tools using single solvent solubility data and first-principle models were used to identify potential solvent systems to improve crystallization performance. Over 85 solvent combinations were prepared using an automated platform for solubility measurement and morphology screening, and several synergistic solubility behaviors were observed. Experimental solubility trends conformed well with in-silico predictions, indicating these tools can be a beneficial first screen at identifying synergistic solvents to explore. The presentation will delve into the utility of an integrated approach comprising of the solubility modeling and automated experimental screening platform in the efficient data generation to enable timely crystallization process development.

All authors are employees of AbbVie and may own AbbVie stock. AbbVie sponsored and funded the study; contributed to the design; participated in the collection, analysis, and interpretation of data, and in writing, reviewing, and approval of the final publication.

606112 Utilization of Quality-By-Control for Rapid Process Design of Agrochemical Crystallization

Abstract:

Crystallization is a key separation process that is used in the agrochemical industry to separate the agrochemical actives from impure solutions. A poorly designed crystallization step can lead to poor yield, low purity, and long filtration time. Traditionally, a series of design of experiments led by the Quality-by-Design (QbD) approach is used to optimize operating profile of the crystallization process (i.e. temperature profile, solvent/antisolvent addition profile, pH profile).[1] An exhaustive list of experiments covering all factors is needed to explore the whole design space experimentally. To minimize the number of experiments and personnel exposure to toxic chemicals, Quality-by-Control (QbC) utilizes control strategies implemented for the target critical quality attributes (CQAs) to determine the operating profile of the process.[2] For rapid process design, direct design or model free approaches can be used to quickly determine an operating profile of a process leading the system to the desired CQAs. Two direct design approaches, direct nucleation control (DNC) and supersaturation control (SSC), are used in this work to control the crystallization of a model agrochemical compound. Both DNC and SSC utilize process analytical technology (PAT) tools to acquire data and control the process. DNC utilizes closed-loop feedback control approach with particle measurements from focused beam reflectance measurement (FBRM) to generate temperature cycles.[3] On the other hand, SSC uses concentration measurement via UV/Vis detector in a closed-loop feedback control approach to control the concentration by manipulating temperature.[4]

In this work, both direct design approaches were implemented to improve the particle shape, length, and filtration time of needle shaped particles. Preliminary results have indicated significant improvement in the overall crystallization-filtration process performance. Using a DNC-based direct design approach, the filtration time was reduced by a factor of four compared to the standard recipe. The improved procedure not only reduces the unit operation cycle times, but also improves the particle shape for better downstream operations (i.e. drying, transport). The analysis of the product crystals using Ultra Performance Liquid Chromatography (UPLC) indicates that the impurity profile of the agrochemical compound was not compromised with the thermocycles. With SSC-based direct design approach, the concentration of the system was controlled via a set point of the absolute supersaturation. The system was able to perform a cooling profile and generate uniform needle particles. The QbC-based direct design approaches were also evaluated in larger scale (5 L) crystallization systems indicating that the feedback control based rapid design approach can lead to fast and robust scale-up of agrochemical crystallization processes.

References:

(1) Bondi, R. W.; Drennen, J. K. *Quality by Design and the Importance of PAT in QbD*; Academic Press, 2011; Vol. 10.

(2) Su, Q.; Ganesh, S.; Moreno, M.; Bommireddy, Y.; Gonzalez, M.; Reklaitis, G. V.; Nagy, Z. K. A Perspective on Quality-by-Control (QbC) in Pharmaceutical Continuous Manufacturing. *Comput. Chem. Eng.* **2019**, *125*, 216–231.

(3) Bakar, M. R. A.; Nagy, Z. K.; Saleemi, A. N.; Rielly, C. D. The Impact of Direct Nucleation Control on Crystal Size Distribution in Pharmaceutical Crystallization Processes. *Cryst. Growth Des.* **2009**, *9* (3), 1378–1384.

(4) Saleemi, A. N.; Rielly, C. D.; Nagy, Z. K. Comparative Investigation of Supersaturation and Automated Direct Nucleation Control of Crystal Size Distributions Using ATR-UV/Vis Spectroscopy and FBRM. *Cryst. Growth Des.* **2012**, *12* (4), 1792–1807.

609295 Dual Chamber and Membrane Embedded Continuous Microfluidic Device for MOFs Screening

Abstract:

Metal Organic Frameworks (MOFs) are a class of three-dimensional materials comprised of organic and inorganic building unit cells. The self-assembly of metal ions and organic linker through coordinate bonds results in synthesis of infinite and diverse micro-porous structures. High porosity and ultra-high surface to volume ratio of MOFs along with their fascinating chemical and design flexibility have prompt to numerous applications. Despite of the ubiquitous applications of MOFs, great challenges still exist for developing a relatively cheap, safe and readily scalable process that is sought after by industries. Here we have developed a continuous-flow synthesis of MOFs by modifying the design of the previously developed microfluidic mixer to accommodate for the separated nucleation and growth chambers to induce synthesis of the uniform nanosized particles at constant supersaturation. Both chambers have identical designs, yet the growth chamber has a membrane that separates inlets and outlets. Nucleation chamber is initially filled with premixed solutions of the aqueous solutions of the metal ligand and organic ligand while keeping its outlet open. After running the system for few minutes to ensure all the air has been moved out and system is filled with the solution, the outlet is closed, and circulation is started. Using an optical system, the nucleation of the crystals can be observed within the nucleation chamber. By that time, outlet is opened, and nucleated crystals will move to the growth chamber. Crystals growing in the growth chamber are kept inside and extracted for the further analysis. Using this microfluidic platform, parametric screening is performed and the effect of the ratio of metal ligand to the organic linker on the morphology, growth rate and yield of the synthesized MOFs are measured.

598093 In Situ Polymorph Transformation of a Pharmaceutical Intermediate

Abstract:

We discovered that the polymorph of a pharmaceutical intermediate transformed *in situ* during crystallization from solution. Initially, we noted an anomaly in the signal recorded during a crystallization experiment using a Mettler-Toledo Particle Track G400 instrument, an unexplained decrease followed by an increase in the average chord length of the particles. At roughly the same time, we discovered that there is a clear change in the rheology of the suspension. Rapid crystallization early in the process produces a suspension with a yield-stress that flows and mixes poorly. This gradually transforms during subsequent antisolvent addition to a free-flowing suspension without yield stress. We determined that the phenomenon causing both the observed change in particle size and rheology is crystal polymorph transformation accompanied by a dramatic change in habit. Our first indication of the transformation was obtained observing suspension sampled early in the crystallization under an optical microscope. We observed a clear habit change with time. We then further explored this transformation using simultaneous *in situ* Raman spectroscopy (Kaiser RXN1-785) and microscopy (Mettler-Toledo PVM and EasyViewer). We verified that a polymorph transformation occurs during the antisolvent addition phase of crystallization, accompanied by a change in crystal habit. The initial form grows rapidly as acicular crystals, followed by a gradual transformation to a second, more stable form characterized by roughly equant crystal aggregates. This transformation is responsible for the change in suspension rheology and the observed change in particle size. The dependence of the transformation on process conditions and seeding, as well as the kinetics of the transformation, have been determined. Knowledge of this transformation allowed us to understand the experimental observations and enabled confident scale-up.

608146 Detection of Morphology and Modality Changes during Crystallization Using in Situ Chord Length Monitoring and Inference System Model.

Abstract:

A new framework has been recently introduced to correlate in situ chord length distribution (CLD) measurements and offline particle size distribution (PSD) data (Irizarry et al, 2017) and applied to different particle morphologies (Schoell et al, 2019). In this work we are expanding this modelling scheme to consider multimodal distribution measurements (Irizarry et al, 2020). This work also applies the proposed approach to predict if the particle morphology or distribution of different APIs can be predicted from a CLD signal. To accomplish this a data-driven inference system and a more robust CLD2PSD model was built for the studied pharmaceutical compounds since the differences in optical properties and particle morphology require a unique model for each individual substance. The inference system is trained on PSDs annotated by modality and morphology, and the individual models have been combined with online solute concentration data to allow for the estimation of crystal growth, secondary nucleation and agglomeration kinetics, depending on the supersaturation level encountered during the process. The validity of this approach is studied in terms of classification and prediction error rates on test sets. The main advantage of the proposed approach is that all data is generated online and in situ, thus avoiding large sampling errors. Also, it can operate in areas of high concentrations and on small particle sizes where PVM pictures are more difficult to resolve.

References

- Irizarry, R. et al., 2017. Data-driven model and model paradigm to predict 1D and 2D particle size distribution from measured chord-length distribution. Chemical Engineering Science, 164, pp. 202-218.
- Schoell, J. et al., 2019. Determining particle-size distributions from chord length measurements for different particle morphologies. AIChE Journal, 65, e16560
- Irizarry, R. et al., 2020. CLD-to-PSD model to predict bimodal distributions and changes in modality and particle morphology. Submitted to Chemical Engineering Science.

611603 Industrial Scale Downstream Process Development for Low Concentration Amino Acid Product Using PAT

Abstract:

This work utilizes PAT to develop a downstream purification process on an industrial scale for a low yielding synthetic amino acid from a dilute chemical reaction product. Cooling crystallization optimized by use of inline IR measurement was used to remove 90% of the highest impurity paving the way for efficient use of continuous chromatography to remove lower concentration impurities. The steps leading up to the final design of the Downstream Process (DSP) process are outlined in this work.

The food grade product, an amino acid (AA) is used globally for animal feed as well as human consumption. Production is targeted to be on several MT/year scale. The process of synthesis of the AA involves several reactions involving organic and inorganic compounds.

The resulting reaction product is dilute, over 50% water and contains a higher molar ratio of impurities in the form of organic and inorganics compounds than the product AA. The salts in the reaction product are metal sulfate and sulfite, both at concentrations higher than the AA. Ion exclusion chromatography pulse test results were promising with AA eluting after all impurities. However, scale up to continuous chromatographic separation using Simulated Moving Bed (SMB) chromatography failed due to pressure build up in the columns with the sulfates precipitating out and high ionic load causing osmotic pressure to build up, shattering the resins.

To allow the use of SMB chromatography, precipitation of sulfates prior to chromatography was imperative. Various methods of sulfate salt crystallization including evaporative, cooling and antisolvent crystallization were studied. Evaporative crystallization lead to formation of anhydrous sulfates while cooling crystallization forms decahydrate sulftates, removing considerable amounts of solvent in the process. In line ReactIR® was used to determine the end point for cooling crystallization when the sulfate concentration in the ML plateaued. Given the translucent nature of decahydrate crystals Inline particle track FBRM® was not capable of accurate crystal count measurement. Nucleation and growth kinetics were estimated by off line microscopy as well as real time solution concentration measurement from ReactIR. Use of PAT allowed for faster optimization of the salt precipitation step with 90% of sulfates removed prior to SMB. Salt precipitation lead to some product loss which required an additional cake wash step to recover AA and recycle back into the process with the feed. Permeability of decahydrates crystals were considerably better than anhydrous crystals allowing for better wash and product loss reduction in salt precipitation step.

A second round of crystallization process is required for purification of the AA in the SMB extract from 95% purity to commercial grade product. Inline particle track FBRM(R) was utilized for kinetic measurements in that process.

605492 Hybrid Modeling Approach Integrating PLS Models with First Principles Knowledge

Abstract:

Projection to latent spaces (PLS) is a widely used statistical modeling approach and has found excellent applicability in modeling, monitoring and control [1-2] of complex batch processes. In batch processes, the quality of the final product is of utmost importance and predicting the quality of products well in advance is essential to move the process towards the desired objective. The simple PLS models built with data from previous batches have demonstrated great success in this regard. However, use of process knowledge in developing these models have not been explored in sufficient detail. Motivated by these considerations, in the present work, a hybrid PLS model built with recorded measurement data and data based on first principles model is proposed. A batch crystallization process is chosen as the motivating example where the particle size distribution at batch end is our desired quality variable.

Measuring quality related variables, such as particle size, in a lab during batch runs is an expensive and time intensive task thus motivating the use of good models to predict time evolution of the process variable trajectories and the final quality attributes at the end of the batch. Detailed first principles models of these processes are often available which can predict such final quality attributes with reasonable precision. However, to do so they need to be calibrated to existing plant data by estimating key parameters in the model. Furthermore, to be used in on-line monitoring and control, where disturbances are present, these models need to be run in the form of an observer, Extended Kalman Filter (EKF) or incorporated into a Moving Horizon Estimator (MHE), all of which require considerable expertise and effort to build and maintain. As a result, fundamental models rarely find application in on-line monitoring, optimization and control.. However, these first principles models can provide much useful information that can augment on-line process data to potentially provide improved final quality attribute predictions.

This work addresses the problem of integrating fundamental process knowledge with measurement data to build hybrid PLS models with better predictive ability. This is done by considering the process measurements of historical batches as inputs to the first principles model to predict batch trajectories of variables which are not directly measured. The X block matrix is then augmented with this valuable information once the data of these variables is generated. The first principles models do not have to be calibrated to plant data. Rather, they only need to help in predicting changes in the direction of the process during its duration. The hybrid batch PLS models will accommodate for any biases in the trajectory predictions of the incorporated first principles model, and, in an on-line monitoring mode, will provide the time varying adaptation of the predictions that might be afforded by an EKF or MHE implementation of a first principles only model.

This hybrid PLS methodology approach is used to illustrate this straightforward but powerful approach to predict the final crystal size distribution for the batch crystallization process and is compared with the standard PLS approach based only on plant data, and to subspace based quality models. The simulation results show the improved predictive capability of the proposed approach over the other modeling techniques considered.

[1] Paul Nomikos, John F. MacGregor, Multi-way partial least squares in monitoring batch processes, Chemometrics and Intelligent Laboratory Systems, Volume 30, Issue 1, 1995, Pages 97-108, ISSN 0169-7439

[2] Latent variable MPC for trajectory tracking in batch processes, Journal of Process Control, Volume 15, Issue 6, 2005, Pages 651-663, ISSN 0959-1524

[3] Dan Shi, Nael H. El-Farra, Mingheng Li, Prashant Mhaskar, Panagiotis D. Christofides, Predictive control of particle size distribution in particulate processes, Chemical Engineering Science, Volume 61, Issue 1, 2006, Pages 268-281, ISSN 0009-2509

607712 Indirect Parameter Estimation of Dynamic Mechanistic Models via Hybrid and Neural Differential Equations.

Abstract:

Continued improvement to data-driven algorithms has led to a growing list of data-driven-driven solutions to real-world problems, including applications for material characterization, image recognition and discovery of novel chemical compounds. [1, 2] Many of these applications can be characterized by 'static' conditions. However, in process systems engineering the applications are more often 'dynamic', requiring models to accurately predict process performance under time-varying conditions. For such applications, data-driven approaches have been less robust due to their limited interpretability and failure to observe known physical laws, especially for conditions not considered in model training. Thus, modeling efforts in these domains rely primarily on building mechanistic models based on first principles, which can be a time and labor-intensive process. A union of mechanistic and data-driven modeling techniques (hybrid modeling, HM), has been proposed that can enhance model interpretability while greatly accelerating the model-development process. [3] Unlike purely data-driven methods, HM has yet to become a standard tool for many who could benefit from its accelerated workflow. This is partially due to the lack of automated, computationally efficient tools, enabling practitioners to use hybrid methods at a high level. Moreover, several competing frameworks make identifying which framework is optimal for a given system difficult.

In this presentation a framework is proposed for integrating data-driven models with mechanistic constraints for problems with the following characteristics: (1) nonlinear dynamic systems without steady state assumptions, (2) unknown mechanisms, and (3) incomplete and noisy data. We demonstrate how open-source software for automated parameter estimation, such as checkpointing, automated differentiation and adaptivity enable faster development of HMs. We weigh the merits of approaches that estimate parameters of the data-driven coupled with and without physical constraints. For methods coupling physical constraints, a formulation based on forward sensitivity equations is often used. [4] In contrast, in light of recent work [5], we investigate a backward adjoint-based formulation for enforcing physical constraints during model fitting. Through this comparison we demonstrate how the choice of numerical formulation plays an important role in the size and number of relationships that can be modeled via a hybrid framework. Advantages and limitations of hybrid frameworks will be identified with the aim of enabling practitioners to integrate hybrid workflows to answer domain-specific questions for applications common in (bio)chemical process engineering, such as reaction analysis, model-based optimization, and adaptive design of experiments.

References

1. Himanen, L., et al., *Data-Driven Materials Science: Status, Challenges, and Perspectives.* Advanced Science, 2019. **6**(21): p. 1900808.
2. Venkatasubramanian, V., *The promise of artificial intelligence in chemical engineering: Is it here, finally?* AIChE Journal, 2019. **65**(2): p. 466-478.
3. von Stosch, M., et al., *Hybrid semi-parametric modeling in process systems engineering: Past, present and future.* Computers & Chemical Engineering, 2014. **60**: p. 86-101.
4. Oliveira, R., *Combining first principles modelling and artificial neural networks: a general framework.* Computers & Chemical Engineering, 2004. **28**(5): p. 755-766.
5. Rackauckas, C., et al. *Universal Differential Equations for Scientific Machine Learning.* arXiv e-prints, 2020. arXiv:2001.04385.

610892 Bayesian Learning from Hybrid Models and Their Applications in Thermodynamics and Reaction Engineering

Abstract:

So-called glass (white)-box[1] models are built from scientific theories to capture expert knowledge and intuition for observed phenomena. Black-box (surrogate)[2] models on the other hand employ statistical and machine learning paradigms to reveal trends in (large) volumes of data. At the intersection of these paradigms, hybrid (grey-box) models fuse glass-box models with black-box statistical machine learning components to correct for phenomena not accounted for by the foundational scientific theories. In this talk, we explore efficient decision-making (optimization) under uncertainty with hybrid models. We advocate for a Bayesian approach and use joint posterior distribution from model training to generate scenarios for stochastic programming. We demonstrate how this workflow enables model-form uncertainty quantification and propagation using a ballistic firing example[3] and two additional case studies.

Epistemic or model form uncertainty arises from model inadequacies due to simplifying assumptions or neglected control variables. Aleatoric uncertainty is a consequence of random phenomena such as experimental variability and noisy observations. The seminal work by Kennedy and O'Hagan[4] introduced a Bayesian framework for the simultaneous quantification of aleatoric and epistemic uncertainty inherent in all mathematical models and is shown in Eq (1) (see attached image).

Experimental observations for experiment i is modeled using three components. The glass-box model $\eta(.,.)$ contains physically meaningful global parameters θ and inputs x_i. The stochastic discrepancy function $\delta(.)$ counteracts systematic bias in the glass-box model. Observation error ε_i is modeled as uncorrelated white noise, i.e., $N(0,\sigma^2 I)$. Outputs of the Gaussian Process (GP) discrepancy function $\delta_i \sim GP(\mu(.),k(.,.))$ follow a conditional normal distribution with the mean and covariance fully specified by $\mu(.)$, and $k(.,.)$, respectively. The parameters of the kernel function, along with those of the white noise model (i.e. σ) are denoted as hyperparameters φ.[5]

The model parameters θ and hyperparameters φ are determined using Bayesian calibration, shown in Eq. (2) (see attached image). Bayesian model calibration interprets parameters ω as random variables. The standard workflow is to encode one's current belief in a prior probability distribution, observe data D, and apply Bayes rule, Eq. (2), to obtain the posterior probability distribution $p(\omega|D)$.[6] In our workflow, we use the posterior distribution from the calibration of the hybrid model as the uncertainty set in a stochastic program to minimize the uncertainty in the model predictions.

Using ballistic firing as an illustration,[1] we demonstrate how Bayesian hybrid models overcome systematic bias from missing physics by simulating computer experiments to hit a target using a single-stage stochastic program. We highlight how the hybrid model outperforms a pure machine-learning (data-driven, GP) model when the amount of data available is low (5 to 6 experiments). We argue that Bayesian hybrid models are an emerging paradigm for data-informed decision-making under uncertainty by discussing two case studies from chemical engineering. In the first case study, we use our workflow to design a continuously stirred tank reactor (CSTR) by calibrating a reaction kinetics model with oversimplified assumptions using Bayesian hybrid models. In the second case study, we use Bayesian hybrid models to fit thermodynamic data to equations of state and propagate epistemic uncertainty in using our workflow in designing a flash process. We contrast the merits of our technique against a pure machine-learning (data-driven) GP based workflows and conclude by advocating for a fully Bayesian decision-making methodology which provides probability distributions necessary for stochastic programming and enables iterative (sequential) design of experiments.

References:

[1] Eason, J. P. & Biegler, L. T. Advanced trust region optimization strategies for glass box/blackbox models. *AIChE Journal* **64**, 3934-3943 (2018). doi.org/10.1002/aic.16364.

[2] Jones, M., Forero-Hernandez, H., Zubov, A., Sarup, B. & Sin, G. Superstructure optimization of oleochemical processes with surrogate models. In *Computer Aided Chemical Engineering*, **44**, 277-282. 10.1016/B978-0-444-64241-7.50041-0.

[3] Eugene, E. A., Gao, X. & Dowling, A. W. Learning and optimization with Bayesian hybrid models. In *Proceedings of the 2020 American Controls Conference*, Accepted (2020). arXiv:1912.06269.

[4] Kennedy, M. C. & O'Hagan, A. Bayesian calibration of computer models. *Journal of the Royal Statistical Society: Series B (Statistical Methodology)* **63**, 425-464 (2001). 10.1111/1467-9868.00294.

[5] Bishop, C. M. *Pattern Recognition and Machine Learning* (Springer, 2006).

[6] Higdon, D., Kennedy, M., Cavendish, J. C., Cafeo, J. A. & Ryne, R. D. Combining field data and computer simulations for calibration and prediction. *SIAM Journal on Scientific Computing* **26**, 448-466 (2004). 10.1137/S1064827503426693.

$$\underbrace{y_i}_{observation} = \underbrace{\eta(x_i,\theta)}_{glass-box\ model} + \underbrace{\delta(x_i,\theta)}_{discrepancy\ model} + \underbrace{\epsilon_i}_{observation\ error} \tag{1}$$

$$\underbrace{p(\omega|\mathcal{D})}_{posterior} = \frac{\overbrace{p\,(\mathcal{D}|\omega)}^{likelihood}\ \overbrace{p\,(\omega)}^{prior}}{\underbrace{\int p\,(\mathcal{D}|\omega)\,p\,(\omega)\,d\omega}_{evidence}} \tag{2}$$

598462 Development of a Hybrid First Principles – Artificial Intelligence Approach for Dynamic Modeling of Complex Systems

Abstract:

First-principles models can provide very good prediction even for cases when there are no data at all, or data are limited in certain range of operating conditions or for cases where data collection is infeasible. However, construction of these models can be time consuming, computationally expensive, and intractable for online adaptation. Furthermore, it can be difficult, if not impossible, to develop accurate models for some complex phenomena, that are poorly understood. On the other hand, artificial intelligence (AI) models can be simpler to develop even for complex and ill-defined dynamic systems. These models are typically computationally inexpensive, and therefore can be adapted online. However, development of AI models requires large amount of data[1] so can be infeasible where data collection is expensive or collection of certain type of data is practically impossible given the current state of the measurement technology. Furthermore, AI models may not be predictive especially when they are extrapolated and/or if the data used for developing the AI models suffer from information gap. This work develops an approach where the first-principles models can be synergistically coupled with the AI models thus exploiting their key strengths.

Various types of AI modeling techniques such as neural network (NN), deep learning, expert systems and fuzzy logic have been employed for process synthesis, design and modelling[2]. More recently, hybrid first-principle AI approaches have been proposed[3] for reactive systems by using a linear dynamic model that is obtained through linearization of a non-linear dynamic model followed by a static neural network. That approach fails to account for the nonlinearity in the first-principles model and dynamics in the nonlinear AI model. Furthermore, neither the first-principles model nor the NN model is adaptive. If both the first-principles model and AI model are dynamic, nonlinear, and adaptive, it becomes considerably challenging to optimally synthesize such hybrid models with due consideration of complexity, computational expense, and accuracy.

A hybrid first principles-AI modeling technique has been developed where not only both the first principles and AI models are dynamic and nonlinear but they also interact with each other leading to a time-varying model. Instead of a series structure, we propose a hybrid series-parallel structure where, instead of propagation of information from the first- principles model through the AI model in the series structure mentioned above, the AI model learns the residual error from the first-principles model stemming from the unmodeled and/or unknown phenomena. The AI model is desired to be dynamic, nonlinear, and adaptive. While existing adaptive NN models such as adaptive bidirectional associative memory (ABAM)[4] and transversal/recursive filters[5,6] have found applications in signal processing and communication, they are inadequate for many chemical engineering applications where learning must be accomplished with data that may be noisy, limited, and non-informative. In this work, we propose a hybrid static-dynamic NN structure that can quickly learn to equilibrate to a minimum energy state. The developed gradient descent algorithm for learning can efficiently deal with blowing and vanishing gradients by developing a modified batch normalization approach. The algorithm also quantifies optimality gaps at every iteration reflecting the tradeoff between the computational expense and accuracy. For making the learning process computationally efficient for highly parameterized system, the algorithm down selects weights using a sensitivity-based approach.

The proposed algorithm is applied to the Van de Vusse reactor model as well as to a supercritical boiler system where complex dynamics associated with reactive-diffusive processes leading to oxide scale formation in the superheater tube banks coupled with mass and heat transfer makes it a challenging system. Our work shows the tradeoff between computational expense and accuracy. It is observed that the stability and speed of the learning algorithm are critical for success of the proposed algorithm for a real-life application.

References

1. Zendehboudi, S., Rezaei, N. & Lohi, A. Applications of hybrid models in chemical, petroleum, and energy systems: A systematic review. *Appl. Energy***228**, 2539–2566 (2018).
2. Venkatasubramanian, V. The promise of artificial intelligence in chemical engineering: Is it here, finally? *AIChE J.***65**, 466–478 (2019).
3. Chen, L., Hontoir, Y., Huang, D., Zhang, J. & Morris, A. J. Combining first principles with black-box techniques for reaction systems. *Control Eng. Pract.***12**, 819–826 (2004).
4. Zupan, J. & Gasteiger, J. Neural networks: A new method for solving chemical problems or just a passing phase? *Anal. Chim. Acta***248**, 1–30 (1991).
5. Roussel-Ragot, P., Personnaz, L., Dreyfus, G. & Marcos, S. Communicated by Steven Nowlan Neural Networks and Nonlinear Adaptive Filtering: Unifying Concepts and New Algorithms 0. Nerrand. **199**, 165–199 (1993).
6. Schädler, K. & Wysotzki, F. Comparing structures using a Hopfield-style neural network. *Appl. Intell.***11**, 15–30 (1999).

605298 A Hybrid Simulation-Based and Mathematical Optimization Framework: Demonstration on a Carbon Capture System Optimization

Abstract:

This work presents an automated framework for optimization which integrates a modified trust region approach, generation of surrogate models, and a direct link with commercial simulators and algebraic modeling systems. The effectiveness of the proposed approach is demonstrated with two case studies. First, a simple flash model case study is proposed. Secondly, an MEA-based carbon capture system [1] where the reboiler duty was minimized while achieving a 90% capture rate. A comparison with available derivative free optimization methods shows improved solutions in both time and accuracy.

Two main approaches for optimization have been used in carbon capture: 1) pure mathematical optimization and 2) simulation-based optimization. Pure mathematical optimization directly leverages the equations describing the physical system to be optimized. This approach encounters challenges, however, where large sets of PDE's and highly complex, nonlinear representations are required to sufficiently characterize the process of interest. Simulation-based optimization, on the other hand, considers the system model to be a black box and is based on a heuristic algorithm that uses the results from process simulations to obtain the relationship between the relevant system input and output variables. Although this approach can be used to obtain satisfactory results for large scale, complex systems, it can often be computationally expensive [2]. Hybrid approaches have been proposed in literature [3]; however, their general implementation often requires a very large effort. Considering the advantages and disadvantages of these approaches, a streamlined hybrid framework is introduced to perform efficient optimization of process systems, in terms of solution accuracy, and time.

This work is based on a surrogate model (SM)-based optimization algorithm that involves generating a simplified approximation of the rigorous process model (i.e. built using advanced commercial simulators like ASPEN, gPROMs, Python, etc.) that is more amenable to gradient-based optimization methods [4] and nonlinear programming (NLP) solvers. The approach can overcome the difficulties associated with complex process models (for example black box models or rigorous first principle models), without significantly compromising solution quality and speed, provided that the surrogate modeling method is accurate. The proposed work consists of a modified trust region method, which generates complex surrogate models in order to obtain the optimal solution [5]. The proposed framework has been implemented as a capability within NETL's Framework for the Optimization and Quantification of Uncertainty and Surrogates tool (FOQUS) [6]. The algorithm for the SM-based optimization is highlighted in Figure 1. The algorithm is described in five steps.

Step 1: The algorithm begins with the user providing a process simulation, input and output variables, an initial data set, upper and lower bounds for the optimization variables, and termination conditions. Then, an initial algebraic surrogate model is developed for each output variable for the full space problem (initial data set, set of candidate basis functions, and type of surrogate modeler are selected by the user).

Step 2: Using the initial set of surrogate models and user input (bounds, objective function, ad hoc constraints, etc.), the software builds and solves the mathematical optimization model. The first optimization problem consists of the full-scale problem (original upper and lower bounds of decision variables). Using a multi-start approach, it proceeds to solve multiple problems at different initial conditions, reporting the optimal solution x* and y.

Step 3: The original simulator model is solved at the optimal solution (x*) obtained by the optimization problem. This solution is input into the black box model to determine y* which corresponds to the solution obtained from the advanced process simulator model.

Step 4: This is followed by a set of termination conditions to check whether the optimal solution of the surrogate model is in good agreement with the simulator model (y*). If this condition is satisfied, and the constraints are satisfied, the algorithm is terminated. However, if the termination conditions are not satisfied, the trust region is updated.

Step 5: The trust region is updated by reducing the upper and lower bounds on the decision variables, generating a new data set using Latin Hypercube sampling method, re-running the advanced process simulation for the new data set, and generating a new set of surrogate models.

The algorithm will repeat steps 2 to 5 until convergence. A summary of each algorithm iteration and the final optimal solution are provided, along with a file containing the surrogate models built at each iteration, and a python file containing the parity plots for surrogate model validation.

The algorithm performance is compared against the derivative free optimizers available within FOQUS using two case studies. The first case study entails optimizing a simple flash model simulated in Aspen Plus [7]. The process model includes 59 variables and equations that get solved for simulation in Aspen Plus. Here, the revenue ($/hr) from selling two products is maximized, subject to a CO_2 purity constraint on the vapor stream. In this example, the DFO (based on NLOpt) and SM algorithms find the same optimal solution, though the SM-algorithm converges far more quickly (128 vs. 358 CPU-seconds).

The second case study entails optimizing an MEA carbon capture system model simulated in Aspen Plus. The process model includes 46071 variables and equations that get solved for simulation in Aspen Plus. In the second case study, an MEA carbon capture simulation is solved to optimality more quickly using the SM-algorithm compared with DFO solver (48 vs. 65 CPU-minutes). Additionally, the SM-algorithm finds a better optimal solution, which would translate in large savings during a year of operation.

Compared with DFO's, the algorithm presented is expected to show better performance (time and solution accuracy) when larger problems are studied. Finally, the proposed approach presents a generic framework for rigorous optimization of large-scale nonlinear systems.

References

[1] Morgan, J.C., Soares, C.A., Omell, B., Bhattacharyya, D., Tong, C., Miller, D.C. (2018). Development of a rigorous modeling framework for solvent-based CO2 capture. Part 2: Steady-state validation and uncertainty quantification with pilot plant data. Industrial & Engineering Chemistry Research 57: 10464-10481.

[2] Larson, J., Menickelly, M., Wild S. M. Derivative-Free Optimization Methods. Optimization Online. http://www.optimization-online.org/DB_FILE/2019/04/7153.pdf

[3] Lee, U., Burre, J., Caspari, A., Kleinekorte, J., Schweidtmann, A.M., Mitsos, A. (2016) Techno-economic Optimization of a Green Field Post-Combustion CO2 Capture Process Using Superstructure

and Rate-Based Models. Industrial & Engineering Chemistry Research. 12014-12026.

[4] Cozad, A., Sahinidis, N., Miller, D.C. (2014). Learning Surrogate Models for Simulation-Based Optimization. AIChE Journal 60(6): 2211-2227.

[5] Eason, J. P., Biegler L. T. (2016). A Trust Region Filter Method for Glass Box/Black Box Optimization. AIChE Journal. 3124-3136.

[6] Miller, D. C., Agarwal D. A., Bhattacharyya D., Boverhof J., Cheah Y-W., Chen Y., et al. (2016). Innovative computational tools and models for the design, optimization and control of carbon capture processes. 26th European Symposium on Computer Aided Process Engineering – ESCAPE 26. 2391-2396

[7] Aspen Plus V8.0 – Optimization Tutorial, Merten Morales, Swiss Federal Institute of Technology Zurich (Aug 2013)

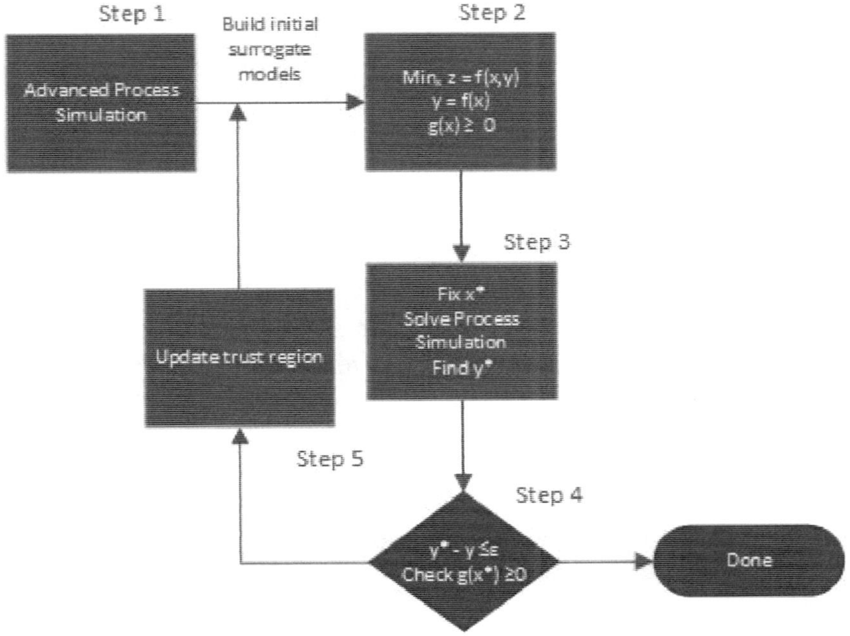

Figure 1: Modified trust region algorithm.

599288 Coupling Mechanistic and Data-Driven Models By Means of Neural Differential Equations to Incorporate Unmodeled Dynamics

Abstract:

Good predictive models are essential in modern chemical industry for control and optimization. Mechanistic models, incorporating physical and empirical knowledge, are dominant. However, the incorporated knowledge is inherently a simplification of reality, resulting in model structure uncertainty. Moreover, gathering the necessary knowledge is time-consuming and might be unfeasible for complex poorly-understood processes. With an ever-increasing amount of data available, data-driven methods are becoming more attractive. In these models, the structure is not explicitly specified, but rather determined by searching for relationships in the available data. Given sufficient and representative data, these models can make highly accurate predictions, unconstrained by any assumptions made. Therefore, they are particularly powerful for learning complex and poorly understood dynamics. The downside is their complete lack of interpretability. Moreover, as representative data is needed, they fail to extrapolate into regions not seen before.

Combining both approaches into a hybrid model aims at taking the best of both worlds. The mechanistic component leverages the available knowledge to model the well-understood parts. The data-driven component models the poorly-understood dynamics to fill in the knowledge gaps. Combinations of both approaches are usually achieved through an uncoupled approach, whereby first a mechanistic model is set-up and calibrated and subsequently the data-driven model is calibrated on the residual error of the mechanistic model (i.e., the difference between the model prediction and data). The data-driven component thus corrects the errors of the mechanistic component. This so-called parallel architecture leverages knowledge and data, resulting in an increased predictive power at lower data requirements [1]. However, as the data-driven model operates on the states of the system, it does not learn its dynamics (i.e., the differentials of the states), but rather the effects of those dynamics. Therefore, an uncoupled hybrid model does not generalize well [2]. Moreover, as the data-driven model is calibrated on the residual error of the mechanistic model, the parameters of both models are uncoupled. Therefore, no synergies between the data-driven and mechanistic models are possible.

Here, we propose a novel framework for coupled hybrid models based on neural differential equations [3]. Neural differential equations define the right-hand side of a differential equation as a neural network. The neural network is subsequently calibrated using backpropagation on the gradient obtained through automatic differentiation, directly learning the dynamics of the system. We combine these neural differential equations with mechanistic differential equations and calibrate the parameters simultaneously, resulting in a coupled hybrid model.

We show the power of this coupled approach on two case studies with incomplete knowledge of the dynamics of the system. In a first study, a simple reactor is modeled whose kinetics are not completely known. A known forward reaction, converting chemical A into chemical B, is modeled mechanistically. However, unknown to the modeler, a reversible reaction occurs, converting chemical B into chemical A. This reaction is thus not incorporated into the mechanistic model, resulting in a model that does not fully capture the dynamics of the system. We show that the coupled hybrid model successfully retrieves the missing kinetics, resulting in a model with high predictive power. Although this result seems trivial on this toy problem, it highlights the possibilities of applying this approach to larger, more complex systems.

Second, we model an enzyme microreactor with a pulsating feed (e.g., resulting from a positive displacement pump), inducing complex hydrodynamics. We generate synthetic data of this system using a transient CFD solver and log the outlet concentrations as a timeseries. The system is subsequently modelled using a simple mechanistic model that only includes kinetics. As the hydrodynamics are not incorporated, the model does not fully capture the dynamics of the system. Again, we show that the data-driven component of the coupled hybrid model successfully retrieves the hydrodynamics, resulting in a model that truly represents the system.

In conclusion, we introduce an architecture for coupled hybrid models based on neural differential equations. The mechanistic component incorporates all available knowledge of the modeled system, while the data-driven component captures any unmodeled dynamics, including (but not limited to) unknown kinetics and complex hydrodynamics.

602260 Enhancing Symbolic Regression with Gradient Information

Abstract:

Incorporating gradient information during the regression of surrogate models of various processes can greatly improve their representation of the process and help simulation and optimization of processes by providing more reliable gradient behavior [3]. Recently, Cozad and Sahinidis [2] proposed an MINLP formulation of symbolic regression that can be expanded to simultaneously calculate the first and second partial derivatives of key parameters. The purpose of the current paper is to enhance the surrogate model regression procedure with gradient information to improve the bounding and model discrimination by incorporating partial first and second derivatives into the regression framework. Symbolic regression learns both the model structure and parameters to model a data set, unlike traditional regression that limits the scope of the regression to a fixed functional form and has typically been performed with genetic programming [1]. The regression only requires the specification of a set of operators and operands (+, -, *, /, exp(.), log(.), sqr(.), cub(.), √(.), etc.) to flexibly develop new functional forms that accurately represent the data. Recent developments in applying a global deterministic approach have shown improved fitting metrics, such as sum of squared error and other information criterion [2,5]. Further expanding the symbolic regression capabilities by applying chain rule evaluations of the derivatives while simultaneously determining the model structure allows for gradient information to impact the regression process and improve the surrogate representation of the data. We applied this technique to benchmarked equations to demonstrate these new capabilities.

Reference cited:

[1] Koza, J.R.: Genetic Programming: On the Programming of Computers by Means of Natural Selection. MIT Press, Cambridge, MA (1992) 24.

[2] Cozad, A. & Sahinidis, N.V. (2018). A global MINLP approach to symbolic regression. Mathematical Programming, 170, 97-119, 2018.

[3] Matias, J. & Jaschke, J (2019) Using a neural network for estimating plant gradients in real-time optimization with modifier adaptation. IFAC-PapersOnLine, 15, 808-813, 2019.

[4] J. McDermott, D.R. White, S. Luke, L. Manzoni, M. Castelli, L. Vanneschi, W. Jaśkowski, K. Krawiec, R. Harper, K.D. Jong, U.M. O'Reilly, Genetic programming needs better benchmarks, in *Proceedings of the Genetic and Evolutionary Computation Conference (GECCO)* (ACM, Philadelphia, 2012)

[5] Tawarmalani, M.;Sahinidis, N. V. Global optimization of mixed-integer nonlinear programs: A theoretical and computational study, Mathematical Programming, 99, 563-591, 2004.

598636 Mixed-Integer Formulations for Fair Classification

Abstract:

Classification models are typically used to separate datasets based on a set of descriptors; however, such models can also be used to make decisions (e.g., make go/no-go decisions in a project). In this context, fairness becomes a major issue because such decisions tend to affect multiple stakeholders and because descriptors might inadvertently introduce biases [1, 2, 3]. A recent application of fair classification models include allocation of financial loans (e.g., descriptor is credit score). In this work, we show how fair classification models can also be used to tackle problems of interest to the chemical engineering community. For instance, in a manufacturing facility, access to experimental lab equipment is simultaneously requested by different stakeholders to test a variety of products. The decision here is whether a sample should be tested or not (or when) given the importance of the sample (captured by descriptors) and given that there is limited lab equipment and budgets. Such decisions are often based on minimizing a loss function, which is an aggregate of the cost associated with false positives and false negatives. The model output is a threshold value (based on a linear combination of sample features) that activates the binary decision. The inherent degeneracy of such models can lead to decisions that prioritize testing in an unfair manner [4].

State of the art research in fairness in machine learning focuses on developing algorithms that do not violate state or federal anti-discrimination laws. This is achieved by adding either post-processing steps or additional constraints to the mathematical model to achieve properties such as demographic parity, equalized odds, and equal opportunity [1,2]. These conditions can be mutually exclusive and often lack a theoretical basis that connects them to the axiomatic view of fairness. In this work, we will present mixed-integer formulations that address the problem of fair classification from an axiomatic perspective [5]. In the field of game theory, the resource allocation problem has been viewed as a bargaining game between stakeholders. Nash [6] first provided an axiomatic approach to obtain solutions to the bargaining problem. These axioms include Pareto optimality, symmetry, affine invariance, and independence of irrelevant alternatives. Nash also proved that there exists an allocation scheme that satisfies these axioms (what is now known as the Nash solution). We observe that the ultimate goal of a fairness measure such as the Nash solution is to *shape an allocation distribution* in a desirable way. As such, the resource allocation problem can also be interpreted as a stochastic programming problem in which one seeks to find allocations that shape distribution of outcomes (in stochastic programming the outcome distribution is shaped by using a risk measure) [7, 8]. As in the case of fairness measures, axioms have been proposed in the stochastic programming literature to study the selection of suitable risk measures [9].

We present a theoretical analysis of the statistical properties of different fairness criteria introduced in binary classification models. We will also provide a case study of a manufacturing facility where a binary classification model is used to drive the decision of which samples should be tested. We will demonstrate the impact of using Nash solution on the allocation distribution of tests between the equipment and the overall testing efficiency (measured by false positives and false negatives). We will showcase how the Nash solution inherently captures this tradeoff between efficiency and fairness, and selects the optimal solution based on the axiomatic properties.

References:

[1] Gölz, P., Kahng, A. and Procaccia, A.D., 2019. Paradoxes in Fair Machine Learning. In *Advances in Neural Information Processing Systems* (pp. 8340-8350).

[2] Hardt, M., Price, E. and Srebro, N., 2016. Equality of opportunity in supervised learning. In *Advances in neural information processing systems* (pp. 3315-3323).

[3] Wattenberg, M., Viégas, F. and Hardt, M., 2016. Attacking discrimination with smarter machine learning. *Google Research, 17.*

[4] Sampat, A.M. and Zavala, V.M., 2019. Fairness measures for decision-making and conflict resolution. *Optimization and Engineering*, *20*(4), pp.1249-1272.

[5] Moulin, H., 1991. *Axioms of cooperative decision making* (No. 15). Cambridge university press.

[6] Nash Jr, J.F., 1950. The bargaining problem. *Econometrica: Journal of the Econometric Society*, pp.155-162.

[7] Dowling, A.W., Ruiz-Mercado, G. and Zavala, V.M., 2016. A framework for multi-stakeholder decision-making and conflict resolution. *Computers & Chemical Engineering*, *90*, pp.136-150.

[8] Hu, J. and Mehrotra, S., 2012. Robust and stochastically weighted multiobjective optimization models and reformulations. *Operations research*, *60*(4), pp.936-953.

[9] Artzner, P., Delbaen, F., Eber, J.M. and Heath, D., 1999. Coherent measures of risk. *Mathematical finance*, *9*(3), pp.203-228.

602264 Modeling and Nonlinear State Estimation for Advanced Process Control of the Enzymatic Conversion of Lactose into Value-Added Products

Abstract:

Due to increased concerns over sustainability of chemical processes, significant research has been conducted to minimize the amount of process waste generated. Despite advances in efficiency, selectivity, and conversion, waste management is still a concern for all processes, especially biological. The most effective method for waste disposal would be its conversion into a value-added product through profitable means. Many industries are attempting to achieve this goal, but it is challenging to design a novel catalyst, enzyme, or process. In particular, recent advances in enzymatic process modeling have created an opportunity for small-scale cheese manufactures to convert their waste into value-added products. The manufacturing of cheese results in the formation of lactose rich whey permeate that cannot be easily utilized and thus is often disposed of at cost to manufacturers [1].

A recent advance in enzymatic modeling corresponds to a new kinetic model to convert lactose waste into value-added galactooligosaccharide (GOS) products using β-galactosidase [2]. This specific reaction has been studied by several researchers in the past, but each proposed model presented challenges such as mass balance issues and overfitting due to the excessive number of parameters [3]. The proposed kinetic model contains no mass balance errors, while minimizing the number of parameters to accurately depict the process characteristics. Furthermore, this model uniquely distinguishes between purely galactose derived disaccharides, trisaccharides, and tetrasaccharides and those containing a glucose saccharide group. Due to the similarity of glucose and galactose, there are no online measuring techniques that can discern between the specific di, tri, or tetrasaccharides stereoisomers. A High Performance Liquid Chromatography (HPLC) method coupled with an ion-exchange column and refractive index detector can distinguish between glucose and galactose, but has no resolution for other isomers such as disaccharide, trisaccharide, and tetrasaccharide stereoisomers. Without the ability to measure specific component concentrations, state estimation must be used to predict the individual states using the online measurements and the proposed kinetic model.

This work addresses the state estimation and advanced control employing the developed enzymatic model for the first time. Unique estimation challenges are posed and addressed, with the specific novelty of using nonlinear state estimation to find stereoisomer concentrations from available online measurements and the nonlinear kinetic model. Both an Extended Kalman Filter (EKF) and a Moving Horizon Estimator (MHE) are analyzed for process implementation. The EKF is an unconstrained estimator that allows for rapid estimation but does not guarantee feasible and optimal results. The MHE is a constrained estimator that guarantees feasibility at the cost of increased computational time [4] [5]. Using these estimation techniques, a Biologically-Inspired Optimal Control Strategy (BIO-CS) algorithm is applied to optimize the feeding of lactose and enzyme into a semi-batch reactor [6][7]. The proposed estimation and advanced control framework allows the feasibility of this process to be examined and provides a guide for other bioprocesses to follow.

References:

[1] Illanes, A. (2011). Whey Upgrading by Enzyme Biocatalysis. *Electronic Journal of Biotechnology*.

[2] Schultz, G., Alexander, R., Lima, F.V., Giordano, R., & Ribeiro, M. (2019). Kinetic Model for the Enzymatic Synthesis of Galacto-Oligosaccharides: Describing Galactobiose Formation. *AIChE Annual Meeting*.

[3] Vera, C., Guerrero, C., Illanes, A., & Conejeros, R. (2011). A Pseudo Steady-State Model for Galacto-Oligosaccharides Synthesis With ß-Galactosidase From *Aspergillus oryzae*. *Biotechnology and Bioengineering*.2270-2279.

[4] Campani, G., Ribeiro, M., Zangirolami, T., Lima, F.V. (2019). A Hierarchical State Estimation and Control Framework for Monitoring and Dissolved Oxygen Regulation in Bioprocesses. *Bioprocess and Biosystems Engineering*.1467-1481.

[5] Lima, F.V., Rawlings, J. (2011). Nonlinear Stochastic Modeling to Improve State Estimation in Process Monitoring and Control. *Process Systems Engineering*.996-1007.

[6] Mirlekar, G., Gebreslassie, B., Diwekar, U., Lima, F.V. (2018). Biomimetic Model-Based Advanced Control Strategy Integrated with Multi-Agent Optimization for Nonlinear Chemical Processes. *Chemical Engineering Research and Design*.229-240.

[7] Mirlekar, G., Al-Sinbol, G., Perhinschi, M., Lima, F.V. (2018). A Biologically-Inspired Approach for Adaptive Control of Advanced Energy Systems. *Computers and Chemical Engineering*.378-390.

604108 Designing Data-Based and Model-Based Methods for Process Monitoring and Equipment Degradation Tracking

Abstract:

A wide variety of industrial processes utilize monitoring methods in order to ensure that safety and quality is maintained. Conventional model-based methods include dimensionality reduction techniques such as Principal Component Analysis (PCA) are often employed as they are computationally simple, and easy to implement in practice [1]. Unfortunately, these techniques assume that the process data is Gaussian, uncorrelated, and only contain a moderate level of noise. This work will demonstrate how wavelet-based representation of data can be utilized in order to efficiently handle data that violate these assumptions. Statistical hypothesis testing methods such as the generalized likelihood ratio chart have shown promise with respect to fault detection, and thus will be used to enhance monitoring capabilities [2], [3].

Additionally, this work will demonstrate the necessity to track deviations in the process models themselves in order to identify process drifts or equipment degradation. When a process operates under control, a controller continuously adjusts the level of manipulated variables in order to ensure that the controlled variables are being maintained within specific limits. Unfortunately, this may be at the expense of increased operating and additional costs. Therefore, this work will utilize a dynamic contour-based algorithm in order to illustrate how equipment degradation can be efficiently tracked in multiple operating regimes [4].

For both algorithms, illustrative examples using simulated synthetic data, and real data from different applications will be utilized in order to highlight the effectiveness of the developed algorithms.

References

[1] I. T. Joliffe, *Principal Component Analysis*, 2nd ed. New York, NY: Springer-Verlag, 2002.

[2] M. R. Reynolds and J. Y. Lou, "An Evaluation of a GLR Control Chart for Monitoring the Process Mean," *J. Qual. Technol.*, vol. 42, no. 3, pp. 287–310, 2010.

[3] M. Z. Sheriff, M. Mansouri, M. N. Karim, H. Nounou, and M. Nounou, "Fault detection using multiscale PCA-based moving window GLRT," *J. Process Control*, vol. 54, 2017, doi: 10.1016/j.jprocont.2017.03.004.

[4] M. Z. Sheriff, H. Nounou, M. Nounou, and M. N. Karim, "Monitoring process degradation through operating regime based process monitoring," in *AIChE Spring Meeting and Global Congress on Process Safety: Process Control Monitoring and Analytics*, 2019.

610824 Optimal Sensor Network Design and Nonlinear State Estimation for in-Situ Corrosion Monitoring in Coal-Fired Boilers

Abstract:

Existing coal fired power plants are increasingly being subjected to significant load-following due to the increased penetration of intermittent renewables to the grid. During load following operations maintaining the boiler environment at the desired conditions is significantly difficult. In particular, higher flue gas temperature than expected and higher concentration of corrosive sulfur species in the flue gas are causing external sulfidic corrosion especially in the waterwall (WW) tubes, which is one of the primary causes for the WW failure [1]. Experimental studies on corrosion rates of Ni-Cr alloys suggest that corrosion rate is a function of alloy composition, temperature, SO_2 and O_2 concentrations in the flue gas [2-3]. The effect of metal temperature on the corrosion rate is highly nonlinear with a "bell" shaped response between 600°C and 1000°C for most of the alloys used in the WW section. Unfortunately, as of now, there is no online commercial sensor for real-time measurement of corrosion rate due to the high operating temperature and challenging environment in the WW section. Recently a novel type of electrochemical sensor is being developed in-house for in-situ measurement of the corrosion rate [4]. As the operating temperature and concentration of sulfur species greatly vary spatially and temporally in the WW, the corrosion rate can vary significantly with space and time. However, it is infeasible to install such sensors in the entire WW. Motivated by these challenges, an optimal sensor network is designed, and a nonlinear state estimation technique is developed for real-time monitoring of corrosion rate at all desired locations inside the WW.

A first-principles dynamic process model of the WW is developed for calculating the spatial and temporal variation in the temperature and concentration of sulfur species. Models are also developed for the corrosion rate. Electrochemical sensor model has been previously developed at West Virginia University [4]. This is a high nonlinear differential algebraic equation (DAE) system. Therefore, an unscented Kalman filter algorithm for DAE system is developed for state estimation, particularly for estimating the corrosion rate given the data from the electrochemical sensor.

Typically, process information which can be obtained from a given sensor network and the reliability of sensor network improves as the number and/or the accuracy of the installed sensor increases. However, it is neither feasible nor necessary to install sensors at all desired locations. An optimal sensor network is designed by minimizing the posterior error covariance of the optimal estimator. The algorithm provides the optimal number, location and type of sensors.

The optimal sensor network is used to estimate the corrosion rate for a coal-fired boiler. A number of operating scenarios are simulated by changing the temperature in the WW and concentration of the sulfur species. Our work shows that even in the presence of large mismatch between the process and the model and high measurement uncertainty, the sensor network provides satisfactory spatial and temporal resolution of the corrosion rate under aggressive load-following operation.

References

1. North American Electric Reliability Council (NERC), "State of Reliability," 2017.

o

2. L. Luthra and D. A. Shores, "Mechanism of Na_2SO_4 Induced Corrosion at 600-900 C," *J. Electrochem Soc.*, no. 1971, pp. 2202–2210, 1978.
3. Viswanathan and C. J. Spengler, "Corrosion of 85 Ni-15 Cr Alloy at 1600 F in Controlled Atmospheres Containing O_2, SO_2, SO_3, H_2S, and N_2," *Corrosion*, vol. 26, no. 1, pp. 29–41, 1970.
4. N. Aung and X. Liu, "High temperature electrochemical sensor for in situ monitoring of hot corrosion," *Corros. Sci.*, vol. 65, pp. 1–4, 2012.
5. Kumari, S. K. Das, and P. K. Srivastava, "Modeling fireside corrosion rate in a coal fired boiler using adaptive neural network formalism," *Port. Electrochim. Acta*, vol. 34, no. 1, pp. 23–38, 2016.

609445 CO_2 Injection Monitoring with a Scalable, Automated Sparse Seismic Array

Abstract:

Successful applications of the 3D surface land seismic method to monitor CO_2 injection into a reservoir have been found in the literature. Reducing the operational effort to acquire 3D data and the turnaround time for the processing and interpretation of field data are highly desirable for more efficient seismic monitoring. The scalable, automated sparse seismic array (SASSA) method has been ideated as a cost-effective alternative or complementary to conventional 3D seismic methods. Based on a sparse array of receivers deployed strategically on Earth's surface and one or more fixed seismic sources, the SASSA method can be used to study specific portions of the subsurface to detect the movement of CO2 in a reservoir with lower environmental footprint and shorter turnaround time to deliver results than conventional 3D seismic methods.

The Energy & Environmental Research Center (EERC) carried out seismic monitoring of an oilfield undergoing CO_2 enhanced oil recovery in Montana, USA, using the SASSA method. Initial efforts in collaboration with the field operator were invested in designing an optimum distribution of two sources and 96 receivers not only to honor the subsurface geologic conditions and CO_2 injection patterns but also to minimize the impact of cultural noise sources. The signal-to-noise ratio was further improved in data acquisition by use of two fixed surface orbital vibrators that were remotely controlled by the EERC team in North Dakota. Weekly data acquisition was conducted on weekends from May 2019 through March 2020. Data-processing workflows were developed to preserve the true seismic amplitudes representing the time-lapse changes of the reservoir due to CO_2 injection and maximize the representation of the reservoir in the seismic sections from the distribution of sources and receivers on the surface. Encouraging results were obtained using the SASSA method in areas adjacent to the CO_2 injection wells. Two validation methods were considered: a time-lapse 2D seismic survey acquired at the beginning and the end of the monitoring activities and reservoir simulation.

615040 Continuous Revolutions: Distributed Manufacturing to Re-Shore the Pharmaceutical Supply Chain

Abstract:

Coronavirus has demonstrated the life and death circumstances related to critical shortages of essential medical items. The U.S. dependence on foreign drug supply has intensified already serious drug shortages, and questions remain over the safety of the drugs we are able to procure. Pandemic-driven drug shortages due to supply chain disruption are ongoing and will only intensify as we struggle with burgeoning hospitalizations and critical care patients.

A sophisticated distributed manufacturing platform could be an answer to rectify supply chain disruption. This platform, coined 'Pharmacy on Demand' (POD) is a miniaturized manufacturing unit the size of a household refrigerator that features proprietary micro-reactors and continuous flow synthetic chemical processes to make active pharmaceutical ingredients (API) and final formulated medicines. POD can make medicine from American-made chemical inputs to a finished product of a pill or liquid, resulting in complete medicine self-reliance. The flexible and distributed manufacturing capability will mitigate drug shortages during both peacetime and crisis, reducing U.S. dependence on foreign manufacturers.

614847 Building Resilient Supply Chains through the Manufacturing USA Institute Network

Abstract:

Over the last few decades, optimization for cost reductions in production across industry segments has been prioritized over maintaining flexible and resilient supply chains. Off-shoring of pharmaceutical production is an important example of this evolution, and the COVID-19 pandemic has exposed the U.S. population and economy to vulnerabilities created by limited domestic control over strategically important supply chains. The Manufacturing USA institutes are working collectively to craft a different future for critical U.S. supply chains by (1) mapping existing supply chains to understand critical gaps in U.S. manufacturing capabilities and infrastructure, (2) identifying critical feed stock limitations, materials and process development needs, and opportunities to refine current and design new technologies for the optimized system design, and (3) coordinating and managing the work to build a more resilient and flexible manufacturing infrastructure in the U.S. In particular, the RAPID Manufacturing Institute is partnering with its members to understand the current supply chains for several critical medicines and develop flexible, modular process options for cost competitive domestic production of these therapeutics.

614848 Change Creates Opportunity and Disruptive Organizational Innovation

Abstract:

The novel coronavirus has brought significant change to our lives, and our workplaces. While some changes are temporary, the need to work remotely has significantly accelerated the adoption of new technology and business practices that will remain in place long after vaccines are available. Travel restrictions have forced tech-transfer teams to collaborate in new ways, requiring remote support of manufacturing suites, and democratizing decision-making through virtual teams. The need to work from home as put a spotlight on the importance of solid data-management infrastructure and enabled a culture change toward "default-to-open" data security practices. Crises serve to highlight the most critical work, and allow a fresh-eyed re-evaluation of organizational priorities. Change creates opportunity, and major change has created disruptive organizational innovation.

614849 Responses to COVID-19 from Dow Life Sciences

Abstract:

Dow Life Sciences is a strategic market segment in Dow. The goal of this market segment is to deliver high quality product and technical solutions to Human Health and Human Comfort. There are 5 sub-platforms that we are participating in: API and Manufacturing, Topical Drugs, Oral Drugs, Biopharmaceutical, and medical devices. Dow products sold in this area include Carbowax Sentry PEGs, Isopropanol, defoamers etc. Under this challenging COVID-19 scenario, our Life Sciences global team, quickly responded by identifying unmet needs from the market and customers. Due to the high demand of isopropanol (IPA), the first response was to address the supply issue. This resulted in a capacity expansion of IPA. Meanwhile, we also coordinated the IPA supply among various pharmaceutical customers and continued supporting our strategic customers to launch drug products. In addition to IPA, our global team also captured opportunities related to other Dow products such as amines, chelants, and PEGs to support our global pharma customers in formulated hand sanitizers. Meanwhile, some disruptive opportunities were also identified and investigated due to a great effort and collaboration across Dow multiple functions.

614850 Social Distancing Via Telepresence for Product Development and Manufacturing

Abstract:

Product Development a Bristol Myers Squibb has an unwavering dedication to developing innovative medicines that transform patients' lives. Transferring oral and parenteral drug product manufacturing processes for clinical or commercial manufacturing is an essential deliverable from Product Development to maintain delivery of innovative medicines to patients. In the time of SARS-CoV-2/Covid-19 and our fight to combat the pandemic through social distancing and working remotely, hands on manufacturing processes - particularly when transferring the process between organizations - can introduce risk of viral transmission. Our team has employed augmented reality, remote monitoring protocols, and other digital technology to maximize the ability to collaborate remotely, collect data, and transfer information and know-how while mitigating the risk of viral transmission. Several projects at different stages have been de-risked, troubleshot, and otherwise enabled through application of these innovative digital technologies.

614851 Innovation in Challenging Times

Abstract:

Innovation is uniquely dependent on supply chains as new vendors and new components need to be designed, built, prototyped, sourced. Impacts can come at us in unexpected ways ranging from government mandated staff reductions that affect productivity, communication hurdles, and challenges when multiple disciplines need to work together on-site but staff limitations preclude collaborative work environments.

Flight reductions can result in limited cargo capacity leading to schedule delays and cost escalation. Vendors may no-longer be allowed on site, or if they do require extensive, complex planning to get approval to be on-site to accomplish a previously simple 4 hr service call. Internally the team responsible for implementation innovations can become fractured in their communications, their perception of priorities and goals.

Overcoming these challenges become drivers for yet other innovations in how we accomplish work.

598416 (Invited Plenary Talk) Wearable Electrochemical Sensors

Abstract:

Wearable sensors have received a major recent attention owing to their considerable promise for monitoring the wearer's health and wellness. The medical interest for wearable systems arises from the need for monitoring patients over long periods of time. These devices have the potential to continuously collect vital health information from a person's body and provide this information to them or their healthcare provider in a timely fashion. Such sensing platforms provide new avenues to continuously and non invasively monitor individuals and can thus tender crucial real-time information regarding a wearer's health. Our research over the past decade has aimed toward filling the gaps toward providing biochemical information, beyond that given by common wrist-watch mobility trackers. This presentation will discuss recent developments in the field of wearable electrochemical sensors integrated directly on the epidermis or within the mouth for various non-invasive biomedical monitoring applications. Particular attention will be given to non-invasive and minimally-invasive monitoring of metabolites and electrolytes using skin-worn and microneedle amperometric and potentiometric sensors, respectively, along with related materials, energy and integration considerations. The preparation and characterization of such wearable electrochemical sensors will be described, along with their current status and future prospects and challenges.

610185 (Invited Plenary Talk) an Integrated Paradigm for Precision Exposure to Airborne Chemical and Biological Stressors Based on Personal Sensing

Abstract:

Introduction

The current manuscript outlines the fundamentals of a methodological framework for precision exposure assessment and gives concrete examples of its application in different European cities to assess the health risk associated with exposure to fine and ultra fine particles of chemical and biological origin. The work described herein involves a multi-city personal sensor campaign aiming to refine PM exposure using low-cost portable sensor data, exposure and human respiratory tract deposition modelling to estimate the biologically effective internal daily intake on an individual basis and agent-based modelling to estimate exposure and consequent health risk to the community.

Methodology

A prototype monitoring device (IcaruSense) was developed for measuring three size fractions of PM (1, 2.5 and 10 μm), enabling direct assessment of personal exposure. The device is based on an Arduino microcontroller where miniaturized electrochemical sensor-modules are connected. In order to improve sensor performance, correction factors were used to account for diverse environmental conditions and they were calibrated with the help of a well-validated light scattering optical sensor. Study participants wore a physical activity wristband (Garmin Vivosmart 3) that records steps, distance, type of activity, heartbeat and sleeping patterns. Finally, participant positions were recorded using a GPS sensor, integrated into the PM sensor kit. After calibration and validation, the IcaruSense devices were used to capture daily variability of PM exposure. Exposure was further refined by estimating inhalation adjusted intake rate from the continuous time-activity data collected from the physical activity and the GPS sensors. Using the inhalation adjusted intake rate PM deposition across the human respiratory tract (HRT) was readily estimated using the Multiple Path Particle Deposition (MPPD) model.

A city scale agent-based model (ABM) was developed to support this work by expanding the precision exposure estimates from the individual to the community level. The model feeds into population-based exposure assessment without imposing prior bias, basing its estimations onto emergent properties of the behaviour of the computerised autonomous decision makers (agents) that compose the city-system. Population statistics, road and buildings network data were transformed into human, road and building agents, respectively. Informed by literature-extracted relationships between Socio-Economic Status (SES) and human behavioural patterns, the established ABM simulates societal dynamics. Time-use survey outputs were associated with human agent behavioural rules, aiming to model representative to real-world routines. Moreover, real time-geography of exposure data, extracted from the precision sensors campaigns in nine European cities were used to parametrise and enhance the model.

Mobility behaviour and eventually exposure to pollutants change as a result of a human agent-specific decision making. Virtual individuals of different sociodemographic backgrounds use different means of transportation and follow different sequence and types of activities. Behaviours that were not explicitly programmed, arise through the human agents' interactions, enabling the examination of emergent behaviours from the bottom-up. Spatiotemporal trajectories are coupled with spatially resolved Particulate Matter (PM) levels, enabling the assessment of personal and inhalation adjusted exposure, based on type of location and intensity of enrolled activities.

Results and Discussion

The above methodology was applied in the personal sensors campaign of the HORIZON2020 EU Project ICARUS, where exposure and intake to PM of almost 100 individuals in each out of nine European cities were monitored. The integrated hybrid approach outlined above, which couples multi-sensor datasets on an individual basis with advanced respiratory tract modeling, allowed us to significantly differentiate the actual intake of the participants, highlighting larger differences than the ones attributed to the spatial differentiation of air pollution based on fixed stations (difference of ambient PM levels of 50% were translated in intake differences up to 110%). These differences are the result of the differences in PM size fractions that are captured by the sensors and the capability of the HRT model to translate these differences in PM size distribution accounting for variation in respiratory tract physiology among the participants, which is ultimately reflected in intake estimates.

ABM findings indicated that inhalation adjusted exposure differences between 2 individuals living in the same neighbourhood can vary by as much as 87%, due to radically different spatiotemporal behaviours. This major difference was observed for cases where one human agent has a full-time job and often exercises in outdoor courts whereas his/her neighbour is a retired homemaker. For the same reason, ABM runs confirmed that even people who reside in the same dwelling might not necessarily be exposed to similar levels of pollution. Model results indicated that PM inhalation adjusted exposure between flatmates can differ by 61%. Evidence was, also, provided for vulnerable subgroups of population that are disproportionately subjected to high levels of exposure. ABM-retrieved population exposure findings indicated that subgroup median exposure for children, adult males with a full-time job and low income and human agents of a lower educational level was notably higher than the total population median value. The validated model can assess exposure for the entire sociodemographic spectrum of a simulated region, without the constant need for repeated individual monitoring which is often not feasible due to high costs, ethical constraints and complex organisation that is associated with such measurements. By modelling the heterogeneous routines of human agents, the ABM produces detailed information related to the societal system examined and generates data that could be used to fill in the gaps that exist in traditional datasets.

Conclusions

This study presents a new methodological approach to estimate the external exposome, which encompasses the totality of human environmental exposures at an individual and community level. This method helps to integrate substantive considerations of individual status into long-term urban planning and management; it would also facilitate the identification of both physically and socially practical means for reducing life-threatening exposure levels. The methodological paradigm for precision assessment of exposure at the individual and community levels laid out in this work can be used for evaluating the impacts of different public health policies prior to implementation.

601270 High Throughput Process Development for CEX Chromatography

Abstract:

Chromatographic operations are the workhorse for protein purification in the biopharmaceutical industry due to their unparalleled separation efficiency and capacity. Typically, chromatographic processes consist of multiple polishing steps, at least one of which is operated in a bind and elute mode. Although bind and elute chromatography is very effective at removing product variants, it suffers from limited capacity and long processing times. In this work, we developed efficient cation exchange chromatography (CEX) flowthrough steps for mAbs using a combination of high throughput screens and lab scale experiments. CEX resins were screened across parameter space using high throughput screening techniques. Conditions which improved process performance as compared to a bind and elute operation were identified. Following high throughput screening, this process was evaluated at the lab scale for further development and refinement. Comparable product quality and process consistency between the Tecan and the lab scale experiments suggest an efficient and reliable high-throughput process development workflow for CEX flowthrough operations.

605951 The Use of Tangential Flow Filtration Systems in Synthetic Peptide Purification

Abstract:

Tangential flow filtration (TFF) is a pressure and shear driven separation process that uses membranes to separate components in a liquid solution based primarily on size differences. TFF processes such as microfiltration and ultrafiltration are widely used in the pharmaceutical industry for concentration and buffer exchange in biologics. Due to the nature of TFF membrane materials, these unit operations have been primarily limited to aqueous or low-solvent processes. Example TFF applications include monoclonal antibodies and microbially derived proteins and peptides.

Recent advances in membrane technology have enabled the fabrication of membranes with smaller pore sizes and composed of materials with improved compatibility to organic solvents. For larger and more complex synthetic peptides, ultrafiltration may provide a superior alternative to distillation and/or chromatography for concentration and solvent exchange.

This presentation will highlight the use of ultrafiltration as part of a synthetic peptide process to concentrate purified peptide solution and to perform a solvent exchange in preparation for precipitation of the pure API. This work represents a novel transfer of TFF technology from the biologic/microbial platform to the synthetic peptide purification platform.

608933 In-Situ Monitoring during Mixing for Protein Crystallisation

Abstract:

Bio-active peptides are the next generation of pharmaceuticals which lead to a significant increase in demand over the last years [1]. Remarkable improvements in the upstream processing of peptides has been achieved, leading to a shift in the bottleneck to downstream purification [2]. Recently, crystallisation has emerged as a novel technology to replace expensive downstream purification techniques in the production line of bio-active peptides in order to increase economic efficiency [3, 4]. Due to their high molecular weight, low diffusivity, and soft and fragile nature, peptides require a gentle, but intensive mixing in order to achieve the same level of crystallisation reproducibility at larger scales than microlitres. The lack of suitable mixing strategies for large-scale peptide crystallisation, while guaranteeing high crystallisation throughput and reproducibility of the crystal qualities, has been identified as one of the main limitations that has to be overcome in order to apply crystallisation as a sustainable downstream purification technique

In this study, the importance of mixing on the controllability and reproducibility of the crystallisation process in large-scale batch crystallisers will be discussed. Instead of commonly used mixing strategies, such as over-head stirring, that usually leads to a more uncontrolled crystallisation with a poor crystallisation reproducibility, the impact of a gentler mixing strategy by using mechanical, oscillatory shaking will be reported. Furthermore, a newly developed image-analysis tool specific to peptide crystals will be presented. It is used to determine crystal sizes and distribution in-situ by processing images of crystals to evaluate the impact of mixing. A simple model peptide with a low solubility and molecular weight, which is known to crystallise, is used to investigate the importance of mechanical mixing on achieving a reproducible peptide crystallisation process. A tubular shaped batch crystalliser with volumes from the millilitres to the tens of millilitres along with different geometries was utilised. The crystallisation condition, such as pH, concentration, shaker speed, templating additives, and the orientation of the crystalliser to the shaking plane have been optimised. The impact on the crystallisation process is evaluated by means of measuring induction time, crystal yield, crystal sizes, and size distribution.

The induction time was determined from peptide concentration by UV-vis spectroscopy every 20-30 minutes and compared to optical microscopy images. These images were analysed using a novel automated evaluation-routine we developed to evaluate the crystallisation behaviour and crystal properties. This new computer-based image evaluation-routine enables the autonomous processing of hundreds of peptide crystal images. Thereby, commonly applied image processing steps, such as background noise filtering and image-gradient based edge determination for obtaining the crystal contours, have been utilised to determine the macro-scale crystal properties, such as crystal size distribution. This novel evaluation-routine is less expensive, able to calculate the macro-scale crystal properties within a couple of minutes, and only requires tens of microlitres of the crystal suspension instead of tens of milligrams of solid crystals, which is required in commonly used analytical techniques such as dynamic light scattering. Furthermore, it enables the evaluation of hundreds of crystals with sizes from the micrometre to the tens of centimetres scale. This makes the evaluation-routine suitable for studying macro-scale crystal properties for expensive and poorly soluble peptides in-situ, allowing to optimise crystallisation parameters during the crystallisation process. Comparison of crystal size distribution obtained with the new routine and manually obtained crystal sizes was used to optimise and identify problems within the image processing steps.

Monitoring the peptide's concentration gradient in crystallisers with volumes of up to 15 millilitres found that an increase in oscillatory shaking speed from 50 to 150 rpm leads to a reproduceable higher homogeneity of the peptide concentration within the crystalliser resulting in a higher consistency of the crystallisation processes. Additionally, higher shaking speed lead to a faster decrease in peptide concentration implying a reduce in induction time and increase in crystal growth rate. These results can be supported by investigation of the crystal size distribution, obtained with the evaluation-routine, showing that a shaking speed of 150 rpm results in a more uniform crystal size distribution with a smaller span whereas a shaking speed of 50 rpm results in a wider spread size distribution. Furthermore, the reproducibility of controlling nucleation with templating additives, such as amino acids, can be enhanced significantly when a faster shaking speed is utilised, leading to a more homogeneous distribution of the additives. Mounting the tubular crystalliser in different orientations to the mixing plane found that if the axial direction of the crystalliser is aligned to the mixing plane, higher mixing length scales can be achieved resulting in an increase of mixing efficiency while keeping the energy input constant.

This study demonstrates the importance of mixing to achieve a high reproducibility of the crystallisation process in crystalliser volumes up to the tens of millilitres for peptides with low solubilities and molecular weights. Higher shaker speed enhances the mass transfer leading to a faster crystallisation process and a narrower crystal size distribution. To study the impact of mixing on the macro-scale crystal properties in-situ, hundreds of crystal images in combination with a novel developed evaluation-routine were evaluated.

Bibliography

1. Kastin, A.J., *Handbook of Biologically Active Peptides*. 2 ed. 2013: Academic Press. 2032.
2. Guillemot-Potelle, C., et al., *Cost of Goods Modeling and Quality by Design for Developing Cost-Effective Processes*. BioPharm International, 2010. 23(6): p. 26-35.
3. Yang, H., et al., *Development and Workflow of a Continuous Protein Crystallization Process: A Case of Lysozyme.* Crystal Growth & Design, 2019. 19(2): p. 983-991.
4. Roque, A.C.A., et al., *Anything but Conventional Chromatography Approaches in Bioseparation.*Biotechnology Journal, 2020.

609025 The Use of Nanofiltration Systems in Synthetic Peptide Processes

Abstract:

Tangential flow filtration (TFF) process such as microfiltration and ultrafiltration has been widely used in the pharmaceutical industry for biologics to remove smaller molecular weight process waste from the proteins. Due to the nature of these membranes, this unit operation has been limited to aqueous systems and larger molecular weight assets. Recent advances in membrane technology have enabled the fabrication of membranes with significantly smaller pore sizes and composed of materials with markedly improved compatibility to organic solvents. These changes make TFF systems a useful alternative to precipitation, distillation and solvent exchange operations in synthetic chemistry steps in small molecules and peptide processes, and in particular for amorphous or thermally labile materials.

This presentation will highlight the use of nanofiltration as part of a synthetic peptide process to remove small molecular weight by-products, to concentrate reaction solution at low temperatures and to perform a solvent exchange between chemical transformation steps.

603722 Design Space Via Symbolic Computation

Abstract:

Design space (DS) is a very important concept in the pharmaceutical industry. The importance of DS definition lies in the assurance of quality, a key goal of Quality by Design (QbD), and in the broadening from an acceptable operating setpoint to a collection of tolerable operating regions [1]. Defining proper limits for the design space is vital to provide manufacturing flexibility and quality assurance. The design space is determined by the uncertainty in the model parameters along with the allowable set of conditions in the process parameter space [2]. Flexibility index is an important measure to evaluate the operability of a design model with uncertainties [3], and it also provides a way to describe the design space; for example, the design space can be approximately regarded as a largest inscribed hyperrectangle or hyperelliptic. In addition, if the nominal conditions are not given, the flexibility index can be extended to the design centering problem [4], which is another important problem of DS definition, where it might be desirable to obtain optimal nominal conditions that maximize the feasible region of operation. The ultimate goal of flexibility analysis is to accurately and explicitly describe the design space, regardless of the convex and nonconvex nature of the design space.

Design space description, flexibility index, and design centering are viewed as three different issues in the DS field, and they are commonly formulated as multi-level optimization models. In general, these issues can be handled by numerical computation methods; for example, through the KKT conditions and complementarity conditions, the multi-level optimization models can be transformed into MILP/MINLP models [5]. However, it is well known that it is difficult to find the global solutions of an MINLP model. Moreover, the numerical methods can only estimate the outer envelope of the design space, rather than providing an accurate description, especially for the nonconvex cases. Mathematically, the expressions with variables can be obtained through the symbolic computation methods, which is very different from the numerical methods.

As we will show in this presentation, the original design model can be obtained directly in the space of the uncertain parameters through the proposed symbolic computation method; thus, it is unnecessary to solve numerical optimization problems. In order to provide an accurate description of the design space, we propose to reformulate the traditional flexibility analysis problem as an existential quantifier model. Then, we introduce a symbolic computation method, cylindrical algebraic decomposition (CAD) [6, 7] to eliminate the quantifiers and transform the flexibility analysis formulation into a series of explicitly triangular formulas. Each of these formulas represents a subspace of the design space, and the logical combination of them constitutes the complete symbolic representation of the design space projected in the space of the uncertain parameters. We show that with this symbolic computation method, the nonconvex design space can also be derived effectively.

Symbolic computation provides a novel way to deal with the characterization of a design space. Moreover, the proposed unified symbolic framework can integrate the three approaches described above for design space definition. These three problems have a similar symbolic model, and the only difference is that the dimension is increased successively, and the corresponding computational complexity is also increased successively. For the design space description problem, through the CAD method, the original high-dimensional feasible space can be projected on the space of the uncertain parameters. For the flexibility index problem, the feasible space can be projected onto the one-dimensional space of flexibility index, and all of the candidate values of flexibility index can be generated at the same time. For the design centering problem, the feasible space can be projected onto the space of the flexibility index and the uncertain parameters, and a rule to locate the final nominal point is developed, which only requires a series of simple numerical operations. In summary, the proposed method for design space has four properties, (1) it is a unified symbolic framework; (2) it can avoid solving numerical optimization problems; (3) it can guarantee finding the optimal flexibility index or nominal points; (4) it can provide explicit algebraic expressions of the design space. A design space may be constructed for single unit operation, multiple unit operations, or for the entire process. However, as the scale of the design model increases, the complexity is greatly increased. For reducing the complexity of symbolic computation, we propose to use the adaptive sampling method to construct the surrogate model and eliminate the equalities of the model. Then, the CAD method can be adapted to obtain the explicit expressions of the design space, and the boundary of the design space can be validated and revised iteratively to satisfy a specified accuracy. Using several examples from the pharmaceutical industry, we illustrate the application of the proposed symbolic computation method to show its novelty and value.

Keywords: Design space, flexibility analysis, design centering, cylindrical algebraic decomposition, symbolic computation.

Reference:

1. Pramod K, Tahir MA, Charoo NA, et al. Pharmaceutical product development: A quality by design approach. *Int J Pharm Investig*.2016;6(3):129.
2. Rooney WC, Biegler LT. Optimal process design with model parameter uncertainty and process variability. *AIChE J*. 2003;49(2):438-449.
3. Swaney RE, Grossmann IE. An index for operational flexibility in chemical process design. Part I: Formulation and theory. *AIChE J*. 1985;31(4):621-630.
4. Harwood SM, Barton PI. How to solve a design centering problem. *Math Methods Oper Res*. 2017;86(1):215-254.
5. Grossmann IE, Floudas CA. Active constraint strategy for flexibility analysis in chemical processes. *Comput Chem Eng*. 1987;11(6):675-693.
6. Collins GE. Quantifier elimination for real closed fields by cylindrical algebraic decomposition. *Lec Notes Comp Sci*. 1975;33:134-183.
7. Arnon DS, Collins GE, McCallum S. Cylindrical algebraic decomposition I: the basic algorithm. *SIAM J Comput*. 1984;13:865-877.

602090 A Multi-Scale Hybrid Modeling Approach for Continuous Pharmaceutical Unit Operations

Abstract:

A prominent research focus for systematic process development in the continuous pharmaceutical industry is the implementation of Quality-By-Design [1] (QbD) approach. This approach focuses on detailed system understanding from a mechanistic perspective, using high-fidelity multi-physics simulation tools like computational fluid dynamics (CFD) [2] and discrete element modeling (DEM) [3]. These tools can accurately replicate the process dynamics within a unit operation, providing insight into complex processes, such as powder mixing and flow. However, due to huge computational costs associated with such simulations, direct implementation of these tools is prohibitive for quick process evaluations. On the other hand, flowsheet modeling focuses on empirical, low-order models using simplified representations of process dynamics. Though it provides quick assessment, it fails to capture important process complexities, such as powder mechanics. These underlying differences create a need for a synergistic integration of high-fidelity simulation models for capturing complex process dynamics with predictive flowsheet models for fast evaluations.

In this study, the proposed integration strategy is demonstrated for continuous blending unit operation using a sequential hybridization approach [4]. Under the proposed approach, particle-scale information [5,6] from DEM simulations (using Simcenter STAR_CCM+ software) is obtained to develop a data-driven model. This data-driven model is then combined with a flowsheet model (developed in PSE gFORMULATE modeling platform [7]) to provide fast predictions for the entire production line. The combination of DEM simulation with flowsheet modeling via a data-driven approach is performed such that any process changes occurring in the process flowsheet are recorded and replicated in the DEM simulation. This allows DEM simulation to capture dynamic effects of process changes on particle mechanics, which is then transferred to update the data-driven model linked to the flowsheet. This approach allows the flowsheet to incorporate effects of real-time process dynamics on powder behavior. The proposed study focuses on the application of this approach for evaluation and prediction of cross-sectional blend uniformity within the blender. Given the difficulties encountered in experimental measurement of cross-sectional blend uniformity for continuous blenders [8,9], the proposed approach provides an alternative for accurate evaluation of blend uniformity with detailed insights into the effects of process design and operating parameters. Blend uniformity is a critical quality attribute for tablet manufacturing and its accurate evaluation can avoid non-conformity of the final product.

In conclusion, the proposed work focuses on incorporation of detailed particle-scale information obtained from high-fidelity simulations within flowsheet models using hybrid data-driven approaches. This increases the mechanistic understanding of powder systems while improving the predictive ability of flowsheet simulations. The integrated approach presented in this work for powder blending systems, can be extended for other unit operations for the construction of robust flowsheet models and allow the implementation of QbD for continuous pharmaceutical process development.

Reference:

[1] L.X. Yu, G. Amidon, M.A. Khan, S.W. Hoag, J. Polli, G.K. Raju, et al., Understanding Pharmaceutical Quality by Design, Aaps J. 16 (2014) 771–783.

[2] J.D. Anderson, J. Wendt, Computational fluid dynamics, Springer Berlin Heidelberg, Berlin, Heidelberg, 1995.

[3] P.A. Cundall, O.D.L. Strack, A discrete numerical model for granular assemblies, Géotechnique. 30 (1980) 331–336.

[4] M. von Stosch, R. Oliveria, J. Peres, S.F. de Azevedo, Hybrid semi-parametric modeling in process systems engineering: Past, present and future, Computers & Chemical Engineering. 60 (2014) 86–101.

[5] N. Metta, M. Verstraeten, M. Ghijs, A. Kumar, E. Schafer, R. Singh, et al., Model development and prediction of particle size distribution, density and friability of a comilling operation in a continuous pharmaceutical manufacturing process, International Journal of Pharmaceutics. 549 (2018) 271–282.

[6] D. Barrasso, A. Tamrakar, R. Ramachandran, Model Order Reduction of a Multi-scale PBM-DEM Description of a Wet Granulation Process via ANN, Procedia Engineering. 102 (2015) 1295–1304.

[7] Z. Wang, M.S. Escotet-Espinoza, M. Ierapetritou, Process Analysis and optimization of continuous pharmaceutical manufacturing using flowsheet models, Computers and Chemical Engineering. 107 (2017) 77–91.

[8] P.M. Portillo, M. Ierapetritou, F.J. Muzzio, Characterization of continuous convective powder mixing processes, Powder Technology. 182 (2008) 368–378.

[9] A.U. Vanarase, F.J. Muzzio, Effect of operating conditions and design parameters in a continuous powder mixer, 208 (2011) 26–36.

607035 Integrated Use of Digital Twin-Based Framework for the Development of a Lomustine Manufacturing Pathway

Abstract:

Traditionally, batch-wise operation has accounted for a large portion of small- and large-scale industrial pharmaceutical processes. However, the development of newer technologies through intensified unit operations for pharmaceutical manufacturing, as well as more precise modeling and online control, has accelerated over the past 10 – 20 years. In addition to these advancements, interest in implementing more continuous pharmaceutical manufacturing techniques in industry has also increased. With the adoption of new unit operations and operational modes, quality guarantees through initiatives such as Quality by Design [1] (QbD) and Quality by Control (QbC) must be maintained to ensure both consumer and operational safety through good manufacturing and modeling practices. In this work, we present a case study utilizing a start-to-finish framework for the full analysis of a high value, low volume pharmaceutical drug pathway with simulation-based alternative flowsheet comparisons. Our group is developing an end-to-end simulation and optimization framework for generating and comparing flowsheet alternatives for pharmaceutical drug manufacturing. The framework includes a custom model library and supports estimation of model parameters from experimental data.

In this presentation, the pharmaceutical drug is analyzed under various operational modes, i.e. batch and continuous for various unit operations. Process development was performed starting with parameter estimation from experiments. For example, crystallization kinetic parameters were estimated with metastable width zone determination at the lab scale in batch crystallizers [2] and nucleation monitoring in microfluidic nitrogen-segmented droplets [3]. We simultaneously developed a digital twin of the process using gPROMS FormulatedProducts to analyze the response of critical quality attributes to process disturbances and better characterize process dynamics under uncertainty. Subsequently, flowsheet simulation and dynamic process analysis via the digital twin was performed. The framework utilizes many computational platforms for analysis by exploiting strengths of different software environments. We utilize Python for parameter estimation and flowsheet simulation, Pyomo for unit operation and flowsheet optimization, and gPROMS FormulateProducts for the digital twin of the process. Thus, the fidelity between gPROMS FormulatedProducts and the custom model library was analyzed as well. The start-to- finish framework is shown to be robust in analyzing different operational modes and conditions in pharmaceutical processing. To demonstrate this capability, we analyze in detail one batch flowsheet, one continuous flowsheet, and one hybrid flowsheet. Here, we compare these flowsheet alternatives manually for proof-of-concept; however, this framework is ultimately to become an end-to-end optimization framework which will algorithmically generate and compare flowsheet alternatives, specifically for the comparison of batch, continuous, and hybrid operational modes.

1. Food and Drug Administration. *Pharmaceutical cGMPs for the 21st Century—A Risk-Based Approach*; Technical Report; U.S. Department of Health and Human Services, Food and Drug Administration, Center for Drug Evaluation and Research (CDER): Rockville, MD, USA, 2004.
2. Nagy, ZK; Fujiwara, M; Woo, XY; Braatz, RD; Determination of the kinetic parameters for the crystallization of paracetamol from water using metastable zone width experiments. *Industrial and Engineering Chemistry Research*. **2008**, 47 (4), 1245-1252.
3. Lu, J; Litster, JD; Nagy, ZK. Nucleation studies of active pharmaceutical ingredients in an air-segmented microfluidic drop-based crystallizer. *Crystal Growth & Design*. **2015**, 15 (8). 3645-3651

603536 Digital Design of End-to-End Manufacturing Process for Mefenamic Acid Using Mechanistic Modeling

Abstract:

Digital twins built using mechanistic models are playing an increasingly significant role in helping pharmaceutical industries develop robust and more economically efficient manufacturing processes. Building a digital twin for an end-to-end drug manufacturing process allows exploration of the individual and combined effects of numerous process parameters during the active ingredient and drug product manufacturing stages on the final drug product performance. This allows for the development of a more robust drug manufacturing process and greater assurance of product quality while reducing process development timelines and resources.

This work will outline the quantitative analysis performed on the manufacturing process for mefenamic acid by building the end-to-end mechanistic flowsheet model Each element of the drug substance production step, including synthesis step, crystallization and separation are validated using process data individually. Once each unit operation step in the production step is validated, a systems model was configured to provide a quantitative representation of the end-to-end production process including drug substance, drug product and product performance elements. This end-to-end model was subsequently utilized to develop a quantitative understanding of the effect of process disturbances, raw material variability, process parameters, formulation parameters and model uncertainty on CQAs (e.g. tablet properties) and manufacturability. Along with the analysis, the relative influence of those factors (sensitivity indices) on the desired CQAs will also be presented. Furthermore, the impact of batch and continuous operational elements of the process will also be probed. This work helps to identify any existing gaps in the scientific understanding of the effects of some of the API attributes on the desired KPIs during this end-to-end flowsheet analysis will also be discussed.

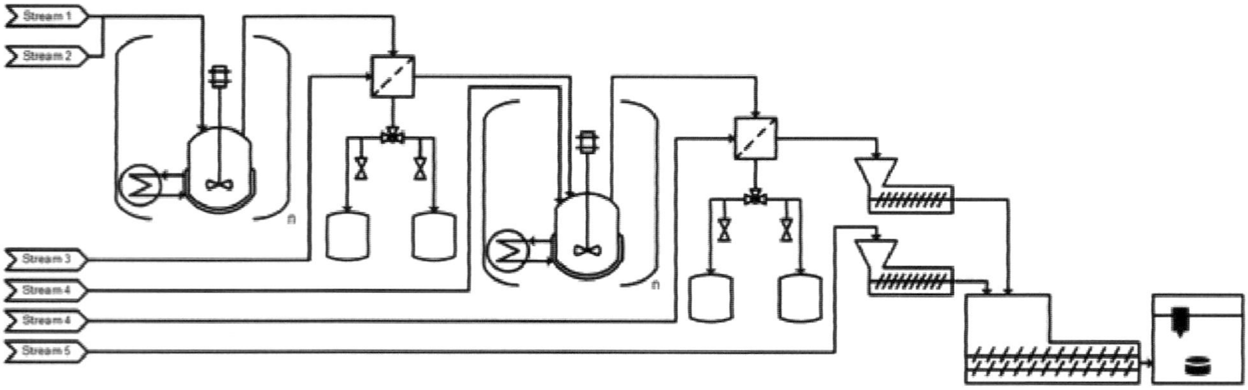

609946 Digital Design of an End-to-End Drug Manufacturing Process Using Mechanistic Modeling

Abstract:

Digital twins built using mechanistic models are playing an increasingly significant role in helping pharmaceutical industries develop robust and more economically efficient manufacturing processes. Building a digital twin for an end-to-end drug manufacturing process allows exploration of individual and combined effects of numerous process parameters and material attributes within the end-to-end process. The end-to-end process includes the active pharmaceutical ingredient (API) and drug product manufacturing stages as well as the final drug product performance. This can allow for the development of a more robust drug manufacturing process and greater assurance of product quality while reducing process development timelines and resources.

This work details an actual application and analysis performed on an existing UCB drug product manufacturing process by building the end-to-end flowsheet model in gPROMS FormulatedProducts, shown in Figure 1. This analysis includes qualitative understanding of the effect of process disturbances, API variability, process parameters, formulation parameters and model uncertainty on tablet properties, and manufacturability key performance indicators (KPIs). Along with this analysis, the relative influence of those factors (sensitivity indices) on the desired tablet properties are presented.

Critical gaps were identified in the scientific understanding of the effects of some of the API and granule attributes on the desired tablet properties and KPIs (e.g. mechanistic link of the particle size distribution to the bulk density of the powder impacting on cohesive forces of powder during roller compaction operation, critical fine fraction generated during the particle forming steps leading to segregation of the powder during powder charge operation and impact on content uniformity, etc.). The importance of data integrity, breaking silos across departments and ensuring correct operating ranges used in the analysis are some of the lessons learnt when developing a digital twin.

This work acts as a starting point for UCB to begin utilizing the end-to-end approach with gPROMS FormulatedProducts as a platform for future drug development pipelines. It also opens up new avenues for scientific understanding in the mechanistic models employed to address the gaps identified during the project (e.g. availability of critical material characterization information for fully exploiting the early adoption of these digital twins in a model risk assessment approach and analysis the feasibility and manufacturability of the drug product production processes). As regards gaps in science, these will be addressed either through improving mechanistic understanding or incorporating data driven elements in the overall mechanistic end-to-end process model.

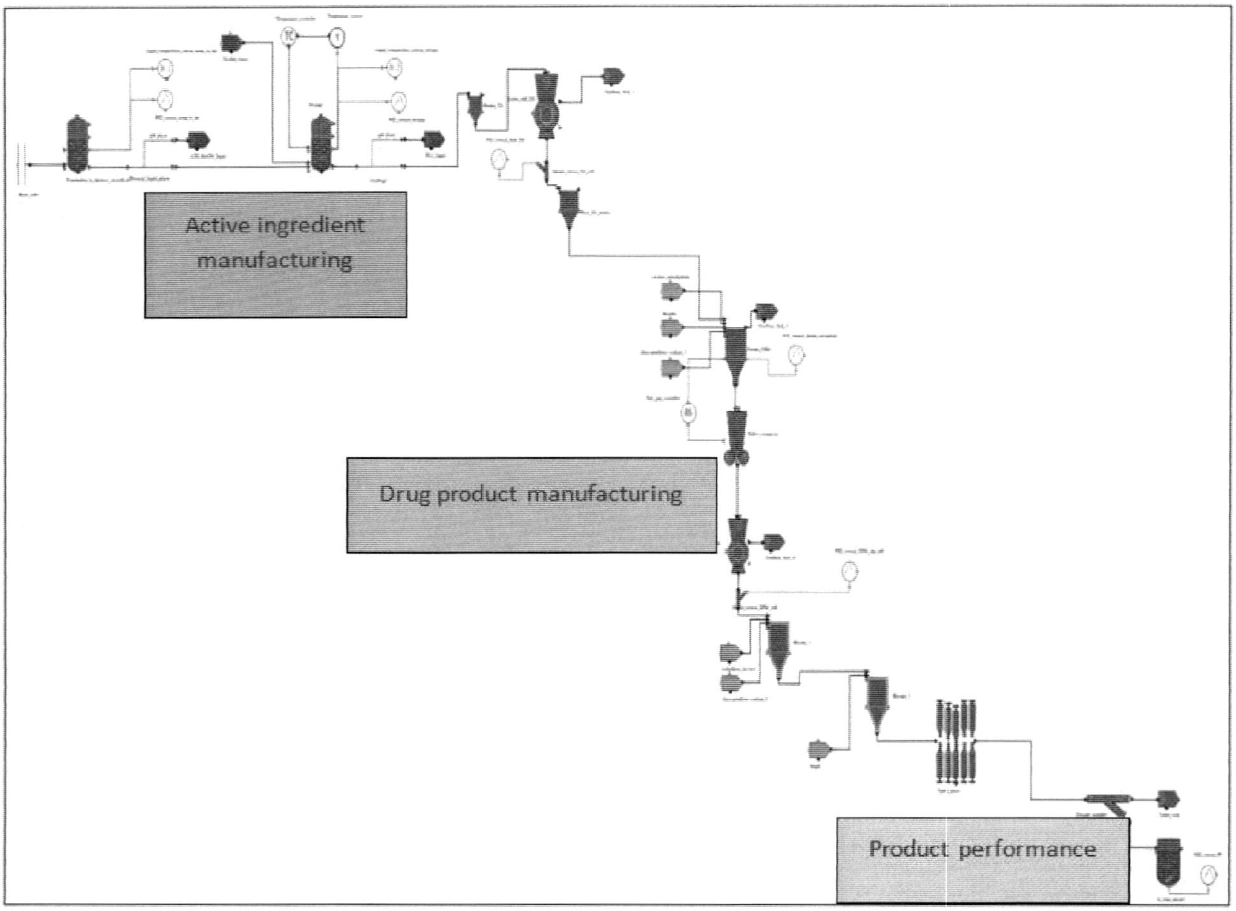
Active ingredient manufacturing
Drug product manufacturing
Product performance

602450 Digital Design of a Dry Granulation-Based Drug Product through Integrated Manufacturing Process and Product Performance Modeling

Abstract:

Motivated by aims to improve R&D efficiency, reduce time to market, and develop robust drug products, the pharmaceutical industry has seen increased application of mechanistic-based system models. Such approach establishes science-based understanding of relationships between various process parameters and Quality Attributes (QAs) with reduced experimentation requirements.

The primary objective of this study is to demonstrate system model of a dry granulation-based drug product manufacturing process with *in-vitro* dissolution. This approach is used to determine the effect of drug product manufacturing process on tablet (API) dissolution. The system model, which was developed in gPROMS FormulatedProducts, couples multiple unit operations together, namely roller compaction, screen milling, lubrication, tablet compression, and *in vitro* dissolution. The roller compactor and the tablet press were calibrated by estimating the respective compaction-related model parameters against experimental data for ribbon and tablet properties. In order to capture the effect of the disintegration of the tablet solid matrix on the delayed API dissolution, a hierarchical description of the tablet and a porosity-dependent disintegration scheme was adopted. The *in-vitro* dissolution was calibrated by estimating the tablet and granule disintegration kinetic parameters in order to capture the dissolution profile of the API.

Following model development activities, the system model (Figure 1) was used to assess the impact of the drug product manufacturing process parameters on API dissolution. Monte Carlo and variance-based analyses were performed using gPROMS FormulatedProducts' Global System Analysis framework to understand the relationship between drug product process parameters and API dissolution. Such framework can be effectively used in the workflow of pharmaceutical product development to define the input vs output relationships between process parameters and product performance. Furthermore, it has the potential to significantly contribute to QbD framework.

Figure: An integrated drug product and *in-vitro* dissolution model in gPROMS FormulatedProducts

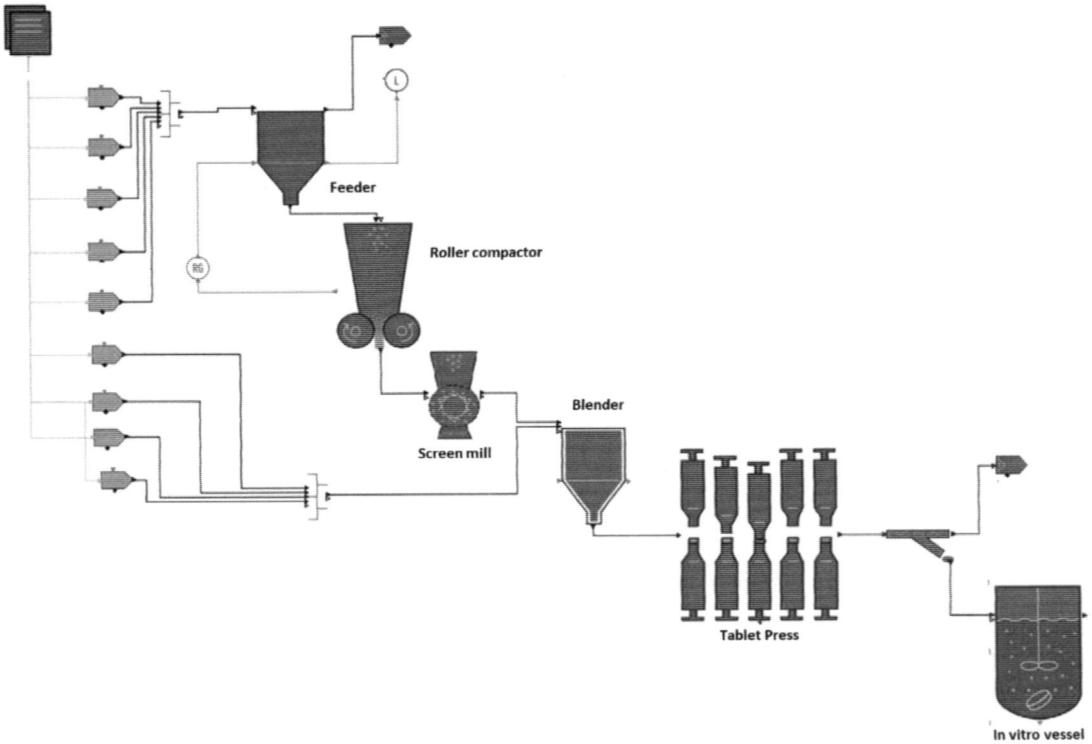

597932 A Mechanistic Model-Based Digitalisation Framework for Continuous Manufacturing Processes

Abstract:

Due to the improvement of manufacturing technologies and publicised industrial cases realising the potential value of adopting continuous manufacturing in pharmaceutical production, much of the research focus has been realigned to the understanding of integrating batch-wise operations into a single production line. However, there is little regulatory instruction and good model-based principles to follow. From a modelling point of view there is a lack of coherence in approach due to the profound lack of understanding of the process dynamics. Despite a significant adoption of particulate system modelling approaches, it is apparent that very few existing, science-based models from academia are fully explored and implemented for industrial practice. Furthermore, while a generic modelling platform for integrated modelling of an industrial process could be an enabling technology, it can be formidable for an industrial practitioner to apply such digital tools.

This paper aims to develop a systematic framework for approaching digital design of a continuous manufacturing process by transferring state-of-the-art innovations from the modelling world into industrial practice through a mechanistic model-based digitalisation platform. The digitalisation framework is built upon gPROMS using a defined case explored in the DiPP (Diamond Pilot Plant), a continuous powder processing in the University of Sheffield, to demonstrate the usability and expandability of the mechanistic model approach. The DiPP of a Consigma^{TM}-25 manufacturing line from powder to tablet consists of twin screw granulation, fluid bed dryer, cone mill, vertical blender and tableting compaction. Global System Analysis of the DiPP with sound engineering principles defined in underlying mechanistic models is carried out to enable a comprehensive exploration of the systematic responses and the development of a robust operating space. The systematic modelling framework will enable the pharmaceutical industry to effectively and efficiently design and implement continuous pharmaceutical manufacturing processes, leading to safer and faster design and operating decisions.

612384 Data-Rich Experimentation at Merck. a Retrospective on a 6year Journey.

Abstract:

Since its inception in early 2014, the Process Development Intensification Laboratory (iLab) within Small Molecule Process Research & Development at Merck has been focused on the development of tools and techniques to enable acquisition of deeper, more fundamental process understanding in a resource sparing manner. This is predominantly centered around the concept of data intensification, specifically, around enabling more efficient data capture, data reduction and data analysis. Since 2014, we have grown to call these activities Data-Rich Experimentation, or DRE, and these investments have spawned a significantly broader strategic initiative, across large and small molecule Process R&D, which we eloquently call "the DRE initiative".

In this submission, we will look back on the development, demonstration, and perhaps more importantly, the deployment of novel tools designed to increase the information density of the experiments we conduct, and the efficiency of the chemists and engineers who conduct them. We will highlight the installation of novel, enabling technologies such as process modeling, bench-top automation, medium and high-throughput experimentation, process analytical technologies and statistically designed experimentation. Furthermore, this submission will emphasize the importance, as well as demonstrate the utility, of application of such tools across scales typically encountered in a process R&D setting. We will also provide perspective on future directions the team is moving.

604842 Embedding Automation in the Pharma Research Laboratory to Enable Drug Substance and Product Development

Abstract:

Automation and High Throughput Screening are helping to drive innovation in drug development. Advances in technology and collaboration have opened the door to a wide range of physical and digital automation solutions to produce higher quality data at a much faster pace. Modular Solid and Liquid handlers along with a range of integrated analytical equipment offer both flexibility in experiment design as well as coordination among different users. Solubility and stability screens create robust synthesis and isolation processes on an accelerated timeline. Automated parallel reactor systems allow for controlled pressurized reactions and timed sampling, enabling kinetic reaction screens and Design of Experiment (DoE). Data tools such as Spotfire give options to better organize, analyze, and visualize data, enabling faster recall of information and better decision making. How these technologies have been implemented and the benefits they have on research timelines, data quality, and productivity will be presented.

609402 Development a Workup for Removing Triphenylphosphine Oxide from a Mitsunobu Coupling

Abstract:

Development a Workup for Removing Triphenylphosphine Oxide from a Mitsunobu Coupling

Eric G. Moschetta, Benoit Cardinal-David, and Moiz Diwan

AbbVie, Inc. Process R&D 1401 Sheridan Road North Chicago, IL 60064, USA

Triphenylphosphine is a common and useful reagent in many organic reactions, such as the Wittig reduction, the Mitsunobu coupling, and the Staudinger reaction, among others. Additionally, triphenylphosphine is used often as a stoichiometric reagent in these reactions, resulting in a stoichiometric amount of triphenylphosphine oxide, the main byproduct of such reactions. Triphenylphosphine oxide (TPPO) is notoriously difficult to remove from reaction mixtures, especially at such large concentrations. Chromatography works to purify reactions in the laboratory, but is impractical for the larger scales required to manufacture drug substance. Extraction is not a viable option, either. TPPO is poorly water soluble and would lead to a solvent inefficient process (nonpolar solvents are common media for Mitsunobu couplings). Process chemists in pharmaceutical development often avoid implementing reactions where TPPO will be a byproduct because of the challenges associated with removing it from the reaction mixture. Bypassing such reactions limits the synthetic tools available for API process development and may lead to the implementation of longer synthetic routes, thereby creating more wasteful processes and requiring more time and effort to develop the process. As such, developing a scalable workup to remove TPPO from such reactions presents a valuable opportunity for pharmaceutical process development to enable reactions where triphenylphosphine is used as a stoichiometric reagent.

This presentation describes the development of a procedure to remove TPPO by reacting it with solid magnesium chloride. This protocol was developed on a Mitsunobu coupling in toluene. Magnesium chloride and the resulting magnesium chloride-TPPO complex are virtually insoluble in toluene, creating an intrinsic biphasic system. This presentation will detail the key features of the solid-liquid reaction and the factors that control the rate of reaction as observed with on-line IR spectroscopy. The rate of reaction was observed to be very fast upon addition of solid magnesium chloride, with a sharp decrease in the rate of reaction seconds after addition. Based on this observation, it was hypothesized that employing a surface regenerating technique to expose fresh magnesium chloride for reaction with TPPO would increase the rate of reaction. Including surface regeneration significantly increased the rate of reaction compared to conditions without. The protocol was then implemented on a model Mitsunobu coupling to demonstrate that the workup had no impact on the impurity profile and could effectively remove TPPO from the reaction mixture.

Eric Moschetta, Benoit Cardinal-David, and Moiz Diwan are employees of AbbVie and may own AbbVie stock. AbbVie sponsored and funded the study; contributed to the design; participated in the collection, analysis, and interpretation of data, and in writing, reviewing, and approval of the final publication.

600699 Drying Process Characterization of a Novel Continuous Vacuum Drying Technology

Abstract:

In the last years continuous technologies have emerged in the pharmaceutical industry, but still some steps are not sufficiently developed. One of those technologies is the last step in the active pharmaceutical ingredient (API) production, drying. Suitable, continuous equipment is not available, especially for the case of cohesive, poorly flowable and thermosensitive materials, as many APIs are. Such a technology would be promising for a truly continuous primary manufacturing route and with attractive particle properties. Additionally, the time to market can be reduced significantly, and scale-up and scale-down considerations can be neglected.

Drying is an especially challenging process, as it involves simultaneous mass and heat transfer and can have challenging consequences for the particle structure and morphology which eventually control powder behavior. In addition, agglomeration and attrition compete. This is an undesired effect, as during crystallization the particle properties are usually fixed in an optimal way for the intended use of the API. Often, the drying process changes the tailored particle properties, as particle size distribution (PSD) and morphology.

This presentation will show the concept of a continuous drying process, using a newly developed dryer design which intends to overcome these challenges by a unique approach of balancing forced feed, powder bed motion and residence time. The results indicate that the particle properties are maintained to a large extent. The influence of several process parameters (mass flow, air flow, rotational speed, temperature, inlet moisture) is initially evaluated, and the product morphology and PSD is investigated.

As further improvement to the dryer, a vacuum-tight design was implemented, to enable the dryer to dry off organic solvents and thermosensitive materials. This enables the dryer to continuously operate at the intended scale, in an industrially relevant way. The governing drying kinetics are investigated under vacuum and the drying efficiency and limitations of the systems are discussed.

As main test substance Ibuprofen was used, with varying inlet moisture levels (10 to 50 wt. %). It was assured that the technology is able to dry cohesive powders with a small particle size (<100 μm) continuously, in the range of 0.5 to 2.0 kg/h, dry basis. In this range, the PSD was maintained and the residual moistures of the product for some process configurations could be reduced to below 1 wt. %. Additionally, the residence time distribution (RTD) was investigated throughout the process and scanning electron microscopy (SEM) was used to classify the morphology prior and after the drying step.

In summary, the presented continuous technology gives a robust drying process, suitable to dry temperature-sensitive cohesive particles at low mass flows under vacuum.

597739 Predicting Optimal Salt Forms for Active Pharmaceutical Ingredients Using the SAFT-γ Mie Equation of State

Abstract:

The pharmaceutical industry is facing a constantly rising demand for drugs of growing complexity and more efficient manufacturing processes. The solubility of pharmaceuticals is a key property during the drug formulation and the subsequent design of the processes involved in the manufacturing of the drug. Common challenges in these manufacturing processes include: the large number of experiments required and the extremely low solubilities of active pharmaceutical ingredients (APIs) in water for which experiments are difficult to perform. Computer-aided approaches provide an attractive alternative to performing time-consuming and costly experiments. In this context, molecular modelling approaches can deliver physical properties predictively are key enabling tools.

The SAFT-γ Mie group contribution (GC) equation of state (EoS) [1, 2] is such a predictive thermodynamic modelling technique. In the SAFT-γ Mie framework, molecules are modelled as heteronuclear chains formed from fused spherical segments, which represent the distinct functional groups comprising the molecule. In this framework, it is assumed that the properties of a molecule or a mixture can be determined from the weighted contributions of the functional groups present in the system of interest, with the assumption that the parameters characterizing the functional groups are fully transferable across molecules.

We demonstrate the validity of the SAFT-γ Mie EoS in the prediction of thermodynamic properties of ionizable APIs. The pH-solubility profiles and the complete phase diagrams of the APIs in water, outlining the solid-liquid, liquid-liquid and vapour-liquid equilibria for these mixtures, are successfully predicted. It is well known that the bioavailability of a drug is improved by salt formulation, which is the preferred method to enhance the solubility of ionizable drugs. The SAFT-γ Mie EoS accounts for the complex speciation phenomena that take place under pH changes including partially ionised (weak electrolytes) systems. We demonstrate the effect on the pH-solubility profile of ibuprofen of multiple inorganic counter ions including sodium, potassium, lithium, calcium and magnesium. This is followed by studying the impact of organic counter ions, particularly the amines, on the pH-solubility profile of ibuprofen and the selection of its optimal salt form.

[1] Papaioannou, V. *et al. J. Chem. Phys.* **140**, (2014).

[2] Dufal, S. *et al. J. Chem. Eng. Data* **59**, (2014).

610442 Elucidation of Local Drug Release Dynamics in Complex Formulations with Multi-Nuclear NMR Micro-Imaging

Abstract:

Purpose: Concurrent visualization of liquid transport, drug mobilization and release process through complex polymer structures is of great scientific significance and challenge in the development of advanced small molecule formulations. Such elucidation of the local, non-linear dynamics of drug release not only can support formulation design and optimization, but also reveals potential risk factors for drug stability that is influenced by local transient phenomena. This work discusses the advancement of quantitative multi-nuclear (^{1}H and ^{19}F) magnetic resonance imaging (MRI) and spectroscopic techniques, applied to revealing drug dissolution and mobilization dynamics inside a long-acting implant formulation in order to support formulation design and risk mitigation.

Methods: Long-acting implant samples, consisted of an active drug, MK-A, and ethylene-vinyl acetate (EVA) copolymer, were placed in dissolution media under controlled conditions and imaged periodically with a 400 MHz NMR Bruker Avance spectrometer. At each time point, ^{1}H NMR imaging (at 20 mm pixel resolution) and spectroscopy were used to observe water concentration and mobility; ^{19}F pulsed-field gradient (PFG) NMR spectroscopy was also used to observe drug mass transport, via its molecular self-diffusion (D^{self}), within the polymer matrix.

Results: Clear signals of the dissolution media were observed via ^{1}H MRI, showing the ingress of the media into the implant via a porous polymer structure created by dissolution of the drug. Analysis of the ^{1}H images reveal a dissolution medium *transport diffusivity*, D^{T}, of around 10^{-13} m^2s^{-1}.The pores are initially created as the drug dissolves into the ingressing dissolution media and thereafter provides a pathway for the drug to leave the implant. Concurrently, ^{19}F NMR spectroscopy reveals that the solid drug dissolves into the media with a square root time dependence within the pores. D^{self} of the dissolution media inside the pores was found to be 2 orders of magnitude higher than the transport diffusivity, D^{T}. ^{19}F PFG-NMR measurements of the dissolved drug molecules within the porous structure exhibit a similar self-diffusivity as the dissolution media, suggesting the concurrent diffusion of these mobile species in the porous system. We tentatively assert that both the transport diffusivity, D^{T}, of the dissolution media ingress into the polymer matrix and the self-diffusivity, D^{self}, of the drug are key descriptors that govern the overall drug release rate into the surrounding dissolution medium. The characterization of such drug mobility provides a mechanistic understanding of the slow drug release as a design feature of such implants. In addition, changes in the ^{19}F signal from within the implant with extended media exposure are indicative of recrystallization of the drug, pointing to the potential stability risks that such a slow drug release can carry.

Conclusions: The concurrent elucidation of the dynamics of both active drug and the dissolution media, made possible by advanced multi-nuclear NMR micro-imaging technology, provides unique perspectives of the local release dynamics inside the implant formulation, and greatly empowers formulation scientists to optimize the formulation design and manufacturing process based on detailed mechanistic understandings.

610889 Pharmacy on Demand: An End-to-End Approach to Developing a High-Dose, Poorly Flowing API into a Robust Continuous Manufactured Formulation

Abstract:

The Pharmacy on Demand (PoD) project enables an end-to-end continuous manufacturing (CM) of API using four refrigerator-sized, portable units; synthesis, purification, crystallization and drug product. High-dose, poorly flowing API always represents a challenge for developing a robust continuous manufacturing (CM) process for solid dosage forms. We explored the nature of powder flow of Ciprofloxacin HCl and investigated multiple techniques in order to maximizing flow properties. Multiple feeders were studied, and an optimum design was selected to facilitate formulation of the targeted batch size of 1000 doses *per* day. Since our goal is to produce ciprofloxacin using a continuous process on a mobile platform, a simple formulation was implemented. It consisted of a filler, disintegrant, glidant and a lubricant. We investigated the effect of excipient types and levels on the formulation manufacturability and performance. Disintegration, *in-vitro* dissolution, tablet mass variability, hardness and assay were used as performance measures which are required to pass the United States Pharmacopeia (USP) standards. Moreover, the dissolution profiles for the tested formulations were compared to the reference listed drug. The chosen formulation passed the USP standards criteria for tablets friability <1%, % label claim (90-110%), and an *in-vitro* dissolution profile reaching 85% dissolved within 30 minutes. The selected formulation robustness and design space were tested by varying the input ciprofloxacin properties in terms of water content and particle size distribution (PSD) and evaluated by testing drug product (DP) performance. The selected formulation showed robustness across all the tested levels of variables and passed appearance and depicted great control of the module over online tablet masses within less than ±5% standard deviation of the target tablet mass (500 mg). In the near future, the developed formulation will be manufactured in a GMP facility for a clinical bioequivalence study required for an abbreviated New Drug Application (aNDA) filling.

599048 Preparation of Solid Dispersions By Spray Drying and Electrostatic Precipitation

Abstract:

Introduction

The search for new active pharmaceutical ingredients (API) has gone through an important development over the last few decades. Especially by using technologies like high throughput screening and combinatorial chemistry, improving solubility and thereby oral bioavailability of drugs in development has become a main challenge for pharmaceutical development [1]. About 40 % of the drugs in the market are considered to be poorly soluble, but regarding the substances in development, the percentage of BCS class II and IV drugs is expected to be much higher [2]. Therefore it is necessary to develop new formulation techniques to enhance solubility and thereby bioavailability of these drug substances.

In this work submicron API particles are produced with ultrasonic atomization technique in a specially designed aerosol generator. As nanoparticular aerosols create a hazard to the environment, and frequently show reduced wettability, the generated particles are charged and deflected into a carrier melt in the melt electrostatic precipitator (MESP) [3]. Thereby the hazardous material is handed safely on the one hand. On the other hand the drug particles are embedded in a water soluble carrier and show an enhanced dissolution in water. The concept of the MESP and first experiments have already been discussed in previous work [3]. The aim of the present work is to characterize the MESP with a new designed aerosol generator in order to obtain a product with a higher drug load than it has been achieved before with a spray drying plant with aerosol conditioning [4].

Materials and Methods

Materials

Phenytoin (Recordati Pharma GmbH, Ulm, Germany) was dissolved in acetone (Merck KGaA, Darmstadt, Germany) for the spray drying process. The sugar alcohol xylitol (Xylisorb 300, Roquette Pharma, Lestrem, France) was chosen as carrier matrix. Drug loads were determined by UV-Vis spectroscopy in a mixture of isopropyl alcohol (Carl Roth GmbH & Co. KG, Karlsruhe, Germany) and water (1:1).

Methods

Particle size and shape were investigated with scanning electron microscopy (SEM) (H-S4500 FEG, Hitachi, Krefeld, Germany) and laser diffraction measurement (Spraytec, Malvern Panalytical, Malvern, UK) right after spray drying.

The drug load in the xylitol melt was analyzed. The chilled and recrystallized melt was dissolved in a mixture of water and isopropyl alcohol (1:1). The phenytoin content was measured with UV-Vis (Jenway 7305, Stone, UK) at a wavelength of 212 nm. The loading time was kept constant at 5 min, while the applied voltage was varied above the corona inception voltage and below the corona breakdown voltage.

Dissolution experiments were conducted in United States Pharmacopeia (USP) Apparatus 2 (DT 6, Erweka, Heusenstamm, Germany) in demineralized water of 37 °C at a stirrer speed of 50 rpm. The drug-laden xylitol was crushed and sieved with a mesh size of 400 µm. For comparison a physical mixture of xylitol and phenytoin was prepared using a turbula mixer (Turbula T10B, W.A. Bachofen AG, Muttenz, Switzerland).

Results

The Process

A spray drying plant for the production of submicron particles and the use of organic solvents was designed. A piezo crystal (WHQ 3005/1530-12N, Siansonic Technology, Bejing, China) operating at a high frequency of 3 MHz, was used to produce an aerosol of small, uniform droplets. The drug solution was pumped constantly out of a feed tank into the aerosol chamber in order to cover the piezo ceramic. The liquid level was kept constant throughout the nebulization process. To separate larger droplets out by a vortex and to transport small, gas-borne droplets upwards, carbon dioxide was let in tangentially. A particularly low volume flow was chosen, as the droplet size is independent of the volume flow, but increases the separation efficiency in the melt electrostatic precipitator (MESP). The droplets were transported upwards through a dip pipe into a drying stage, where the solvent was evaporated and removed from the aerosol by condensation. The particles were collected on a glass surface for investigation in scanning electron microscope (SEM) (Figure 1).

The particles were transported into the separation stage, the MESP. The concept of the MESP has been described in a previous study [3]. It consists of two stages. In the first stage the entering particles were charged by applying high voltage of 5 to 9 kV in order to cause a corona discharge. In the second stage the particles were separated out of the gas stream by deflecting them towards the collecting electrode. The collecting electrode was covered with a melt of xylitol, so that drug particles were embedded (Figure 1) [3].

Particles Size and Shape

The spray dried API particles were characterized by scanning electron microscope (SEM) and laser diffraction measurement. One representative SEM picture is shown in Figure 2. The particles are rectangular or rhombic shaped and thereby indicate a crystalline state [5,6] (Figure 2a). Laser Diffraction measurements were conducted in three measurements and no relevant differences were seen. The particle size distribution is narrow, with a d_{50} of 300 nm (Figure 2b).

Electrostatic Precipitation

The drug load increased with increasing applied voltage hence the particles are charged more and can be separated more efficiently [7]. The drug load could be increased 25 times compared to a previous study [3] (Figure 3). The volume flow of the transporting gas was set at 3.6 L/h compared to 120 L/h. This led to a longer residence time of the particles in the loading area of the MESP and thereby to more efficient precipitation. The aim to increase the drug load could be achieved by the introduction of a new spray drying plant. Nevertheless the drug load is still rather low. During the process a particle layer on the inner walls of the MESP was built up. It can be explained by the phenomenon of back corona. Therefore, an optimization of the MESP is required.

In vitro dissolution

In vitro dissolution experiments were conducted with a solid dispersion with a high drug load and a physical mixture. The mean dissolution times (MDT) were calculated and indicated a significant (α=0.05) shorter mean dissolution time of the solid dispersion (MDT= 125 ± 12.5 s) compared to the physical mixture (MDT= 143 ± 6.9 s). Nevertheless a faster dissolution was expected, so that dissolution experiments with solid dispersions with a low drug load were conducted. A significant reduction of the mean dissolution time (MDT= 92 ± 7.9 s) was observed (Figure 4). It is assumed that agglomeration of the drug particles is a reason for this observation, since agglomeration tendency increases with a higher particle load.

Conclusion

In this study a spray drying plant for the production of submicron particles was built and tested. The design was adapted to the requirements of a combined use with a previously designed Melt Electrostatic Precipitator. The gas volume flow could be set as low as it was needed for an efficient electrostatic separation without influencing the small size of the generated droplets. With this adaption of the spray drying section the aim to obtain higher drug loads in the product was achieved.

The in vitro dissolution behavior was improved, nevertheless an influence of the drug load was observed. An agglomeration tendency with increasing drug load is assumed. Thereby further improvements concerning the MESP design are required in order to reduce particle agglomeration during the process.

Acknowledgements

The authors thank Volker Brandt (TU Dortmund University) for taking the SEM pictures and are grateful for the material support from Roquette Pharma.

References

1. Lipinski, C. Drug-like properties and the causes of poor solubility and poor permeability. *Journal of Pharmacological and Toxicological Methods* **2000**, *44*, 235-249, doi:10.1016/S1056-8719(00)00107-6.
2. Ku, M.; Dulin, W. A biopharmaceutical classification-based Right-First-Time formulation approach to reduce human pharmacokinetic variability and project cycle time from First-In-Human to clinical Proof-Of-Concept. *Pharmaceutical Development and Technology* **2012**, *17*, 285-302, doi:10.3109/10837450.2010.535826.

3. Dobrowolski, A.; Pieloth, D.; Wiggers, H.; Thommes, M. Electrostatic Precipitation of Submicron Particles in a Molten Carrier. *Pharmaceutics* **2019**, *11*, doi:10.3390/pharmaceutics11060276.
4. Strob, R.; Dobrowolski, A.; Schaldach, G.; Walzel, P.; Thommes, M. Preparation of spray dried submicron particles: Part A - Particle generation by aerosol conditioning. *International Journal of Pharmaceutics* **2018**, *548*, 423-430, doi:10.1016/j.ijpharm.2018.06.067.
5. Muhrer, G.; Meier, U.; Fusaro, F.; Albano, S.; Mazzotti, M. Use of compressed gas precipitation to enhance the dissolution behavior of a poorly water-soluble drug: Generation of drug microparticles and drug-polymer solid dispersions. *International Journal of Pharmaceutics* **2006**, *308*, 69-83, doi:10.1016/j.ijpharm.2005.10.026.
6. Nokhodchi, A.; Bolourtchian, N.; Dinarvand, R. Crystal modification of phenytoin using different solvents and crystallization conditions. *International Journal of Pharmaceutics* **2003**, *250*, 85-97, doi:10.1016/S0378-5173(02)00488-X.
7. White, H.J. Particle Charging in Electrostatic Precipitation. *Transactions of the American Institute of Electrical Engineers* **1951**, *70*, 1186-1191, doi:<b _ngcontent-c10=""> 10.1109/T-AIEE.1951.5060545.

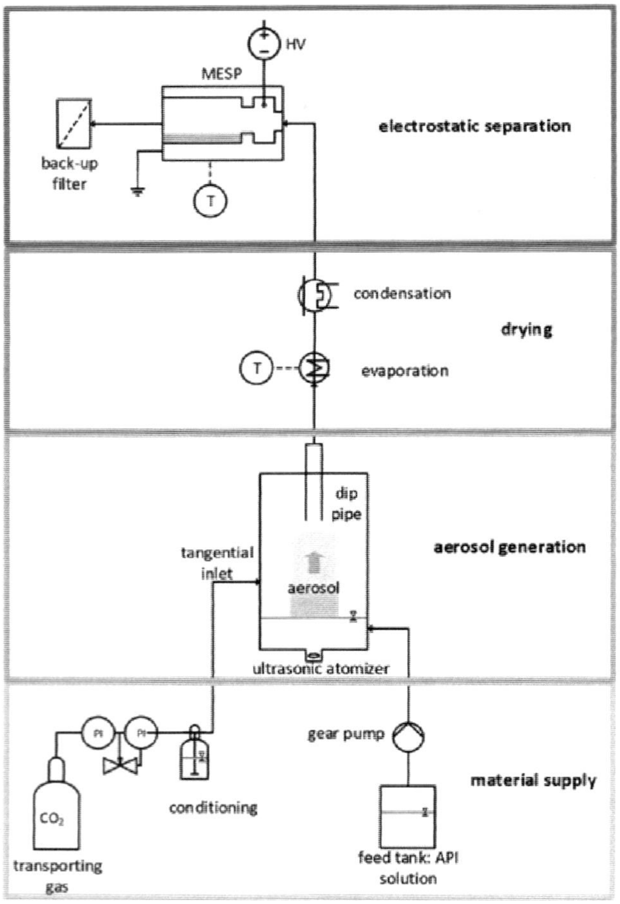

Figure 1: Experimental setup of the aerosol-generator combined with the MESP.

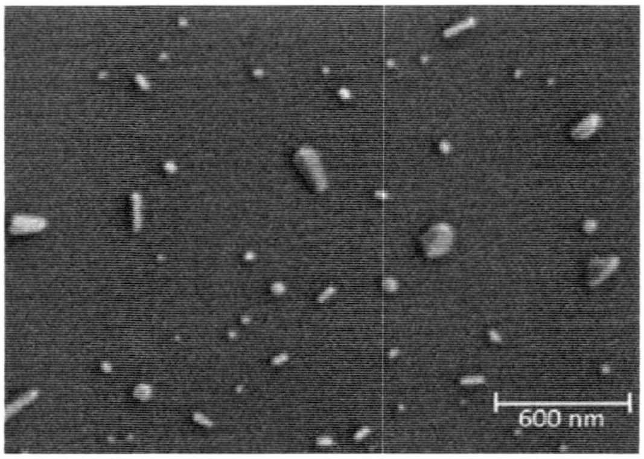

Figure 2 (a): SEM Image of spray dried phenytoin particles

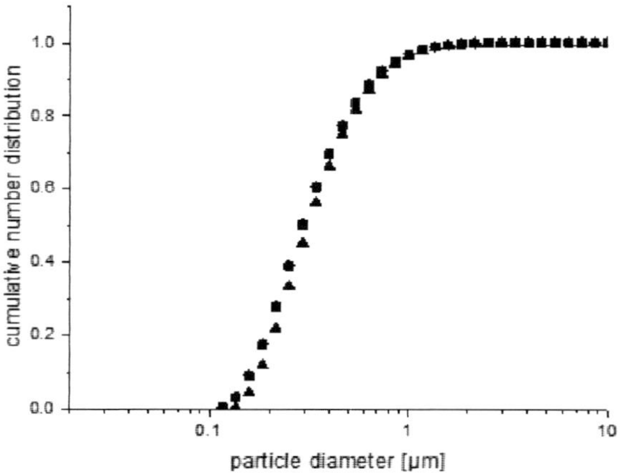

Figure 2 (b): particle size distribution of spray dried phenytoin particles, n=3.

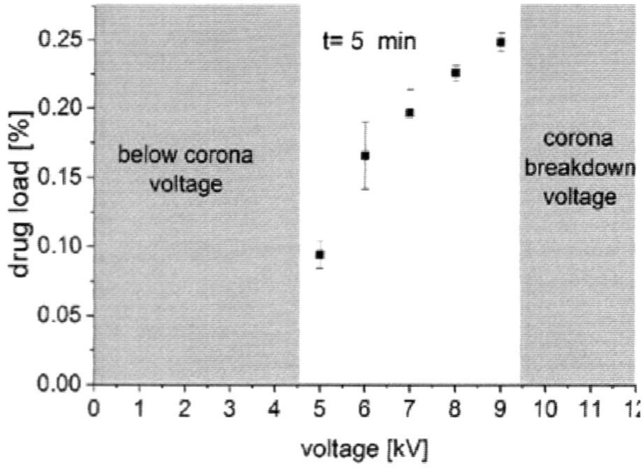

Figure 3: Influence of the corona voltage on the drug load at constant loading time (av±s,n=3).

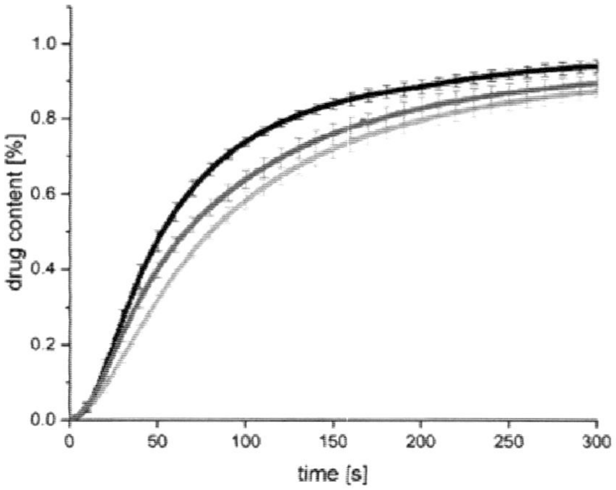

Figure 4: Dissolution profiles of physical mixture (black), solid dispersion with high drug load (dark grey) and solid dispersion with low drug load (light grey) (av±CI).

601798 Modeling and Simulation of Magnetophoresis of Nanoparticles for Magnetic Targeting Applications

Abstract:

Despite numerous studies on drug delivery to the brain there still remain some questions on how can blood brain barrier be overcome in the process of drug transport, and how can drug be targeted and confine within a particular portion of the brain. Drug loaded paramagnetic particle seems to have potential to eliminate the challenges encountered during drug delivery to the central nervous system. Paramagnetic particles, under the effect of external magnetic field, can be used as carriers to effectively transport drugs through cerebrospinal fluid and deliver to the targets without being disturbed by the blood brain barrier and brownian motion. Since the in vivo experiments are expensive and difficult to conduct for such systems, a mechanistic model that can predict the dynamics and distribution of paramagnetic particles is necessary for the optimal design of drug delivery systems. The existing models are applicable to dilute solutions of paramagnetic particles and cannot be extended to concentrated solution, which is generally created near the magnetic targets. Here we present a framework to model magnetophoretic diffusion and convection of paramagnetic particles in a concentrated solution. An analytical solution to the magnetic field-driven transport of particles in 1-D and 3D were developed with application to drug delivery through cerebrospinal fluid. This solution gives rise to concentration profiles of the paramagnetic nanoparticle. The effect of activity coefficient was investigated. Invitro experimental study of paramagnetic nanoparticle transport in a tube was used to back up our model prediction. The result obtained from our model and experiment show that paramagnetic nanoparticle can be confined in a particular domain of the central nervous system (CNS) which will aid the delivery and targeting of drug to the domain.

603765 Rational Design and Mathematical Modeling of Intrathecal Therapeutics: Biodispersion and Pharmacokinetics

Abstract:

Intrathecal (IT) drug delivery can efficiently target regions in the central nervous system without hindrance of the blood brain barrier. This delivery mode is especially promising for novel therapeutics such as enzyme replacement, gene or antisense oligonucleotide (ASO) therapies for the treatment of neurodegenerative diseases including Alzheimer's, Huntington's disease, and stroke.
However, the development of dosing guidelines for IT delivery requires intricate knowledge of the interplay between cerebrospinal fluid (CSF) dynamics, patient anatomy and pharmacokinetics of drug molecules. Fluid mechanical and chemical principles governing IT delivery are difficult to observe in vivo, and experimentation in small animals (mice, rats) does not translate to human applications. In order to fill the knowledge gap concerning IT biodispersion, we propose to use an anatomically accurate replica of the entire human central nervous system. This model faithfully matches spinal and cranial CSF spaces, and also operates with oscillatory CSF motion similar to physiological CSF flow conditions obtained with in vivo cine phase contrast MRI. The in vitro human CNS replica served to study CSF motions and mixing patterns that occur in acute and continuous infusion via a drug pump. The relationships between infusion volumes, injection impulse, catheter choices, and pump settings were systematically studied. Moreover, experimental data informed a novel distributed pharmacokinetic model to predict drug biodispersion for different infusion modes, drug properties, and subject specific physiology. A 1-D mathematical model was developed to replicate both the deformation and dispersion processes as experienced in the experiment, by solving continuity equation (for deformation) and transport equation (for dispersion). This mathematical model gives us a scalable method to predict the biodispersion of drug in the human spine with computer simulation. The presentation demonstrates a workflow to determine optimal infusion guidelines for novel drug leads based on anatomically accurate experimentation, optimal integration of animal data, and computer simulations using the developed mathematical model. The proposed experimental method avoids unnecessary animal experimentation, cuts costs and time effort needed to develop dosing guidelines for new drug leads, which potentially shortens time to market for new therapies. The proposed mathematical model enables a predictive method to quantify the effect of deformation and diffusion on the distribution of drugs.

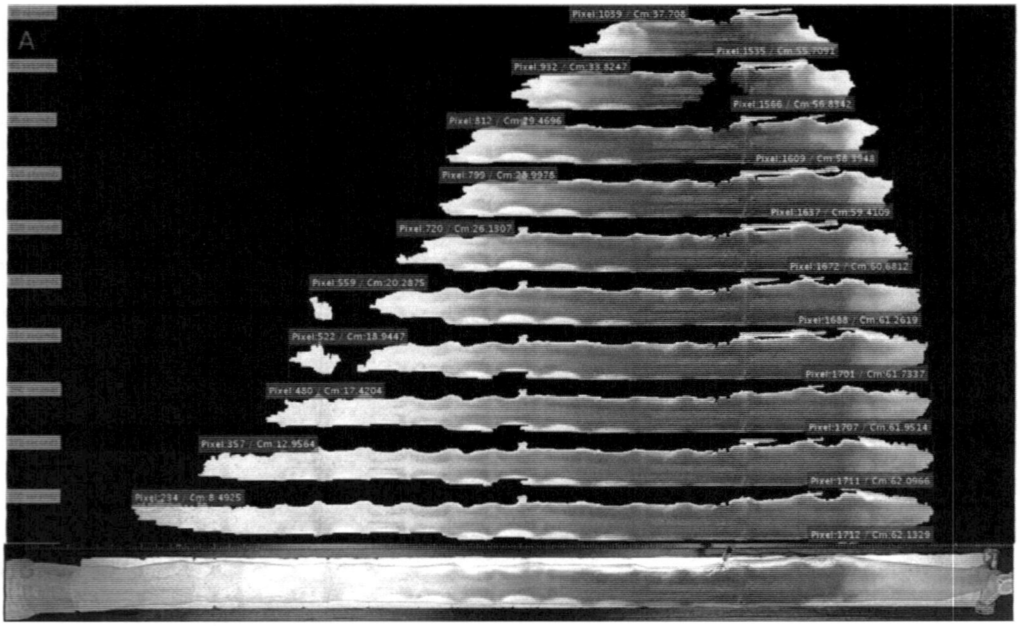

604172 Computer Aided Drug Formulation Design with Image-Based Microstructure Modeling

Abstract:

In pharmaceutical formulation development, microstructures, such as porosity and the distribution of active ingredient particles, play an increasingly important role. Three-dimensional microscopic imaging, specifically X-Ray Microscopy (XRM) and focused ion beam scanning electron microscopy (FIB-SEM), provides insight into these microstructures at unprecedented resolution. However, acquiring such images can be challenged by cost, time, and the availability of high-end imaging facilities and required expertise. Even when the images are acquired at desirable quality, there can be uncertainty that the acquired data is a representative elementary volume (REV).

An image-based microstructure modeling framework will be presented in this talk. XRM images, FIB-SEM images, or correlative images using both, are processed with an artificial intelligence-based image analysis engine using supervised machine learning and generative adversarial networks. A matrix of quantitative descriptors such as volume fractions, particle size, particle dispersion, porosity, and transport properties are computed. A computer aided drug formulation (CADF) model can be numerically generated that matches these image-based descriptors. From the CADF model, full 3D microstructures corresponding to variants of the formulation can be numerically generated.

Figure 1 demonstrates one example CADF model for an amorphous drug formulation. Two dimensional SEM images, Figure 1a, were acquired and segmented with a hybrid supervised machine learning and deep learning method (Figure 1b). Mathematical descriptors are extracted and divided into three groups. The first group is used as the generator input in the GAN model, and the second group as the discriminator input in the GAN model. The third group serves as the validation control group to drive the GAN model iteratively toward a microstructure model with desirable properties, Figure 1c. Once the model is validated, three dimensional microstructures of new formulations can be generated numerically. For example, Figure 1d-1f illustrates five new formulations with porosity 10%, 20%, 30%, 40% and 50%. A broader porosity range, at desirable intervals, can be covered. All quantitative descriptors can be computed and correlated with release behavior without making a new sample.

The CADF model provides a numerical framework to study the sensitivity of these quantitative descriptors to drug formulation parameters (such as drug loading, particle size of the active pharmaceutical ingredients, milling conditions, particle orientation alignment, and more) and under various processing conditions (such as hot melt extrusion temperature, polymer properties, compaction force, and more). The numerical model allows a formulation scientist to rapidly traverse multi-variable parameter space, and narrow down to lead formulation parameters and optimal processing conditions. When *in vitro* release test or dissolution data is available, release simulations can then be conducted on the CADF model to validate, before applying the CADF model to iteratively predict release profiles under different parameters.

The proposed framework has broad applications in formulation development and solid dosage forms, and in material science in general, where microstructures play a critical role.

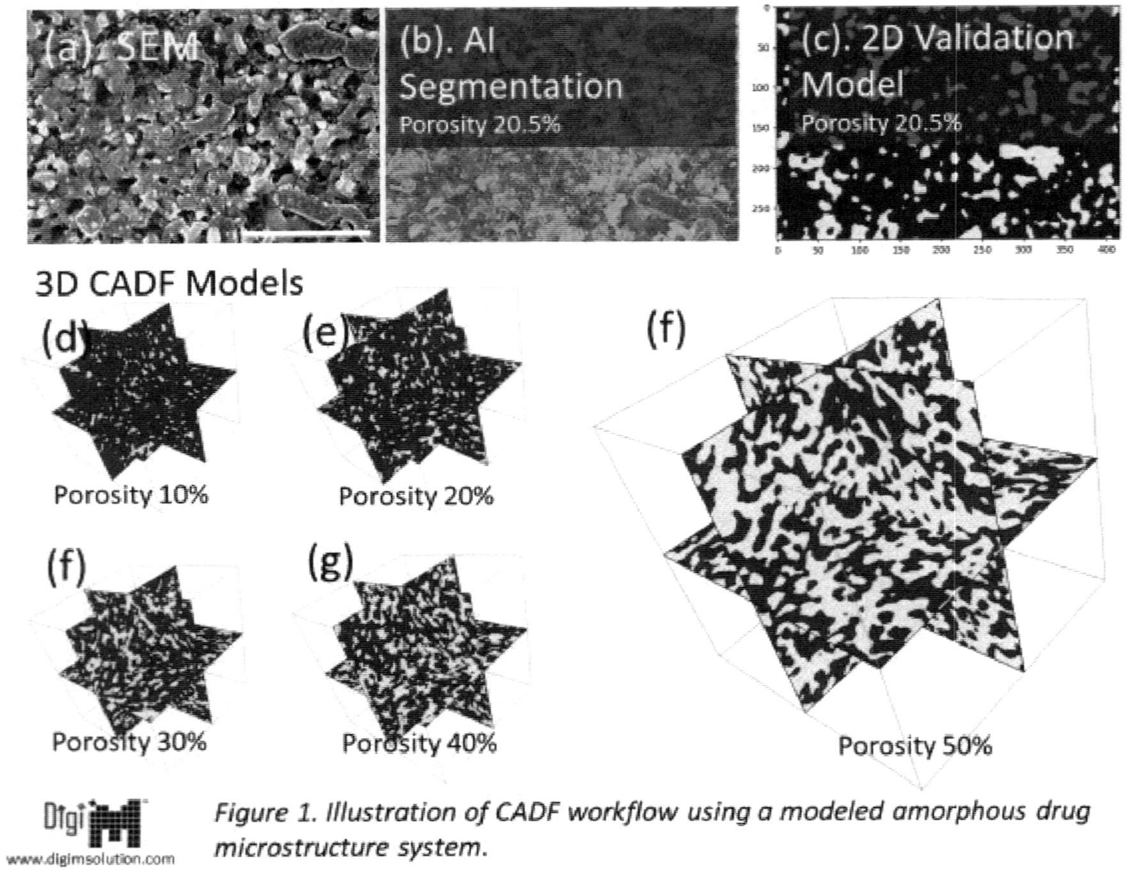

Figure 1. Illustration of CADF workflow using a modeled amorphous drug microstructure system.

1

609668 Mechanistic Mass Transport Analysis of Predictive Improved Dissolution and Bioavailability Using pH Modifier in Drug Formulations

Abstract:

Oral solid dosage formulations require first to be dissolved in the gastrointestinal (GI) tract before being absorbed through the epithelial cell membrane. In vivo drug dissolution depends on the physiological variables in the GI tract, pH, residence time, intestinal motility, and drug properties. The pH-dependent dissolution profile of an ionizable drug may benefit from the manipulation of in vivo variables such as the environmental pH using pH-modifying agents incorporated into the dosage form. The dissolution of dipyridamole, a BCS class II (high permeability, low solubility) weak-base drug formulation in the presence of pH-modifier (tartaric acid) was investigated in this study. A comprehensive mass-transport model was developed that takes into account the effect of physical and chemical properties of drug and pH modifiers such as pKa, solubility, particle size distribution. Besides, the model considered dissolution media conditions such as the hydrodynamical parameters in the dissolution media (e.g., shear rate, convection, confinement, and diffusion) and the dynamic changes in the bulk pH due to the drug dissolution. Good agreement between the mass-transport analysis prediction and in vivo clinical data lends the credibility to apply this model for predicting the performance of a range of weak-base drugs with different solubilities, pKa, and particle size by considering the effect of gastric emptying rate and pH under different conditions. Furthermore, sensitivity analysis quantifies the impact of physiological parameters and drug-pH modifier properties. This analysis provides an improved procedure for formulation design using the pH-modifiers by minimizing the experimental iterations.

594141 Hybrid Nanocrystal-Amorphous Solid Dispersions (HyNASDs): A New Class of Materials for Bioavailability Enhancement of Poorly Soluble Drugs

Abstract:

Drug nanoparticle-laden microparticles (nanocomposites) and amorphous solid dispersions (ASDs) have been commonly used to deliver poorly soluble drugs and enhance their solubility, dissolution rate, and bioavailability. A major shortcoming of drug nanocomposites as compared with drug ASDs is their limited supersaturation capability in dissolution, whereas ASDs have various issues about physical stability of the metastable form of the drug (amorphous). Here, we prepared drug hybrid nanocrystal–amorphous solid dispersions (HyNASDs) and compared their performance to ASDs. A wet-milled griseofulvin (GF, model poorly soluble drug) nanosuspension and a GF solution, both containing the same dissolved polymer–surfactant (SDS: sodium dodecyl sulfate) with 1:1, 1:3, and 1:5 GF:polymer mass ratios, were spray-dried. Hydroxypropyl cellulose (HPC) and Soluplus (Sol) were used as matrix-forming polymers. XRPD, DSC, and Raman spectroscopy reveal that ASDs were formed upon spray-drying the solution-based feed, whereas nanocomposites and nanocomposites with >10% amorphous content, HyNASDs, were formed with the nanosuspension-based feed. Sol provided higher GF relative supersaturation in the dissolution tests than HPC owing to Sol's stronger intermolecular interactions and miscibility with GF and its recrystallization inhibition. Besides the higher kinetic solubility of GF in Sol, presence of GF nanoparticles vs. micron-sized particles in the nanocomposites enabled fast supersaturation. SDS provided enhanced wettability, allowing for fast supersaturation from Soluplus-based HyNASDs (~300% within 20 min), while higher Soluplus loading led to higher supersaturation. This study demonstrates successful preparation of fast supersaturating HyNASDs, which renders nanoparticle formulations competitive to ASDs in bioavailability enhancement of poorly soluble drugs.

608921 Leveraging Hollow Fibre Membranes and High Boiling Point Organic Solvents in the Production of Enteric Amorphous Solid Dispersions

- Download/View Extended Abstract

Abstract:

Leveraging hollow fibre membranes and high boiling point organic solvents in the production of enteric amorphous solid dispersions

Rute Mota, Rui C. Silva, João Vicente

In recent years up to 90 % of the drugs in the discovery pipeline are poorly soluble molecules which poses challenges for bioavailability and dosage form development. Furthermore, the physical and chemical properties of these candidate compounds result in poor solubility in aqueous and volatile organic solvents which hinders the development of amorphous solid dispersion (ASD) using spray drying (SD), while at the same time possessing melting points too high to use hot-melt extrusion (HME): such compounds are typically designated as “brick dust” [1].

In the literature several technologies can be found to deal with poorly soluble APIs: nanomilling or complexation with cyclodextrins fall within the non-ASD alternatives; co-precipitation; supercritical CO2 and KinetiSol [2], as an alternative ASD technology. One such strategy has been developed where nano or micro particles of API-polymer composite precipitate are produced via co-precipitation process using microfluidization. This technology offers the advantage of precise control of particle morphology and generates particles with surface areas that can be 10 times greater than those obtained using HME or SD [3]. Although this technology circumvents the low solubility aspect of the “brick dust” molecules it has some the following drawbacks: a) a suspension with very low solids load and, thus, longer drying process times; b) with “brick dusts” finding a solvent for both API and polymer in amounts for economically viable throughput usually means using organic polar solvents with high boiling points, such as dimethylacetamide (DMA) or dimethylformamide (DMF) or dimethyl sulfoxide (DMSO) which poses challenges to spray dry [3]. These drawbacks amount to longer process times with low throughput which may not be financially appealing in early stage development.

The work presented herein consists on a co-precipitation process with the inclusion of a cross-filtration hollow fibre membranes unit as a straightforward and scalable technology focusing on producing amorphous solid dispersions with “brick dust” compounds. The cross-filtration operation consists in a two-step approach: 1) the first step, concentration of the suspension, the removal of excess antisolvent to increase the solids load; 2) the second step, a solvent swap, where the presence of high boiling point organic solvents in the suspension is minimized by replacing it with an aqueous based solvent. This approach widens the co-amorphous technology to include high boiling points organic solvents best suited for “brick dust” while circumventing the need for secondary drying and reducing cycle time.

References

[1] Temtem M, The Coming of Age of Amorphous Solid Dispersions, Pharma's Almanac, 26

October, 2018.

[2] Jermain SV, Brough C, Williams RO 3rd, Amorphous Solid Dispersions and Nanocrystal Technologies for Poorly Water-Soluble Drug Delivery – An Update, Int J Pharm. 2018 Jan 15; 535 (1-2): 379-392.

[3] Duarte Í, Corvo ML, Serôdio P, Vicente J, Pinto JF, Temtem M, Production of nano-solid dispersions using a novel solvent-controlled precipitation process – benchmarking their in vivo performance with an amorphous micro-sized solid dispersion produced by spray drying, Eur J Pharm Sci. 2016 Oct 10;93:203-14.

Figure 1 – Co-Precipitation Process Schematics with cross-filtration membranes setup

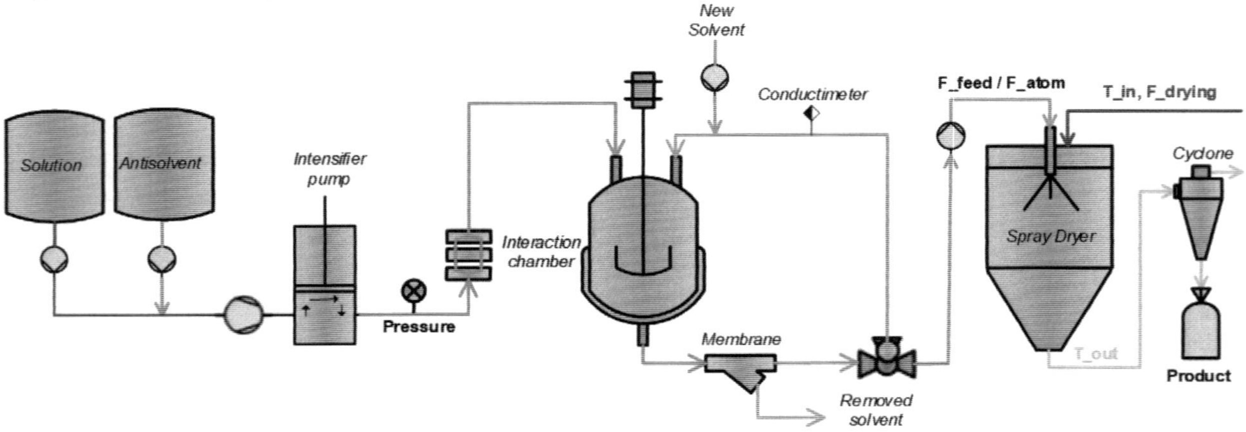

610310 Insight into Amorphous Solid Dispersion Dissolution Behavior: Laser-Diffraction and Raman Spectroscopy Holding Hands

Abstract:

Introduction

Formulation of Biopharmaceutics Classification System (BCS) class II and IV drugs into amorphous solid dispersions (ASDs) is a promising strategy to increase both the apparent aqueous solubility and bioavailability of these drugs. Although drug release from ASDs is still poorly understood, it has been suggested that drug release from ASDs is polymer-controlled at low drug loadings, whereby both drug and polymer release at the same rate (congruently) [1]. However, in contrast, when a higher drug loading is reached, which differs for different drug-polymer combinations, an abrupt decrease in drug release rate may be observed which may be explained by a switch to drug-controlled release, whereby release of drug and polymer is incongruent [2].

In the present work, it is intended to further understand the dissolution performance of different ASD formulations by following the dissolution rate of ASD particles using a laser-diffraction methodology. Furthermore, it will be also studied the possible drug enrichment on the partially dissolved ASD particles by Raman. Intrinsic dissolution was also applied to measure the dissolution rate of the API from the ASD matrix. For this study several ASDs were prepared with two grades of HPMC-AS having different ratios of succinoyl:acetyl groups (L and M) by spray-drying. A BCS class II drug will be used as a model drug.

Materials and methods

Production of ASDs

Amorphous solid dispersions were produced by spray-drying using a mixture of methanol and dichloromethane at 1:4 ratio (w/w), respectively, as solvents. The ASDs were produced at different Drug:HPMC-AS grades (L and M) and at different drug loads (15% and 35 % w/w). For these experiments, a Buchi Mini Spray Dryer B-290 was used with a two-fluid nozzle in closed loop. The spray-dried samples were further dried to remove any residual solvents in a vacuum drying oven at 40 °C.

Physical characterization of ASD

Amorphous solid dispersions were characterized by X-Ray Powder Diffraction (XRPD) to confirm their amorphous nature. Morphology was studied by scanning electron microscopy (SEM) and particle size distribution was evaluated by Laser Diffraction (Sympatec).

ASDs and API dissolution performance

To get an insight into ASDs disintegration/dissolution kinetics, these ASDs particles were added to the dispersion unit of Mastersizer 2000 containing 150 mL of FaSSIF (biorelevant dissolution media). 86 mg and 200 mg of each ASD with 35 wt.% and 15 wt.% drug loads, respectively, were added to this dispersion unit to obtain 0.2 mg/mL as a drug target concentration. Particle size of ASDs was monitored for 4 hours. The reduction of particle size of ASDs during dissolution and colloidal formation was compared with the dissolution rate of the API released from these ASD particles. The drug concentration over time was measured by HPLC.

Results

ASD prepared by spray-drying showed an amorphous nature with no diffraction peaks from crystalline molecule. The particle morphology of the produced ASD was also analyzed by SEM. All the ASDs presented a shriveled morphology (raisin-like particles) which is characteristic of slower spray-drying kinetics. The particle size distribution was similar between different amorphous solid dispersions.

In order to monitor the dissolution performance of the four ASD, the different formulations were suspended in FaSSIF (pH 6.5) in the dispersion unit of the Mastersizer 2000.

In all ASD formulations it was possible to see the appearance of a colloidal population with a submicron size (< 600 nm) after the 240 minutes (Figure 1, A). ASD dissolution showed to be slower in formulations containing higher API loading, since after 60 minutes it is still possible to observe ASD particles in suspension. It's worth mentioning that for formulation - API:HPMC-AS M (35:65) - a full disintegration of ASD did not occur even after 240 minutes.

Raman imaging was performed to the API:HPMC-AS M (35:65) ASD before and after being submitted to the experiment in the Mastersizer equipment. Since part of this sample was not dissolved after 240 minutes, an aliquot was taken from the dispersion unit and centrifuged prior Raman analysis. The intensity of one specific Raman band of the API (1605 cm^{-1}) and the Polymer (2943 cm^{-1}) were monitored in ASD particles before and after the experiment. Results showed that the ratio of 2943 cm^{-1}/1605 cm^{-1} bands is above 1 (red spectra, Figure 1, B) in this ASD particles before being suspended into FaSSIF. However, this ratio showed to invert after the experiment, i.e, the polymer specific band at 2943 cm^{-1} seems to reduce faster than API specific band suggesting an absence of a concomitant dissolution of both elements when this ASD is suspended in FaSSIF. This is consistent with a surface drug enrichment phenomenon [2], suggesting faster dissolution of the polymer in relation to the API.

Intrinsic dissolution confirms that the dissolution rate of ASDs is aligned with the dissolution of ASDs particles observed in laser-diffraction experiment. It is possible to note that formulations containing 35% of API present slower dissolution rates than the ones containing 15% of API. Again, API:HPMC-AS M (35:65) formulation showed to have the lowest performance (Figure 1, C), which seem to be related with a drug enrichment of ASD particles in this particular formulation.

Conclusions

The results presented in this work indicate that dissolution of ASD particles (by laser-diffraction) give an important insight on the API dissolution rate. In detail, it was interesting to note that formulations which showed to disintegrate/dissolve faster leading to a stable colloidal suspension in laser-diffraction test were also the formulations with the highest and fastest API dissolution rate. This proves that their improved performance is related with the ASD dissolution mechanism. The faster dissolution of ASD particles in 15% drug load formulations suggests possible polymer-controlled dissolution which is associated with low drug loadings formulations [2]. The Raman analysis of API:HPMC-AS M (35:65) particles after being suspended/stirred in FaSSIF for 240 min suggested a faster dissolution of the polymer in comparison to the API, leading to particles enriched in API. This explains the slowest dissolution rate observed in this formulation. Therefore, it can be concluded that API dissolution performance is impacted not only by the drug load but also by the HPMC-AS grade. As a future work, the different hydrogen bonding between the API and the 2 different grades of HPMC-AS with varying ratios of succinoyl:acetyl groups will be explored. This study also shows that laser diffraction and Raman spectroscopy pose as orthogonal techniques to understand the dissolution mechanisms of ASDs.

REFERENCES

Tres F, Posada MM, Hall SD, Mohutsky MA, Taylor LS. Mechanistic understanding of the phase behavior of supersaturated solutions of poorly water-soluble drugs. *Int J Pharm*. 543(1-2): 2018

Saboo S, Kestur US, Flaherty DP, Taylor LS. Congruent release of drug and polymer from amorphous solid dispersions: insights into the role of drug-polymer hydrogen bonding, surface crystallization and glass transition. *Mol Pharm*. 17(4): 2020.

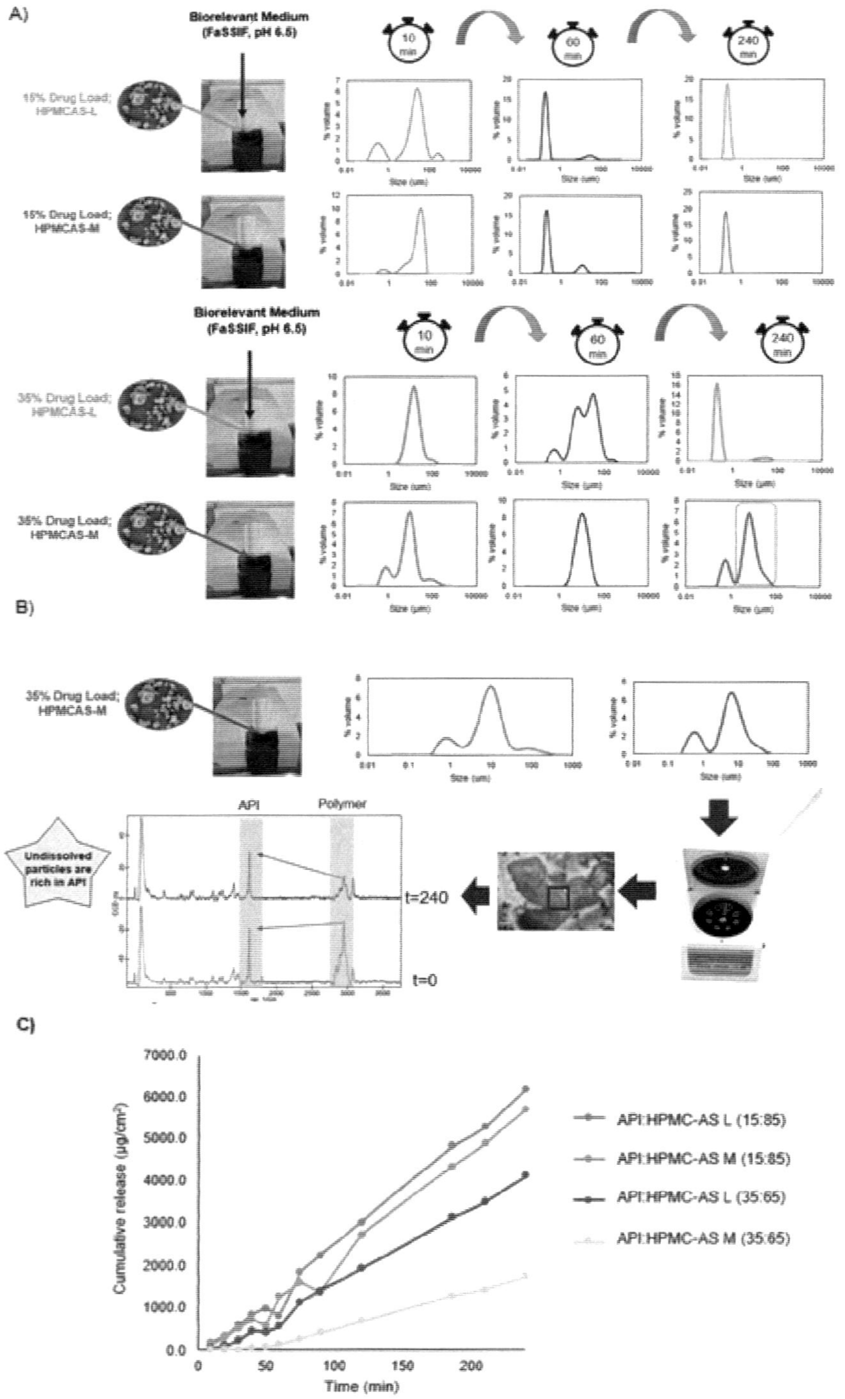

Figure 1:

A) Particle size distribution as function of time when ASDs were suspended in FaSSIF (pH=6.5). From top to bottom 1) API:HPMC-AS L (15:85); 2) API:HPMC-AS M (15:85); 3) API:HPMC-AS L (35:65); 4) API:HPMC-AS M (35:65).

B) Raman spectra of API:HPMC-AS M (35:65) ASD particles before and after being suspended into FaSSIF media in Mastersizer 2000 equipment.

C) Surface normalized dissolution rates of API from 4 different ASD formulations in FaSSIF (pH=6.5).

602798 Multi-Factor Modeling to Predict Significant Factors in Tablet Disintegration and Dissolution Using the FBRM

Abstract:

We have previously demonstrated the application of Focused Beam Reflectance Measurement (FBRM) for analysis during immediate-release tablet disintegration to evaluate changes in particle size and count. This approach could distinguish changes in tablet particle size during disintegration due to varying properties, such as tablet hardness. Here we present a proof-of-concept study demonstrating the generation of a predictive statistical model of a drug product design of experiments (DoE) using parameterization of FBRM measurements taken during tablet disintegration. The FBRM was used to monitor the disintegration of tablets generated in an 18-run DoE evaluating five variables at three levels. The DoE evaluated API properties as well as dry granulation and tableting conditions. Statistical models were generated using the DoE input variables fit to predict FBRM measurement parameters. Robust, multi-factor models identified that several FBRM measures such as total particle count, time of maximum particle count, mean chord (particle) length, or bounded particle size ranges were dependent on several DoE inputs. Furthermore, these factors identified via FBRM disintegration corresponded with the significant factors of dissolution prediction models. In contrast, a USP disintegration method failed to provide a predictive model with multiple significant factors. The FBRM approach to monitoring disintegration allows for a more comprehensive understanding of tablet performance and provides greater insight into the impact of upstream material properties or processing parameters. Material properties and process parameters ultimately impacting dissolution could be predicted earlier and in a material-sparing way. It further offers an orthogonal measurement to gain a better understanding of tablet dissolution explained in a framework of underlying particle behavior.

608277 Melt-Coating of APIs with Surfactant for Drug Dissolution Enhancement

Abstract:

For a solid dosage form, medicine is dispensed based on the dissolving of the active ingredient in the gastrointestinal tract. The significance of the dissolution rate on drug clinical performance has been recognized for a long time. One of the most critical quality attributes was assessed depends on its dissolution behavior. Unfortunately, significant numbers of FDA approved drugs and developmental drugs are hydrophobic and poorly soluble. It has been documented that approximately 40% of approved drugs and 90% of drugs under development are categorized as molecules of low water-solubility and low bioavailability. Due to discovering new drugs is increasing difficulty and become a non-effective way to approach therapeutic requirements, enhancing the solubility of these available drugs is considerably investigated. The prevalent approach for dissolution improvement can be ionizing the drug molecules, producing salt from, micronizing API particles, and using solid dispersion techniques. However, these techniques may require complex unit operations, create unstable solid forms, or generate APIs of poor manufacturability. In this work, we will propose a novel approach in which a small amount of surfactant (< 10% of API weight) is coated on the surface of API particles by applying shearing and heat simultaneously. In this way, we are able to enhance the dissolution of hydrophobic and insoluble API without changing its original solid state.

In our work, three poorly soluble APIs (Ibuprofen, Carbamazepine, and Fenofibrate) and two low-melting-point surfactants (Poloxamer407 and Cetylpyridinium Chloride) are selected. Three specific case studies are performed. In each case study, a twin-screw extruder is used to perform melt-coating because it can apply shearing and heat simultaneously. At a certain temperature, the surfactant can partially melt, then smears and coats on the surface of API particles. Then, the hydrophobic APIs presents a new hydrophilic surface. The particle morphology is photographed using SEM, which shows that the surfactant successfully coats on API particles.

Then, the dissolution test is performed for both treated powder and finished product (tablet and capsule) and compared with the physical mixture and no-surfactant case. Results show that the melt-coating of the surfactant considerably enhances the dissolution rate. The treated product can release drugs 2-5 times faster than the physically mixed product. In contrast, the physical blending surfactant has a marginal effect on dissolution in comparison to the no-surfactant case.

608802 Cocrystallization Process Development of Benzoic Acid – Apremilast Cocrystal

Abstract:

In this work, we have designed several new cocrystals and solvates of a pharmaceutical molecule apremilast. Out of these new multicomponent forms, the cocrystal with benzoic acid is particularly promising due to its improved intrinsic dissolution rate (approximately 400%) and comparable thermal properties to marketed apremilast B. The crystal structure of this cocrystal was solved using single-crystal X-ray diffraction to provide further insight into the cocrystal formation. Cocrystallization process was explored to simplify further scale up. Utilizing the ternary phase diagram, constructed by the combination of specific experiments and a solid-liquid equilibria model, we were able to significantly decrease the experimental efforts compared to traditional approaches such as discontinuous isoperibolic thermal analysis (DITA). Methylehtyl ketone was chosen as a solvent for the cocrystallization experiment. This was performed in a 100ml reactor, were we investigated the impact of process parameters included cooling rate, stirring speed, impeller, concentration and seeding on the size and morphology of the apremilast-benzoic acid cocrystal. Obtained results confirmed strong impact of operating conditions on the crystals sizes and shapes. These crystals were consequently tested during dissolution and it was observed that significant improvements in the dissolution rate can be achieved by varying the crystal sizes and well as shapes.

603493 A Systematic Study of State-of-the-Art Methods in Crystal Structure Prediction for Organic Hydrates

- **Download/View Extended Abstract**

Abstract:

Hydrates are co-crystalline materials containing water as one of the molecules in the crystal lattice. The incorporation of water into the crystal lattice produces a unit cell different from that of the anhydrate and, consequently, the physical properties of the hydrate can differ significantly from those of the anhydrate. The existence and stability of hydrates is an important consideration in the development of pharmaceutical products: the prevalence of water during manufacturing and storage can mean that neat forms of an active pharmaceutical ingredient can undergo a phase transition to hydrate form, impacting the effectiveness of the drug. Crystal structure prediction (CSP) methods can in principle be useful in identifying likely hydrates, by undertaking searches for all polymorphs of water and one or more given compounds for a given co-crystal stoichiometry. Minimal information is needed, typically just the chemical connectivity diagram[1], to search for the low lattice energy arrangements of the constituent atoms in space.

Applications of CSP to hydrates have resulted in mixed success so far. In the fifth blind test[2] organised by Cambridge Crystallographic Data Centre, one of the targets was a hydrate but none of the 10 groups that attempted to predict its structure put forward the correct structure within their shortlist. In the sixth blind test[3], only 8 groups submitted predicted structures for the hydrate target, and only one group generated the experimental structure within their shortlist.

In order to gain a better understanding of the challenges that make CSP for hydrates difficult, we present a systematic evaluation of a CSP state-of-the-art method for organic hydrates, in which the lattice energy is partitioned into intramolecular and intermolecular contributions. Intramolecular interactions are modelled via quantum mechanical calculations[4], and intermolecular interactions are divided into electrostatics, modelled using *ab initio* derived distributed multipoles[5,6,7], and repulsion/dispersion interactions modelled using a semi empirical potential. A total number of 107 hydrates extracted from the Crystal Structure Database are minimized locally using the *CrystalOptimizer* algorithm[8] with six models of different levels of sophistication (functional, basis set, use of continuum polarizability). The geometric differences between experimental structures and the corresponding minimization outputs are compared in terms of root mean-squared deviation and a CPU time required. Five of the six models are found to give a good degree of accuracy in more than 95% of cases, but with varying computational costs. A further assessment of the proposed models is undertaken by determining relative energy rankings, which are critical in generating reliable polymorphic energy landscapes.

References:

(1) Pantelides, C. C.; Adjiman, C. S.; Kazantsev, A. V. General Computational Algorithms for Ab Initio Crystal Structure Prediction for Organic Molecules. In *Prediction and Calculation of Crystal Structures Methods and Applications. Topics in Current Chemistry*; Atahan-Evrenk, S.; Aspuru-Guzik, A., Eds.; Springer, **2014**; Vol. 345, pp. 25–58.

(2) Bardwell, David A., Claire S. Adjiman, Yelena A. Arnautova, Ekaterina Bartashevich, Stephan XM Boerrigter, Doris E. Braun, Aurora J. Cruz-Cabeza et al. Towards crystal structure prediction of complex organic compounds–a report on the fifth blind test. *Acta Crystallographica Sect. B: Struct. Sci. Cryst. Eng. Mater.* **2011**, 535-551.

(3) Reilly, A. M.; Cooper, R. I.; Adjiman, C. S.; Bhattacharya, S.; Boese, A. D.; Brandenburg, J. G.; Bygrave, P. J.; Bylsma, R.; Campbell, J. E.; Car, R.; *et al.* Report on the Sixth Blind Test of Organic Crystal Structure Prediction Methods. *Acta Crystallogr. Sect. B Struct. Sci. Cryst. Eng. Mater.* **2016**, *72*, 439–459.

(4) Gaussian 09, Revision A.02, M. J. Frisch, G. W. Trucks, H. B. Schlegel, G. E. Scuseria, M. A. Robb, J. R. Cheeseman, G. Scalmani, V. Barone, G. A. Petersson, H. Nakatsuji, X. Li, M. Caricato, A. Marenich, J. Bloino, B. G. Janesko, R. Gomperts, B. Mennucci, H. P. Hratchian, J. V. Ortiz, A. F. Izmaylov, J. L. Sonnenberg, D. Williams-Young, F. Ding, F. Lipparini, F. Egidi, J. Goings, B. Peng, A. Petrone, T. Henderson, D. Ranasinghe, V. G. Zakrzewski, J. Gao, N. Rega, G. Zheng, W. Liang, M. Hada, M. Ehara, K. Toyota, R. Fukuda, J. Hasegawa, M. Ishida, T. Nakajima, Y. Honda, O. Kitao, H. Nakai, T. Vreven, K. Throssell, J. A. Montgomery, Jr., J. E. Peralta, F. Ogliaro, M. Bearpark, J. J. Heyd, E. Brothers, K. N. Kudin, V. N. Staroverov, T. Keith, R. Kobayashi, J. Normand, K. Raghavachari, A. Rendell, J. C. Burant, S. S. Iyengar, J. Tomasi, M. Cossi, J. M. Millam, M. Klene, C. Adamo, R. Cammi, J. W. Ochterski, R. L. Martin, K. Morokuma, O. Farkas, J. B. Foresman, and D. J. Fox, Gaussian, Inc., Wallingford CT, **2016**.

(5) Nyman, J.; Day, G. M. Static and Lattice Vibrational Energy Differences between Polymorphs. *CrystEngComm* **2015**, *17*, 5154–5165.

(6) Price, S. L.; Leslie, M.; Welch, G. W. A.; Habgood, M.; Price, L. S.; Karamertzanis, P. G.; Day, G. M. Modelling Organic Crystal Structures Using Distributed Multipole and Polarizability-Based Model Intermolecular Potentials. *Phys. Chem. Chem. Phys.* **2010**, *12*, 8466.

(7) Stone, A. J. Distributed Multipole Analysis, or How to Describe a Molecular Charge Distribution. *Chem. Phys. Lett.* **1981**, *83*, 233–239.

(8) Kazantsev, A. V.; P. G. Karamertzanis; C. S. Adjiman; C. C. Pantelides. Efficient handling of molecular flexibility in lattice energy minimization of organic crystals. *J. Chem. Theory Comput.* **2011**, *7*, 1998-2016.

597238 Effects of Inorganic Salts and Volume Reduction on Glycine Polymorphism

Abstract:

Studies have demonstrated that inorganic salts can drive polymorphic selectivity of glycine from the metastable α-polymorph to the stable γ-polymorph. Here, we explore the effect of NaCl and $CaCl_2$ together with volume reduction on glycine nucleation. The salts were added to supersaturated aqueous solutions of glycine at a 2:1 Gly:NaCl/$CaCl_2$ ratio at volumes from 1000 μL to 50 μL. Analysis showed that a primary γ-glycine nucleation event occurs for Gly:NaCl systems, with observed increase in induction times as volume was decreased. However, β-glycine primary nucleation was also detected for Gly:NaCl systems. Interestingly, it was observed that β-glycine nucleation is preferred over the salt-directed γ-glycine nucleation at a certain volume. The increase in induction times combined with a decrease in probability for γ-glycine nucleation may have allowed preferred nucleation of the β-polymorph. On the other hand, no nucleation events were observed for Gly:$CaCl_2$ systems at a time span of 48 hours. These results provide insight on how polymorph modifiers act in confined environments.

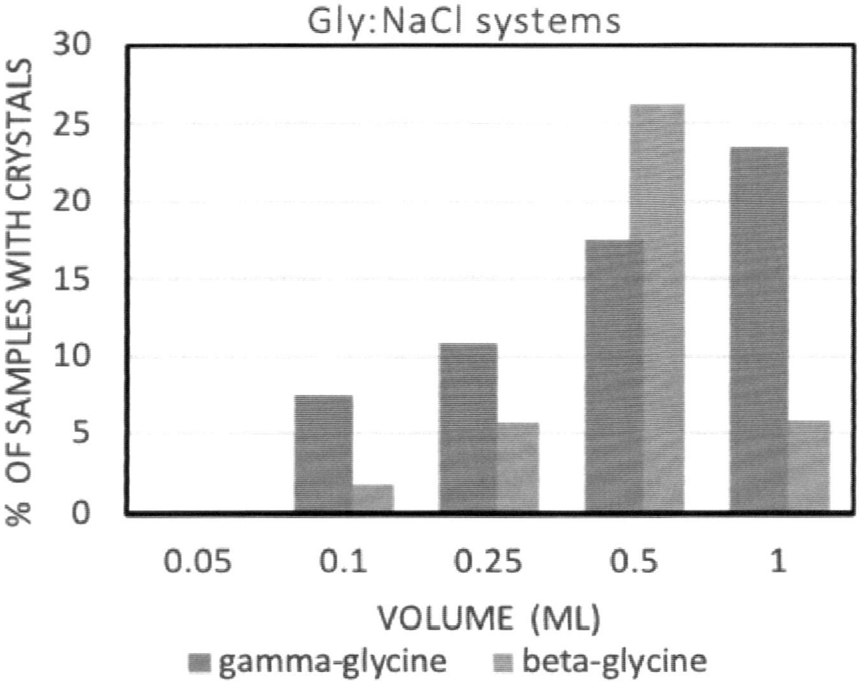

609279 Using Fluid Dynamics and Interface Engineering to Control Polymorphism and Morphology of Organic Molecules.

Abstract:

Crystallization of organic materials as thin films is an area of interest in multiple fields, such as in the pharmaceutical industry, organic electronics and for the multiple applications afforded by metal organic frameworks (MOFs). Over the past decade, significant work has been accomplished in terms of understanding how polymorphism of organic molecules can be enhanced through crystallization as a thin film, through innovative techniques such as confinement, polymer additives, and control of deposition conditions such as the use of electric fields. A deeper understanding of how these polymorphs are formed is critically missing from literature.

In this work, we use a mix of experimental and computational work to study the formation of thin film polymorphs of pharmaceutical compounds and organic semiconductors. The use of interface engineering can influence both the morphology and the polymorphism of acetaminophen, a pharmaceutical compound, and TIPS-pentacene, an organic semiconductor. We also show that changing the flow coating parameters can tune the nucleation and growth rates of organic semiconductors, which then impact the molecule polymorph, crystal size, and crystal orientation. In-situ X-ray diffraction data shows the changes in crystal growth conditions due to changing the fluid dynamics conditions and provides valuable insight in the mechanism of crystal nucleation. Controlling the interface precisely to obtain the desired polymorphs and crystal morphologies will be discussed as well.

609384 Acetaminophen Polymorphism and Texture Control Using Novel Coating Regimes

Abstract:

Spray coating and antisolvent addition have emerged as viable techniques that for use in the continuous processing of pharmaceutical molecules. Through controlling supersaturation during crystallization, these techniques are able to create particle distributions. In this work, meniscus guided flow coating is explored as an alternative to spray coating that creates high surface area crystals for use in pharmaceutical formulations. We have observed the formation of film morphology similar to that obtained using a spray coating technique, and unique from what has been previously described using a flow coating method. Acetaminophen is used as a model compound in this study, where films were fabricated and characterized using polarized optical microscopy and grazing incidence x-ray diffraction (GIXD) to determine particle sizes and crystal orientation on the substrate. Results suggest that the coating regime deviates from traditional evaporative or Landau-Levich coating regimes described in literature. Here we present our current hypothesis for the mechanism of crystallization that produces the observed particulate morphology.

594770 Ternary Phase Diagram and Phase Transformation Behaviors of Various Creatine Phosphate Sodium Hydrates

Abstract:

The thermodynamic stability of creatine phosphate sodium (CPS) in water–methanol system was investigated by the determination of CPS–water–methanol ternary phase diagram. The involved transformation behaviors were explored in solid-state, solutions, and vapor environment by VT-PXRD, Raman, FTIR spectroscopy and DVS. Results show that CPS can exist in three hydrate forms, namely, heptahydrates, tetrahydrates, and dihydrates, in the ternary phase diagram. These three crystal forms become the stable form in turn as the water content of the solvent mixture increases, with no anhydrate form appearing, which was seldom found in current reports. The transition points move towards higher water activity as the temperature increases. Besides, solution-mediated phase transformation could happen between tetrahydrates and dihydrates, which is influenced by solvent composition, temperature, and solid loading. While solid-state phase transformation of them couldn't occur when heating. Heptahydrates and tetrahydrates could transform into each other through adsorption or desorption of the water vapor. The study provides a guide for the control of CPS hydrate forms in theory and in the production to obtain the commercial crystal form with high purity.

602459 Towards the Development of Digital Twin Framework for Continuous Pharmaceutical Manufacturing: Real-Time Model Maintenance and System Analyses

Abstract:

Emerging digitization technologies within Industry 4.0 framework are bringing revolutionary transformations in the chemical manufacturing sector, including pharmaceutical manufacturing. Industry 4.0 in pharmaceutical manufacturing comprises efficient data collection and implementation strategies through the use of process digitization, Internet of Things (IoT), advanced data analytics using big data, machine learning, cyber-physical systems and cloud computing. These data-intensive strategies highlighted in Industry 4.0 are recently being applied towards the development of a digital twin framework [1,2]. For continuous pharmaceutical manufacturing, a data-centric digital-twin framework consists of a physical manufacturing plant inter-connected with its virtual, model-based representation, allowing seamless communication between the two entities. Within this framework, real-time data pertaining to process operation are recorded from the physical plant and sent to update the virtual model for advanced system analyses, and to update the plant operation when needed.

For the efficient implementation of digital-twin framework, real-time modeling strategies concerning virtual model and plant updates need to be developed. The presented work focuses on developing these strategies, where the major aspects involve: 1) model maintenance based on real-time data obtained from plant operation; 2) algorithms for real-time system analyses; 3) determination of frequency criterion for model and plant updates. Model maintenance of the virtual plant is performed using model adaptation strategies, including recursive adaptation and ensemble-based methods [3,4]. Unlike commonly used methods where models are updated and maintained off-line, these strategies allow real-time model updates in the virtual plant during manufacturing plant operation. After developing real-time model updating strategies within the digital twin framework, the model updates are used for dynamic system analyses, like sensitivity and feasibility analysis [5-7], providing uninterrupted monitoring of process performance. Based on these analyses, corrective process improvements are suggested for real plant operation, in situations where, the current process settings can potentially lead to violations of drug's quality specifications. Along with this capability of real-time update of virtual and real plant, frequency criterion for these updates need to be investigated for an efficient update strategy in the digital twin framework. These updates are generally performed at the same frequency associated with plant measurements [8] but can be modified based on a user-defined plant-model mismatch criterion [9] to reduce required computational resources.

In this presentation, the results of such a framework for a continuous direct compaction line, together with Industry 4.0 will be illustrated. This will allow for a computationally efficient and robust implementation of digital-twin framework in pharmaceutical manufacturing for scientific, risk-based regulation compliant process operation.

Reference:

[1] M. Grieves, J. Vickers, Digital Twin: Mitigating Unpredictable, Undesirable Emergent Behavior in Complex Systems, in: Transdisciplinary Perspectives on Complex Systems, Springer, Cham, 2017: pp. 85–113.

[2] A. Belhadi, K. Zkik, A. Cherrafi, S.M. Yusof, S. El Fezazi, Understanding Big Data Analytics for

Manufacturing Processes: Insights from Literature Review and Multiple Case Studies, Computers & Industrial Engineering. 137 (2019).

[3] P. Kadlec, R. Grbic, B. Gabrys, Review of adaptation mechanisms for data-driven soft sensors, Computers & Chemical Engineering. 35 (2011) 1–24.

[4] K. Hazama, M. Kano, Covariance-based locally weighted partial least squares for high-performance adaptive modeling, Chemometrics and Intelligent Laboratory Systems. 146 (2015) 55–62.

[5] G. Bano, Z. Wang, P. Facco, F. Bezzo, M. Barolo, M. Ierapetritou, A novel and systematic approach to identify the design space of pharmaceutical processes, Computers & Chemical Engineering. 115 (2018) 309–322.

[6] Z. Wang, M.S. Escotet-Espinoza, M. Ierapetritou, Process Analysis and optimization of continuous pharmaceutical manufacturing using flowsheet models, Computers and Chemical Engineering. 107 (2017) 77–91.

[7] F. Boukouvala, V. Niotis, R. Ramachandran, F.J. Muzzio, M. Ierapetritou, An integrated approach for dynamic flowsheet modeling and sensitivity analysis of a continuous tablet manufacturing process, Computers & Chemical Engineering. 42 (2012) 30–47.

[8] G. Bano, P. Facco, M. Ierapetritou, F. Bezzo, M. Barolo, Design space maintenance by online model adaptation in pharmaceutical manufacturing, Computers & Chemical Engineering. 127 (2019) 254–271.

[9] A.S. Badwe, R.D. Gudi, R.S. Patwardhan, S.L. Shah, S.C. Patwardhan, Detection of model-plant mismatch in MPC applications, Journal of Process Control. 19 (2009) 1305–1313.

610062 Digital Transformation of Unit Operations Laboratory

Abstract:

The Unit Operations Laboratory at Rose-Hulman Institute of Technology recently underwent its digital transformation and the facilities are now being used in multiple courses as well as research projects. The fundamental shift to allowing open and efficient access to data has allowed this rich resource to influence not only the traditional Unit Operations lab course, but to supplement other core courses, serve as the backbone for our new *Process Analytics* course, and act as a hub for multidisciplinary research projects.

603143 Development of a Joint Infrastructure for Chemical Engineering Applications Related to Industry 4.0 and Digital Transformation

Abstract:

This work provides an overview on the opportunities created when a developed Industry 4.0 infrastructure is applied to chemical engineering processes. Industry 4.0 defines a technology framework designed to revolutionize the world process industries. In this context, research groups have been exploring such a revolution by bringing to this environment improvements to link chemical engineering research, applications and teaching[1,2,3]. In particular, in this presentation, the cooperation between research groups at West Virginia University (WVU, USA) and Federal University of Campina Grande (UFCG, Brazil) in this area is described. Such cooperation aims to lead to a complete IIoT (Industrial Internet of Things) infrastructure connected to several dynamic process simulators (e.g., Aspen, CHEMCAD, SimCentral). With data availability from this infrastructure (at different data request points for a simulated process and in real time), Industry 4.0 aspects can be explored, such as: digital twin, machine learning, on-line data analysis, event frames and notifications, as well as topics directly linked to automatic and statistical process control, process optimization and economic analysis.

The current infrastructure envisioned for analysis involves the use of a simulation platform that represents a chemical or manufacturing plant, data processing/storage and data request points from client devices and applications. The particularly investigated simulations include a subcritical coal-fired power plant model (from WVU), an industrial process from a large Latin America Refining Company (from UFCG), as well as chemical processes from the literature. Central to this infrastructure is a database to store real-time simulation data that represent plant measurements. The database used corresponds to the OSIsoft PI historian (which is part of the PI System)[1]. A Supervisory Control and Data Acquisition (SCADA) system is employed to remotely monitor and control the studied simulation through Open Platform Communications (OPC). In an analogy between the developed infrastructure and the Automation Pyramid, levels 0 (sensors and actuators) and 1 (unit automation) are represented by the dynamic simulations, level 2 (SCADA) provides the remote access to process variables and specifications by customized HMIs (Human Machine Interfaces), and level 3 (Plant Information Management System – PIMS) contextualizes the data from an asset standpoint allowing the management of analytics, event frames, notifications and system integrations with a tree view of the plant structure. The PIMS makes the data available for custom applications, thus interpreting the information and adding value to process related techniques such as predictive process control, optimization and economics analysis. Therefore, this infrastructure can provide a more realistic environment for investigating process automation and implementations under development, using desktop or web-based custom reports and dashboards for decision making at all different levels (operators, engineers or executives).

This work thus explores aspects of Industry 4.0 such as Digital Twins, Simulation, System Integration and Data Analytics using a developed infrastructure at WVU and UFCG that can be used for operator training, research and engineering classes. Additionally, it corresponds to the starting point for future research on more robust data analytics, machine learning, big data and business intelligence studies.

References

[1] E. Ruiz-Ramos, J. M. Romero-García, F. Espínola, I. Romero, V. Hernández and E. Castro, "Learning and researching based on local experience and simulation software for graduate and undergraduate courses in chemical and environmental engineering", Education for Chemical Engineers, 21, pp 50 – 61, 2017;

[2] J. Uhlemann, R. Costa and J. Cl. Charpentier, "Product design and engineering — past, present, future trends in teaching, research and practices: academic and industry points of view", Current Opinion in Chemical Engineering, 27, pp 10–21, 2020;

[3] M. Teles dos Santos, A. S. Vianna Jr. and G. A. C. Le Roux, "Programming skills in the industry 4.0: are chemical engineering students able to face new problems?", Education for Chemical Engineers, 22, pp 69–76, 2018.

[4] PI Server. (2018 SP3). San Leandro, CA, USA. OSIsoft.

602636 Robust Large-Scale State-Estimate Prediction

Abstract:

Effective monitoring and control of a process require online measurements of all state variables of the process. Measurement of all state variables are typically not available due to the lack of reliable online sensors and/or high costs of online sensors. These unmeasured variables are often directly related to product properties, which are important to monitor and control. Their estimates are usually obtained using state estimators. As a result of the need for and the importance of state estimation, the development of accurate state estimators that are robust with respect to model uncertainties and unmeasured inputs, has been the subject of research for decades [1-5]. There have been a few reports on the design of large-scale state estimators [6-11]. Although the implementation of centralized estimators for large-scale systems seems to be an optimal approach, these estimators are not scaleable to complex, large-scale systems, as they suffer from the curse of dimensionality.

In this paper, we address the problem of large-scale robust state-estimate prediction; that is, the prediction of future values of state estimates robustly at every time instant in large-scale processes. To this end, we decompose a large-scale state-estimate prediction problem into a set of smaller-scale state-estimate prediction problems so that the resulting set of the smaller-scale state-estimate predictors are more robust, easier to design and implement, and more computationally efficient. We also study tuning of the smaller-scale state-estimate predictors to maximize the robustness of the predictors set while ensuring the stability of the error dynamics of the entire set. In addition, we address the problem of implementing the smaller-scale state-estimate predictors in parallel (using a computer with parallel processors) to improve computational efficiency of the state estimate prediction while preserving the accuracy and robustness of state-estimate predictions.

References

[1] J.M. Ali, N.H. Hoang, M.A. Hussain, D. Dochain, Review and classification of recent observers applied in chemical process systems, Computers & Chemical Engineering, 76 (2015) 27-41.

[2] A.K. Jana, A nonlinear exponential observer for a batch distillation, in: 2010 11th International Conference on Control Automation Robotics & Vision, IEEE, 2010, pp. 1393-1396.

[3] M. Soroush, State and parameter estimations and their applications in process control, Computers & Chemical Engineering, 23 (1998) 229-245.

[4] S. Tatiraju, M. Soroush, Nonlinear state estimation in a polymerization reactor, Industrial & engineering chemistry research, 36 (1997) 2679-2690.

[5] N. Zambare, M. Soroush, B.A. Ogunnaike, Robustness improvement in multi-rate state estimation, in: Proceedings of the 2001 American Control Conference.(Cat. No. 01CH37148), IEEE, 2001, pp. 993-998.

[6] N. Abdel-Jabbar, C. Kravaris, B. Carnahan, A partially decentralized state observer and its parallel computer implementation, Industrial & engineering chemistry research, 37 (1998) 2741-2760.

[7] N. Abdel-Jabbar, C. Kravaris, B. Carnahan, Structural analysis and partitioning of dynamic process models for parallel state estimation, in: Proceedings of the 1998 American Control

Conference. ACC (IEEE Cat. No. 98CH36207), IEEE, 1998, pp. 3170-3176.

[8] J.B. Carvalho, F.M. Barbosa, Parallel and distributed processing in state estimation of power system energy, in: MELECON'98. 9th Mediterranean Electrotechnical Conference. Proceedings (Cat. No. 98CH36056), IEEE, 1998, pp. 969-973.

[9] R. Ebrahimian, R. Baldick, State estimation distributed processing [for power systems], IEEE Transactions on Power Systems, 15 (2000) 1240-1246.

[10] D.M. Falcao, F.F. Wu, L. Murphy, Parallel and distributed state estimation, IEEE Transactions on Power Systems, 10 (1995) 724-730.

[11] D.B. Pourkargar, M. Moharir, A. Almansoori, P. Daoutidis, Distributed estimation and nonlinear model predictive control using community detection, Industrial & Engineering Chemistry Research, 58 (2019) 13495-13507.

614872 Is Global Harmonization Achievable – an Industry Perspective

Abstract:

ICH introduced risk based approaches and enhanced scientific understanding as an opportunity to encourage continuous process improvement facilitate the introduction of newer technology. In addition, Quality by Design (QbD) promised to improve confidence in quality through the lifecycle of pharmaceutical products. A primary incentive for industry remains the prospect of global regulatory concordance for new applications and post approval changes, and ICHQ12 holds much promise. However, during the last decade, the industry has experienced a proliferation of regulatory divergence regarding the interpretation of ICH guidelines (and control strategies) across geographic regions, including "anxiety" for adoption and implementing ICHQ12. This presentation will illustrate some examples of divergence in control strategies while exploring opportunities for improved harmonization.

614880 Quality By Design and Expedited Development Programs

Abstract:

There is significant pressure on the CMC team to complete the product development process in ever shorter timeframes. Several new molecules have had rapid clinical development programs that have shortened Phase 3 or have had no Phase 3 in the case of oncology therapeutics. Gone are the days when a multiyear Phase 3 clinical program allowed for the CMC team to complete the drug development cycle in a standard timeframe. CMC activities are being pushed forward in the development cycle. The CMC team is more reliant on company platform technology and Quality by Design (QbD) principles that can allow the CMC team to complete development faster and create a robust control strategy.

The regulatory dossier must capture elements of the development process, it must show how our clinical drug product links to the commercial drug product and it must outline our control strategies for commercial manufacturing. This presentation will highlight the successes and challenges of rapid drug development and how our global regulatory dossier has evolved to deal with these complexities.

Kevin Fitzpatrick is an employee of AbbVie and may own AbbVie stock. AbbVie sponsored and funded the study; contributed to the design; participated in the collection, analysis, and interpretation of data, and in writing, reviewing, and approval of the final publication.

614874 QbD in Development and Commercial Manufacture

Abstract:

Quality by Design (QbD) elements are an integral component of product development and regulatory submissions, enabling continual improvement throughout the product lifecycle. QbD can provide a systematic approach in which prior knowledge from platform technologies is leveraged to focus resources on understanding variability critical to a specific product and/ or process. Companies decide on the extent to which to invest in QbD based on the stage of development and product and/or process risk and expand the use of QbD principles during the development process. Design spaces can be established for processes initially identified as high risk as that knowledge is used to establish appropriate process and input material controls for commercial operations. Raw material monitoring plans that support continual improvement can support expanded material attribute ranges based on a robust understanding of their interaction and impact on product quality. Process analytical technology (PAT) is often deployed primarily during development to rapidly understand sensitivity to variability and establish appropriate controls that are subsequently used during production, however PAT has been successfully implemented globally for release, in limited cases.

Regulatory flexibility as a result of QbD development and registrations has not been fully realized, however with continuous manufacturing this is expected to change. Design spaces across integrated unit operations can be established on commercial equipment without requiring extensive input materials and time. This provides a comprehensive understanding of the process sensitivities, the appropriateness of the control strategy and the justification for regulatory and operational flexibility. Furthermore, the shift towards performance-based approaches for analytical methods and models where the focus is on control of the output rather than the input can provide additional flexibility.

The objective of this presentation is to provide examples of how QbD elements and PAT are used to develop a robust process and how continuous manufacturing can fully utilize QbD in development and commercial manufacture allowing for potential regulatory flexibility.

614879 The Handle and the Briefcase: Model Quality for Quality By Design

Abstract:

QbD is based on a fundamental understanding of process performance and product quality, which can be derived from first principles modeling methods. Model development and use routinely suffers from peer-to-peer variability that undermines the credibility of model-based evidence. In this talk we discuss Amgen's approach to model quality and modeling lifecycle management to enable model-informed decisions and promote model-based filing activities. Several case studies will be presented.

614876 QbD at Vertex: Back to the Future

Abstract:

This talk focuses on the lessons learned in implementing the QbD approach in DS regulatory filings for the Vertex CF portfolio ranging from Kalydeco (2012) to Trikafta (2019). Case studies are used to address the evolution of approaches taken for risk assessments, experiment design and analysis, nonlinear modeling, design space limits and other areas. These lessons learned are fueling further development as we prepare for future regulatory filings. Further discussion during the panel forum will be welcome.

614875 QbD in Mobile Production Scenarios: Progress Toward a Multistage PAT Strategy for Pharmacy on Demand

Abstract:

Quality by Design strategies in the area of API production largely focus on definition of design space for reactions for development and commercial processes followed by traditional isolation, sampling and off-line release. PAT monitoring utilized in development has generally been removed when processes enter the commercial manufacturing facilities or is used "for information only" purposes. Continuous, on-demand production strategies require the PAT monitoring and quality control of intermediate as well as crude API solutions in order to release materials for purification. In addition PAT strategies can be envisioned which could eventually remove some end-product testing for API, thus allowing on-demand release to accompany the on-demand production. The talk will focus on potential PAT and control strategy opportunities that will facilitate the eventual move toward RTRT of CQAs for API.

614873 QbD (Quality by Design) an Enabler for Pharmaceutical Product Development: FDA Perspective

Abstract:

Quality by Design can be defined as systematic approach to development that begins with predefined objectives and emphasizes product/process understanding and process control, based on sound science and quality risk management. Implementation of QbD approach leads to improved robustness of a firm's quality system and better assurance of product quality. The US Food and Drug Administration (FDA) encourages risk-based approaches and the adoption of QbD principles in drug product development and manufacturing. This presentation will address impact of QbD primarily as an enabler that supports patient focused drug development, allows leveraging of flexible regulatory filing strategies based on enhanced product/process understanding and supports faster speed to market of novel products and/or manufacturing techniques. There will also be a discussion of how CDER is implementing a science and risk based approach for CMC assessment of regulatory submissions and for site inspections.

AUTHOR INDEX

Abdel-Latif, Kameel 334
Abdelrahman, Omar 188, 216
Abdi, Mesfin 372, 524, 583
Abild-Pedersen, Frank 13
Aboelela, Sarah 595
Abolhasani, Milad 334
Aboulmouna, Lina 206
Abraham, Nathan 99
Abramov, Yuriy 101
Abrams, Vatallio 428
Aburub, Aktham 138
Acharya, Shreyas 176
Achenie, Luke 16, 39
Acosta, Carlos Manuel Ramirez 149
Adam, Molly 69, 71
Adams, Donald 564
Adjiman, Claire S. 390, 511, 593, 616
Agblevor, Foster 39
Agbleze, Selorme 625
Aguda, Remil 268
Agustinus, Yohanes 131
Ahmad, Maaz 301
Ahmed, Bilal 424, 485
Ahsan, Syed 407
Aigner, Isabella 592
Akiti, Otute 127
Akseli, Ilgaz 427, 509
Alamani, Bryan G. 111, 313, 618
Alba-Rubio, Ana 366
Albericio, Fernando 63
Alexander, Ronald 559
Alhindi, Norah Sulaiman 145
Aliado, Michael 428
Aljama, Hassan 333
Allen, Karen 361
Altmeyer, Andreas 412
Alvarado-Hernández, Bárbara B. 513
Alwakwak, Abdo-Alslam 181
Amar, Vinod S. 364
Amidon, Gregory E. 605
Amini, Hossein 139, 427, 509
Andersson, Martin Peter 382
Anifowoshe, Ismail 412
Anselmo, Aaron C. 85
Apfelbaum, Rachel 587
Araujo, Willy 625
Arbogast, Jeffrey E. 627
Ardagh, M. Alexander 174
Arevalo, Faustin 606
Arjmandi-Tash, Omid 424
Armaou, Antonios 379
Armenante, Piero M. 326
Armstrong, Madeleine 537
Arntzen, Julius 281
Arora, Akhil 381
Artamonova, Anastasia 296, 403
Arunachalam, Naveen 37
Assaf, Georges 128
Ataíde, Filipe 128, 359
Austin, Lauren A. 89
Avendano, Carlos 327
Avraamidou, Styliani 168
Ayansiji, Ayankola 601, 602
Aziz, Hossain 407
Babaei, Masoud 327
Bachawala, Sunidhi 346
Bade, Rachel S. 122
Badr, Sara 57
Bajdich, Michal 42
Baker, Tania A. 205
Bakh, Naveed 360
Bakhshi, Santanu 208
Baldovino-Medrano, Victor 214
Balwinski, Karen M. 532
Banerji, Aditya 212
Bao, Zhenan 93
Barajas-Olalde, Cesar 564
Baranwal, Yukteshwar 344
Barolo, Massimiliano 350, 372
Barone, Paul W. 296, 403
Barrasso, Dana 173, 585, 587
Barrett, Lena A. 234
Bauer, Hannes 401, 409
Baumann, John 69, 71
Baumann, Michael 530
Bazaz, Sajad Razavi 58
Belekar, Viraj Vilas 134, 408
Bell, Alexis T. 333
Bell, Tristan A. 205
Bellinghausen, Stefan 430
Belowich, Matthew 184
Bende, Elizabeth M. Cummings 296, 403
Bennett, Charmaine 184
Benítez, Laura Andrade 534
Beran, Gregory 100
Berchielli, Alfred 142, 412

Beretta, Michela 28
Berger, Adam G. 251
Bermingham, Sean 173, 585
Bernhard, Stefan 10
Bertalan, Tom S. 307
Berthiaume, François 276
Bezzo, Fabrizio 350
Bhadriraju, Bhavana 303
Bhagat, Mihir 184
Bhalode, Pooja 232, 581, 622
Bharadwaj, Rahul 134
Bhattacharyya, Debangsu 549, 562
Bhawnani, Rajan 119, 494
Bhebe, Prince 297
Bibiceva, Svetlana 426
Biegler, Lorenz 507
Bilgili, Ecevit 139, 422, 606
Bio, Matthew 61
Biscans, Béatrice 113
Biserni, Stefano 534
Bispo, Heleno 625
Biswal, Prayag 240
Blackwood, Daniel O. 280, 398
Blakely, Logan 51
Bligaard, Thomas 42
Bliznyuk, Valery 95
Bobba, Venkata 427, 509
Bock, Daniel 90
Bodák, Brigitta 496
Bolton, Glen 574
Bommarius, Andreas 20, 58, 120, 136, 369, 433, 538
Bommarius, Bettina 433
Bommireddy, Yasasvi 346
Bonageri, Shrilakshmi 266, 336
Bond, Jesse 216
Borchardt, Christian 26
Born, Stephen 298
Bosco, Marcello 129
Bosetti, Luca 5
Bosshart, Nicholas 564
Botros, Lucy L. 298
Boukerche, Moussa 491
Boukouvala, Fani 51, 194, 546
Bozinovski, Dragana 296, 403
Braatz, Richard 210, 296, 403
Bradford, Eric 387
Bradley, William 194, 546
Brandon, Jenna 280, 398
Breen, Christopher P. 61, 432
Breshamer, Tom 530
Briggs, Naomi 595
Brindle, Joseph 18
Broadbelt, Linda 184
Brown, Cameron 374, 584
Brown, Robert C. 208
Brown, Solomon F. 430
Brueggemeier, Shawn 358
Brunelli, Nicholas 250
Brutto, Patrick Lo 128
Bu, Wei 267
Bullard, Joseph W. 429, 613
Burcham, Christopher 59, 173, 317
Burgess, Diane 2
Burnett, Annessa 595
Bustillos, Christian G. 227
Buttini, Francesca 534
Calado, Filipe 317, 585
Callaghan, Ryann 85
Campos, Rui Pina 65
Canalejas, Laura 534
Caparco, Adam A. 433
Capece, Maxx 26
Capecelatro, Jesse 414
Capellades, Gerard 111, 313, 618
Capozza, Rosario 417
Cardenas, Ilya Orson Sandoval 387
Cardinal-David, Benoit 591
Carmody, Alan 143
Carragher, Adelia 132
Carrasquillo-Flores, Ronald 358
Carrière, David 112
Casas-Orozco, Daniel M. 372, 583
Casati, Federica 298
Cash, Kevin J. 367
Cassayre, Laurent 113
Castellanos, Claudia 149
Castillo, Anya 51
Castillo, Ivan 305
Castro, Jerome 431
Castro-Rodriguez, Daniel 505
Cejas, Pedro J. 89
Cha, Seunghyeok 178
Chaaya, Elie 62, 132
Chakrabarti, Rajshree 118, 137, 182
Champion, Julie 433
Chan, Gabriela Ruphuy 292
Chang, Chao 101
Chanona, E. Antonio Del Rio 384, 387
Chanussot, Lowik 41
Chapizanis, Dimitrios 572
Chappelow, Chris 540
Charitopoulos, Vasileios 384
Charton, Sophie 112
Chatterjee, Sharmista 635

Chaudhuri, Bodhisattwa 2, 407
Chaudhuri, Santanu 119, 494
Chen, Dabing 157
Chen, Feiyang 421
Chen, Hongbo 35
Chen, Ida(Xue) 568
Chen, Jie 539
Chen, Kaitong 23
Chen, Liang 87
Chen, Mingfu 361
Chen, Po-Yen 342
Chen, Qiushi 421
Chen, Ruipeng 323
Chen, Weifeng 507
Chen, Wenqian 123, 164, 320, 489
Chen, Yingjie 622
Chen, Yu 22
Cheng, Emily 197
Chew, Alex K. 237
Chi-Hwa, Wang 148
Chiang, Leo H. 305
Chierici, Veronica 534
Cho, Patricia 309
Choi, Michael 287
Choksi, Tej 13
Chong, Magdalene 584
Choudhary, Madhuresh K. 255
Chow, Shing Fung 323
Chowdhury, Ananya 585, 587
Christofides, Panagiotis D. 49, 386
Chung, Eun Ji 144
Cifuentes, Javier F 149
Claes, Thomas 474
Clark, Sarah 138
Claussen, Jonathan 97
Clay, Mackenzie 362
Close, Edward 585, 587
Coblyn, Matthew 523
Cocchini, Ugo 128, 585
Coelho, Alexander 606
Cohen, Benjamin M. 295
Coleman, Simon 322
Coliaie, Paria 126, 199, 321, 541
Comisso, Davide 127
Conley, Ashley 619
Conrad, Jacinta C. 160
Contreras, Lourdes 143
Conway, Sandra 412
Conway, Shawn 66
Coombs, John Ryan 309
Coppens, Karen 532
Coppey, Nicolas 113
Cordiner, Joan 91, 92
Corzo, Carolina 528
Costa, Antonio 2
Costello, Mark 429
Cremaschi, Selen 376
Cruz, Isaac Jerome C. Dela 111, 313, 618
Cruz, Juan C 149
Crvenjak, Đana 528
Cui, Jianming 110
Cui, Tianqi 307
Curtis, Jennifer S. 415
Dai, Jiayu 621
Daldosso, Matteo 127
Dale, Steven 429
Dallin, Bradley C. 254
Darge, Abenazer W. 95
Das, Abhishek 41
Das, Tapas K. 521
Dauenhauer, Paul 174, 188, 221
David, Alexa A. 367
Davé, Rajesh 24, 87, 422
De Baets, Bernard 554
De Beer, Thomas 410
Dean, James 15
Dean, Stacey L. 532
Del Pozo, David Fernandes 554
Delhommelle, Jerome 265
Deo, Shyam 196
Desai, Parind 587
Deshmukh, Smeet 353
Deshpande, Anuja 551
Deshpande, Siddharth 11
Destrac, Philippe 113
Destro, Francesco 350, 372
Devol, Timothy A. 95
Dhillon, Pritpal Singh 179
Dicaudo, David 276
Dickey, David S. 357
Dighe, Anish V. 159, 601
Dimitrakopoulos, Panagiotis 8, 252
Ding, Yangyao 49
Diwan, Moiz 62, 166, 591
Dobrowolski, Adrian 596
Doerr, Frederik 584
Doherty, Michael F. 102
Donta, Maxsam 97
Dosch, Hope 364
Doshi, Pankaj 142, 280, 398, 412
Douieb, Sélim 585
Dowding, Peter J. 1
Dowling, Alexander 547
Doyle, Patrick S. 68

Du, Hui ... 537
Duan, Chenru ... 37
Duanghathaipornsuk, Surachet ... 366
Dubey, Atul ... 298
Duephans, Vanessa ... 281
Dunn, Robert ... 84
Duran, Tibo ... 2
Durelle, Maxime ... 112
Dutta, Arnab ... 54
Dybeck, Eric ... 99, 389
Dziubla, Thomas D. ... 138
Eckhoff, Julia ... 353
Ellison, Christopher J. ... 212
Elsaidi, Sameh ... 224
Elsayed, Mohamed ... 138
Emaminejad, Sam ... 94
Engle, Marissa ... 556
Epps, Robert ... 334
Eren, Ayse ... 487
Esterhuizen, Jacques ... 12
Eugene, Elvis ... 547
Eydenberg, Michael ... 51
Facco, Pierantonio ... 350
Fakhrabadi, Ehsan Akbari ... 279
Fan, Andy ... 361
Fatemi, Naghmeh ... 310
Feng, Jie ... 537
Feng, Xin ... 372, 524, 583
Fenster, Michaël ... 371
Ferguson, Heidi ... 89
Ferguson, Steven ... 399, 482
Fernández, Verónica Daza ... 534
Fichthorn, Kristen ... 110, 277
Fiedler, Daniela ... 405
Filios, Mike ... 633
Fischer, Kerri ... 173
Fisher, Cyrus ... 575
Fitzpatrick, Kevin ... 630
Flamm, Matthew H. ... 133, 352, 353
Florence, Alastair J. ... 485, 584
Flores, Raul A. ... 42
Florit, Federico ... 432
Folestad, Staffan ... 27
Forgber, Thomas ... 143
Forzano, Anna ... 595
Foster, James C. ... 95
Foulet, Cédric ... 113
Frank, Gregory ... 570
Franklin, Robert D. ... 433
Fraunhoffer, Kenneth ... 371
Freiberg, Lucas ... 523
Fuchsbichler, Agnes ... 528
Furst, Ariel ... 96
Futamura, Haruka ... 406
Gaffney, Piers ... 63
Galagan, James ... 361
Galindo, Amparo ... 390, 511, 593
Galvani, Marina ... 127
Galvanin, Federico ... 384, 387
Ganesh, Sudarshan ... 170
Gao, Hanyu ... 477
Gao, Tora ... 145
Gao, Xian ... 547
Gao, Yijie ... 407
Gao, Yi ... 158
Garcia, Andrew R. ... 163
García-Losada, Pablo ... 501
García-Muñoz, Salvador ... 183, 286, 350, 507, 510, 579
Gbadago, Dela Quarme ... 177
Gehris, Amie ... 530
Gehrke, Daniel ... 602
Geng, Xi ... 160
George, Neil ... 1
Gerceker, Duygu ... 62
Gerlza, Tanja ... 405
Gernaey, Krist V. ... 382
Gerogiorgis, Dimitrios I. ... 505
Ghanekar, Pushkar ... 272
Ghose, Sanjit ... 522
Ghosh, Debanjan ... 545
Ghosh, Deepanjan ... 276
Ghribi, Othman ... 265
Giri, Gaurav ... 619, 620
Glennon, Brian ... 482
Glotzer, Sharon C. ... 233
Goda, Amit ... 585
Goldsmith, Bryan R. ... 12
Goncalves, Ricardo ... 31
Gong, Kai ... 537
Gonzalez, Federico Lora ... 358
Gonzalez, Marcial ... 346
Gonçalves, Elsa ... 359
Gopeesingh, Joshua ... 216
Goreke, Dilara ... 557
Gorle, Ravi Kumar ... 315
Goss, Charles A. ... 542
Goyal, Siddharth ... 41
Grabow, Lars C. ... 178
Grady, Michael C. ... 264
Gramadnikova, Ekaterina ... 374
Greeley, Jeffrey ... 11, 13, 272
Green, William H. ... 392
Grieco, William ... 566
Griego, Charles ... 17

Grieme, Timothy 491
Griffin, Daniel 74
Grinstaff, Mark 361
Grosser, Shane T. 589
Grossmann, Ignacio 299, 579
Grover, Martha A. 20, 58, 120, 136, 369, 538
Gryczke, Andreas 166
Gröger, Harald 525
Gröls, Jan 484
Gu, Wenyu 19
Guan, Yi-Xin 514
Gump, Christopher 522
Guo, Mingxia 164
Guo, Yuan 421
Gupta, Anant 361
Gupta, Rishabh 306
Guthrie, Stephanie 619, 620
Guzmán, Elisa A. Torrico 155
Hadjiev, Viktor 137
Hafezisefat, Parinaz 208
Haider, Shozeb 480
Hamling, John A. 564
Hammersmith, Gregory 595
Hammond, Paula T. 251
Hammond, Robert B. 1
Han, Songi 253
Hanley, Kevin J. 417
Hanley, Thomas R. 213
Hanson, Joshua 183, 286
Harding, Matthew 399
Harold, Michael P. 179
Harper, Daniel 37
Harper, Kaid 62
Harris, J. Keith 532
Harris, J. Kirk 367
Harris, Patrick R. 20, 120, 369, 538
Harrison, Cristian 633
Hart, Travis 432
Hartl, Sonja 405
Hartmann III, Carl 73, 285
Hasan, M M Faruque 381
Haslam, Andrew J. 593
Hassanaly, Malik 518
Hassanpour, Ali 426
Hausner, Douglas 344
He, Fengjiao 537
He, Q. Peter 52
He, Xiaorong 139, 156, 157
Head-Gordon, Martin 333
Healy, Anne-Marie 399
Heindel, Theodore J. 408
Helmbrecht, Hawley 339
Heltzel, Jacob 122
Heng, Jerry Y. Y. 123, 124, 164, 320, 489, 576
Henriques, João 140, 609
Henthorn, David 624
Heras-Domingo, Javier 41
Heraud, Lea 492
Hermant, Paul 298
Hernandez-Betancur, Jose D. 50
Hessel, Volker 525
Hickey, Matthew 371
Hicks, William 327
Hilden, Jon 73, 285
Hill, Priscilla J. 420
Ho, Caleb 41
Hock, Marcus 99
Hodgins, Naomi 412
Holbrook, Debora L. 532
Holloway, Buck 131
Hondred, John 97
Hong, Moo Sun 296, 403
Hong, Yu-Tong 278
Horn, Martin 349
Hou, Xiao Serenity 132
Houson, Ian 485
Howard, Maylin 251
Hsiao, Wen-Kai 28, 349
Huang, Jack Chung-Jr 297
Huang, Yan-Shu 346
Huang, Zuyi (Jacky) 480
Hubbs, Christian D. 299
Huggins, Seth 74
Hur, Inyoung 372
Husson, Scott M. 95
Hwang, Sungwon 177
Ierapetritou, Marianthi 330, 581, 622
Ikeda, Craig 133
Irdam, Erwin 502
Irizarry, Roberto 495, 543
Irwin, Nicola 117
Isayev, Olexandr 395
Jackson, Alexander 7
Jackson, George 390, 511, 593
Jackson, Kyle 97
Jacob, Nitya 574
Jain, Ankit 42
Jain, Ashwinkumar 280, 398
Jain, Karnesh 265
Jain, Rishabh 255
Jajcevic, Dalibor 143, 398
Jamison, Timothy 61, 432
Janet, Jon Paul 37
Janik, Michael 196

Jarmer, Daniel J. 294
Jawlekar, Suhas 315
Jayaraman, Arul 337
Jayaraman, Arun 26
Jayjock, Eric 504
Jensen, Klavs F. 432, 477, 501
Jewell, Megan 367
Jiang, Mo 319
Jiang, Shuhong 537
Jiang, Zhihua 52
Jimeno, Guillermo 492, 584
Jiménez, Ana Alejandra Barrera 410
Jin, Can 525
Jin, Lu 564
Jin, Yingdi 101
Jinschek, Joerg 366
Jirát, Jan 615
Johanson, Kerry 416
Johnson, Brad 183, 286
Johnson, Luke 235
Johnson, Martin 501
Jonuzaj, Suela 390, 511
Jordan, Christopher 131
Joseph, Andrea 145
Joseph, Kristeen 174
Jovanovic, Goran 523
Jung, Juhyung 97
Justen, Anna 596
Kagan, Michael 90
Kalal, Zbynek 353
Kalantar, Thomas H. 532
Kamyar, Reza 412
Kape, Adrian 143
Kappatou, Chrysoula D. 77
Karakitsios, Spyros 572
Karim, M. Nazmul 561
Karimi, Iftekhar A. 54, 79, 301
Kasat, Gopal 412
Kaskiewicz, Peter L. 1
Katragadda, Apoorva 79
Kawasaki, Koji 406
Kayode, Gbolade O. 40
Keith, John A. 17
Kelkar, Manish S. 126, 321, 491, 539, 541
Kermenidou, Marianthi 572
Kerr, Mark S. 294
Kevrekidis, Ioannis G. 307
Khan, Faisal 303
Khan, M. Arif 138
Khan, Saif A. 68
Khandare, Aditya 176
Khinast, Johannes G. 143, 349, 398, 401, 405, 409, 592
Khurshid, Madiha 297
Kienle, Daniel 271
Kilbourne, Jacquelyn 276
Kilwein, Zachary 51
Kim, Dong-Shik 366
Kim, Doyoung 117
Kim, Sangah 24, 87
Kim, Seulhee 515
Kimber, James 280, 398
Kitchin, John R. 10
Kitching, Victoria R. 424
Klapperich, Catherine M. 361
Klimsa, Vojtech 292
Kline, Billie 633
Klinger, Jordan 421
Klitzing, Nicholas 578
Kobler, Mirjam 534
Kong, Hyunjoon 278
Kontny, Tanner 67
Kontoravdi, Cleo 168
Koolivand, Abdollah 8
Kopp, Nathaniel D. 371
Kotecki, Brian 62
Kotidis, Pavlos 168
Koynov, Athanas 122, 603
Koç, Deniz 505
Kravaris, Costas 561
Krumme, Markus 592
Kshirsagar, Shivani 59
Kucuk, Gulsad 139
Kuhn, Simon 310, 473
Kulesza, Tim 432
Kulik, Heather J. 37
Kumar, Dinesh 273
Kumar, Gaurav 188
Kumar, Ramya 341
Kumar, Sumit 499
Kumar, Suneel 276
Kuminek, Gislaine 605
Kungl, Andreas 405
Kunnath, Kuriakose 87, 220
Kuo, Robert 353
Kuzmanovic, Uros 361
Kwofie, Francis 133
Kwon, Joseph 303, 337
La Zara, Damiano 27
Ladika, Mladen 532
Lagare, Rexonni 170
Lagerman, Colton 369
Lai, Pin-Kuang 88
Laird, Carl D. 51

Laird, Carl ... 583
Lakerveld, Richard ... 323
Laky, Daniel ... 583
Lalloo, Anita ... 631
Lamarche, Keirnan ... 504
Landis, Steven ... 102
Lang, Matthew J. ... 205
Larsen, Paul A. ... 540
Lavril, Thibaut ... 41
Le, Ngoc ... 341
Leblebici, M. Enis ... 473
Leblebici, Mumin E. ... 474
Leckband, Deborah E. ... 269
Lee, Dongheon ... 337
Lee, Jong Min ... 80
Lee, Joshua ... 368
Lee, Kai ... 280, 398
Lee, Michael A. ... 360
Lee, Su-Lin ... 372, 524, 583
Lee, Tim Boung-Wook ... 587
Lentner, Matthew ... 97
Levons, Jaquan ... 569
Li, Cheng-Han ... 366
Li, Jiali ... 342
Li, Meng ... 43
Li, Shuyun ... 50
Li, Sizhu ... 101
Li, Wei ... 584
Li, Xiaoyu ... 123, 320
Li, You ... 86
Liao, Jianshan ... 193
Liao, Rick ... 152
Liberatore, Matthew ... 279
Lill, Sten Nilsson ... 103
Lim, Kaizhuo ... 342
Lima, Fernando V. ... 50, 559, 625
Lin, Binhua ... 267
Lin, Huizhuan ... 537
Lin, Kuan-Han ... 507
Lin, Zhixing ... 87
Linic, Suljo ... 12
Link, Frederik J. ... 124, 489, 576
Linninger, Andreas ... 601, 602
Lino, Paulo ... 607
Lippelt, Christopher ... 66
Litster, James D. ... 34, 424, 430, 431, 588
Liu, Bin ... 275
Liu, Chang ... 267
Liu, Chun-Yen ... 43
Liu, Fang ... 37
Liu, Peiyuan ... 412
Liu, Tianyu ... 233
Liu, Xiaoguang ... 515
Liu, Xiaohui ... 396, 397
Liu, Xiaoli ... 342
Liu, Xingbo ... 562
Liu, Yanhong ... 480
Liu, Ying ... 267
Liu, Zhanjie ... 614
Livingston, Andrew G. ... 63
Lochmann, Dirk ... 528
Lomeo, Joshua ... 332, 516, 603
Looker, Adam ... 633
Lopes, Maria Inês ... 65
Lu, Amos E. ... 296, 403
Luciani, Carla ... 63
Luo, Yu ... 231
Ma, Lisa ... 332
Macdonald, Bruce C. ... 142, 412
Macgregor, John ... 545
Mach, Eric ... 97
Macht, Betsy ... 287
Mackey, Jaron ... 524, 583
Macmillan, Andie L. ... 138
Maddipudi, Bharath ... 364
Madison, Tyree ... 428
Madlmeir, Stefan ... 143
Mahadevan, Meera ... 129, 396, 397
Mahanthappa, Mahesh K. ... 35, 212
Mahesan, Bahi ... 530
Majumdar, Paulami ... 13
Majumdar, Sudip ... 61
Maliekkal, Vineet ... 221
Malladi, Jaya ... 427, 509
Mallapragada, Surya K. ... 97
Maloney, Andrew J. ... 270, 296, 403
Mani, Sriramvignesh ... 228
Manjrekar, Onkar ... 62
Manning, Harris W. ... 205
Mannschott, Thomas ... 592
Mansouri, Seyed Soheil ... 382
Mantle, Mick ... 594
Marek, James C. ... 491
Margelefsky, Eric ... 122
Marks, Brian ... 133
Markwalter, Chester E. ... 537
Marlowe, Justin ... 176
Marsac, Patrick ... 138
Marshall, Grant D. ... 120, 136
Marta, Teresa ... 508
Martin, John ... 251
Martinelli, Francesco ... 127
Martinez, Daniel ... 43
Marton, Christopher ... 131

Martín, Mariano 50
Martínez-Cartagena, Pedro A. 513
Martínez-Ortega, Fernando 214
Mascaro, Tracey 133
Mascia, Salvatore 298
Masooleh, Leila Samandari 627
Mateos, Carlos 501
Mateus, Elisabete 128
Mathews, Jane 431
Matic, Josip 401, 409
Matsunami, Kensaku 517
Matuszewski, Michael S. 551
Maus, Martin 139
Mazal, Tobias 162
Mazzotti, Marco 5, 496
Mbachu, Victoria 309
McBride, Kevin 507
McCallen, Rose 415
McDonald, Matthew A. 58, 120, 136, 369, 538
McGinty, John 584
McKee, Laura 132
McManus, Simon A. 537
McMullan, Jason 175
McMullen, Jonathan 368
McReady, Laura 530
Medina-González, Sergio 346
Mehra, Punit 360
Meir, Glen 473
Melkeri, Yash 587
Mendez, Rafael 513
Mendonça, Ricardo 359
Meng, Qingwei 475
Mercandelli, Alberto 534
Metzler, Catherine 613
Meyer, Jonathan 585
Mhamdi, Adel 77
Mhaskar, Prashant 545
Miller, David C. 551
Miller, Jeffrey T. 11
Millstone, Jill E. 10
Milton, Robert A. 430
Milton, Ross D. 19
Miranda, Michael 191
Mitchell, Niall 129, 317, 396, 397, 489, 492, 584, 585
Mitsos, Alexander 77
Mittal, Sachin 353
Miura, Tomohiro 517
Mo, Yiming 501
Mockus, Linas 45
Mohammed, Yakoob Sardar 315
Mohan, Anne 368
Monbouquette, Harold G. 362
Moniri, Saman 161
Monnier, Olivier 492
Monos, Timothy M. 432
Monteil, Dominique 90
Monteiro, Ana 359
Monteiro, Pedro 508
Montemore, Matthew M. 40
Moon, David 633
Moon, Hyunjin 253
Moore, Jessica 117
Morais, Ruth 65
Moreira, Joao 413
Morley, Wayne 127
Morrin, Gregory T. 271
Morris, Aaron 4
Morris, Collin 491
Morton, Colin 1
Moschetta, Eric G. 591
Mota, Rute F. 607
Mou, Mingyao 319
Mourão, Marta 128
Mpourmpakis, Giannis 15
Muanpaopong, Nontawat 422
Mudie, Deanna 71
Mufti, Ahmed 524, 583
Mukherjee, Angan 549
Mukherjee, Samrat 166
Mukhopadhyay, Ahana 196
Mummudi, Mothivel 134, 412
Murnen, Hannah 61
Murphy, Eric 134, 166
Muzika, Frantisek 21
Muzzio, Fernando 171, 344, 503, 614
Myers, Steven 317
Myerson, Allan S. 111, 313, 618
Nagy, Zoltan K. 59, 170, 346, 372, 487, 524, 540, 583
Naik, Ankita 174
Nakanishi, Hayao 291
Nambiar, Anirudh M. K. 432
Nance, Elizabeth 145, 152, 339
Nande, Inês 140
Nandiwale, Kakasaheb 501
Nandy, Aditya 37
Narsimhan, Karthik 368
Nascimento, Susana 65, 359
Nataraj, Akshaya 543
Nazemzadeh, Nima 382
Nelson, Arif Z. 68
Nelson, Celeste M. 234
Nelson, Patrick 18

Nere, Nandkishor K. 126, 134, 321, 408, 491, 541
Neurock, Matthew 221
Ngelale, Randy O. 227
Nguyen, Chanh 415
Nguyen, Freddy T. 360
Nguyen, Sonbinh 184
Nielsen, Caroline 203
Nielsen, Rasmus Fjordbak 382
Nieves-Remacha, Maria Jose 501
Nigra, Michael 18
Nijhuis, Job 525
Nikolakopoulou, Anastasia 210
Nilsson, Mikael 227
Nockett, Anthony J. 542
Nopens, Ingmar 410, 554
Nordstrom, Fredrik 314
Notestein, Justin 184, 197
Nounou, Hazem 561
Nounou, Mohamed 561
Noël, Timothy 525
Nunes, Patricia 609
Nunes, Patrícia D. 140
Nunn, Christopher 590
Nwaokorie, Chukwudi F. 40
O'Toole, Rebecca 522
Odille, Fabrice 128
Oh, Tae Hoon 80
Ojakovo, Peter 429
Oktem, Ulku 627
Oliva, Joseph 59
Omar, Chalak 588
Omell, Benjamin P. 551
Omidvar, Noushin 16
Otten, Pieter 66
Ottoboni, Sara 374, 511, 584
Ou, Jianfa 90
Ou, Rui Wen 296, 403
Overhoff, Kirk 155
Paisana, Maria 31, 609
Paiva, Mafalda 140, 609
Pajaro, Estefany 214
Palahnuk, Hannah 509
Palizhati, Aini 41
Palmer, Jeremy C. 118, 137, 182
Pan, Chuanji 342
Pan, Da 83
Pandya, Gneya Harish 613
Panmai, Santipharp 537
Pantelides, Constantinos C. 616
Papageorgiou, Charles D. 485, 487
Papathanasiou, Maria M. 168
Pappas, Iosif 168, 168
Parga, Catalina 482
Parikh, Devi 41
Parisini, Irene 127
Parmar, Deependra 178
Parrish, Christine 61
Parsons, Anna 82, 327
Parulkar, Aamena 250
Patel, Alpa 343
Patel, Samarthaben J. 187
Patience, Daniel 502
Patton, Jasson T. 540
Paudel, Amrit 28, 401, 534
Paul, Shubhajit 35, 103
Peck, Wesley D. 564
Peeters, Michiel 410
Peeva, Ludmila G. 63
Peng, Hu-Hong 514
Pereira, Simao 317
Perez, Hector 299
Perez, Jem Valerie D. 111, 313, 618
Perini, Giulio 327
Perkins, Daniel 167
Perona, Joseph 84, 131
Persits, Nili 296
Petersson, Gunilla 27
Petridis, Ioannis 572
Petsagkourakis, Panagiotis 384, 387
Pham, Tiep Hoang 267
Phenix, Brian 633
Pieloth, Damian 596
Pietz, Mark A. 294
Pinto, Joana T. 28
Pinto, João F. 140
Pistikopoulos, Efstratios N. 168
Pitt, Kate 34, 424
Plata, Paola Leon 267
Politi, Maria 266, 336
Porter, Stuart C. 287
Portugal, Catarina M. 607
Pozzo, Lilo 266, 336
Pradhan, Shankali 429
Prasad, Elke 584
Price, Chris 374, 511
Probst, Susana Ecenarro 534
Procopio, Adam 353
Prud'Homme, Robert K. 537
Psarellis, Yorgos M. 307
Ptak, Aaron 518
Purdum, Geoffrey E. 295
Putz, Isabella 281
Qi, Yijun 263
Qiu, Yihong 154

Quach, Amy L. 133
Quaghebeur, Ward 554
Quayle, Mike J. 27
Quon, Justin L. 485, 487
Rabbani, Arash 327
Rafat, Neda 363
Ragazzo, Kira 131, 358
Raghuraman, Arjun 184
Rahman, Mahbubur 606
Raj, Rishi 522
Rajat, Ghosh 179
Rajniak, Pavol 139, 413
Ram, Rajeev J 296
Ramachandran, Rohit 171, 344
Ramachandruni, Hanu 537
Ramakrishnan, Srividya 315
Ramirez, Julia E. Vela 89
Ranadive, Pinaki 250
Rapp, Kersten 595
Raudenbush, Katherine 352
Rawal, Aditya 537
Reeves, Sheena 428
Rege, Kaushal 276
Rehrl, Jakob 349
Reineke, Theresa M. 341
Reizman, Brandon 294, 396, 397, 501
Rekhi, Lavie 13
Reklaitis, Gintaras V. 45, 170, 346, 583
Ren, Yi Ming 49
Rendall, Ricardo 305
Reutzel-Edens, Susan M. 616
Reyer, Sebastian 528
Reyes, Kristopher G. 334
Reyes, Luis H. 149
Reynolds, Michael A. 160
Rezaei, Fateme 181
Riazi, Hossein 264
Ribeiro, Ana 128
Ribeiro, Fabio H. 11
Ribeiro, Marcelo P. A. 559
Ricardo, Ana-Aguiar 607
Ricart, Brendon G. 567
Riccitelli, Molly 132
Ridvan, Ludek 615
Rielly, Chris 584
Rimer, Jeffrey D. 117, 160, 178, 255
Rincon, David 386
Rincón, Juan Antonio 501
Rioux, Robert 196
Ristroph, Kurt D. 537
Riviere, Morgane 41
Roberts, Kevin J. 1, 7
Robertson, Murray 584
Roblegg, Eva 405
Roday, Setu 155
Rodriguez Jr., Jaime 266, 336
Rogers, Amanda 295
Rogers, Luke 565
Rolandi, Pablo 632
Romagnoli, Jose A. 305
Romañach, Rodolfo 513
Román-Ospino, Andrés D. 344
Roper, Thomas D. 595, 634
Rosa, Leonor 129
Rosbottom, Ian 1, 123, 164, 320, 489, 576
Rossi, Francesco 45
Rousseau, Ronald 20, 58, 120, 136, 369, 538
Rover, Marge 208
Rownaghi, Ali 181
Ruggles, Alex 166
Ruiz-Mercado, Gerardo 50
Ruth, John C. 19
Ryan, Michael 633
Saccani, Martina 127
Saccomano, Samuel C. 367
Sacher, Stephan 349
Sahinidis, Nick 183, 286, 299, 556
Saklatvala, Rob 89
Salami, Hossein 58, 120, 136, 369, 538
Salar-Behzadi, Sharareh 528
Salari, Kambiz 415
Salavati-Fard, Taha 178
Salehi, Niloufar 605
Salman, Agba D. 431, 588
Sampat, Apoorva 557
Sampath, Janani 163
Sangerman, Jose 296, 403
Sannidhi, Abhinav 213
Santhosh, Shruti 85
Santos, Andrew P. 238
Sarafinas, Aaron 355
Saravanan, Karthikeyan 17
Sardinha, João 359
Sarigiannis, Dimosthenis 572
Sarkar, Avik 143
Sarwar, Owais 299
Satoh, Paul 363
Sauer, Robert T. 205
Saurer, Eric M. 309
Scaravelli, Federico 127
Schaldach, Gerhard 596
Scheiwiller, Sage 266, 336
Schoell, Jochen 495, 543
Schreiber, Igor 21

Schreiberova, Lenka 21
Schroeder, Charles M. 273
Schroeder, Sven 426
Schröder, Daniela 139
Schröttner, Hartmuth 592
Schulte, Kevin L. 518
Schultz, Guilhermina 559
Schultz, Kelly 147
Schultz, Leon 139
Schwabe, Robert 139
Schwartz, Daniel K. 269, 271
Schwartz, Kyra 133
Scott, Lewis 419
Scott, Susannah L. 253
Seemakurthi, Ranga Rohit 11
Sefcik, Jan 374, 511
Seibert, Kevin D. 575, 578
Seider, Warren 627
Sen, Maitraye 183, 286, 413
Senftle, Thomas P. 43
Sensharma, Prerana 361
Shah, Pranali 574
Shahani, Ashwin J. 161
Shamsuzzaman, Farooq 54
Shangguan, Ning 61
Shao, Hui 532
Sharpe, Sonja 308
Shekhar, Shashank 62
Shen, Baoxing 365
Shende, Anuradha 364
Shende, Rajesh 364
Shepard, Kimberly B. 69, 71
Sheriff, M. Ziyan 230, 561
Sheskey, Paul J. 532
Shetty, Manish 174
Shi, Baimei 101
Shi, Jiafu 22
Shi, Zhongping 358
Shields, Richard 354
Shirts, Michael R. 99
Short, Michael 507
Shu, Nicolas S. 361
Shuaibi, Muhammed 41
Shukla, Diwakar 476
Si, Yingnan 515
Siddique, Humera 485
Siegmann, Eva 398
Sierra-Vega, Nobel O. 513
Silva, Rui C. 607
Simo, Georges Motchoffo 145
Simon, John 518
Singh, Meenesh R. 119, 126, 159, 321, 494, 541, 601, 602
Singh, Ravendra 171, 503
Sinha, Kushal 134, 166, 408
Sinha, Parth 412
Sinskey, Anthony J. 296, 403
Sipple, Patrick 371
Sirasitthichoke, Chadakarn 326
Sirk, Kevin 122
Sirota, Eric 368
Sitaraman, Hariswaran 518
Slade, David 183, 286, 588
Smith, Alexander 47
Smith, Rachel M. 34, 424
Sokolowsky, Katie 155
Soliman, Mahmoud 428
Solomon, Michael J. 233
Somasundar, Ambika 218
Somayajula, Chandra Sekhar 562
Son, Manki 360
Song, Zhen 525
Soos, Miroslav 615
Soroush, Masoud 264, 627
Sosa, Ricardo D. 160
Sousa, Lucia Volta E 128
Sousa, Luis 31
Sowrirajan, Koushik 343
Spangler, Jamie B. 478
Spanos, Alexander 250
Splawn, Heather 518
Spormann, Alfred M. 19
Springer, Kirsten 539
Springs, Stacy 296, 403
Stamatis, Stephen D. 579
Stankovic-Brandl, Milica 534
Stinson, Benjamin M. 205
Stocker, Michael 399
Storer, Greg 62
Strano, Michael 360
Streather, David 82
Streibel, Verena 13
Stueber, Dirk 122
Su, Qinglin 298
Su, Zhaoqian 269
Subramaniam, Shankar 134, 408
Sugden, Isaac J. 616
Sugiyama, Hirokazu 57, 291, 406, 517
Sun, Changquan 35, 103
Sun, David 103
Sun, Feilong 27
Sun, Guangxu 101
Sun, Weike 296
Sun, Yiying 22

Sundmacher, Kai 525
Sunil, Vishnu 148
Susarla, Ramana 315
Suthar, Kerul 52
Szilagyi, Botond 487
Tabar, Iman Beheshti 544
Tabora, Jose E. 371
Takizawa, Bayan 298, 298
Tan, Zhe 341
Tanabe, Shuichi 517
Tandogan, Nil 578
Tararina, Margarita 361
Taussig, Michael 544
Taylor, Michael 37
Tew, Jonathan D. 34, 424
Thamizhchelvan, Anbu Mozhi 148
Thierry, David 507
Thomas, John A. 325
Thomas, Stephen 509
Thommes, Markus 281, 596
Thompson, Connor J. 269
Thompson, David H. 524
Tian, Chang 537
Tirumkudulu, Mahesh S. 142
Tocce, Elizabeth 530
Todd, Paul W. 213
Tomar, Bhawana S. 142
Torfs, Elena 554
Torres, Frank Torres 578
Toson, Peter 398
Touma, Jad 523
Traeger, Jeremiah 271
Tran, Kevin 41
Trogrlic, Martina 143
Trojacher, Christina 405
Tsapatsis, Michael 188
Tseng, Yni-Chao 139
Tsilomelekis, George 176
Udipabu, Pompon Mputu 252
Udwary, Kevin 518
Ulissi, Zachary 10, 41
Urie, Russell 276
Uz, Metin 97
Vaideeswaran, Karthik 138
Valla, Clemence 492
Valleau, Stephanie 38
Van Der Wel, Peter 592
Van Gerven, Tom 310, 473, 474
Van Hauwermeiren, Daan 410
Van Hest, Jan 525
Van Ommen, J. Ruud 27
Vansaders, Bryan 233
Vargason, Ava M. 85
Varsolona, Richard 122
Vasenkov, Sergey 163
Vecchiarello, Nicholas 574
Vekilov, Peter G. 118, 137, 182
Velaga, Jagannadharao 315
Vellara, Anand 267
Venkatachalam, Manohar 315
Verma, Chinmay 18
Verma, Lakshmanji 118, 137, 182
Vermeire, Florence H. 392
Vetter, Thomas 135, 327, 542
Vicente, João 607
Villa, Patricia Marce 584
Viswanath, Shekhar K. 138
Vlachos, Dionisios G. 44
Vojvodic, Aleksandra 235
Volk, Amanda A. 334
Vu, Vinh H. 269
Waldo, Michael 308
Wallace, Tanya 127
Wang, Bo 94
Wang, Chenguang 103
Wang, Di 179
Wang, Jin 52
Wang, Joseph 571
Wang, Kunlin 35
Wang, Leon Z. 69, 537
Wang, Li Ge 588
Wang, Linxi 196
Wang, Mei 9
Wang, Qinggang 295
Wang, Sheryl 251
Wang, Shih-Han 16
Wang, Shiyan 204
Wang, Siwen 16
Wang, Tony 297
Wang, Xiaonan 79, 342
Wang, Xueying 22
Wang, Zhenyu 305
Wang, Zi-Dan 514
Wang, Zilong 412
Wangsanuwat, Chatarin 239
Warkiani, Majid Ebrahimi 58
Warren, Nicholas J. 1
Warzecha, Monika 118, 137, 182
Watson, Oliver L. 390, 511
Watson, Thomas 530
Watson, Timothy J. 629
Wegiel, Lindsay 413
Wehbe, Malak 593
Weimer, Alan W. 522

Weltz, James S. 271
Weng, Jingwen 323
Werlang, Caroline 198
White, Claire 537
Whitley, Joshua A. 433
Wiggers, Helmut 596
Wijethunga, Tharanga 308
Wilbert, Christopher 131
Wilkinson, Sam 129, 396, 397
Willcox, David 633
Williams, Bianca 376
Williams, Katerina 575
Williams, Kindle 201
Wilsdon, David 280, 398
Wilson, Chick 584
Winther, Kirsten T. 42
Woehl, Taylor J. 9
Wolfrum, Jacqueline 296, 403
Wong, Qing 90
Wong, Renato 585
Wood, Erin 372, 524, 583
Wood, Geoffrey 389
Worden, Robert M. 363
Wright, Dillon 71
Wu, Huiquan 343
Wu, Nan 147
Wu, Richard 297
Wu, Wei-Lee 59, 540
Wu, Wei 298
Wu, Yinghao 269
Wu, Zhe 386
Wutscher, Thomas 534
Xia, Han 396
Xia, Yidong 421
Xin, Hongliang 16
Xu, Hongyun 35
Xu, Jiayi 275
Xu, Xiaoming 2
Xu, Yinan 11
Xue, Yao 379
Yabuta, Keisho 406
Yaginuma, Keita 517
Yamada, Masahiro 291
Yan, Tianyu 277
Yang, Haitao 342
Yang, Huaiyu 320
Yang, Jiankai 537
Yang, Jiecheng 415
Yang, Ou 195, 330
Yang, Shang-Tian 86, 479
Yang, Yihui 485
Yang, Zhuocen 101
Yao, Shan-Jing 514
Yao, Yuan 414
Yaron, Jordan 276
Yeardley, Aaron S. 430
Yeo, Jet 63
Yin, Qiuxiang 621
Yoon, Bola 522
Yoon, Junwoong 41
Yoshioka, Yasuyuki 291
Yu, Miao 295
Yu, Mingzhe 431
Yu, Weili 142
Yu, Ying 184
Yuan, Changxia 295
Zacuto, Michael 84
Zajac, Paul 530
Zamani, Siavash 4
Zamarripa, Miguel A. 551
Zang, Jia 371
Zavala, Victor M. 47, 557
Zellnitz, Sarah 534
Zemanick, Edith 367
Zettl, Manuel 592
Zhai, Tianhua 480
Zhang, Chenyue 525
Zhang, Dongda 387
Zhang, Fangyuan 480
Zhang, Fengli 479
Zhang, Fuweng 27
Zhang, Li 297
Zhang, Pin 267
Zhang, Qi 306
Zhang, Shawn 332, 516, 603
Zhang, Shijia 43
Zhang, Yeli 530
Zhang, Yichi 49
Zhang, Yingyue 537
Zhang, Yizu 616
Zhao, Fei 579
Zhao, Jingnan 475
Zhao, Jin 530
Zhao, Lingyan 17
Zhao, Yang 22
Zhao, Yichao 94
Zhao, Yongsheng 102
Zheng, Kai 87
Zhou, Ling 621
Zhou, Liping 516
Zhou, Lufang 515
Zhou, Maotang 348
Zhou, Teng 525
Zhu, Guanghui 295

Zhu, Wenbo 305
Zidan, Ahmed 75
Ziegler, Kirk J. 163
Zielinski, Margot 113
Ziff, Robert M. 605
Zimmer, Andreas 528
Zimmerman, William B. 92
Zitnick, C. Lawrence 41
Zuber, Adam 176
Zvonícek, Vít 615
Zürcher, Philipp 57

AIChE
120 Wall Street, FL 23
New York, NY 10005-4020

ISBN 978-1-7138-2319-3

9 781713 823193

CURRAN ASSOCIATES INC.
proceedings.com